D1476536

Handbook of Aviation Human Factors

HUMAN FACTORS IN TRANSPORTATION
A Series of Volumes Edited by
Barry H. Kantowitz

Bainbridge • *Complex Cognition and the Implications for Design*

Barfield/Dingus • *Human Factors in Intelligent Transportation Systems*

Billings • *Aviation Automation: The Search for a Human-Centered Approach*

Garland/Wise/Hopkin • *Handbook of Aviation Human Factors*

Noy • *Ergonomics and Safety of Intelligent Driver Interfaces*

Parasuraman/Mouloua • *Automation and Human Performance: Theory and Application*

Handbook of Aviation Human Factors

Edited by

Daniel J. Garland
John A. Wise
V. David Hopkin
Embry-Riddle Aeronautical University

LEA LAWRENCE ERLBAUM ASSOCIATES, PUBLISHERS
1999 Mahwah, New Jersey London

Lawrence Erlbaum Associates, Inc., Publishers
10 Industrial Avenue
Mahwah, NJ 07430

Cover design by Kathryn Houghtaling Lacey

Library of Congress Cataloging-in-Publication Data

Handbook of aviation human factors / edited by Daniel J. Garland, John A. Wise, V. David Hopkin.
p. cm.
Includes bibliographical references and indexes.
ISBN 0-8058-1680-1
1. Aeronautics—Human factors. I. Garland, Daniel J. II. Wise, John A., 1944– . III. Hopkin, V. David.
TL553.6.A95 1998
629.13—dc21 98-17058
CIP

Books published by Lawrence Erlbaum Associates are printed on acid-free paper, and their bindings are chosen for strength and durability.

10 9 8 7 6 5 4 3 2 1

To Linda, Danny, and Brianna
—Daniel J. Garland

To my family
—John A. Wise

To Betty
—V. David Hopkin

* * *

Dedicated to those early pioneers
of aviation human factors,
who made this book possible.

Contents

Series Foreword

Barry H. Kantowitz
Battelle Human Factors Transportation Center

The domain of transportation is important for both practical and theoretical reasons. All of us are users of transportation systems as operators, passengers, and consumers. From a scientific viewpoint, the transportation domain offers an opportunity to create and test sophisticated models of human behavior and cognition. This series covers both practical and theoretical aspects of human factors in transportation, with an emphasis on their interaction.

The series is intended as a forum for researchers and engineers interested in how people function within transportation systems. All modes of transportation are relevant, and all human factors and ergonomic efforts that have explicit implications for transportation systems fall within the series purview. Analytic efforts are important to link theory and data. The level of analysis can be as small as one person, or international in scope. Empirical data can be from a broad range of methodologies, including laboratory research, simulator studies, test tracks, operational tests, fieldwork, design reviews, or surveys. This broad scope is intended to maximize the utility of the series for readers with diverse backgrounds.

I expect the series to be useful for professionals in the disciplines of human factors, ergonomics, transportation engineering, experimental psychology, cognitive science, sociology, and safety engineering. It is intended to appeal to the transportation specialist in industry, government, or academia, as well as the researcher in need of a testbed for new ideas about the interface between people and complex systems.

This volume is focused on the aviation domain. It combines theoretical views of human performance, methodological issues, and practical implications, thus achieving a major goal of the series, which is to demonstrate the interaction between practical and theoretical aspects of human factors. Section I introduces aviation human factors, discussing history, methodology, and organizational factors. Section II reviews theoretical underpinnings as related to individual and crew performance. Section III focuses on parameters related to the aircraft itself, while Section IV considers air traffic control from a systems perspective. Section V covers aviation operations and design, including avionics, maintenance, security, and accident investigation. Forthcoming books in this series will continue this blend of practical and theoretical perspectives on transportation human factors.

Preface

Aviation is expanding, and this expansion is expected to continue for the foreseeable future. More people want to fly, and aircraft types are increasing in number. Technical innovations and automation introduce changes in the control of aircraft as vehicles and as traffic, and in the numerous human roles and jobs in the air and on the ground that support aviation. The domain of human factors as a discipline is also expanding, independently of aviation. It now embraces more topics and applications than it formerly did, and employs a greater variety of techniques. Aviation human factors is therefore expanding in two distinct ways: the range of applications of human factors within aviation has increased and so the titles of the chapters in this text cover more applications than previous texts did, but the range of topics within each chapter has also increased because the boundaries of human factors as a discipline have been extended to encompass additional themes and approaches. These developments have both led to a net expansion of the subject matter of aviation human factors, because scarcely any applications or topics have been dropped. Currently, aviation human factors is a developing and dynamic force.

The rapid and sometimes revolutionary developments in technology, aviation, and human factors, and the complex interdisciplinary interactions between them, lead to a requirement to review and appraise progress from time to time by taking stock of what has happened and by peering into the future. This text has this objective. It attempts to employ recently developed concepts and explanatory frameworks where they are appropriate. Authors were urged to consider what is known now, and encouraged to develop a point of view. Any disagreements between them have not been edited out, but indicate where there is room for authoritative views to differ, because knowledge is not yet firm or standardized or because divergent practices have evolved. The international perspective sometimes adopted is broadly representative of aviation itself and of its human factors issues.

Aviation has often been among the first contexts to apply new technologies safely and successfully. It has therefore also been among the first to encounter and resolve the human factors issues associated with them. This has required their suitable matching with human capabilities and knowledge, and the devising of appropriate training. The full range of the human factors consequences of new technologies has not always been foreseen, and may not always have been predictable while the technologies were being developed or when they were first introduced, but at last there seems to be a growing recognition of what the full range of these consequences is likely to be and a greater willingness to take account of them during development. This text attempts to represent this perspective.

There is now an extensive body of human factors knowledge and experience but still there remain some differences in its application, though perhaps fewer in aviation than in some other contexts. The broadening range of applications emphasizes the practical issue of the generalizability of human factors evidence. Where similar problems apparently occur in different applications of human factors within aviation (such as in cockpits, helicopters, and air traffic control), the criteria for determining whether they have similar solutions inevitably come to the fore. As the boundaries of human factors are extended and more applications of human factors within aviation are introduced, such circumstances must arise more often. The effective and efficient deployment of scarce human factors resources depends on reliable guidelines about when existing human factors evidence requires independent confirmation for a new application and when it does not. Human factors evidence ranges from fundamental truths, which remain valid for all applications because they are based on unchanging human attributes, to findings that are specific to, and valid for, a single application only. Some human factors recommendations are enduring, but others date quickly as they succumb to technological or other advances. Which human factors problems are unique, and do they require unique solutions? These are real human factors issues in aviation, with major practical consequences for the effective deployment of human factors effort.

Typically, in aviation and elsewhere, there are not sufficient human factors resources to accomplish all that needs to be done. Some of the most crucial human factors decisions therefore concern the optimum allocation of the limited resources available. This entails costing the actual or putative benefits of human factors contributions in relation to the costs of providing them, though the tools for doing so may be rudimentary. Necessary progress towards better costing is now being accomplished, and the quest is for tangible benefits and practical improvements that can be quantified and costed. Human factors contributions to aviation cannot ignore costs. If a technological innovation requires the installation of new airborne or ground-based equipment, strong evidence is needed beforehand that the outlay will be worthwhile and the costs justified. Such evidence includes significant human factors data to provide reassurance about efficiency, performance, safety, and user acceptability. The potential benefits of standardization in a cost-conscious era may extend to improved safety and performance, better transfer of training, fewer human errors, and enhanced prospects for the international usage and marketing of systems.

The chapters in this text are grouped into five broad sections. The introductory section sets the scene by describing the evolution of aviation human factors, by considering the supporting role of research and development, by tackling issues of measurement, by discussing the problems of ensuring the safety and correctness of what is done, and by demonstrating the relevance of organizational and management settings to what can be achieved.

Part II presents a succession of themes dealing with the accommodation of human capabilities and attributes within aviation settings. Mental processes underlying human performance are synthesized and integrated into a hierarchy that shows the interrelationships among them. The progressive automation that characterizes much of aviation has to foster and be compatible with human task performance. Many aviation activities are accomplished through the collaborative efforts of a whole team of people, and these have to be managed as a resource. People in aviation working irregular hours encounter problems of fatigue and the disruption of diurnal rhythms. Aviation tasks are accomplished by relating existing knowledge and the understanding of current situations to previous and future scenarios. People capable of doing aviation jobs have to be selected and appropriately trained.

Part III deals with aircraft. At one time, work on aircraft constituted almost the whole of aviation human factors, and it is still the predominant application. The pilot's performance in achieving the flight objectives in the environment of the cockpit is studied, and the cockpit is considered as a workspace with tasks, information sources, communications, and controls. Simulation is appraised as a tool to represent selected features and functioning of the cockpit for various objectives. Human factors studies now embrace aspects of the design and accessibility of aircraft cabins. Helicopter flight raises some specific human factors issues.

The fourth part covers human factors work on air traffic control. A comparatively modest amount of such work has been done for a long time, but it has become accepted as part of aviation human factors only quite recently. Principles of air traffic control are described. The main themes of human performance and the impact of automation are addressed in relation to the air traffic controller. The introduction of data links will change greatly the communications between air and ground, and raises human factors problems of integration for both the pilot and the controller.

Part V presents a compendium of recent and extended applications of human factors to aviation. Developments in avionics and in their associated human–machine interfaces provide more sophisticated support for many human aviation activities. Meteorology has also been affected by automation, and developments in the presentation of weather information are viewed in relation to aviation activities. Comparatively recently, human factors principles have been applied more extensively to maintenance procedures in aviation. The increased emphasis on security and its job attributes in aviation prompts examination of its procedures and effectiveness in human factors terms. Systematic efforts to learn human factors lessons by studying accidents and incidents should benefit future flight safety. Human factors roles and evidence at judicial and other formal procedures and investigations related to aviation are described.

Collectively, these chapters include a broad range of human factors expertise, all of it relevant to aviation in some way. It is probably no longer feasible for most human factors specialists in aviation to broaden their expertise sufficiently to possess detailed, up-to-date, professional knowledge of every facet of aviation human factors covered in this text. There would be far too much to learn. A comprehensive human factors audit of any workspace in aviation may usually require the expertise of more than one specialist in the future, because so many diverse influences are relevant, with a large body of evidence on each. Perhaps all texts on aviation human factors must now be written by quite a large group of authoritative specialists in their own spheres. Full coverage of aviation human factors may always need more than one editor or more than one teacher.

The editors would like to thank all the authors who accepted the invitation to contribute to this text, sometimes after a little coaxing, as they are all busy people with many commitments. The effort made by Lloyd Hitchcock, who completed his chapter on time despite failing health, was particularly appreciated. Sadly, Lloyd died during the production stages of the book, and it is hoped that his chapter, his last publication, will serve as a fitting tribute to his professionalism and integrity. The editors also acknowledge the encouragement and constructive advice of Barry Kantowitz, the Series Editor, and assistance at Embry-Riddle Aeronautical University from Elaine Gardner, Sue Buman, and Heather Duckett.

I

INTRODUCTION

1

A Historical Overview of Human Factors in Aviation

Jefferson M. Koonce
University of Central Florida

Human factors in aviation are involved with the study of the human's capabilities, limitations, and behaviors and the integration of that knowledge into the systems we design for them with the goals of enhancing safety, performance, and the general well-being of the operators of the systems (Koonce, 1979).

THE EARLY DAYS: PRE-WORLD WAR I (CUTTING THEIR TEETH)

The role of human factors in aviation has its roots in the earliest days of aviation. The pioneers of aviation had their concerns for the welfare of those who flew their aircraft (particularly themselves), and as the capabilities of the vehicles were expanded, the aircraft rapidly exceeded human capability of directly sensing and responding to the vehicle and the environment to effectively exert sufficient control to ensure the optimum outcome, safety of flight. The first flight of only 12 sec. in which Orville Wright flew 540 ft. was on Thursday, December 17, 1903. The fourth and final flight of that day was made by Wilbur for 59 sec. and traversed 825 ft!

The purposes of aviation were principally adventure and discovery. To see an airplane fly was indeed unique; to actually fly an airplane was a daring feat! The early pioneers did not take it lightly, for to do so meant flirting with death in these fragile unstable craft. Thus, the earliest aviation was restricted to relatively straight and level flight and fairly level turns. The flights were performed under visual conditions in places carefully selected for elevation, clear surroundings, and certain breeze advantages to get the craft into the air sooner and land at the slowest possible ground speed.

The greatest problems with early flight were the reliability of the propulsion system and the strength and stability of the airframe. Many accidents and some fatalities occurred because of the structural failure of an airplane component or the failure of the engine to continue to produce power.

Although human factors were not identified as a scientific discipline at this time, there were serious human factors problems in the early stages of flight. The protection of the pilot from the elements, as he sat out in his chair facing them head on, was merely a transfer of technology from bicycles and automobiles. The pilots wore goggles, topcoats, and gloves similar to those used when driving the automobiles of that period.

The improvement of the human–machine interface was largely an undertaking of the designers, builders, and fliers of the machines (the pilots themselves). They needed some critical information to ensure proper control of their craft and some feedback about the power plant. At first, the aircraft did not have instrumentation. The operators directly sensed the attitude, altitude, and velocity of the vehicle and made their inputs to the control system to achieve certain desired goals. But in the 2 years following the first flight, the Wright brothers spent considerable effort trying to provide the pilot with information that would aid in keeping the airplane coordinated, especially in turning flight where the lack of coordinated flight was most hazardous. Soon these early craft had a piece of yarn or other string trailing from one of the struts of the airplane to provide yaw information as an aid in avoiding the turn-spin threat, and the Wright brothers came up with the incidence meter, a rudimentary angle of attack or flight path angle indicator.

As the altitude capabilities and range of operational velocities increased, the ability of the human to accurately sense the critical differences did not commensurately increase. Thus, early instrumentation was devised to aid the operator in determining the velocity of the vehicle and the altitude above the ground. The magnetic compass and barometric altimeter, pioneered by balloonists, soon found their way into the airplanes. Additionally, the highly unreliable engines of early aviation seemed to be the source of demise to many aviators. Either mechanical failure of the engine or propeller or interruption of the flow of fuel to the engine from contaminants or mechanical problems led to the introduction of the tachometer, to provide engine speed to the pilot, and of gauges about critical temperatures and pressures of the engine's oil and coolant.

WORLD WAR I (DARING KNIGHTS IN THEIR AERIAL STEEDS)

The advantages of an aerial view and ability to drop bombs on ground troops from above gave the airplane a unique role in the First World War. Although still in its infancy, the airplane made a significant contribution to the war on both sides, and it became an object of wonder while thousands of our nation's youth aspired to become aviators. The roles of the airplane were principally those of observation, attack of ground installations and troops, and air-to-air aerial combat. The aircraft themselves were strengthened to take the increased G-loads imposed by the combat maneuvering and the increased weight of ordinance payloads.

Pilots had to possess special abilities to sustain themselves in this arena. The human factors problems in the selection of pilot candidates had begun to emerge. Originally, family background, character traits, athletic prowess, and recommendations from significant persons secured an individual applicant a position in pilot training. Being a good hunter indicated an ability to lead and shoot at other moving targets, and strong physique and endurance gave promise of ability to endure the rigors of altitude, heat and cold, and the forces of aerial combat. Additionally, the applicant needed to be brave and show courage.

Later, psychologists began to take a more systematic and scientific approach to the classification of individuals and assignment to various military specialties. The aviation medicine people became concerned about the pilots' abilities to perform under extreme climatic conditions (the airplanes were open cockpits without heaters) and the effects of altitude on performance. During this period, industrial engineers began to utilize knowledge about human abilities and performance to improve factory productivity in the face of significant changes in the composition of the work force. Women began to play a major role in this area. Frank Gilbreth, an industrial engineer, and his wife Lillian, a psychologist, teamed up to solve many questions about the improvement of human performance in the work place, and the knowledge gained was useful in industry as well as to the armed forces.

Early in the war, it became apparent that the Allied forces were losing far more pilots to accidents than to combat. In fact, two thirds of all of the aviation casualties were not due to engagement in combat. The failure of the airframes or engines, mid-air collisions, and weather-related accidents (geographical or spatial disorientation) took the greater toll. However, the performance of individuals also contributed significantly to the number of accidents. Fortunately, with the slower airspeeds of the airplanes at that time and the light crushable structure of the airframe itself, many aviators in initial flight training who crashed and totaled an airplane or two still walked away from the crash(es) and later earned their wings. Of course, with the cost of today's airplanes that would hardly be the case.

The major human factors problems of the World War I era were the selection and classification of personnel, the physiological stresses upon the pilots, and the design of the equipment to ensure mission effectiveness and safety. The higher altitude operations of these airplanes, especially the bombers, pushed the development of liquid oxygen converters, regulators, and breathing masks. Also, due to the size and weight of these oxygen systems, they were not utilized in the fighter aircraft. Cold weather flying gear, flight goggles, and rudimentary instruments were just as important as improving the reliability of the engines and the strength and crash-worthiness of the airframes. To protect the pilots from the cold, leather flight jackets or large heavy flying coats, leather gloves, and leather boots, some fur lined, were used. While wearing all of this heavy clothing, the thoughts of wearing a parachute were out. In fact, many pilots thought that it was not sporting to wear a parachute, and such technologies were not well developed.

The experience of the British was somewhat different from other reported statistics of World War I: "The British found that of every 100 aviation deaths, two were by enemy action, eight by defective airplanes, and 90 for individual defects, 60 of which were a combination of physical defects and improper training" (Engle & Lott, 1979, p. 151). One explanation offered is that, of these 60, many had been disabled in France or Flanders before going to England and joining the Royal Air Corps.

BARNSTORMING ERA (THE THRILL OF IT ALL)

After the war, these aerial cavalrymen came home to the admiration of the public. Stories of great heroism and aerial combat skills had preceded them so that they came home to a public eagerly awaiting the opportunity to talk to these aviators and see demonstrations of their aerial daring. From this setting came the post-World War I barnstorming era.

The airplanes themselves changed in that more of them had enclosed cabins for passengers and often the pilot's cockpit was also enclosed. Instead of the variations on the box kite theme of the earliest airplanes, those after World War I were more aerodynamic, more rounded in design than box-like. Radial engines became the more popular means of propulsion, and they were air cooled as opposed to the earlier heavy water-cooled engines. With greater power-to-weight ratios these airplanes were more maneuverable and could fly higher, faster, and farther than their predecessors.

Flying became an exhibitionist activity, a novelty, and a source of entertainment. Others had visions of it as a serious means of transportation. The concept of transportation of persons and mail via air was in its infancy, and this brought many new challenges to the aviators. The commercial goals of aviation came along when the airplanes became more reliable and capable of staying aloft for longer durations, making distant places within easier, but relatively uncomfortable, reach. The major challenges were the weather and navigation under unfavorable conditions of marginal visibility.

Navigation over great distances over unfamiliar terrain became a real problem. Much of the western United States and some parts of the central and southern states were not well charted. In the older days where one flew around one's own barnyard or local town, getting lost was not a big concern. However, to fly hundreds of miles away from home, pilots used very rudimentary maps or hand-sketched instructions and attempted to follow roads, rivers, and railway tracks. Getting lost was indeed a problem. IFR flying in those days probably meant I Follow Roadways instead of Instrument Flight Rules!

Writing on water towers, the roofs of barns, municipal buildings, hospitals, or airport hangars identified the cities. As pilots tried to navigate at night natural landmarks and writing on buildings became less useful and tower beacons came into being to "light the way" for the aviator. The federal government had an extensive program for the development of lighted airways for the mail and passenger carriers. The color of the lights and the flashing of codes on the beacons were the means of identifying the particular airway that one was following. In the higher, drier southwestern United States some of the lighted airway beacons continued to stand into the 1950s. Runway lighting replaced the use of automobile headlights or brush fires to identify the limits of a runway at night. Under the low visibility of fog, haze, and clouds, however, even these lighted airways and runways became less useful, and new means of navigation had to be provided to guide the aviators to the airfields.

Of course, weather was still a severe limitation to safe flight. Protection from icing conditions, thunderstorms, and low ceilings and fog were still problems. From the developments resulting from the war effort, there were improved meteorological measurement, plotting, forecasting, and dissemination of weather information. In the 1920s, many expected that "real pilots" were capable of flying at night and into clouds without the aid of any special instruments. But, there were too many instances of pilots flying into clouds or at night without visual reference to the horizon, which resulted in them entering a spiraling dive (graveyard spiral) or spinning out of the clouds too late to recover before impacting the ever-waiting earth. In 1929, Lt. James Doolittle managed to take off, maneuver, and land his airplane solely by reference to instruments inside the airplane's cockpit. This demonstrated the importance of basic attitude, altitude, and turn information to maintaining the airplane right-side-up when inside clouds or in other situations where a distinct external world reference to the horizon is not available.

Much research had been performed on the effects of high altitude on humans (Engle & Lott, 1979), going back as early as the 1780s when the English surgeon Dr. John

Sheldon studied the effects of altitude upon himself in balloon ascents. In the 1860s, the French physician Dr. Paul Bert, later known as the "father of aviation medicine," performed altitude research on a variety of animals and himself in altitude chambers that he designed. During this post-World War I era, airplanes were capable of flying well over 150 miles per hour and at altitudes of nearly 20,000 ft., but little, other than oxygen breathing bags and warm clothing, had been provided to ensure safety at high altitudes. Respiratory physiologists and engineers worked hard to develop a pressurized suit that would enable pilots to maintain flight at very high altitudes. These technologies were "spinoffs" from the deep sea diving industry. On August 28, 1934, in his supercharged Lockheed Vega *Winne Mae*, Wiley Post became the first to fly an airplane while wearing a pressure suit. He made at least 10 subsequent flights and attained an unofficial altitude of approximately 50,000 ft. In September 1936, Squadron Leader F. D. R. Swain set an altitude record of 49,967 ft. Later, in June 1937, Flight Lt. M. J. Adam set a new record of 53,937 ft.

Long endurance and speed records were attempted, one after another. Problems of how to perform air-to-air refueling were faced as well as the stresses that long-duration flight imposed upon the engines and the operators. In the late 1920s, airplanes managed to fly over the North and South Poles and across both the Atlantic and Pacific Oceans. From the endurance flights came the development of the early autopilots in the 1930s. Of course, these required reliable electrical systems on the aircraft and imposed certain weight increases that were generally manageable on the larger multiengine airplanes. This is thought of as being the first introduction of automation into airplanes, which continues today.

In 1939, the Medical Research Council requested that Professor Sir Frederic Bartlett, head of the Cambridge University Psychological Laboratory, develop and execute a program of research on human performance. This resulted in numerous studies on selection, human performance, the effects of physiological and psychological stressors on performance, vigilance, and visual and vestibular systems.

THE WORLD WAR II ERA (SERIOUS BUSINESS)

Despite the heyday of the barnstorming era, military aviation had shrunk after the United States had won "the war to end all wars." The wars in Europe in the late 1930s stimulated American aircraft designers to plan ahead, advancing the engine and airframe technologies for the development of airplanes with capabilities far superior to those that were left over from the First World War.

The "necessities" of World War II resulted in airplanes capable of airspeeds four times as fast as those of World War I and with the shifted impellers and turbochargers altitude capabilities exceeded 30,000 ft. With the newer engines and airframes came much greater payload and range capabilities. The environmental extremes of high altitude, heat, and cold became the challenges to designers for the safety and performance of aircrew members. Also, land-based radio transmitters greatly improved cross-country navigation and instrument landing capabilities, as well as communications between airplanes and between the airplane and persons on the ground responsible for aircraft control. Ground-based radar was developed to alert the Allied forces to incoming enemy aircraft and as an aid to guide aircraft to their airfields. Also, radar was installed in aircraft to navigate to their targets when the weather prevented visual "acquisition" of the targets.

The rapid expansion of technologies brought far many more problems than ever imagined. Although the equipment had advanced, the humans who were being selected and trained to operate this equipment had not significantly changed. Individuals who had not moved faster than 30 miles per hour in their lifetime were soon being trained to operate vehicles capable of speeds 10 times as fast and far more complex than anything they had experienced. The art and science of selection and classification of individuals from the general population to meet the responsibilities of maintaining and piloting the new aircraft had to undergo significant changes. In the process of having to process hundreds of thousands of individuals, the selection and classification centers became a source of great amounts of data about human skills, capabilities, and limitations. Much of these data are documented in a series of 17 "blue books" of the U.S. Army Air Force Aviation Psychology Program (Flanagan, 1947). Another, more broad, source of information on the selection of aviators is the North and Griffin (1977) *Aviator Selection 1917–1977.*

A great deal of effort was put forth in the gathering of data about the capabilities and limitations of humans and the development of guidelines for the design of displays and controls, environmental systems, equipment, and communication systems. Following the war, *Lectures on Men and Machines: An Introduction to Human Engineering* by Chapanis, Garner, Morgan, and Sanford (1947), Paul Fitts' "blue book" on *Psychological Research on Equipment Design* (1947), and the *Handbook of Human Engineering Data for Design Engineers* prepared by the Tufts College Institute for Applied Experimental Psychology and published by the Naval Special Devices Center (1949) helped to disseminate the vast knowledge regarding human performance and equipment design that had been developed by the early human factors psychologists and engineers (Moroney, 1995).

Stevens (1946), in his article "Machines Cannot Fight Alone," wrote of the development of radar during the war. "With radar it was a continuous frantic race to throw a better and better radio beam farther and farther out, and to get back a reflection which could be displayed as a meaningful pattern before the eyes of an operator" (p. 391). As soon as technology would make a step forward a human limitation would be encountered or the enemy would devise some means of degrading the reflecting signal so that it would be virtually useless. Often weather conditions would result in reflections from the moisture in the air which would reduce the likelihood of detecting a target. In addition to the psychophysical problems of detecting signals in the presence of "noise" there was the well-known problem that humans are not very good at vigilance tasks.

Without pressurization, the airplanes of World War II were very noisy, and speech communications were most difficult in the early stages. At the beginning of the war, the oxygen masks did not have microphones built in them, so throat microphones were utilized, making speech virtually unintelligible. The headphones that provided information to the pilots were "leftovers" from the World War I era and did little to shield out the ambient noise of the airplane cockpit.

In addition to the noise problem, as one might expect, there was a great deal of vibration, which contributed to apparent pilot fatigue. Stevens (1946) mentioned that a seat was suspended such that it "floated in rubber" to dampen the transmission of vibrations from the aircraft to the pilot. Although technically successful, the seat was not acceptable to the pilots because it isolated them from a sense of feel of the airplane.

Protecting the human operator while still permitting a maximum degree of flexibility to move about and perform tasks was also a major problem (Benford, 1979). The necessity to protect aviators from anti-aircraft fire from below was initially met with the installation of seat protectors—plates of steel built under the pilot's seat to deflect rounds coming up from below. For protection from other than below, B. Gen. Malcolm C. Grow, surgeon of the 8th Air Force, got the Wilkinson Sword Company, designer of early suits of armor, to make body armor for B-17 aircrew members. By 1944, B-17 crews with body armor had a 60% reduction in men wounded.

Dr. W. R. Franks developed a rubber suit with a nonstretchable outer layer to counter the effects high G-forces on the pilot. The Franks flying suit was worn over the pilot's underwear and was filled with water. As the G-forces would increase, they would also pull the water down around the lower extremities of the pilot's body, exerting pressure to help prevent the pooling of blood. In November 1942, this was the first G-suit worn in actual air operations. Because of the discomfort and thermal buildup in wearing the Franks suit, pneumatic anti-G suits soon followed. One manufacturer of these pneumatic G-suits, David Clark Co. of Worcester, Massachusetts, later became involved with the production of microphones and headsets. The Gradient Pressure Flying suit, Type NS-9 or G-1 suit, was used by the Air Force in the European theater in 1944.

The training of aviators to fly airplanes soon integrated flight simulators into the program. Although flight simulation began as early as 1916, the electromechanical beginning of the modern flight simulator was the invention of E. A. Link in 1929 (Valverde, 1968). The Link Trainer, affectionately known as the "Blue Box," was used extensively during World War II, particularly in the training of pilots to fly under instrument conditions.

Although the developments in aviation were focused principally on military applications during this time, civilian aviation was quietly advancing in parallel to the military initiatives. Some of the cargo and bomber aircraft proposed and built for the military were also modified for civilian air transportation. The DC-3, one of the most popular civil air transport aircraft prior to the war, was the "workhorse" of World War II, used for the transportation of cargo and troops around the world. After the war, the commercial airlines found that they had a large experienced population from which they could select airline pilots. However, there were few standards to guide them in the selection of the more appropriate pilots for the tasks of commercial airline piloting: passenger comfort, safety, and service. McFarland (1953), in *Human Factors in Air Transportation*, provided a good review of the status of the commercial airline pilots selection, training, and performance evaluation, as well as aviation medicine, physiology, and human engineering design. Gordon (1949) noted the lack of selection criteria to discriminate between airline pilots who were successful (currently employed) and those who were released from the airlines for lack of flying proficiency.

The problems of air traffic control in the civilian sector were not unlike those in the operational theater. As radar was developed and used for military purposes, it later became integrated into the civilian air traffic control structure. There were the customary problems of ground clutter, precipitation attenuating the radar signals, and the detection of targets. Advances in communications between the ground controllers and the airplanes as well as communications between ground control sites greatly facilitated the development of the airways infrastructure and procedures that continue on to this date. Hopkins (1995) provided an interesting and rather complete review of the history of human factors in air traffic control.

Following the war, universities got into the act with the institution of aviation psychology research programs sponsored by the government (Koonce, 1984). In 1945, the National Research Council's Committee on Selection and Training of Aircraft Pilots gave a grant to the Ohio State University to establish the Midwest Institute of Aviation. In 1946, Alexander C. Williams founded the Aviation Psychology Research Laboratory at the University of Illinois, and Paul M. Fitts opened the Ohio State University's Aviation Psychology Laboratory in 1949. These and other university research programs in aviation psychology and human engineering attracted veterans returning from the war to use the G.I. Bill to go to college, advance their education, and work in the area of human factors psychology and engineering.

Although developed under the blanket of secrecy, toward the end of World War II jet aircraft made their debut in actual combat. These jet airplanes gave our imaginations a glimpse of what was to come in terms of aircraft altitude and airspeed capabilities of military and civilian aircraft in the near future.

THE JET ERA (NEW HORIZONS)

The military airplanes developed after World War II were principally jet fighters and bombers. The inventory was "mixed" with many of the piston engine airplanes being left over, but as the United States approached the Korean War the jet aircraft became the prominent factor in military aviation. Just before World War II, Igor Sikorsky developed a successful helicopter. During the Korean War, the helicopters found widespread service. These unique flying machines were successful, but tended to have a rather high incidence of mechanical problems, which were attributed to the reciprocating engines that powered them. The refinement of the jet engine and use of it in the helicopters made them much more reliable and in more demand, both within the armed forces as well as in the civilian sector.

Selection and classification of individuals brought into the military changed little from the advances that had been made during the pressure of World War II. Moving into the jet era of aviation did not have a significant effect on selection and classification procedures until the advent of personal computers. Commercial air carriers typically sought their pilots from those who had been selected and trained by the armed forces. These pilots had been through rigorous selection and training criteria, were very standardized, had good leadership skills, and generally possessed a large number of flight hours.

Boyne (1987) described the early entry of the jet airplanes into commercial air travel. In the United States, aircraft manufacturers were trying to develop the replacement for the fabled DC-3 in the form of various two- and four-radial-engine propeller airplanes. There were advances in the range that the airplanes could fly without refueling, the speed was increased, and most of the airplanes soon had pressurization for passenger safety and comfort. In the meantime, Great Britain's Vickers-Armstrong came out with the Vicount in 1950, a four-engine turboprop airplane that provided much faster, quieter, and smoother flight. Soon thereafter, in 1952, the deHavilland Comet 1A entered commercial service. The Comet was an innovative full jet airliner capable of carrying 36 passengers at 500 miles per hour between London and Johannesburg. These entries into the jet era had a significant impact on America's long-standing prominence in airline manufacturing. After two in-flight breakups of Comets in 1954, deHavilland had difficulty promoting any airplane with the name Comet. With that, the focus of interest in airliner production shifted back to the United States, where Boeing, which had experience in developing and building the B-47 and B-52 jet bombers, made its entry into the commercial jet airplane market. In 1954, the Boeing 367-80 prototype of the resulting Boeing 707 made its debut. The Boeing 707 could economically fly close to Mach 1 and was very reliable but expensive. Later, Convair came out with its model 880 and Douglas made its DC-9, both closely resembling the Boeing 707 (Josephy, 1962).

The introduction of jet airplanes brought varied responses from the pilots. A number of the pilots who had served many years flying airplanes with reciprocating engines and propellers exhibited some "difficulties" in transitioning to the jet airplanes. The jet airplanes had fewer engine instruments for the pilots to monitor, fewer controls for the setting and management of the jet engines, and, with the advancement of technology, more simplistic systems to control. However, the feedback to the pilot was different between piston propeller and jet airplanes. The time to accelerate (spool-up time) with the advance of power was significantly slower in the jet airplanes, and the time with which the airplane transited distances was significantly decreased. Commercial airlines became concerned with the human problems in transition training from propeller to jet airplanes, wondering if there should be an age limit for pilots transitioning to jet airplanes. Today, that "problem" seems to be no longer an issue. With the advent of highly sophisticated flight simulators and other training systems and jet engines that build up their thrust more rapidly, there has been little written recently about difficulties of transition training from propeller airplanes to jet airplanes.

Eventually, the jet era resulted in reductions in the size of the flight crews required to manage the airplanes. In the "old days" the transoceanic airliners required a pilot, a copilot, a flight engineer, a radio operator, and a navigator; the jet airliners required only a pilot, copilot, and in some instances a flight engineer. With the aid of computers and improved systems engineering, many of the jet airplanes that previously had three flight crew members eliminated the need for a flight engineer and now require only two pilots.

The earlier aircraft with many crew members, sometimes dispersed and out of visual contact with each other, required good communication and coordination skills, which were "trained" in such programs as crew coordination training (CCT). But with the reduction in the number of crew members and placing them all within hand's reach of each other, lack of "good" crew coordination, communication, and utilization of available resources became a real problem in the jet airline industry. The tasks of interfacing with the on-board computer systems, through the Flight Management System (FMS), changed the manner in which the flight crewmembers interact. Reviews of accident data and reports into the Aviation Safety Reporting System (ASRS) (Foushee, 1984; Foushee & Manos, 1981) revealed crew coordination as a "new" problem. Since the mid-1980s, much has been written about crew resource management (CRM; Weiner, Kanki, & Helmreich, 1994), and the Federal Aviation Administration (FAA) has issued an Advisory Circular 120-51B (FAA, 1995) for commercial air carriers to develop CRM training. Despite over 10 years of research, programs, and monies, there still seems to be a significant problem with a lack of good CRM behaviors in the cockpits.

The jet engines have proven to be much more reliable than the piston engines of the past. This has resulted in a reliance on their safety and sometimes a level of complacency and disbelief when things go wrong. With highly automatized systems and reliable equipment the flight crew physical workload has been significantly reduced; however, there seems to have been a resultant increase in the cognitive workload.

THE NEW TECHNOLOGY ERA (THE COMPUTER IN THE COCKPIT)

Here we are in the 1990s, and although many things have changed in aviation, many other things have not. The selection of pilots for the armed forces is still about as accurate as it has been for the past 40 years. There are new opportunities and challenges in selection and classification since women are now permitted to be pilots in the military, and they are not restricted from combat aircraft. The selection and classification tests developed and refined over the past 40 years on males might not select the females with the greatest likelihood of successfully performing as pilots (McCloy & Koonce, 1982). Human factors engineers will have to reconsider the design of aircraft cockpits for a wider range of anthropometric dimensions, and the development of personal protective and life support equipment to include females is a pressing need.

With the advent of the microcomputers and flat panel display technologies, the aircraft cockpits of the modern airplanes are vastly different from those of the past. The navigational systems are extremely precise, and they are integrated with the autopilot systems for the capability of fully automated flight from just after takeoff to after the airplane touches down on the runway. The pilot is becoming a passive observer of the airplane's systems while the automation does the flying. A challenge for the designers is what to do with the pilot during the highly automated flight (Mouloua & Koonce, 1997).

Recently there has been a great amount of attention to the concept of situation awareness in the advanced airplanes (Garland & Endsley, 1995). Accidents have occurred in which the flight crew members were not aware of their location with respect to dangerous terrain or were unaware of the current status of the airplane's systems when that knowledge was essential for correct decision making. A large amount of basic research has been initiated to learn more about individual differences in situation awareness, the potential for selection of individuals with that capability, and the techniques for improving one's situation awareness. In retrospect, much of the literature is reminiscent of the earlier research on attention and decision making.

Looking forward, human factors practitioners will have numerous challenges, from the effects of advanced display technologies and automation at all levels of aviation, right down to the general aviation recreational pilot. Efforts to invigorate general aviation so that it will be more affordable and thus attractive to a larger portion of the public will bring issues of selection and training down to the private pilot level, where, historically, a basic flight physical and a source of funds were all that were necessary to get into pilot training.

Economics are pushing a restructuring of the way in which the airspace system works (Garland & Wise, 1993; Hopkin, 1995). Concepts such as data links between controlling agencies and the aircraft they control, free flight to optimize flight efficiency, comfort and safety, automation of weather observation and dissemination, and modernization of the air traffic controllers workstations will all require significant inputs from aviation human factors practitioners in the near future.

The future supersonic aircraft, to reduce drag and weight costs, might not provide windows for forward visibility, but might provide an enhanced or synthetic visual environment which the pilots can "see" to maneuver and land their airplanes. Other challenges might include the handling of passenger loads of 500 to 600 persons in one airplane, the design of the terminal facility to handle such airplanes, waiting and loading facilities for the passengers, and the systems for handling the great quantity of luggage and associated cargo. Of course, the future problems in airport security will have to be faced by planners and design teams including human factors practitioners.

REFERENCES

Benford, R. J. (1979). *The heritage of aviation medicine.* Washington, DC: The Aerospace Medical Association.
Boyne, W. J. (1987). *The Smithsonian book of flight.* Washington, DC: Smithsonian Books.
Chapanis, A., Gardner, W. R., Morgan, C. T., & Sanford, F. H. (1947). *Lectures on men and machines: An introduction to human engineering.* Baltimore, MD: Systems Research Laboratory.
Engle, E., & Lott, A. S. (1979). *Man in flight: Biomedical achievements in aerospace.* Annapolis, MD: Leeward.
Federal Aviation Administration. (1995, March). *Crew resource management training* (Advisory Circular AC 120-51B). Washington, DC: Author.
Fitts, P. M. (1947). *Psychological research on equipment design* (Research Rep. No. 17). Washington, DC: Army Air Forces Aviation Psychology Program.
Flanagan, J. C. (1947). *The Aviation Psychology Program in the Army Air Force* (Research Rep. No. 1). Washington, DC: Army Air Forces Aviation Psychology Program.
Foushee, C. J. (1984). Dyads and triads at 35,000 feet. *American Psychologist, 39,* 885–893.
Foushee, C. J., & Manos, K. L. (1981). Information transfer within the cockpit: Problems in intracockpit communications. In C. E. Billings & E. S. Cheaney (Eds.), *Information transfer problems in the aviation system* (NASA Rep. No. TP-1875; pp. 63–71). Moffett Field, CA: NASA-Ames Research Center.
Garland, D. J., & Endsley, M. R. (Eds.). (1995). *Experimental analysis and measurement of situation awareness.* Daytona Beach, FL: Embry-Riddle Aeronautical University Press.
Garland, D. J., & Wise, J. A. (Eds.). (1993). *Human factors and advanced aviation technologies.* Daytona Beach, FL: Embry-Riddle Aeronautical University Press.
Gordon, T. (1949). The airline pilot: A survey of the critical requirements of his job and of pilot evaluation and selection procedures. *Journal of Applied Psychology, 33,* 122–131.
Hopkin, V. D. (1995). *Human factors in air traffic control.* London: Taylor & Francis.
Josephy, A. M., Jr. (Ed.). (1962). *The american heritage history of flight.* New York: American Heritage.
Koonce, J. M. (1979, September). Aviation psychology in the U.S.A.: Present and future. In F. Fehler (Ed.), *Aviation psychology research.* Brussels, Belgium: Western European Association for Aviation Psychology.
Koonce, J. M. (1984). A brief history of aviation psychology. *Human Factors, 26*(5), 499–506.
McCloy, T. M., & Koonce, J. M. (1982). Sex as a moderator variable in the selection and training of persons for a skilled task. *Journal of Aviation, Space, and Environmental Medicine, 53*(12), 1170–1173.
McFarland, R. A. (1953). *Human factors in air transportation.* New York: McGraw-Hill.
Moroney, W. F. (1995). The evolution of human engineering: A selected review. In J. Weimer (Ed.), *Research techniques in human engineering* (pp. 1–19). Englewood Cliffs, NJ: Prentice-Hall.
Mouloua, M., & Koonce, J. M. (Eds.). (1997). *Human–automation interaction: Research and practice.* Mahwah, NJ: Lawrence Erlbaum Associates.
Naval Special Devices Center. (1949, December). *Handbook of human engineering data for design engineers.* Prepared by the Tufts College Institute for Applied Experimental Psychology (NavExos P-643, Human Engineering Rep. No. SDC 199-1-2a). Port Washington, NY: Author.
North, R. A., & Griffin, G. R. (1977). *Aviator selection 1917–1977* (Technical Rep. No. SP-77-2). Pensacola, FL: Naval Aerospace Medical Research Laboratory.
Stevens, S. S. (1946). Machines cannot fight alone. *American Scientist, 334,* 389–400.
Valverde, H. H. (1968, July). *Flight simulators: A review of the research and development* (Technical Rep. No. AMRL-TR-68-97). Wright–Patterson Air Force Base, OH: Aerospace Medical Research Laboratory.
Weiner, E., Kanki, B. G., & Helmreich, R. (Eds.). (1993). *Cockpit resource management.* New York: Academic Press.

2

Aviation Research and Development: A Framework for the Effective Practice of Human Factors, or "What Your Mentor Never Told You About a Career in Human Factors . . ."

John E. Deaton
CHI Systems, Inc., Orlando, FL

Jeffrey G. Morrison
Space Warfare Systems Center, San Diego, CA

THE ROLE OF HUMAN FACTORS RESEARCH IN AVIATION

Since its meager beginnings in the chaos of the second world war, human factors has come to fill a substantial role in aviation. In fact, it is arguably in this domain that human factors has received its greatest acceptance as an essential part of the research, development, test, and evaluation cycle. This acceptance has come from the critical role humans, notably pilots, play in these person–machine systems, the unique problems and challenges these systems place on human perception, physiology, and cognition, and the dire consequences of human error in these systems. As a result, there have been numerous opportunities for the development of a science of human factors that has contributed significantly to the safety and growth of aviation.

Times are changing, and with the end of the Cold War, funding for human factors research and development is shrinking along with military spending. Being a successful practitioner in the field of human factors will require considerable skills that are beyond those traditionally taught as part of a graduate curriculum in human factors. New challenges are being presented that will require a closer strategic attention to what we do, how we do it, and what benefits accrue as a result of our efforts. This chapter offers snippets of what, we hope, passes for wisdom based on the authors' experience in the practice of human factors. It describes questions and issues that the successful practitioner of human factors must bear in mind to conduct research, development, testing, and engineering (RDT&E) in any domain. The bulk of the authors' experience is with the Department of Defense (DoD), and this is the basis of our discussion. Nonetheless, the lessons learned and advice should be applicable across other endeavors related to the science of human factors.

Focus Levels of RDT&E

An important part of succeeding as a human factors practitioner is recognizing the type of research being funded, and the expectancies a sponsor is likely to have for the work being performed. The DoD identifies four general categories of RDT&E and has specific categories of funding for each of these categories.[1] These categories of research are identified as 6.1 through 6.4, where the first digit refers to research dollars and the second digit refers to the type of work being done (Table 2.1). DoD sponsors are typically very concerned with the work being performed, as Congress mandates what will be done with the different categories of funding and has mechanisms in place to audit how it is spent. This issue is also relevant to the non-DoD practitioner as well, because regardless of the source of RDT&E funding, understanding the expectations that are attached to it is critical to successfully concluding a project. Therefore, the successful practitioner will understand how their projects are funded and the types of products expected for that funding.

Basic research is that typically thought of as being performed in an academic setting. Typically, a researcher has an idea that he or she feels would be of some utility to a sponsor, and obtains funding to try to explore the idea further. Alternatively, the work performed may be derived from existing theory but represents a novel implication of that theory. Human factors work at the 6.1 level will typically be done with artificial tasks and naive subjects, such as a university laboratory with undergraduate students as subjects. Products of such work may be theoretical development, a unique model or theory, and the work typically entails empirical research to validate the theory. This work is generally not focused with a particular application or problem, although it may be inspired by a real-world problem and may utilize a problem domain to facilitate the research. However, this research is not generally driven by a specific operational need; its utility for a specific application may only be speculated on. Examples of this type of research might ask questions such as:

- How do we model strategic decision making?
- How is the human visual perception process affected by the presence of artificial lighting?
- What impact do shared mental models have on team performance?

Applied research is still very much at the research end of the research–development spectrum; however, it is typically where an operational need or requirement first comes into the picture in a significant way. This research can be characterized as taking established theories or models shown to have some scientific validity, and exploring their use to solve a specific problem. Because of its applied flavor, it is common and advisable to have some degree of subject matter expertise involved with the project, and to utilize tasks that have at least a theoretical relationship to those of the envisaged

---

[1]In point of fact, these categories are being redefined as part of the downsizing and redefinition of the DoD procurement proces. For instance, there was until the early 1990s a distinction in the 6.3 funding between core-funded prototype demonstrations (6.3a) and actual field demonstrations (6.3b) that received specific funding from the Congressional budget. This distinction has been eliminated. The authors were unable to locate a specific set of definitions that are current as of the time this chapter was written. Therefore these definitions are based on the authors' current understanding of the DoD procurement system based on current practice rather than an official set of definitions.

TABLE 2.1
Types and Characteristics of DoD Research and Development

Number	*Type*	*Definition*	*Research Questions*	*Products*
6.1	Basic research	Research done to develop a novel theory or model, or extend existing theory into new domains. The work may be funded in interest in solving a specific problem; however, there is typically, no single application of the research that drives the work.	Can we take an idea and turn it into a testable theory? Can we assess the utility of a theory in understanding a problem?	Theoretical papers, describing empirical studies, mathematical models, recommendations for continued research, and discussion of potential applications.
6.2	Applied research	Research done to take an existing theory, model or approach, and apply it to a specific problem.	Can we take this theory/model and apply it to this problem to come up with a useful solution?	Rudimentary demonstrations, theoretical papers describing empirical studies, recommendations for further development.
6.3	Advanced development	Move from research to development of a prototype system to solve a specific problem.	Can we demonstrate the utility of technology in solving a real-world need? What are the implications of a proposed technology? Is the technology operationally viable?	Working demonstrations in operationally relevant environments. Assessment with intended users of the system. Technical papers assessing the operational requirements for the proposed system/technology.
6.4	Engineering development	Take a mature technology and develop a fieldable system.	Can we integrate and validate the new technology into existing systems? What will it cost? How will it be maintained?	The products of this stage of development would be a matured, tested system ready for procurement—notably, detailed specifications and performance criteria, life-cycle cost estimates, etc.
6.5	System procurement	Go out and support the actual buying, installation, and maintenance of the system.	Does it work as per specification? How do we fix the problems?	Deficiency reports and recommended fixes.

application being developed. Questions asked by this type of human factors research might include:

- How is command-level decision making in tactical commanders affected by time stress and ambiguous information?
- How should we use advanced automation in a tactical cockpit?
- How do we improve command-level decision making of Navy command and control staff?
- How can synthetic 3-D audio be used to enhance operator detection of sonar targets?

Advanced development is where the work starts moving away from research and toward development. Although demonstrations are often done as part of 6.2 and even 6.1 research, there is an implicit understanding that these demonstrations are not of fieldable systems to be used by specific operators. A major product of 6.3 R&D, however, is typically a demonstration of a fairly well thought out system in an operationally relevant test environment with the intended users of the proposed system. As a result, this type of research is typically more expensive than that which takes place at 6.1 or 6.2, and often involves contractors with experience and necessitates the involvement of subjects and subject-matter experts with operational experience related to the development that is going to take place. Research questions in advanced development are typically more concerned with demonstrating meaningful performance gains and the feasibility of transferring the underlying technology to fielded systems. Representative questions in 6.3 human factors research might include:

- Is the use of a decision support system feasible and empirically validated for tactical engagements?
- What are the technical requirements for deploying the proposed system in terms of training, integration with existing systems, and so on?
- What are the expected performance gains from the proposed technology and what are the implications for manning requirements based on those gains?

As the procurement process for a technology or system moves beyond 6.3, human factors will typically play a less and less dominant role. This is not to say that it should not be, or that it is not, necessary for human factors to have continued involvement in the RDT&E process. It is just that at the 6.4 level, most of the critical human factors issues are typically solved, and the mechanics of constructing and implementing technology tend to be the dominant issue. It becomes harder and harder (as well as more and more expensive) to implement changes as the system matures. As a result only critical shortcomings will be addressed by program managers in later stages of technology development. If we, as human factors practitioners, have been contributing appropriately through the procurement process, our relative involvement at this stage is not going to be problematic and will naturally be less prominent than it was earlier in the RDT&E process. Human factors issues still need to be addressed to ensure that the persons in the person–machine systems are not neglected. Typically at this stage of the procurement process we are concerned with testing issues such as compliance and verification. The questions asked become more related to testing and evaluation of the developed person–machine interfaces, documenting the final system, and development of required training curriculum. Thus, although it is imperative that human factors professionals continue to have a role, there are in fact few dedicated research and development dollars for them at the 6.4 and 6.5 stages. What funding is received for human factors at this stage typically comes from the project itself and is at the discretion of project management. Research done at these levels might ask questions related to:

- Can the persons in the system read the displays?
- What training curriculum is required for the people in the system to ensure adequate performance?
- What criteria should be used in selecting the individuals to work in this system?

DEVELOPMENT OF AN EFFECTIVE R&D PROGRAM

The R&D process is similar no matter what the application domain. Unfortunately, R&D managers often lose track of the real purpose of behavioral research: solving a problem. More to the point—sponsors want, and deserve to have, products that make their investment worthwhile. They (and you) need to know where you are and where you are going and have a pretty good sense of how you are going to get there. Keeping these issues in the forefront of your mind as a program manager or principle investigator will likely result in further support in the near future. Having said that, what makes an R&D program successful? One quickly comes to the realization that successful programs require the necessary resources, and that there is a "critical mass" of personnel, facilities, and equipment resources that must be available to be effective. It also is intuitively obvious that proper program management, including a realistic funding base, is crucial if research is to be conducted in an effective manner. But what factors do we often neglect to attend to, that may play a deciding role in defining the eventual outcome of the research program? What does one do when the resources do not match the magnitude of task required to get to the end goal?

You must understand your customers and their requirements. Often, particularly in the DoD domain, there are multiple customers with different, sometimes competing, and sometimes directly conflicting agendas. You must understand these customers and their needs and find a way to give them not only what they ask for or expect, but what they need. The successful practitioner will understand what they need, and sometimes will have to understand their needs better than they do if the project is to succeed. Needless to say this can be something of an art rather than a science, and often requires significant diplomatic skills. For example, in the DoD model there are typically two customers: the sponsors, or the people responsible for the money being spent in support of RDT&E, and the users, or those who will make use of the products of this effort. In the Navy, the former is typically the Office of Naval Research (ONR) and the latter is the Fleet. ONR is typically interested in the theory and science underlying the RDT&E process, and is interested in an audit trail whereby it can show (a) that quality science is being performed as measured by meaningful research studies and theoretical papers, and (b) the successful transition of the science through the various levels of the RDT&E process. The Fleet is also interested in transition, but it is interested in the applicability of the developed technology to solving its real-world needs in the near future. Thus the users are interested in getting the useful products out to the ships (or airplanes or whatever), and are less interested in the underlying science. The competing needs of these two requirements are often one of the most challenging aspects of managing a human factors project, and failure to manage them effectively is often a significant factor in the project's failure. One must understand where one's technology/research is in the RDT&E process and where it needs to go in order to be successful, and do whatever one can to facilitate its getting on to the next stage in the procurement process. Understanding this process and knowing what questions to ask from a management perspective are key to meeting your own objectives as a researcher/practitioner, as well as those of your sponsors/customers. How might this be accomplished?

First, we suggest that the successful human factors practitioner should emphasize providing information that best fits the nature of the problem and the environment in which it is to be applied. In other words, providing a theoretical treatment of an issue when the real problem involves an operational solution will not be met with overwhelming support. There has to be a fit between theory and application. That is not to say that theory does not have an important role to play in aviation human factors. The problems arise when researchers (usually more comfortable in describing issues conceptually) are faced with sponsors who want the "bottom line" and they want it now, not tomorrow. Those of us in academics may not be comfortable with this mindset. The solution is to become familiar with the operational issues involved, and how best to translate your input to the sponsor so that the sponsor can, in turn, communicate such information into something that can be meaningful to the user group in question.

Second, the most common reason that research programs get into trouble is that they propose to do more than is feasible given the resources available. Initially, one will get approving gestures from the sponsors, but what happens a year or two down the road when it becomes evident that initial goals were far too ambitious? Successful R&D efforts are underscored by their ability to meet project goals on time and within specified funding levels. Promising and not delivering is not a strategy that can be repeated twice! Therefore it is critical that the program manager keep track of where the program is, where it is committed to going, and the available resources available and required to get there. When there is a mismatch between available and required resources, the program manager must be proactive in redefining objectives, rescoping the project and/or obtaining additional resources. It is far better to meet the most critical of your research objectives, and have a few fall to the wayside (for good reason), than to have the entire project be seen as a failure. In recent years, many programs have been jeopardized less by reductions in funding than by the inability or unwillingness of program management to realistically deal with the effects of those cuts.

Third, and perhaps most important (certainly to the sponsor), is how you measure the effectiveness of a new system or technology that you have developed. This issue is often referred to as "exit criteria" and deals with the question: How do you know when you are done? This is by no means a trivial task, and can be critical to the success of obtaining and maintaining funding. Many projects are perceived as failing by sponsors not because they aren't doing good work, but because there is no clear sense as to when it will pay off. Measures of effectiveness (MOEs) to assess these exit criteria are often elusive and problematic. However, they do provide a method for assessing the efficacy of a new system. Determining the criteria that will be used to evaluate the usefulness of a system is a process that needs to be up front during the developmental stage. That way there are no "surprises" at the end of the road, where the system (theory) does wonderful things, but the customer doesn't understand why he or she should want it. A colleague once told me that the best he could imagine was a situation where there were no surprises at the end of a research project. It is interesting to note that such a statement runs against the grain of what one is taught in doing academic research. In academic research, we prize the unexpected discovery and are taught to focus on the identification of additional research. This is often the last thing that a user wants to hear; users want answers—not more questions. One of the most important things learned by novice practitioners is how to reconcile the needs of the customer with their research training.

Fourth, it is to your advantage to make personal contact (i.e., face to face) with your sponsor and supporting individuals. The people whose money you are spending will almost universally appreciate getting "warm fuzzies" that can only come from one-on-one contacts. New developments in the areas of communications (i.e., teleconferencing, e-mail, etc.) are no substitute for close contact with individuals supporting your efforts. As you become a proficient practitioner of human factors you will learn that there is no better way to sense what aspects of a project are of greatest interest to your customers, and which are problematic, than to engage in an informal discussion with them. Further, your value to the customer will be significantly increased if you are aware of hidden agendas and their priorities. Although many times these will not be directly relevant to you or your project, your sensitivity to them will make you much more effective as a practitioner. This will become painfully obvious when things go wrong. Your credibility is, in part, established through initial contact.

Fifth, do you have outside endorsements for the kind of work you are attempting? In other words, who really cares what you are doing? Generating high-level support from the intended users of your effort is indispensable in convincing sponsors that there is a need for such work. In the military environment, this process is de facto mandatory. Few projects receive continued funding unless they have the support of specific program offices within the DoD. Operational relevancy and need must be demonstrated if funding is to be secured, and defended in the face of funding cuts.

Sixth, interagency coordination and cooperation will undoubtedly enhance the probability of a successful research program. Your credibility as a qualified and responsible researcher depends on being aware of ongoing related work elsewhere and its relevance to the issues going on in your project. Generally, efforts made to leverage off this ongoing work to avoid duplication of effort have become more and more critical in this era of limited research and development resources. The lack of senior-level support and ineffective coordination among outside research organizations will in fact be a significant impediment to executing program goals. Through the use of coordinating and advisory committees, working groups, cooperative research agreements, and widespread dissemination of plans and products, duplication of effort should be minimized.

Finally, you must be prepared to discuss where your research will go subsequent to the conclusion of the project. What transition opportunities are available in both the civilian and military sectors? Describe the applicability of your work to other domains, particularly those of interest to your sponsors and customers. This is critical to building on any success achieved in a particular research project, and maintaining your credibility. Will there be additional follow-up work required? What other sponsors/customers would be interested in your findings/products? Who could most benefit from the results of your work? Extracting the critical information from your project and demonstrating how this will assist other work is often neglected once a project has been finished. The successful practitioner will not entirely walk away from an area once a particular project is finished, but will track its transitions, both planned and unplanned. An excellent way to build credibility and develop new contracts and funding opportunities is to contact those people whose work you are building on to (a) advise them of the utility of their work and (b) make them aware of your expertise and capability. Not only are these people generally flattered by the interest, but they may advocate you as a resource when they meet colleagues with similar interest.

SOME WORDS OF WISDOM REGARDING DEALING WITH THE SPONSOR, MANAGEMENT, AND USER

Be honest. Don't tell them what you think they want to hear—unless that bears some resemblance to reality. Be honest with yourself as well. There is nothing more dangerous to a project or an organization than someone who doesn't know what he or she is talking about. Trying to bluff your way through a discussion will only damage your credibility, and that of your cause, particularly if you are with people who do know what they are talking about. Colleagues and sponsors generally will not confront you with your ignorance, but they will be impressed by it—negatively. If you are not sure of something, the best bet is to ask an intelligent, appropriate question of an appropriate person, at the appropriate time and in the appropriate place. You can use this strategy to turn a potentially negative situation into a positive one by displaying your sensitivity, judgment, and wisdom despite your possible lack of technical knowledge.

Management really does not want to hear about your problems. If you must present a problem, it behooves you to have identified prospective solutions and to present the recommended solution with underlying rationale and implications for the decision. Deal with problems at the lowest level of management you can. Don't jump the chain of command. Tell higher levels of management about successes—not problems. When in doubt, document everything. It is in everyone's best interests in the midst of turbulence to document discussions, alternatives, and recommended solutions. This way, if the problem becomes terminal to your efforts, you have the ammunition to fend off accusations and blame, and to potentially demonstrate your widom and foresight.

If the problem being discussed is threatening to one's project or career, document this situation in the form of memos distributed to an appropriate group of individuals. Generally this will be to all the affected parties, with copies to supervisory personnel if that seems necessary. (Note that this is almost never appropriate for the first memorandum.) Memos of this nature must be well written and self-explanatory. Assume the reader knows nothing, particularly if you are going to use one of the most powerful features of a memo—the courtesy copies (cc:) routing. This is one of the best tools available to ensure that you have covered your backside, and that management recognizes that you appreciate the significance of problems in your project, your skills in dealing with them at an appropriate level, and the consequences of not dealing with the problems effectively. The tone of such memoranda is critical to their effectiveness. Never be vindictive, accusatory, or in any way judgmental in a memorandum. State the facts (as you see them) and be objective. Describe in a clear, concise manner what has been done and when, as well as what needs to be done by when, and, if appropriate, by whom. One of the most effective techniques in writing such a memorandum is to demonstrate awareness of the constraints and factors creating your problem and limiting yourself and the other relevant parties from getting the problem solved. Again, such a strategy will demonstrate your appreciation of conflicting agendas and send the message that you wish to work around them by building bridges to the other parties involved.

DEVELOPING A LONG-TERM RESEARCH STRATEGY

It has been the authors' experience that the most successful and interesting research program is not in fact a single program at all, but related programs operating at several levels of the RDT&E process in parallel. This is an effective strategy for a variety of reasons. First, it offers built-in transition from basic through applied research and on into advanced development. Second, it provides a vehicle by which to address interesting, important and often unexpected problems that may appear in more advanced R&D at more basic levels of R&D when appropriate resources might not be available to explore the problem at the higher level of research. Third, it provides a basis for leveraging of resources (people, laboratory development and maintenance costs, etc.) across a variety of projects. This will make you more effective, efficient, and, of particular importance in this era of down-sizing, cost-effective. Further, such efforts go a long way toward establishing the critical mass of talent necessary to do quality research on a regular basis. Finally, a multithrust strategy provides a necessary buffer when one or another line of funding comes to an end.

Figure 2.1 shows how such a strategy would be laid out over time. Note that the lower levels of research tend to cycle more rapidly than do projects performing advanced development. In addition, the further along the project is in the R&D process, the more expensive and resource-intensive it tends to be. New problems and ideas for additional research are shown as being inspired by the needs of ongoing applied research. The products of each level of research are shown as feeding down into the next available cycle of more developmental research. Note also that the products of one level of research need not necessarily flow to the next level of research. They may jump across levels of research or even spawn entirely new research efforts within the same line of funding.

CRITICAL TECHNOLOGY CHALLENGES IN AVIATION RESEARCH

Several excellent sources are available that will assist in developing a realistic perspective regarding future opportunities in aviation research. For example, the recent National Plan for Civil Aviation Human Factors developed by the Federal Aviation Administration

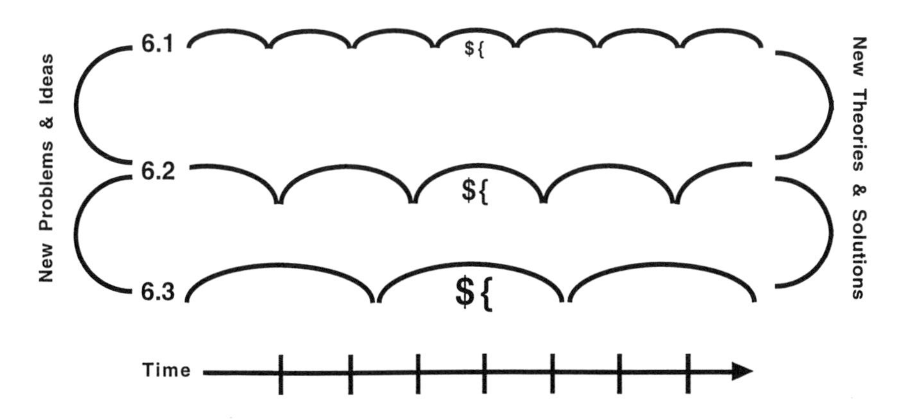

FIG. 2.1. Representation of ideal R&D investment strategy.

(FAA, March 1995) supports several critical areas within aviation. This initiative describes the goals, objectives, progress, and challenges for both the long- and short-term future of human factors research and application in civil aviation. More specifically, the FAA plan identifies the following five research thrusts: (a) human-centered automation, (b) selection and training, (c) human performance assessment, (d) information management and display, and (e) bioaeronautics. The primary issues in each of the first four thrust areas are summarized in Tables 2.2 to 2.5. These issues certainly exemplify the challenges facing human factors specialists in the upcoming years. These are the areas that will most certainly receive sponsorship support, as they have been deemed to be impacting the rate of human error-related incidents and accidents.

It behooves researchers to be aware of several changes within the R&D environment in the last few years that will have significant influence on new initiatives. These changes will substantially change the role of human factors researchers conducting aviation research. First, there has been an increased awareness and sensitivity to the critical importance of the human element in safety. With this increased understanding we should see a renewed interest on safety, even if that results in less funding for

TABLE 2.2
Issues in Human-Centered Automation

Workload	1. Too little workload in some phases of flight and parts of ATC operations to maintain adequate vigilance and awareness of systems status. 2. Too much workload associated with reprogramming when flight plans or clearances change. 3. Transitioning between different levels of workload, automation-induced complacency, lack of vigilance, and boredom for flight deck, ATC, and monitoring of system and service performance.
Operational situation awareness and system mode awareness	1. The ability of operators to revert to manual control when the advanced automation equipment fails. 2. An inadequate "cognitive map," or "situational awareness" of what the system is doing. 3. Problematic recovery from automation failures. 4. The potential for substantially increased head down time. 5. Difficulty and errors in managing complex modes.
Automation dependencies and skill retention	1. The potential for controllers, pilots, and others to over-rely on computer generated solutions (e.g., in air traffic management and flight decisions). 2. Hesitancy of humans to take over from an automated air traffic and flight deck system. 3. Difficulty in maintaining infrequently used basic and critical skills. 4. Capitalizing on automation-generated alternatives and solutions. 5. Monitoring and evaluating pilot and controller skills where computer-formulated solutions disguise skill weaknesses. 6. Supporting diagnostic skills with the advent of systems that are more reliable and feature built-in-self-diagnostics (e.g., those in "glass cockpit" systems and fully automated monitoring systems).
Interface alternatives	1. Major system design issues that bridge all aviation operations include selecting and presenting information for effective human–computer interface. 2. Devising optimal person–machine interfaces for advanced ATC systems and for flight deck avionics. 3. Devising strategies for transitioning to new automation technologies without degrading individual or contemporary system performance.

TABLE 2.3
Issues in Selection and Training

New equipment training strategies	1. Training pilots, controllers, security personnel, and systems management specialists to transition to new technologies and the associated tasks for new equipment. 2. New training concepts for flight crews, controller teams, security staffs, and system management teams. 3. Measuring and training for the performance of new tasks associated with equipment predictive capabilities (vs. reactive type tasks) for pilots and air traffic controllers. 4. Methods to train personnel in the use of computer decision-aiding systems for air and ground operations. 5. Improved strategies for providing the required student throughput within training resource constraints on centralized training facilities, training devices, and simulation.
Selection criteria and methods	1. Evaluation of individual and aggregate impacts on personnel selection policies of changing requirements in knowledge, abilities, skills, and other characteristics for flight crew, controller, and airway facilities operations associated with planned and potential changes in the national airspace system (NAS). 2. Expanded selection criteria for pilots, controllers, technicians, and inspectors from general abilities to include both more complex problem-solving, diagnostic, and meta-cognitive abilities as well as the social attributes, personality traits, cultural orientation, and background biographical factors that significantly influence operational performance in a highly automated national airspace system (NAS). 3. Development of measures to evaluate these more complex individual and team-related abilities in relation to job/task performance.

TABLE 2.4
Issues in Human Performance Assessment

Human capabilities and limitations	Determining the measures and impacts of (a) cognitive factors underlying successful performance in planning, task/workload management, communication, and leadership; (b) the ways in which skilled individuals and teams prevent and counteract errors; (c) ways to reduce the effects of fatigue and circadian dysrhythmia on controllers, mechanics, and flight deck and cabin crews; (d) baseline performance characteristics of controllers so that the impact of automation can be assessed; and (e) qualifying the relationship between age and skilled performance.
Environmental impacts (external and internal)	1. Assessing the influence of "culture" on human performance, including the impact of different organizational and ethnic cultures, management philosophies and structures, and procedural styles. 2. Determining methods to accommodate mixed corporate, regional, and national views of authority, communication, and discipline. 3. Addressing variations in aviation equipment design philosophies and training approaches. 4. Understanding population stereotypical responses in aviation operations.
Methods for measurement	Devising effective aviation system monitoring capabilities with emphasis upon: (a) expansion of the collection, usage, and utility of human performance data and databases; (b) standardization and improved awareness of critical human factors variables for improved collection, classification, and use of reliable human performance data; (c) standardization of classification schemes for describing human factors problems in person–machine systems; (d) better methods and parameters to assess team (vs. individual) performance parameters for flight and maintenance crews, air traffic controllers, security and aviation operations personnel; and (e) improved understanding of relationship between actual performance and digital data measurement methodologies for the flight deck to predict future air crew performance based on trend data.

TABLE 2.5
Issues in Information Management and Display

Information exchange between people	1. Identify requirements for access to critical NAS communications for analysis purposes. 2. Determine the effects of pilot response delays in controller situation awareness and controller/pilot coordination (particularly with regard to delayed "unable" responses). 3. Set standards for flight crew response to messages. 4. Assess the changes in pilot/controller roles. 5. Enhance the communication training for pilots and controllers. 6. Identify sources, types, and consequences of error as a result of cultural differences. 7. Develop system design and procedural solutions for error avoidance, detection, and recovery.
Information exchange between people and systems	1. Assess and resolve the effects of data communications on pilots/controllers situational awareness. 2. Determine the best display surfaces, types, and locations for supporting communication functions in the cockpit, at the ATC workstation, and at monitoring and system maintenance control centers. 3. Identify sources, types, and consequences of error, and error avoidance, detection, and recovery strategies. 4. Establish requirements and set standards for alerting crew, controller, and system management personnel to messages of varying importance.
Information displays	1. Establish policies for operationally suitable communication protocols and procedures. 2. Set standards for display content, format, menu design, message displacement, control and interaction of functions, and sharing. 3. Assess the reliability and validity of information coding procedures. 4. Provide design guidelines for message composition, delivery, and recall. 5. Prescribe the most effective documentation and display of maintenance information. 6. Prototype technical information management concepts and automated demonstration hardware to address and improve the content, usability, and availability of information in flight deck, controller, aircraft, maintenance, security, AF system management, and aviation operations.
Communications processes	1. Devise methods of reconstructing the situational context needed to aid in the analysis of communications. 2. Analyze relationships between workload factors and errors in communication. 3. Evaluate changes in information transfer practices. 4. Set standards and procedures for negotiations and modifications to clearances. 5. Establish procedures for message prioritization and response facilitation. 6. Set standards for allocation of functions and responsibilities between pilots, controllers, and automated systems. 7. Provide guidelines on the distribution of data to, and integration with, other cockpit systems. 8. Prescribe communication policies related to flight phases and airspace, such as use in terminal area and at low altitudes. 9. Determine the impact of data communications on crew and controller voice communications proficiency.

non-safety-related research. Second, programmatic changes within organizations, such as increased NASA emphasis on aeronautics, and DoD technology transfer programs will very likely generate cooperative agreements between agencies that heretofore had not considered sharing technological advances. Moreover, the emphasis away from strictly military applications is obviously one of the "dividends" resulting from the end of the Cold War and the draw-down of the military complex. Finally, technological changes in the design and development of aviation systems continue at an increasing level of effort. Systems are becoming more complex, requiring modifications to training regimens. Advances in the development of aircraft structures have surpassed the capabilities of the operator to withstand the environmental forces impinging upon him or her. These new developments will certainly stimulate innovative efforts to investigate how to enhance the capabilities of the human operator, given the operator's physiological limitations. What this all means is that we in the human factors field need to be aware of what these changes are, and, more importantly, of how we can be more responsive to the needs of both civilian and military research agencies.

In regard to these ongoing and future challenges, there are several driving factors that contribute to the role aviation human factors will play in the near future. Some of these drivers are: (a) technology, (b) demographics, (c) cultural, and (d) economic. Each one of these drivers is next discussed in light of its impact on the direction of future aviation research efforts.

Technology. With the advent of new aircraft and future changes to air traffic control systems, we will see even higher levels of automation and complexity. How these changes impact operator performance and how system design should be modified to accommodate and minimize human error need to be determined. A blend of the best of computer and human capabilities should result in some type of human–computer interaction designed to minimize errors.

Demographics. With the military draw-down a reality, there will be fewer and fewer pilots trained by military sources. Changing skill levels and other work-related demographics will likely affect personnel selection and training of pilots as well as ancillary personnel, that is, controllers, maintenance, and operations. How these changes drive the development of new standards and regulations remains to be seen. We have already seen a change from strict adherence to military specifications in DoD system acquisition requirements to industrial standards. Not only is the "leaner, meaner" workforce the hallmark of the new military, but also it gives justification to support further developments in the area of personnel training. The acquisition of additional weapon systems will most likely decrease, resulting in a redoubling of our efforts to train existing personnel to operate the current generation of weapon systems to a more optimal and efficient level.

Cultural. Opportunities to collaborate with our foreign counterparts will increase as organizations become increasingly international. The development of aviation standards and practices will take into account incompatible cultural expectations that could lead to increased human errors and unsafe conditions. We have already seen these developments in the area of air traffic control, and we will certainly see analogous efforts in other areas as well in the near future.

Economic. Economic factors have vastly affected the aerospace industry. Available funding to continue R&D efforts has steadily decreased. Under this kind of austere environment, competition for limited research dollars is fierce. Many agencies, especially the military, are cutting back on the development of new systems and are now refocusing on improving training programs to assure a high-level skill base given the reduction in available personnel.

The role the human factors field plays in aviation research is no different than the role it plays in any research endeavor. The methods, for the most part, remain the same. The difference lies in the impact it has on our everyday lives. In its infancy, human factors focused on the "knobs and dials" issues surrounding aircraft and aircraft design. Today we are faced with more complex issues, compounded by an environment that is driving scarce resources into areas that go beyond theoretical pursuits to that of practical, applied areas of concentration. This is not to say that this area is not vital, progressive, or increasing in scope and value. It merely means that we, as professionals working in the field of aviation human factors, have to be aware of the technology gaps and how best to satisfy the needs of our customers. This can be accomplished, but it requires a certain kind of flexibility and visionary research acumen to anticipate what these problems are and how best to solve them.

MAJOR FUNDING SOURCES FOR AVIATION RESEARCH

In the past, many educational institutions manually searched a selection of sources, from the *Commerce Business Daily* and the *Federal Register*, to periodicals and agency program directories and indexes that were updated on a regular basis. Today, much of this search can be done online, electronically. An array of available technologies can significantly improve the ease of retrieval of information in areas such as funding opportunities, announcements, forms, and sponsor guidelines. If you have an Internet connection of some type, you can find federal opportunities through *FEDIX*, an online database retrieval service of government information for college, universities, and other organizations. The following agencies are included in the *FEDIX* database:

1. Department of Energy
2. Office of Naval Research
3. NASA
4. FAA
5. Department of Commerce
6. Department of Education
7. National Science Foundation
8. National Security Agency
9. Department of Housing and Urban Development
10. Agency for International Development
11. Air Force Office of Scientific Research

A user's guide is available from FEDIX that includes complete information on getting started, including an appendix of program titles and a list of keywords by agency.

All government agencies can also be accessed through the Internet. Most colleges and universities provide Internet access. Individuals who must provide their own service need to subscribe to an Internet provider such as America Online or CompuServe. Generally, you pay a subscription service fee that may include a specified number of free minutes per month.

In addition to online searches, you may wish to make direct contact with one of many federal sources of research support. The Department of Defense has typically funded many human factors programs. Behavioral and social science research and development are referred to as manpower, personnel, training, and human factors R&D in the DoD.

Although it is beyond the scope of this chapter to review each and every government funding source, the following sources would be of particular interest to those conducting aviation human factors research. Contact these agencies directly for further information.

U.S. Air Force

Air Force Office of Scientific Research
Life Sciences Directorate
Building 410
Bolling Air Force Base
Washington, DC 20332

Armstrong Laboratory
Human Resources Directorate (AL/HR)
7909 Lindbergh Drive
Brooks AFB, TX 78235-5340

Armstrong Laboratory
Crew Systems Directorate (AL/CF)
2610 7th Street
Wright-Patterson AFB, OH 45433-7901

USAF School of Aerospace Medicine
USAFSAM/EDB
Aerospace Physiology Branch
Education Division
USAF School of Aerospace Medicine
Brooks AFB, TX 78235-5301

U.S. Army

Army Research Institute for the Behavioral and Social Sciences
5001 Eisenhower Avenue
Alexandria, VA 22233

U.S. Army Research Laboratory
Human Research & Engineering
Directorate ATTN: AMSRL-HR
Aberdeen Proving Ground, MD 21005-5001

U.S. Army Research Institute of Environmental Medicine
Commander
U.S. Army Natick RD&E Center
Building 42
Natick, MA 01760

Walter Reed Army Institute of Research
ATTN: Information Office
Washington, DC 20307-5100
U.S. Army Aeronautical Research Laboratory
P.O. Box 577
Fort Rucker, AL 36362-5000

U.S. Navy

Office of Naval Research
800 North Quincy Street
Arlington, VA 22217-5000

Space Warfare Systems Center
Code D44
53560 Hull Street
San Diego, CA 92152-5001

Naval Air Warfare Center, Aircraft Division
Crew Systems
NAS Patuxent River, MD 20670-5304

Naval Air Warfare Center, Training Systems Division
Human Systems Integration
12350 Research Parkway
Orlando, FL 32826-3224

Naval Air Warfare Center, Weapons Division
Crew Interface Systems
NAS China Lake, CA 93555-6000

Naval Health Research Center
Chief Scientist
P.O. Box 85122
San Diego, CA 92138-9174

Naval Aerospace Medical Research Laboratory
NAS Pensacola, FL 32508-5700

Naval Biodynamics Laboratory
Commanding Officer
P.O. Box 29407
New Orleans, LA 70189-0407

Miscellaneous

National Science Foundation
4201 Wilson Boulevard
Arlington, VA 22230

Federal Aviation Administration Technical Center
Office of Research and Technology Applications
Building 270, Room B115
Atlantic City International Airport, NJ 08405

3

Measurement in Aviation Systems

David Meister
Human Factors Consultant

A LITTLE HISTORY[1]

One cannot understand measurement in aviation human factors (HF) without knowing a little of its history, which goes back to World War I and even earlier. In that war new aircraft were tested at flight shows and selected in part on the basis of pilot opinion. The test pilots were the great fighter aces, men like Guynemer and von Richtoffen. Such tests were not tests of pilot performance as such, but the pilot and his reactions to the aircraft were a necessary part of the test.

Between the wars human factors participation in aviation system research continued (Dempsey, 1985). The emphasis in the Army Air Force was primarily medical/physiological. For example, researchers using both animals and men studied the effects of altitude and acceleration on human performance. "Angular accelerations were produced by a 20 ft-diameter centrifuge, while a swing was used to produce linear acceleration" (Moroney, 1995). Work on anthropometry in relation to aircraft design began in 1935. As early as 1937 a primitive G-suit had been developed. This was also the period when Edwin Link marketed his flight simulator (which became the grandfather of all later flight simulators) as a coin-operatored amusement device.

In World War II efforts in aircrew personnel selection led to the Air-Crew Classification Test Battery to predict success in training and combat (Taylor & Alluisi, 1993). Human factors specialists were also involved in a wide variety of activities, including determining human tolerance limits for high-altitude bailout, automatic parachute opening devices, cabin pressurization schedules, pressure breathing equipment and protective clothing for use at high altitudes, airborne medevac facilities, and ejection seats (Van Patten, 1994). Probably the best known of the World War II researchers was Paul Fitts, who worked with his collaborators on aircraft controls and displays (Fitts & Jones, 1947).

[1] I am indebted for parts of this historical review to W. F. Moroney (1995).

During the 1950s and 1960s, HF personnel contributed to the accommodation of men in jet and rocket-propelled aircraft. Under the prodding of the new U.S. Air Force all engineering companies bidding on the development of military aircraft had to enlarge their staffs to include HF specialists, and major research projects like the Air Force Personnel and Training Research Center were initiated. Although the range of HF investigations in these early days may have been limited, a later section of this chapter shows that these have expanded widely.

THE DISTINCTIVENESS OF AVIATION HF MEASUREMENT

Despite this relatively long history the question comes to mind: Is there anything that specifically differentiates aviation HF measurement from measurement of other types of systems, such as surface ships, submarines, railroads, tanks, or automobiles? Asked in this way, the answer to this question must be: Except for a very small number of specific environment-related topics, no, there is not. Except for physiological areas such as the topics mentioned in the previous historical section, every topic addressed in aviation HF research is addressed also in connection with other systems.

For example, questions of workload, stress, and fatigue are raised with other transportation and even with nontransportation systems. Questions dealing with such present-day "hot" topics in aviation research as situational awareness (addressed in chapter 11 of this book) and those dealing with the effects of increasing automation (see chapter 7) are also raised in connection with such widely different systems as nuclear power plants.

So why a chapter on measurement in a text on aviation HF? Because, although the questions and the methods are much the same as in other fields, the aircraft is a distinctive system functioning in a very special environment. It is that environment that makes aviation HF measurement important. Because of that environment general behavioral principles and knowledge cannot automatically be generalized to the aircraft. Aviation HF measurement emphasizes the *context* in which its methods are employed.

This chapter is therefore not one on general psychological measurement, and only enough description of the methods employed is provided to enable the reader to understand the way in which the methods were used. We mention statistics and experimental design, but go no further.

Even so constrained, the scope of aviation HF measurement is very wide; almost every type of method and measure that one finds in the general behavioral literature has been used in researching aviation issues. That measurement is largely research oriented, because, although there is nonresearch measurement in aircraft development and testing, it is rarely reported in the literature.

MAJOR MEASUREMENT TOPICS

One of the first questions one asks about measurement is: What topics does this measurement encompass? Given the broad range of aviation HF research, the list that follows cannot be all-inclusive, but it includes the major questions addressed. Space does not permit a detailed description of what is included in each category, although many of these topics are subject matter for subsequent chapters. They are not listed in any particular order of importance. References to illustrative research are appended. Of course, each individual study may investigate more than one topic.

1. The effect of automation on crew proficiency (e.g., the "glass cockpit"; McClumpha, James, Green, & Belyavin, 1991).
2. Effects of and methods of predicting pilot workload, stress, and fatigue (Selcon, Taylor, & Koritsas, 1991).
3. Training, training devices, training effectiveness evaluation, transfer of training to operational flight (Goetti, 1993).
4. Perceptual cues used by flight personnel (Battiste & Delzell, 1991).
5. Design and use of simulators (Kleiss, 1993).
6. Measurement in system development, for example, selection among alternative designs and evaluation of system adequacy (Barthelemy, Reising, & Hartsock, 1991).
7. Evaluation of crew proficiency (McDaniel & Rankin, 1991).
8. Amount of and reasons for pilot error (Pawlik, Simon, & Dunn, 1991).
9. Factors involved in aircraft accidents and accident investigation (Schwirzke & Bennett, 1991).
10. Aircrew selection, such as determination of factors predicting pilot performance (Fassbender, 1991).
11. Evaluation of the human engineering characteristics of aircraft equipment, such as varying displays and helmets (Aretz, 1991).
12. Situational awareness (see chapter 11).
13. Effects of environmental factors (e.g., noise, vibration, acceleration, lighting) on crew performance (Reynolds & Drury, 1993).
14. Effects of drugs and alcohol on pilot performance (Gawron, Schiflett, Miller, Slater, & Ball, 1990).
15. Factors leading to more effective crew coordination and communication (Conley, Cano, & Bryant, 1991).
16. Pilot personality characteristics (Orasanu, 1991).
17. Pilot decision making and information processing: flight planning; pilot's mental model (Orasanu, Dismukes, & Fischer, 1993).
18. Checklists and map formats; manuals (Degani & Wiener, 1993).
19. Cockpit display and control relationships (Seidler & Wickens, 1992).
20. Methods of measuring pilot performance (Bowers, Salas, Prince, & Brannick, 1992).
21. Air traffic control (Guidi & Merkle, 1993).
22. Crew health factors, age, experience, and sex differences (Guide & Gibson, 1991).

PERFORMANCE MEASURES AND METHODS

Aviation HF measurement can be categorized under four method/measure headings: flight performance, nonflight performance, physiological, and subjective. Before describing each category, it is useful to say something about how to select among them.

For convenience we refer to all methods and measures as *metrics*, although there is a sharp distinction between them. Any individual method like the experiment can be utilized with many different measures.

A large number of metric selection criteria exist, the most prominent ones being *validity* (how well does the metric measure and predict operational performance) and *reliability* (the degree to which a metric reproduces the same performance under the same measurement conditions consistently). Others include *detail* (does it reflect performance with sufficient detail to permit meaningful analysis); *sensitivity* (does it reflect significant variations in performance caused by task demands or environment); *diagnosticity* (does it discriminate among different operator capacities); *intrusiveness* (does it cause degradation in task performance); *requirements* (what does it require in system resources to use it); and personnel *acceptance* (will test personnel tolerate it). Obviously one would prefer a metric that, all other things being equal, was *objective* (was not mediated by a human observer) and *quantitative* (capable of being recorded in numerical format). *Cost* is always a significant factor.

It is not possible to make unequivocal judgements of any metric outside of the measurement context in which it will be used. Certain generalizations can, however, be made: All other things being equal, one would prefer objective to subjective and nonphysiological to physiological metrics (because the latter often require expensive and intrusive instrumentation and in most cases have only an indirect relationship to performance), although if one is concerned with physiological variables, they cannot be avoided. Any metric is preferable that can be embedded in the operator's task and does not degrade task performance. The cheaper the metric (less time to collect and analyze data), the better. Again, all other factors being equal, data gathered in operational flight or in the operational environment are to be preferred to data collected nonoperationally.

Flight Performance Metrics

The following is indebted in part to Hubbard, Rockway, and Waag (1989).

Because pilot and aircraft are so closely interrelated as a system, aircraft state can be used as an indirect measure of how the pilot performs in controlling the aircraft. In state-of-the-art simulators and to a slightly lesser extent in modern aircraft, it is possible to obtain automatically measures of aircraft state, such as altitude, deviation from glide slope, pitch roll and yaw rates, airspeed, bank angle, and so forth. In a simulator it is possible to sample these parameters at designated intervals, such as fractions of a second. The resultant time-series plot is extremely useful in presenting a total picture of what happens to the pilot/aircraft system. This is not a direct measurement of the pilot's arm or hand actions or perceptual performance, but is mediated through the aircraft's instrumentation. Measurement of arm and hand motions or the pilot's visual glances would, however, be perhaps a little too molecular and probably would not be measured except under highly controlled laboratory conditions. The reader is referred to chapter 14, which discusses the capabilities of the simulator in measurement of aircrew performance. Measurement within the operational aircraft has been much expanded as aircraft, such as the F-16, have become highly computer controlled.

Because the pilot is controlling the aircraft directly, it is assumed that deviations from specified flight performance requirements (e.g., a given altitude, a required glide slope) represent errors directly attributable to the pilot, although one is obviously not measuring pilot behavior (e.g., hand tremor) directly. This assumes that the aircraft has no physical malfunctions that would impact the pilot's performance.

Where the pilot is supposed to react to a stimulus (e.g., a topographic landmark) appearing during the flight scenario, the length of time the pilot takes to respond to that stimulus is also indicative of the pilot's skill. Reaction time and response duration measures are also valuable in measuring pilot performance.

The time-series plot referred to previously will resemble a curve with time represented horizontally, aircraft state vertically. Such a plot is useful in determining when and for how long a particular parameter was out of bounds. Such plots can be very useful when, in a simulator, a stimulus condition like wind gust or aircraft malfunction is presented; the plot indicates how the pilot has responded. In pilot training these plots can be utilized as feedback for debriefing students.

In research on flight performance, researchers usually compute summary measures based on data that have been sampled in the course of the flight. This is necessary because large amounts of data must be reduced to a number that can be more readily handled. For the same reason the flight course is characteristically broken up into segments based on the tasks to be performed, such as straight and level portions, ridge crossings, turns, and so on. One then summarizes the pilot performance within the designated segment of the course.

One of the most common summary metrics is root mean square error (RMSE), which is computed by taking the square root of the average of the squared error or deviation scores as shown:

$$\mathrm{RMSE} = \frac{\sqrt{\Sigma e_i^2}}{N} \tag{1}$$

where e_i is the individual error score and N is the total number of error scores sampled. A limitation of RMSE is that position information is lost. Notwithstanding that, this metric is often used.

Two other summary metrics are the mean of the error scores (ME) and the standard deviation of those scores (SDE). ME is computed as:

$$\mathrm{ME} = \frac{\Sigma e_i}{N} - e \tag{2}$$

SDE is computed as:

$$\mathrm{SDE} = S_e = \frac{\Sigma(e_i - e)^2}{N} \tag{3}$$

RMSE is completely defined by ME and SDE, and according to Hubbard et al. (1989) the latter are to be preferred because RMSE is less sensitive to differences between conditions and more sensitive to measurement bias.

There are many ways to summarize pilot performance, depending on the individual mission goals and pilot tasks. In air combat maneuvering, for example, number of hits and misses of the target and miss distance would follow from the nature of the mission. The method and measure selected are determined by the questions the investigator asks. However, it is possible, as Stein (1984) did, to develop a general-purpose pilot performance index. This was based on subject matter experts and revised to eliminate those

TABLE 3.1
Pilot Performance Index Variable List

Takeoff	Climb	En route
Pitch angle	Heading	Altitude
	Airspeed	Pitch angle
		Heading
		Course deviation indicator
		Omni bearing sensor
Descent	**Initial approach**	**Final approach**
Heading	Heading	Heading
Airspeed	Manifold left	Gear position
Bank angle	Manifold right	Flap position
Course deviation indicator	Bank angle	Course deviation indicator
		Omni bearing sensor

measures that did not differentiate experienced from novice pilots. Variables possible are shown in Table 3.1. Refer to Berger (1977) and Brictson (1969) for examples of studies in which flight parameters were used as measures to differentiate different conditions.

At the present time the U.S. Air Force with the assistance of the Arvin/Calspan Corporation is attempting to develop a standard system to assess the operability of the aircraft crew station as a part of flight testing. The following is based on a paper by Cohen, Gawron, Mummaw, and Turner (1993).

Currently the crew station evaluation process is not standardized, with a variety of metrics and procedures being used. Data from one flight test are often not comparable to those of another. The research effort underway endeavors to develop a structured process for crew station evaluations.

This system (the Test Planning, Analysis and Evaluation System or Test PAES) provides various computerized tools to guide evaluation personnel, who in many cases are not measurement specialists. The system includes materials to plan and conduct ground or flight tests. The tools available include word processing, spreadsheet, graphics, statistics, and networking, as well as a model to predict system performance in the field based on simulation and test data, tools to administer crew questionnaires, and to help present and analyze the data.

When the Test PAES user clicks on the current step in the process, the appropriate applications open on the screen automatically. After the data have been collected, they will be downloaded to the system through either network, tape, or disk. Videotapes will be digitized and the system will collect debrief data from the crew. The statistical analysis package and the integrated data display can be used to evaluate the data.

Nonflight Performance Metrics

Certain performances are not reflected in aircraft state. For example, the aircrew may be required to communicate upon takeoff or landing with air traffic control, or to use a radar display or direct visualization to detect possible obstacles, or to perform contingency planning in the event of an emergency.

Each such nonflight task generates its own metric. Examples are content analysis of communications or speed of target detection/acquisition or number of correct target identifications.

All flight performance metrics must be collected during an operational or a simulator flight; nonflight metrics can be utilized at any time during an operational or simulated flight, following that flight (on the ground), or can be used in a nonflight environment, such as a laboratory.

Some nonflight metrics are related to flight but do not measure a specific flight. An example is a summary measure of effectiveness such as the number of flights or other actions performed by the pilot to achieve some sort of criterion (mostly in training). In the study of map displays or performance of map-of-the earth helicopter flight the pilot may be asked to draw a map or may be asked to make time or velocity estimates. Researchers have developed extensive lists of measures (Meister, 1985) from which one can select those that seems appropriate for the task to be measured. Review of the papers in the literature of aviation psychology (see the references at the end of this chapter) will suggest others.

The metrics referred to so far are an integral part of the flight task, but there are also those that are not, that are utilized purely for research purposes, and therefore are somewhat artificial. The recent emphasis on pilot workload studies, for example, has created a great number of subjective workload metrics (see chapter 7). Besides the well-known scales such as SWAT or TLX (Vidulich & Tsang, 1985), which require the pilot to rate his or her own performance, there are others that demand that pilots perform a second task (in addition to those required for flight), such as sort cards, solve problems, make a choice reaction, or detect a specific stimulus event. The problem that one faces with secondary tasks is that in the actual flight situation they may cause deteroriation of performance in the primary flight task, which could be dangerous. This objection might not be pertinent in a flight simulator. In general, any secondary task that distracts the pilot from flight performance is undesirable in actual flight.

Performance measures taken after the flight is completed, or where a copilot takes the controls while the pilot performs a research task, are of course acceptable. Measurement of flight performance variables is usually accomplished by sensors linked to a computerized data collection system. Such instrumentation is of course not available for measurement of nonflight performance variables. The following is a description of the instrumentation that could be particularly useful for aviation HF variables.

Although there are many instruments that can measure human performance variables and the measurement environment (e.g., photometer, thermometer, sound level meter, vibration meter and analyzer; American Institute for Aerospace and Aeronautics, 1992, describes these in more detail), two are of particular interest for us. The accelerometer is a device, such as a strain gauge or piezoelectric force transducer, that measures acceleration along one or more axes. Obviously such a device would be necessary for any study of G-forces.

More commonly used, however, is the videotape recorder, which is becoming increasingly popular for providing records of visual and audio operator performance for posttest analysis. A complete system includes a television camera, a videotape recorder (VTR), and a television monitor. Crites (1980) described how the instrumentation should be used.

1. Complex tasks should be recorded simultaneously from various aspect angles.
2. Zoom lenses with focal-length ratios of at least 8:1 should be used. This allows the framing of the subject without having to move the camera. Remote-controlled focus and zoom are recommended. For data recording inside compartments a wide-angle lens of 5 mm or higher is recommended.

TABLE 3.2
Recommended Data-Collection Media and Formats

Data Type	*Recommended Media and Format*
Videotapes	Standard VHS-formatted tape
Magnetic tapes	Reel-to-reel: nine-track tape at 1600 bits-per-inch density Helical scan: 8-mm cassettes
Magnetic disks	Preferred: 3.5-inch disks in 1.4 MB
Sound tapes	Preferred: digital audio tapes (DAT) Alternative: standard audio tapes
Photography	35-mm film
Graphical data	TIFF on IBM PC-compatible PICT format on Macintosh-compatible Encapsulated Postscript HPGL (Hewlett-Packard Graphics Language)

3. Small hand-held cameras are needed for work in aircraft cockpits.
4. A video/audio junction box facilitates the selection of audio and video inputs.
5. The VTR should have a conventional counter to locate scenes on tape; a timer should also be included to time the recorded performance.
6. Extensive lighting, perhaps beyond common floodlights and spotlights, is required to secure good detail. Cameras should have good low-light level imaging characteristics.

If the VTR does not have a time display or a more accurate one is required, a time-code generator can add a time code on the videotape, with a resolution of up to 0.001 sec. Recommendations from the AIAA (1992) are presented in Table 3.2.

Physiological Measures

Only a relative small percentage of aviation HF studies make use of physiological instrumentation and measures. That is because such measures are useful only when the variables being studied involve a physiological component. In particular, studies involve acceleration (McCloskey, Tripp, Chelette, & Popper, 1992), hypoxia, noise level, fatigue (Krueger, Armstrong, & Cisco, 1985), alcohol, drugs, and workload.

Table 3.3 from Meister (1985) lists physiological measures associated with the major bodily systems. Heart rate and heart-rate variability have been the most commonly used physiological assessment methods, primarily because they are relatively nonintrusive, and because portable devices for recording these data are available. These metrics have been employed in a number of in-flight studies involving workload (Hart & Hauser, 1987; Hughes, Hassoun, Ward, & Rueb, 1990; Wilson & Fullenkamp, 1991; Wilson, Purvis, Skelly, Fullenkamp, & Davis, 1987). Itoh, Hayashi, Tsukui, and Saito (1989) and Shively et al. (1987) have demonstrated that heart-rate variability can discriminate differences in workload imposed by flight tasks.

TABLE 3.3
Physiological Measures of Workload

System	*Measure*
Cardiovascular system	* Heart rate
	* Heart-rate variability (sinus arrhythmia)
	* Blood pressure
	Peripheral blood flow
	* Electrical changes in skin
Respiratory system	* Respiration rate
	Ventilation
	Oxygen consumption
	Carbon dioxide estimation
Nervous system	* Brain activity
	* Muscle tension
	* Pupil size
	Finger tremor
	Voice changes
	Blink rate
Biochemistry	* Catecholamines

Note. Those measures most commonly utilized have been indicated by an asterisk.

Nevertheless, all of these metrics have disadvantages. Many of them require intrusive instrumentation, which may not be acceptable in an actual flight environment. However, they are more supportable in a simulator. For example, in a simulator or study of helicopter crew performance, stress, and fatigue over a week-long flight schedule Krueger et al. (1985) had three electrocardiogram chest electrodes wired to a monitoring system to assess heart rate and heart-rate variability as indicators of alertness. Oral temperatures were taken at approximately 4-hr intervals; urine specimens (for catecholamines) were provided at 2-hr intervals between flights. Illustrative descriptions of physiological studies in the flight simulator are also provided by Morris (1985), Armstrong (1985), and Lindholm and Sisson (1985).

Unfortunately, the evidence for the relationship between physiological and performance indices is ambiguous at best. Often the meaning of such a relationship, even when it is documented, is unclear. Moreover, the sensitivity of these metrics to possible contaminating conditions, for example, ambient temperature, is very high.

Subjective Measures

Subjective measures (whatever one may think about their validity and reliability) have always been and still are an integral part of aviation HF measurement. As mentioned previously, in World War I ace fighter pilots like Guynemer and von Richtoffen were used to evaluate the handling qualities of prototype aircraft. Ever since the first aviation school was established, expert pilots have been used not only to train, but also to evaluate the performance of their students. Even today, with all the sophisticated, computerized instrumentation available in test aircraft, the pilot is routinely asked to evaluate handling qualities. Automated performance measurement methods, although highly desirable, cannot entirely replace subjective techniques (Vreuls & Obermayer, 1985).

Muckler (1977) pointed out that all measurement is subjective at some point in test development; the objective/subjective distinction is a false issue. Given this, the problem is to find ways of enhancing the adequacy of the subjective techniques. There is need for more research to develop more adequate methods to train and calibrate expert observers.

The subjective techniques described in the research literature include interviews, questionnaire surveys, ratings and rankings, categorization, and communications analyses. Subjective data, particularly ratings, are characteristically used to indicate pilot preference, performance evaluations, task difficulty, estimates of distance traveled or velocity, and, in particular, workload, which is one of the "hot" topics in aviation HF research.

Because of the variability in these subjective techniques, efforts have been made to systematize them quantitatively in scales of various sorts (for a discussion of scales, see Meister, 1985). The Likert 5-point scale (e.g., none, some, much, very much, all) is a very common scale that can be created in moments even by someone who is not a psychometrician. However, the validity of such self-created scales may be suspect. Development of valid and reliable scales requires prior research on the dimensions of the scale, and empirical testing and analysis of test results. Most complex phenomena cannot be scaled solely on the basis of a single dimension, because most behavior of any complexity is multidimensional. The interest in measurement of workload, for example, has created a number of multidimensional scales: SWAT, which has been used extensively in simulated and actual flight (see AIAA, 1992, pp. 86–87), has three scalar dimensions, Time Load, Mental Effort Load, and Psychological Stress. Scales, either individually or as part of questionnaire surveys, have probably been used more frequently as a subjective measurement device than any other technique. That is because it is difficult to quantize interviews, except as part of formal surveys, in which case they turn into rating/ranking scales.

CHARACTERISTICS OF AVIATION HF RESEARCH

What has been described so far is somewhat abstract and only illustrative. One can ask, how can one describe the aviation HF measurement literature as a whole?

One way to answer this question is to review the recent literature in this area. The author examined the *Proceedings of the Human Factors and Ergonomics Society* (HFES) for 1990, 1991, 1992, and 1993, and the journal that society publishes, *Human Factors*, for the same period, for all studies of aviation HF variables. As a check on the representativeness of these two sources the 1991 *Proceedings of the International Symposium on Aviation Psychology*, sponsored by Ohio State University (OSU), were examined. One hundred and forty-four relevant papers were found in the HFES *Proceedings* and the journal, and 87 papers were found in the OSU *Proceedings*. Only papers that described specific measurement were included in the sample. Those that were reviews of previous measurement research or described prospective research were excluded. Those papers selected as being relevant were content analyzed by applying seven taxonomies:

1. General topic, such as flight, navigation, design, workload
2. Specific topic, such as situational awareness
3. Measures employed, such as tracking error, reaction time
4. Measurement venue, such as laboratory, simulator, operational flight
5. Type of subject, such as pilot, air traffic controllers, nonflying personnel

6. Methodology, such as experiment, questionnaire, observation, incident reports
7. Statistical analysis employed, such as analysis of significance of differences, correlation, factor analysis

Space does not permit a listing of all the taxonomic categories employed, because of their large number. Categories were developed on the basis of the individual papers themselves. The number of categories by topic are: general topic (47); specific topic (71); measures (44); measurement venue (8); subject type (12); methodology (16); statistical analysis (16). The categories were not mutually exclusive. Every category that could describe a particular paper was counted. For example, if a paper dealt with instrument scanning, and in the process described the visual factors involved in the scanning, both categories were counted. Thus categories might tend to overlap, but the procedure employed made the measurement picture that resulted more detailed than would otherwise be the case.

Only those categories that described 5% or more of the total number of papers are listed in the following tables. Because the number of these categories is small compared to the total number of categories reported, it is apparent that although aviation HF measurement is extensive in its subject matter and its tools, it is not very intensive, except in relatively few areas. These presumably are the areas that most excite funding agencies and individual researchers.

An analysis was performed to ensure that the two data sources (HFES and OSU) were not so different that they could not be combined. Roughly the same data patterns could be discerned (broad but not intensive), although there were some differences of note. For example, the OSU sample dealt much more than HFES with flight-related topics (OSU 72%, HFES 35%). Such differences could be expected, because the two sources draw from different venues (e.g., OSU is international, HFES almost exclusively American; OSU preselects its topic areas, HFES does not). In the event, the differences were not considered sufficient to make combination impossible.

Of the 47 categories under "general topic," 13 met the 5% criterion. These are listed in Table 3.4, which indicates that most research was basic. The term means that the research dealt with general principles rather than specific applications. Applied research was only 11% of the total. Both basic and applied research total 91%. The fact that the figures do not add to 100% simply means that a small number of papers, although dealing with measurement, did not involve empirical research. The second point is

TABLE 3.4
General Topic Categories

Category	Percentage
1. Military or commercial flight	50% (113 papers)
2. Design	10% (23 papers)
3. Workload/stress	8% (17 papers)
4. Air traffic control	10% (23 papers)
5. Training	14% (32 papers)
6. Automation	8% (18 papers)
7. Basic research	80% (189 papers)
8. Instrument scanning	7% (16 papers)
9. Visual factors	9% (20 papers)
10. Evaluation	6% (13 papers)
11. Accidents	6% (14 papers)
12. Applied research	11% (25 papers)
13. Pilot personality	5% (12 papers)

TABLE 3.5
Specific Topics

1. Display design/differences	21% (50 papers)
2. Transfer of training	5% (11 papers)
3. Personnel error	6% (14 papers)
4. Personnel demographics	5% (12 papers)
5. Perceptual cues	16% (36 papers)
6. Decision making	6% (13 papers)
7. Workload	14% (33 papers)
8. Communications	6% (14 papers)
9. Coding	5% (11 papers)
10. Tracking	9% (21 papers)
11. Crew coordination	5% (12 papers)
12. Incidents	6% (14 papers)
13. HUD/HMD	5% (12 papers)
14. Mental model	8% (17 papers)
15. Dual tasks	6% (13 papers)
16. Cognition	6% (13 papers)

that only half the papers presented dealt directly with flight-related topics; the others involved activities incident to or supportive of flight, but not flight directly. For example, 10% of the papers dealt with air traffic control, which is of course necessary for aviation, but which has its own problems.

Table 3.5 lists the 15 specific topics that were most descriptive of the papers reviewed. As one can see, only 16 categories out of 71 met the 5% criterion. Although the table reveals a wide assortment of research interests, only 3, display design/differences, perceptual cues (related to display design), and workload, described a relatively large number of papers.

Table 3.6 describes the measures employed by researchers. Of the 44 measures found, only 10 satisfied the 5% criterion. Of course, many studies included more than one type of measure. Error and time are of course the most common measures. The frequency and percentage of measures was the most common statistical treatment of these measures. The relatively large number of ratings of, for example, attributes, performance, preferences, similarity, difficulty, and so on attest to the importance of subjective measures, particularly when these are utilized in a workload measurement context (e.g., SWAT, TLX).

TABLE 3.6
Measures Employed

1. Reaction time	13% (31 papers)
2. Response duration	16% (48 papers)
3. Response error	33% (76 papers)
4. Tracking error	12% (29 papers)
5. Frequency, percentage	33% (80 papers)
6. Ratings	30% (66 papers)
7. Interview data	5% (11 papers)
8. Workload measure	8% (18 papers)
9. Flight performance variables	10% (22 papers)
10. Categorization	8% (17 papers)

TABLE 3.7
Measurement Venue

1. Laboratory (not simulator)	16% (36 papers)
2. Full-scale simulator	23% (52 papers)
3. Part-task simulator or simulated displays	27% (63 papers)
4. Operational flight	11% (26 papers)
5. Irrelevant	16% (46 papers)

Table 3.7 describes where measurements took place. Of the nine categories, five met the 5% criterion. Some explanations are necessary. A laboratory does not simulate any of the characteristics of flight; a full-scale simulator, with at least two degrees of motion, does. A part-task simulator or simulated display reproduces some part of the cockpit environment. Some measures were taken in flight. Where the measurement venue is unimportant, the situation was usually one in which questionnaire surveys were administered by mail or elsewhere.

There is apparently great reliance on flight simulators, both full-scale and part-task, but in many cases there is no flight relationship at all (e.g., the laboratory). The fact that only 26 of the 231 papers dealt with the actual flight environment in the air is somewhat surprising, because measurements taken outside that environment are inevitably artificial to a greater or lesser extent.

Of the 12 categories describing the type of subject used in these studies, only three were significant: 60% of the subjects were pilots (140 papers), 33% (75 papers) of the subjects were nonflying personnel (college students, government workers, the general public), and 9% (20 papers) were air traffic controllers. That the largest proportion of subjects is pilots is not at all surprising, but the relatively large number of nonflying personnel is somewhat daunting.

Nine of the 16 categories under the heading of methodology (Table 3.8) met the 5% criterion. As one would expect, more than half the papers published were experimental in nature. What is somewhat less expected is the large number of studies that are not experimental, although there is some overlap here, because some of the experimental studies did make use of nonexperimental methodology in addition to the experiment. There is heavy reliance on subjective techniques, observation, questionnaires, interviews, and self-report scales. Pilot opinion is, as it has always been, extremely important in aviation.

Of the 16 statistical analysis categories, 4 were most frequently employed (Table 3.9). Again, as one would expect, tests of the significance of differences between conditions or groups accounted for the great majority of analyses. The percentage

TABLE 3.8
Methodology

1. Experiment	54% (126 papers)
2. Observation	12% (29 papers)
3. Questionnaire survey	16% (48 papers)
4. Rating/ranking scale	30% (65 papers)
5. Performance measurement (general)	21% (50 papers)
6. Interviews	10% (22 papers)
7. Physical/physiological data recording	8% (17 papers)
8. Analysis of incident reports	8% (17 papers)
9. Verbal protocol analysis	5% (11 papers)

TABLE 3.9
Statistical Analysis

1. Tests of significance of differences	67% (155 papers)
2. Correlation	10% (22 papers)
3. Frequency, percentage	24% (56 papers)
4. None	5% (12 papers)

would be even greater if one included such tests as multiple regression, discriminant analysis, or factor analysis in this category. Although the categories in this content area tend to overlap, the relatively large number of studies in which the analysis stopped at frequency and percentage should be noted.

What does this review tell us about the nature of aviation HF research? The large number of topic areas, both general and specific, ranging from information processing to geographical orientation to electroencephalography to pilot attitudes (to note only a few topics taken at random), indicates that many areas are being mined, but very few intensively. The major concerns are basic research as it relates to flight and displays. In spite of the fact that presumably automation (the "glass cockpit"), situational analysis, and workload are all "hot" topics in the aviation research community, they receive only a modest degree of attention. If one adds up all the topics that deal with sophisticated mental processes (e.g., decision making, mental models, cognition) and add to these crew coordination, a fair bit of attention is being paid to higher order behavioral functions. This represents some change from earlier research areas.

Most of the behavioral research in aviation is conducted on the ground, for which there are obvious reasons: nonavailability of aircraft and cost of flights. Another reason is perhaps that much of the research deals with cockpit or display variables, which may not require actual flight. Reliance on opinion expressed in questionnaires and on incident/accident reports and on full-scale simulators diminishes the need to measure in actual flight. It may also reflect the fact that behavioral research in general (not only in aviation) rarely takes place in the operational environment, which is not conducive to sophisticated experimental designs and instrumentation. This leaves us, however, with the question of whether results achieved on the ground (even with a high degree of simulation) are actually valid for flight conditions. The problem is compounded by the fact that a third of all subjects employed in these studies were not flying personnel.

HF research in aviation is not completely wedded to an experimental format; only half the studies reported were of this type. It is remarkable that with a system whose technology is so advanced, there is so much reliance on nonexperimental techniques and subjective data.

SUMMARY APPRAISAL

What this review of the aviation HF literature suggests is that research in the future should endeavor to concentrate on key issues to a greater extent than in the past. "Broad but shallow" is not a phrase one would wish to describe that research in general. One of the key issues in aviation HF research (as it should be in general behavioral research as well) is that of the effects of automation on human performance. It seems inevitable that technological sophistication will increase in the coming century and that some of that sophistication will be represented on the flight deck. Its effects are not uniformly positive, so the match between human and the computer in the air must be explored more intensively.

Another recommendation based on the literature review is that results achieved in the simulator should be validated in the air. Simulators have become highly realistic, but they may well lack certain features that can be found only in flight. The frequency with which part-task simulators and laboratories are used in aviation HF research makes one wonder whether precisely the same effects will be found in flight. It is true that there is in behavioral research as a whole little validation in the operational context of effects found in the laboratory, but flight represents a critically distinct environment from that in which most aviation behavioral studies are conducted.

A similar recommendation refers to test subjects. Although it is true that the majority of subjects in the studies reviewed were pilots, it is somewhat disturbing to see the large number of nonflying personnel who are used for this purpose. True, almost all nonpilots were used as subjects in nonflight studies, such as studies of displays, but if one believes that the experience of piloting is a distinctive one, it is possible that that experience generalizes to and subtly modifies nonpiloting activities. In any event, this matter should be looked into.

Finally, we note that the highest percentage of studies dealt with flight variables, and this is quite appropriate. The comparative indifference to other aviation aspects is somewhat disturbing, however. There has been in recent years increasing attention in aviation research to ground maintenance, but proportionately this area, although critical to flight safety, is underrepresented. By comparison, air traffic control (ATC) receives more attention, probably because of the immediacy of the relationships between ATC personnel and pilots. We would recommend a more intensive examination of how well ground maintainers function and the factors that affect that efficiency. A little more attention to passengers, too, would not be be amiss. The role of the passenger in flight is a very passive one, but on long-distance flights particularly the constraints involved in being a passenger are very evident.

REFERENCES

American Institute of Aeronautics and Astronautics. (1992). *Guide to human performance measurement* (Rep. No. BSR/AIAA, G-035-1992). New York: Author.

Aretz, A. J. (1991). The design of electronic map displays. *Human Factors, 33*, 85–101.

Armstrong, G. C. (1985). Computer-aided analysis of in-flight physiological measurement. *Behavior Research Methods, Instruments, & Computers, 17*, 183–185.

Barthelemy, K. K., Reising, J. M., & Hartsock, D. C. (1991, September). Target designation in a perspective view, 3-D map using a joystick, hand tracker, or voice. *Proceedings of the Human Factors and Engineering Society* (pp. 97–101). San Francisco, CA.

Battiste, V., & Delzell, S. (1991, June). Visual cues to geographical orientation during low-level flight. *Proceedings of the Symposium on Aviation Psychology* (pp. 566–571). Columbus, OH: Ohio State University.

Berger, I. R. (1977, March). Flight performance and pilot workload in helicopter flight under simulated IMC employing a forward looking sensor (Rep. No. AGARD-CP-240). *Proceedings of the Guidance and Control Design Considerations for Low-Altitude and Terminal-Area Flight.* Neuilly-sur-Seine, France: AGARD.

Bowers, C., Salas, E., Prince, C., & Brannick, M. (1992). Games teams play: A method for investigating team coordination and performance. *Behavior Research Methods, Instruments, & Computers, 24*, 503–506.

Brictson, C. A. (1969, November). Operational measures of pilot performance during final approach to carrier landing (Rep. No. AGARD-CP-56). *Proceedings of the Measurement of Aircrew Performance—The Flight Deck Workload and its Relation to Pilot Performance.* Neuilly-sur-Seine, France: AGARD.

Cohen, J. B., Gawron, V. J., Mummaw, D. A., & Turner, A. D. (1993, June). Test planning, analysis and evaluation system (Test PAES), a process and tool to evaluate cockpit design during flight test. *Proceedings of the Symposium on Aviation Psychology* (pp. 871–876). Columbus, OH: Ohio State University.

Conley, S., Cano, Y., & Bryant, D. (1991, June). Coordination strategies of crew managment. *Proceedings of the Symposium on Aviation Psychology* (pp. 260–265). Columbus, OH: Ohio State University.

Crites, D. C. (1980). Using the videotape method. In *Air Force systems command design handbook DH-1-3*, Part 2, Series 1-0, General Human Factors Engineering, Chapter 7, Section DN 7E3 (pp. 1–6). Washington, DC: U.S. Government Printing Office.

Degani, U., & Wiener, E. L. (1993). Cockpit checklists: Concept, design, and use. *Human Factors, 35*, 345–359.

Dempsey, C. A. (1985). *50 years of research on man in flight*. Wright-Patterson AFB, OH: U.S. Air Force.

Fassbender, C. (1991, June). Culture-fairness of test methods: Problems in the selection of aviation personnel. *Proceedings of the Symposium on Aviation Psychology* (pp. 1160–1168). Columbus, OH: Ohio State University.

Fitts, P. M., & Jones, R. E. (1947). *Psychological aspects of instrument display. I. Analysis of 270 "pilot-error" experiences in reading and interpreting aircraft instruments* (Rep. No. TSEAA-694-12A). Dayton, OH: Aeromedical Laboratory, Air Materiel Command.

Gawron, V. J., Schiflett, S. G., Miller, J. C., Slater, T., & Ball, J. F. (1990). Effects of pyridostigmine bromide on in-flight aircrew performance. *Human Factors, 32*, 79–94.

Goetti, B. P. (1993, October). Analysis of skill on a flight simulator: Implications for training. *Proceedings of the Human Factors Society* (pp. 1257–1261). Seattle, WA.

Guide, P. C., & Gibson, R. S. (1991, September). An analytical study of the effects of age and experience on flight safety. *Proceedings of the Human Factors Society* (pp. 180–183). San Francisco, CA.

Guidi, M. A., & Merkle, M. (1993, October). Comparison of test methodologies for air traffic control systems. *Proceedings of the Human Factors Society* (pp. 1196–1200). Seattle, WA.

Hart, S. G., & Hauser, J. R. (1987). Inflight applications of three pilot workload measurement techniques. *Aviation, Space and Environmental Medicine, 58*, 402–410.

Hubbard, D. C., Rockway, M. R., & Waag, W. L. (1989). Aircrew performance assessment. In R. S. Jensen (Ed.), *Aviation psychology* (pp. 342–377). Brookfield: Gower Technical.

Hughes, R. R., Hassoun, J. A., Ward, G. F., & Rueb, J. D. (1990). *An assessment of selected workload and situation awareness metrics in a part-mission simulation* (Rep. No. ASD-TR-90-5009). Wright-Patterson AFB, OH: Aeronautical Systems Division, Air Force Systems Command.

Itoh, Y., Hayashi, Y., Tsukui, I., & Saito, S. (1989). Heart rate variability and subjective mental workload in flight task validity of mental workload measurement using H.R.V. method. In M. J. Smith & G. Salvendy (Eds.), *Work with computers: Organizational, management stress and health aspects* (pp. 209–216). Amsterdam: Elsevier.

Kleiss, J. A. (1993, October). Properties of computer-generated scenes important for simulating low-altitude flight. *Proceedings of the Human Factors Society* (pp. 98–102). Seattle, WA.

Krueger, G. P., Armstrong, R. N., & Cisco, R. R. (1985). Aviator performance in week-long extended flight operations in a helicopter simulator. *Behavior Research Methods, Instruments, & Computers, 17*, 68–74.

Lindholm, E., & Sisson, N. (1985). Physiological assessment of pilot workload in simulated and actual flight environments. *Behavior Research Methods, Instruments, & Computers, 17*, 191–194.

McCloskey, K. A., Tripp, L. D., Chelette, T. L., & Popper, S. E. (1992). Test and evaluation metrics for use in sustained acceleration research. *Human Factors, 34*, 409–428.

McClumpha, A. J., James, M., Green, R. C., & Belyavin, A. J. (1991, September). Pilots's attitudes to cockpit automation. *Proceedings of the Human Factors Society* (pp. 107–111). San Francisco, CA.

McDaniel, W. C., & Rankin, W. C. (1991). Determining flight task proficiency of students: A mathematical decision aid. *Human Factors, 33*, 293–308.

Meister, D. (1985). *Behavioral analysis and measurement methods*. New York: Wiley.

Moroney, W. F. (1995). Evolution of human engineering: A selected review. In J. Weimer (Ed.), *Research techniques in human factors*. Englewood Cliffs, NJ: Prentice-Hall.

Morris, T. L. (1985). Electroocculographic indices of changes in simulated flying performance. *Behavior Research Methods, Instruments, & Computers, 17*, 176–182.

Muckler, F. A. (1977). Selecting performance measures: "Objective" versus "subjective" measurement. In L. T. Pope & D. Meister (Eds.), *Productivity enhancement: Personnel performance assessment in Navy systems* (pp. 169–178). San Diego: Naval Personnel Research and Development Center.

Orasanu, J. (1991, September). Individual differences in airline captains' personalities, communication strategies, and crew performance. *Proceedings of the Human Factors Society* (pp. 991–995). San Francisco, CA.

Orasanu, J., Dismukes, R. K., & Fischer, U. (1993, October). Decision errors in the cockpit. *Proceedings of the Human Factors Society* (pp. 363–367). Seattle, WA.

Pawlik, E. A., Sr., Simon, R., & Dunn, D. J. (1991, June). Aircrew coordination for Army helicopters: Improved procedures for accident investigation. *Proceedings of the Symposium on Aviation Psychology* (pp. 320–325). Columbus, OH: Ohio State University.

Reynolds, J. L., & Drury, C. G. (1993, October). An evaluation of the visual environment in aircraft inspection. *Proceedings of the Human Factors Society* (pp. 34–38). Seattle, WA.

Schwirzke, M. F. J., & Bennett, C. T. (1991, June). A re-analysis of the causes of Boeing 727 "black hole landing" crashes. *Proceedings of the Symposium on Aviation Psychology* (pp. 572–576). Columbus, OH: Ohio State University.

Seidler, K. S., & Wickens, C. D. (1992). Distance and organization in multifunction displays. *Human Factors, 34,* 555–569.

Selcon, S. J., Taylor, R. M., & Koritsas, E. (1991, September). Workload or situational awareness?: TLX vs. SART for aerospace systems design evaluation. *Proceedings of the Human Factors Society* (pp. 62–66). San Francisco, CA.

Shively, R., Battiste, V., Matsumoto, J., Pepitone, D., Bortolussi, M., & Hart, S. G. (1987, June). Inflight evaluation of pilot workload measures for rotorcraft research. *Proceedings of the Symposium on Aviation Psychology* (pp. 637–643). Columbus, OH: Ohio State University.

Stein, E. S. (1984). *The measurement of pilot performance: A master-journeyman approach* (Rep. No. DOT/FAA/CT-83/15). Atlantic City, NJ: Federal Aviation Administration Technical Center.

Taylor, H. L., & Alluisi, E. A. (1993). Military psychology. In V. S. Ramachandran (Ed.), *Encyclopedia of human behavior* (pp. 503–542). San Diego: Academic Press.

Van Patten, R. E. (1994). *A history of developments in aircrew life support equipment, 1910–1994.* Dayton, OH: SAFE—Wright Brothers Chapter.

Vidulich, M. A., & Tsang, P. S. (1985, September). Assessing subjective workload assessement: A comparison of SWAT and the NASA-bipolar methods. *Proceedings of the Human Factors Society* (pp. 71–75). Baltimore, MD.

Vreuls, D., & Obermayer, R. W. (1985). Human–system performance measurement in training simulators. *Human Factors, 27,* 241–250.

Wilson, G. F., & Fullenkamp, F. T. (1991). A comparison of pilot and WSO workload during training missions using psychophysical data. *Proceedings of the Western European Association for Aviation Psychology, II,* 27–34.

Wilson, G. F., Purvis, B., Skelly, J., Fullenkamp, F. T., & Davis, I. (1987, October). Physiological data used to measure pilot workload in actual and simulator conditions. *Proceedings of the Human Factors Society* (pp. 779–783). New York, NY.

4

Underpinnings of System Evaluation

David W. Abbott
Mark A. Wise
University of Central Florida

John A. Wise
Embry-Riddle Aeronautical University

BACKGROUND

Rapid advances in software and hardware have provided the capability to develop very complex systems that have highly interrelated components. Although this has permitted significant increases in system efficiency and has allowed the development and operation of systems that were previously impossible (e.g., negative stability aircraft), it has also brought the danger of system-induced catastrophes. Perrow (1984) argued that complex systems that are highly coupled (i.e., have highly interdependent components) are unstable and have a disposition toward massive failure. This potential instability makes human-factors-based evaluation more important than it has been in the past, whereas the component coupling makes the traditional modular evaluation methods obsolete.

Systems that are highly coupled can create new types of failures. The coupling of components that were previously independent can result in unpredicted failures. As systems become more coupled, interdisciplinary issues will become more critical. For example, it is possible that new problems could reside in the human–machine interface where disciplines meet and interact. It is in these intellectual intersections that new compromises and cross-discipline trade-offs will be made. And it will be in these areas that new and unanticipated human-factors-based failures may emerge.

As systems grow in complexity and intradependence the cost of performing adequate testing is rapidly approaching a critical level. The cost of certification in aviation has been a significant cost driver. The popular aviation press is continually carrying articles on an aviation part (e.g., an alternator) that is exactly the same as an automobile part (i.e., comes off exactly the same assembly line) but that costs two to three times as much because of the aviation certification costs. Human-factors-based verification, validation, and certification methods must thus not only be effective, they must also be cost-effective.

"Technically adequate" human factors testing may not even be sufficient or even relevant to a system becoming safely operational. The political and emotional issues associated with the acceptance of some technically adequate systems (e.g., nuclear power, totally automatic public transportation systems) must also be considered. For many systems the human factors evaluation must answer questions beyond safety and reliability, such as "What type of evaluation will be acceptable to the users and the public?," "How much will the public be willing to spend to test the system?," and "What level of security and reliability will they demand from the system?"

In spite of the fact that the importance of human-factors-based evaluation of the complex systems is increasing, the processes by which it is accomplished may be the most overlooked aspect of the system development. Although a considerable amount has been written about the design and development process, very little organized information is available on how to verify and validate high-complexity and highly coupled dynamic systems. In fact, the inability to adequately evaluate such systems may become the limiting factor in society's ability to employ systems that our technology and knowledge will allow us to design.

This chapter is intended to address issues related to human factors underpinnings of system evaluation. To accomplish this goal it addresses two general areas. The first section addresses the basic philosophical underpinnings of verification, validation, and certification. The second is a simple description of the basic behavioral science statistical methods. The purpose of this section is to provide the statistically naive reader with a very basic understanding of the interpretation of results using those tools.

DEFINITIONS

Verification and validation are very basic concepts in science, design, and evaluation. They form the foundation of the success or failure of each. Verification and validation should both be thought of as processes. In science *verification* is usually thought of as the process of determining the truth or correctness of a hypothesis. When dealing with system design, Carroll and Campbell (1989) argued that verification should also include determining the truth of conclusions, recommendations, practices, and procedures. And Hopkin (1994) suggested that one may need to extend the definition of verification to explore major system artifacts, such as software, hardware, and interfaces.

Validation is defined broadly by Reber (1985) as the process of determining the formal logical correctness of some proposition or conclusion. In the human factors context it may be seen as the process of assessing the degree to which a system or component does what it purports to do.

In aviation human factors an example of verification and validation might involve testing an interface for a flight management system (FMS). As a type of in-cockpit computer, the FMS provides ways for the pilot to enter data into the FMS and to read information from it. The design guidelines for a particular FMS might call for the input of information to be carried out through a variety of cryptic commands, using a number of different modes. If these guidelines are carried out exactly, then we have a system that is verifiable. However, if the system proves to be unusable, because of the difficult nature of the commands or poor legibility of the display output, then it may not be valid (assuming that one of the design goals was to be usable).

Hopkin (1994) suggested that:

- Verification and validation tend to be serial rather than parallel processes.
- Verification normally precedes validation.

- Usually both verification and validation occur.
- Each should be planned considering the other.
- The two should be treated as complementary and mutually supportive.

CERTIFICATION

Certification can be thought as the legal aspect of verification and validation: that is, it is verification and validation carried out such that a regulatory body agrees with the conclusion and provides some "certificate" to that effect. The concept of the certification of aircraft and their pilots is not new. For many years the engineering and mechanical aspects of aviation systems have had to meet certain criteria of strength, durability, and reliability before they could be certified as airworthy. Additionally, pilots of aircraft have to be certificated (a certification process) on their flight skills and must meet certain medical criteria. However, these components (the machine and the human) are the tangible aspects of the flying system, and there remains one more, a less readily quantifiable entity—the interface between human and machine (Birmingham & Taylor, 1954).

Why Human Factors Certification?

Why conduct human factors certification of aviation systems? On the surface this may seem like a fairly easy question to answer. Society demands safety. Society expects that the things it buys, the planes it uses and the cars it drives, are safe. Western society has traditionally looked to the government to ensure that safety by establishing laws and holding individuals or companies responsible when they are negligent. It is therefore no surprise that there should be a collective societal cry for the human factors of an aviation system to be certified. After all, the plane itself has to be certified, and the pilot has to be technically and medically certified. Does it not seem that the interface between these two should also be the focus of examination to assure that a "safe" pilot can in fact safely operate the "safe" engineering system?

If the intended goal of human factors certification is to insure the safety and efficiency of systems, then one might consider the following questions about certification. Would the process of human factors certification improve system safety by itself? Or would the threat of a human factors audit merely provide the impetus for human factors considerations in system development? Or would the fact that a design passed a human factors certification process inhibit further research and development for the system? Or would the fact that something was not explicitly included in the process, cause it to be neglected? Would it inhibit the development of new approaches and technologies so as to decrease the cost of certification? (One can see the effects of the last question in the area of general aviation where 30- to 50-year-old designs predominate.)

As mentioned earlier, the viability of human factors certification is not necessarily accepted by everyone as a proper or effective means for increasing the safety of complex systems. Another way in which to look at certification, accepting that this might be true, is as a "Machiavellian certification." In his discourse on ethics, *The Prince*, Machiavelli described a scenario in which a young man uses unethical means to obtain a political position with which he will perform many altruistic and socially benefiting programs. Machiavelli concluded that in such a scenario, the ends justified the means.

Could the same be true for certification, where a questionable means (certification) leads to a desirable end (safety and efficiency)?

Similarly, Endsley (1994) argued that the certification process may be not unlike a university examination. Most exams do not claim to be valid reflections of a student's knowledge of the course material; however, by merely imposing an exam upon the students they are forced to study the material, thus learning it. Certification can be viewed the same way—that is, certification, in and of itself, may not cause good human factors design. However, the threat of failing a product or system for poor human factors will "scare" the designers into considering the user from the beginning.

A formal human factors certification process may or may not be a feasible reality. It may be that an institutionalized certification process will not in actuality improve system safety or efficiency by any significant amount, but instead would merely be "a palliative and an anodyne to society" (Hancock, 1994). However, one could also point to many examples of certificated pilots fumbling the interface with certificated systems (capable of safe operation), thus leading to actual unsafe operation resulting in actual loss of life.

It is not the purpose of this chapter to address the legal issues associated with human factors certification of aviation (or any other type of system). Rather, this chapter addresses the technical and philosophical issues that would underpin the potential technical evaluation. For simplicity's sake, however, the word *evaluation* is used to imply verification, validation, and certification processes.

UNDERPINNINGS

Effective evaluation of large human–machine systems will always be difficult. The complexity and integration of such systems require techniques that seek consistent or describable relationships among several independent variables, with covariation among the dependent variables according to some pattern that can be described quantitatively. It cannot rely on tools that identify simple relationships between an independent variable and a single dependent measure, which one normally uses in classical experimental psychology research. Hopkin (1994) warned however, that although more complex multivariate procedures can be devised in principle, caution is required because the sheer complexity can ultimately defeat meaningful interpretation of the findings, even where the methodology is orthodox.

Hopkin (1994) went further to suggest that the following data sources can contribute to the evaluation process of new systems:

- Theories and constructs that provide a basis and rationale for generalization
- Data representative of the original data, but that may be at a different level (e.g., theories vs. laboratory studies)
- Similar data from another application, context, or discipline
- Operational experience relevant to expectations and predictions
- Expert opinion compared to the preceding items
- Users' comments based on their knowledge and experience
- Case histories, incidents, and experience with the operational system

This list is not intended to be all-inclusive but rather to be a model of the sort of data that should be considered.

Another basic issue that needs to be addressed early has to do with the selection of both the appropriate measures and data that will be relevant to the evaluation of the system to be tested. Experience has shown all too often that data are gathered based on the gut instincts without considering how the data are related and how they contribute to the evaluation process. Often, one comes to the data analysis phase only to find that key data are missing.

When Should Human Factors Evaluation Be Done?

When evaluation should take place during the cycle of a system's development has been touched upon briefly already. The temporal location of the evaluation will also affect the type of evaluation that can be applied. There are three different types, or times, of evaluation: a priori, ad hoc, and post hoc.

A priori evaluation includes the consideration of human factors requirements during the formation of initial design to criteria. This would require human factors input at the time when the design specifications are being laid out. Ad hoc evaluation would be evaluation that takes place throughout the production of the system. Post hoc evaluation would be the evaluation of the entire completed system. This would include the hardware, software, and human, and most importantly their interactions.

The issue in the "real" world is always going to be somehow related to cost. And given that the costs of making a change tend to increase geometrically as one moves away from conceptual design, cost considerations alone argue for a priori and ad hoc approaches, where a human factors evaluation process is done in a manner that allows needed changes to be made when the cost impact is low. This is particularly true of software intensive systems, where the loss of the original programmer or a loss of a the original programmer's memory for how the code was developed could result in the requirement to completely rewrite the code.

In the best-case scenario, human factors evaluation of complex aviation systems would require human factors consultation throughout the predesign, design, and implementation process. The involvement of a human factors practitioner during the process would guarantee consideration of the users' needs and insure an optimal degree of usability. However, the best-case scenario is unfortunately not always reality. In fact, as mentioned earlier, it may be that evaluation is only a hovering threat that forces designers to consider the human element, even though the process itself has no demonstrated operational validity. It may also be, as suggested by recent actual operational experience, that the cost of lives from operational failure of poor human factors design can easily justify the cost of developing and evaluating optimal human factors design throughout the design process. This could include simulation evaluation of pilot performance under various designs to validate the certified systems.

How Should Human Factors Evaluation Be Done?

Current standards and guidelines, such as the various military standards, provide a basis for evaluation of products. These standards can be useful for checking workspace design; however, the conclusions gained from "passing" these guidelines should be taken with a grain of salt.

Evaluation should be based on more than traditional design standards (e.g., Mil-Specs). Hopkin (1994) used the design of the three-pointer altimeter as an example of this point. If the task were to ensure that a three-pointer altimeter followed good human factors standards (good pointer design, proper contrast, text readability, etc.), it could be concluded that the altimeter was in fact certifiable. However, research has shown that the three-pointer altimeter is a poor way of presenting this type of information. In fact, errors of up to 10,000 ft are not uncommon (Hawkins, 1987). So, by approving the three-pointer altimeter based on basic design standards, a poorly designed instrument could be certified. Principle-based evaluation would have noted that a three-pointer altimeter was inappropriate even if it does meet the most stringent human factors standards. Principle-based evaluation would recommend a different type of altimeter altogether.

Wise and Wise (1994) argued that there are two general approaches to the human factors evaluation of systems: (a) the top-down or systems approach, and (b) the bottom-up or monadical approach. The top-down approach is built on the assumption that evaluation can be best served by examining the systems as a whole (its goals, objectives, operating environment, etc.), then examining the individual subsystems or components.

> In an aircraft cockpit, this would be accomplished by first examining what the aircraft is supposed to do (e.g., fighter, general aviation, commercial carrier), identify its operating environment (IFR, VFR, IMC, VMC, combat, etc.) and looking at the entire working systems which includes the hardware, software, liveware (operators) and their interactions; then evaluative measures can be applied to the subsystems (e.g., individual instruments, CRT displays, controls). (Wise & Wise, 1994, p. 15)

Top-down, or the systems approach to evaluation, is valuable in that it requires an examination of the systems as a whole. This includes the relationship between the human and the machine—the interface.

On the other hand, the bottom-up approaches look at the system as a series of individual parts, monads, that can be examined and certified individually. Using this method, individual instruments and equipment would be tested against human factors guidelines. Then the certified components would be integrated into the system. The bottom-up approach is very molar; that is, it tries to break down the whole into its component parts. The benefit of this method is that the smaller parts are more manageable and lend themselves to controlled testing and evaluation. It is obviously much easier to certify that a bolt holding a tier in place is sound, than to certify the entire mechanical system.

However, the simplicity and apparent thoroughness of this approach are somewhat counteracted by the tendency to lose sight of the big picture, such as what the thing is supposed to do. For a given purpose, a weak bolt in a given location may be acceptable; in another it may not be. Unless the purpose is known, one may end up with a grossly overengineered (i.e., overpriced) system.

Additionally, the sum of the parts does not equal the whole. A set of well-designed and well-engineered parts may all do their individual jobs well (verification) but may not work together to perform the overall task they were intended to perform (validation). A good example of this outside the world of aviation can be found in the art of music. Molecularly, a melody is simply made up of a string of individual notes; however, the ability to recognize and play the notes individually does not give sufficient cause for believing that the melody will in fact be produced. Thus individual subcomponents may individually function as designed, but not be capable of supporting an integrated performance in actual operational settings.

Human factors evaluation of an aviation system interface would be difficult, to say the least. However, it has been argued that the top-down evaluation produces the most operationally valid conclusions about the overall workability of a system. Perhaps only full system evaluation within high-fidelity operational-relevant simulation settings should be utilized.

HUMAN FACTORS EVALUATION AND STATISTICAL TOOLS

A basic and extremely important issue raised in the first half of this chapter is the reasoning that would underpin any of the proposed answers. That is, how does one scientifically demonstrate that a generated answer is indeed acceptable given a certain set of circumstances? The traditional method of evaluating the "truth" of a hypothesis (the most basic function in the evaluation process) in behavioral science and human factors has been the experimental paradigm. The basic guarantor of this paradigm is the statistical methods that support the experimental designs and that establish whether the results are meaningful or "truthful." Thus, an understanding of the basic concepts of statistics is necessary for anyone who even reviews one of the processes. To examine the results of a evaluation process, without understanding the capabilities and limits of statistics, would be like reviewing a book written in a language one could not read.

Unfortunately, there are a number of common misunderstandings about the nature of statistics and the real meaning or value of the various classes of statistical tools. Although it is impossible to provide a reader with adequate tools in part of a chapter, a chapter, or probably even a complete book, the goal of the following section is to provide:

- Awareness of the basic types of statistical tools
- Basic description of their assumptions and uses
- Simple understanding of their interpretation and limits

Anyone who is serious about this topic should prepare to undertake a reasonable period of study. A good place to start would be Shavelson (1996).

Introduction to Traditional Statistical Methods

Reaching valid conclusions about complex human–machine performance can be difficult. Research approaches and statistical techniques have been developed specifically to aid researchers in the acquisition of such knowledge. Familiarity with the logical necessity for various research designs, the need for statistical analysis, and the associated language used is a help in understanding research reports in the behavioral science and human factors areas.

This section is written to help the statistics-naive reader better understand and interpret the basic statistics used in behavioral science and human factors research. It addresses the following issues:

- Estimates of population values
- Relationships between factors
- Differences between groups

It is not intended as a "how to" chapter, as that is far beyond the scope of this undertaking. Rather, the hope is that the statistics-naive reader will be able to better understand and evaluate human factors and behavioral science research that utilizes the basic techniques covered in this text.

Estimates of Population Values

To understand or evaluate studies of human performance, one can begin with the most basic kind of research question. What is typical of this population? This would be the situation where a researcher is interested in knowing the behavior or characteristics that are typical of a large defined group of people (the population), but is able to study only a smaller subgroup (a sample) to make judgments. What is the problem here? A researcher who wanted to discover the typical number of legs that human beings have could pick a few and note that there is no person-to-person variability in number of legs; people all have two. People do not vary in number of legs from one to the other, so how many people a researcher selects for his sample, which ones are selected, how they are selected, and so on makes very little difference. The problem for researchers using human behavior and many human characteristics as the object of study is that virtually all nontrivial human behavior varies widely from person to person. Consider the researcher who wants some demographic information and skill level information regarding operators of FMS-equipped aircraft. The research involves selecting a subset (sample) of people from the entire defined group (population) and measuring the demographic and performance items of interest. How does a researcher select the sample? A researcher who seeks findings that will be applicable to the entire population will have to select the people to be studied in a way that will not give an unrepresentative, biased sample but will instead give a sample that is typical of the whole group and that will allow the researcher to state to what extent the sample findings might differ from the entire group. The correct selection techniques involve some method of random sampling. This simply means that all members of the population have an equal chance of being included in the sample. Not only does this technique avoid having a biased nonrepresentative sample, but researchers are able to calculate the range of probable margin of error that the sample findings might have from population truth. For example, it might be possible to state that the sample mean age is 40.5 years and that there is a 95% chance that this value is within 1.0 year of the true population value. If the researcher gathered this type of information without using a random sample—for example, by measuring only those pilots who fly for the researcher's friend Joe—the researcher might get a "sample" mean of 25 if Joe has a new, underfunded flight department, or of 54 if Joe has an older, stable flight department. In either case the researcher would not know how representative these group means are of the population of interest and would not know how much error might be present in the calculation.

Random sampling gives an approximate representation of the population, without any systematic bias, and allows one to determine how large an error may be present in the sample findings. This sort of research design is called a survey, or sample survey. It can take the form of a mailed questionnaire sent to the sample, or personal interviews with the selected sample, or obtaining archival data on the selected sample. In all cases the degree of likely error between the sample findings and the population values is determined by the person-to-person variability in the population and the size of the sample. If the population members have little individual difference on a particular characteristic, then the "luck of the draw" in selecting the random sample will not produce a sample that differs from the population. For example, in assessing the number of arms our pilot population has, because all have the same amount (i.e., "0" variability in the population) the sample mean will be identical to the population mean (i.e., both will be "2") no matter how the researcher selects the sample, with no error in the sample value. For characteristics on which pilots do differ, the greater is the variability in the individuals in the population, the greater will be the probable difference between any random sample mean and the population truth. This difference is called *sampling error* and is also influenced by the size of the sample selected. The larger the sample, the smaller is the sampling error. Consider a sample of 999 pilots from the entire population of 1,000 pilots. Obviously this sample will have a mean on any characteristic that is very close to the actual population value. Because only one score is omitted from any selected sample, the sample will not be much influenced by the "luck" of who is included. The other extreme in sample size would be to take a sample of only one pilot. Obviously here the sample-to-sample fluctuation of "mean" would be equal to the individual variability in the measured characteristic that exists in the population. Very large sampling error would exist because our sample mean could literally be take on any value from the lowest to the highest individual population score value.

Thus, the design issues in sample surveys are to be sure to obtain a random (thus unbiased) sample and to have a large enough sample size for the inherent variability in the population being studied so that the sample value will be close to the population truth.

There are two additional research questions that are frequently asked in behavioral research. One is, within one group of people, do scores on two variables change with each other in some systematic way? That is, do people with increasing amounts of one variable (e.g., age) also have increasing (or decreasing) amounts of some other variable (e.g., time to react to a warning display)? The second type of research question that is asked is, for two or more groups that differ in some way (e.g., type of altimeter display used), do they also have different average performances (e.g., accuracy in maintaining assigned altitude) on some other dimension? Let's look in more detail at these two questions and their research design and statistical analysis issues.

Questions of Relationships

In question of relationships researchers are interested in describing the degree to which increases (or decreases) in one variable go along with increased or decreased scores on a second variable. For example, is visual acuity related to flying skill? Is the number of aircraft previously flown related to time required to train to proficiency in a new type? Is time since last meal related to reaction time or visual perception? These example questions can all be studied as relationships between variables within a single group of research participants.

The statistical index that is used to describe such relationships is a Pearson correlation coefficient, r. This statistic describes the degree and direction of straight line relationship between values of the two variables or scores. The absolute size of the statistic varies from 0 to 1.0, where 0 indicates there is no systematic variation in one score dimension related to the increase or decrease in the other score dimension. A value of 1.0 would indicate that as one variable increases there is an exact and constant amount of change in the other score so that a plot of the data points for the two variables would all fall along a straight line. The direction of the relationship is indicated by the algebraic sign of the coefficient, with a minus sign indicating that as values on one dimension increase, values on the other decrease, forming a negative relationship. A plus sign indicates a positive relationship, with increases in one dimension going along with increases on the other.

In order to study such questions of relationships one must have a representative sample from the population of interest and two scores for each member of the sample, one on each variable.

Once the degree and direction of linear relationship have been calculated with the Pearson r, it is then necessary to consider if the described relationship in our sample came about because such a relationship actually exists in the population or came about by having sampling variability provide some nonrepresentative members in our sample who show such a relationship even though the true population situation is that no such relationship exists. Unfortunately, it is possible to have a relationship in a sample when none exists in the general population.

Was the result obtained because of taking a sample from a population where this relationship also exists, or was the sample relationship a result of a sampling error that gave a sample relationship when the population has no such relationship? Fortunately this apparent dilemma is easy to solve with statistical knowledge of sampling variability involved in random selection of correlational relationships, just as it is possible to calculate random sampling variability for sample means. A typical method for deciding if the observed correlation is real (exists in the population) or is simply due to non-representative sampling error is to calculate the probability of sampling error producing the observed size of sample correlation from a population where there is zero correlation. Thus if a researcher found an observed $r(n = 50) = .34$, $p = .02$; the p (probability) of .02 tells us that the chance of having sampling error produce a sample r of .34 when the population r is 0.0 is only 2 times in 100. As a general rule in the behavioral sciences, when sampling error has a probability as small as 5 in 100, or less, of producing our observed r, we conclude that our observed r came from a population that really has such a relationship rather than having come about by this sampling error from a population with zero correlation. This decision is reported by saying we have a *statistically significant*, or simply a *significant*, correlation. We are really concluding that our sample correlation is too big to have come just from sampling luck and thus there is a real relationship in the population.

> A random sample of corporate pilots showed a significant degree of relationship between total flying hours and time required to learn the new flight management system, $r(98) = -.40$, $p < .01$.

The interpretation of this standard results report is that the more flying hours corporate pilots have, the less time it takes them to learn a new FMS. The relationship within the sample of pilots is substantial enough that the researcher can conclude the relationship also exists among corporate pilots in general, because the chance of a nonrepresentative sample with this relationship being selected from a population not having this relationship is less than 1 in 100.

The researcher who finds a significant degree of relationship between the two variables may then want to calculate an index of the *effect size*, which will give an interpretable meaning to the question of how much relationship exists. This can be easily accomplished with the correlation relationship by squaring the r value to obtain the coefficient of determination, r^2. The coefficient of determination indicates the proportion of variability in one variable that is related to variation in the other variable. For example, an $r = .60$ between years of experience and flying skill would lead to an r^2 of .36. Thus it could said that 36% of the variability in pilot skill is related to individual differences in pilot experience. Obviously, 64% of variation in pilot skill is related to something(s) other than experience. It is this effect size index, r^2, and not the size of the observed p value that gives us information on the size or importance of the relationship. Although size of relationship does have some influence on the p value, it is only one of several factors. The p value is also influenced by sample size, and by variability in the population, so that no direct conclusion of effect size can be obtained by looking at the p value. The coefficient of determination, r^2, is needed.

What interpretation can be made about the relationship between two variables when a significant r is found? Is it possible to conclude that one variable influences the other, or is the researcher limited to only concluding that performance on one variable is related to (goes along with) the other variable without knowing why? The distinction between these two types of valid conclusion of significant research findings may seem minor, but in fact this is a major and important distinction. This is particularly true for any application of our results. What can be concluded from the significant ($r = .60$, $p = .012$) correlation between pilot experience (hours flown) and pilot skill (total simulation proficiency score)? There are essentially two options. The decision on what is a legitimate interpretation is based on the way in which the research study has been conducted. One possibility is to select a representative random sample of pilots from our population of interest and obtain scores on the two variables from all pilots in our sample. The second possibility would be to start again with a random sample, but this time the sample would have to come from initial pilots with low experience. A researcher would than randomly assign the sample pilots to receive a certain amount of experience, and after waiting for the experience to be obtained, skill measurements would be taken.

What is the difference in legitimate interpretation of the two studies? In the first approach, simply measuring experience and skill, it is not possible to know why the more experienced pilots have more skill. It could be that experience develops skill. It could be that pilots who have high skill get the opportunity to acquire flight time experience. It could also be that highly motivated pilots work hard to acquire both skill and experience. In short, the data show that experience and skill go together, but it cannot show if it is because experience develops skill, or skill leads to experience, or both follow from some other unmeasured factor.

For pilot selection applications of this study, this may be all that is needed. If a company selects more experienced pilots, they will on average be more skillful even if it is not known why. For training applications, however, sufficient information is not available from this study; that is, it cannot suggest that obtaining experience will lead to improved skill. This type of research design is called a post facto study.

Researchers simply find people who have already been exposed to or elected to be exposed to some amount of one variable and evaluate the relationship of scores on that variable to another aspect of behavior. Such designs only permit relatedness interpretations. No cause-and-effect interpretation or conclusion that the first variable actually influences the behavior is justified. A causal influence may or may not exist—one simply can not decide from this type of design. If it does exist, its direction (which is cause and which is effect, or are both variables "effects" of some other cause) is unknown. The researcher observes a relationship after the fact of the research participants having been exposed to different amounts of the variable of interest. Thus, if a statistically significant post facto relationship between the two variables is found, it will show the relationship does exist in the population but it will be impossible to determine why.

Questions of Group Difference

This approach to design involves creating groups of research participants that differ on one variable and then evaluating statistically to see if these different groups also differ significantly on the behavior of interest. The goal of the research may be either to find out if one variable is simply related to another (post facto study), or the goal may be to establish if one variable actually influences another (true experiment). With either goal, the question being asked using this method is whether or not groups differ, as opposed to the previous correlational design, which asked if scores were related for a single group.

If the groups are formed by noting the amount of one variable the participants currently have and assigning them to the appropriate group, it is a post facto design. If there is a significant group difference on the behavior performance, the interpretation will still be simply that the group difference variable and behavior are related without knowing why. The information obtained from a post facto group difference study is the same as would be obtained from the correlational relationship post facto study described earlier.

The statistical evaluation for "significance" would not be based on a correlation coefficient, but would use something like a *t*-test or an analysis of variance (ANOVA) procedure. These two techniques simply allow a researcher to calculate the probability of obtaining the observed differences in mean values (assuming random sampling) if in fact the populations are not different. That is, it is possible that the samples have different means when their populations do not have different means. Sampling variability can certainly lead to this situation. Random samples do not necessarily match the population exactly, so two samples can easily differ when their populations do not. If observed groups have mean values that have a very low probability (.05 or smaller) of coming from equal populations, that is, differing due to sampling error only, it is possible to conclude that the group variable being studied and behavior are truly related in the population, not just for the sample studied.

This is the same as the result from a post facto relationship question evaluated with a correlation coefficient described in the previous section. The legitimate interpretation of a post facto study will be the same whether the researcher evaluates the result as a relationship question with a correlation coefficient, or evaluates it as a group difference question with a test for significant differences between means. If the more powerful interpretation that a variable actually influences the behavior is required, then the researcher would need to conduct a *true experiment*. Although it is possible to conduct a true experiment as a relationship question evaluated with a correlation coefficient, this is very rare in practice. True experiments producing information on one variable actually influencing performance on another are almost always conducted as a question of group differences and evaluated for statistical significance with something other than a correlation coefficient.

To obtain cause-and-effect information, a research design where only the group difference variable could have lead to the observed difference in group performance is required. This research would begin by creating two or more groups that do not initially differ on the group difference variable, or anything else that might influence performance on the behavior variable; for example, research participants do not decide which group to join, the top or lowest performers are not placed in "groups," and existing intact groups are not used. Instead, equal groups are actively formed by the researcher, and controls are imposed to keep unwanted factors from influencing the behavior performance. Experimental controls are then imposed to be sure the groups are treated equally throughout the experiment. The only factor that is allowed to differ between the groups is the amount of the group difference variable that the participants experience. Thus, the true experiment starts with equal groups and imposes differences on the groups to see if a second set of differences result.

In this way it is possible to determine if the imposed group difference really influences the performance, because all alternate logical possibilities for why the groups differ on the behavior of interest are eliminated. In practice, the equal groups are formed either by randomly assigning an existing pool of research participants into equal groups, or by selecting several equal random samples from a large population of research participants. In either procedure, the groups are formed so that the groups are equal on all factors, known and unknown, that have any relationship to or potential influence on the behavior performance. The researcher then imposes the research variable difference on the groups and later measures the individuals and compares group means on the behavior performance.

As discussed earlier, random sampling or random assignment might have assigned people to groups in a way that did not produce exact equality. Thus the researcher needs to know if the resulting group differences are greater than the initial inequality random chance might have produced. This is easily evaluated with a test for statistical significance. If the test statistic value has a probability of .05 associated with it, then the sampling variability alone only has a 5/100 chance of producing as large a group mean difference as the one found. Again, for any observed result that has a probability of being produced by sampling luck alone that is as small as or smaller than 5/100 one will conclude that the difference came about from something other than this unlikely source and is "statistically significant." In this case, the researcher would conclude that the reason the groups have different behavior performance means is that the imposed group difference variable created these performance differences, and, if these differences are imposed on other groups, one would expect to reliably find similar performance differences.

Effect size indices similar to the r^2 in correlation coefficients are available here, also. The eta square statistic is frequently used for group difference designs, and it reports the proportion of performance variability that is due to the imposed variable influence (true experiment) or related to our observed group variable difference (post facto). The eta square index for differences between group means is analogous to the r^2 for relationships between variables.

Examples

As an example of a group difference true experiment versus a group difference post facto study, consider an attempt to find out if unusual attitude training influences pilot performance in recovering from an uncommanded 135-degree roll. Researcher A investigates this by locating 30 pilots in his company who have had unusual attitude training within the past 6 months and who volunteer for such a study. He compares their simulator performance to that of a group of 30 pilots from the company who have never had such training and have expressed no interest in participating in the study. A statistical comparison of the performance of the two groups in recovering from the uncommanded 135-degree roll indicates the mean performances for pilots who were or were not trained in unusual attitude recovery were 69.6 and 52.8, respectively. These means do differ significantly with $t(38) = 3.45$, $p = .009$.

With such a design one can conclude that the performance means for the population of trained and untrained pilots do differ in the indicated direction. The chance of obtaining nonrepresentative samples with means this different (from populations without mean differences) is less than 1 in 100. However, because this is a post facto study it is impossible to know whether the training or other pilot characteristics are responsible for the difference in the means.

Because Researcher A used a post facto study—that is, did not start with equal groups and did not impose the group difference variable (i.e., having or not having unusual attitude training) on the groups—there are many possible reasons that the trained group performed better. For example, the more skilled pilots sought out such training and thus perform any flight test better because of their inherent skill, not because of the training. Allowing the pilots to self-select the training created groups that differ in ways other than the training variable under study.

It is, of course, also possible that the attitude training is the real active ingredient leading to the roll recovery performance, but one can not know from Researcher A's study. It is only possible to know that seeking and obtaining attitude training *is related* to better roll recovery. Is it because better pilots seek such training, or because such training produces increased skill? It is impossible to know. Is this difference in interpretations relevant? If one is selecting pilots to hire, perhaps not. Simply hire those who have obtained such training, and they will (based on group averages) be more skilled. If one is trying to decide whether to purchase unusual attitude training for a company's pilots and the cost of such training is expensive, then one would want to know if such training actually leads to improved skill in pilots in general. If the relationship between attitude training and performance is due to the fact that only high-skill pilots have historically sought out such training, providing it to all may be a waste of time and money.

Researcher B has a better design for this research. Sixty pilots are identified in the company who have not had unusual attitude training. They are randomly assigned to one of two equal groups, either a group that is then given such training or to a group that gets an equal amount of additional standard training. Again the mean performance of the two groups differs significantly with a $p = .003$.

This time the research provides much better information from the significant difference. It is now possible to conclude that the training produced the performance difference and would reliably produce improved performance if imposed on all of the company's pilots. The pilot's average performance on unusual attitude recovery would be better because of the training. How much better could be indicated by looking at our effect size index. If eta squared equaled .15 we would know that the training leads to 15% of the variability among pilots on the performance being measured.

Frequently these questions of group difference are addressed with a research design involving more than two groups in the same study. For example, a researcher might randomly assign research participants to one of three groups and then impose a different amount of training or a different type of training on each group. One could then use a statistical analysis called analysis of variance (ANOVA) to see if the three amounts or types differ in their influence on performance. This is a very typical design and analysis in behavioral science studies. Such research can be either a true experiment (as described earlier) or a post facto study. The question of significance is answered with an F statistic rather than the t in a two-group study, but eta squared is still used to indicate the amount or size of the treatment effect.

For example, unusual attitude recovery was evaluated with three random samples of pilots using a normal attitude indicator, a two-dimensional outside-in heads-up display (HUD), or a three-dimensional HUD. Mean times to recovery were 16.3, 12.4, and 9.8 sec, respectively. The means do differ significantly with a one-way ANOVA, $F(2, 27) = 4.54$, $p < .01$. An eta squared of .37 indicates that 37% of pilot variability in attitude recovery is due to the type of display used. One can conclude that the three methods would produce differences among pilots in general, because the probability of finding such large sample differences just from random assignment effects rather than training effects is less than 1 in 100. Further, the display effects produce 37% of the individual pilot variability in time to recover. The analysis of variance (ANOVA) has established that the variance among the means came from display effects, not from random assignment differences in who was assigned to which group. This ANOVA statistical procedure is very typical for the analysis of data from research designs involving multiple groups.

Statistical Methods Summary

These are the basics of design and statistical procedures used in human factors research. These basics can be expanded in several dimensions, but the basics remain. Questions are asked about what is typical of a group, about relationships between variables for a group, and about how groups that differ on one variable differ on some behavior. More than one group difference can be introduced in a single study, and more than one behavior can be evaluated. Questions can be asked about group frequencies of some behavior such as pass/fail rather than average scores. Rank order of performance rather than actual score can be evaluated. Statistical options are numerous, but all answer the same question, that is, is the observed relationship or difference real or simply sampling variability?

Throughout all simple or elaborate designs and statistical approaches the basics are the same. The question being answered may be either of relationships between variables or of differences between groups. The design either may be post facto yielding only relatedness information or may be a true experiment with information on the influence a variable has on behavior. If one takes group differences as they are found and sees if they differ in other behaviors, this is a post facto design and it determines if the two differences are related, but not why. If the design starts with equal groups and then imposes a difference, it is a true experiment and such a design can determine if the imposed difference creates a behavior difference.

In reviewing or conducting research on the effects of design evaluation on system operational safety the research "evidence" needs to be interpreted in light of these statistical guidelines. Has an adequate sample size been used to assure representative information for the effect studied? Did the research design allow a legitimate cause and effect interpretation (true experiment), or was it only post facto information about relatedness? Were the sample results evaluated for statistical significance?

HOW WOULD WE KNOW EVALUATION WAS SUCCESSFUL?

One of the arguments against all types of evaluation is that evaluation drives up cost dramatically, whereas it adds little increase in safety. This is especially true for aviation systems, which have proven to have fewer accidents and incidents than any other type of transportation system (Endsley, 1994; Hancock, 1994). However, if society accepts fewer aviation accidents than it accepts in other modes of transportation, designers working in aviation must accept this judgment and work toward improved safety. Fewer operator errors in a simulator for certified systems than for poorly designed systems would be a better design evaluator than waiting for infrequent fatal accidents in actual operation.

A second problem inherent within this one is deciding when the evaluation process should stop. In a test of system (interface) reliability, there will always be some occurrences of a mistake. How few mistakes should the evaluation strive for? The problem is that the answer goes on and on and never is completely done. The challenge is to find how "reliable" a system needs to be before the cost of additional evaluation overcomes its benefits. Rather than slipping into this philosophical morass, perhaps the evaluation question should be: Does this certified system produce significantly fewer operational errors than other currently available systems? From a purely economic basis, insurance costs for aviation accidents are probably always cheaper than good aviation human factors evaluation design. This should not be an acceptable reason to settle for a first "best guess" at design. Rather, the best possible evaluation with human factors consultation and evaluation at the predesign, design, and implementation stages should be utilized.

REFERENCES

Birmingham, H. P., & Taylor, F. V. (1954). A design philosophy for man-machine control systems. *Proceedings of the IRE, 42*, 1748–1758.

Carroll, J. M., & Campbell, R. L. (1989). Artifacts as psychological theories: The case of human–computer interaction. *Behaviour and Information Technology, 8*, 247–256.

Endsley, M. R. (1994). Aviation system certification: Challenges and opportunities. In J. A. Wise, V. D. Hopkin, & D. J. Garland (Eds.), *Human factors certification of advanced aviation technologies* (pp. 9–12). Daytona Beach, FL: Embry-Riddle Aeronautical University Press.

Hancock, P. A. (1994). Certification and legislation. In J. A. Wise, V. D. Hopkin, & D. J. Garland (Eds.), *Human factors certification of advanced aviation technologies* (pp. 35–38). Daytona Beach, FL: Embry-Riddle Aeronautical University Press.

Hawkins, F. H. (1987). *Human factors in flight.* Aldershot, Hampshire: Gower.

Hopkin, V. D. (1994). Optimizing human factors contributions. In J. A. Wise, V. D. Hopkin, & D. J. Garland (Eds.), *Human factors certification of advanced aviation technologies* (pp. 3–8). Daytona Beach, FL: Embry-Riddle Aeronautical University Press.

Perrow, C. (1984). *Normal accidents: Living with high-risk technologies.* New York: Basic Books.

Reber, A. S. (1985). *The Penguin dictionary of psychology.* Harmondsworth, Middlesex: Penguin Books.

Shavelson, R. J. (1996). *Statistical reasoning for the behavioral sciences* (3rd ed.). Needham Heights, MA: Allyn & Bacon.

Wise, J. A., & Wise, M. A. (1994). On the use of the systems approach to certify advanced aviation technologies. In J. A. Wise, V. D. Hopkin, & D. J. Garland (Eds.), *Human factors certification of advanced aviation technologies* (pp. 15–23). Daytona Beach, FL: Embry-Riddle Aeronautical University Press.

5

Organizational Factors Associated With Safety and Mission Success in Aviation Environments

Ron Westrum
Anthony J. Adamski
Eastern Michigan University

This chapter examines the organizational factors in aviation safety and mission success. The organizations involved cover the entire range of aviation organizations, from airline operations departments to airports, manufacturing organizations, air traffic control, and corporate flight departments. Organizational factors include such things as organizational structure, management, corporate culture, training, and recruitment. Although the greater part of this chapter is focused on civil aviation, we also devote some attention to space and military issues. We also use examples from other high-tech systems for illustration of key points. Obviously, full treatment of a field so broad could result in a publication the size of this book. So we concentrate on key organizational processes involved in recent studies and major accidents. These will open the general issues.

The authors have tried to integrate empirical studies within a broader framework, a model of effective operation. We believe failures occur when various features of this model are not present. In choosing any model we risk leaving out some critical factors. This is a calculated risk. We believe that further discussion will proceed best with such an integrative framework.

HIGH INTEGRITY

The underlying basis for this chapter is a model of *high integrity* for the development and operation of equipment and people. The model is guided by adapting a principle stated by Arthur Squires. Squires was concerned about the integrity of the engineering design process in large systems. Considering several major failures, Squires (1986) proposed this criterion: "An applied scientist or engineer shall display utter probity toward the engineered object, from the moment of its conception through its commissioning for use" (p. 10). Following Squires' idea, we propose to state the principle in this way:

The organization shall display utter probity toward the design, operation, and maintenance of aviation and aerospace systems.

Thus, organizations with "utter probity" will get the best equipment for the job, use it with intelligence, and maintain it carefully (Fig. 5.1). In addition, they will display an honesty and sense of responsibility appropriate to a profession with a high public calling. Organizations that embody this principle are "high-integrity" organizations. These organizations can be expected to do the best job they can with the resources available. The concept unites two related emphases, both common in the organization literature: high reliability and high performance.

High Reliability. The high-reliability organization concentrates on having few incidents and accidents. Organizations of this kind typically have systems in which the consequences of errors are particularly grave.

For example, operations on the decks of aircraft carriers involve one of the most tightly coupled systems in aviation. During the Vietnam war, for instance, two serious carrier fires, each with high loss of life, war materiel, and efficiency, were caused when minor errors led to chains of fire and explosion (Gillchrist, 1995, pp. 24–26). Today aircraft carrier landings are one of the archetypical "high-reliability" systems (Roberts & Weick, 1993).

High Performance. The high-performance organization concentrates on high effectiveness. Here, instead of the multifaceted approach of the high-reliability organization there is often a single measure that is critical. "Winning" may be more important than flawless operation. The emphasis is on getting the job done (e.g., beating an adversary) rather than error-free operation.

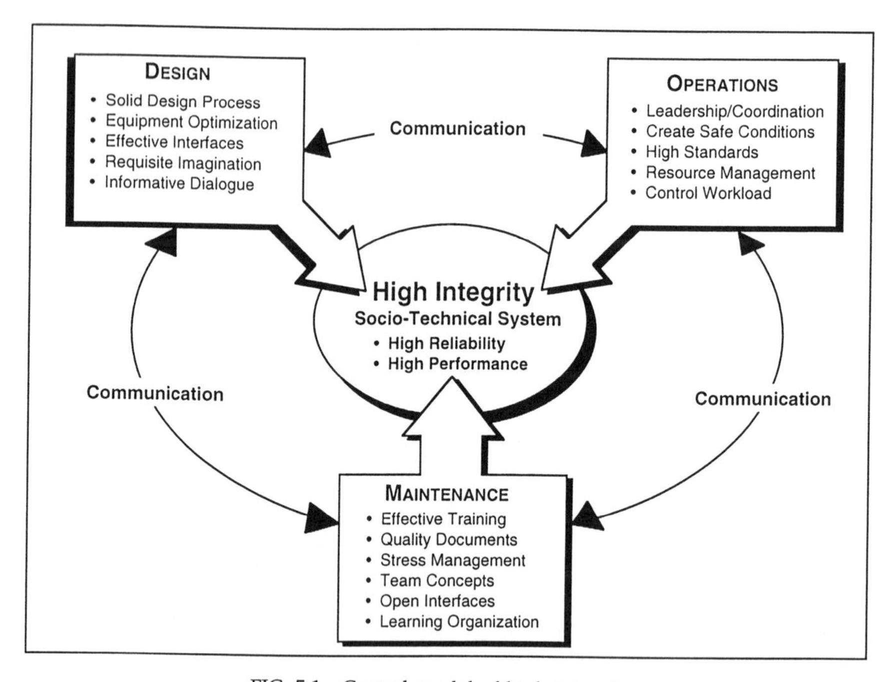

FIG. 5.1. Central model of high integrity.

For example, during the Korean conflict the Naval Ordnance Test Station at China Lake designed and produced an anti-tank rocket, the RAM, in 29 days. The need for this weapon was so critical that safety measures usually observed were suspended. The Station's Michelson Laboratory was turned into a factory at night, and the production line ran down the main corridor of the laboratory. Wives came into the laboratory to work alongside their husbands to produce the weapon. The RAM was an outstanding success, but its production was a calculated risk.

A suggestive hypothesis is that in high-performance situations there is a more masculine emphasis on winning, on being an "ace," and individual achievement, whereas high-reliability situations put an emphasis on balanced objectives and team effort. The context will determine which of these two emphases is more critical to the situation at hand. Usually in civilian operations high reliability is the stronger emphasis, whereas in a military context high performance would be more important than error-free operation. Because organizations may face situations with differing performance requirements, effective leadership may shift emphasis from one of these orientations to the other. We believe, however, that high-integrity operation implies protection of critical information flows. Maintaining utter probity is possible only when information is freely shared and accurately targeted. Thus high-integrity organizations would have certain common features involving information including the following:

1. All decisions are taken on the best information available.
2. The processes that lead to or underlie decisions are open and available for scrutiny.
3. Personnel are placed in an environment that promotes good decision making and encourages critical thought.
4. Every effort is made to train and develop personnel who can and will carry out the mission as intended.
5. Only those persons who are in a fit state to carry out the mission are made responsible to do so.
6. Ingenuity and imagination are encouraged in finding ways to fulfill the organization's objectives.

The rest of this chapter is concerned with the development of organizations that exhibit these performance characteristics. We believe that these features allow high-integrity systems to operate with safety and effectiveness. Conversely, organizations where incidents or accidents are likely to occur are organizations where one or more of these principles are compromised. The authors believe that every movement away from these principles is a movement away from high integrity and toward failure of the system (cf. Maurino, Reason, Johnston, & Lee, 1995).

BUILDING A HIGH-INTEGRITY HUMAN ENVELOPE

Around every complex operation there is a human envelope that develops, operates, maintains, interfaces, and evaluates the functioning of the sociotechnical system. The system depends on the integrity of this envelope, on its thickness and strength. Compromises to its strength and integrity uncover the system's weakness and make it

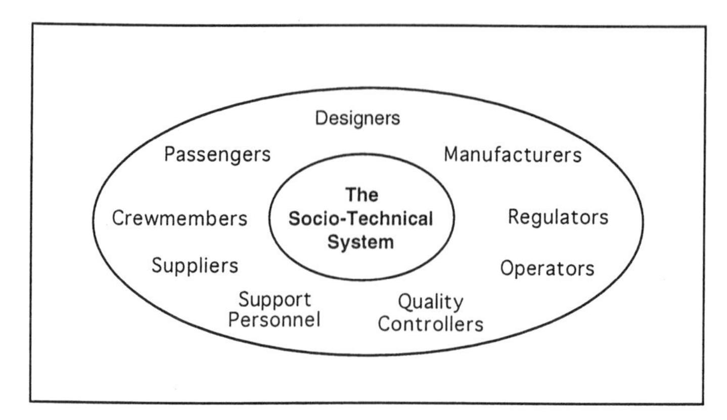

FIG. 5.2. Members of the human envelope.

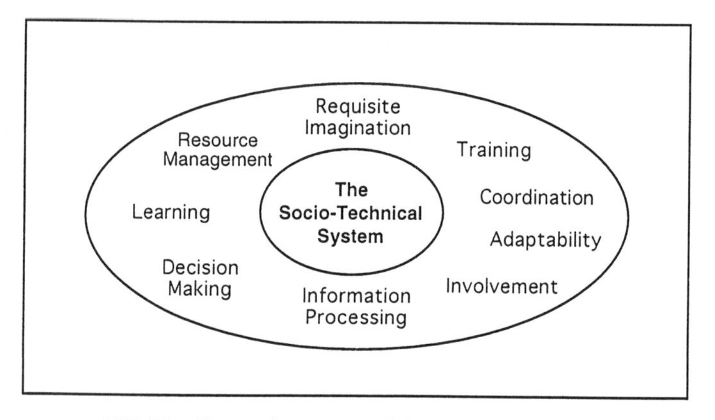

FIG. 5.3. Essential activities of the human envelope.

vulnerable. Accordingly, an aviation organization that nurtures this envelope will be strong. One that weakens it is asking for trouble (Figs. 5.2 and 5.3).

In the sections that follow, we examine the activities that provide the high-integrity human envelope including:

1. Getting the right equipment,
2. Operating the equipment,
3. Growing a high-integrity culture,
4. Maintaining human assets,
5. Managing the interfaces, and
6. Evaluation and learning.

THE RIGHT STUFF: GETTING PROPER EQUIPMENT

Design: Using Requisite Imagination

The focus of this section is the design process, and subsequent interactions over design, rather than technical aspects of the designs themselves. It may seem strange to begin with the design of equipment, because in many cases aviation organizations take it for granted.

However, getting proper equipment is essential to high-integrity functioning. The organization that uses bad equipment will have to work harder to achieve success than the one that starts out with the proper equipment. The equipment the organization uses should be both adequate to insure a reasonable level of safety and the best available for the job—within the constraints of cost. This principle suggests that no aviation organization can afford to be indifferent to the equipment that it uses to its development, manufacture, and current state of functioning. It should systematically search out the best equipment it can afford to match mission requirements, test it carefully, and endeavor to use it with close attention to its strengths and weaknesses.

Design serves human purpose, and should do so in an economical and safe way. But system design, particularly on a large scale, often fails through lack of foresight. In designing big systems, mistakes in conception can lead to large and costly foul-ups, or even system failure. (Collingridge, 1994). This seems to be particularly true of software problems. About 75% of major software projects actually get put into operation; the other 25% are canceled (Gibbs, 1994). Many large systems, furthermore, may need considerable local adjustment, as has happened with the ARTS III software used by the Federal Aviation Administration (FAA) to manage major airport traffic control (Westrum, 1994).

Recent years have provided many examples of compromised designs that affected safety. The destruction of the Challenger, the Hyatt Regency disaster, and the B-1 and B-2 bombers provide some major instances. In each case designers did not think through the design or executed it badly.

Another major example of design failure is the Hubble Space Telescope. Hubble failed because neither the National Aeronautics and Space Administration (NASA) nor the contractor insisted on carrying out all the tests necessary to determine if the system was functioning correctly. Instead, overreliance on a single line of testing, failure to use outside critical resources, and rationalization of anomalies ruled the day. When the telescope was launched, there was already ample evidence that the system had problems, but this evidence had been ignored. In spite of the many indications that the telescope was flawed, none were pursued. Critical cross-checks were omitted, inquiry was stifled, and in the end, a flawed system was launched, at great public cost (Caspars & Lipton, 1991). The failure of the Hubble Space Telescope was a failure of the design process.

An equally flagrant example is the Denver Airport automated baggage handling system. Here an unproven system for moving passengers' luggage was to be a key interface between parts of the airport. The concept cried out for a careful scale-up, but none was carried out. When the airport opened, the automated baggage system did not work, and a manual backup was used instead, at great cost (Hughes, 1994).

The Hubble telescope and Denver Airport cases were mechanical failures. In others the equipment works mechanically, but does not interface well with people. This can happen through poor interface design (such as error-encouraging features), or because unusual or costly operations are necessary to maintain the equipment (cf. Bureau of Safety, 1967).

Recently a group of French researchers carried out a major study of French pilots' attitudes about automation (Gras, Morocco, Poirot-Delpech, & Scardigli, 1994). One of the most striking findings from this study was pilots' concern about lack of dialogue with the engineers who designed their equipment. Not only did pilots feel that there was insufficient attention to their needs, but they also felt that designers and even test pilots had a poor grasp of the realities that pilots faced. Although attitudes toward automation were varied, pilots expressed very strong sentiments that more effort was needed to get designers in dialogue with pilots before equipment features were finalized.

One of the key skills of a project manager is an ability to anticipate what might go wrong, and test for that when the system is developed. Westrum (1993) has called this "requisite imagination" (cf. Petroski, 1994). Requisite imagination often indicates the direction from which trouble is likely to arrive. Understanding the ways in which things can go wrong often allows one to test to make sure that they have not. As Petroski (1994) showed, great designers are more likely to ask deeper and more probing questions, and to consider a wider range of potential problems.

Although foresight is valuable, however, it cannot be perfect. Even the best systems design strategy (Petroski, 1994; Rechtin, 1992) cannot forsee everything. So once the system is designed and produced, monitoring needs to continue, even if nothing appears to be wrong. If things begin to go wrong, a vigilant system will catch the problems sooner. The Comet and Electra airliners, for instance, needed this high level of vigilance, because each had built-in problems that were unanticipated (Schlager, 1994, pp. 26–32, 39–45). Such examples show that engineering is seldom so far advanced that all problems can be anticipated beforehand. Even maestros (discussed later) do not anticipate everything. Joseph Shea, a fine systems engineer, blamed himself for the fire that killed three of the Apollo astronauts. Yet Shea had done far more than most managers to anticipate and correct problems (Murray & Cox, 1989).

Getting the Knowledge as Well as the Hardware

No equipment comes without an intellectual toolkit. This toolkit includes, but is not limited to, the written manuals. Kmetz (1984), for instance, noted that the written documentation for the F-14 Tomcat fighter requires 300,000 pages. But these abundant materials often are deficient in both clarity and usability. We have observed that the creators of many operational documents—that is, checklists, operational manuals, training manuals, and so on—assume that their message is transparent and crystal clear. Often the message is anything but transparent and clear. It's faults can include documents that are difficult to use, and therefore are not used; complex procedures that encourage procedural bypasses and workarounds; and difficult-to-understand documents, composed by writers who have not considered the needs of the end users. The writers of such documents unwittingly set up future failures.

Manuals always leave things out. All equipment is surrounded by a body of tacit knowledge regarding the fine points of its operation. Getting this tacit knowledge along with the formal communication may be vital. Tacit knowledge may include matters that are difficult to put into words or unusual modes of the equipment that for liability or other reasons are difficult to commit to print. Sometimes manuals are not updated for reasons of organizational politics (e.g., Gillchrist, 1995, pp. 124–125). What goes into manuals may involve erroneous assumptions about what people would "naturally" do. For instance, during an investigation into two Boeing 737 accidents, an FAA team discovered that the designers assumed that pilots would respond to certain malfunctions by taking actions that were not in the written manual for the 737. Among other assumptions, the designers believed that if one hydraulic system jammed, pilots then would turn off both hydraulic systems and crank the landing gear down by hand. Of course, if the plane were on landing approach, there might not be time to do this. Although the hydraulic device failure is rare in the landing situation, the key point is that expected pilot actions were not communicated in the manual (Wald, 1995). The Boeing 737 is one of the safest jets in current use, yet this example illustrates that not all information regarding the equipment is expressed in the manual. Nor should it be; there are lots of things that one need not know. But sometimes critical things can get left out. In accepting a new airliner, a used airliner, or any other piece of machinery, care needs to be taken to discover this tacit knowledge.

The designers may not be the only holders of this tacit knowledge. Sometimes other pilots, operators of air traffic control equipment, or mechanics may hold this not-written-down knowledge. A study of Xerox copier repair people, for instance, showed that much of the key information about the machines was transmitted orally through scenario exchange between repair people (Brown & Duguid, 1991). Similarly, process operators in paper pulp plants often solved problems through such scenario exchange (Zuboff, 1984). Kmetz (1984) found that unofficial procedures ("workarounds") were committed only to the notebooks of expert technicians working on avionics repair. Sensitivity to such off-the-record information, stories, and tacit knowledge is important. It is often just such knowlege that gets lost in layoffs, personnel transfers, and reshuffling (cf. Franzen, 1994).

The use of automation especially requires intensive training in the operation and the quirks of the automated system. But training requires constant updates. Some key problems may be pinpointed only with field experience of the hardware. Failure of the organization to collect and transmit information about quirks in a timely way could well lead to failure of the equipment, death, and injury.

For instance, on December 12, 1991, an Evergreen Air Lines 747 over Thunder Bay in Canada ran into trouble with its autopilot. The autopilot, without notifying the pilots, began to tip the plane over to the right, at first slowly, then more rapidly. The pilots did not notice the motion because it was slow. Finally with the right wing dipping radically, the plane lost lift, and began plummeting downward. After much struggle, the pilots succeeded in regaining control, and landed in Duluth, Minnesota. An FAA investigation revealed that over the years similar problems had occurred with 747 autopilots used by other airlines. Particularly intriquing, however, was the discovery that the Evergreen plane's roll computer had previously been installed in two other planes in which it also had caused uncommanded rolls. The exact cause of the problem in the roll computer remains unknown (Carley, 1993).

Because automation problems are more fully covered elsewhere in this book (see chaps. 6, 7, and 20), we do not discuss them further here. However, it is worth noting that hardware and software testing can in principle never be exhaustive (Littlewood & Strigini, 1992) and that therefore the price of safety is constant vigilance and rapid diffusion of knowledge about equipment problems.

Sustaining Dialogues About Key Equipment

For aviation organizations, we should think about information in terms of a constant dialogue rather than a single transmission. Once a system is turned over to the users, the design process does not stop, it simply scales down. And around each piece of key equipment in the aviation organization, a small or large dialogue may be needed. This dialogue includes manufacturers, operators, and regulators as the most obvious participants. Obviously, aircraft themselves and their engines are particularly important subjects of such dialogue, but other items of equipment also need consideration. When dialogue is lacking, unpleasant things can happen.

Take, for instance, the disastrous fire on a Boeing 737 at Ringway Airport near Manchester in the United Kingdom, on August 22, 1985. The fire involved an engine "combustion can" that fractured, puncturing a fuel tank. The can had been repaired by a welding method that had met British CAA standards, but was not what the manufacturer called for in the manual it issued to British Airways. This accident was the most dramatic of a series of problems with the cans. Earlier problems had been written off as improper repairs, but this masked a key breakdown. One sentence in the accident report highlighted this key breakdown in communication between the operators (British Airways) and the engine makers (Pratt & Whitney):

> It has become evident from the *complete absence of dialogue* between British Airways and Pratt & Whitney on the subject of combustion can potential failures that, on the one hand, the manufacturer believed that his messages were being understood and acted upon, and on the other that the airline interpreted these messages as largely inapplicable to them at the time. (cited in Prince, 1990, p. 140)

It was management's responsibility to notice and eliminate the discrepancy between what the manual called for and what was asked of the maintenance technicians. Obviously, the bad practices continued only through management's willingness to allow variance from the recommended practice.

It should therefore be obvious that the security of an airplane is shaped—in part—by the quality of dialogue between maker and user. The combustion can problems were evidently a case of the "encapsulation" response (explained later), in which the system did not pay attention to the fact that it was having a problem.

A particularly important study was conducted by L. Homer Mouden (1992, p. 141) for the Aviation Research and Education Foundation to determine the most significant factors in preventing airline accidents. Mouden's study included personal interviews with senior airline executives, middle management personnel, and airline safety officers to determine what actions by management they considered the most effective for accident prevention. Several of those interviewed indicated that they thought complete safety was probably an unattainable goal. Many also indicated that risk management was part of every manager's responsibilities. The factors most often mentioned by managers as having strong influence on safety were effective communication, training, and standard operating procedures.

Mouden's study demonstrates the need for a sensitivity to the communication channels in the organization. He noted that sometimes the designated communication channels in the organization are less effective than they are believed to be, but their failure is discovered only after some unpleasant event has occurred. Thus latent failures may accumulate unseen (cf. Reason, 1990). Mouden presented a series of case studies that show these problems with communication. While the organization chart emphasized vertical communication, Mouden discovered that managers at virtually all levels considered lateral communication more effective than vertical.

Customizing the Equipment

Equipment in constant use does not long stay unchanged. Through use, repair, and on-the-spot redesign, its form mutates. Customizing equipment can lead to two situations, each of which is worth consideration:

1. Enhancements May Improve Safety. Changes may provide substantial advantages by improving the ease, efficiency of operations, or aesthetic qualities for the local users.

Eric Von Hippel, in studies of "lead users," found that lead users are more likely to customize their equipment (Peters, 1992, pp. 83–85). Often in the changes that lead users make are the secrets for improving equipment, which, if carefully studied, will provide better manufactured products in the future. This certainly appeared to be true of the ARTS-III traffic control software, developed by the FAA. A considerable number of "patches" had to be made to the software to allow for local conditions. These patches, furthermore, were more likely to be spotted and transmitted face to face, rather than through any official channels. Many of the patches were tested late at night, when traffic was light, before being officially submitted for approval. The FAA, however, seemed slow to pick up on these changes (Westrum, 1994).

There has been intense interest in the "high-performance team" ever since Peter Vaill wrote his 1978 article. We can define a high-performance team as one operating beyond ordinary expectations for the situation in which the group finds itself. Just as the ace or the virtuoso embodies unusual individual performance, the "crack" team shows a group peforming at a virtuoso level. This does not mean simply a group of virtuosos, but rather a group whose interactions allow performance of the task at a high effectiveness level. Although the literature on high reliability seems to have ignored Vaill's work, it is evident that high reliability shares many of the same characteristics as high performance. In any case, high-integrity teams get more out of their equipment. It is a common observation that such teams can get the same equipment that may turn out a lackluster performance for others to perform "like a Stradivarius" for them. There are two reasons for this.

First, these teams know their equipment better. High-integrity teams or organizations take little for granted and make few assumptions. The equipment is carefully studied, and its strengths and limitations are recognized (Wetterhahn, 1997, p. 64). The team checks out and understands what it has been given, then "tunes it up" for optimal performance. High-performance teams will often go beyond the usual boundaries to discover useful or dangerous features. When the "Top Gun" air combat maneuvering school was formed, the characteristics of the F-4 Phantom were carefully studied, and so the team was able to optimize its use in combat (Wilcox, 1990). Similarly, in the Falklands war, one of the two British Harrier squadrons, the 801, carefully studied learned how to use its Blue Fox radar, whereas the companion 800 squadron considered the Blue Fox unreliable and of limited value. The combat performance of the two groups strongly reflected this difference, with the 801 outperforming the other. Captain Sharkey Ward, Officer in Charge of the 801, summed up what he learned from the conflict: "I have no hesitation in presenting the following as the most important lessons of the Falklands air war. The two main lessons must be: Know your weapons platforms, their systems and operational capabilities; then employ them accordingly and to best effect" (Ward, 1992, p. 355). Thus, it is not just discovering the "edge of the envelope" that is important for high performance teams, but also training to exploit exactly the features discovered.

High-integrity teams will sometimes even reject the equipment they have been given. If what they have been given is not good enough, they may go outside channels to get the equipment they need. They are also natural "tinkerers." In a study of nuclear power plants and their "incident" rates, Marcus and Fox (1988) noted that the teams that carefully worked over their equipment were likely to have lower incident rates. Peters (1988, p. 166) also remarks that high-performance R&D teams customize their equipment more.

Often the procedures of high-integrity teams skirt or violate official policy. Sometimes this can affect safety. High-level policies are sometimes shaped by forces that have little to do with either mission success or safety. So when high performance is the number one criterion for the front line, policy may get violated. In Vietnam, when Air Force Falcon missiles did not work, they were replaced by Sidewinder missiles that did (Wetterhahn, 1997, p. 69). In a study of the use of the VAST avionics check-out system in aircraft carrier shops, for instance, Kmetz found that many procedures that were not official policy were used to get the job done (Kmetz, 1984). Similarly, in Vietnam American technicians often used "hangar queens," contrary to official policy (Trotti, 1984). Knowing when to intervene in such situations and when to leave them untouched is the essence of managerial judgment.

2. ***Safety-Degrading Changes.*** Wherever there is choice, there is danger as well as opportunity. A failure to think through actions with equipment may lead to human factors glitches. One example was United Airlines' new color scheme, dark gray above and dark blue below, which some employees called a "stealth" look. The poor visibility created for both planes and airport vehicles in matching colors evidently was not considered. It apparently led to a number of airport "fender benders" (Quintanilla, 1994). Similarly, methods for saving time, money, or hassles with equipment can often lead into the danger zone. Some airliners, for instance, may "fly better" with certain circuit breakers pulled. Although it is good to know such things, overuse of this inside knowledge can encourage carelessness and cause incidents.

Bad maintenance or repairs may cause equipment failures almost as dramatic as the use of substandard parts. In the Manchester fire case, there would have been no problem if the manufacturer's instructions for maintenance had been followed.

Yet it may be almost as bad to accept the equipment "as delivered," and "hope for the best" along with manuals and supportive documentation. Cultural barriers that impede or impair information search or active questioning may be one reason that this happens. Unwillingness to question may be particularly strong when the providers of the hardware are a powerful technical culture (e.g., the United States) and the recipients do not have a strong indigenous technical culture of their own. Airliners delivered to some developing countries may thus arrive with inadequate dialogue.

The organization receiving the equipment may cause further problems by dividing up the information involved and using it in adversarial ways. In fact, for groups with low team skills or internal conflicts, equipment may become a center for organizational struggle. Different subgroups may assert their prerogatives, hiding knowledge from the other, instead of working in a cooperative manner. Stephen Barley studied two hospital groups using computer tomography (CT) scanners, and found that cooperation between doctors and technicians was difficult to achieve (Barley, 1986). When such knowledge is divided between groups that do not communicate well, the best use of equipment is not possible.

MANAGING OPERATIONS: COORDINATION OF HIGH-TECH OPERATIONS

Creating Optimal Conditions

One of the key functions for all levels of management in an aviation system is creating optimum human factors situations in which others will operate. This means making sure that all the human factors environments in the aviation organization provide contexts and personnel that will result in a safe accomplishment of the job. In high-integrity organizations pilots, flight attendants, maintenance personnel, and dispatchers are more likely to find themselves in situations where they can operate successfully, when they have received the appropriate training for the activity, and where they get an adequate flow of information to do the job correctly.

Environmental design is thus a management responsibility. At the root of many accidents is a failure to manage the working environment. For instance, on March 1, 1994, the crew of a Boeing 747-251B in a landing rollout at Narita Airport found one of its engines dragging (NTSB, 1994). The reason, it seemed, was that pin retainers for a diagonal engine brace lug had not been reinstalled during the "C" check in St. Paul, Minnesota. In looking into the accident, the National Transportation Safety Board (NTSB) found that conditions in the Northwest Airlines Service Facility in St. Paul consituted an error-prone environment. Mechanics' understanding of procedures was inconsistent, training was not systematically carried out, and the layout of the inspection operations was inefficient, causing stress to the inspectors. Clearly, these were conditions that management needed to identify and improve.

James Reason, in introducing his well-known theory of accidents, noted that errors and mistakes by the operators at "the sharp end" are often promoted as the "cause" of accidents, when actions by management have actually created unsafe conditions in the first place. These management actions create situations that Reason termed *latent pathogens*—accident-prone or damage-intensifying conditions (Reason, 1990). It is important, therefore, to be aware of the potential for putting personnel in situations where they should never be in the first place. A reluctance to create hazardous situations needs to go hand in hand, however, with a willingness to deal with them when they appear.

For instance, both British airlines and the British pilots union, BALPA, were reluctant to admit that pilot fatigue was a problem. Fatigue is a proven killer, yet a good many senior managers used a "public relations" strategy (discussed later) to paper over the problem (Prince, 1990, pp. 111–129). A latent pathogen existed, but the organizations steadfastly hid it from sight. Unhappily, the problem did not go away, just its visibility.

Similarly, when fire broke out on a grounded Saudi Arabian Airlines flight in Riyadh on August 19, 1980, the three Saudi Arabian Airlines pilots involved failed to take critical actions in a timely way. Their casualness and inaction apparently caused the entire complement of flight SV 163—301 persons—to die needlessly. All three pilots had records that indicated severe problems (Prince, 1990, p. 130). So who put these pilots at the controls? It would appear a serious failure for management at any airline to place such men at the controls of a Lockheed L-1011.

Planning and Teamwork

Emphasis on planning is a strong indicator of high integrity. High-integrity organizations do not just "let it happen." More of their activities and decisions are under conscious and positive control. A popular bumper sticker in the United States announces that "Shit Happens." The implication is that bad things happen in ways that are difficult to predict or control. This expresses a common working-class attitude about the level of control of the person over his or her life—that is to say, very little. The "shit happens" philosophy of life is at the opposite pole from that of the high-reliability team. Very little "shit" is allowed to happen in a high-integrity organization, and what does is carefully noted, and, if possible, designed out of the next operation.

High-integrity organizations often appear to have mastered disciplines that others have not, and thus are able to do things that other organizations consider outside their realm of control. In civilian operations, this has meant a higher degree of safety; for the military, it has meant higher mission success rates.

A remarkable picture of a high-reliability team is given in Aviel's article (1994) on the tire repair shop at United Airlines' San Francisco maintenance facility. High integrity is evident in the small team's self-recruitment, self-organization, high morale, excellent skills, customized layout, and obvious comprehensive planning. We would all like to know how to build such teams in the first place. But to not interfere with them is something that every management group can learn. Aviel points out that United was willing to give up some apparent economies to keep the team together.

Some high-integrity teams require extensive practice. But what should be done when the crew—such as an airliner flight deck team—needs to be a team temporarily? It appears that high-integrity characteristics may form even in a short space of time with the right leadership, the right standard operating procedures, and proper training. The captain, in the preflight briefing, shapes the crew atmosphere, and this in turn shapes interactions during the flight (Ginnett, 1993). A cockpit with a crew resource management (CRM) atmosphere therefore can be created (or destroyed) rapidly.

One instance of excellent CRM skills took place on United Airlines flight 811, flying from New York to New Zealand. Flight 811 was a Boeing 747. The front cargo door blew out, killing several passengers, and a 50% power loss was experienced. Company policy in such a situation was to lower the landing gear. After considerable discussion, however, the crew decided not to lower the gear, because they didn't really know the state of the equipment. This decision, it later turned out, saved their lives. United Airlines Captain Ken Thomas associates this deliberative behavior with the intense CRM training United uses (K. Thomas, personal communication, October 20, 1994).

Intellectual Resource Management

High-integrity organizations are marked by intelligent use of intellectual resources. Because crew resource management is covered in detail in chapter 9 by Captain Daniel Maurino, we concentrate here only on the more general application of the same principles. The wise use of intellectual resources is critical to all aviation operations inside, outside, and beyond the aircraft. There are basically three principles.

1. *Use the full brainpower of the organization.* Coordinate leadership is vital to this principle. Coordinate leadership consists in allowing the person best able to make a decision to take control—temporarily. Coordinate leadership is basic to aviation. In flying the plane, for instance, control on the flight deck will shift back and forth between the left- and right-hand seats, even though the pilot retains ultimate authority. We would like to suggest, however, that coordination has wider implications that need to be examined.

For instance, General Chuck Yeager, in command of a Tactical Air Command squadron of F-100 Supersabres, managed to cross the Atlantic and deploy his planes to Europe without any failures. His perfect deployment was widely considered exemplary. Yet one of the keys to this accomplishment was Gen. Yeager's insistence on allowing his maintenance staff to decide whether the airplanes were fit to fly. Yeager had been in maintenance himself, but his basic attitude was that the maintenance people knew best whether the equipment was ready to fly.

> I never applied pressure to keep all of our airplanes in the air; if two or three were being serviced, we just lived with an inconvenience, rather than risking our lives with aircraft slapdashed onto the flight line. I wouldn't allow an officer-pilot to countermand a crew chief-sergeant's decision about grounding an unsafe airplane. A pilot faced with not flying

> wasn't always the best judge about the risks he was willing to take to get his wheels off the ground. And it paid off. My pilots flew confident, knowing that their equipment was safe. (Yeager & Janos, 1985, p. 315)

Yeager's example shows that great leadership may include emphasis on high reliability as well as winning. This might seem surprising in view of Yeager's overall "ace" qualities. When coordinate leadership does not take place, problems occur. In the BAC One-Eleven windscreen accident on June 10, 1990 (Birmingham, United Kingdom), a windscreen detached at 17,300 ft because it had been badly attached, nearly ejecting the pilot with it. A maintenance supervisor had done the job himself, due to a shortage of personnel. Because the supervisor did the job in a hurry, he installed the wrong bolts. No one else was present. He needed to have someone else check his work, but instead he became lost in the task (Maurino et al., 1995, pp. 86–101). Failure to coordinate leadership can thus overload the person in charge.

2. *Get the information to the person who needs it.* The information on which decisions are made should be the best available, and information possessed by one member of the organization has to be available in principle to anybody who needs it. Probably no better example of intellectual resource management can be cited than the Apollo moon flights. The organization was able to concentrate the needed intellectual resources to design systems and to solve problems. Apollo 13's emergency and recovery took place at the apogee of NASA's high-integrity culture (Murray & Cox, 1989, pp. 387–449). In fact, one might use this criterion for cognitive efficiency of the organization: "The organization is able to make use of information, observations or ideas, wherever they exist within the system, without regard for the location or status of the person or group originating such information, observations or ideas" (Westrum, 1993). We will see later in this chapter that an organization's cognitive adequacy can be assessed by just how closely it observes this principle.

3. *Keep track of what is happening, who is doing what, and who knows what.* The ability to secure appropriate vigilance and attention for all the organization's tasks, so that someone is watching everything that needs watching, is critical to safety. We are all familiar with the concept of mental workload, from studies of pilots and other operators of complex machinery. Yet often the most important workload is that shouldered by top management. If "situational awareness" is important for the pilot or flight deck crew, "having the bubble" is what top management needs (Roberts & Rousseau, 1989). The importance of management keeping track cannot be underestimated. Management's having "too much on their minds" was implicated in the Clapham Junction railroad accident (Hidden, 1989), but it is a common problem in aviation as well. John H. Enders, vice chairman and past president of the Flight Safety Foundation, stated that the distribution of contributing causes for the last decade's fatal accidents included "perhaps 60–80 percent management or supervisory inattention at all levels" (Enders, 1992).

Maestros

A key feature promoting high integrity in any aviation organization is the standards set by leaders. The most powerful standards are likely to be those set by *maestros,* who believe that the organization should operate in a manner consistent with their own high expectations (Vaill, 1982). In these organizations, persons of high technical virtuosity, with broad attention spans, high energy levels, and an ability to ask key questions, shape the culture. The maestro's high standards, coupled with the other personal features, force awareness and compliance with these standards on the rest of the organization. Arthur Squires, in his book on failed engineering projects, notes that major technical projects without a maestro present often founder (Squires, 1986).

The absence of a maestro may cause standards to slip or critical functions not to be performed. Such failures can be devastating to aerospace projects. An excellent example of such a project is the Hubble Space Telescope. Although the telescope's primary mirror design and adjustment were critical for the mission, the mirror had no maestro. No single person was charged with the responsibility of making the system work (Caspars and Lipton, 1991). Likewise, historical analysis might well show that safety in the American space program was associated with the presence or absence of maestros. During the palmy days of Apollo, NASA fairly bristled with maestros (see Murray & Cox, 1989). Michael Collins, an astronaut, made this comment about NASA Flight Directors:

> I never knew a "Flight" who could be considered typical, but they did have some unifying characteristics. They were all strong, quick, and certain. [For instance] Eugene Kranz, as fine a specimen of the species as any, and the leader of the team during the first lunar landing. A former fighter pilot . . . he looked like a drill sergeant in some especially bloodthirsty branch of the armed forces. Mr. Kranz and the other Flights—Christopher C. Kraft, Jr., John Hodge, Glynn Lunney, Clifford Charlesworth, Peter Frank, deserve a great deal of the praise usually reserved for the astronauts, although their methods might not have passed muster at the Harvard Business School. For example, during practice sessions not only were mistakes not tolerated, but miscreants were immediately called to task. As one participant recalls, "If you were sitting down in Australia, and you screwed up, Mr. Kraft, or Mr. Kranz, or Mr. Hodge would get on the line and commence to tell you how stupid you were, and you knew that every switching center . . . ships at sea, everybody and his mother, everybody in the world was listening. And you sat there and took it. There was no mercy in those days." (Collins, 1989, p. 29)

And there could hardly afford to be. Space travel is even less forgiving than air travel when it comes to mistakes. This maestro-driven environment defined the atmosphere for Project Apollo. By the days of the Space Shuttle, maestros were much harder to find. When NASA standards weakened, so did safety (Cooper, 1986; McCurdy, 1993).

Maestros shape climates by setting high standards for aviation organizations. Consider Gen. Yeager's description of Colonel Albert G. Boyd in 1946. Colonel Boyd was then head of the Flight Test Division at Wright Field:

> Think of the toughest person you've ever known, then multiply by ten, and you're close to the kind of guy that the old man was. His bark was never worse than his bite: he'd tear your ass off if you screwed up. Everyone respected him, but was scared to death of him. He looked mean, and he was.
>
> And he was one helluva pilot. He flew practically everything that was being tested at Wright, all the bombers, cargo planes, and fighters. If a test pilot had a problem you could bet Colonel Boyd would get in that cockpit and see for himself what was wrong. He held the three-kilometer low altitude world speed record of 624 mph, in a specially built Shooting Star. So, he knew all about piloting, and all about us, and if we got out of line, you had the feeling that the old man would be more than happy to take you behind the hangar and straighten you out. (Yeager & Janos, 1985, p. 113)

But standards are not strong only because they have penalties attached. They must be intelligently designed, clear, well understood, and consistently applied. Not all maestros are commanding personalities. Some maintain standards through more subtle means. Leighton I. Davis, Commanding Officer of Holloman Air Force Missile Development Center in the 1950s, managed to elicit a fierce loyalty from his officers, so much so that many of them worked 50- or 60-hour weeks so as not to let him down. He got this loyalty by providing a highly supportive environment for research and testing (Lt. Col. Thomas McElmurry, personal communication, August 15, 1993).

Maestros protect integrity through insistence on honest and free-flowing communications. Maestro systems exhibit a high degree of openness. Decisions have an open, available quality as opposed to a secretive or political one. Maestros may also be critical for organizational change. A maestro at United Airlines, Edward Carroll, a vice president, acted as the champion who sponsored United's original program "Command, Leadership, and Resource Management," that organization's version of CRM. Carroll responded to the Portland, Oregon, crash of 1978 by promoting understanding of the root causes and devising a comprehensive solution (K. Thomas, personal communication, October 20, 1994).

Communities of Good Judgment

We speculate that a high-integrity organization must constitute a "community of good judgment." Good judgment is different from technical competence. Although technical knowledge is objective and universal, judgment pertains to the immediate present. Judgment is the ability to make sound decisions in real situations, which often involve ambiguity, uncertainty, and risk. Good judgment includes knowledge of how to get things done, who can be counted on to do what, and usually reflects deep experience. Maestros exemplify good judgment.

High integrity demands a culture of respect. When good judgment is compromised, respect is impossible. In communities of good judgment, the individual's position in the system is proportional to recognized mastery. Each higher level in the system fosters an environment below it that encourages sound decisions. Individual capabilities are carefully tracked, and often knowledge of individuals' abilities will not be confined to the next level up, but will go two levels higher in the system, thus allowing higher-ups to evaluate the abilities of those below them to assess their own personnel. Furthermore, a knowledge of organizational tasks runs parallel to the knowledge of people. In other words, there is awareness not only of what people can do, but also of what they are supposed to do. This knowledge allows a high degree of empowerment. But it is also demanding.

If this speculation is correct, then the most critical feature may well be this: that respect is given to the practice of good judgment, wherever it occurs in the organization, rather than to hierarchical position. And this observation leads to an interesting puzzle: If the organization is to operate on the best judgment, how does it know what the best judgment is?

ORGANIZATIONAL CULTURE

Corporate Cultural Features That Promote or Degrade High Integrity

Organizational Culture. Organizations move to a common rhythm. What ties together the diverse strands of people, decisions, and orientations is the organization's microculture. This organizational culture is an ensemble of patterns of thought, feeling, and behavior that guide the actions of the organization's members. The closest analogy one can make is to the personality or character of an individual. The ensemble of patterns is a historical product, and it may reflect the organization's experiences over a surprisingly long span of time (Trice & Beyer, 1993). It is also strongly shaped by external forces, such as national cultures and regional differences. Finally it is shaped by conscious decisions about structure, strategy, and policy taken by top management (cf. Schein, 1992).

Organizational culture has powerful effects on the individual, but it influences rather than determines individual actions. An organization's norms, for instance, constrain action, by rewarding or punishing certain kinds of acts. But individuals can violate both informal norms and explicit policy. Furthermore, some organizational cultures are stronger than others, and have a greater influence on the organization's members. For the individual, the norms constrain only to the extent that the organization is aware of what the individual is doing, and the individual in turn may decide to "buy into" or may remain aloof from the norms.

Organizational culture is an organic, growing thing, and changes over time—and of course sometimes it changes more rapidly than at other times. Different parts of the organization will reflect variations of the culture, sometimes very substantial variations, due to different backgrounds, varying experiences, local conditions, and different leaders.

Aspects of Culture. Anthropologists, sociologists, and psychologists (including human factors specialists) have addressed organizational culture from the perspectives of their respective disciplines. Because culture has several facets, some researchers have emphasized one, some another, or some combination of these facets. Three of the facets are cognitive systems, values, and behavior.

Culture exists as a shared *cognitive system* of ideas, symbols, and meanings. This view was emphasized by Trice and Beyer (1993), who saw ideologies as the substance of organizational culture. Similarly, Schein, in his discussion (1992) of culture, discussed organizational *assumptions*. An organization's assumptions are the tacit beliefs that members hold about themselves and others, shaping what is seen as real, reasonable, and possible. Schein saw assumptions as "the essence of a culture" (Schein, 1992, p. 26), and maintained that a culture is (in part) "a pattern of shared basic assumptions that the group learned as it solved its problems of external adaptation and internal integration, that has worked well enough to be considered valid and, therefore, to be taught to new members as the correct way to perceive, think, and feel in relation to those problems" (p. 12).

Assumptions are also similar to what others have called "theories-in-use." Argyris, Putnam, and Smith (1985) and Schon (1983) distinguished between espoused theory and theory-in-use. The former is what the group presents itself as believing; the latter is what it really believes. Espoused theory is easy to discuss, but changing it will not change behavior. Theory-in-use may be hard to bring to the surface.

Values reflect judgments about what is right and wrong in an organization. They may be translated into specific norms, but norms may not always be consistent with values, especially those openly espoused. For instance, Denison (1990, p. 32) defined perspectives as "the socially shared rules and norms applicable to a given context." The rules and norms may be viewed as the solutions to problems encountered by organizational members; they influence how members interpret situations and prescribe the bounds of acceptable behavior. But the values held by an organization may be very difficult to decipher, since what is openly proclaimed may not in fact be what is enforced (Schein,

1992, p. 17). Espoused values (Argyris et al., 1985) may reflect what people may say in a variety of situations but not what they do. Many participants in unsuccessful "quality" programs have found out too late that quality is a concept supported by management only as an espoused value, not a value-in-use. This separation is parallel to the differences in "theory" mentioned earlier. In any case, values may be different for different subgroups, regions, and levels of responsibilty. Sometimes constellations of values are described as an organization's climate. Dodd (1991), for instance, defines organizational culture as the communication climate rooted in a common set of norms and interpretive schemes about phenomena that occur as people work toward a predetermined goal. The climate shapes how organizations think about what they do, and thus how they get things done. Although some aviation organizations may have a strong common vision and we-feeling (e.g. Southwest Airlines), others may represent an amalgam of competing values, loyalties, and visions. Lautman and Gallimore (1987) found that management pilots in 12 major carriers thought that standards were set at the top of the organization, but in-depth studies to confirm this assertion are so far lacking.

Finally, culture is a pattern of observable *behavior*. This view is dominant in Allport's theory (1955) of social structure. Allport argued that social events involve observable patterns that coalesce into structures. He explored patterns that defined social structures and he implied that examining the impact of single variables in complex systems should be replaced with examining the ongoing structure of interacting events. Although Allport did not define structures as cultures, his research provides a basis for the study of organizational culture. Similarly, Lineberry and Carleton (1992) cited Burke and Litwin regarding organizational culture as "the way we do things around here" (p. 234). Emphasizing behavior suggests that cultures can be discovered by watching what people do.

These definitions and orientations constitute only a handful of those available. While they are intellectually stimulating, none has been compelling enough to gain general acceptance. Even the outstanding survey of the literature by Trice and Beyer (1993) is short of a synthesis. Thus, no one has yet evolved a complete and intellectually satisfying approach to organizational culture. While this basic task is being accomplished, however, incidents and accidents occur, and lives and money are being lost. So some researchers have tried to focus on specific cultural forms that affect safety. For instance:

- Pidgeon and O'Leary (1994) defined safety culture "as the set of beliefs, norms, attitudes, roles and social and technical practices within an organization which are concerned with minimizing the exposure of individuals, both within and outside an organization, to conditions considered to be dangerous" (p. 32).
- Lauber (1993) maintained that safe corporate culture requires clear and concise orders, discipline, attention to all matters affecting safety, effective communications, and a clear and firm management and command structure.
- Wood (1993) stated that culture, taken literally, is what we grow things in. He maintained:

> The culture itself is analogous to the soil and water and heat and light needed to grow anything. If we establish the culture first, the safety committee, the audit program and the safety newsletter will grow. If we try to grow things, such as safety programs, without the proper culture—they will die. (p. 26)

- Westrum suggested that the critical feature of organizational culture for safety is information flow. He defined three types of climates for information flow: the pathological, the bureaucratic, and the generative (Westrum, 1993). Because these types bear directly on the concept of high integrity, we will elaborate on them in the section that follows.

Communications Flow and the Human Envelope

Using his well-known model, Reason (1990) suggested that accidents occur when latent pathogens (undetected failures) are associated with active failures and failed defenses by operators at "the sharp end" (Fig. 5.4). Ordinarily, this is represented by a "Swiss cheese model" in which accidents occur when enough "holes" in the Swiss cheese slices overlap. However, this can also be represented by the "human envelope" model proposed earlier. Each of Westrum's organization types, because of its communication patterns, represents a different situation vis-à-vis the buildup of latent pathogens in the human envelope. Effective communication is vital for identifying and removing these latent pathogens. We can represent each one in terms of both a diagram (Fig. 5.5) and typical behaviors:

1. The *pathological* organization typically chooses to handle anomalies by using suppression or encapsulation. The person who spots a problem is silenced or driven into a corner. This does not make the problem go away, just the message about it. Such organizations constantly generate "latent pathogens," since internal political forces act without concern for integrity. Pathogens are also likely to remain undetected, because detectors may be punished. The pathological organization may be rare in aviation, but it is dangerous where it exists.

2. *Bureaucratic* organizations tend to be good at routine or predictable problems. They do not actively create pathogens at the rate of pathological organizations, but they are not very good at spotting or fixing them. They sometimes make light of problems or only address those immediately presenting themselves. Underlying causes

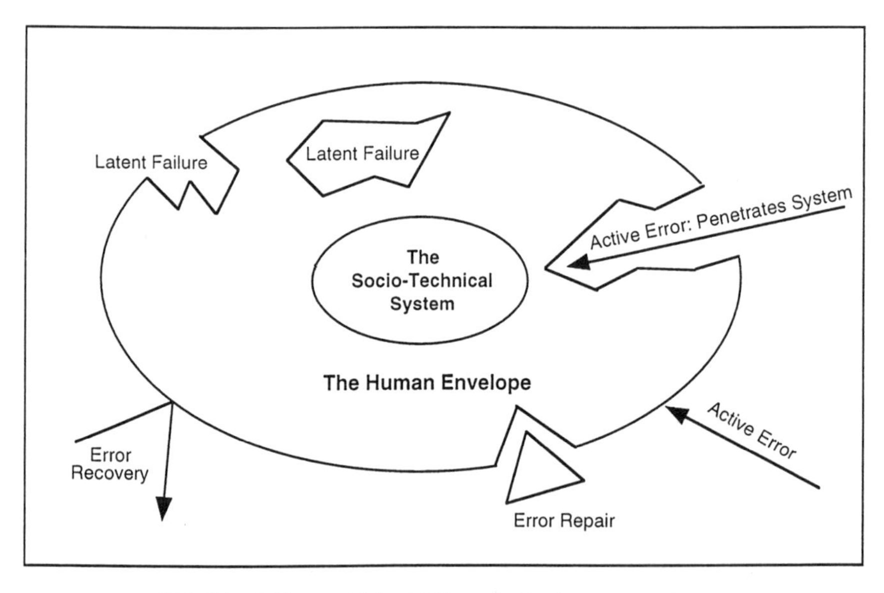

FIG. 5.4. Active and latent failures in the human envelope.

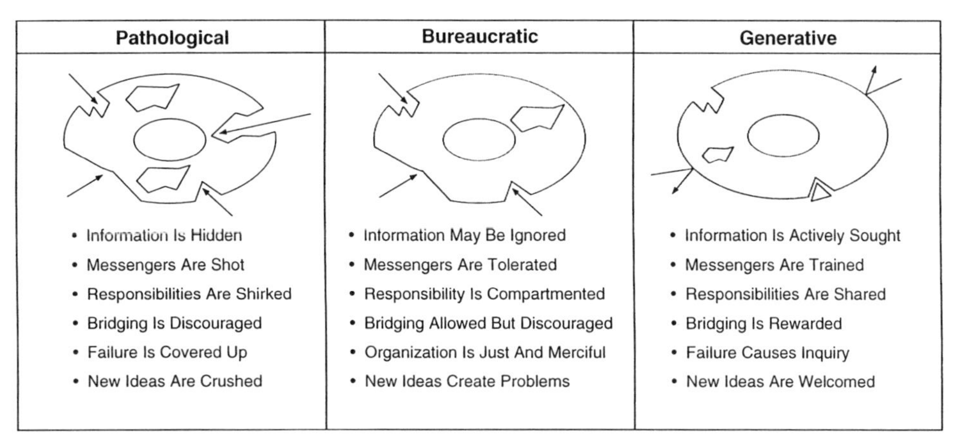

FIG. 5.5. How organizational cultures treat information.

may be left untouched. When an emergency occurs, they find themselves unable to react in an adaptive way.

3. The last type of organization is the *generative* organization. It encourages communication, as well as self-organization. There exists a culture of conscious inquiry that tends to root out and solve problems that are not immediately apparent. The generative organization possesses a high degree of integrity, a human envelope in depth that protects the sociotechnical system. When the system occasionally generates a latent pathogen, the problem is likely to be quickly spotted and fixed.

Although Westrum's schema is intuitive and is well known in the aviation community, it has yet to be shown by quantitative studies that in fact "generativity" correlates with safety.

Subcultures. As if coping with organizational cultures were not enough in itself, the problem is compounded by the existence of subcultures within the aviation organization. Over time, any social unit that produces subunits will produce subcultures. So as organizations grow and mature, subcultures arise (Schein, 1992). In most cases, the subcultures are shaped by the tasks each performs. Differing tasks and backgrounds lead to different assumptions. Within aviation organizations subcultures have been identified primarily by job positions. For example, distinctive subcultures may exist among corporate management, pilots, mechanics, flight attendants, dispatch, and ground handling. Furthermore, these subcultures may show further internal differentiation such as maintenance technicians versus avionics technicians, male flight attendants versus female flight attendants, sales versus marketing personnel, day-shift versus night-shift dispatch, and baggage versus fuel handlers.

Subcultural differences can become important through differing assumptions. Dunn (1995) reported on five factors identified at the NASA Ames Research Center that led to differences between the cabin crew and cockpit crew. Four of the five factors were rooted in assumptions that each group held about the other. Dunn reported:

- The historical background of each group influences the attitudes they hold about each other.

- The physical separation of the groups' crew stations leads to a serious lack of awareness of each groups' duties and responsibilities.
- Psychological isolation of each group from the other leads to personality differences, misunderstanding of motivations, pilot skepticism, and flight attendant ambivalence regarding the chain of command.
- Organizational factors such as administrative segregation and differences in training and scheduling create group differences.
- Regulatory factors lead to confusion over sterile cockpit procedures and licensing requirements.

Dunn argued that often the subcultures, evolving from shared assumptions, are not in harmony with each other—nor do they always resonate with the overall organizational culture. These groups are very clearly separated in most companies. The groups work for different branches of the company, have different workplace conditions, power, and perspectives. This lack of harmony can erode the integrity of the human envelope. Dunn provided a number of examples that depict hazardous situations that can result from differences between the cockpit crew and the flight attendant crew. She noted that a Human Factor Team that investigated the 1989 Dryden accident found that such separation was a contributing factor to the accident. These problems were further confirmed in an important study by Chute and Wiener (1995). Chute and Wiener document the safety problems caused by lack of common training, physical separation, and ambiguous directives—such as the sterile cockpit rule. When emergencies arise, the resulting lack of coordination can have lethal consequences (Chute & Wiener, 1996).

Schein (1992) proposed that in some cases the communication barriers between subcultures are so strong that organizations have to invent new boundary-spanning functions or processes. One example of such efforts is the recent intiative by the FAA and some industry groups calling for joint training programs between pilots and flight attendants. Such joint training can be very effective. Some years ago one of us (Adamski) spoke about pilot and flight attendant relationships with a close friend, a captain with a major U.S. airline. The captain said that he had just attended his first joint training session between pilots and flight attendants since his employment with the airline. With some amazement, he said that previously he never had any idea about the problems or procedures dealt with by the cabin crew. This joint training was the airline's first attempt to provide a bridge between the two subcultures. Joint training efforts have often produced positive results (Chute & Wiener, 1996).

Major Empirical Studies. Much research has been conducted to explore the many facets of organizational culture in the aviation community and related high-tech industries. Improving safety and reliability has been the primary purpose in most of this research. Although the findings are valuable, generally they have been advanced without a previously articulated theory.

One of the earliest and most interesting examples of subtle creation of a safety culture in an aviation operation was related by Patterson (1955), who managed to shift attitudes about accidents at a remote airbase in World War II at the same time that he accomplished cross-functional cooperation. Patterson's approach would later become well known as "sociotechnical systems theory" and under the leadership of Eric Trist and others it has accumulated an imposing body of knowledge (e.g., Pasmore, 1988). CRM concepts and sociotechnical ideas have a great deal in common. Yet although sociotechnical systems (STS) theory would be enormously helpful in aviation, it has yet to move out of the industrial environment that spawned it. Current aviation research has focused instead on the organizational antecents of "systems accidents" and on CRM-related attitude and behavior studies.

The work on "systems accidents" was begun by Turner (1978) and Perrow (1984), with major contributions by Reason (1984, 1990) and others. Turner and Perrow showed that accidents were "man-made disasters" and that the dynamics of the organizations routinely generated the conditions for these unhappy events. Reason traced the psychological and managerial lapses leading to these accidents in more detail. Reason noted that in accident investigations, blame was often placed on the operators at the "sharp end," whereas the conditions leading up to the accident (the "soft end") are given less emphasis. But in fact more probing has shown management actions are strongly implicated in accidents. For instance, although the Dryden, Ontario, accident (1989), was initially dismissed as pilot error, investigation showed it to be rooted in problems far beyond the cockpit (Maurino et al., 1995, pp. 57–85). Similarly, in the controlled-flight-into-terrain accident on Mt. Saint-Odile, near Strasbourg, January 20, 1992, a critical deficiency was the lack of a ground proximity warning system (Paries, 1994). The reasons for the lack of such a system reached far beyond the pilots, to management and national regulation.

Climates for Cooperation

In a parallel development there has been some outstanding ethnographic work by the "high-reliability" group at the University of California, Berkeley. In contrast to Perrow, the Berkeley group decided to find out why some organizations could routinely and safely carry out hazardous operations. Gene Rochlin, Todd LaPorte, Karlene Roberts, and other members of the "high-reliability group" carried out detailed ethnographic studies of aircraft carriers, nuclear power plants, and air traffic control to determine why the accident rates for some of these operations were as low as they were found to be. These studies suggested some of the underlying principles for safe operation of large, complex systems, including:

1. "Heedful interaction" and other forms of complex cooperation.
2. Emphasis on cooperation instead of hierarchy for task accomplishment. Higher levels monitor lower ones instead of direct supervision in times of crisis.
3. Emphasis on accountability and responsibility and avoidance of immature or risky behavior.
4. High awareness of hazards and events leading to them.
5. Forms of informal learning and self-organization embedded in organizational culture.

The richness of the Berkeley studies is impressive, yet they remain to be synthesized. A book by Sagan (1993) sought to compare and test the Perrow and Berkeley approaches, but after much discussion (*Journal of Contingencies and Crisis Management*, 1994) by the parties involved, many issues remain unresolved.

Meanwhile, another approach has developed out of work on crew resource management (see Maurino, chap. 9, this volume). Robert Helmreich and his colleagues developed and tested materials for scoring actions and attitudes indicative of effective crew resource management. Originally, these materials grew out of the practical task of evaluating pilots' CRM attitudes, but have since been developed and extended to be used as measures of organizational attributes as well—for example, the presence of safety-supportive cultures in organizations. The more recent work has been strongly influenced by scales developed by Hofstede (1980) for studying differences in the work cultures of nations (discussed later). Using the Flight Management Attitudes Questionnaire, Merritt and Helmreich (1995) made some interesting observations about safety-supportive attitudes in airlines. For instance, they observe that national cultures differed on some attitudes relevant to safety (see Figs. 5.6 through 5.8).

The data in Figs. 5.6 through 5.8 require some discussion. It is evident, for instance, that there are differences both between nations and within nations. In terms of differences between nations, one might expect "Anglo" (U.S./northern European) cultures

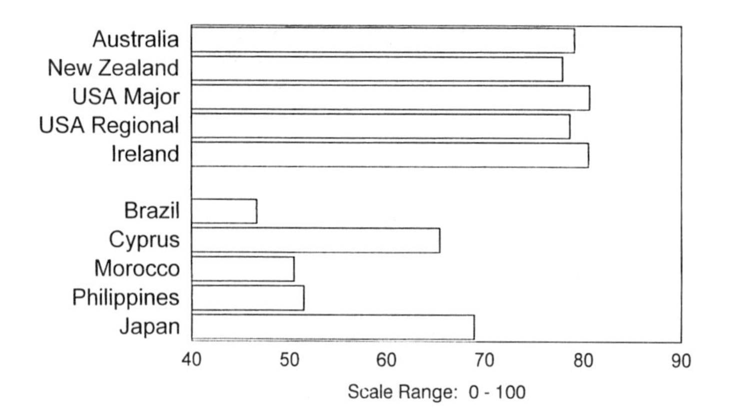

FIG. 5.6. Support for a flattened command structure among pilots. (Data are from the NASA/University of Texas/FAA Crew Resource Project.)

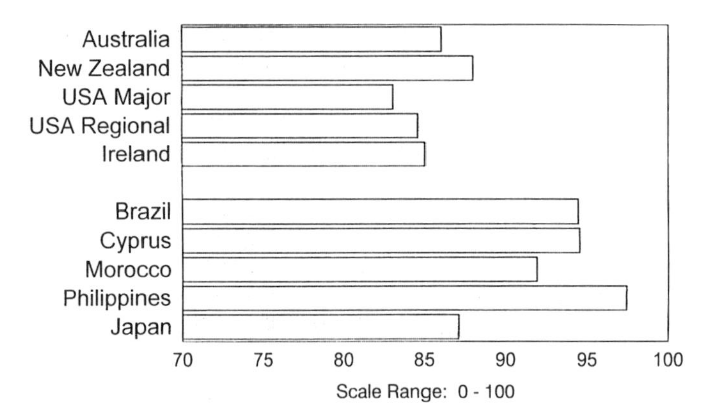

FIG. 5.7. Support for information sharing among pilots. (Data are from the NASA/University of Texas/FAA Crew Resource Project.)

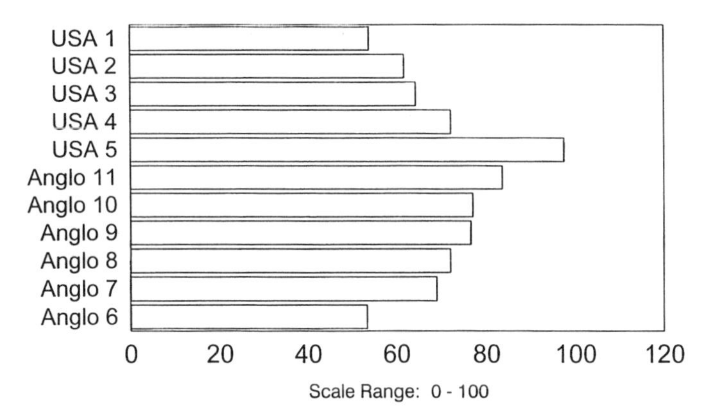

FIG. 5.8. Positive organizational culture in 11 airlines. (Data are from the NASA/University of Texas/FAA Crew Resource Project.)

to have features that support better information flow. So it is not surprising to find that pilots in Anglo cultures seem more willing to support a flattened command structure (Fig. 5.6). However, pilots from more authoritarian cultures apparently support a higher degree of information sharing than their Anglo counterparts (Fig. 5.7)! According to Merritt and Helmreich, in authoritarian cultures, precisely because of the large status differences in command, information sharing needs to be especially emphasized. Most interesting, however, are the dramatic differences between airlines from the same nation on positive organizational culture (Fig. 5.8). Positive organizational culture reflects questions about positive attitudes toward one's job and one's company. The airline designated USA 1 has a culture in the doldrums, compared with the remarkable showing for USA 5, especially considering that these are averaged scores for the organization's members. One can only ponder the impacts that these organizational attitudes have on safety, because the airlines in the study are anonymous.

In a related paper, Law and Wilhelm (1995) show that there are equally remarkable *behavioral* differences between airlines. Using the Line/LOS Checklist developed by the NASA/University of Texas/FAA Aerospace Crew Project, raters observed and scored 1,300 pilots. Figure 5.9 shows the results for two airlines identified only as "1" and "2." These assessments of behavioral markers show even greater variations in safety-related behavior than the attitudes studied by Merritt and Helmreich. Law and Wilhelm (1995) showed also that there are differences as well between the fleets of the same airline in CRM (Fig. 5.10). What underlying features (history, recruitment, leadership, etc.) account for these differences is unknown. However, both sets of data provide very strong evidence that that organizational cultures are related to safety.

National Differences in Work Cultures

Aviation operates in a global community. Some aviation organizations are monocultural: They operate within a specific area of the world and employ people largely from that same national culture. These aviation organizations manifest many of the features

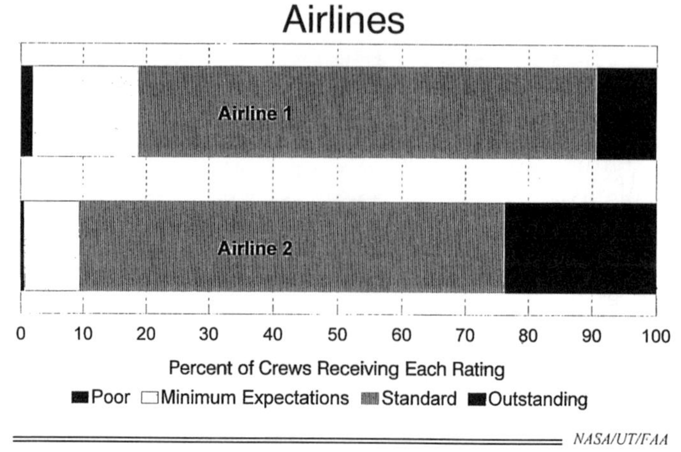

FIG. 5.9. Overall crew effectiveness ratings in two airlines.

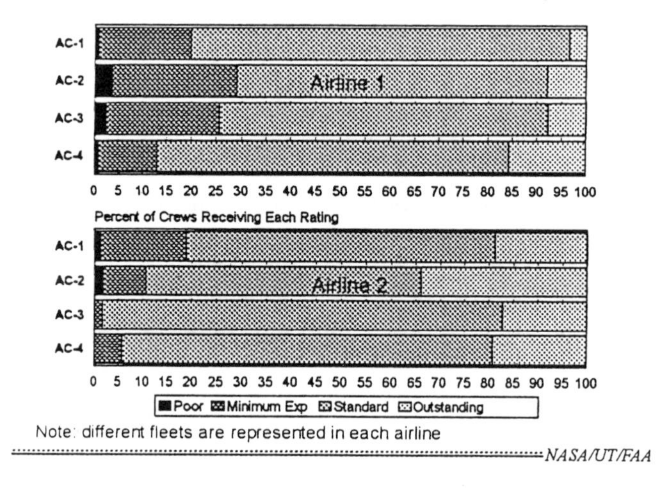

FIG. 5.10. Ratings of crew effectiveness by fleet in two airlines.

of the national cultures out of which they arose. Others are multicultural: They have facilities throughout the world that employ people from a variety of national cultures. Multicultural crews represent a particular challenge. Recently a physical struggle over the controls of an Airbus 300 broke out on a Korean Airlines flight deck as a Canadian captain and a Korean first officer struggled over how the landing should be managed. The first officer's command of English was insufficient to express his concerns, so he simply grabbed the wheel. Finally the plane crash-landed and then burned; fortunately there were no casualties (Glain, 1994). Obviously getting multicultural groups to work well together will be one of the key tasks the aviation community has to face in the next decade.

As anthropologists such as Hall (1959) have pointed out for many years, each society provides its members with a "mental program" that specifies not only general orientations but also minute details of action, expression, and use of space. Travelers are often taken aback when foreigners act in ways that seem incomprehensible at home. On a flight deck or in a control tower, however, these differences can have serious consequences.

One useful framework for sorting out differences in organization-relevant values between cultures was developed by Geert Hofstede (1980). He identified four dimensions of national culture: power distance, uncertainty avoidance, individualism/collectivism, and masculinity.

Power distance is the degree to which members of a culture will accept differences in power between superiors and subordinates. An unequal distribution of power over action is common to aviation organizations. It provides one means by which organizations focus control and responsibility. The power distance, however, varies considerably. In some cultures the "gradient" is far steeper than others. As we have seen in the Helmreich and Merritt data, discussed earlier, this trait shows strong variations, especially between Anglo and non-Anglo cultures.

The second dimension Hofstede identified is uncertainty avoidance. This is the tolerance a culture holds toward the uncertainty of the future, which includes the elements of time and anxiety. Cultures cope with this uncertainty through the use of technology, law, and religion; organizations cope using technology, rules, and rituals. Organizations reduce internal uncertainty caused by the unpredictable behavior of their members by establishing rules and regulations. According to Hofstede (1980, p. 116) organizational rituals are nonrational, and their major purpose is to avoid uncertainty. Training and employee development programs may also be used to reduce uncertainty. Because technology creates short-term predictability, it too can be used to prevent uncertainty. One way in which this takes place is through over reliance on flight management systems (FMS) as opposed to CRM. Sherman and Helmreich (1995) found a stronger reliance on automation, for instance, in cultures with high power distance and strong uncertainty avoidance.

Individualism/collectivism, the third dimension, expresses the relationship between a member of a culture and his or her group. It is reflected in the way people live together and are linked with societal norms, and it affects members' mental programming, structure, and functioning of organizations. The norm prevalent within a given society regarding the loyalty expected from its members obviously shapes how people relate to their organizations. Members of collectivistic societies have a greater emotional dependence on their organizations. Organizations may emphasize individual achievement or the welfare of the group. The level of collectivism affects willingness of an organization's members to comply with organizational requirements. Willingness to "go one's own way" is at one pole of the continuum. At the other pole is a willingness to keep silent and go along with the group—often a fatal response in an emergency.

How different societies cope with masculinity/femininity is the fourth dimension identified by Hofstede (1980, p. 176). Although masculine and feminine roles are associated in many societies with the roles for males and females, respectively, how polarized the sexes are on this dimension varies a great deal. This dimension is obviously important for aviation. The "macho" attitude so often complained about in CRM seminars reflects a high masculinity orientation. "Task leadership" versus "socioemotional leadership" is associated with this dimension (Bales, 1965). Similarly, some cultures may value masculine roles more highly than feminine ones. Recently it was reported by the *Chicago Sun Times* that 20 Indian Airline flights were canceled because the pilots were upset that some senior flight attendants were getting more pay than they were. The article stated that the pilots sat at their seats with arms folded and refused to fly if certain flight attendants were on board. The flight attendants retaliated by refusing to serve the pilots tea.

Helmreich (1994) made a convincing argument that three of Hofstede's four variables were important in the crash of Avianca 052, which ran out of fuel in a holding pattern over Long Island on January 25, 1990. The pilots failed to communicate successfully with each other and with the ground, allowing a worsening situation to go unrecognized by air traffic control. Many of the CRM failures Helmreich identified as present during the flight seem associated with the high power-distance, collectivist, and uncertainty-avoiding features of the pilots' Colombian culture.

Johnston (1995) speculated that differences in cultural orientations might affect response to and acceptance of CRM. CRM itself a value system; it may or may not collate with local value systems. But it is dangerous, as Johnston points out, to assume that regional differences in accident rates reflect CRM orientations. He cites a paper by Weener (1990) that shows that although small aircraft accident rates vary strongly by region, accident rates for intercontinental aircraft are similar for developed and developing nations. The reason, Johnston suggests, is that international airports are more likely to operate on a world standard, whereas differences in infrastructure show up more strongly in general accident rates. So economic differences may be just as important as culture in understanding accident rates. Culture may thus be an important explanatory variable, but other differences between nations need to be taken into account.

MAINTAINING HUMAN ASSETS

Training, Experience, and Work Stress

Maintaining the human assets of an organization is critical to high integrity. Yet human assets are often neglected. Accident and incident reports are filled with descriptions of inadequate training, inappropriate tasking, fatigue, job-related stress, boredom, and burnout.

Big differences can be found in the approaches organizations take to their members. Although high-integrity organizations are careful with their people, obviously many others are not. High-performance teams, for instance, are anything but passive in their attitude toward the people who are members. They show special care in hiring, in making sure their people get trained correctly, in giving personnel appropriate tasks, and in monitoring how they are doing. New members are carefully vetted and "checked out" to see what their capabilities are. Previous training is not taken for granted. Rather, new recruits are given a variety of formal and informal tests to assess their abilities.

Evaluating new members is not enough. Once skills have been certified, personnel have to join the team psychologically as well as legally. Aviation systems are often tightly coupled (Perrow, 1984). This means that all personnel need to be considered part of the system, because a failure by any one of them may cause grave problems. Yet often higher managers fail to secure "buy in" by the organization's less visible members. Resulting disaffection by the "invisibles" can be costly. For example, maintenance personnel often have important roles in protecting safety, but seldom receive anything like the attention lavished on the flight deck crew by management, the public, and academics (Shepherd, 1994). Securing "buy in" by this group will be difficult, because while their failures recieve swift attention, their successes are seldom so visible.

In a high-integrity organization human assets are carefully maintained and assigned, and the experience of the operators matches the requirements of the task. If inexperienced or stressed workers are present, they are put under careful supervision. In the study by Mouden (1992) previously mentioned, he thought that frequent high-quality training may be the most important means of preventing accidents within aviation organizations. But training, especially high-quality training, is expensive. Organizations on the economic margin or in the process of rapid change or expansion frequently have neither the money nor the time to engage in the training needed. In these organizations, integrity is often compromised by economic pressures.

One way to lower integrity is to hire managers who allow standards to slip. This appears to have been the case at Continental Express previous to the stabilizer detachment accident (discussed later). NTSB Board Member John Lauber, in a minority opinion, noted that:

> The multitude of lapses and failures committed by many employees of Continental Express discovered in this investigation is not consistent with the notion that the accident originated from isolated, as opposed to systematic, factors. It is clear based on this [accident] record alone, that the series of failures that led to the accident were not the result of an aberration, but rather resulted from the normal, accepted way of doing business at Continental Express (NTSB, 1992, p. 53)

In an Addendum to this report Brenner further explored the probability that two managers in particular, the subsidiary's president and its senior director of maintenance and engineering, allowed the airline's maintenance standards to deteriorate (NTSB, 1992, Addendum). Continental's president had been an executive for Eastern Airlines, and during this period had made positive statements about the quality of maintenance during his watch that did not accord with Eastern practices as discovered by investigators. The maintenance director had earlier been director of quality control at Aloha Airlines when one of its planes suffered a preventable structural failure, resulting in the detachment of a fuselage upper lobe. Putting such people in critical positions in an airline suggests that higher management at Continental did not put high integrity foremost.

Another way to create hazardous conditions is to turn operations over to undertrained or temporary personnel. It is well known that training flights, for instance, have unusually high accident rates. And the accident literature describes many major blunders, sometimes fatal, that have taken place because inexperienced people are at the controls of the airplane, the bridge of the ship, the chemical or nuclear reactor, and so on (cf. Schneider, 1991). Having such people in control often causes incidents or accidents because:

1. They make decisions based on lack of knowledge, incorrect mental models, or fragmentary information. For instance, they may not have an adequate idea what a lapse on their part may mean for another part of the operation.
2. Newcomers or temporaries may not be part of the constant dialogue, and may intentionally be excluded from participation in informal briefings, story-swapping, and so on.
3. Those who need watching by supervisors increase the latters' mental workloads and thus distract them.

4. Newcomers and temporary workers may have little commitment to the organization's standards, values, and welfare.
5. If they make errors, or get in trouble, they are less likely to get the problem fixed rapidly, for fear of getting into trouble.

Even trained people can become risks if they are overstressed or tired. All too often, moreover, economic pressures during highly competitive times or periods of expansion will encourage dubious use of human assets. This can happen even in the best firms. For instance, in 1988 users of Boeing 737s and 767s found that some of the fire extinguishers on these planes had crossed connections—that is, when one side was called for, the other side's sprinklers came on. Although the crossed connections were not implicated in an accident, the possbility was present. An even more serious problem with engine overheat wiring was discovered on a Boeing 747 by Japan Air Lines. Investigation showed that hoses as well as wires were misconnected, and that the problem was widespread. Ninety-eight instances of plumbing or wiring errors were found on Boeing aircraft in 1988 alone. FAA inspections in the Boeing plant at Everett, Washington, showed that quality control had slipped. Even the maintenance manual for the 757 was found to be incorrect, showing the connections reversed. A possible explanation for these various problems was the sudden brisk demand for Boeing products. Boeing's response may have been to use its assets outside the envelope of safe operation. According to one engineer:

> . . . a too ambitious schedule for the new 747-400 aircraft has caused wiring errors so extensive that a prototype had to be completely rewired last year, a $1 million job. . . . The Boeing employee also said the long hours some employees were working last year [1988] on the 747-400 production line—12-hour days for seven days a week, including Thanksgiving, Christmas, and New Year's Day—had turned them into zombies. (Fitzgerald, 1989, p. 34)

Such high-stress situations are likely to make errors easier to commit and harder to spot, thus creating latent pathogens.

Layoffs of experienced people, whether due to strikes, downsizing, or retirement policies, are likely to endanger integrity, in aviation organizations and elsewhere. When the Chicago Post Office retired large numbers of its senior, experienced personnel, it shortly encountered severe problems: mail piled up, was put in trash baskets, or even burned. The senior managers were badly needed to keep the system running, and the effects of their retirement were both unexpected and damaging to the integrity of the Post Office operations (Franzen, 1994). Similarly, when the PATCO strike led to large numbers of experienced air traffic controllers being fired, extreme measures were needed to keep the system running. In fact, the air traffic control system experienced many anxious moments. Although the feared increase in accidents did not take place, the stress experienced by many managers and others who took the place of the fired controllers in control towers was evident.

Major changes of any kind are likely to cause stress. Such changes include mergers, expansions, downsizing, or moving to new facilities. One of the most severe impacts to safety was the deregulation of U.S. airlines in 1978. Deregulation put additional pressures on many marginal operators, and led to mergers that brought together incompatible cultures. A study of one unstable and two stable airlines by Little, Gaffney, Rosen, and Bender (1990) showed that pilots in the unstable airline showed significantly more stress than those in the stable airlines. This supports what common sense suggests:

A pilot's workload will increase with worries about the company. The Dryden, Ontario, accident also took place in the wake of a merger between Air Ontario and Austin Airways Limited. Investigation showed that the merger resulted in unresolved problems, such as unfilled or overburdened management roles, minimal flight following, and incompatible operations manuals (Maurino et al., 1995, pp. 57–85).

Pilot worries about companies in trouble may be well founded. A company in economic trouble may encourage pilots to engage in hazardous behavior, may confront the pilot with irritable supervisors, or may skimp on maintenance or training. It may be tempting to operate on the edge of the "safe region." An investigation of the airline U.S. Air by the *New York Times* showed that a climate existed in which fuel levels might not be carefully checked, resulting in some cases in planes leaving the airport with less fuel than they should have had (Frantz & Blumenthal, 1994).

Government organizations are not immune from economic pressures, either. The American Federal Aviation Administration often uses undertrained inspectors to carry out its critical role of monitoring the safety of air carriers. It has a huge workload, and a relatively small staff to do the job. It should come as no surprise that inspections are often perfunctory and sometimes overlook serious problems (Bryant, 1995b).

These examples suggest that while human assets may be expensive to maintain, not maintaining them may well prove more expensive.

MANAGING THE INTERFACES

Working at the Interface

One of the biggest problems faced by aviation organizations is handling transactions across the boundaries of organizational units. This includes subsystems of the organization as well as the organization's relations with external bodies, ranging from unions to regulators. It is in the interfaces that things frequently go wrong.

One interface problem is hand-offs. When there is a failure to communicate across interfaces, the breakdown can set up some of the most dangerous situations in aviation. As an airplane is handed off from one set of controllers to another by air traffic control, as a plane is turned over from one maintenance crew to another, and as initiative on the flight deck is passed back and forth, loss of information and situational awareness can occur. It is essential that the two spheres of consciousness, that of the relinquisher and that of the accepter, intersect long enough to transfer all the essential facts.

The loss of a commuter aircraft, Embraer-120RT on September 11, 1991, belonging to Continental Express (Flight 2574), took place when the leading edge of the left horizontal stabilizer detached during flight. The aircraft crashed, killing all aboard. Investigation showed that the de-icer boot bolts had been removed by one maintenance shift, but were not replaced by the succeeding one, due to faulty communications. The accident report (NTSB, 1992) commented that management was a contributing factor in setting up the conditions that led to confusion at the interface.

Another common problem is the failure to put together disparate pieces of information to get a picture of the whole situation. This apparently was one of the problems that led to the shoot-down of two U.S. Navy helicopters by Air Force fighters in Iraq. Inside the AWACS aircraft monitoring the airspace, radarmen at different positions each had a piece of the puzzle; they failed to compare notes, however. The failure in crew coordination led to the helicopters being identified as unfriendly, and they were shot down (Morrocco, 1994).

When two organizations are jointly responsible for action at an interface, neither may assume responsiblity. We have already noted the breakdown of an interface in the Manchester fire of 1985. Note also this comment by John Nance on the source of the de-icing failure that led to the Air Florida (Potomac) Crash of 1982:

> There were rules to be followed, inspections to be made, and guidelines to be met, and someone was supposed to be supervising to make certain it was all accomplished according to plan. But neither Air Florida's maintenance representative nor American's personnel had any idea whose responsibility it was to know which rules applied and who should supervise them. So no rules were applied at all and no one supervised anything. They just more or less played it by ear. (Nance, 1986, p. 255)

By contrast with this catch-as-catch-can approach, high-integrity organizations carefully control what comes into the organization and what goes out. An excellent example of such management of an interface is Boeing's use of customer information to provide better design criteria for the 777. Airlines were actively involved in the design process, providing input not only about layout, but also about factors that affected inspection and repair (O'Lone, 1992). By contrast, the Airbus 320 development seems to have made many French pilots, at least, feel that dialogue between them and the designers was unsatisfactory (Gras et al., 1994)

The best interfaces include overlapping spheres of consciousness. We can think of the individual "bubble," or field of attention, as a circle or sphere. (In reality an octopus or a star might be a better model.) The worst situation would be if such spheres do not overlap at all; in this case there would be isolation, and the various parties would not communicate. The best situation would be if overlap was substantial, so each would have some degree of awareness of the other's activities. Sometimes, however, the spheres only touch at a single tangent point. In this case there is a "single-thread" design, a fragile communication system. Single-thread designs are vulnerable to disruption, because the single link is likely to fail. For this reason, redundant channels of communication and cross-checking characterize high-integrity teams. Unfortunately, some individuals do not want to share information, as it would entail sharing power. This is one of the reasons that pathological organizations are so vulnerable to accidents: In such organizations there are few overlapping information pathways.

External Pressures

Another problem for the aviation community is coping with external forces. Aviation organizations are located in interorganizational "fields of force," and are affected by social pressures. These fields of force often interfere with integrity. The actions of organizations are often shaped by political, social, and economic forces. These forces include airlines, airports, regulators, and others. One air charter organization, B & L Aviation, experienced a crash in a snowstorm in South Dakota. The crash was blamed on pilot error. But after the crash, questions were raised about regulatory agencies' oversight of B & L's safety policies. One agency, the FAA, had previously given the flying organization a clean bill of health, but the Forest Service, which also carries out aviation inspections, described it instead as having chronic safety problems. Investigations, furthermore, disclosed that a U.S. Senator and his wife (an FAA official) had tried to limit the Forest Service's power and even eliminate it from inspecting B & L (Gerth & Lewis, 1994). The Federal Aviation Administration generally is caught in such fields of local political and economic forces, and some have questioned its ability to function as a regulator due to conflicting pressures and goals (e.g. Adamski & Doyle, 1994; Hedges, Newman, & Carey, 1995). Similarly, groups monitoring the safety of space shuttles (Vaughn, 1990) and the Hubble Space Telescope (Lerner, 1991) were subtly disempowered, leading to major failures.

Other individuals and groups formally "outside" the aviation organization may have a powerful impact on its functioning. Terrorists are an obvious example, but there are many others. Airport maintenance and construction crews, for instance, can cause enormous damage when they are careless. In May 1994, a worker in Islip, New York, knocked over a ladder and smashed a glass box, turning on an emergency power button; aircraft in three states were grounded for a half-hour (Pearl, 1994). In September 1994, a worker caused a short circuit that snarled air traffic throughout the Chicago region (Pearl, 1994). On January 9, 1995, power to Newark International Airport was shut down when a construction crew drove pilings through both the main and auxiliary power cables for the airport (Hanley, 1995).

EVALUATION AND LEARNING

Organizational Learning

All aviation organizations learn from experience. How well they learn is another matter. In the aviation community learning from mistakes is critical because failure of even a subsystem can be fatal. Because aircraft parts are mass-produced, what is wrong with one plane may be wrong with others. Systematic error, then, must be detected soon and rooted out quickly. Compared to other transport systems, aviation seems to have a good system for making such errors known and corrected quickly (Perrow, 1984). For instance, when two rudders on Boeing 737s malfunctioned, all units were checked that had been modified by the procedure thought to have caused the problem (Bryant, 1995a). Similarly, when some propellers manufactured by Hamilton Standard proved defective, the FAA insisted that some 400 commuter planes be checked and defective propellers replaced (Karr, 1995). This form of "global fix" is typical of, and somewhat unique to, the aviation industry. But many other problems are not dealt with so readily.

It may be useful to classify the cognitive responses of aviation organizations to anomalies into a rough spectrum such as the one in Fig. 5.11 (based on Westrum 1986):

Suppression and Encapsulation

These two responses are likely to take place when political pressures or resistance to change is intense. In suppression, the person raising questions is punished or eliminated. *Encapsulation* happens when the individuals or group raising questions are

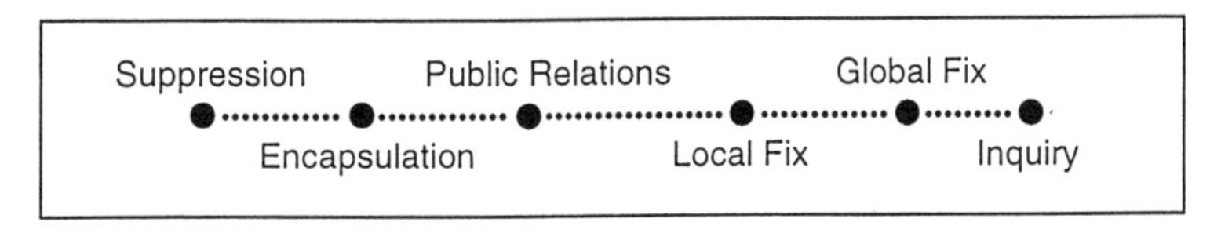

FIG. 5.11. Organizational response to anomaly.

isolated by management. For instance, an Air Force lieutenant colonel at Fairchild Air Force Base, in Washington state, showed a long-term pattern of risky flying behavior that climaxed in the spectacular crash of a B-52. Although similar risky behavior continued over a period of years, and must have been evident to a series of commanding officers, none prevented the officer from flying, and in fact he was put in charge of evaluating all B-52 pilots at the base (Kern, 1995). When this case and others were highlighted in a report by Allan Diehl, the Air Force's top safety official, Diehl was transferred from the Air Force Safety Agency in Albuquerque, New Mexico, to a nearby Air Force testing job (Thompson, 1995). Fixing the messengers instead of the problems is typical of pathological organizations. Cover-ups and isolation of whistle-blowers are not, of course, a monopoly of the U.S. Air Force.

Public Relations and Local Fixes

Organizational inertia often interferes with learning. It makes many organizations respond to failure primarily as a political problem. Failure to learn from the individual event can often take place when failures are explained away through public relations or when the problem solved is seen as a personal defect or a random glitch in the system. For instance, even though the Falklands air war was largely won by the Royal Navy, public relations presented the victory as a triumph for the Royal Air Force (Ward, 1992, pp. 337–351). The public relations campaign obscured the many RAF failures, some of which should have forced a reexamination of doctrine. Similarly, it has been argued that problems with Boeing 737-200s pitching up needed more attention than the situation got, even after the Potomac crash of an Air Florida jet (Nance, 1986, pp. 265–279). Previously Boeing had responded to the problem with local fixes, but without the global reach that Boeing could easily have brought to bear. When Mr. Justice Moshansky was investigating the Dryden, Ontario accident, legal counsel for both the carrier and the regulatory body sought to limit the scope of the inquiry and its access to evidence. Fortunately both these attempts were resisted, and the inquiry had far-reaching effects (Maurino et al., 1995, Foreword).

Global Fix and Reflective Inquiry

In a high-integrity organization, failures are considered occasions for inquiry, not blame and punishment (cf. Johnston, 1993). Aviation organizations frequently use global fixes (e.g., airworthiness directives) to solve common problems. But the aviation community also has a large amount of "reflective inquiry" (Schon, 1983), in which particular events trigger more general investigations, leading to far-reaching action. A comprehensive system of inquiry is typical of a community of good judgment, and it is this system that spots and removes the "latent pathogens." This system gives each person in the system a "license to think" and thus empowers anyone anywhere in it to identify problems and suggest solutions. Such a system actively cultivates maestros, idea champions, and internal critics. The Dryden, Ontario, accident inquiry and the United Airlines Portland, Oregon (1978), accident were both used as occasions for "system learning" far beyond the scope of the individual accident.

One can see in this spectrum an obvious relationship to the three types of organizational cultures discussed earlier. Pathological organizations are more likely to choose responses from the left side of the spectrum, generative organizations from the right side. We would also expect that organizations with strong CRM skills would favor responses toward the right. We believe that a study would show that higher mission success and lower accident rates are more typical of organizations choosing responses toward the right of this distribution. Although anecdotal evidence supports the relationship, such a study remains to be done.

Pop-Out Programs

One of the features of reflective inquiry is a willingness to bring otherwise hidden problems into view. These problems may be "hidden events" to management, suppressed because of unwritten rules or political influence (cf. Wilson & Carlson, 1996). Nonetheless, in high-integrity organizations considerable effort may be exerted to make such invisible events visible so that action can be taken on them. A "pop-out program" brings into the organization's consciousness what otherwise would have remained unknown. For instance, a factor in United Airlines developing its Command, Leadership, and Resources (CLR) program was a survey among United's pilots that brought to the surface a number of serious unreported incidents. With this expanded database, management became ready to take stronger actions than it might otherwise have done (Sams, 1987, p. 30).

Similarly, the use of anonymous reporting to third parties was critical in the development of the Aviation Safety Reporting System (ASRS) in the United States. Through ASRS, information on a wide variety of incidents is obtained through confidential communications from pilots and others (Reynard, Billings, Cheaney, & Hardy, 1986). The ability to get information that would otherwise be withheld allows decision making from a broader base of information, and also allows hidden events to become evident. The ASRS does not confer complete immunity on those who report to it, however, and some critics have noted that key information can be withheld (Nance, 1986).

Putting the right information together is sometimes the key to getting hazards to stand out. Information not seen as relevant for cultural or other reasons is sometimes ignored. Disaster may follow such a lapse. Information relevant to icing problems on a small commuter plane called the ATR-72 was ignored by the FAA (Engelberg & Bryant, 1995b). In particular, European information was not collated with U.S. information (Engelberg & Bryant, 1995a). Failure to collate this foreign evidence—in part, due to political pressures—about the design's hazards meant that the FAA did not arrange the information so that the failure pattern stood out (Frederick, 1996). Similarly, failure of the Space Shuttle Challenger occurred in part because statistics pointing clearly to a problem with low temperatures were not assembled in such a way that the pattern linking temperature and blow-by was evident (Bell & Esch, 1989; Tufte, 1997, pp. 38–53).

A famous example of the encouragement for pop-out is Wernher von Braun's reaction to the loss of a Redstone missile prototype. After a prototype went off course for no obvious reason, von Braun's group at Huntsville tried to analyze what might have gone wrong. When this analysis was fruitless, the group faced an expensive redesign to solve the still unknown problem. At this point an engineer came forward and told von Braun that he might inadvertently have caused the problem through creating a short circuit. He had been testing a circuit before launch, and his screwdriver had caused a spark. Although the circuit seemed fine, obviously the launch had not gone well. Investigation showed that the engineer's action was indeed at fault. Rather than punishing the engineer, von Braun sent him a bottle of champagne (von Braun, 1956).

Cognition and Action

Recognizing problems, of course, is not enough. Organizations have to do something about them. It must be remarked that although high-performance teams often have error-tolerant systems, the teams themselves are not tolerant of error, do not accept error as "the cost of doing business," and constantly try to eliminate it. High-performance teams spend a lot of time going over past successes and failures, trying to understand the reasons why. Then they fix the problems.

But many organizations do not always follow through after the recognition of problems. Politically influenced systems may respond with glacial slowness while key problems remain, as with the systems used to carry out air traffic control in the United States (Wald, 1996). Many of the computers used to direct traffic at U.S. airports can otherwise be found only in computer museums. At other times aviation organizations are caught up in political pressures that influence them to act prematurely. New equipment may be installed (as in the case of the new Denver Airport) before it has been thoroughly tested or put through an intelligent development process (Paul, 1979).

Sometimes aviation organizations seem to need disaster as a spur to action. Old habits provide a climate for complacency while problems go untreated (Janis, 1972). In other cases, the political community simply will not provide the resources or the mandate for change unless the electorate demands it and is willing to pay the price. Often it can require a horrendous event to unleash the will to act. For instance, the collision of two planes over the Grand Canyon in 1956 was a major stimulus to providing more en route traffic control in the United States (Adamski & Doyle, 1994, pp. 4–6; Nance, 1986, pp. 89–107). When FAA chief scientist Robert Machol warned of the danger of Boeing 757-generated vortices for small following aircraft, the FAA did not budge until two accidents with small planes killed 13 people (Anonymous, 1994). After the accident, following distance was changed from 3 to 4 miles. It is possible to trace the progress of the aviation system in the United States, for instance, through the accidents that brought specific problems to public attention. Learning from mistakes is a costly strategy, no matter how efficient the action is following the catastrophe. The organization that waits for a disaster to act is inviting one to happen.

CONCLUSION

"Human factors" has moved beyond the individual and even the group. Human factors is now seen to include the nature of the organizations that design, manufacture, operate, and evaluate aviation systems. Yet although recent accident reports acknowledge the key roles that organizations play in shaping human factors, this area is usually brought in only as an afterthought. It needs to be placed on an equal footing with other human factors concerns. We recognize that "organizational factors" is a field just opening up. Nonetheless, we hope to have raised some questions that further investigations can now proceed to answer.

On one point we are sure. High integrity is difficult to attain, as its rarity in the literature attests. Nonetheless, it is important to study those instances where it exists and understand what makes it operate successfully there. We have attempted here to show that "high-integrity" attitudes and behaviors form a coherent pattern. Those airlines, airports, corporate and commuter operations, government agencies, and manufacturers that have open communication systems, high standards, and climates supporting inquiry know things that the rest of the industry could learn. Civilians could learn from the military and vice versa. Out of such inquiries and exchanges we may learn to design sociotechnical systems more likely to get us safely to our destinations.

ACKNOWLEDGMENTS

The authors acknowledge the kind assistance rendered by Timothy J. Doyle and Ashleigh Merritt in writing this chapter.

REFERENCES

Adamski, A. J., & Doyle, T. J. (1994). *Introduction to the aviation regulatory process* (2nd ed.). Westland, MI: Hayden McNeil.

Allport, F. H. (1955). *Theories of perception and the concept of structure: A review and critical analysis with an introduction to a dynamic-structural theory of behavior.* New York: John Wiley & Sons.

Argyris, C., Putnam, R., & Smith, D. M. (1985). *Action science.* San Francisco: Jossey-Bass.

Aviel, D. (1994, November 14). Flying high on auto-pilot. *Wall Street Journal*, p. A10.

Bales, R. F. (1965). The equilibrium problem in small groups. In A. P. Hare, E. F. Borgatta, & R. F. Bales (Eds.), *Small groups: Studies in social interaction* (pp. 444–476). New York: Alfred A. Knopf.

Barley, S. (1986). Technology as an occasion for structuring: Evidence from the observation of CT scanners and the social order of radiology departments. *Administrative Science Quarterly, 31*(1), 78–108.

Bell, T., & Esch, K. (1989). The space shuttle: A case of subjective engineering. *I.E.E.E. Spectrum*, 42–46.

Brown, J. S., & Duguid, P. (1991). Organizational learning and communities of practice: Toward a unified view of working, learning, and innovation. *Organization Science, 2*(1).

Bryant, A., (1995a, March 15). FAA orders rudder checks on all 737s. *New York Times.*

Bryant, A. (1995b, October 15). Poor training and discipline at FAA linked to six crashes. *New York Times*, pp. 1, 16.

Bureau of Safety. (1967, July). *Aircraft design-induced pilot error.* Washington, DC: Civil Aeronautics Board, Department of Transportation.

Carley, W. M. (1993, April 26). Mystery in the sky: Jet's near-crash shows 747s may be at risk of autopilot failure. *Wall Street Journal*, pp. A1, A6.

Caspars, R. S., & Lipton, E. (1991, March 31–April 3). Hubble error: Time, money, and millionths of an inch. *Hartford Courant.*

Chute, R. D., & Wiener, E. L. (1995). Cockpit–cabin communications I: A tale of two cultures. *International Journal of Aviation Psychology, 5*(3), 257–276.

Chute, R. D., & Wiener, E. L. (1996). Cockpit–cabin communications II: Shall we tell the pilots? *International Journal of Aviation Psychology, 6*(3), 211–231.

Collingridge, D. (1992). *The management of scale: Big organizations, big decisions, big mistakes.* London: Routledge.

Collins, M. (1989, July 16). Review of Murray and Cox, *Apollo: The race to the moon. New York Times Book Review*, pp. 28–29.

Cooper, H. S. F. (1986, November 10). Letter from the space center. *New Yorker*, pp. 83–114.

Denison, D. R. (1990). *Corporate culture and organizational effectiveness.* New York: John Wiley & Sons.

Dodd, C. H. (1991). *Dynamics of intercultural communications.* Dubuque, IA: Wm. C. Brown.

Dunn, B. (1995). Communication: Fact or fiction. In N. Johnston, R. Fuller, & N. McDonald (Eds.), *Aviation psychology: Training and selection* (Vol. 2, pp. 67–74). Aldershot, England: Avebury Aviation.

Enders, J. (1992, February). Management inattention greatest aircraft accident cause, not pilots, says Enders. *Flight Safey Foundation News, 33*(2), 1–15.

Engelberg, S., & Bryant, A. (1995a, February 26). Lost chances in making a commuter plane safer. *New York Times*, pp. 1, 14, 15.

Engelberg, S., & Bryant, A. (1995b, March 12). Since 1981, federal experts warned of problems with rules for icy weather flying. *New York Times*, pp. 1, 12.

Fitzgerald, K. (1989). Probing Boeing's crossed connections. *I.E.E.E. Spectrum*, pp. 30–35.

Frantz, D., & Blumenthal, R. (1994, November 13). Troubles at USAir: Coincidence or more? *New York Times*, pp. 1, 18, 19.

Franzen, J. (1994, October 24). Lost in the mail. *New Yorker*, pp. 62–77.

Frederick, S. A. (1996). *Unheeded warning: The inside story of American Eagle flight 4184.* New York: McGraw-Hill.

Gerth, J., & Lewis, N. A. (1994, October 16). Senator's bill to consolidate air inspection is questioned. *New York Times*, pp. 1, 14.

Gibbs, W. W. (1994, September). Software's chronic crisis. *Scientific American*, pp. 86–95.

Gillchrist, P. T. (1995). *Crusader! Last of the gunfighters.* Atglen, PA: Shiffer.

Ginnett, R. C. (1993). Crews as groups: Their formation and their leaderhsip. In E. L. Wiener, B. G. Kanki, & R. L. Helmreich (Eds.), *Cockpit resource management.* New York: Academic Press.

Glain, S. (1994, October 4). Language barrier proves dangerous in Korea's skies. *Wall Street Journal*, pp. B1, B4.

Gras, A. C., Morocco, S., Poirot-Delpech, L., & Scardigli, V. (1994). *Faced with automation: The pilot, the controller, and the engineer.* Paris: Publications de la Sorbonne.

Hall, E. (1959). *The silent language.* New York: Doubleday.

Hanley, R. (1995, January 12). Blackout at Newark Airport leads to study of cable rules. *New York Times.*

Hedges, S. J., Newman, R. J., & Cary, P. (1995, June 26). What's wrong with the FAA? *U.S. News and World Report*, pp. 29–37.

Helmreich, R. (1994). Anatomy of a system accident: The crash of Avianca 052. *International Journal of Aviation Psychology, 4*(3), 265–284.

Hidden, A. (1989). *Investigation into the Clapham Junction Railway accident.* London: H.M.S.O.

Hofstede, G. (1980). *Culture's consequences: International differences in work-related values.* Beverly Hills, CA: Sage.

Hughes, D. (1994, August 8). Denver Airport still months from opening. *Aviation Week & Space Technology*, pp. 30–31.

Janis, I. L. (1972). *Victims of groupthink: A psychological study of foreign-policy decisions and fiascoes* Boston: Houghton Mifflin.

Johnston, N. (1993, October). Managing risk and apportioning blame. *I.A.T.A. 22nd Technical Conference*, Montreal.

Johnston, N. (1995). CRM: Cross-cultural perspectives. In E. Wiener, B. G. Kanki, & R. L. Helmreich (Eds.), *Cockpit resource management* (pp. 367–398). San Diego: Academic Press.

Journal of Contingencies and Crisis Management. (1994). [Special issue] 2 (4).

Karr, A. R. (1995, August 28). Propeller-blade inspection set on small planes. *Wall Street Journal*, p. A34.

Kern, T. (1995). *Darker shades of blue: A case study of failed leadership.* Published by the author. Colorado Springs: United States Air Force Academy.

Kmetz, J. L. (1984). An information-processing study of a complex workflow in aircraft electronics repair. *Administrative Science Quarterly, 29*(2), 255–280.

Lauber, J. K. (1993, April). A safety culture perspective. *Proceedings of the Flight Safety Foundation 38th Annual Corporate Aviation Safety Seminar* (pp. 11–17).

Lautman, L. G., & Gallimore, P. L. (1987, June). The crew-caused accident. *Flight Safety Foundation Flight Safety Digest*, pp. 1–8.

Law, J. R. & Willhelm, J. A. (1995, April). *Ratings of CRM skill markers in domestic and international operations: A first look.* Symposium conducted at the 8th International Symposium on Aviation Psychology, Columbus, OH.

Lerner, E. (1991, February). What happened to Hubble? *Aerospace America*, pp. 18–23.

Lineberry, C., & Carleton, J. R. (1992). Culture change. In H. D. Stolovitch & E. J. Keeps (Eds.), *Handbook of human performance technology* (pp. 233–246). San Francisco: Jossey-Bass.

Little, L., Gaffney, I. C., Rosen, K. H., & Bender, M. (1990, November). Corporate instability is related to airline pilots' stress symptoms. *Aviation, Space, and Environmental Medicine, 61*(11), 977–982.

Littlewood, B., & Strigini, L. (1992, November). The risks of software. *Scientific American*, pp. 62–75.

Marcus, A., & Fox, I. (1988, December). Lessons learned about communicating safety. *Related concerns to industry: The Nuclear Regulatory Commission after Three Mile Island.* Paper presented at the Symposium on Science Communication: Environmental and Health Research, University of Southern California, Los Angeles.

Maurino, D. E., Reason, J., Johnston, N., & Lee, R. (1995). *Beyond aviation human factors.* Aldershot, England: Avebury Aviation.

McCurdy, H. E. (1993). *Inside NASA: High technology and organizational change in the u.s. space program* Baltimore: Johns Hopkins Press.

Merritt, A. C., & Helmreich, R. L. (1995). *Culture in the cockpit: A multi-airline study of pilot attitudes and values.* Paper presented at the 1995 papers: The NASA/University of Texas/FAA aerospace crew research project: VIIIth International Symposium on Aviation Psychology, Ohio State University.

Morrocco, J. D. (1994, July 18). Fratricide investigation spurs U.S. training review. *Aviation Week & Space Technology*, pp. 23–24.

Mouden, L. H. (1992, April). *Management's influence on accident prevention.* Paper presented at the Flight Safety Foundation 37th Corporate Aviation Safety Seminar: The Management of Safety, Baltimore, MD.

Murray, C., & Cox, C. B. (1989). *Apollo: The race to the moon.* New York: Simon & Schuster.

Nance, J. J. (1986). *Blind trust.* New York: William Morrow.

National Transportation Safety Board. (1992). *Aircraft accident report: Britt Airways, Inc., d/b/a Continental Express Flight 2574 in-flight structural breakup, EMB-120RT, N33701, Eagle Lake, Texas, September 11, 1991.* Washington, DC: Author.

National Transportation Safety Board. (1994). *Special investigation report of maintenance anomaly resulting in dragged engine during landing rollout of Northwest Airlines Flight 18, Boeing 747-251B, N637US, New Tokyo International Airport, Narita, Japan, March 1, 1994.* Washington, DC: Author.

O'Lone, R. G. (1992, October 12). 777 design shows benefits of early input from airlines. *Aviation Week and Space Technology.*

Paries, J. (1994, July/August). Investigation probed root causes of CFIT accident involving a new-generation transport. *I.C.A.O. Journal*, pp. 37–41.

Pasmore, W. A. (1988). *Designing efficient organizations: The sociotechnical systems perspective.* New York: John Wiley.

Patterson, T. T. (1955). *Morale in war and work.* London: Max Parrish.

Paul, L. (1979, October). *How can we learn from our mistakes if we never make any?* Paper presented at 24th Annual Air Traffic Control Association Fall Conference, Atlantic City, NJ.

Pearl, D. (1994, September 15). A power outage snarls air traffic in Chicago region. *Wall Street Journal*, p. 5.

Perrow, C. (1984). *Normal accidents: Living with high-risk technologies.* New York: Basic Books.

Peters, T. (1988). *Thriving on chaos: Handbook for a management revolution.* New York: Alfred A. Knopf.

Peters, T. (1992). *Liberation management: Necessary disorganization for the nanosecond nineties.* New York: Alfred A. Knopf.

Petroski, H. (1994). *Design paradigms: Case histories of error and judgment in engineering.* New York: Cambridge University Press.

Pidgeon, N., & O'Leary, M. (1994). Organizational safety culture: Implications for aviation practice. In N. Johnston, N. McDonald, & R. Fuller (Eds.), *Aviation psychology in practice* (pp. 21–43). Aldershot, England: Avebury Technical.

Prince, M. (1990). *Crash course: The world of air safety.* London: Collins.

Quintanilla, C. (1994, November 21). United Airlines goes for the stealth look in coloring its planes. *Wall Street Journal*, pp. A1, A4.

Reason, J. (1984). Little slips and big accidents. *Interdisciplinary Sciences Reviews, 11*(2), pp. 179–189.

Reason, J. (1990). *Human error.* New York: Cambridge University Press.

Rechtin, E. (1992, October). The art of system architecting. *I.E.E.E. Spectrum*, pp. 66–69.

Reynard, W. D., Billings, C. E., Cheaney, E. S., & Hardy, R. (1986). *The development of the NASA aviation safety reporting system* (NASA Reference Publication 1114). Moffett Field, California: NASA Ames.

Roberts, K. H. (Ed.). (1993). *New challenges to understanding organizations.* New York: Macmillan.

Roberts, K. H., & Rousseau, D. M. (1989, May). Research in nearly failure-free, high-reliability organizations: Having the bubble. *I.E.E.E. Transactions on Engineering Management, 36*(2), 132–139.

Roberts, K. H., & Weick, K. (1993, September). Group mind: Heedful interaction on aircraft carrier flight decks. *Administrative Science Quarterly, 38*(3).

Sagan, S. D. (1993). *The limits of safety: Organizations, accidents, and nuclear weapons.* Princeton, NJ: Princeton University Press.

Sams, T. L. (1987, December). *Cockpit resource management concepts and training strategies.* Unpublished doctoral dissertation, East Texas State University.

Schein, E. H. (1992). *Organizational culture and leadership* (2nd ed.). San Francisco: Jossey-Bass.

Schlager, N. (Ed.). (1994). *When technology fails: Significant technological disasters, accidents, and failures of the twentieth century.* Detroit: Gale Research.

Schon, D. A. (1983). *The reflective practitioner: How professionals think in action.* New York: Basic Books.

Schneider, K. (1991, July 30). Study finds link between chemical plant acccidents and contract workers. *New York Times*, p. A10

Shepherd, W. T. (1994, February 1). *Aircraft maintenance human factors.* Presentation at International Maintenance Symposium, San Diego.

Sherman, P. J., & Helmreich, R. L. (1995). *Attitudes toward automation: The effect of national culture.* Paper presented at the 1995 Papers: The NASA/University of Texas/FAA Aerospace Crew Research Project. VIIIth International Symposium on Aviation Psychology, Ohio State University.

Squires, A. (1986). *The tender ship: Government management of technological change.* Boston: Birkhauser.

Thompson, M. (1995, May 29). Way out in the wild blue yonder. *Time*, pp. 32–33.

Trice, H., & Beyer, J. M. (1993). *The cultures of work organizations.* Englewood Cliffs, NJ: Prentice-Hall.

Trotti, J. (1984). *Phantom in Vietnam.* Novato, CA: Presidio.

Tufte, E. R. (1997). *Visual explanations.* Cheshire, CT: Graphics Press.

Turner, B. A. (1978). *Man made disasters.* London: Wykeham.

Vaill, P. B. (1978). Toward a behavioral description of high-performing systems. In M. McCall & M. Lombardo (Eds.), *Leadership: Where else can we go?* Durham, NC: Duke University Press.

Vaill, P. B. (1982). The purposing of high-performing systems. *Organizational Dynamics, 11*(2), 23–39.

Vaughn, D. (1990, June). Autonomy, interdependence, and social control: NASA and the Space Shuttle Challenger. *Administrative Science Quarterly, 35*(2), pp. 225–257.

von Braun, W. (1956, October). Teamwork: Key to success in guided missiles. *Missiles and Rockets*, pp. 38–43.

Wald, M. (1995, May 7). A new look at pilots' role in emergency. *New York Times*, p. 12.

Wald, M. (1996, January 29). Ambitious update of air navigation becomes a fiasco. *New York Times*, pp. 1, 11.

Ward, Cmdr. S. (1992). *Sea Harrier over the Falklands.* London: Orion.

Weener, E. F. (1990). *Control of crew-caused accidents: The sequal* (Boeing flight operations regional seminar: New Orleans). Seattle: Boeing Commercial Aircraft Company.

Westrum, R. (1986, October). *Organizational and inter-organizational thought.* Paper presented at the World Bank Conference on Safety and Risk Management.

Westrum, R. (1991). *Technologies and society: The shaping of people and things.* Belmont, CA: Wadsworth.

Westrum, R. (1993). Cultures with requisite imagination. In J. Wise & D. Hopkin (Eds.), *Verification and validation: Human factors aspects* (pp. 401–416). New York: Springer.

Westrum, R. (1994). Is there a role for a "test controller" in the development of new ATC equipment? In J. Wise, V. D. Hopkin, & D. Garland (Eds.), *Human factors certification of new aviation technologies.* New York: Springer.

Wetterhahn, R. F. (1997, August). Change of command. *Air and Space*, pp. 62–69.

Wilcox, R. K. (1990). *Scream of eagles: Top Gun and the American aerial victory in Vietnam.* New York: John Wiley.

Wilson, G. C., & Carlson, P. (1996, January 1–7). The ultimate stealth plane. *Washington Post National Weekly Edition*, pp. 4–9.

Wood, R. C. (1993). *Elements of a safety culture.* Proceedings of the Flight Safety Foundation 38th Annual Corporate Aviation Safety Seminar, 26–29.

Yeager, Gen. C., & Janos, L. (1985). *Yeager: An autobiography.* New York: Bantam.

Zuboff, S. (1984). *In the age of the smart machine: The future of work and power.* New York: Basic Books.

II

HUMAN CAPABILITIES AND PERFORMANCE

6

Processes Underlying Human Performance

Lisanne Bainbridge
University College London

Two decades ago, a chapter on aviation with this title might have focused on physical aspects of human performance, on representing the control processes involved in flying. There has been such a fundamental change in our knowledge and techniques that this chapter focuses almost exclusively on cognitive processes. The main aims are to show that relatively few general principles underlie the huge amount of information relevant to interface design, and that context is a key concept in understanding human behavior.

Classical interface human factors/ergonomics consists of a collection of useful but mainly disparate facts and a simple model of the cognitive processes underlying behavior—that these processes consist of independent information–decision–action or if–then units. (I use the combined term *human factors/ergonomics*, shortened to HF/E, because these terms have different shades of meaning in different countries. *Cognitive* processing is the unobservable processing between arrival of stimuli at the senses and initiating an action.) Classic HF/E tools are powerful aids for interface design, but they make an inadequate basis for designing to support complex tasks. Pilots and air traffic controllers are highly trained and able people. Their behavior is organized and goal-directed, and they add knowledge to the information given on an interface in two main cognitive activities: understanding what is happening, and working out what to do about it.

As the simple models of cognitive processes used in classic HF/E do not contain reminders about all the cognitive aspects of complex tasks, they do not provide a sufficient basis for supporting HF/E for these tasks. The aim of this chapter is to present simple concepts that could account for behavior in complex dynamic tasks and provide the basis for designing to support people doing these tasks. As the range of topics and data that could be covered is huge, the strategy is to indicate key principles by giving typical examples, rather than attempting completeness. This chapter does not present a detailed model for the cognitive processes suggested, or survey HF/E techniques, and it does not discuss collective work. The chapter offers three main sections, on simple use of interfaces; understanding, planning, and multitasking; and learning, workload, and errors. The conclusion outlines how the fundamental nature of human cognitive processes underlies the difficulties met by HF/E practitioners.

USING THE INTERFACE, CLASSIC HF/E

This chapter distinguishes between cognitive functions or goals, what is to be done, and cognitive processes, how these are done. This section starts with simple cognitive functions and processes underlying the use of displays and controls, on the interface between a person and the device the person is using. More complex functions of understanding and planning are discussed in the next main section.

I take the view that simple operations are affected by the context in which they are done. Someone does not press a button in isolation. For example, a pilot keys in a radio frequency as part of contacting air traffic control, as part of navigation, which is multi-tasked with checking for aircraft safety, and so on. From this point of view, an account of cognitive processes should start with complex tasks. However that is just too difficult. Here, I have started with the simple tasks involved in using an interface, and point out how even simple processes are affected by a wider context. The next main section builds up from this to discuss more complex tasks.

Five main cognitive functions are involved in using an interface:

- Discriminating a stimulus from a background, or from other possible stimuli. The process usually used for this is decision making.
- Perceiving "wholes." The main process here is integrating together parts of the sensory input.
- Naming.
- Choosing an action. The cognitive process by which the functions of naming, and choosing an action, are done (in simple tasks) is recoding, that is, translating from one representation to another, such as (shape → name), or (display → related control).
- Comparison, which may be done by a range of processes from simple to complex.

Because discriminating and integrating stimuli are usually done as the basis for naming or for choosing an action, it is often assumed that the processes for carrying out these functions are independent, input driven, and done in sequence. However, the discussion shows that these processes are not necessarily distinct, or done in sequence, and that they all involve use of context and knowledge.

This section does not discuss displays and controls separately, as both involve all the functions and processing types. Getting information may involve making a movement such as visual search or accessing a computer display format, whereas making a movement involves getting information about it. The four subsections are on detecting and discriminating; visual integration; naming and simple action choices; and action execution.

Detecting and Discriminating

It might be thought, because the sense organs are separate from the brain, that at least basic sensory effectiveness, the initial reception of signals by the sense organs, would be a simple starting point, before considering the complexities that the brain can introduce such as naming a stimulus or choosing an action. However, sensing processes turn out not to be simple: There can be a large contribution of prior knowledge and present context.

This part of the chapter is in four subsections, on detecting, discriminating one signal from others that are present, or not present (absolute judgment), and sensory decisions. It is artificial to distinguish between sensory detection and discrimination, although they are discussed separately here, because they both involve (unconscious) decision making about what a stimulus is. In many real tasks, other factors have more effect on performance than any basic limits to sensory abilities. Nevertheless, it is useful to understand these sensory and perceptual processes, because they raise points that are general to all cognitive processing.

Detecting. *Detection* is one of those words that may be used to refer to different things. In this section I use it to mean sensing the presence of a stimulus against a blank background. Detecting the presence of light is an example. A human eye has the ultimate sensitivity to detect one photon of electromagnetic energy in the visible wavelengths. However, we can only detect at this level of sensitivity if we have been in complete darkness for about half an hour (Fig. 6.1). The eyes adapt so they are sensitive to a range of light intensities around the average (Fig. 6.2); this adaptation takes time. Adaptation allows the eyes to deal efficiently with a wide range of stimulus conditions, but it means that sensing is relative rather than absolute.

The two curves on the dark adaptation graph (Fig. 6.1) indicate that the eyes have two different sensing systems, one primarily for use at high, and the other for use at low, light intensities. These two systems have different properties. At higher levels of illumination the sensing cells are sensitive to color. There is one small area of the retina (the sensory surface inside the eye) that is best able to discriminate between spatial positions, and best able to detect stationary objects. The rest of the sensory surface (the periphery) is better at detecting moving than stationary objects. At lower levels of illumination intensity, the eyes see mainly in black and white, and peripheral vision is more sensitive for detecting position.

Therefore it is not possible to make a simple statement that "the sensitivity of the eyes is . . . ." The sensitivity of the eyes depends on the environment (e.g., the average level of illumination) and on the stimulus (e.g., its movement, relative position, or color). The sensitivity of sense organs adapts to the environment and the task, so

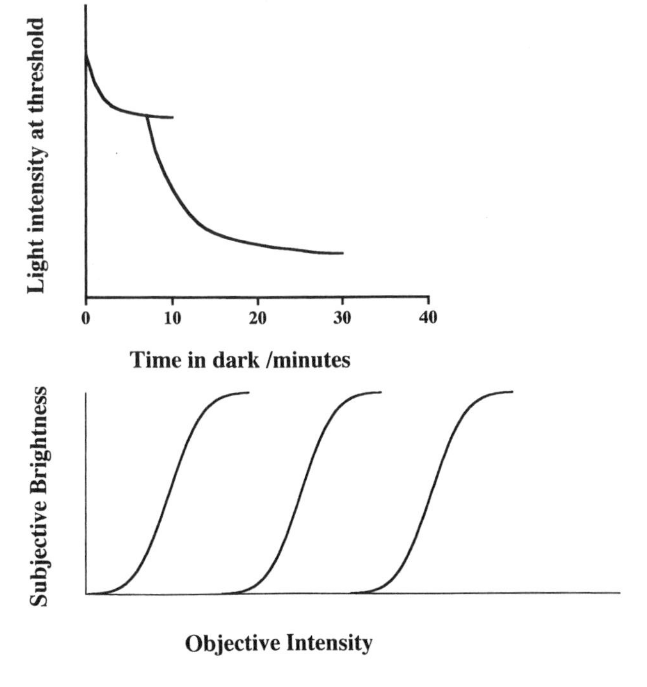

FIG. 6.1. Increasing sensitivity to light after time in darkness ("dark adaptation").

FIG. 6.2. The sensitivity of the eye when adapted to three different levels of average illumination. At each adaptation level, the eye is good at discriminating between intensities around that level.

sensitivity does not have an absolute value independent of these influences. This means it is difficult to make numerical predictions about sensory performance in particular circumstances, without testing directly.

However, it is possible to draw practical implications from the general trends in sensitivity. For example, it is important to design to support both visual sensing systems in tasks that may be done in both high and low levels of illumination, such as flying. It is also sensible to design so that the most easily detected stimuli (the most "salient") are used for the most important signals. Visual salience depends not only on intensity but also on the color, movement, and position of the stimulus. Very salient stimuli attract attention; they override the usual mechanism for directing attention (see next main section). This means that very salient signals can be either useful as warning signals, or a nuisance as irrelevant distractions that interrupt the main task thinking.

Discriminating Between Stimuli. In this section I use the word *discrimination* to mean distinguishing between two (or more) stimuli. As with detection, the limits to our ability to discriminate between stimulus intensities are relative rather than absolute. The just noticeable difference between two stimuli is a ratio of the stimulus intensities. (There is a sophisticated modern debate about this, but it is not important for most practical applications.) The ratio is called the Weber fraction. Again, the size of this ratio depends on the environmental and task context. For example, in visual intensity discriminations, the amount of contrast needed to distinguish between two stimuli depends on the size of the object (more contrast is needed to see smaller objects) and on the level of background illumination (more contrast is needed to see objects in lower levels of background illumination).

The Weber fraction describes the difference between stimuli that can just be discriminated. When stimuli differ by larger amounts, the time needed to make the discrimination is affected by the same factors: Finer discriminations take longer, and visual discriminations can be made more quickly in higher levels of background illumination.

Touch and feel (muscle and joint receptor) discriminations are made when using a control. For example, a person using a knob with tapered sides may make three times more positioning errors than when using a knob with parallel sides (Hunt & Warrick, 1957). Neither of the sides of a tapered knob actually points in the direction of the knob, so touch information from the sides is ambiguous.

Resistance in a control affects how easy it is to discriminate by feel between positions of the control. Performance in a tracking task, using controls with various types of resistance, shows that inertia makes performance worse, whereas elastic resistance can give the best results. This is because inertia is the same whatever the size of movement made, so it does not help in discriminating between movements. Elastic resistance, in contrast, varies with the size of movement, so gives additional information about the movements being made (Howland & Noble, 1955).

Absolute Judgment. The Weber fraction describes the limit to our abilities to discriminate between two stimuli when they are both present. When two stimuli are next to each other we can, at least visually, make very fine discriminations in the right circumstances. However, our ability to distinguish between stimuli when only one of them is present is much more limited. This process is called absolute judgment. The judgment limits to our sensory abilities are known in general, for many senses and dimensions (Miller, 1956). These limits can be affected by several aspects of the task situation, such as the range of possible stimuli that may occur (Helson, 1964).

When only one stimulus is present, distinguishing it from others must be done by comparing it with mental representations of the other possible stimuli. So absolute judgment must involve knowledge and/or working memory. This is an example of a sensory discrimination process that has some processing characteristics in common with what are usually considered much more complex cognitive functions. There is not always a clear distinction between simple and complex tasks in the aspects of processing involved.

Although our ability to make absolute judgments is limited, it can be useful. For example, we can discriminate among eight different positions within a linear interval. This means that visual clutter on scale-and-pointer displays can be reduced; it is only necessary to place a scale marker at every 5 units that need to be distinguished. But our ability is not good enough to distinguish between 10 scale units without the help of an explicit marker.

In other cases, the limitations need to be taken into account in design. For example, we can only distinguish among 11 different color hues by absolute judgment. As we are very good at distinguishing between colors when they are next to each other, it can be easy to forget that color discrimination is limited when one color is seen alone. For example, a color display might use green-blue to represent one meaning (e.g., main water supply) and purple-blue with another meaning (e.g., emergency water supply). It might be possible to discriminate between these colors, and so use them as a basis for identifying meaning, when the colors are seen together, but not when they are seen alone. (Some discussion of meaning is given later.)

Again, discrimination is a process in which the task context, in this case whether or not the stimuli occur together for comparison, has a strong effect on the cognitive processes involved and on our ability to make the discriminations.

Sensory Decision Making. Detections and discriminations involve decisions about whether the evidence reaching the brain is sufficient to justify deciding that a stimulus (difference) is present. For example, detection on a raw radar screen involves deciding whether a particular radar trace is a "blip" representing an aircraft, or something else that reflects radar waves. A particular trace may only be more or less likely to indicate an aircraft, so a decision has to be made in conditions of uncertainty. This sort of decision can be modeled by signal detection or statistical decision theory. Different techniques are now used in psychology, but this approach is convenient here because it distinguishes between the quality of the evidence and the observer's prior biases about decision outcomes.

Suppose that radar decisions are based on intensity, and that the frequencies with which different intensities have appeared on the radar screen when there was no aircraft present have been as shown in Fig. 6.3a at the top, whereas the intensities that have appeared when an aircraft was present are shown in Fig. 6.3a at the bottom. There is a range of intensities that occurred only when an aircraft was not present, a range of intensities that occurred only when an aircraft was present, and an intermediate range of intensities that occurred both when an aircraft was present and when it was not (Fig. 6.3b). How can someone make a decision when one of the intermediate intensities occurs? The decision is made on the basis of signal likelihood. The height of the curve above a particular intensity indicates how likely that intensity was to occur when there was or was not an aircraft. At the midpoint between the two frequency distributions,

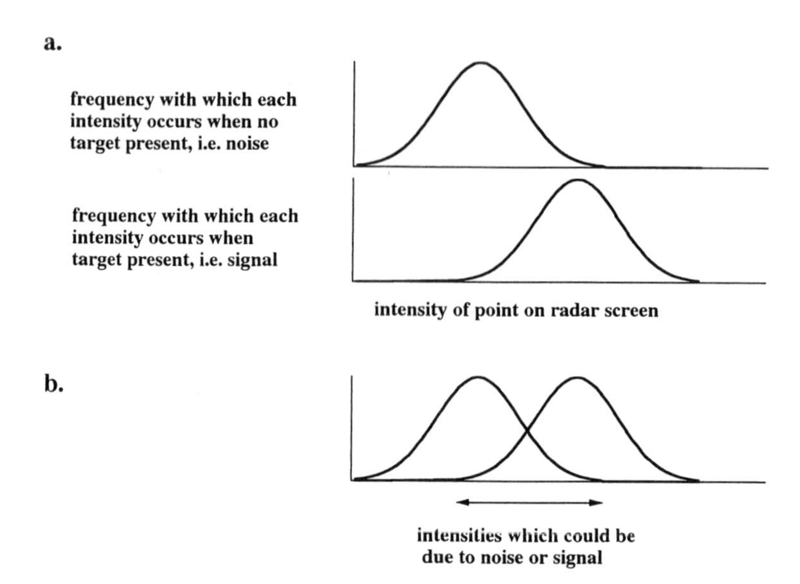

FIG. 6.3. Knowledge about the occurrence of intensities. Decision making uses knowledge about the alternatives, based on previous experience.

both possibilities are equally likely. Intensities less than this midpoint are more likely not to come from an aircraft, and intensities greater than this midpoint are more likely to come from an aircraft.

Note that when a stimulus is in this intermediate range, it is not always possible to be right about a decision. A person can decide a trace is not an aircraft when it actually is (a "miss"), or can decide it is an aircraft when it is not (a "false alarm"). These ways of being wrong are not called *errors*, because it is not mathematically possible always to be right when making uncertain decisions. The number of wrong decisions and the time to make the decision increase when signals are more similar (overlap more).

Note that when the radar operator is making the decision, there is only one stimulus actually present, with one intensity. The two frequency distributions, against which this intensity is compared to make the decision, must be supplied from the operator's previous experience of radar signals, stored in the operator's knowledge base. Decisions are made by comparing the input stimulus ("bottom-up") with stored knowledge about the possibilities ("top-down").

In addition to the uncertainty due to similarity between possible interpretations of a stimulus, the second major factor in this type of decision making is the importance or costs of the alternative outcomes. In the example just given, the person's decision criterion, the intensity at which the person changes from deciding "yes" to deciding "no," was the point at which both possibilities are equally likely. But suppose it is very important not to miss a signal—for instance, when radar watch keeping in an early warning system. Then it might be sensible to use the decision criterion in Fig. 6.4. This would increase the number of hits. It would also increase the number of false alarms, but this might be considered a small price to pay compared with the price of missing a detection. Alternatively, imagine people doing a job in which when they detect a signal they have to do a lot of work, and they are feeling lazy and not committed to their job. Then they might move their decision criterion in the other direction, to minimize the number of hits.

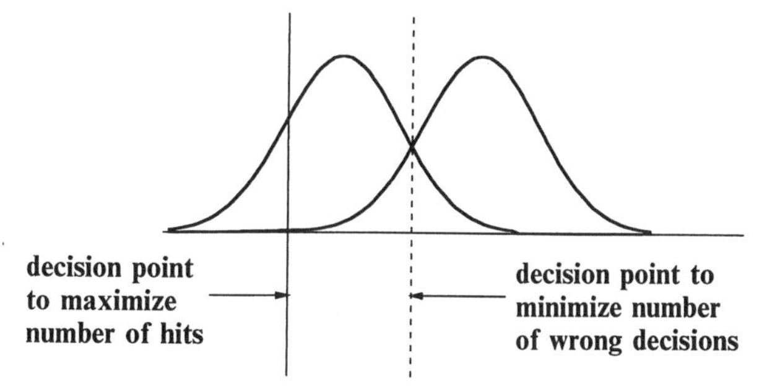

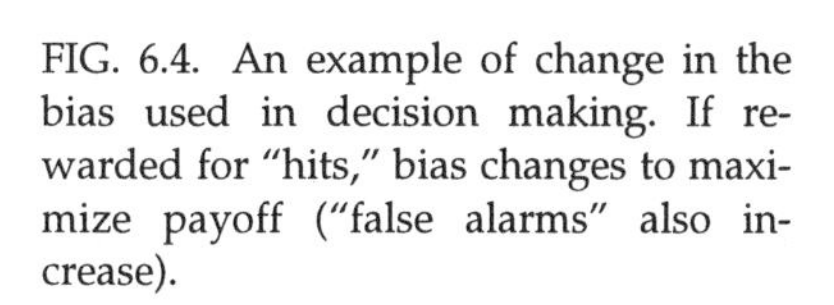
FIG. 6.4. An example of change in the bias used in decision making. If rewarded for "hits," bias changes to maximize payoff ("false alarms" also increase).

This shift in decision criterion is called bias. Decision bias can be affected by probabilities and costs. The person's knowledge of the situation provides the task and personal expectations/probabilities and costs that are used in setting the biases, so again top-down processing influences sensory decisions. There are limits to human ability to assess biases (Kahneman, Slovic, & Tversky, 1992). At extreme probabilities we tend to substitute determinacy for probability. We may think something is sure to happen, when it is just highly likely. Some accidents happen because people see what they expect to see, rather than what is actually there (e.g., Davis, 1966). Inversely, we may think something will never happen, when it is objectively of very low probability. For example, when signals are very unlikely, then it is difficult for a human being to continue to direct attention to watching for them (the "vigilance" effect).

Visual Integration

The effects of knowledge and context are even more evident in multidimensional aspects of visual perception, such as color, shape, size, and movement, in which what is seen is an inference from combined evidence. This discussion is in subsections on movement, size, and color; grouping processes; and shape. (There are also interesting auditory integrations, much involved in music perception, but these are not discussed here.)

Movement, Size, and Color Constancies. It is actually quite odd that we perceive a stable external world, given that we and other objects move, and the wavelength of the environmental light we see by changes, so the size, position, shape, and wavelength of light reflected from objects onto the retina all change. As we do perceive a stable world, this suggests our perception is relative rather than absolute: We do not see what is projected on the retina, but a construction based on this projection, made by combining evidence from different aspects of our sensory experience. The processes by which a wide variety of stimuli falling on the retina are perceived as the same are called *constancies*.

When we turn our heads the stimulation on the retina also moves. However, we do not see the world as moving, because information from the turning receptors in the ear is used to counteract the evidence of movement from the retina. The changes on the retina are perceived in the context of changes in the head rotation receptors. When the turning receptors are diseased, or the turning movements are too extreme for the receptors to be able to interpret quickly, then the person may perceive movement that is not actually occurring, as in some flying illusions.

There is also constancy in size perception. As someone walks away from us, we do not see them becoming smaller and smaller, although there are large changes in the size of the image of that person that falls on the retina. In interpreting the size of objects, we take into account all the objects that are at the same distance from the eye, and then perceive them according to their relative size. Size constancy is more difficult to account for than movement constancy, as it involves distance perception, itself a complex process (Gibson, 1950). Distance is perceived by combining evidence about texture, perspective, changes in color of light with distance, and overlapping (itself a construct, discussed later). Information from the whole visual field is used in developing a percept that makes best overall sense of the combination of inputs. Cognitive psychology uses the concept that different aspects of stimulus processing are done simultaneously, unless an aspect is difficult and slows processing down. Each aspect of processing communicates its "results so far" to the other aspects via a "blackboard," and all aspects work together to produce a conclusion (Rumelhart, 1977).

Color perception is also an integrative process that shows constancy. Research on the color-receptive cells in the retina suggests that there are only three types of cell, which respond to red, green, and blue light wavelengths. The other colors we "see" are constructed by the brain, based on combinations of stimulus intensities at these three receptors. The eyes are more sensitive to some colors, so if a person looks at two lights of the same physical intensity but different wavelengths, the lights may be of different experienced intensity (brightness). The effectiveness of the color construction process is such that there are some visual demonstrations in which people see a range of colors, even though the display consists only of black and white plus one color. This constructive process also deals with color constancy. The wavelength of ambient lighting can change quite considerably, so the light reflected from objects also changes in wavelength, but objects are perceived as having stable color. The wavelengths of light from all the objects change in the same way, and color is perceived from the relative combinations of wavelengths, not the actual wavelengths. This constancy process is useful for perceiving a stable world despite transient and irrelevant changes in stimuli, but it does make designing color displays more difficult. As with our response to stimulus intensity, our perception of color is not a fixed quantity that can easily be defined and predicted. Instead, it depends on the interaction of several factors in the environment and task contexts, so it may be necessary to make color perception tests for a particular situation.

Grouping Processes. Another type of perceptual integration occurs when several constituents of a display are grouped together and perceived as a "whole." The Gestalt psychologists in the 1920s first described these grouping processes, which can be at several levels of complexity.

1. Separate elements can be seen as linked into a line or lines. There are four ways in which this can happen: when the elements are close together, are similar, lie on a line, or define a contour. The grouping processes of proximity and similarity can be used in the layout of displays and controls on a conventional interface, to show which items go together.
2. When separate elements move together they are seen as making a whole. This grouping process is more effective if the elements are also similar. This is used in the design of head-up displays and predictor displays, as in Fig. 6.5.

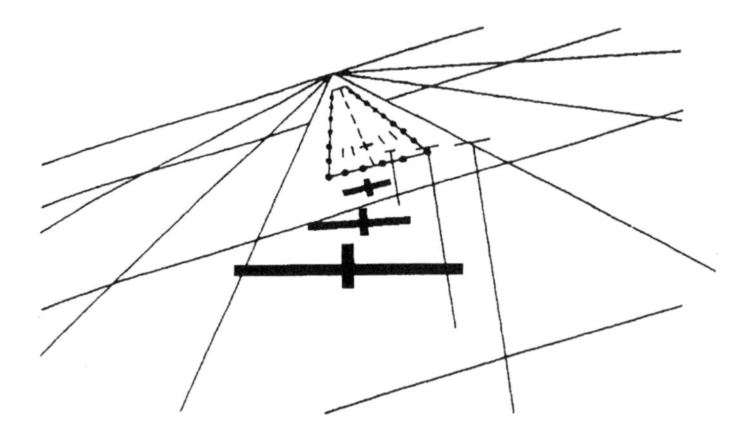

FIG. 6.5. Gestalt grouping processes relate together the elements of a predictor landing display. (Reprinted with permission from *Human Factors*, Vol. 19, No. 6, 1977. Copyright © 1977 by the Human Factors and Ergonomics Society. All rights reserved.)

3. Something that has uniform color or a connected contour is seen as a "whole"—for example, the four sides of a square are seen as a single square, not as four separate elements.

4. The strongest grouping process occurs when the connected contour has a "good" form, that is, a simple shape. For example, a pull-down menu on a computer screen is seen as a distinct unit in front of other material, because it is a simple shape, and the elements within the shape are similar and (usually) different from the elements on the rest of the screen. When the visual projections of two objects are touching, then the one with the simplest shape is usually seen as in front of (overlapping) the other.

The visual processes by which shapes and unities are formed suggest recommendations for the design of symbols and icons that are easy to see (Easterby, 1970).

Shape Constancy. Visual integrative processes ensure that we see a unity when there is an area of the same color, or a continuous contour. The shape we see depends on the angles of the contour lines (there are retinal cells which sense angle of line). Again there are constancy processes. The shape perceived is a construction, taking into account various aspects of the context, rather than a simple mapping of what is projected from the object onto the retina. Figure 6.6 shows a perspective drawing of a cube, with the same ellipse placed on each side. The ellipse on the front appears as an ellipse on a vertical surface. The ellipse on the top appears to be wider and sloping at the same angle as the top. The ellipse on the side is ambiguous—is it rotated, or not part of the

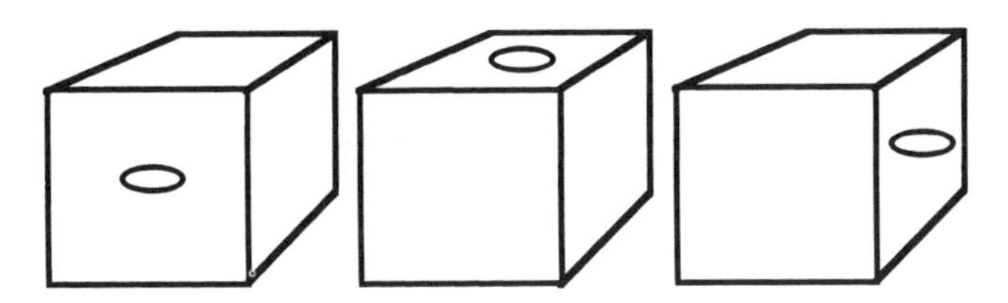

FIG. 6.6. Shape and size "constancy": the same cube with the same ellipse in three different positions. The ellipses are computer-generated duplicates.

FIG. 6.7. Ambiguous "wife/mother-in-law" figure. The same stimulus can be given different interpretations.

cube at all? The ellipse on the top illustrates shape "constancy." It is perceived according to knowledge about how shapes look narrower when they are parallel to the line of sight, so a flat narrow shape is inferred to be wider. Again, the constancy process shows that the surrounding context (in this case the upper quadrilateral) affects how particular stimuli are seen.

The Gestalt psychologists provided dramatic examples of the effects of these inference processes, in their reversible figures as in Fig. 6.7. The overall interpretation that is given to this drawing affects how particular elements of it are grouped together and named—for example, whether they are seen as parts of the body or pieces of clothing. It is not possible to see both interpretations at the same time, but it is possible to change quickly from one to the other. As the interpretation given to an object affects how parts of it are perceived, this can cause difficulty with the interpretation of low-quality visual displays, for example, from infrared cameras or on-board radar.

Naming, and Simple Action Choices

The next functions to consider are identifying name, status, or size, and choosing the nature and size of actions. These cognitive functions may be met by a process of recoding (association) from one form of representation to another, such as:

Shape	→ name
Color	→ level of danger
Spatial position of display	→ name of variable displayed
Name of variable	→ spatial position of its control
Length of line	→ size of variable
Display	→ related control
Size of distance from target	→ size of action needed

Identifications and action choices that involve more complex processing than this recoding are discussed in the section on complex tasks. This section discusses interdependence of the processes and functions; identifying name and status—shape, color, and location codes; size → size codes; and recoding/reaction times. Computer displays have led to the increased use of alphanumeric codes, which are not discussed here (see Bailey, 1989).

Interdependence of the Functions

Perceiving a stimulus, naming it, and choosing an action are not necessarily independent. Figure 6.7 shows that identification can affect perception. This section gives three examples that illustrate other HF/E issues.

Naming difficulties can be based on discrimination difficulties. Figure 6.8 shows the signal/noise ratio needed to hear a word against background noise. The person listening not only has to detect a word against the noise background, but also has to discriminate it from other possible words. The more alternatives there are to distinguish, the better the signal/noise ratio needs to be. This is the reason for using a minimum number of standard messages in speech communication systems, and for designing these messages to maximize the differences between them, as in the International Phonetic alphabet, and standard air-traffic control language (Bailey, 1989).

An important aspect of maximizing differences between signals can be illustrated by a visual example. Figure 6.9 shows some data on reading errors with different digit designs. Errors can be up to twice as high with design A than with design C. At a quick glance, these digit designs do not look very different, but each digit in C has been designed to maximize its difference from the others. Digit reading is a naming task based on a discrimination task, and the discriminations are based on differences between the straight and curved elements of the digits. It is not possible to design an 8 that can be read easily, without considering the need to discriminate it from 3, 5, 6, and 9, which have elements in common. As a general principle, design for discrimination depends on knowing the ensemble of alternatives to be discriminated, and maximizing the differences between them.

However ease of detection/discrimination does not necessarily make naming easy. Figure 6.10 shows an *iconic* display. Each axis displays a different variable, and when all eight variables are on target, the shape is symmetrical. It is easy to detect a distortion in the shape, to detect that a variable is off target. However, studies show that people have difficulty with discriminating one distorted pattern from another by memory, and with

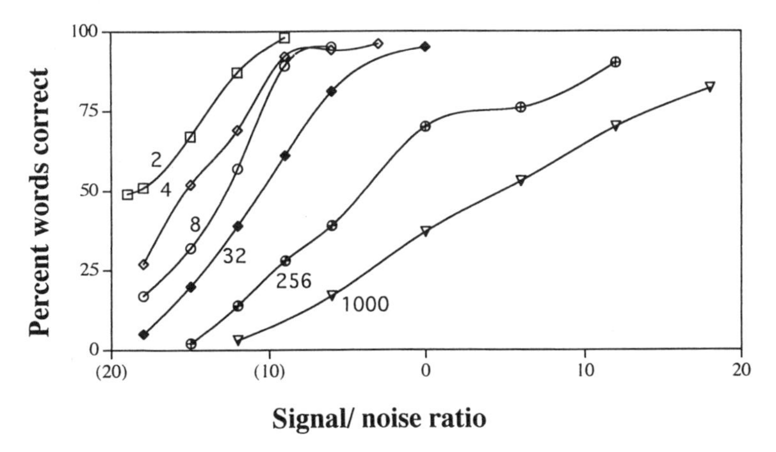

FIG. 6.8. Percentage of words heard correctly in noise, as a function of the number of different words that might occur (Miller, Heise, & Lichten, 1951).

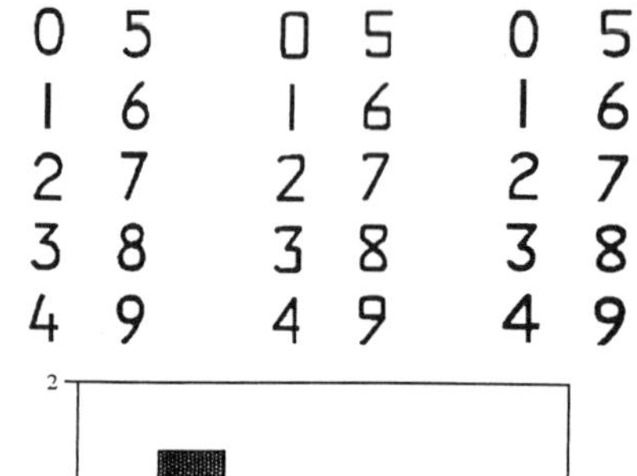

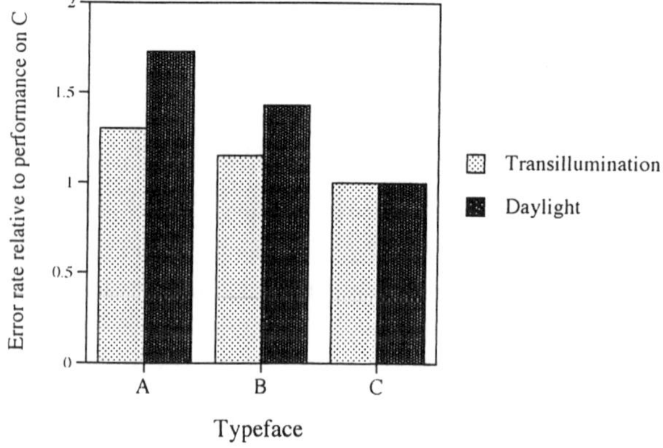

FIG. 6.9. Reading errors with three different digit designs (Atkinson, Crumley, & Willis, 1952). Errors are fewest with the design which minimizes the number of elements which the alternatives have in common.

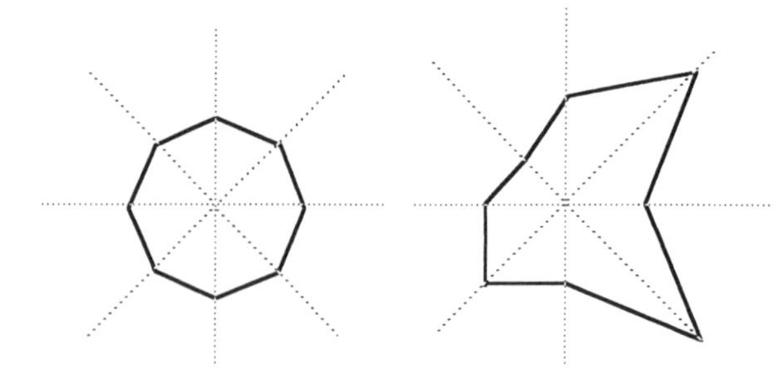

FIG. 6.10. "Iconic" display: Eight variables are displayed, measured outward from the center. When all eight variables are on target, the display has an octagon shape.

identifying which pattern is associated with which problem. This display supports detection, but not discrimination or naming. It is important in task analysis to note which of the cognitive functions are needed, and that the display design supports them.

Shape, Color, and Location Codes for Name and Status

Conventional interfaces all too often consist of a sea of displays or controls that are identical both to sight and touch. The only way of discriminating between and identifying them is to read the label or learn the position. Even if labels have well-designed typeface, abbreviations, and position, they are not ideal. What is needed is an easily seen "code" for the name or status, which is easy to recode into its meaning. The codes used most frequently are shape, color, and location. (Felt texture can be an important code in the design of controls.) The codes need to be designed for ease of discrimination, and for ease of making the translation from code to meaning.

Shape Codes. Good shape codes are "good" figures in the Gestalt sense, and also have features that make the alternatives easy to discriminate. However, ease of discrimination is not the primary criterion in good shape code design. Figure 6.11

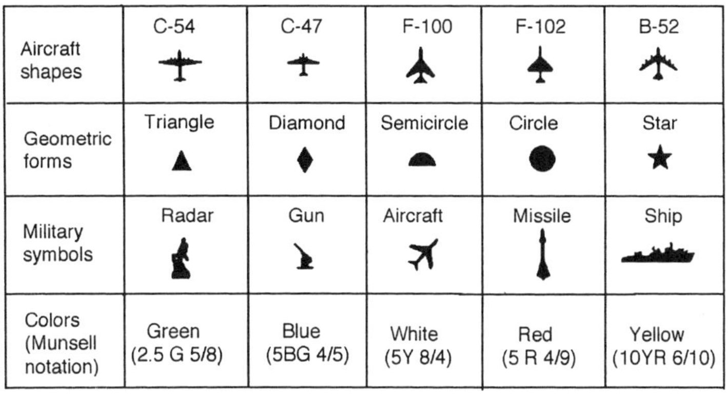

Aircraft shapes	C-54	C-47	F-100	F-102	B-52
Geometric forms	Triangle ▲	Diamond ♦	Semicircle	Circle ●	Star ★
Military symbols	Radar	Gun	Aircraft	Missile	Ship
Colors (Munsell notation)	Green (2.5 G 5/8)	Blue (5BG 4/5)	White (5Y 8/4)	Red (5 R 4/9)	Yellow (10YR 6/10)

FIG. 6.11. Symbols used in discrimination tests (Smith & Thomas, 1964; copyright © 1964 by the American Psychological Association. Reproduced with permission).

shows the materials used in discrimination tests between sets of colors, military look-alike shapes, geometric forms, and aircraft look-alike shapes. Color discrimination is easiest, military symbols are easier to distinguish than aircraft symbols because they have more different features, and geometric forms are discriminated more easily than aircraft shapes. (Geometric forms are not necessarily easier to discriminate. For example, the results would be different if the shapes included a octagon as well as a circle.) The results from naming tests rather than discrimination tests would be different if geometric shapes or colors had to be given a military or aircraft name. Naming tests favor look-alike shapes, as look-alike shapes can be more obvious in meaning.

Nevertheless, using a look-alike shape (symbol or icon) does not guarantee obviousness of meaning. That people make the correct link from shape to meaning needs to be tested carefully. People can be asked, for each possible shape, what they think it is a picture of; what further meaning, such as an action, they think it represents; and, given a list of possible meanings, which of these meanings they choose as the meaning of the shape. To minimize confusions when using shape codes, it is important not to include in the coding vocabulary any shape that is assigned several meanings, or several shapes that could all be assigned the same meaning. Otherwise there could be high error rates in learning and using the shape codes. It is also important to test these meanings on the appropriate users, naive or expert people, or an international population. For example, in Britain a favored symbol for "delete" would be a picture of a space villain from a children's TV series, but this is not understood by people from other European countries!

As well as the potential obviousness of their meaning, look-alike shapes have other advantages over geometric shapes. They can act as a cue to a whole range of remembered knowledge about this type of object (see later discussion on knowledge). Look-alike shapes can also vary widely, whereas the number of alternative geometric shapes which are easy to discriminate is small. An interface designer using geometric shape as a code runs out of different shapes quite quickly, and may have to use the same shape with several meanings. The result of this is that a person interpreting these shapes has to notice when the context has changed to one in which a different shape → meaning translation is used, and then to remember this different translation, before the person can work out what a given shape means. This multistage process can be error prone, particularly under stress. Some computer-based displays have the same shape used with different meanings in different areas of the same display. A person using such a display has to remember to change the coding translation used every time the person makes an eye movement.

Color Codes. Using color as a code poses the same problems as using geometric shape. Except for certain culture-based meanings, such as red → danger, the meanings of colors have to be learned specifically, rather than being obvious. And only a limited number of colors can be discriminated by absolute judgment. The result is that a designer who thinks color is easy to see rapidly runs out of different colors, and has to use the same color with several meanings. There are computer-based displays on which color is used simultaneously with many different types of meaning, such as:

Color → substance (steam, oil, etc.)
Color → status of item (e.g., on/off)
Color → function of item
Color → subsystem item belongs to
Color → level of danger
Color → attend to this item
Color → click here for more information
Color → click here to make an action

A user has to remember which of these coding translations is relevant to a particular point on the screen, with a high possibility of confusion errors.

Location Codes. The location of an item can be used as a basis both for identifying an item and for indicating its links with other items.

People can learn where a given item is located on an interface, and then look or reach to it automatically, without searching. This increases the efficiency of behavior. But this learning is effective only if the location → identity mapping remains constant; otherwise there can be a high error rate. For example, Fitts and Jones (1947/1961a), in their study of pilot errors, found that 50% of errors in operating aircraft controls were choosing the wrong control. The layout of controls on three of the aircraft used at that time shows why it was easy to be confused:

	Position of Control		
Aircraft	*Left*	*Center*	*Right*
B-25	Throttle	Prop	Mixture
C-47	Prop	Throttle	Mixture
C-82	Mixture	Throttle	Prop

Suppose a pilot had flown a B-25 sufficiently frequently to be able to reach to the correct control without thinking or looking. If he then transferred to a C-47, two thirds of his automatic reaches would be wrong, and if to a C-82, all of them. As with other types of coding, location → identity translations need to be consistent and unambiguous. Locations will be easier to learn if related items are grouped together, such as items from the same part of the device, with the same function, or the same urgency of meaning.

Locations can sometimes have a realistic meaning, rather than an arbitrary learned one. Items on one side in the real world should be on the same side when represented on an interface. (Ambiguity about the location of left/right displays could have contributed to the Kegworth air crash; Green, 1990.) Another approach is to put items in meaningful relative positions. For example, in a mimic/schematic diagram or an electrical wiring diagram, the links between items represent actual flows from one part of the device to another. On a cause–effect diagram, links between the nodes of the diagram represent causal links in the device. On such diagrams relative position is meaningful, and inferences can be drawn from the links portrayed (see later discussion on knowledge).

Relative location can also be used to indicate which control goes with which display. When there is a one-to-one relation between displays and controls, then choice of control is a recoding that can be made more or less obvious, consistent, and unambiguous by the use of spatial layout. Gestalt proximity processes link items together if they are next to each other. But the link to make can be ambiguous, such as in the layout: O O O O X X X X. Which X goes with which O? People bring expectations about code meanings to their use of an interface. If these expectations are consistent among a particular group of people, the expectations are called *population stereotypes*. If an interface uses codings that are not compatible with a person's expectations, then the person is likely to make errors.

If two layouts to be linked together are not the same, then studies show that reversed but regular links are easier to deal with than random links (Fig. 6.12). This suggests recoding may be done, not by learning individual pairings, but by having a general rule from which one can work out the linkage.

In multiplexed computer-based display systems, in which several alternative display formats may appear on the same screen, there are at least two problems with location coding. One is that each format may have a different layout of items. We do not know whether people can learn locations on more than one screen format sufficiently well to be able to find items on each format by automatic eye movements rather than by visual search. If people have to search a format for the item they need, studies suggest this could take at least 25 sec. This means that every time the display format is changed, performance will be slowed down while this search process interrupts the thinking about the main task (see also later discussion on short-term memory). It may not be

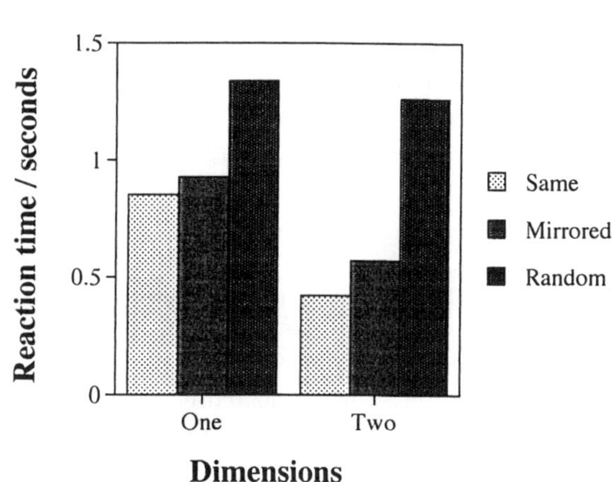

FIG. 6.12. Effect of relative spatial layout of signals and responses on response time (Fitts & Deininger, 1954).

possible to put items in the same absolute position on each display format, but one way of reducing the problems caused by inconsistent locations is to locate items in the same relative positions on different formats.

The second location problem in multiplexed display systems is that people need to know the search "space" of alternative formats available, where they currently are in it, and how to get to other formats. It takes ingenuity to design so that the user of a computer-based interface can use the same sort of "automatic" search skills for obtaining information that are possible with a conventional interface.

In fact, there can be problems with maximizing the consistency and reducing the ambiguity of all types of coding used on multiple display formats (Bainbridge, 1991). Several of the coding vocabularies and coding translations used may change between and within each format (beware the codes used in figures in this chapter). The cues a person uses to recognize which coding translations are relevant need to be learned, and are also often not consistent. A display format may have been designed so the codes are obvious in meaning for a particular subtask, when the display format and the subtask are tested in isolation. But when this display is used in the real task, before and after other formats used for other subtasks, each of which uses different coding translations, then a task-specific display may not reduce either the cognitive processing required or the error rates.

Size → Size Codes

On an analogue interface, length of line is usually used to represent the size of a variable. The following arguments apply both to display scales and to the way control settings are shown. There are three aspects: the ratio of the size on the interface to the size of the actual variable; the way comparisons between sizes are made; and the meaning of the direction of a change in size.

Interface Size: Actual Size Ratio. An example of the interface size to actual size ratio is that, when using an analogue control (such as a throttle), a given size of action has a given size of effect. Once people have learned this ratio, they can make actions without having to check their effect, which gives increased efficiency (see later discussion).

The size ratio and direction of movement are again codes used with meanings that need to be consistent. Size ratios can cause display reading confusions if many displays are used, which all look the same but differ in the scaling ratio used. If many controls that are similar in appearance and feel are used with different control ratios, then it may be difficult to learn automatic skills in using them to make actions of the correct size. This confusion could be increased by using one multipurpose control, such as a mouse or tracker ball, for several different actions each with a different ratio.

A comparison of alternative altimeter designs is an example that also raises some general HF/E points. The designs were tested for reading speed and accuracy (Fig. 6.13). The digital display gives the best performance, and the three-pointer design (A) is one of the worst. The three-pointer altimeter poses several coding problems for someone reading it. The three pointers are not clearly discriminable. Each pointer is read against the same scale using a different scale ratio, and the size of pointer and size of scale ratio are inversely related (the smallest pointer indicates the largest scale, 10,000s, the largest pointer 100s).

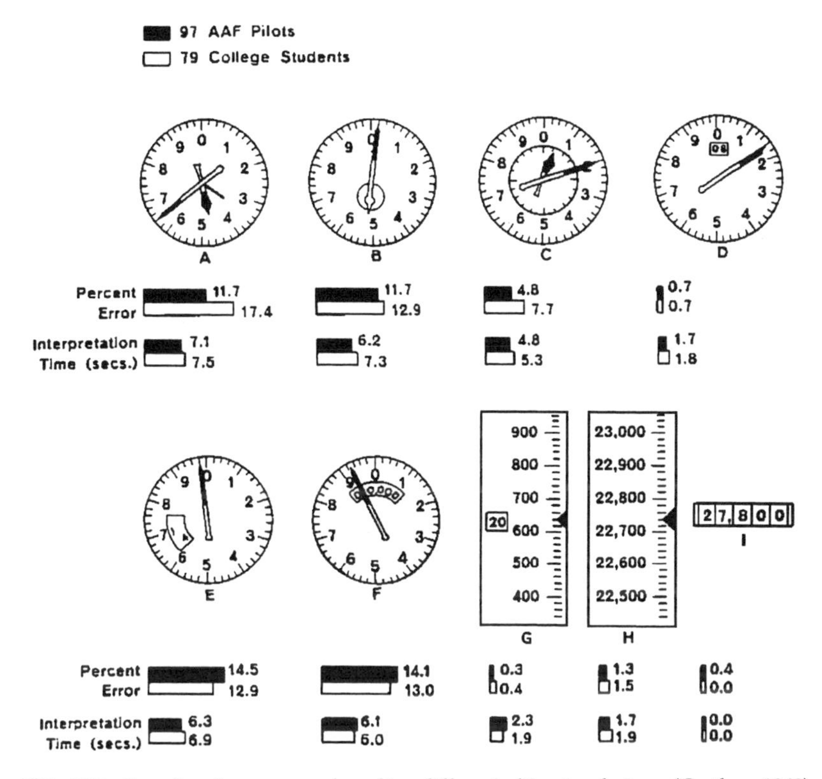

FIG. 6.13. Speed and accuracy of reading different altimeter designs (Grether, 1949).

Despite these results, a digital display is not now used. A static reading test is not a good reflection of the real flying task. In the real task, altitude changes rapidly so a digital display would be unreadable. And the user also needs to identify rate of change, for which angle of line is an effective display. Unambiguous combination altimeter displays are now used, with a pointer for rapidly changing small numbers, and a digital display for slowly changing large numbers (D). Before this change, many hundreds of deaths were attributed to misreadings of the three-pointer altimeter, yet the display design was not changed until these comparative tests were repeated two decades later. This delay occurred for two reasons, which illustrate that HF/E decisions are made in several wider contexts. First was the technology: In the 1940s, digital instrument design was very much more unreliable than the unreliability of the pilot's instrument readings. Second, cultural factors influence the attribution of responsibility for error. There is a recurring swing in attitudes, between saying that a user can read the instrument correctly so the user is responsible for incorrect readings, to saying that if a designer gives users an instrument that it is humanly impossible to read reliably, then the responsibility for misreading errors lies with the designer.

Making Comparisons Between Sizes. There are two important comparisons in control tasks: Is the variable value acceptable/within tolerance (a check reading), and if not, how big is the error? These comparisons can both usually be done more easily on an analogue display. Check readings can be made automatically (i.e., without processing that uses cognitive capacity) if the pointer on a scale is in an easily recognizable position when the value is correct. And linking the size of error to the size of action needed to correct it can be done easily if both are coded by length of line.

An example shows why it is useful to distinguish cognitive functions from the cognitive processes used to meet them. Comparison is a cognitive function that may be done either by simple recoding or by a great deal of cognitive processing, depending on the display design. Consider the horizontal bars in Fig. 6.13 as a display from which an HF/E designer must get information about the relative effectiveness of the altimeter designs. The cognitive processes needed involve searching for the shortest performance bar by comparing each of the performance bar lines, probably using iconic (visual) memory, and storing the result in working memory, then repeating to find the next smallest, and so on. Visual and working memory are used as temporary working spaces while making the comparisons; working memory is also used to maintain the list of decision results. This figure is not the most effective way of conveying a message about alternative designs, because most people do not bother to do all this mental work. The same results are presented in Fig. 6.14. For a person who is familiar with graphs, the comparisons are inherent in this representation. A person looking at this does not have

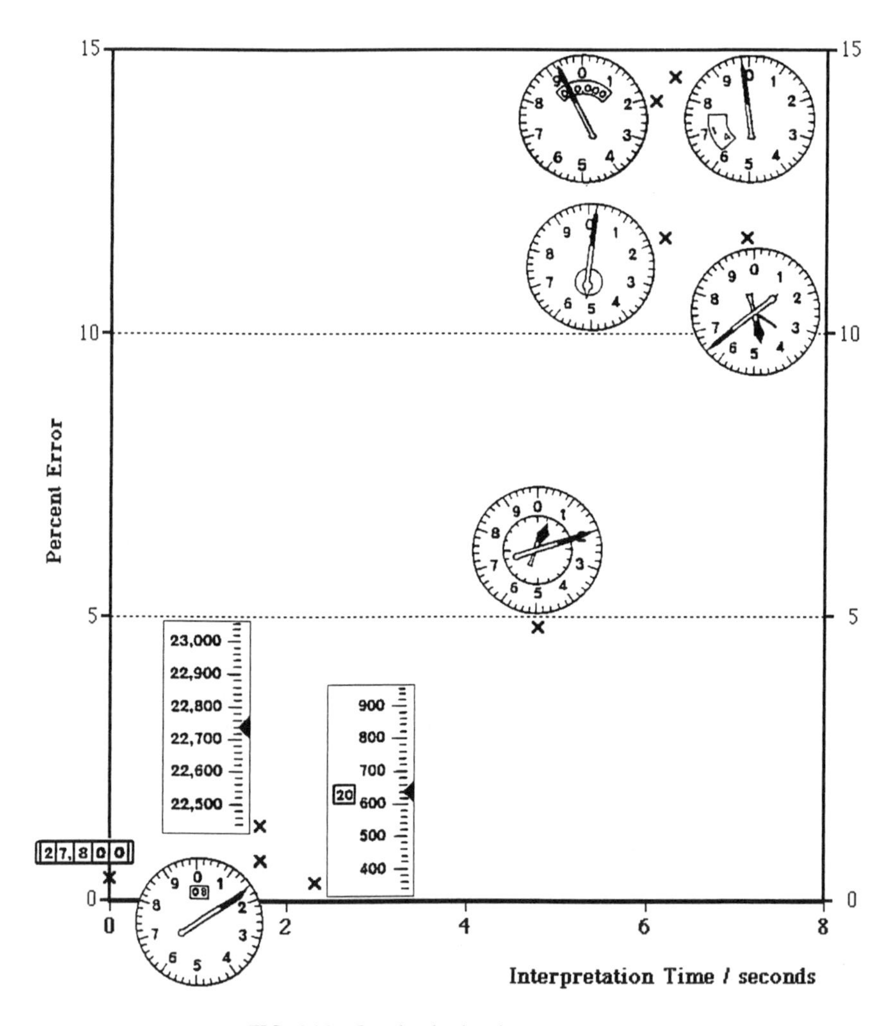

FIG. 6.14. Graph of pilot data in Fig. 6.13.

to do cognitive processing that uses processing capacity and is unrelated to and interrupts the main task of thinking about choice of displays. (See later discussion for more on memory interruption and on processing capacity.) This point applies in general to analogue and digital displays. For many comparison tasks, digital displays require more use of cognitive processing and working memory.

Direction of Movement → Meaning. The second aspect to be learned about interface sizes is the meaning of the direction of a change in size. Cultural learning is involved here, and can be quite context specific. For example, people in technological cultures know that clockwise movement on a display indicates increase, but on a tap or valve control means closure, therefore decrease. Again there can be population stereotypes in the expectations people bring to a situation, and if linkages are not compatible with these assumptions, error rates may be at least doubled.

Directions of movements are often paired. For example, making an control action to correct a displayed error involves two directions of movement, on the display and on the control. It can be straightforward to make the two movements compatible in direction if both are linear, or both are circular.

It is in combining three or more movements that it is easy to get into difficulties with compatibility. One classic example is the aircraft attitude indicator. In the Fitts and Jones (1947/1961b) study of pilots' instrument reading errors, 22% of errors were either reversed spatial interpretations, or attitude illusions. In the design of the attitude indicator, four movements are involved: of the external world, of the display, of the control, and of the pilot's turning receptors (see Fig. 6.15). The attitude instrument can show a moving aircraft, in which case the display movement is the same as the joystick control movement but opposite to the movement of the external world. Or the instrument can show a moving horizon, which is compatible with the view of the external world but not with the movement of the joystick. There is no solution in which all three movements are the same, so some performance errors or delays are inevitable. Similar problems arise in the design of moving scales and of remote-control manipulation devices.

Reaction Times

The evidence quoted so far about recoding has focused on error rates. The time taken to translate from one code representation to another also gives interesting information. Teichner and Krebs (1974) reviewed the results of reaction time studies. Figure

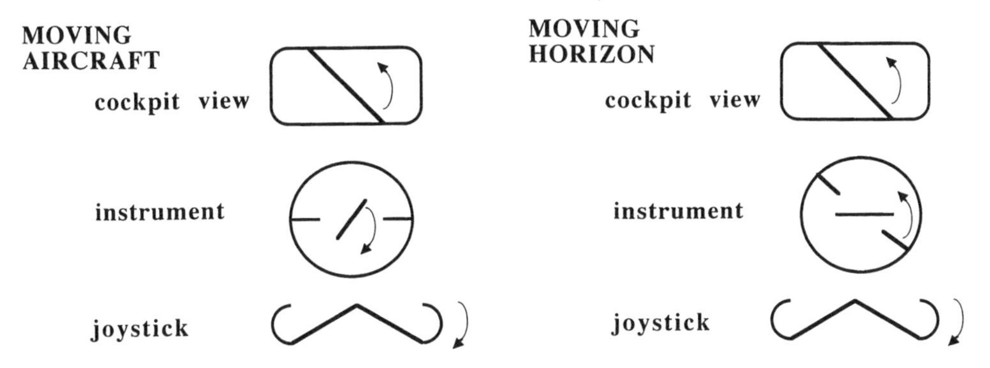

FIG. 6.15. Two designs for the attitude indicator, showing incompatible movements.

6.16 shows the effect of the number of alternative items and the nature of the recoding. The effect of spatial layout was illustrated in Fig. 6.12. Teichner and Krebs also reviewed evidence that, although unpracticed reaction times are affected by the number of alternatives to choose between, after large amounts of practice this effect disappears and all choices are made equally quickly. This suggests that response choice has become automatic; it no longer requires processing capacity.

The results show the effect of different code translations: using spatial locations of signals and responses (light, key) or symbolic ones (visually presented digit, spoken digit, i.e., voice). The time taken to make a digit → voice translation is constant, but this is already a highly practiced response for the people tested. Otherwise, making a spatial link (light → key) is quickest. Making a link that involves a change of code type, between spatial and symbolic (digit → key, or light → voice), takes longer. (So these data show it can be quicker to locate than to name.) This coding time difference may arise because spatial and symbolic processes are handled by different areas of the brain, and it takes time to transmit information from one part of the brain to another. The brain does a large number of different types of coding translation (e.g., Barnard, 1987).

The findings presented so far come from studies of reacting to signals that are independent and occur one at a time. Giving advance information about the responses that will be required, which allows people to anticipate and prepare their responses, reduces response times. There are two ways of doing this, illustrated in Fig. 6.17. One is to give a preview, allowing people to see in advance the responses needed. This can more than halve reaction time. The second method is to have sequential relations in the material to be responded to. Figure 6.16 showed that reaction time is affected by the number of alternatives; the general effect underlying this is that reaction time depends on the probabilities of the alternatives. Sequential effects change the probabilities of items. One way of introducing sequential relations is to have meaningful sequences in the items, such as prose rather than random letters.

Reaction time and error rate are interrelated. Figure 6.18 shows that when someone reacts very quickly, the person chooses a response at random. As the person takes a longer time, and can take in more information before initiating a response, there is a trade-off between time and error rate. At longer reaction times there is a basic error rate that depends on the equipment used.

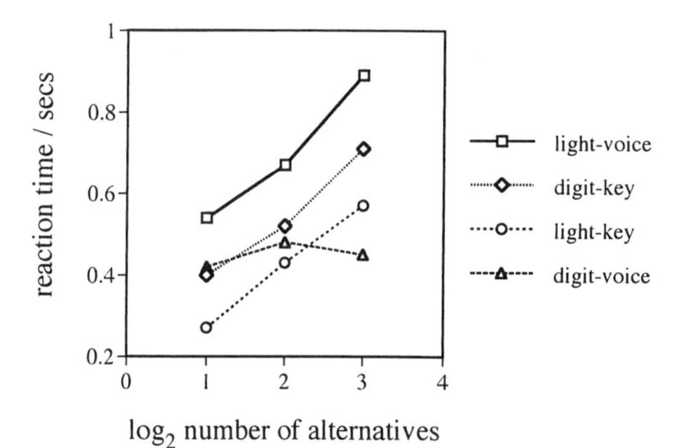

FIG. 6.16. Response times are affected by the number of alternatives to be responded to, the nature of the "code" linking the signal and response, and the amount of practice (Teichner & Krebs, 1974; Copyright © 1974 by the American Psychological Association. Reproduced with permission).

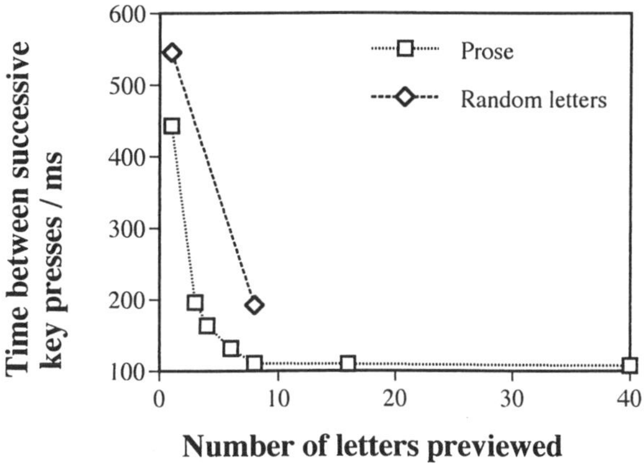

FIG. 6.17. Effect of preview, and predictability of material, on response time (based on data in Shaffer, 1973; Copyright © 1973 Academic Press. Reproduced with permission).

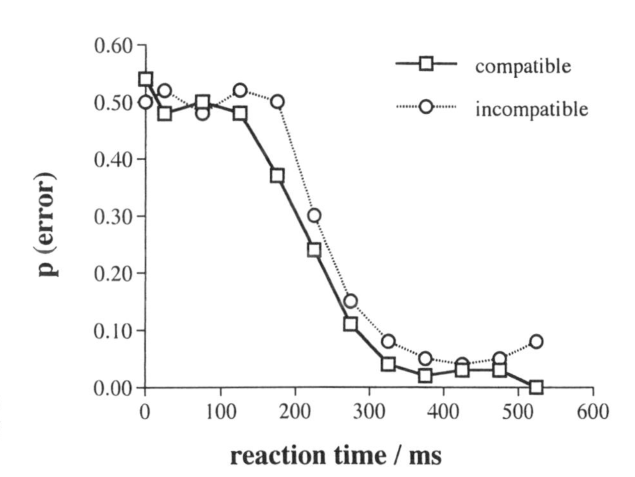

FIG. 6.18. Speed–accuracy tradeoff in two-choice reactions, and the effect of stimulus–response compatibility.

Action Execution

This chapter does not focus on physical activity, but this section makes some points about cognitive aspects of action execution. The section is in two parts, on acquisition movements, and on continuous control or tracking movements.

The speed, accuracy, and power a person can exert in a movement depend on its direction relative to the body position. Human biomechanics and its effects on physical performance, and the implications for workplace design, are large topics, which are not reviewed here (Pheasant, 1991). Only one point is made. Workplace design affects the amount of physical effort needed to make an action, and the amount of postural stress a person is under. These both affect whether a person is willing to make a particular action or to do a particular job. So workplace design can affect performance in cognitive tasks. Factors that affect what a person is or is not willing to do are discussed more in the section on workload.

Acquisition Movements. When someone reaches to something, or puts something in place, this is an *acquisition* movement. Reaching a particular endpoint or target is more important than the process of getting there. The relation between the speed and accuracy of these movements can be described by Fitts's Law (Fitts, 1954), in which movement time depends on the ratio of movement length to target width. However, detailed studies show that movements with the same ratio are not all carried out in the same way. Figure 6.19 shows that an 80/10 movement is made with a single pulse of velocity. A 20/2.5 movement has a second velocity pulse, suggesting the person has sent a second instruction to his or her hand about how to move. Someone making a movement gives an initial instruction to his or her muscles about the direction, force, and duration needed, then monitors how the movement is being carried out, by vision and/or feel. If necessary the person sends a corrected instruction to the muscles to improve the performance, and so on. This monitoring and revision is called using feedback. A finer movement involves feedback to and a new instruction from the brain. A less accurate movement can be made with one instruction to the hand, without needing to revise it. An unrevised movement ("open-loop" or "ballistic") probably involves feedback within the muscles and spinal cord, but not visual feedback to and a new instruction from the brain.

Movements that are consistently made the same way can be done without visual feedback, once learned, as mentioned in the section on location coding. Figure 6.20 shows the double use of feedback in this learning. A person chooses an action instruction that he or she expects will have the effect wanted. If the result turns out not to be as intended, then the person needs to adjust knowledge about the expected effect of an action. This revision continues each time the person makes an action, until the expected result is the same as the actual result. Then the person can make an action with minimal need to check that it is being carried out effectively. This reduces the amount of processing effort needed to make the movement. Knowledge about expected results is a type of meta-knowledge. Meta-knowledge is important in activity choice, and is discussed again later.

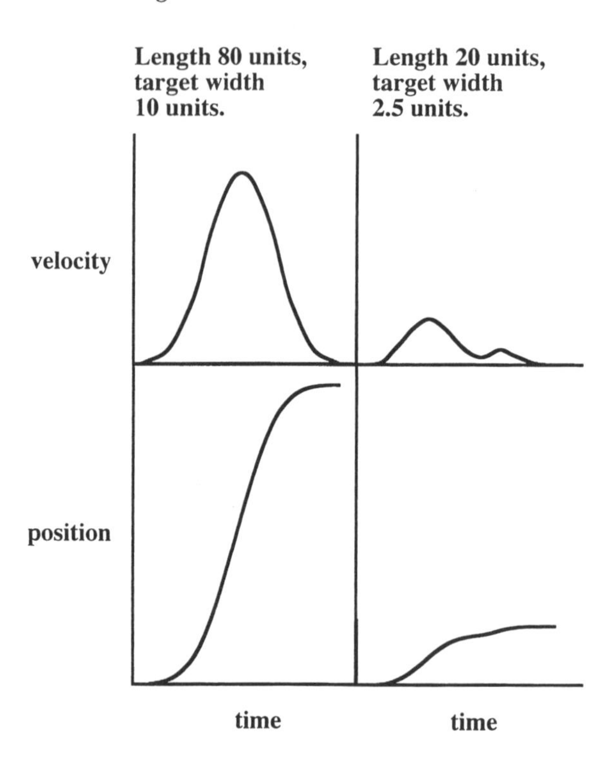

FIG. 6.19. Execution of movements of different sizes (Crossman & Goodeve, 1963).

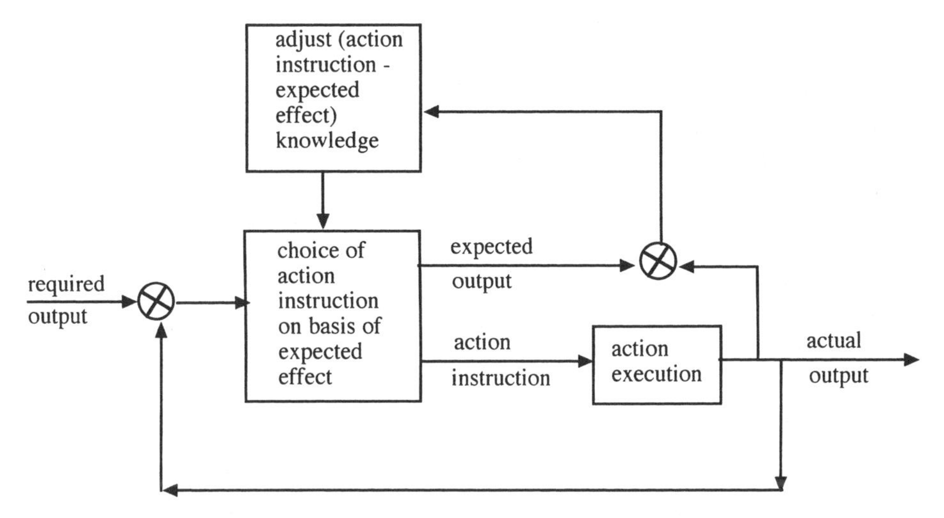

FIG. 6.20. Double use of feedback in learning to make movements.

Control or Tracking Movements. Control movements are ones in which someone makes frequent adjustments, with the aim of keeping some part of the external world within required limits. They might be controlling the output of an industrial process, or keeping an aircraft straight and level. In industrial processes, the time lag between making an action and its full effect in the process may be anything from minutes to hours, so there is usually time to think about what to do. By contrast, in flying, events can happen very quickly, and human reaction time plus neuromuscular lag, adding up to half a second or more, can have a considerable effect on performance. So different factors may be important in the two types of control task.

There are two ways of reducing the human response lag (cf. Fig. 6.17). Preview allows someone to prepare actions in advance and therefore to overcome the effect of the lag. People can also learn something about the behavior of the track they are following, and can then use this knowledge to anticipate what the track will do and so prepare their actions.

There are two ways of displaying a tracking task. In a pursuit display, the moving target and the person's movements are displayed separately. A compensatory display system computes the difference between the target and the person's movements, and displays this difference relative to a fixed point. Many studies show human performance is better with a pursuit display, as in Fig. 6.21. As mentioned earlier, people can learn about the effects of their actions, and about target movements, and both types of learning lead to improved performance. On the pursuit display, the target and human movements are displayed separately, so a person using this display can do both types of learning. In contrast, the compensatory display only shows the difference between the two movements. It is not possible for the viewer to tell which part of a displayed change is due to target movements and which is due to the viewer's own movements, so these are difficult to learn.

A great deal is known about human fast tracking performance (Rouse, 1980; Sheridan & Ferrell, 1974). A person doing a tracking task is acting as a controller. Control theory provides tools for describing some aspects of the track to be followed and how a device responds to inputs. This has resulted in the development of a "human transfer function," a description of a human controller as if the person were an engineered control device.

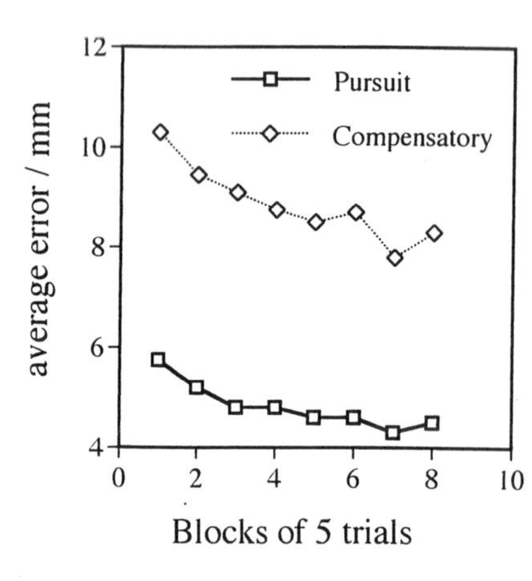

FIG. 6.21. Errors in tracking performance using pursuit and compensatory displays (Briggs & Rockway, 1966; copyright © 1966 by the American Psychological Association. Reproduced with permission).

The transfer function contains some components that describe human performance limits, and some that partially describe human ability to adapt to the properties of the device the person is controlling. This function can be used to predict combined pilot–aircraft performance. This is a powerful technique with considerable economic benefits. However, it is not relevant to this chapter as it describes performance, not the underlying processes, and it only describes human performance in compensatory tracking tasks. It also focuses attention on an aspect of human performance that can be poorer than that of fairly simple control devices. This encourages the idea of removing the person from the system, rather than appreciating what people can actively contribute, and designing support systems to overcome their limitations.

Summary and Implications

Theory. The cognitive processes underlying classic HF/E can be relatively simple, but not so simple that they can be ignored. Cognitive processing is carried out to meet cognitive functions. Five functions were discussed in this section: distinguishing between stimuli; building up a percept of an external world containing independent entities with stable properties; naming; choosing an action; and comparison.

This section suggests these functions could be met in simple tasks by three main cognitive processes. (What happens when these processes are not sufficient has been mentioned briefly and is discussed in the next main section.) The three processes are: deciding between alternative interpretations of the evidence; integrating data from all sensory sources, together with knowledge about the possibilities, into an inferred percept that makes best sense of all the information; and recoding, that is, translating from one type of code to another.

Five other key aspects of cognitive processing have been introduced:

1. Sensory processing is relative rather than absolute.
2. The cognitive functions are not necessarily met by processes in a clearly distinct sequence. Processes that are "automated" may be done in parallel. The processes communicate with each other via a common "blackboard," which provides the context within which each process works, as summarized in Fig. 6.22.

As processing is affected by the context in which it is done, behavior is adaptive. However, for HF/E practitioners this has the disadvantage that the answer to any HF/E question is always, "it depends."

3. The processing is not simply input driven: All types of processing involve the use of knowledge relevant in the context. (It can therefore be misleading to use the term *knowledge-based* to refer to one particular mode of processing.)
4. Preview and anticipation can improve performance.
5. Actions have associated meta-knowledge about their effects, which improves with learning.

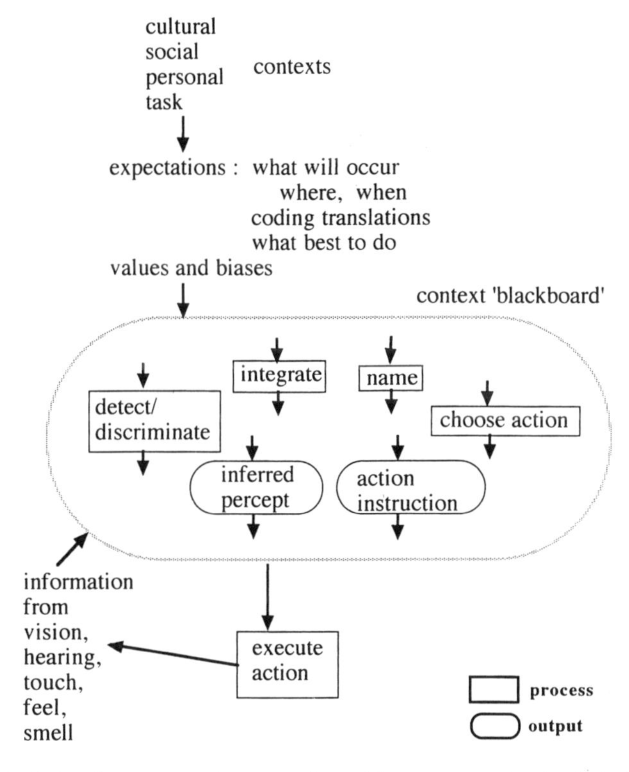

FIG. 6.22. The contextual nature of cognitive processes in simple tasks.

Practical Aspects. The primary aim of classic HF/E has been to minimize unnecessary physical effort. The points made here emphasize the need to minimize unnecessary cognitive effort.

Task analysis should not only note which displays and controls are needed, but might also ask such questions as: What cognitive functions need to be carried out? By what processes? Is the information used in these processes salient?

In discrimination and integration: What is the ensemble of alternatives to be distinguished? Are the items designed to maximize the differences between them? What are the probabilities and costs of the alternatives? How does the user learn these?

In recoding: What coding vocabularies are used (shape, color, location, size, direction, alphanumeric) in each subtask? In the task as a whole? Are the translations unambiguous, unique, consistent, and if possible obvious? Do reaction times limit performance, and if so can preview or anticipation be provided?

COMPLEX TASKS

Using an interface for a simple task entails the functions of distinguishing between stimuli, integrating stimuli, naming, comparing, and choosing and making simple actions. When the interface is well designed, these functions can be carried out by decision making, integration, and recoding processes. These processes use knowledge about the alternatives that may occur, their distinguishing features, probabilities, and costs, and the translations to be made.

Doing a more complex task uses more complex knowledge in more complex functions and processes. For example, suppose an air traffic controller is given the two flight strips in Fig. 6.23. Commercial aircraft fly from one fix point to another. These two aircraft are flying at the same level (31,000 ft) from fix OTK to fix LEESE7. DAL1152 is estimated to arrive at LEESE7 at 2 min after AAL419 (18 – 16), and is traveling faster (783 > 746). So DAL1152 is closing relatively fast and the controller needs to take immediate action, to tell one of the aircraft to change flight level. The person telling the aircraft to change level is doing more than simply recoding the given information. The person uses strategies for searching the displays and for comparing the data about the two aircraft, plus a simple dynamic model of how an aircraft changes position in time, to build up a mental picture of the relative positions of the aircraft, with one overtaking the other so a collision is possible. The person then uses a strategy for optimizing the choice of which aircraft to instruct to change.

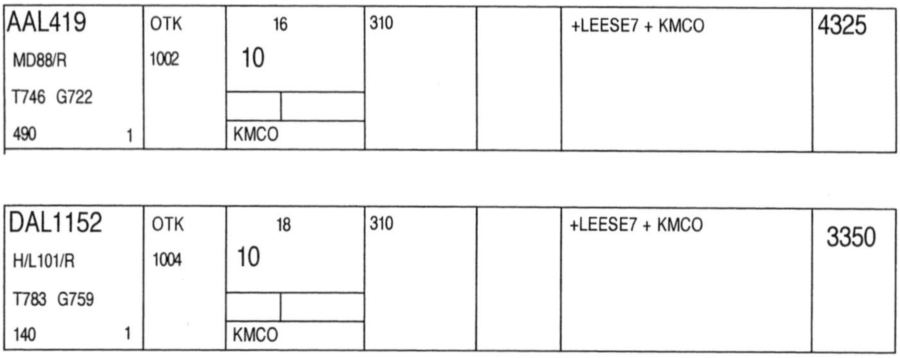

AAL419	OTK	16	310		+LEESE7 + KMCO	4325
MD88/R	1002	10				
T746 G722						
490 1		KMCO				

DAL1152	OTK	18	310		+LEESE7 + KMCO	3350
H/L101/R	1004	10				
T783 G759						
140 1		KMCO				

FIG. 6.23. Two flight strips, each describing one aircraft. Column 1: (top) aircraft identification; (bottom) true airspeed/knots. Column 2: (top) previous fix. Column 3: (top) estimated time over next fix. Column 4: flight level (i.e., altitude in 100s of ft). Column 6: next fix.

The overall cognitive functions or goals are to understand what is happening and to plan what to do about it. In complex dynamic tasks these two main cognitive needs are met by subsidiary cognitive functions such as:

- Infer/review present state.
- Predict/review future changes/events.
- Review/predict task performance criteria.

- Evaluate acceptability of present or future state.
- Define subtasks (task goals) to improve acceptability.
- Review available resources/actions, and their effects.
- Define possible (sequences of) actions (and enabling actions) and predict their effects.
- Choose action/plan.
- Formulate execution of action plan (including monitor the effects of actions, which may involve repeating all the preceding).

These cognitive functions are interdependent. They are not carried out in a fixed order but are used as necessary. Lower level cognitive functions implement higher level ones. At the lowest levels, the functions are fulfilled by cognitive processes such as searching for the information needed, discrimination, integration, and recoding. The processing is organized within the structure of cognitive goals/functions.

An overview is built up in working storage by carrying out these functions. This overview represents the person's understanding of the current state of the task and the person's thinking about it. The overview provides the data the person uses in later thinking, as well as the criteria for what best to do next and how best to do it. There is a cycle: Processing builds up the overview, which determines the next processing, which updates the overview, and so on (see Fig. 6.24). Figure 6.22 showed an alternative representation of context, as nested rather than cyclic. (For more about this mechanism, see Bainbridge 1993a.)

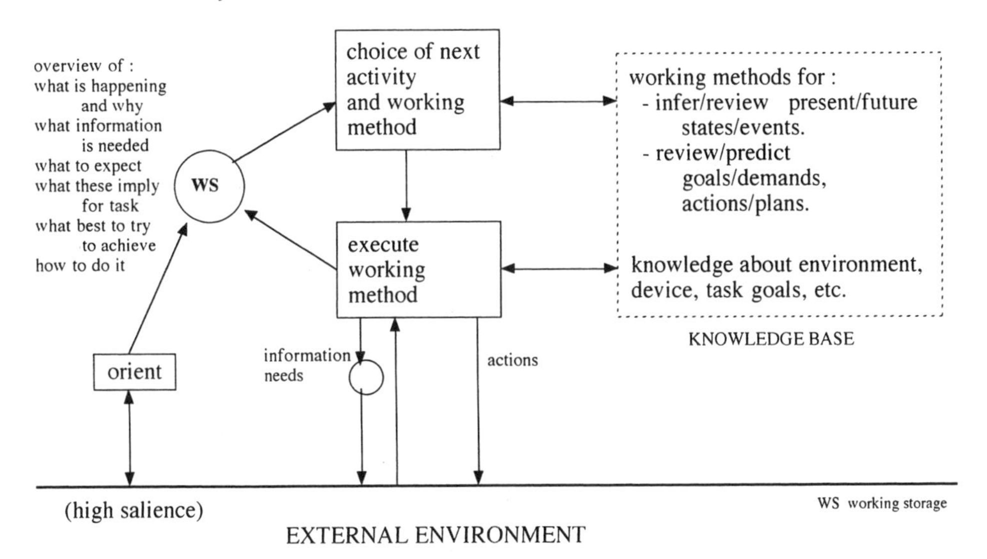

FIG. 6.24. A sketch of the contextual cycle in relation to the knowledge base and the external environment.

The main cognitive processes discussed in the previous section were decision making, integrating stimuli, and recoding. Additional modes of processing are needed in complex tasks, such as:

- Carrying out a sequence of recoding transformations, and temporarily storing intermediate results in working memory.

- Building up a structure of inference, an overview of the current state of understanding and plans, in working storage, using a familiar working method.
- Using working storage to mentally simulate carrying out a cognitive or physical strategy.
- Deciding between alternative working methods on the basis of meta-knowledge.
- Planning and multitasking.
- Developing new working methods.

These complex cognitive processes are not directly observable. The classic experimental psychology method, which aims to control all except one or two measured variables, and to vary one or two variables so their effects can be studied, is well suited to investigating discrimination and recoding processes. It is not well suited to investigating cognitive activities in which many interrelated processes may occur without any observable behavior. Studying these tasks involves special techniques: case studies, videos, verbal protocols, or distorting the task in some way, perhaps slowing it down or making the person do extra actions to get information (Wilson & Corlett, 1995). Both setting up and analyzing the results of such studies can take years of effort. The results tend to be as complex as the processes studied, so they are difficult to publish in the usual formats. Such studies do not fit well into the conventions about how research is done, so there are unfortunately not many of this type. However, the rest of this section gives some evidence about the nature of complex cognitive processes, to support the general claims made so far. The subsections are on sequences; language understanding; inference and diagnosis; working storage; planning, multitasking, and problem solving; and knowledge.

Sequences of Transforms

After decision making, integrating, and recoding, the next level of complexity in cognitive processing is carrying out a sequence of recoding translations or transforms. The result of one step in the sequence acts as the input to the next step, so has to be kept temporarily in working memory. Here the notion of recoding needs to be expanded to include transforms such as simple calculations and comparisons, and conditions leading to alternative sequences. Note that in this type of processing the goal of the behavior, the reason for doing it, is not included in the description of how it is done. Some people call this type of processing *rule-based*. There are two typical working situations in which behavior is not structured relative to goals.

When a person is following instructions that do not give them any reason for why the person has to do each action, then the person is using this type of processing. This is usually not a good way of presenting instructions, as if anything goes wrong, the person has no reference point for identifying how to correct the problem.

The second case can arise in a stable environment, in which behavior can be done in the same way each time. If a person has practiced often, the behavior may be done without needing to check it, or to think out what to do or how to do it (see later discussion). Such overlearned sequences give a very efficient way of behaving, in the sense of using minimal cognitive effort. But if the environment does change, then overlearning is maladaptive and can lead to errors (see later discussion on learning and errors).

Language Processing

This section covers two issues: using language to convey information and instructions, and the processes involved in language understanding. Although language understanding is not the primary task of either pilot or air traffic controller, it does provide simple examples of some key concepts in complex cognitive processing.

Written Instructions. Providing written instructions is often thought of as a way of making a task easy, but this is not guaranteed. Reading instructions involves interpreting the words in order to build up a plan of action. The way the instructions are written may make this processing more or less difficult. Videorecorder operating manuals are notorious for this.

Various techniques have been used for measuring the difficulty of processing different sentence types. Some typical results are (Savin & Perchonock, 1965):

Sentence Type	*Example*	*% Drop in Performance*
Kernel	The pilot flew the plane.	0
Negative	The pilot did not fly the plane.	–16
Passive	The plane was flown by the pilot.	–14
Negative passive	The plane was not flown by the pilot.	–34

Such data suggest that understanding negatives and passives involves two extra and separate processes. This suggests it is best in general to use active positive forms of sentence. But when a negative or restriction is the important message, it should be the most salient and come first. For example, "No smoking" is more effective than "Smoking is not permitted." And using a simple form of sentence does not guarantee that a message makes good sense. I recently enjoyed staying in a hotel room with a notice on which the large letters said:

> **Do not** use the elevator during a fire.
>
> **Read this notice carefully.**

Connected prose is not necessarily the best format for showing alternatives in written instructions. Spatial layout can be used to show the groupings and relations between phrases, by putting each phrase on a separate line, by indenting to show items at the same level, and by using flow diagrams to show the effect of choice between alternatives (e.g., Oborne, 1995, chapter 4). When spatial layout is used to convey meaning in written instructions, it is a code and should be used consistently, as discussed earlier.

Instructions also need to be written from the point of view of the reader: "If you want to achieve this, then do this." Instruction books are often written the other way round: "If you do this, then this happens." The second approach requires from the reader much more understanding, searching, and planning to work out what to do. Note that the effective way of writing instructions is goal oriented. In complex tasks, methods of working are in general best organized in terms of what is to be achieved; this is discussed again later.

Language Understanding. In complex tasks, many of the cognitive processes and knowledge used are only possible because the person has considerable experience of the task. Language understanding is the chief complex task studied by experimental psychologists (e.g., Ellis, 1993), because it is easy to find experts to test. When someone is listening to or reading language, each word evokes learned expectations. For example:

The
can only be followed by
—a descriptor, or
—a noun.

The pilot
depending on the context, either:
(a) *will be followed by the word "study" or:*
(b) *—evokes general knowledge (scenarios) about aircraft or ship pilots.*
—can be followed by:
—a descriptive clause, containing items relevant to living things/animals/human beings/ pilots, or
—a verb, describing possible actions by pilots.

Each word leads to expectations about what will come next; each constrains the syntax (grammar) and semantics (meaning) of the possible next words. To understand the language, a person needs to know the possible grammatical sequences, the semantic constraints on what words can be applied to what types of item, and the scenarios. During understanding, a person's working storage contains the general continuing scenario, the structure of understanding built up from the words received so far, and the momentary expectations about what will come next. (Many jokes depend on not meeting these expectations.)

The overall context built up by a sequence of phrases can be used to disambiguate alternative meanings, such as:

The Inquiry investigated why
the pilot turned into a mountain.

or

In this fantasy story
the pilot turned into a mountain.

The knowledge base/scenario is also used to infer missing information. For example:

The flight went to Moscow.
The stewardess brought her fur hat.

Answering the question "Why did she bring her fur hat?" involves knowing that stewardesses go on flights and about the need for and materials used in protective clothing, which are not explicitly mentioned in the information given.

Understanding language does not necessarily depend on the information being presented in a particular sequence. Although it requires more effort, we can understand someone whose first language uses a different word order from English, such as:

The stewardess her fur hat brought.

We do this by having a general concept that a sentence consists of several types of unit (noun phrases, verb phrases, and so on) and we make sense of the input by matching it to the possible types of unit. This type of processing can be represented as being organized by a "frame with slots," where the frame coordinates the slots for the types of item expected, which are then instantiated in a particular case, as in:

Noun phrase	Verb	Noun phrase
The stewardess	**brought**	**her fur hat.**

(As language has many alternative sequences, this is by no means a simple operation; Winograd, 1972.)

The understanding processes used in complex control and operation tasks show the same features that are found in language processing. The information obtained evokes both general scenarios and specific moment-to-moment expectations. The general context, and additional information, can be used to decide between alternative interpretations of the given information. A structure of understanding is built up in working storage. Frames or working methods suggest the types of information the person needs to look for to complete their understanding. These items can be obtained in a flexible sequence. And knowledge is used to infer whatever is needed to complete the understanding but is not supplied by the input information. There is an important addition in control/operation tasks, which is that the structure of understanding is built up in order to influence the state of the external world, to try to get it to behave in a particular way.

Inference and Diagnosis

To illustrate these cognitive processes in an aviation example, this section uses an imaginary example so the presentation can be short. Later sections describe real evidence on pilot and air traffic controller behavior, which justifies the claims made here.

Suppose that an aircraft is in flight and the "engine oil low" light goes on. What might be the pilot's thoughts? The pilot needs to *infer the present state* of the aircraft (cognitive functions are indicated by italics). This involves considering alternative hypotheses that could explain the light, such as that there is an instrument fault, or there genuinely is an engine fault, and then choosing between the hypotheses according to their probability (based on previous experience of this or other aircraft) or by looking for other evidence that would confirm or disprove the possibilities. The pilot could *predict the future changes* that will occur as a result of the chosen explanation of events. Experienced people's behavior in many dynamic tasks is future oriented. A person takes anticipatory action, not to correct the present situation, but to ensure that predicted unacceptable states or events do not occur. Before *evaluating the predictions for their acceptability*, the pilot needs to *review the task performance criteria*, such as the relative importance of arriving at the original destination quickly, safely, or cheaply. The result of comparing the predictions with the criteria will be to *define the performance needs to be met*. It is necessary to *review the available resources*, such as the state of the other engines or the availability of alternative landing strips. The pilot can then *define possible alternative action sequences and predict their outcomes*. A *review of action choice criteria*, which includes the task performance criteria plus others such as the difficulty of the proposed procedures, is needed as a basis for *choosing an action sequence/plan*, before beginning to *implement the plan*. Many of these cognitive functions must be based on incomplete evidence, for example, about future events or the effects of actions, so risky decision making is involved.

A pilot who has frequently practiced these cognitive functions may be able to carry them out "automatically," without being aware of the need for intermediate thought. And an experienced pilot may not be aware of thinking about the functions in separate stages; for example, (predict + review criteria + evaluation) may be done together.

Two modes of processing have been used in this example: "automatic" processing (i.e., recoding), and using a known working method that specifies what thinking to carry out. Other modes of processing are suggested later. The mode of processing needed to carry out a function depends on the task situation and the person's experience (see later discussion on learning). An experienced person's knowledge of the situation may enable the person to reduce the amount of thinking to do, even when the person does need to think things out explicitly. For example, it may be clear early in the process of predicting the effects of possible actions that some will be not acceptable and so need not be explored further (see later discussion on planning).

Nearly all the functions and processing mentioned have been supplied from the pilot's knowledge base. The warning light evoked working methods for explaining the event and for choosing an action plan, as well as knowledge about the alternative explanations of events and suggestions of relevant information to look for. The combination of (working method + knowledge referred to in using this method + mental models for predicting events) is the scenario. Specific scenarios may be evoked by particular events, or by particular phases of the task (phases of the flight).

This account of the cognitive processes is goal oriented. The cognitive functions or goals are the means by which the task goals are met, but are not the same as them. Task and personal goals act as constraints on what it is appropriate and useful to think about when fulfilling the cognitive goals.

The cognitive functions and processing build up a structure of data (in working storage) that describes the present state and the reasons for it, predicted future changes, task performance and action choice criteria, resources available, the possible actions, the evaluations of the alternatives, and the chosen action plan. This data structure is an overview that represents the results of the thinking and deciding done so far, and provides the data and context for later thinking. As an example, the result of reviewing task performance criteria is not only an input to evaluation; it could also affect what is focused on in inferring the present state, or in reviewing resources, or in action choice. The overview ensures that behavior is adapted to its context.

The simple example just given described reaction to a single unexpected event. Normally, flying and air traffic control are ongoing tasks. For example, at the beginning of shift an air traffic controller has to build up an understanding of what is happening and what actions are necessary, from scratch. After this, each new aircraft that arrives is fitted into the controller's ongoing mental picture of what is happening in the airspace; the thinking processes do not start again from the beginning. Aircraft usually arrive according to schedule and are expected, but the overview needs to be updated and adapted to changing circumstances (see later discussion on planning and multitasking).

There are two groups of practical implications of these points. One is that cognitive task analysis should focus on the cognitive functions involved in a task, rather than simply prespecifying the cognitive processes by which they are met. The second is that designing specific displays for individual cognitive functions may be unhelpful. A person doing a complex task meets each function within an overall context, the functions are interdependent, and the person may not think about them in a prespecified sequence. Giving independent interface support to each cognitive function, or subtask within a function, could make it more difficult for the person to build up an overview that interrelates the different aspects of the person's thinking.

Diagnosis. The most difficult cases of inferring what underlies the given evidence may occur during fault diagnosis. A fault may be indicated by a warning light or, for an experienced person, by a device not behaving according to expectations. Like any other inference, fault diagnosis can be done by several modes of cognitive processing, depending on the circumstances. If a fault occurs frequently, and has unique symptoms, it may be possible to diagnose the fault by visual pattern recognition, that is, pattern on interface → fault identity (e.g., Marshall, Scanlon, Shepherd, & Duncan, 1981). This is a type of recoding. But diagnosis can also pose the most difficult issues of inference, for example, by reasoning based on the physical or functional structure of the device (e.g., Hukki & Norros, 1993).

In-flight diagnosis may need to be done at speed. Experienced people can work rapidly using *recognition-primed decisions,* in which situations are assigned to a known category with a known response, on the basis of similarity. The processes involved in this are discussed by Klein (1989). The need for rapid processing emphasizes the importance of training for fault diagnosis.

Amalberti (1992, Expt. 4) studied fault diagnosis by pilots. Two groups of pilots were tested: Pilots in one group were experts on the Airbus, and those in the other group were experienced pilots beginning their training on the Airbus. They were asked to diagnose two faults specific to the Airbus, and two general problems. In 80% of responses, the pilots gave only one or two possible explanations. This is compatible with the need for rapid diagnosis. Diagnostic performance was better on the Airbus faults, which the pilots had been specifically trained to watch out for, than on the more general faults. One of the general problems was a windshear on take-off. More American than European pilots diagnosed this successfully. American pilots are more used to windshear as a problem, so are more likely to think of this as a probable explanation of an event. People's previous experience is the basis for the explanatory hypotheses they suggest.

In the second general fault there had been an engine fire on take-off, during which the crew forgot to retract the landing gear, which made the aircraft unstable when climbing. Most of the hypotheses suggested by the pilots to explain this instability were general problems with the aircraft, or were related to the climb phase. Amalberti suggested that when the aircraft changed the phase of flight, from take-off to climb, the pilots changed their scenario that provides the appropriate events, procedures, mental models, and performance criteria for use in thinking. Their knowledge about the previous phase of flight became less accessible, and so was not used in explaining the fault.

Working Storage

The inference processes build up the contextual overview or situation awareness in working storage. This is not the same as short-term memory, but short-term memory is an important limit to performance and is discussed first.

Short-Term Memory

Figure 6.25 shows some typical data on how much is retained in short-term memory after various time intervals. Memory decays over about 30 sec, and is worse if the person has to do another cognitive task before being tested on what the person can remember.

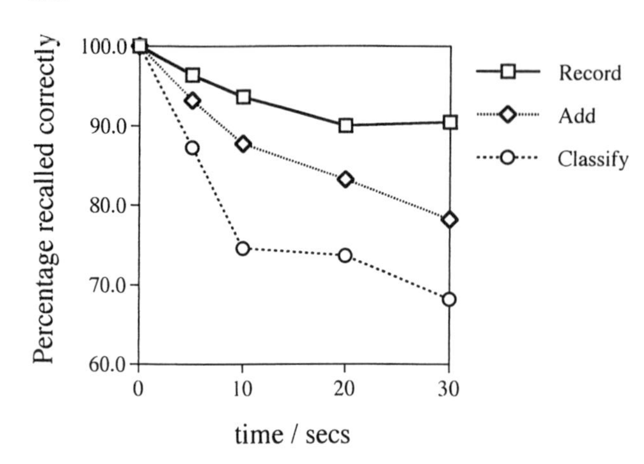

FIG. 6.25. Decrease in recall after a time interval, with different tasks during the retention interval (Posner & Rossman, 1965; copyright © 1965 by the American Psychological Association. Reproduced with permission).

This memory decay is important in the design of computer-based display systems in which different display formats are called up in sequence on a screen. Suppose the user has to remember an item from one display, for use with an item on a second display. Suppose that the second display format is not familiar, so the person has to search for the second item: This search may take about 25 sec. The first item must then be recalled after doing the cognitive processes involved in calling up the second display and searching it. The memory data suggest that the person will have forgotten the first item on 30% of occasions.

The practical implication is that, to avoid this source of errors, it is necessary to have sufficient display area so that all the items used in any given cognitive processing can be displayed simultaneously. Minimizing non-task-related cognitive processes is a general HF/E aim, to increase processing efficiency. In this case it is also necessary in order to reduce errors. This requirement emphasizes the need to identify what display items are used together, in a cognitive task analysis.

The Overview in Working Storage

Although there are good reasons to argue that the cognitive processes in complex dynamic tasks build up a contextual overview of the person's present understanding and plans (Bainbridge 1993a), not much is known about this overview. This section makes some points about its capacity, its content, and the way items are stored.

Capacity. Bisseret (1970) asked air traffic area controllers, after an hour of work, what they remembered about the aircraft they had been controlling. Three groups of people were tested: trainee controllers, people who had just completed their training, and people who had worked as controllers for several years. Figure 6.26 shows the number of items recalled. The experienced controllers could remember on average 33 items. This is a much larger figure than the 7 ± 2 chunk capacity for static short-term memory (Miller, 1956) or the 2 items capacity of running memory for arbitrary material (Yntema & Mueser, 1962). Evidently a person's memory capacity is improved by doing a meaningful task and by experience. A possible reason for this is given later.

Content. Bisseret also studied which items were remembered. The most frequently remembered items were flight level (33% of items remembered), position (31%), and time at fix (14%). Leplat and Bisseret (1965) had previously identified the strategy the

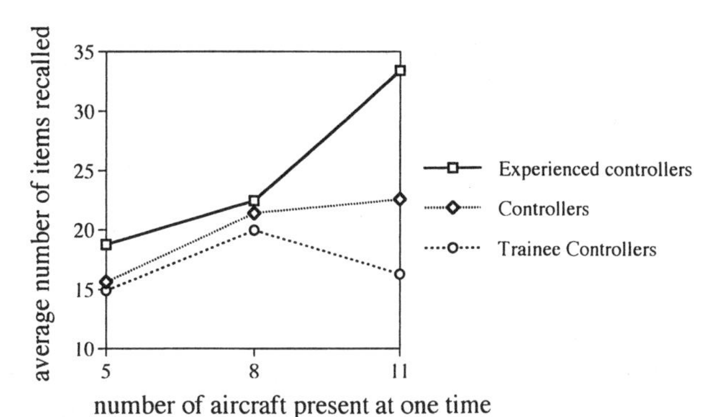

FIG. 6.26. Number of items recalled by air traffic controllers (data from Bisseret, personal communication, based on Bisseret, 1970).

controllers used in conflict identification (checking whether aircraft are a safe distance apart). The frequency with which the items were remembered matches the sequence in which they were thought about: The strategy first compared aircraft flight levels, then position, then time at fix, and so on.

Sperandio (1970) studied another aspect (Fig. 6.27). He found that more items were remembered about aircraft involved in conflict than ones that were not. For nonconflict aircraft, more was remembered about aircraft that had been in radio contact. For conflict aircraft, more was remembered about aircraft on which action had been taken, and most was remembered about aircraft for which an action had been chosen but not yet made.

These results might be explained by two classic memory effects. One is the rehearsal or repetition mechanism by which items are maintained in short-term memory. The more frequently the item or aircraft has been considered by the controllers when identifying potential collisions and acting on them, the more likely it is to be remembered. The findings about aircraft in conflict could be explained by the recency effect, that items that have been rehearsed most recently are more likely to be remembered. These rehearsal and recency mechanisms make good sense as mechanisms for retaining material in real as well as in laboratory tasks.

The Form in Which Material Is Retained. The controllers studied by Bisseret (1970) remembered aircraft in pairs or threes: "There are two flying towards DIJ, one at level 180, the other below at 160," "there are two at level 150, one passed DIJ towards BRY several minutes ago, the other should arrive at X at 22," or "I've got one at level 150 which is about to pass RLP and another at level 170 which is about 10 min behind." The aircraft were not remembered by their absolute positions, but in relation to each other. Information was also remembered relative to the future; many of the errors put the aircraft too far ahead. These sorts of data suggest that, although rehearsal and recency are important factors, the items are not remembered simply by repeating the raw data, as in short-term memory laboratory experiments. What is remembered is the outcome of working through the strategy for comparing aircraft for potential collisions. The aircraft are remembered in terms of the key features that bring them close together—whether they are at the same level, or flying toward the same fix point, and so on.

A second anecdotal piece of evidence is that air traffic controllers talk about "losing the picture" as a whole, not piecemeal. This implies that their mental representation of the situation is an integrated structure. It is possible to suggest that experienced controllers remember more because they have better cognitive skills for recognizing the relations between aircraft, and the integrated structure makes items easier to remember.

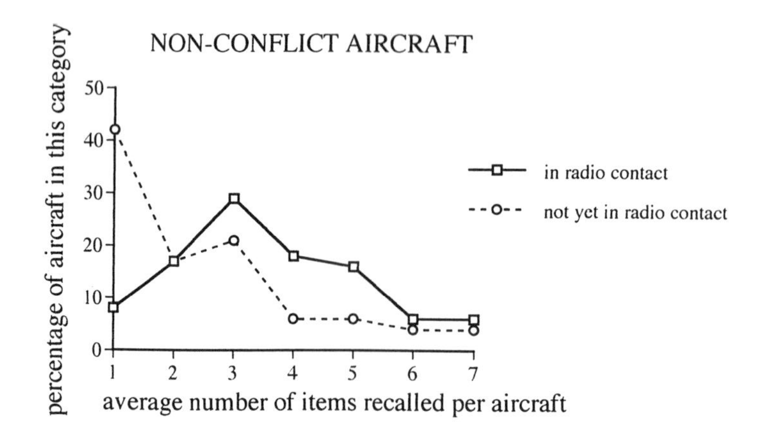

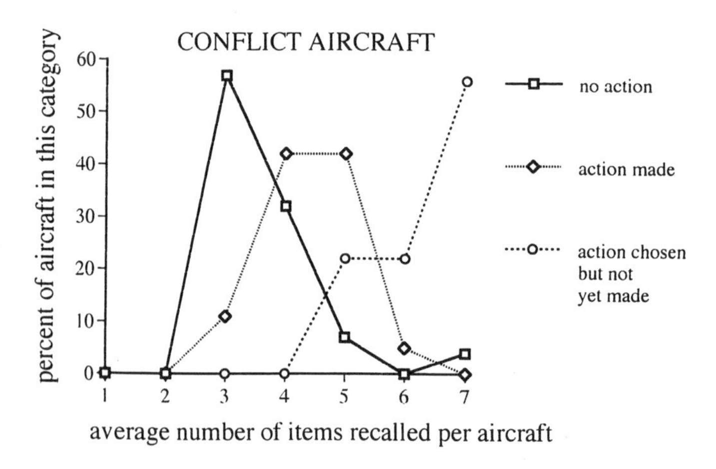

FIG. 6.27. Recall of items about aircraft in different categories (based on data in Sperandio, 1970).

The only problem with this integrated structure is that the understanding, predictions, and plans can form a "whole" that is so integrated and self-consistent that it becomes too strong to be changed. People may then only notice information that is consistent with their expectations, and it may be difficult to change the structure of inference if it turns out to be unsuccessful or inappropriate (this rigidity in thinking is called *perceptual set*).

Some Practical Implications. Some points have already been made about the importance of short-term memory in display systems. The interface also needs to be designed to support the person in developing and maintaining an overview. It is not yet known whether an overview can be obtained directly from an appropriate display, or whether the overview can only be developed by actively understanding and planning the task, with a good display enhancing this processing but not replacing it. It is important, in display systems in which the data needed for the whole task are not all displayed at the same time, to ensure there is a permanent overview display and that it is clear how the other possible displays are related to it.

Both control automation (replacing the human controller) and cognitive automation (replacing the human planner, diagnoser, and decision maker) can cause problems with the person's overview. A person who is expected to take over manual operation or decision making will only be able to make informed decisions about what to do after the person has built up an overview of what is happening. This may take 15–30 min to develop. The system design needs to allow for this sort of delay before a person can take over effectively (Bainbridge, 1983). Also, the data just given show that a person's ability to develop a wide overview depends on experience. This means that, to be able to take over effectively from an automated system, the person needs to practice building up this overview. Practice opportunities should therefore be allowed for in the allocation of functions between computer and person, or in other aspects of the system design such as refresher training.

Planning, Multitasking, and Problem Solving

Actions in complex dynamic tasks are not simple single units. A sequence of actions may be needed, and it may be necessary to deal with several responsibilities at the same time. Organization of behavior is an important cognitive function, which depends on and is part of the overview. This section is in three interrelated parts: planning future sequences of action; multitasking, dealing with several concurrent responsibilities, including sampling; and problem solving, devising a method of working when a suitable one is not known.

Planning

It may be more efficient to think out what to do in advance, if there is a sequence of actions to carry out, or multiple constraints to satisfy, or it would be more effective to anticipate events. Alternative actions can be considered and the optimum ones chosen, and the thinking is not done under time pressure. The planning processes may use working storage, for testing the alternatives by mental simulation, and for holding the plan as part of the overview.

In aviation, an obvious example is preflight planning. Civilian pilots plan their route in relation to predicted weather. Military pilots plan their route relative to possible dangers and the availability of evasive tactics. In high-speed, low-level flight there is not time to think out what to do during the flight, so the possibilities need to be worked out beforehand. The plan then needs to be implemented, and adjusted if changes in circumstances make this necessary. This section is in two parts, on preplanning and online revision of plans.

Preplanning. Figure 6.28 shows results from a study of preflight planning by Amalberti (1992, Expt. 2). Pilots thought out the actions to take at particular times or geographical points. Planning involves thinking about several alternative actions, and choosing the best compromise given several constraints. Some of the constraints the pilots consider are the level of risk of external events, the limits to maneuverability of the aircraft, and their level of expertise to deal with particular situations, as well as the extent to which the plan can be adapted, and what to do if circumstances mean that major changes in plan are needed.

Amalberti studied four novice pilots, who were already qualified but at the beginning of their careers, and four experts. The cognitive aims considered during planning are

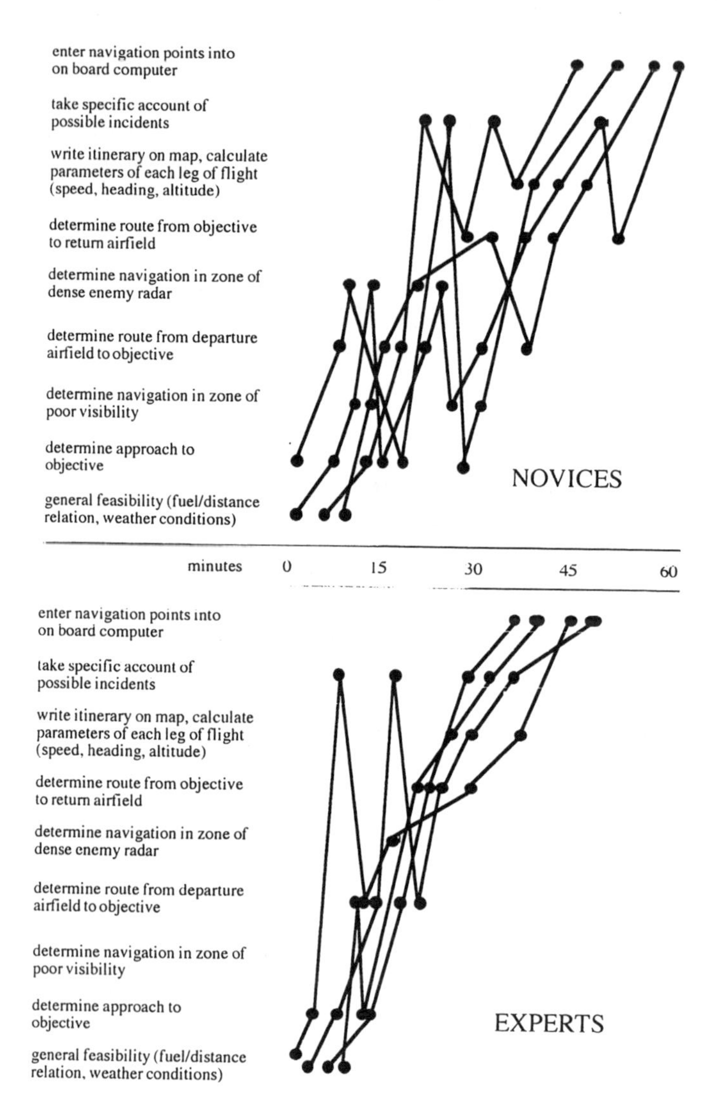

FIG. 6.28. Preflight planning by pilots with different levels of expertise (translated from Amalberti, 1992).

listed on the left of the figure. Each line on the right represents one pilot, and shows the sequence in which he thought about the cognitive functions. The results show that novice pilots took longer to do their planning, and that each of the novice pilots returned to reconsider at least one point he had thought about earlier. Verbal protocols collected during the planning showed that novices spent more time mentally simulating the results of proposed actions to explore their consequences. The experts did not all think about the cognitive functions in the same sequence, but only one of them reconsidered an earlier point. Their verbal protocols showed they prepared fewer responses to possible incidents than the novices.

One of the difficulties with planning is that later in planning the person may think of problems that mean that parts of the plan already devised need to be revised. Planning is an iterative process. The topics are interdependent; for example, the possibility of incidents may affect the best choice of route to or from the objective. What is chosen as the best way of meeting any one of the aims may be affected by, or affect, the best way of meeting other aims. As the topics are interdependent, there is no one optimum sequence for thinking about them. The results suggest that experts have the ability, when thinking about any one aspect of the flight, to take into account its implications for other aspects, so it does not need to be revised later.

The experts have better knowledge about the scenario, about possible incidents and levels of risk. They know more about what is likely to happen, so they need to prepare fewer alternative responses to possible incidents. The experts also know from experience the results of alternative actions, including the effects of actions on other parts of the task, so they do not need to mentally simulate making actions to check their outcomes. They also have more confidence in their own expertise to deal with given situations. All these are aspects of their knowledge about the general properties of the things they can do, how risky these are, how good they are at them, and so on. This meta-knowledge was introduced in the earlier section on actions, and is also central to multitasking and in workload and learning (see later discussion).

Online Adaptation of Plans. In the second part of Amalberti's study, the pilots carried out their mission plan in a high-fidelity simulator. The main flight difficulty was that they were detected by radar. The pilots responded immediately to this. The response had been been preplanned, but had to be adapted to details of the situation when it happened. The novice pilots showed much greater deviations from their original plan than the experts. Some of the young pilots slowed down before the point at which they expected to be detected, as accelerating was the only response they knew for dealing with detection. This acceleration led to a deviation from their planned course, so they found themselves in an unanticipated situation. They then made a sequence of independent, reactive, short-term decisions, because there was not time to consider the wider implications of each move. The experts made much smaller deviations from their original plan, and were able to return to the plan quickly. The reason for this was that they had not only preplanned their response to radar, they had also thought out in advance how to recover from deviations from their original plan. Again experience, and therefore training, plays a large part in effective performance.

In situations in which events happen less quickly, people may be more effective in adapting their plans to changing events at the time. The best model for the way that people adapt their plans to present circumstances is probably the opportunistic planning model of Hayes-Roth and Hayes-Roth (1979; see also Hoc, 1988).

Multitasking

If a person has several concurrent responsibilities, each of which involves a sequence of activities, then interleaving these sequences is called multitasking. Doing this involves an extension of the processes mentioned under planning. Multitasking involves working out in advance what to do, combined with opportunistic response to events and circumstances at the time.

Examples of Multitasking. Amalberti (1992, Expt. 1) studied military pilots during a simulated flight. Figure 6.29 shows part of his analysis, of activities during descent to low-level flight. The bottom line in this figure is a time line. The top part of the figure describes the task as a hierarchy of task goals and subgoals. The parallel double-headed arrows beneath represent the time that the pilot spent on each of the activities. These

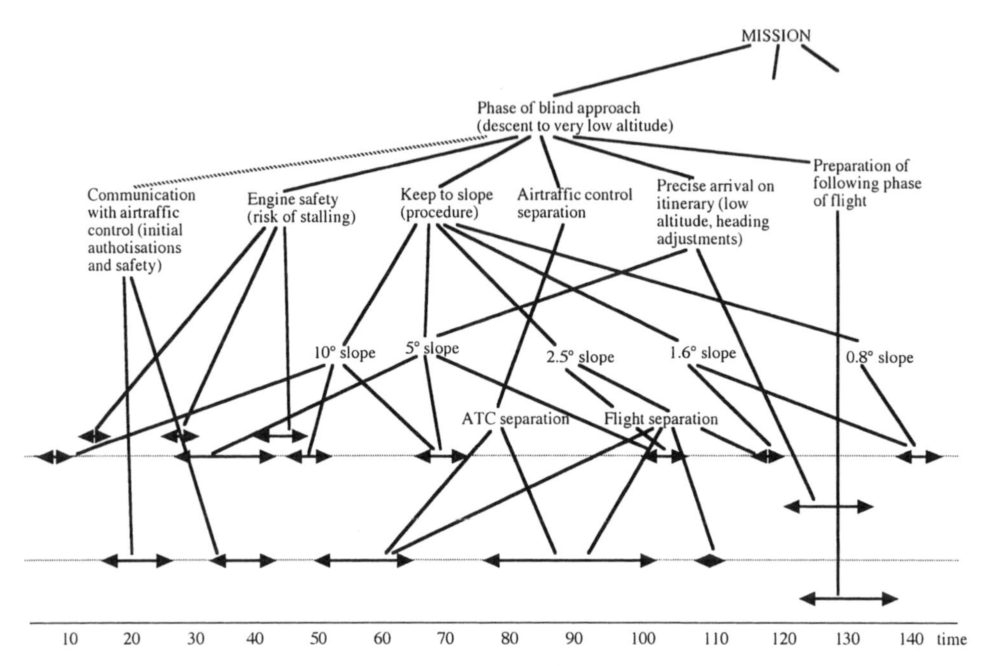

FIG. 6.29. Multitasking by a pilot during one phase of the flight (translated from Amalberti, 1992).

arrows are arranged in five parallel lines that represent the five main tasks in this phase of flight: maintain engine efficiency at minimum speed; control angle of descent; control heading; deal with air traffic control; and prepare for the next phase of flight. Other principal tasks that occurred in other phases of flight were: keep to planned timing of maneuvers; control turns; check safety. Figure 6.29 shows how the pilot allocated his time between the different tasks. Sometimes it is possible to meet two goals with one activity. The pilot does not necessarily complete one subtask before changing to another. Indeed, this is not often not possible in a control task, in which states and events develop over time. Usually the pilot does one thing at a time. However, it is possible for him to do two tasks together when they use different cognitive processing resources. For example, controlling descent, which uses eyes + motor coordination, can be done at the same time as communicating with air traffic control, which uses hearing + speech (see also later discussion on workload).

Some multitasking examples are difficult to describe in a single figure. For example, Reinartz (1989), studying a team of three nuclear power plant operators, found they might work on 9 to 10 different goals at the same time. Other features of multitasking have been observed by Benson (1990):

- Multitasking may be planned ahead (a process operator studied by Beishon, 1974, made plans for up to 1.5 hr ahead). These plans are likely to be partial, and incomplete in terms of timing and detail. Planned changes in activity may be triggered by times or events. When tasks are done frequently, much of the behavior organization may be guided by habit.
- Executing the plan. Interruptions may disrupt planned activity. The preplan is incomplete, and actual execution depends on details of the situation at the time. Some tasks may be done when they are noticed in passing (Beishon, 1974, first

noticed this, and called it serendipity). This is opportunistic behavior. The timing of activities of low importance may not be preplanned, but may be fitted in in spare moments. The remaining spare moments are recognized as spare time.

- Effects of probabilities and costs. In a situation that is very unpredictable, or when the cost of failure is high, people may make the least risky commitment possible. If there is a high or variable workload, people may plan to avoid increasing their workload, and use different strategies in different workload conditions (see later discussion on workload).

A Possible Mechanism. Sampling is a simple example of multitasking in which people have to monitor several displays to keep track of changes on them. Mathematical sampling theory has been used as a model for human attention in these tasks. In the sampling model, the frequency of attending to an information source is related to the frequency of changes on that source. This can be a useful model of how people allocate their attention when changes to be monitored are random, as in straight and level flight, but this model is not sufficient to account for switches in behavior in more complex phases of flight.

Amalberti (1992) made some observations about switching from one task to another. He found that:

- Before changing to a different principal task the pilots review the normality of the situation, by checking that various types of redundant information are compatible with each other.
- Before starting a task that will take some time, they ensure that they are in a safe mode of flight. For example, before analyzing the radar display, they check that they are in the appropriate mode of automatic pilot.
- While waiting for feedback about one part of the task, pilots do part of another task that they know is short enough to fit into the waiting time.
- When doing high-risk, high-workload tasks, pilots are less likely to change to another task.

These findings suggest that, at the end of a subsection of a principal task, the pilots check that everything is all right. They then decide (not necessarily consciously) what next to devote effort to, by combining their preplan with meta-knowledge about the alternative tasks, such as how urgent they are, how safe or predictable they are, how difficult they are, how much workload they involve, and how long they take (see later discussion on workload).

Practical Implications. Multitasking can be preplanned, and involves meta-knowledge about alternative behaviors. Both planning and knowledge develop with experience, which underlines the importance of practice and training.

The nature of multitasking also emphasizes the difficulties that could be caused by task-specific displays. If a separate display was used for each of the tasks combined in multitasking, then the user would have to call up a different display, and perhaps change coding vocabularies, each time the person changed to a different main task. This would require extra cognitive processing and extra memory load, and could make it difficult to build up an overview of the tasks considered together. This suggests an extension to the point made in the section on working storage. All the information used in all the principal tasks that may be interleaved in multitasking needs to be available at the same time, and easily cross-referenced. If this information is not available, then coordination and opportunistic behavior may not be possible.

Problem Solving

A task is familiar to a person who knows the appropriate working methods, plus the associated reference knowledge about the states that can occur, the constraints on allowed behavior, and the scenarios, mental models, and so on that describe the environmental possibilities within which the working methods must be used.

Problem solving is the general term for the cognitive processes a person uses in an unfamiliar situation, which the person does not already have an adequate working method or reference knowledge for dealing with. Planning and multitasking are also types of processing that are able to deal with situations that are not the same each time. However, both take existing working methods as their starting point, and either think about them as applied to the future, or work out how to interleave the working methods used for more than one task. In problem solving, a new working method is needed.

There are several ways of devising a new working method. Some are less formal techniques that do not use much cognitive processing, such as trial and error, or asking for help. There are also techniques that should not need much creativity, such as reading an instruction book. People may otherwise use one of three techniques for suggesting a new working method. Each of these uses working methods recursively; it uses a general working method to build up a specific working method.

1. Categorization. This involves grouping the problem situation with similar situations for which a working method is available. The working method that applies to this category of situation can then be used. This method is also called *recognition-primed decision making*. The nature of "similarity" and the decisions involved are discussed by Klein (1989).
2. Case-based reasoning. This involves thinking of a known event (a *case*) that is similar or analogous to the present one, and adapting the method used then for use in the present situation. This is the reason why stories about unusual events circulate within an industry. They provide people in the industry with exemplars for what they could do themselves if a similar situation arose, or with opportunities to think out for themselves what would be a better solution.
3. Reasoning from basic principles. In the psychological literature, the term *problem solving* may be restricted to a particular type of reasoning in which a person devises a new method of working by building it up from individual components (e.g., Eysenck & Keane, 1990, chapters 11 and 12). This type of processing may be called *knowledge based* by some people.

A general problem-solving strategy consists of a set of general cognitive functions that have much in common with the basic cognitive functions in complex dynamic tasks (see introduction to this section). Problem solving, for example, could involve understanding the problem situation, defining what would be an acceptable solution, and identifying what facilities are available. Meeting each of these cognitive needs can be difficult, because the components need to be chosen for their appropriateness to the situation and then fitted together. This choice could involve: identifying what properties are needed from the behavior; searching for components of behavior that have the right properties (according to the meta-knowledge which the person has about them); and then combining them into a sequence.

The final step in developing a new working method is to test it, either by mental simulation, or by trial and error. This mental simulation could be similar to the techniques used in planning and multitasking. Working storage may thus be used in problem solving in two ways: to hold both the working method for building up a working method and the proposed new method; and to simulate carrying out the proposed working method to test whether its processing requirements and outputs are acceptable.

Knowledge

Knowledge is closely involved in all modes of cognitive processing, It provides the probabilities, utilities, and alternatives considered in decision making, and the translations used in recoding. In complex tasks it provides the working methods and reference knowledge used in thinking about cognitive functions, and the meta-knowledge. Different strategies may use different types of reference knowledge. For example, a strategy for diagnosing faults by searching the physical structure of the device uses one type of knowledge, whereas a strategy that relates symptoms to the functional structure of the device uses another. The reference knowledge may include scenarios, categories, cases, mental models, performance criteria, and other knowledge about the device the person is working with. Some knowledge may be used mainly for answering questions, for explaining why events occur or actions are needed. This basic knowledge may also be used in problem solving.

There are many interesting fundamental questions about how these different aspects of knowledge are structured, interrelated, and accessed (Bainbridge, 1993c), but these issues are not central to this chapter. The main questions here are the relation between the type of knowledge and how it can best be displayed, and what might be an optimum general display format.

Knowledge and Representation. Any display for a complex task can show only a subset of what could be represented. Ideally, the display should make explicit the points that are important for a particular purpose, and provide a framework for thinking. The question of which display format is best for representing what aspect of knowledge has not yet been thoroughly studied, and most of the recommendations about this are assumptions based on experience (Bainbridge, 1988). For example, the following formats are often found:

Aspect of knowledge	*Form of display representation*
Geographical position	Map
Topology, physical structure	Mimic/schematic, wiring diagram
Cause–effect, functional structure	Cause–effect network, mass-flow diagram
Task goals–means structure	Hierarchy
Sequence of events or activities	Flow diagram
Analogue variable values and limits	Scale + pointer display
Evolution of changes over time	Chart recording

Each of these aspects of knowledge might occur at several levels of detail, for example, in components, subsystems, systems, and the complete device. And knowledge can be at several levels of distance from direct relevance; for example, it could be about a specific aircraft, about all aircraft of this model, about aircraft in general, about aerodynamics, or about physics.

Knowledge-display recommendations raise three sorts of question. One arises because each aspect of knowledge is one possible "slice" from the whole body of knowledge. All the types of knowledge are interrelated, but there is not a simple one-to-one relation between them. Figure 6.30 illustrates some links between different aspects of knowledge. Any strategy is unlikely to use only one type of knowledge, or to have no implications for aspects of thinking that use other types of knowledge. It might mislead the user to show different aspects of knowledge with different and separate displays that are difficult to cross-refer between, as this might restrict the thinking about the task. Knowledge about cross-links is difficult to display, and is gained by experience. This emphasizes training.

A second question is concerned with salience. Visual displays emphasise (make more salient) the aspects which can easily be represented visually. (For example, see the discussion at the end of this chapter on the limitations of Figs. 6.22 and 6.24 as models of behavior.) It might be unwise to make some aspects of knowledge easy to take in simply because they are easier to display, rather than because they are important in the task. There are vital types of knowledge that are not easy to display visually, such as the associations used in recoding, or the categories, cases, scenarios, and meta-knowledge used in complex thinking. These are all learned by experience. The main approach to supporting nonvisual knowledge is to provide the user with reminder lists about the alternatives (see later discussion on cued recall). Display design and training are

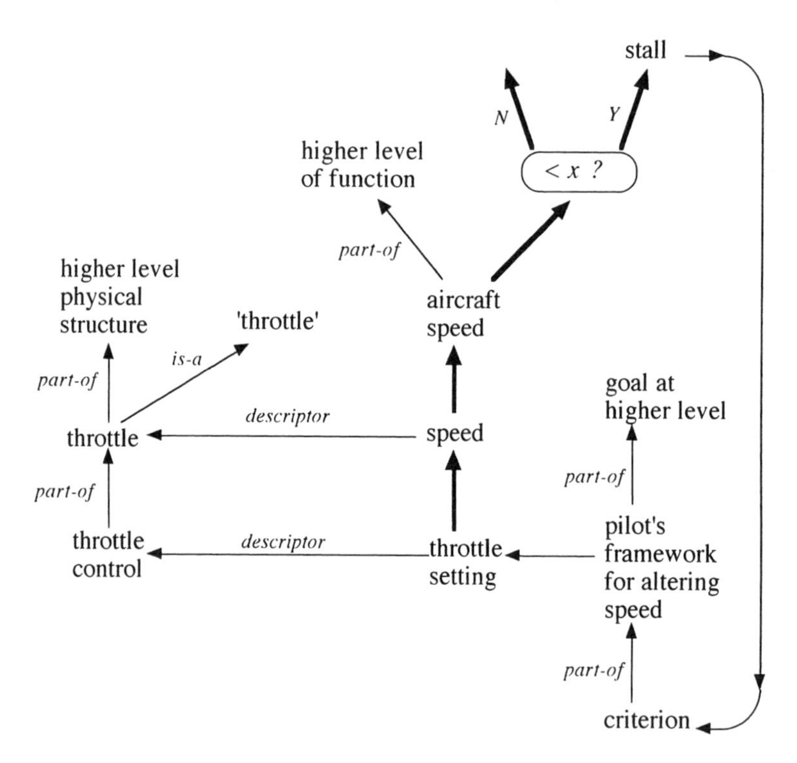

FIG. 6.30. Some of the links in a small part of a pilot's knowledge base (thick arrow indicates cause–effect relation).

interdependent, as they are each effective at providing different types of knowledge. It could be useful to develop task analysis techniques that identify different aspects of knowledge, as well as to do more research on how types of knowledge, and the links between them, can best be presented.

The third issue about all these multiple possible display formats repeats the questions raised previously about efficient use of codes. If a user was given all the possible display types just listed, each of which would use different codes, possibly with different display formats using the same code with different meanings (e.g., a network with nodes could be used to represent physical, functional, or hierarchical relations between items), the different codes might add to the user's difficulties in making cross connections between different aspects of knowledge.

An Optimum Format? These issues suggest the question: Is there one or a small number of formats that subsume or suggest the others? This is a question that has not yet been much studied. A pilot study (Brennan, 1987) asked people to explain an event, given either a mimic or a cause–effect diagram of the physical device involved. The people tested either did or did not already know how the device worked. The results suggested that people who did not know how the device worked were most helped by a cause–effect representation (which does show how it worked), whereas experts were best with the mimic representation. Contextual cues can greatly aid memory performance (e.g., Eysenck & Keane, 1990, chapter 6). A cue is an aid to accessing the items to be recalled. The reason for expert performance with mimic displays might be that the icons and flow links on this type of display not only give direct evidence about the physical structure of the device, but they also act as cues or reminders about other knowledge the person has about the device—they evoke other parts of the scenario. This is an example from only one type of cognitive task, but it does point to the potential use of contextual cued recall in simplifying display systems. However, cued recall can only be effective with experienced people, who can recognize the cues and know what they evoke.

MENTAL WORKLOAD, LEARNING, ERRORS

Workload, learning, and errors are all aspects of the efficiency of cognitive processing. There are limits to human processing capacities, but these are difficult to define, because of the adaptability of human behavior. As a result of learning, processing becomes more efficient and adapted to what is required. As efficiency increases, mental workload may decrease. Error rates can be affected by both expertise and workload, and errors are closely involved in the processes of learning. There is a huge wealth of material that could be discussed, so the aim here is only to give a brief survey.

Mental Workload

There is a large number of issues involved in accounting for mental workload and how it is affected by different aspects of a task. This section mentions three main topics: whether people can only do one task at a time; factors affecting processing capacity; and the ways in which people typically respond to overload.

Single- or Multichannel Processing

Many types of evidence, including the example of multitasking in Fig. 6.29, show that people usually do one task at a time. This section looks at how people attend to one source of stimuli among many, and under what circumstances people can do more than one task at a time. As usual, the findings show how adaptable human beings are, and that there is not yet a full account of the processes involved.

Focused Attention. People have the ability to pick out one message against a background of others, visual or auditory. Studies show, however, that a person does not only process one of the stimulus sources, but takes in enough about the other possible signals to be able to separate them. This chapter has already used the notion of *depth* of processing, as in discrimination, recoding, sequences of recoding, and building up an overview. This notion is also involved here. Separation of two signal sources requires the least processing if they can be discriminated by physical cues, such as listening to a high voice while a low voice also speaks, or reading red lettering against a background of green letters. The sorts of factors discussed earlier on discrimination affect how easy it is to do this separation. If stimuli cannot be distinguished by physical cues, then "deeper" processing may be involved. For example, Gray and Wedderburn (1960) found that messages presented to the ears as:

Left ear	:	mice	5	cheese
Right ear	:	3	eat	4
Were heard as	:	(3 5 4)		(mice eat cheese)

In this case, the words may be grouped by recognizing their semantic category. In some tasks deeper processing for meaning may be needed, that is, building up an overview, as in:

> **It is important that the subject** man **be** car **pushed** house **slightly** boy **beyond** hat **his** shoe **normal** candy **limits** horse **of** tree **competence** pen **for** be **only** in **phone** this **cow** way **book** can **hot** one **tape** be **pin** certain **stand** that **snaps** he **with** is **his** paying **teeth** attention **in** to **the** the **empty** relevant **air** task **and** hat **minimal** shoe **attention** candy **to** horse **the** tree **second or peripheral task.**
> (Lindsay & Norman, 1972)

Note that if the cue used becomes ineffective, this is disconcerting. It then takes time, and a search for clues about what would be effective, before the person can orient to a new cue and continue with the task. There is also an interplay of depths of processing: When the physical cue becomes inadequate for following the message, then the reader uses continuity of meaning as a basis for finding a new physical cue. This account fits in with several points made earlier. The person is using active attention for what the person wants to take in, not passive reception of signals. The task setting provides the cue that can be used to minimize the effort needed to distinguish between signal sources. This cue than acts as a perceptual frame for searching for relevant inputs.

The concept of depth of processing was first introduced by Craik and Lockhart (1972) to explain results in some memory experiments. The word *depth* is here distinguished from depth in the organization of behavior, as in goal/subgoal and so on.

Parallel Processing. The criteria defining whether or not people are able to do two tasks at the same time have so far proved elusive to identify. Figure 6.16 shows that, after high levels of practice, choice time is not affected by number of alternatives. Such tasks are said to be automated, or to require no conscious attention. They can be done at the same time as something else, unless both tasks use the same peripheral resources such as vision or hand movement. Wickens (e.g., 1984) did a series of studies showing that people can use different peripheral resources at the same time. People can also learn to do some motor tasks so that movements are monitored by feel rather than visually; then movements can be made at the same time as looking at or thinking about something else. In practice the possibility of multiple processing means that care is needed in designing tasks. One might, for example, think it would reduce unnecessary effort for an air traffic controller to have the flight strips printed out, rather than expecting the controller to write the strips by hand. However if the controller, while writing, is simultaneously thinking out how the information fits into their overview, then printing the flight strips might deprive him or her of useful attention and thinking time.

Whether or not two tasks that both involve "central" processing can be done at the same time is less clear. This is partly because what is meant by central processing has not been clearly defined. People can do two tasks at the same time if the tasks are processed by different areas of the brain—for example, a music task and a language task (Allport, Antonis, & Reynolds, 1972)—though both tasks need to be simple and perhaps done by recoding. Going to "deeper" levels of processing, there does seem to be a limit to the extent to which people can build up distinct overviews for two different tasks at the same time. Whether or not an overview is needed to do a task may be part of the question. For example, people playing multiple chess games may have very good pattern recognition skills and so react to each game by recognition-primed decisions as they return to it, rather than having to keep in mind a separate and continuing overview for each of the games they are playing. Most experienced drivers can drive and hold a conversation on a different topic at the same time when the driving task is simple, but they stop talking when the driving task becomes more difficult.

This is an area in which it is challenging to identify the limits to performance, and it is probably beyond the competence of HF/E at the moment either to define the concepts or to investigate and measure the processing involved. Fortunately, in practice the issue can often be simplified. When predicting performance, the conservative strategy is to assume that people cannot do two tasks at the same time. This will always be the worst-case performance.

Factors Influencing Processing Capacity

The amount of mental work a person can do in a given time is not a simple quantity to specify. If it is assumed that a person can only do one thing at a time, then every factor that increases the time taken to do a unit task will decrease the number of those tasks that can be done in a given time interval, and so decrease performance capacity. Thus every factor in interface design might affect performance capacity.

Focusing on performance time emphasizes performance measures of workload effects. Other important measures of workload are physiological, such as the rate of secretion of stress chemicals, and subjective, such as moods and attitudes. Any factor could be considered a "stressor" if its effect is that performance levels, stress hormone secretion rates, or subjective feelings deteriorate. The approach in this section is to indicate some key general topics, rather than to attempt a full review. There are reviews of workload topics in the chapters on fatigue and biological rhythms, pilot performance, and controller performance (chapters 10, 13, and 19).

The points made here are concerned with the capacities of different mental processes; extrinsic and intrinsic stressors; individual differences; and practical implications.

Capacities of Different Cognitive Resources. Different aspects of cognitive processing have different capacities. For a review of processing limits see Sage (1981). The capacity of different processes may be affected differently by different factors. Figure 6.31 shows time-of-day effects on performance in four tasks: serial search, verbal reasoning (working memory) speed, immediate retention, and alertness. The different performance trends in these tasks suggest that each task uses a different cognitive resource that responds differently to this stress. It is difficult to make reliable analyses of these differences, but some other tasks in which performance may differ in this way are coding and syllogisms (Folkhard, 1990).

Extrinsic and Intrinsic Stressors. Extrinsic stressors are stressors that apply to any person working in a particular environment, whatever task they are doing. Time-of-day, as in Fig. 6.31, is extrinsic in this sense. Some other extrinsic stressors that can affect performance capacity are noise, temperature, vibration, fumes, fatigue, and organizational culture.

Intrinsic stressors are factors that are local to a particular task. All the HF/E factors that affect performance speed or accuracy come in this category. The effect of task difficulty interacts with motivation. Easy tasks may be done better with high motivation,

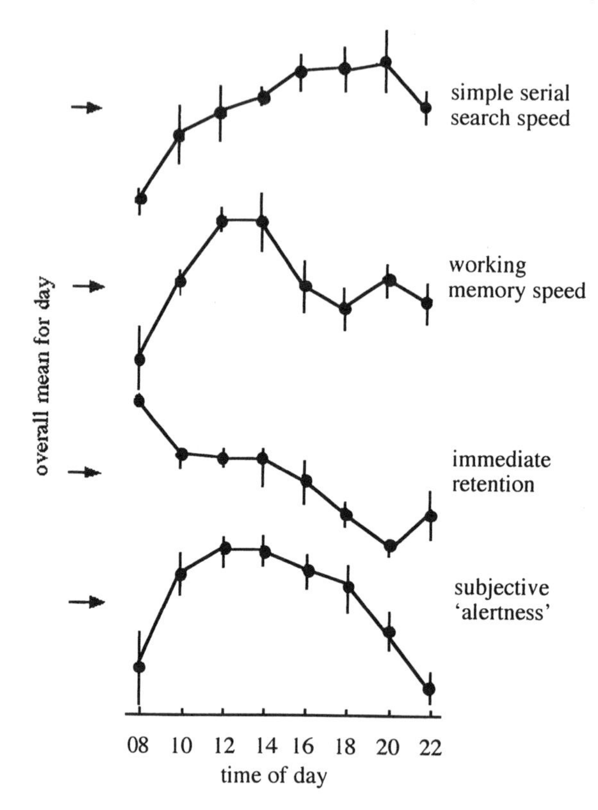

FIG. 6.31. Cognitive processing capacities change during the day. The different patterns of change suggest these capacities have different mechanisms. Reproduced from Folkhard, 1990, Circadian performance rhythms, in Broadbent et al. (Eds.), *Human Factors in Hazardous Situations* (pp. 543–553), copyright © 1990 Clarendon Press, reproduced by permission of Oxford University Press.

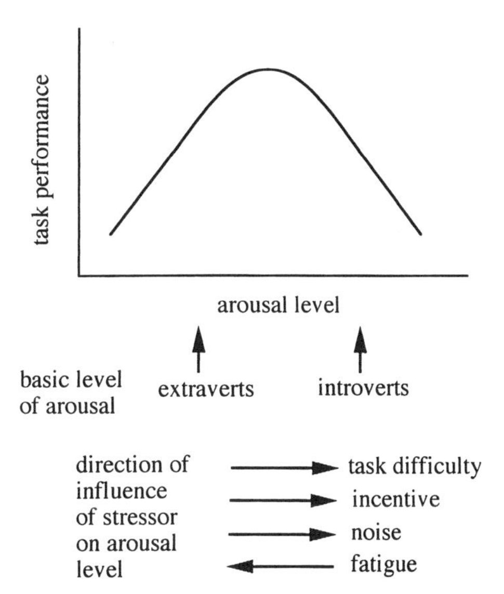

FIG. 6.32. "Inverted U" relation between internal arousal level and performance (it is not possible to account for the effect of all stressors in this way).

whereas difficult tasks are done better at lower levels of motivation. This can be explained by assuming that stressors affect a person's "arousal" level, and that there is an inverted-U relation between arousal level and performance (see Fig. 6.32).

Measures of stress hormones and of workforce attitudes show that several factors to do with the pacing of work, and the amount of control over their work that a person feels he or she has, can be stressors (e.g., Johansson, Aronsson, & Lindström, 1978). Such aspects are of more concern in repetitive manufacturing jobs than in work such as flying or air traffic control.

Individual Differences. Individual differences affect a person's capacity for a task, and the person's willingness to do it. Aspects of individual differences fall into at least five groups.

1. Personality. Many personality dimensions, such as extroversion/introversion, sensitivity to stimuli, need for achievement or fear of success, and preference for facts/ideas or for regularity/flexibility, can affect a person's response to a particular task.

2. Interests and values. A person's interests and values affect the response to various factors in the task and the organizational climate, which influence willingness and commitment to do or learn a given task. People differ in their response to incentives or disincentives such as money, status, or transfer to a job that does not use their skills.

3. Talent. Different people have different primary senses, different cognitive styles, and different basic performance abilities (e.g., Fleishman, 1975). For example, very few of us have the ability to fly high-speed aircraft.

4. Experience. The rest of us may be able to develop higher levels of performance though practice. Even the few who can fly high-speed aircraft have millions spent on their training. The effects of training on cognitive capacities is discussed more in the section on learning.

5. Nonwork stressors. There may be nonwork stressors on an individual that affect the person's ability to cope with work, such as illness, drugs, or home problems.

Practical Implications. There are so many factors affecting the amount of effort any particular individual is able or willing to devote to a particular task at a particular time, that performance prediction might seem impossible. Actually the practical ways of dealing with this variety are familiar. There are two groups of issues, in HF/E design and in performance prediction.

Nearly all HF/E design recommendations are based on measures of performance capacity. Any factor that has a significant effect on performance should be improved, as far as is economically justifiable. Design recommendations could be made about all the intrinsic and extrinsic factors mentioned earlier, and individual differences might be considered in selection.

However, it is easier to predict that a design change will improve performance than to predict the size of the improvement. Numerical performance predictions may be made in order to assess whether a task can be done in the time available, or with the available people, or to identify limits to speed or accuracy on which design investment should best be concentrated. Obviously it is not practical to include all the possible effective factors when making such predictions. Three simplifying factors can reduce the problem. One is that, although smaller performance changes may give important clues about how to optimize design, from the point of view of performance prediction these factors may only be important if they make an order of magnitude difference to performance. Unfortunately, our data relevant to this issue are far from complete. The second point is that only conservative performance predictions are needed. For these purposes it may be valid to extrapolate from performance in simple laboratory tasks in which people with no relevant expertise react to random signals, which is the worst case. To predict minimum levels of performance, it may not be necessary to include the ways in which performance can improve when experienced people do tasks in which they know the redundancies, can anticipate, and so forth. The third point is that, in practice, many of the techniques for performance prediction that have been devised have the modest aim of matching expert judgements about human performance in a technique that can be used by someone less expert, rather than attempting high levels of accuracy or validity.

Response to Overload

If people doing a simple task have too much to do, they only have the options of omitting parts of the task or of accepting a lower level of accuracy in return for higher speed (Fig. 6.18). People doing more complex tasks may have more scope for responding to increased workload while maintaining acceptable task performance. This section discusses increasing efficiency, changing strategy, and practical implications.

Increasing Efficiency. Complex tasks often offer the possibility of increasing the efficiency with which a task is done. For example, Sperandio (1972) studied the radio messages of air traffic approach controllers. He found that when they were controlling one aircraft they spent 18% of their time in radio communication. When there were nine aircraft, they spent 87% of their time on the radio. In simple models of mental workload:

$$\text{Total workload} = \text{workload in one task} \times \text{number of tasks}$$

Evidently that does not apply here, or the controllers would spend 162% of their time on the radio. Sperandio found that the controllers increased the efficiency of their radio messages in several ways. There were fewer pauses between messages. Redundant and unimportant information were omitted. And conversations were more efficient: The average number of conversations per aircraft decreased but the average number of messages per conversation increased, so fewer starting and ending procedures were necessary.

Changing Strategy. The controllers studied by Sperandio (1972) not only altered the efficiency of their messages; the message content also altered. The controllers used two strategies for bringing aircraft into the airport (this is a simplification so the description can be brief). One strategy was to treat each aircraft individually. The other was to standardize the treatment of aircraft by sending them all to a stack at a navigation fix point, from which they could all enter the airport in the same way. When using the individual strategy, the controllers asked an aircraft about its height, speed, and heading. In the standard strategy they more often told an aircraft what height and heading to use. The standard strategy required less cognitive processing for each aircraft. Sperandio found that the controllers changed from using only the individual strategy when there were three or fewer aircraft, to using only the standard strategy when there were eight or more aircraft. Expert controllers changed to the standard strategy at lower levels of workload. Sperandio argued that the controllers change to a strategy which requires less cognitive processing, in order to keep the total amount of cognitive processing within achievable limits (Fig. 6.33A).

The relation between task performance and workload is therefore not the same in mental work as it is in physical work. In physical work, conservation of energy ensures there is a monotonic relation between physical work and task performance. In mental

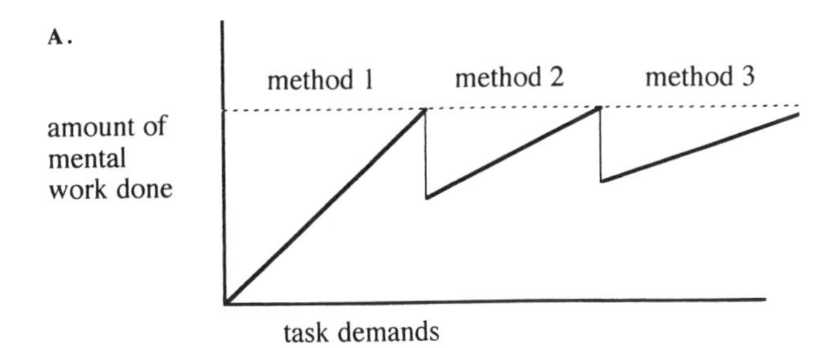

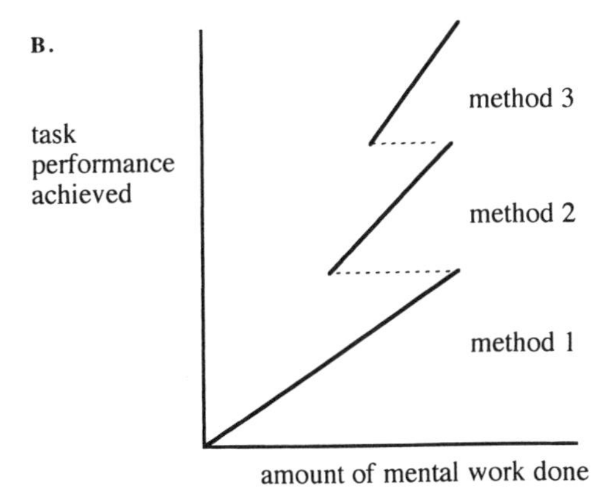

FIG. 6.33. Effect of changing working methods on relation between mental work and task work (Part A from Sperandio, 1972). This figure is a simplification: in practice the use of methods overlaps, so there are not discontinuities.

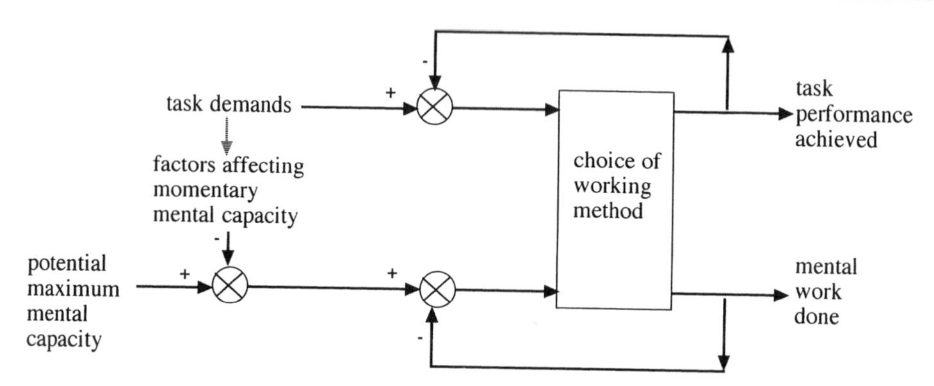

FIG. 6.34. Choice of optimum working method depends on task and personal factors.

workload, if there are alternative working methods for meeting given task demands, there is not necessarily a linear relation between the task performance achieved and the amount of mental work needed to achieve it. By using different methods, the same amount of mental effort can achieve different amounts of task performance (Fig. 6.33B).

In choosing an optimum working method, two adaptations are involved. A person must choose a method that meets the task demands. The person must also choose a method that maintains mental workload at an acceptable level. Whichever method is chosen will affect both the task performance achieved and the mental workload experienced (Fig. 6.34). There needs to be a mechanism for this adaptive choice of working method. This is another contextual effect that could be based on meta-knowledge. Suppose that the person knows, for each method, both how well it meets various task demands and what mental workload demands it poses. The person could then compare this meta-knowledge with the demands of the task and mental context, to choose the best method for the circumstances (Bainbridge, 1978).

Practical Implications. This flexibility of working method has several practical implications. It is not surprising that many studies have found no correlation between task performance and subjective experience of mental workload. There are also problems with predicting mental workload, similar to the problems of predicting performance capacity mentioned earlier.

A person can only use several alternative working methods if the performance criteria do not strictly constrain what method must be used. For example, in air traffic control, safety has much higher priority than the costs of operating the aircraft. Task analysis could check that alternative methods are possible, and perhaps what these methods are (it may not be possible to predefine all methods; see earlier discussion on problem solving and later discussion on learning).

Adaptive use of working methods suggests that strategy specific displays should not be provided, as they could remove the possibility of this flexibility for dealing with varying levels of workload. It could also be useful to train people to be aware of alternative methods and of the use of meta-knowledge in choosing between them.

When decision support systems are introduced with the aim of reducing workload, it is necessary to consider a wider situation. Decision support systems can increase rather than decrease mental workload if the user does not trust the decision support system and so frequently checks what it is doing (Moray, Hiskes, Lee, & Muir, 1995).

Learning

Learning is another potentially huge topic. All the expertise of psychology on learning, of HF/E on training, and of educational psychology on teaching cognitive skills and knowledge could be included. As this chapter focuses on cognitive processes, this section primarily discusses cognitive skill and knowledge. The coverage only attempts a brief mention of some key topics, which indicate how learning interrelates with other aspects of cognitive processing rather than being a separate phase of performance.

This section uses the word *skill* in the sense in which it is used in psychology and in British industry. There are two key features of skilled behavior in this sense. Processing can be done with increased efficiency, either because special task-related abilities have been developed that would not be expected from the average person, or because no unnecessary movements or cognitive processing are used. And behavior is adapted to the circumstances. Choices, about what best to do next and how to do it, are adapted to the task and personal context. In this general sense, any type of behavior and any mode of cognitive processing can be skilled, so it can be confusing to use the word *skill* as the name for one mode of processing.

This section has three main parts: changes in behavior with experience; learning processes; and relations between mode of processing and appropriate training method.

Changes Within a Mode of Processing

This subsection briefly surveys the modes of processing that have formed one framework of this chapter, and indicates the ways in which each can change by introducing new aspects of processing or losing inefficient ones. This is a summary of points made before and is by no means complete. Learning can also lead to changes from one mode of processing to another, as discussed later.

Physical Movement Skills. By carrying out movements in a consistent environment, people can learn:

- Which movement has which effect (i.e., they develop their meta-knowledge about movements; Fig. 6.20). This means they do not need to make exploratory actions, and their movements do not oscillate around the target. People can then act with increased speed, accuracy, and coordination, and can reach to the correct control or make the correct size of action without checking.
- To use kinesthetic rather than visual feedback.
- The behavior of a moving target, so its movements can be anticipated.

Changes in performance may extend over very long periods. For example, Crossman (1959) studied people doing the manually dexterous task of rolling cigars, and found that performance continued to improve until people had made about 5 million items.

Perceptual Skills. These are discriminations and integrations. People learn:

- The discriminations, groupings, and size, shape, and distance inferences to make.
- The probabilities and biases to use in decision making.

- The appropriate items to attend to.
- The eye movements needed to locate given displays.

Recodings. The person connects from one item to another by association, without intermediate reasoning. These associations may need to be learned as independent facts, or there may be some general rule underlying a group of recodings, such as "choose the control with its location opposite to the location of the display." Many people need a large number of repetitions before they can learn arbitrary associations.

Sequence of Recodings. Two aspects of learning may be involved:

- When a sequence is the same each time, so that the output of one recoding and the input of the next recoding are consistent, then a person may learn to "chunk" these recodings together, to carry them out as a single unit without using intermediate working memory.
- When a goal/function can be met in the same way each time, then choosing a working method that is adapted to circumstances is not necessary. A previously flexible working method may then reduce to a sequence of transforms that does not include goals or choice of working method.

Familiar Working Methods. People need to learn:

- Appropriate working method(s).
- The reference knowledge needed while using each method. When this reference knowledge has been learned while using the method, then it may be accessed automatically, without having to think out explicitly what knowledge is needed in a particular situation.
- How to build up an integrated overview.
- Meta-knowledge about each working method, which is used in choosing the best method for a given context.

Planning and Multitasking. People can become more skilled in these activities. They can learn a general method for dealing with a situation, and the subsidiary skills for dealing with parts of it (Samurçay & Rogalski, 1988).

Developing New Working Methods. The process of developing new working methods can itself be more or less effective. Skill here lies in taking an optimum first approach to finding a new working method. There are several possible modes of processing for doing this.

1. Recognition-primed decisions. People can only make recognition primed decisions once they have learned the categories used. Several aspects of learning are involved:

    - The features defining a category, and how to recognize that an instance has these features so is a member of the category.
    - The members of a category, and their properties (e.g., for each category of situation, what to do in it).

- How to adapt a category method to specific circumstances.

2. Case-based reasoning. Cases (or, more distant from a particular task, analogies) provide examples as a basis for developing the knowledge or working method needed. To be able to do this, people need to know:
   - Cases
   - How to recognize which case is appropriate to which circumstances.
   - How to adapt the method used in one case to different circumstances.
3. Reasoning from basic principles. For this sort of reasoning, people need to have acquired an adequate base of knowledge about the task and the device(s) they are using, with associated meta-knowledge. The same type of knowledge may also be used for explaining events and actions.

Learning Processes

Little is known about how changes in processing take place. Similar processes may be involved in developing and maintaining physical and cognitive skills. This section indicates some mechanisms: repetition; meta-knowledge and feedback; independent goals–means; and changing modes of processing.

Repetition. Repetition is crucial for acquiring and maintaining skills. The key aspects are that, each time a person repeats a task, some aspects of the environment are the same as before, and knowledge of results is given. This knowledge of results has two functions: It gives information about how and how well the task was done, and it acts as a reward.

Meta-Knowledge and Feedback. As described in the section on movement execution, learning of motor skills involves learning both how to do an action and meta-knowledge about the action. Actions have associated expectations about their effect (meta-knowledge). Feedback about the actual effect provides information that can be used to refine the choice made next time (Fig. 6.20). Thus, during learning, feedback is used both to revise the present action and to revise the next action.

Choosing an action instruction on the basis of meta-knowledge is similar in process to choosing the working method used to maintain mental workload at an acceptable level. The choice of working method involves checking meta-knowledge about each method, to find which method has the properties best suited to the present situation. A similar process is also involved when developing a new cognitive working method: A person develops a working method, hoping (on the basis of a combination of meta-knowledge and mental simulation) that it will give the required result, and then revises the method on the basis of feedback about the actual effectiveness of what they do.

Independent Goals–Means. In coping with mental workload, and in developing cognitive processes while learning, several working methods may be used for meeting the same function/goal. Also the same behavior may be used to meet several goals. Thus, the link between goal and means must be flexible. The goal and means are independent in principle, although, after learning, particular working methods may become closely linked to particular goals. In the section on workload, the goal–means link was described as a point at which a decision between working methods is made on the basis of meta-knowledge.

It is generally the case (Sherrington, 1906/1947) that behavior at one level of organization transfers information about the goal to be met, and constraints on how it should be met, to the lower levels of behavior organization by which the goal is met, but not detailed instructions about how to meet it. How to carry out the function is decided locally, in the context at the time. As behavior is not dictated from above, but has local flexibility, human beings are not by nature well suited to following standardized procedures.

Changes in the Mode of Processing. Learning does not lead only to changes within a given mode of processing. A person may also change to a different mode of processing. If the task is consistent, than a person can learn to do the task in a more automatic way, that is, by using a simpler mode of processing. Inversely, when there is no fully developed working method or knowledge for meeting a given goal/function, then it is necessary to devise one. Thus the possibility or need for developing a simpler or more complex mode of processing depends both on a person's experience with the task and on the amount and types of regularity in the task. It may be possible through learning to change from any mode of processing to any other mode of processing, but two types of change are most typical: from more complex to simpler processing, or vice versa.

Someone may start a new task by developing a working method. But once the person has had an opportunity to learn the regularities in the task, the processing may become simpler. If the task and environment are sufficiently stable, the person may learn that making a choice between methods to meet a goal, or search for appropriate knowledge, is not necessary. In familiar stable situations, the working method may become so standardized that the person using it is not aware of goals or choices.

Alternatively, someone may start by learning parts of a task, and gradually become able to organize them together into a wider overview, or become able to choose behavior that is compatible with several cognitive functions. These changes depend on changes in processing efficiency. When someone first does a complex task, the person may start at the lowest levels of behavior organization, learning components of the task that will eventually be simple but that at first require all the person's problem-solving, attention and other processing resources. As the processing for doing these subtasks becomes simpler with learning, this releases processing capacity. This capacity can then be used for taking in larger segments of the task at the same time, so the person can learn about larger regularities in the task.

In general any cognitive function, and any subgoal involved in meeting it, may be met by any mode of processing, depending on the person's experience with the task, and the details of the circumstances at the moment. A task can become "automated" or flexible at any level of behavior organization, depending on the repetitions or variety of situations experienced. Thus in some tasks a person may learn to do the perceptual-motor components automatically but have to rethink the task each time at a higher level, as in a professional person using an office computer. In other tasks, "higher" levels of behavior organization such as planning may become automated whereas lower levels remain flexible, as in driving to work by the same route every day. It is not necessarily the case that "higher" levels of behavior organization are only done by more complex modes of processing such as problem solving, or vice versa.

As any of the main cognitive functions in a task could become so standardized that they are done automatically or unconsciously, this is the origin of so-called "shortcuts" in processing. Inversely, at any moment, a change in the task situation, such as a fault, may mean that what could previously be done automatically now has no associated standard working method, so problem solving is needed to find one. At any time, or at any point in the task, there is the potential for a change in the mode of processing. So care is needed, if an interface design strategy is chosen of providing displays that support only one mode of processing.

Some Training Implications

Gagné (e.g., 1977) first suggested the concept that different modes of processing are best developed by different training methods. It is not appropriate to survey these methods here, but some general points link to the general themes of this chapter.

Simple Processes. Training for simple processes needs to:

- Maximize the similarity to the real task (the transfer validity) of discriminations, integrations, and recodings that are learned until they become automatic, by using high-fidelity simulation.
- Minimize the need for changes in mode of processing during learning, by presenting the task in a way that needs little problem solving to understand.
- Ensure that trainees retain a feeling of mastery, as part of their meta-knowledge about the task activities, by avoiding training methods in which errors are difficult to recover from, and by only increasing the difficulty of the task at a rate such that trainees continue to feel in control.

Complex Processes. Tasks that involve building up an overview and using alternative strategies need more than simple repetition if they are to be learned with least effort. The status of errors is different in learning complex tasks. In training for simple discriminations, recodings, and motor tasks, the emphasis is on minimizing the number of errors made, so that wrong responses do not get associated with the inputs. By contrast, when learning a complex task, an "error" can have positive value as a source of information about the nature and limits of the task. Thus in learning complex tasks, the emphasis should be more on exploring the possibilities without negative consequences, in order to develop a variety of working methods and wide knowledge of the task alternatives. Flexibility could be encouraged by giving trainees:

- Guided discovery exercises, in which the aim is to explore the task rather than to achieve given aims.
- Recovery exercises in which people practice recovering from nonoptimal actions.
- Problem solving and planning exercises, with or without real time pressures.
- Opportunities to share with other trainees the discoveries made.
- Practice with considering alternative working methods, and with assessing the criteria for choosing between them.
- Practice with thinking about alternative "hypotheses" for the best explanation of events, or the best action.
- Practice with multitasking.
- Practice with using different methods for developing working methods, and with the case examples and recognition categories used.

A feature of cognitive skill is having a knowledge base that is closely linked to the cognitive processing that uses it, so that the knowledge is appropriately organized and easy to access. This suggests that knowledge is best learned as part of doing the task, not separately.

Training as Part of System Design. This chapter has mentioned several ways in which training needs interact with the solutions chosen for other aspects of the system design:

- The need for training, and the quality of interface or procedure design, may be inversely related.
- Skills are lost if they are not maintained by practice, so the amount of continuing training needed may be related to the extent of automation.

Difficulties and Errors

Errors occur when people are operating at the limits of modes of processing. Errors result from misuse of normally effective processes. The concept of relating error types to modes of processing was first suggested by Rasmussen (1982), although the scheme suggested here is somewhat different.

There are several points that the approach to complex tasks taken in this chapter suggests should be added to most error schemes. First, the notion of *error* needs to be expanded. In some simple tasks such as recoding it is possible to be wrong. But in control tasks and in complex tasks it is useful to think in terms of difficulty, or lowered effectiveness, rather than focusing on being wrong. For example, Amalberti's novice pilots (Fig. 6.28) were already qualified. They completed the task; they just did it less effectively than the more experienced pilots. Thus, as a basis for supporting people doing complex tasks, it is useful to look at factors that make the task more difficult, as well as factors that slow behavior down or increase errors.

Second, many error schemes assume that task behavior can be broken down into small independent units, each of which may be right or wrong. In probabilistic risk assessment (PRA) or human reliability assessment (HRA) techniques, behavior is segmented into separate units. A probability of error is assigned to each unit, and the total probability of human error for the combined units is calculated by addition or multiplication. But this chapter has stressed that human behavior in complex tasks does not consist of independent units. The components of complex behavior are organized into an integrated interdependent structure. This means that, although PRA/HRA techniques are useful for practical purposes, any attempt to increase their fundamental validity while retaining an "independent units" model of behavior is doomed to failure (Hollnagel, 1993).

Third, as the processes of building up and using an overview are often not included in models of human processing, the related errors are also often not discussed, so they are the focus here. This section briefly suggests some of the ways in which performance can be weaker (e.g., see Bainbridge, 1993b).

Discriminations. Decisions made under uncertainty cannot always be right, and are more likely to be wrong if the evidence on which they are based is ambiguous or incomplete. Incorrect expectations about probabilities and incorrect biases about payoffs can also increase error rates. People make errors such as misattributing risk, importance, or urgency; ignoring a warning that is frequently a false alarm; or seeing what they expect to see. Some people when under stress refuse to make decisions involving uncertainty.

Recodings. There are many sorts of error that can be attributed to mistranslations. Sometimes the person does not know the coding involved. People are more likely to make coding errors when they have to remember which specific code translation to use in which circumstances. Difficult codes are often ambiguous or inconsistent. The salience of some stimuli may give improper emphasis to them or to their most obvious meaning.

Sequences. The items that need to be retained in working memory during a sequence of behavior may be forgotten within half a minute, if other task processing distracts or interrupts the rehearsal needed to remember the items.

In an overlearned sequence, monitoring/supervision of parts of the activity may be omitted. This can lead to "slips" in performance, or to rigid behavior that causes difficulties when the environment changes and adaptive behavior is needed.

Overview and Behavior Organization. There may be errors in organizing the search for information. People may only attend to part of the task information, fail to keep up-to-date with changes in the environment, or look at details without taking an overall view. They may not get information which there is a cost on getting. They may only look for information that confirms their present interpretation of the situation ("confirmation bias"). In team work, people may assume without checking that another member of the team, particularly someone with higher status, has done something which needed doing.

There may also be errors in the allocation of time between tasks, which may lead to omissions or repetitions. People may react to events rather than anticipating events and how to deal with them. They may not apply available strategies in a systematic way. They may shift between subtasks, without relating them to the task as a whole ("thematic vagabonding," Doerner, 1987). They may break the task down into sub-problems in an inadequate way, or fail to devise intermediate subgoals, or they may continue to do parts of the task which they know how to do ("encystment," Doerner, 1987). Under high workloads, people may delay decisions in the hope that it will be possible to catch up later, or they may cycle through thinking about the task demands without taking any action.

The overview influences a person's biases about what will happen and what to do about it. If the overview is incorrect, this can lead to inappropriate behavior or expectations. People who have completed a subtask, and so completed a part of their own overview, may fail to tell other members of the team about this. Once people have built up a complete and consistent overview, it may be difficult to change it when it turns out to be inadequate ("perceptual set"). The overview may also be lost completely if a person is interrupted.

Use of Knowledge. People's knowledge of all types may be incomplete or wrong, so they make incorrect inferences or anticipations. There may be problems with assumed shared knowledge in a team if team members change.

A person may have an incorrect or incomplete representation of the device they are using. For example, the person may not know the correct causalities or interactions, or may not be able to represent correctly the development of events over time. Or someone may use an inappropriate category in recognition-primed decisions or case in case-based reasoning.

Knowledge about probabilities may be incorrect, or used wrongly. People may be under- or overconfident. They may have a "halo effect," attributing the same probabilities to unrelated aspects. They may give inappropriate credence to information or instructions from people of higher status. Different social groups—for example, unions, management, and the general public—may have different views on the risks and payoffs of particular scenarios.

This list of human weaknesses should not distract from two important points. One is that people can be good at detecting their errors and recovering from them, if they are given an interface and training that enable them to do this. Therefore design to support recovery should be included in cognitive task analysis.

The second point is that care is needed with the attribution of responsibility for faults. Although it may be a given individual who makes an error, the responsibility for that error may be attributed elsewhere, to poor equipment or system design (training, workload, allocation of function, teamwork, organizational culture).

CONCLUSION

There are several integrative concepts in this chapter.

Cognitive Goals. In complex tasks people use cognitive goals when implementing task goals. A person's cognitive goals are important in organizing the person's behavior, in directing attention to parts of the task, in choosing the best method for meeting a given goal, and in developing new working methods. The cognitive goals might be met in different ways in different circumstances, so the goals and the processes for meeting them can be independent. For example, flying an aircraft involves predicting the weather, and this may be done in different ways before and during the flight.

Contextual Overview. People build up an overview of understanding and planning, which then acts as the context for later activity. The overview provides data, expectations and values, and the criteria for deciding what would be the next best thing to do and how to do it.

Goal–Means Independence and Meta-Knowledge. Meta-knowledge is knowledge about knowledge, such as the likelihood of alternative explanations of what is happening, or the difficulty of carrying out a particular action. Alternative working methods have associated with them meta-knowledge about their properties. Decisions about how best to meet a particular aim are based on meta-knowledge, and are involved in adapting behavior to particular circumstances and in the control of multitasking and mental workload and in learning.

Modes of Processing. As well as using different working methods, people may use different modes of processing, such as knowing the answer by association or thinking out a new working method. The mode of processing used varies from moment to moment, depending on the task and the person's experience.

Modeling Human Behavior

Basing HF/E on an analysis of behavior into small independent units fits well with a "sequential stages" concept of the underlying structure of human behavior. But a sequential stages model does not include many of the key features of complex tasks such as flying and air traffic control. Complex behavior is better described by a contextual model, in which processing builds up an overview that determines what processing is done next and how, which in turn updates the overview, and so on. In this mechanism for behavior organization, choices about what to do and how to do it depend on details of the immediate situation interacting with the individual's nature and previous experience.

The aspects missing from many sequential stages models are:

- The goal-oriented nature of behavior, and the independence of goals from the means by which they are met.
- The continuing overview.
- The flexible sequencing of cognitive activity, and the organization of multitasking.
- The knowledge base, and the resulting predictions, anticipations, and active search for information that are part of top-down processing.

Some of these aspects require a fundamental change in the nature of the model used. The most important aspect to add is the overview, as all cognitive processes are done within the context provided by this overview, and the sequence in which they are done is determined by what is in the overview.

A simple version of a contextual model has been suggested in Figs. 6.22 and 6.24. These figures can act as an aide-mémoire about contextual processing, but any one-page representation can only indicate some features of what could be expressed. These simple figures do not make explicit important aspects such as:

- Risky decision making and the effects of biases.
- Goal orientation of behavior.
- Typical sequences of activity.
- Different modes of processing, including devising new working methods.
- Use of meta-knowledge.

Perhaps the most important disadvantage of the one-page contextual model will be felt by people who are concerned with tasks that are entirely sequential, rather than cyclic as in flying or air traffic control. But I would argue that although dependencies may define the order in which some parts of a task are done, it could still be useful, when designing to support sequential tasks, to consider the task sequence as a frame for structuring the overt behavior, whereas the underlying order of thinking about task aspects may be more varied (cf. Fig. 6.28).

The Difficulty of HF/E

Contextual processing underlies two types of difficulty for HF/E. One group of issues is concerned with HF/E techniques. As indicated earlier, the overview suggests the need for several additions to HF/E techniques:

- Considering the codings used in the task as a whole, rather than for isolated subtasks.
- Orienting cognitive task analysis toward the cognitive goals or functions to be met, as an intermediary between the task goals and the cognitive processing. (Analyzing either goals or working methods alone is necessary but not sufficient.)
- Designing the interface, training, and allocation of function between people and machines, to support the person's development and use of the contextual overview, alternative strategies, and the processes involved in the development of new working methods.
- Extending human error schemes to include difficulties with the overview and with the organization of sequences of behavior.

The second group of issues is concerned with a fundamental complexity problem in human behavior and therefore in HF/E. Human behavior is adapted to the particular circumstances in which it is done. This does not make it impossible to develop a general model of human behavior, but it does make it impossible to predict human behavior in detail. Predicting human behavior is like weather prediction: It is not possible to be right, but it is possible to be useful. Any HF/E answer is always going to be context sensitive. The continuing complaint of HF/E practitioners, that researchers do not provide them with what they need, is a consequence of the fundamental nature of human behavior. Specific tests of what happens in specific circumstances will always be necessary. What models of human behavior can provide is, not the details, but the key issues to focus on when doing such tests or when developing and applying HF/E techniques.

REFERENCES

Allport, D. A., Antonis, B., & Reynolds, P. (1972). On the division of attention: A disproof of the single channel hypothesis. *Quarterly Journal of Experimental Psychology, 24,* 225–235.

Amalberti, R. (1992). *Modèles d'activité en conduite de processus rapides: implications pour l'assistance á la conduite.* Unpublished doctoral thesis, University of Paris.

Atkinson, W. H., Crumley, L. M., & Willis, M. P. (1952). *A study of the requirements for letters, numbers and markings to be used on trans-illuminated aircraft control panels. Part 5: the comparative legibility of three fonts for numerals* (Report No. TED NAM EL-609, part 5). Naval Air Material Center, Aeronautical Medical Equipment Laboratory.

Bailey, R. W. (1989). *Human performance engineering: A guide for system designers* (2nd ed.). London: Prentice Hall.

Bainbridge, L. (1974). Problems in the assessment of mental load. *Le Travail Humain, 37,* 279–302.

Bainbridge, L. (1978). Forgotten alternatives in skill and workload. *Ergonomics, 21,* 169–185.

Bainbridge, L. (1983). Ironies of automation. *Automatica, 19,* 775–779.

Bainbridge, L. (1988). Types of representation. In L. P. Goodstein, H. B. Anderson, & S. E. Olsen (Eds.), *Tasks, errors and mental models* (pp. 70–91). London: Taylor & Francis.

Bainbridge, L. (1991). Multiplexed VDT display systems. In G. R. S. Weir & J. L. Alty (Eds.), *Human-computer interaction and complex systems* (pp. 189–210). London: Academic Press.

Bainbridge, L. (1993a). *Building up behavioural complexity from a cognitive processing element* (p. 95). London: Department of Psychology, University College London.

Bainbridge, L. (1993b). *Difficulties and errors in complex dynamic tasks.* Unpublished manuscript, University College London.

Bainbridge, L. (1993c). Types of hierarchy imply types of model. *Ergonomics, 36,* 1399–1412.

Barnard, P. (1987).Cognitive resources and the learning of human–computer dialogues. In J. M. Carroll (Ed.), *Interfacing thought: Cognitive aspects of human–computer interaction.* Cambridge, MA: MIT Press.

Beishon, R. J. (1974). An analysis and simulation of an operator's behaviour in controlling continuous baking ovens. In E. Edwards & F. P. Lees (Eds.), *The human operator in process control* (pp. 79–90). London: Taylor & Francis.

Benson, J. M. (1990). *The development of a model of multi-tasking behaviour.* Unpublished doctoral thesis, Ergonomics Unit, University College London.

Bisseret, A. (1970). Mémoire opérationelle et structure du travail. *Bulletin de Psychologie, XXIV,* 280–294. English summary in *Ergonomics, 14,* 565–570 (1971).

Brennan, A. C. C. (1987). *Cognitive support for process control: designing system representations.* Unpublished master's thesis, Ergonomics Unit, University College London.

Briggs, G. E., & Rockway, M. R. (1966). Learning and performance as a function of the percentage of pursuit tracking component in a tracking display. *Journal of Experimental Psychology, 71,* 165–169.

Craik, F. I. M., & Lockhart, R. S. (1972). Levels of processing: A framework for memory research. *Journal of Verbal Learning and Verbal Behaviour, 11,* 671–684.

Crossman, E. R. F. W. (1959). A theory of the acquisition of speed-skill. *Ergonomics, 2,* 153–166.

Crossman, E. R. F. W., & Goodeve, P. J. (1963, April) *Feedback control of hand-movement and Fitts' Law.* Communication to the Experimental Psychology Society, University of Oxford.

Davis, D. R. (1966). Railway signals passed at danger: The drivers, circumstances and psychological processes. *Ergonomics, 9,* 211–222.

Doerner, D. (1987). On the difficulties people have in dealing with complexity. In J. Rasmussen, K. D. Duncan, & J. Leplat (Eds.), *New technology and human error* (pp. 97–109). Chichester: Wiley.

Easterby, R. S. (1970). The perception of symbols for machine displays. *Ergonomics, 13,* 149–158.

Ellis, A. W. (1993). *Reading, writing and dyslexia: A cognitive analysis.* Hillsdale NJ: Lawrence Erlbaum Associates.

Eysenck, M. W., & Keane, M. T. (1990). *Cognitive psychology: A student's handbook.* Hillsdale, NJ: Lawrence Erlbaum Associates.

Fitts, P. M. (1954). The information capacity of the human motor system in controlling the amplitude of movement. *Journal of Experimental Psychology, 47,* 381–391.

Fitts, P. M., & Deininger, R. L. (1954). S-R compatibility: Correspondence among paired elements within stimulus and response codes. *Journal of Experimental Psychology, 48,* 483–492.

Fitts, P. M., & Jones, R. E. (1961a). Analysis of factors contributing to 460 "pilot error" experiences in operating aircraft controls (Memorandum Report No. TSEAA-694-12, Aero Medical Laboratory, Air Materiel Command, Wright Patterson Air Force Base, Dayton, Ohio, July 1, 1947). In H. W. Sinaiko (Ed.), *Selected papers on human factors in the design and use of control systems* (pp. 332–358). New York: Dover. (Original work written 1947)

Fitts, P. M., & Jones, R. E. (1961b). Psychological aspects of instrument display. I: Analysis of 270 "pilot error" experiences in reading and interpreting aircraft instruments (Memorandum Report No. TSEAA-694-12A, Aero Medical Laboratory, Air Materiel Command, Wright Patterson Air Force Base, Dayton, Ohio, October 1, 1947). In H. W. Sinaiko (Ed.), *Selected papers on human factors in the design and use of control systems* (pp. 359–396). New York: Dover. (Original work written 1947)

Fleishman, E. A. (1975). Taxonomic problems in human performance research. In W. T. Singleton & P. Spurgeon (Eds.), *Measurement of human resources* (pp. 49–72). London: Taylor & Francis.

Folkhard, S. (1990). Circadian performance rhythms: Some practical and theoretical implications. In D. E. Broadbent, J. Reason, & A. Baddeley (Eds.), *Human factors in hazardous situations* (pp. 543–553). Oxford: Clarendon Press.

Gagné, R. M. (1977). *The conditions of learning* (3rd ed.). New York: Holt, Rinehart & Winston.

Gallaher, P. D., Hunt, R. A., & Williges, R. C. (1977). A regression approach to generate aircraft predictor information. *Human Factors, 19,* 549–555.

Gibson, J. J. (1950). *The perception of the visual world.* Boston: Houghton Mifflin.

Gray, J. A., & Wedderburn, A. A. I. (1960). Grouping strategies with simultaneous stimuli. *Quarterly Journal of Experimental Psychology, 12,* 180–184.

Green, R. (1990). Human error on the flight deck. In D. E. Broadbent, J. Reason, & A. Baddeley (Eds.), *Human factors in hazardous situations* (pp. 55–63). Oxford: Clarendon Press.

Grether, W. F. (1949). The design of long-scale indicators for speed and accuracy of quantitative reading. *Journal of Applied Psychology, 33,* 363–372.

Hayes-Roth, B., & Hayes-Roth, F. (1979). A cognitive model of planning. *Cognitive Science, 3,* 275–310.

Helson, H. (1964). *Adaptation-level theory.* New York: Harper and Row.

Hoc, J.-M. (1988). *Cognitive psychology of planning.* London: Academic Press.

Hollnagel, E. (1993). *Reliability of cognition: Foundations of human reliability analysis.* London: Academic Press.

Howland, D., & Noble, M. (1955). The effect of physical constraints on a control on tracking performance. *Journal of Experimental Psychology, 46*, 353–360.

Hukki, K., & Norros, L. (1993). Diagnostic orientation in control of disturbance situations. *Ergonomics, 35*, 1317–1328.

Hunt, D. P., & Warrick, M. J. (1957). *Accuracy of blind positioning of a rotary control* (USAF WADC Tech. Notes 52-106). U.S. Air Force.

Johansson, G., Aronsson, G., & Lindström, B. G. (1978). Social psychological and neuroendocrine stress reactions in highly mechanised work. *Ergonomics, 21*, 583–599.

Kahneman, D., Slovic, P., & Tversky, A. (Eds.). (1982). *Judgement under uncertainty: Heuristics and biases*. New York: Cambridge University Press.

Klein, G. A. (1989). Recognition-primed decisions. In W. B. Rouse (Ed.), *Advances in man-machine systems research* (Vol. 5, pp. 47–92). Greenwich, CT: JAI Press.

Leplat, J., & Bisseret, A. (1965). Analyse des processus de traitement de l'information chez le controleur de la navigation aérienne. *Bulletin du CERP, XIV*, 51–67. English translation in *Controller: IFACTCA Journal, 5*, 13–22 (1965).

Lindsay, P. H., & Norman, D. A. (1972). *Human information processing*. New York: Academic Press.

Marshall, E. C., Scanlon, K. E., Shepherd, A., & Duncan, K. D. (1981). Panel diagnosis training for major hazard continuous process installations. *Chemical Engineer, 365*, 66–69.

Miller, G. A. (1956). The magical number seven, plus or minus two: Some limits on our capacity for processing information. *Psychological Review, 63*, 81–97.

Miller, G. A., Heise, G. A., & Lichten, W. (1951). The intelligibility of speech as a function of the context of the test materials. *Journal of Experimental Psychology, 41*, 329–335.

Moray, N. P., Hiskes, D., Lee, J., & Muir, B. M. (1995). Trust and human intervention in automated systems. In J.-M. Hoc, P. C. Cacciabue, & E. Hollnagel (Eds.), *Expertise and technology: Cognition and human-computer cooperation* (pp. 183–194). Hillsdale, NJ: Lawrence Erlbaum Associates.

Oborne, D. (1995). *Ergonomics at work* (3rd ed.). Chichester: Wiley.

Pheasant, S. (1991). *Ergonomics: Work and health*. London: Macmillan.

Posner, M. I., & Rossman, E. (1965). Effect of size and location of informational transforms upon short-term retention. *Journal of Experimental Psychology, 70*, 496–505.

Rasmussen, J. (1982). Human errors: A taxonomy for describing human malfunction in industrial installations. *Journal of Occupational Accidents, 4*, 311–335.

Reinartz, S. J. (1989). Analysis of team behaviour during simulated nuclear power plant incidents. In E. D. Megaw (Ed.), *Contemporary ergonomics 1989* (pp. 188–193). London: Taylor & Francis.

Rouse, W. B. (1980). *Systems engineering models of human–machine interaction*. Amsterdam: North Holland.

Rumelhart, D. E. (1977). Towards an interactive model of reading. In S. Dornic (Ed.), *Attention and performance VI*. Hillsdale, NJ: Lawrence Erlbaum Associates.

Sage, A. P. (1981). Behavioural and organizational considerations in the design of information systems and processes for planning and decision support. *IEEE Transactions on Systems, Man and Cybernetics, SMC-11*, 640–678.

Samurçay, R., & Rogalski, J. (1988). Analysis of operator's cognitive activities in learning and using a method for decision making in public safety. In J. Patrick & K. D. Duncan (Eds.), *Training, human decision making and control*. Amsterdam: North-Holland.

Savin, H. B., & Perchonock, E. (1965). Grammatical structure and the immediate recall of English sentences. *Journal of Verbal Learning and Verbal Behaviour, 4*, 348–353.

Shaffer, L. H. (1973). Latency mechanisms in transcription. In S. Kornblum (Ed.), *Attention and performance IV* (pp. 435–446). London: Academic Press.

Sheridan, T. B., & Ferrell, W. R. (1974). *Man-machine systems: Information, control and decision models of human performance*. Cambridge, MA: MIT Press.

Sherrington, C. (1947). *The integrative action of the nervous system*. Cambridge: Cambridge University Press. (Original work published 1906)

Smith, S. L., & Thomas, D. W. (1964). Color versus shape coding in information displays. *Journal of Applied Psychology, 48*, 137–146.

Sperandio, J. C. (1970). *Charge de travail et mémorisation en contrôle d'approche* (Report No. IRIA-CENA, CO 7009, R24). Institut de Recherche en Informatique et Aeronautique, Paris.

Sperandio, J. C. (1972). Charge de travail et régulation des processus opératoires. *Le Travail Humain, 35*, 86–98. English summary in *Ergonomics, 14*, 571–577 (1971).

Teichner, W. H., & Krebs, M. J. (1974). Laws of visual choice reaction time. *Psychological Review, 81*, 75–98.

Wickens, C. D. (1984). Processing resources in attention. In R. Parasuraman & D. R. Davies (Eds.), *Varieties of attention*. London: Academic Press.

Wilson, J. R., & Corlett, E. N. (1995). *Evaluation of human work: A practical ergonomics methodology* (2nd ed.). London: Taylor & Francis.

Winograd, T. (1972). *Understanding natural language*. Edinburgh: Edinburgh University Press. Reprinted from *Cognitive Psychology, 1972, 3,* 1–195.

Yntema, D. B., & Mueser, G. E. (1962). Keeping track of variables that have few or many states. *Journal of Experimental Psychology, 63,* 391–395.

7

Automation in Aviation: A Human Factors Perspective

René R. Amalberti
Val de Grâce Military Hospital, Paris, and Institut de Médecine Aéronautique du Service de Santé des Armées (IMASSA), Brétigny-sur-Orge cedex, France

Aircraft automation is part of a vast movement to improve and control performance and risks in our so-called "advanced societies." Beyond the consequences for crews, automation is incorporated into the global evolution of these advanced societies as a tool providing people with more comfort and happiness, better performance, and fewer problems. But automation is also part of the aviation business and gives rise to permanent national and international competition between manufacturers. Inevitably accidents, incidents, and successes feed this competition and are overexploited by the press.

From a technical point of view, any new technology calls for a period of adaptation to eliminate residual problems and to allow users to adapt to it. This period can take several years, with successive cycles of optimization for design, training, and regulation. This was the case when jets replaced propeller planes. The introduction of aircraft like the B707 or the B727 were major events in aviation. These cycles are invariably fraught with difficulties of all types including accidents. People do not yet know how to optimize complex systems and reach the maximum safety level without field experience. The major reason for this long adaptive process is the need for harmonization between the new design on one hand and the policies, procedures, and moreover the mentalities of the whole aviation system on the other hand. This harmonization goes far beyond the first months or years following the introduction of the new system, both because of the superimposition of old and modern technologies during several years, and because of the natural reluctance of people and systems to change.

This is also true for the automation of cockpits. The current transition phase from classic aircraft to glass cockpits has been marked by a series of major pilot-error-induced incidents and accidents. Some of these human errors have been recognized as facilitated by technical drawbacks in the design, operation, and/or training of automated aircraft.

An important question for the aviation community is to determine what part of these identified design, operations, procedures, and training drawbacks would disappear with appropriate personnel training and regulation but without design change.

The second part of the question is which of the problems are more difficult to treat and will require changes in automation design and policy? However, one should note that there is no precise timetable for adaptation. Policies are often revisited under the pressure of an accident; for example the recent U.S. public reactions to the Valujet accident in Miami led the FAA to considerably strengthen surveillance of airlines in a highly competitive market where seat prices, services, and sometimes safety practices may be relaxed.

Human factors research in cockpit automation has been greatly influenced from this complex and highly emotional climate. There are dozens of human factors research teams in the world involved in the study of automation-induced problems. They have proposed series of diagnostics of automation drawbacks for more than a decade (two decades if the work done in automation within the nuclear industry is included). The concepts of irony of automation (Bainbridge, 1987), clumsy automation (Wiener, 1989), and insufficient situation/mode awareness (Endsley, 1996; Sarter & Woods, 1992) are good examples of these diagnostics. Theorists have also suggested generic solutions to improve the design of automated systems, such as the concepts of human-centered automation (Billings, 1997), user-friendly design, ecological interfaces, ecological design, and ecological safety (Flach, 1990; Rasmussen &Vicente, 1989; Wioland & Amalberti, 1996). This profusion of research and ideas contrasts with the fact that human factors research does not have much application within the industry. It is the goal of this chapter to understand better the reasons for this paradox.

Of course, that is not to say that nothing exists in the industry regarding human factors. First, the growing adhesion of industry to cockpit resource management (CRM) courses is more than a success for psychosociologists. Second, human factors are intensively used by manufacturers and regulators. How can a manufacturer or an operator ignore the end user (the pilot) if it wants to please the customer and sell a product?

However, one must question how effectively human factors are used in design and operations and try to comprehend why these human factors are not the ones suggested by the academics. Regardless of their success, CRM courses have encouraged a relative capsulation of human factors theoretical concepts within flight operations. In the rest of the industry, there is still a belief that human factors relies much more on good sense and past experience than on theory; "We are all human, so we can all do human factors" (Federal Aviation Administration [FAA] human factors report: Abbott et al., 1996, p. 124).

One further step in this analysis should allow us to comprehend what model of end user designers and operations staffs had (and/or still have) in mind. According to the nature of this end user model, we suspect that the design process will be different because the design is oriented toward the replacing of end users' weaknesses.

In sum, speaking about automation is like addressing a very sensitive question that goes beyond the technical and "objective" analysis of automation-induced problems. The problems include those of ethical questions and reflect a persisting deficit in communication between the human factors academy and industry.

Fortunately, this situation is not frozen. Several recent initiatives by regulatory authorities (FAA Human Factors Team, JAA Human Factors Steering Group, OCAI' s International Human Factors Seminar) are casting new light on the problem. For safety reasons, and also because of a growing tendency of managers and politicians to fear the legal consequences and public disgrace if new crashes occur, authorities ask more and more human factors academies and industry staff to be more open-minded and to bridge their disciplines. Note that bridging does not mean unidirectional efforts. Industry should incorporate more academic human factors, but the academic realm should equally improve its technical knowledge, consider industry constraints, and consider the feasibility and cost of suggested solutions.

Because of this big picture, and because numerous (good) reports, books and chapters have already been devoted to the overview of automation-induced problems in commercial and corporate aviation (see, e.g., Amalberti, 1994; Billings, 1997; Funk, Lyall, & Riley, 1995; Parasuraman & Mouloua, 1996; Sarter & Woods, 1992, 1995; Wiener, 1988; Wise et al., 1993), this chapter gives priority to identifying mismatches between industry and the theorists and to suggesting critical assessment of human factors solutions, rather simply listing drawbacks and presenting new solutions as miracles.

The chapter is divided into three sections. The first section describes the current status of aircraft automation, why it has been pushed in this direction, what end user model triggered the choice, how far can the results be considered a success by the aviation technical community, and how drawbacks were explained and controlled. The second section starts from a brief description of automation drawbacks, then turns to the presentation of a causal model based on the mismatch of the pilot's model between industry and academia. The third and last section focuses on solutions suggested by the theorists to improve the human–machine interface or to assist industry in improving jobs. It also considers the relevance, feasibility, and limitations of these suggestions.

AUTOMATION: A SUCCESS STORY FOR AVIATION

Today's automated aircraft have a variety of features ranging from automatons, automated tools and pilot support systems, numerous and complex subsystems, and highly integrated components. Changes in the mode of information display to the pilot have been equally important. Use of cathode ray tube (CRT) screens has made for more succinct information presentation and a better emulation of the outside world through map displays.

These new cockpits have been christened "glass cockpits" to reflect this better representation of the outside world (the similarities between the instrument panel and the outside world) and the use of electronic display panels. Aircraft automation and computerization are now used on all types of two-person crew aircraft.

Premises of Automation in the Early 1970s: Emergence of an End User's Model Based on Common Sense

At the end of the 1960s, aviation was facing a big challenge. The early age of aviation with pilot heroes was over. Economic big growth and promises of a tremendously expanding world market were pushing the industry to demonstrate its capability to carry massive numbers of passengers, safely and routinely.

Human limitations were recognized as potential barriers to the attainment of this goal. Research on the related physiological and psychological human limitations was echoing the feeling that humans were "intelligent but fragile" machines. Among the strongest limitations were the limited capacity of attention and resources, which led to an oversimplified end user model with a single-channel metaphor. For all these reasons, individual and group performance were seen as highly unreliable.

Not surprisingly, the growing demands on performance placed by increasing traffic were resulting in calls for a new definition of the pilot role. Automation and prosthesis (intelligent assistance) were logically developed to bypass human limitations in order to meet the new challenges just described: routinely carrying more passengers, more safely, more efficiently, and more cheaply.

To sum up, designers originally hoped to:

- Reduce workload and difficulty of carrying out the phases of the flight.
- Relieve pilots of having to perform repetitive sequences that are unrewarding and for which human beings in their inconsistency can be at their best or their worst.
- Endow pilots with the gratifying part of their jobs: decision making.

Automation: A Success Story

Automation in aviation has reached all the goals assigned in the 1960s. There is a consensus that nobody should design a modern aircraft with much less automation than the current glass-cockpit generation. The global performance of systems has been multiplied by a significant factor: Cat III landing conditions have been considerably and safely extended, lateral navigation (LNAV) has been tremendously enhanced, computer-support systems aid the pilot in troubleshooting failures, and so forth.

The economic benefits have been equally remarkable: fuel saving, enhanced reliability, ease in maintenance support, reduction in crew complement, and tremendous benefits in the reduced training time. The duration of the transition course for type-rating on modern glass cockpits is at least 1 week shorter than for the old generation and has considerably improved cross-crew qualification among similar types of aircraft. For example, the transition course between A320 and A340 asks for an 11-day course instead of a 22-day regular course. Another benefit relies on the zero-flight training concept, which should aim soon at type-rating pilots on a machine without any flight (flight is of course the most expensive part of training). However, some U.S. airlines have come recently to the conclusion that crews are not learning enough in transition and are considering making the transition course much longer (fallout of Cali accident investigation).

Automation has also contributed greatly to the current level of safety. Safety statistics show that glass cockpits have an accident rate half that of the previous generation of aircraft. The global level of safety now appears to be equal to that of the nuclear industry or railways (in Western Europe). The only apparent concern is that this (remarkable) level of safety has not improved since the 1970s. Of course, one should also note that the current level of safety is not due only to automation, but to the overall improvement of systems, including air traffic control (ATC), airports, and the use of simulators in training. But one has also to acknowledge that the current plateau of safety is not solely a function of automation problems. There is growing evidence that increases in traffic, in the complexity of organizations and management, and in the variety of corporate and national cultures all play significant roles in the plateau and need more consideration in future design and training.

Last, but not least, the regulation of aviation safety relies on the acceptance of a certain risk of accident that compromises between safety costs and safety benefits; this level of accepted risk is close to the one observed at the present time. Therefore manufacturers and managers do not feel guilty and consider that they meet the targets (better efficiency, less cost, with a high level of safety). Objectively, these people are correct. The reader will easily understand that this self-satisfaction does not promote room for dialogue when criticisms of automation are expressed.

AUTOMATION FOR THE BEST AND THE WORST: THE PILOTS' AND THE THEORISTS' PERSPECTIVES

No important technical change can occur without important social change. Because automation is able to carry on most flying tasks, and because the use of automation has been necessary to reach the goal of better performance and safety, it was explicitly assumed by designers that pilots would follow the technical instructions given by computers, and use automation as much as possible. Pilots would have to adapt to their new position, with a noble role in programming the system, monitoring execution by automation, and only overriding in case of system malfunction.

This adaptation has been much more difficult than expected. Pilots' unions rapidly complained about advanced automation. During the 1980s, many pilots refused to transition to the glass cockpit for as long as it was possible in their company. The social reaction to glass has also been exacerbated by pilots' unions because of the reduction of flight-crew members (even though this was only partially true, because the reduction of crew was accomplished in the DC9 and B737). It is probably not a surprise that a series of glass cockpit accidents occurred in a country (France) where the protests and strikes against the introduction of glass had been stronger than anywhere in the world (A320 Habsheim, A320 Mont-Saint-Odile, B747-400 Papeete; see Gras, Moricot, Poirot-Delpech, & Scardigli, 1994).

As noted in the introductory section, it is not the main goal of this chapter to list all the drawbacks of automation. However, a summary of these problems is given next with cross-references to chapters and books already published.

Temporary Problems Due to Change in Habits and Procedures

The human factors problems of pilots flying glass cockpits can be divided into two categories. The first addresses transient difficulties due to changes in habits. First, the transition course from nonglass to glass is very demanding (see Pelegrin & Amalberti, 1993, or Amalberti, 1994, for a complete discussion of the problem). There is a significant handicap of age and experience on multicrew, classic aircraft. Crew coordination calls for even more effort than before (Bowers et al., 1994). Information is in English, and English is far from being the native language of all the pilots in the world. Another series of ab initio problems in the glass cockpit comes with the revolution in flight management from the controlling computer called the FMC. The introduction of the FMC has generated two types of side effects: the consequence of errors has been shifted into the future and aids can turn into traps. Database systems, for example, have a fantastic memory for beacons, airports, and the SIDs and STARs associated with standard takeoffs and landings. There have been several reported incidents where pilots, after a go around or a modification to the flightpath on final approach, displayed the procedure beacon on the screen and persisted erroneously for a fairly long time. The Thai A310 accident near Kathmandu, Nepal, in 1992 and the American Airlines B767 accident near Cali in 1995 are excellent examples of this kind of problem.

In addition, the use of databases is often a source of confusion between points with similar names or incidents related to point coordinate errors within the database itself (Wiener, Kanki, & Helmreich, 1993). Reason (1990) termed the latter *resident pathogen errors* to emphasize that multiple forms of these errors exist in the database but remain undetected until the point is used, just the way that an infection can be lodged in the body for a long time before a sudden outbreak of disease (i.e., its long period of incubation). Most of these drawbacks tend to disappear with experience on glass and with appropriate training, namely, dedicated glass cockpit crew-resource management courses (Wiener et al., 1993).

More Permanent Problems: Poor Situation Awareness

We saw earlier that system planning was both attractive and time-consuming. Familiarization with onboard computers often prompts people to assume that they can and should always use them as intermediaries. Pilots tend to get involved in complex changes to programmed parameters in situations where manual override would be the best way to insure flight safety. Fiddling with the flight management system (FMS) makes pilots lose their awareness of the passing of time and, further, their awareness of the situation and of the flight path (Endsley, 1996; Sarter & Woods, 1991).

This type of problem occurs frequently in final approach with runway changes, and has already been a causal factor in numerous incidents (see excerpts of ASRS quoted in the FAA human factors report, Abbott et al., 1996). These difficulties of reprogramming in real time are indicative of a more serious lack of comprehension that Wiener (1988), and then Woods, Johannesen, Cook, and Sarter (1994) termed "a clumsy system logic," which in any case is a logic that differs greatly from that of a pilot.

The pilot evaluates the computer through his or her own action plan by using what he or she knows about how automated equipment functions. Most pilots lack a fundamental grasp of the internal logic of automation and evaluate the gap between their expectations (governed by what they would do if they were in command) and what the computer does. This analysis is sufficient in most cases, but can rapidly become precarious if the chasm widens between the action plan and the behavior of the machine. This is frequently observed while using autopilot modes that combine or follow one another automatically according to the plane's attitude without any action by the pilot (automatic reversion modes). Sarter and Woods (1991, 1992, 1995), in a series of studies about autopilot mode confusion, have shown the limitations of human understanding of such reversions in dynamic situations.

Several studies of recent incidents show that the human operator responds poorly in such conditions:

- He hesitates about taking over because he will accuse himself of not understanding before accusing the machine of faulty behavior. Trying to do his best, he does not try to reason on the basis of mental procedures, but on his own most efficient capacity (based on logical rules). This change in reasoning is time consuming and resource-depleting (to the detrement of coordination with the human copilot) and often aggravates the situation by postponing action.
- In many cases, lacking knowledge and time, he accepts much greater drift than he would from a human copilot.
- Finally, when he is persuaded that there is an automation problem, the human operator only tends to override the automated procedure that deviates from the goal;

he does not override the others, and he is willing to invest more attention and be less confident in the subsequent phases of the flight (Lee & Moray, 1994).

In short, there are no problems when the automated procedure plan and the pilot's plan coincide. But difficulties surface rapidly as soon as plans differ; different modes of functioning do not facilitate efficient judgment of the difference, and the automation benefits (unjustly) from greater latitude than a human crewmate.

Inaccuracy of Pilot Model in the Minds of Designers

Many of the problems just discussed in coupling human and automated systems come from a poor model of the pilot guiding the design process. This section tries to summarize the weaknesses of this model. We have seen earlier that the reference pilot model in the minds of designers has long been of an "intelligent but fragile and unreliable" partner. Poor performances were long attributed to the limited resources and divided attention (sensitivity to overload). When the cognitive aspects of the flying tasks became more important because of the growing automation and assistance systems, designers added as a characteristic of the pilots model a series of biases related to decision making and incorrect mental representation (leading to poor situation awareness). Note that all the contents of such models are not spontaneous generations. They come from psychologists and human factors specialists, but they are—in this case—extremely oversimplified, like cartoons, and serve as Emperor's new clothes to the technical community for the same idea: Pilots cannot do the job as well as technique can or could do. Pilots need assistance, and when task requirements are too high or too complex, they need to be replaced.

No doubt, these oversimplistic pilots models are extremely incorrect when considering the knowledge existing among theorists. The dynamic control model of cognition, characteristic of pilots' activities—but also of many other situations—has been extensively described in a series of studies not limited to the aviation field (see Cacciabue, 1992; Hoc, 1996; Hollnagel, 1993; Rasmussen, 1986). There is growing evidence that dynamic cognition, regardless of the truth of limitations, is self-protected against the risk of losing control by a series of extremely efficient mechanisms. Moreover, humans are using errors to optimize these mechanisms.

The lessons from such models are twofold: First, dynamic cognition is continuously tuning a compromise between contradictory dimensions: reaching goals with the best objective performance, with minimum resources spent on the job to avoid short-term overload and long-term fatigue and exhaustion. Meta-knowledge and error recovery capacities are at the core of the efficient tuning of this compromise. Human errors are just items among others, such as the feeling of difficulty, the feeling of workload, and the time of error recovery, required to cognitively adapt and control the situation to reach the assigned goals. Second, the design of some automated systems is masking cognitive signals and impoverishing the efficiency of meta-knowledge, therefore causing the potential for new categories of human losses of control.

The following section details this model and explains the negative interaction between automation and dynamic cognition (see also, for extended description, Amalberti, 1996; Wioland & Amalberti, 1996).

An Analogy Between Cognitive Dynamics and a Betting System. A dynamic model of cognition could be seen as a bottleneck in available resources, or a toolset with several solutions to bypass resource limitations. The toolset is composed of perception, action, memory, and reasoning capacities. These capacities are prerequisites of human intelligence, just as tools are prequisites for work. But, because resources are the bottleneck for efficient use of cognitive tools, the very intelligence of cognition relies on solutions that bypass resource limitations. All solutions converge toward a parsimonious use of resources. They are threefold. First, the schematic of mental representation and the capability to use the representation at different levels of abstraction allow humans to oversimplify the world with limited risk (Rasmussen, 1986). Second, planning and anticipation allow humans to reduce uncertainty and to direct the world (proactive position) instead of being directed by the world (reactive position). Third, skills and behavioral automation are natural outcomes of training and notable ways to save resources. These three solutions have two dominant characteristics: They are goal-oriented and based on a series of bets. The subject cannot reduce the universe in order to simplify it without betting on the rightness of his or her comprehension; he or she cannot be proactive without betting on a particular evolution of the situation; he or she cannot drive the system using skill-based behavior without betting on the risk of routine errors. Failures are always possible outcomes of betting, just as errors are logical outcomes of dynamics cognition (see Reason, 1990).

Field experiments in several areas confirm this pattern of behavior. For example, a study considered the activity of fighter pilots on advanced combat aircraft (Amalberti & Deblon, 1992). Because of rapid changes in the short-term situation, fighter pilots prefer to act on their perception of the situation to maintain the short-term situation in safe conditions rather than delaying to evaluate the situation to an optimal understanding.

When an abnormal situation arises, civil and fighter pilots consider very few hypotheses (one to three) which all result from expectation developed during flight preparation or in flight briefings (Amalberti & Deblon, 1992; Plat & Amalberti, in press). These prepared diagnoses are the only ones that the pilot can refer to under high time-pressure. A response procedure is associated with each of these diagnoses, and enables a rapid response. This art relies on anticipation and risk-taking.

Because their resources are limited, pilots need to strike a balance between several conflicting risks. For example, pilots balance an objective risk resulting from the flight context (risk of accident) and a cognitive risk resulting from personal resource management (risk of overload and deterioration of mental performance).

To keep resource management practical, the solution consists in increasing outside risks, simplifying situations, only dealing with a few hypotheses, and schematizing the reality. To keep the outside risk within acceptable limits, the solution is to adjust the perception of reality as much as possible to fit with the simplifications set up during mission preparation. This fit between simplification and reality is the outcome of in-flight anticipation.

However, because of the mental cost of in-flight anticipation, the pilot's final hurdle is to share resources between short-term behavior and long-term anticipation. The tuning between these two activities is accomplished by heuristics that again rely on another sort of personal risk-taking: as soon as the short-term situation is stabilized, pilots invest resources in anticipations and leave the short-term situation under the control of automatic behavior.

Hence, the complete risk-management loop is multifold: Because of task complexity and the need for resource management, pilots plan for risks in flight preparation by simplifying the world. They control these risks by adjusting the flight situation to these simplifications. To do so, they must anticipate. And to anticipate, they must cope with a second risk, that of devoting resources in flight to long-term anticipation to the detriment of short-term monitoring, navigation, and collision avoidance. Thus, pilots accept a continuous high load and high level of preplanned risk to avoid transient overload and/or uncontrolled risk. Any breakdown in this fragile and active equilibrium can result in unprepared situations, where pilots are poor performers.

Protections of Cognitive Betting: Meta-Knowlege, Margins, and Confidence. The findings just given sketch a general contextual control model that serves the subject to achieve a compromise that considers the cost-effectiveness of his performance and maintains accepted risks at an acceptable level. This model is made with a set of control mode parameters (Amalberti, 1996; Hollnagel, 1993). The first characteristic of the model lies in its margins. The art of the pilot is to plan actions that allow him to reach the desired level of performance, and no more, with a comfortable margin. Margins serve to free resources for monitoring the cognitive system, detecting errors, anticipating, and, of course, avoiding fatigue.

The tuning of the mode control depends both on the context and on meta-knowledge and self-confidence. Meta-knowledge allows the subject to keep the plan and the situation within (supposed) known areas, and therefore to bet on reasonable outcomes. The central cue for control-mode reversion is the emerging feeling of difficulty triggered by unexpected contextual cues or change in the rhythms of action, which lead to activation of several heuristics to update the mental representation and to keep situation awareness (comprehension) under (subjectively) satisfactory control. These modes and heuristics control and adapt the level of risk when actions are decided.

A Model of Error Detection and Error Recovery. Controlling risk is not enough. Regardless of the level of control, errors will occur and the operator should be aware of it. He or she develops a series of strategies and heuristics to detect and recover from errors (Rizzo, Ferrente, & Bagnara, 1994; Wioland & Amalberti, 1996).

Experimental results show that subjects detect over 70% of their own errors (Alwood, 1994; Rizzo et al., 1994); this percentage of detection falls to 40% when the subjects are asked to detect the errors of colleagues (Wioland & Doireau, 1995). This lesser performance is because the observer is deprived of the memory of intentions and actions (mental traces of execution), which are very effective cues for the detection of routine errors.

Jumpseat or video observations in civil and military aviation (for an overview see Amalberti, Pariès, Valot, & Wibaux, 1997) shows that error rate and error detection are in complex interactions. The error rate is high at first when task demands are reduced and the subjects are extremely relaxed, then converges toward a long plateau, and finally decreases significantly only when the subjects almost reach their maximum performance level. The detection rate is also stable above 85% during the plateau, then decreases to 55% when the subjects approach their maximum performance, precisely at the moment subjects are making the lowest number of errors.

That figure indicates that error rate and error awareness are cues directly linked to the cognitive control of the situation. The feedback from errors allows the subject to set up a continuous picture of the quality of his own control of the situation. One can note that the best position for the subject does not fit with a total error avoidance, but merely with a total awareness of errors. The combined mechanisms protecting cognitive betting and allowing the subject to recover errors form what is called an ecological safety model of cognition (see Fig. 7.1). These results confirm the hypothesis of the

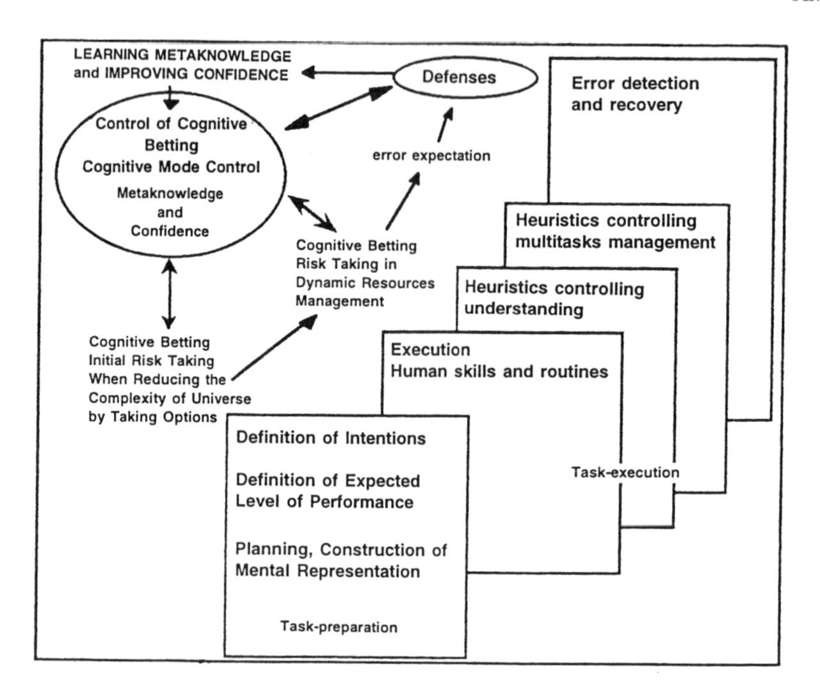

FIG. 7.1. An ecological safety model of cognition. Because operators are resource limited, they have to accept risks in simplifying the universe (setting up a mental representation), using routines to maintain the job under reasonable workload, and giving priorities in checks. Meta-knowledge, confidence, and respect for margins allow the users to maintain these accepted risks within acceptable limits. However, regardless of the value of the balance between accepted risks and risk protection, all subjects will make errors. Error detection and error recovery strategies are the natural complement of this risk of error. The error rate and error recovery time serve also to provide feedback and to help adapt cognitive mode control.

highly dynamic and continuously adapting nature of the ecological safety model. Operators are using a family of cognitive abilities to control risk.

Mapping the Self-Limitation of Performance. All these error results lead to a better understanding of how operators' performance is continuously encapsuled into self-limited values (see Wioland & Amalberti, 1996). Operators manage their performance within a large latitude for action. The performance envelope is usually self-limited so it remains within the safe margins (keeping the situation under control), because of a series of cognitive signals warning the operator when he is approaching the boundaries. This concept of a self-limited performance envelope is close to what Gibson and Cooks (1938) referred to as a desirable region, or "safe field of travel" (for a more recent application to modern technology, see Flach & Dominguez, 1995, or Rasmussen, 1996).

Why Automation Should Represent a Risk for This Cognitive Model

Notwithstanding the good intentions of the aviation industry, the design of many modern support systems (including automation) and the design of associated training courses and safety policies have the potential to interact negatively with the ecological safety model as already described. Some of these designs reduce the end user's cognitive experience of the system and jumble his meta-knowledge, confidence, and protective signals when approaching boundaries. The result is that operators accept new levels of performance, but are not expanding correctly cognitive protections to the new envelope of performance, and therefore are partially out of control of risk management.

This section tries to understand the logic of these negative interactions. The reasons for them are threefold. The first reason is linked to the multiplication of solutions to reach the same goal, the second is the long time needed to stabilize self-confidence and meta-knowledge on the glass cockpit, and the third and last reason corresponds to a certain irony of safety policies and automation design willing to suppress human error, finally resulting in suppressing individual ecological safety capacities.

Expanding the Envelope of Performance Beyond Performance Requirements. The envelope of performance has been considerably extended to reach new standards of performance. Doing so, this envelope has also been extended in side areas that correspond to levels of performance already attainable by pilots. These new areas, which are not necessarily required, have not been carefully analyzed, and create the potential for a series of negative interactions. They often differ from crews' spontaneous behavior (which appears as a heritage from the old generation of aircraft maneuvering), sometimes subtly and sometimes importantly, and therefore demand more pilot knowledge, more attention paid to comprehending how the system works and how to avoid making judgment errors. This is the case for many vertical profile modes on modern autopilots. For example, for the same order of descent given by the crew, the flightpath chosen by the autopilot is often subtly different from the crew's spontaneous manual procedure of descent, and moreover varies following the different types of aircraft. Another example is provided by some modern autoflight systems that allow the pilot to control speed by different means, either by varying the airplane pitch attitude (speed-on-pitch) or by varying the engine thrust level (speed-on-thrust). Even though these same concepts can be used successfully by most crews in manual flight, they still are confused for many crews when using autoflight systems, probably because of subtle differences from crews' spontaneous attitude when selecting and implementing the flight law in service (speed-on-thrust or speed-on-pitch).

Last, but not least, there is a mechanical effect on knowledge when expanding the number of solutions. The expansion of the performance envelope multiplies solutions to reach the same goal for the same level of performance and mechanically reduces the experience of each solution (and the time on training for each of them; see Fig. 7.2, and the next section).

Increasing the Time Needed to Establish Efficient Ecological Safety Defenses. The greater the complexity of assistances and automated systems, the greater is the time needed to stabilize expertise and self-confidence. The self-confidence model can be described as a three-stage model (Amalberti, 1993; Lee & Moray, 1994) with analogy to the Anderson model of expertise acquisition (Anderson, 1985). The first stage is a cognitive stage and corresponds to the type-rating period. During this period, the confidence is based on faith and often obeys an all-or-nothing law. Crews are under- or overconfident. The second stage of expertise is an associative stage and corresponds to the expansion of knowledge through experience. System exploration behaviors (often termed *playing*) are directly related to the acquisition of confidence, because this extends the pilot's knowledge beyond what he normally implements to better understand

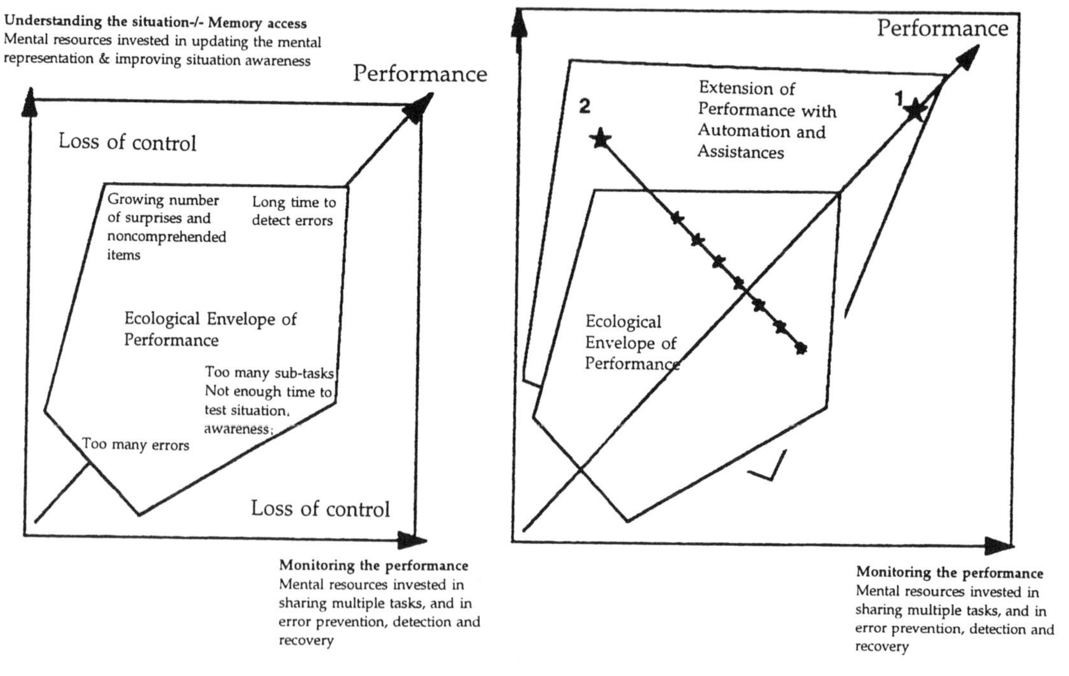

FIG. 7.2. Mapping the limitation of performance and the loss of control. *Left:* The space of possibilities to reach a given level of performance. The performance envelope is normally self-limited, to remain in safe conditions (situation under control), thanks to a series of cognitive signals warning the operator when approaching unstable boundaries. Note that the operator can lose control of the situation whatever the level of performance, but for different reasons. *Right:* A perverse effect of automation. The envelope of cognitive performance has been artificially expanded by assistance and automated systems. This extension of the envelope of performance is an undebatable advantage for Point 1. However, it is much more debatable for Point 2, which represents a level of performance already attainable within the ecological envelope. The expansion of the performance envelope multiplies solutions to reach the same goal for the same level of performance and mechanically reduces the experience of each solution.

system limits (and hence his own limits). Confidence is therefore based on the capacity to comprehend the relationship between system architecture and system behavior. Expertise is stabilized (in general, far from the total possible knowledge that could be learned on the system). This confidence, which was reached in less than 400 hours on the previous generation of transport aircraft, is not stable until after 600 to 1,000 hours of flight time on glass-cockpit aircraft (roughly 2 years' experience on this type), due to the extreme expansion of systems capacities and flight situations.

The drawback of the glass cockpit is not only a matter of training length. The final quality of expertise is also changing. With complex and multimode systems, the human expertise cannot be systematic, and therefore is built upon local explorations. Even after 1,000 flight hours, pilots do not have a complete and homogeneous comprehension of systems; they become specialists in some subareas, such as some FMS functions, and can almost ignore the neighboring areas. When analyzing carefully what pilots think they know (Valot & Amalberti, 1992), it is clear that this heterogeneous structure of expertise is not mirrored accurately by meta-knowlege. Because meta-knowledge is at the core of the control of risk taking, it is not surprising that overconfidence and underconfidence are facilitated. Overconfidence occurs when the system is engaged in areas where the pilot discovers too late that he is beyond his depth; lack of confidence results in delayed decision making or refusal to make decisions. The A320 accidents of Habsheim and Bangalore occurred with pilots precisely at this second stage, transitioning on the glass cockpit with a poor representation not only of the situation but also of their own capacities.

The third stage of expertise corresponds to the autonomous stage of Anderson's model. A retraction to an operative subset of this expertise for daily operations is the last stage of the confidence model. The margins generated by this retraction provide pilots with a good estimate of the level of risk to assign to their own know-how. Again, the heterogeneous structure of expertise provokes some paradoxical retractions. Pilots import from their explorations small personal pieces of procedures that they have found efficient and tricky. The problem is here that these solutions are not part of standard training nor shared by the other crew members. They result in a higher risk of mutual misunderstanding. Last, but not least, it is easy to understand that a little practice in manual procedures does not contribute to confidence, even when these procedures are formally known. The less the practice, the greater the retraction phase. Pilots become increasingly hesitant to switch to a manual procedure, and tend to develop active avoidance of these situations whenever they feel uneasy.

These effects are increased by the masking effect of both the training philosophy and the electronic protections. Because of economic pressure and because of operator and manufacturer competition, training tends to be limited just to the useful and underconsiders the complexity of systems. It is not uncommon that type-rating amplifies the value of the electronic protections to magnify the aircraft, and excludes the learning of areas of techniques, because these areas are not considered as useful or are not used by the company, that is the vertical profile, or some other FMS functions. This approach also favors overconfidence and the heterogeneous structure of expertise and the personal unshared discovery of systems. The experience of cognitive approaches to boundaries increases the efficiency of signals. But in most flying conditions with the glass cockpit, workload is reduced, complexity is hidden by systems, human errors are much more complex and difficult to detect (feedback postponed, error at strategic level rather than execution level), and the result is that operators continue too long to feel at ease. The cognitive signals when approaching boundaries are poorly activated until the sudden situation where all signals are coming in brutally and overwhelming cognition (explosion of workload, misunderstanding, and errors).

Blocking or Hiding Human Errors Could Result in Increasing Risks. It is common sense to assume that the fewer the human errors, the better the safety. Many systems and safety policies have the effect of hiding or suppressing human errors, such as fly-by-wire, electronic flight protections, safety procedures, regulations, and defenses. When considered individually, these defenses and protections are extremely efficient. But their multiplication results in a poor pilot's experience of certain errors. We have seen already in the description of the cognitive model that pilots need the experience of error to set up efficient self-defenses and meta-knowledge. Not surprisingly, the analysis of glass-cockpit incidents/accidents reveals that most (of the very rare) losses of control occur in situations where the pilot experience of error has been extremely reduced in the recent past due to increasing defenses (e.g., noncomprehension of flight protections, AirFrance

B747-400 in Papeete, Tarom A310 in Paris) or where errors are of an absolutely new style (e.g., crews back to automatic flight when lost in comprehension, China Airlines A310 in Nagoya, Tarom A310 in Bucharest). To sum up, although it is obvious that human error should not be encouraged, the efficiency of error reduction techniques (whatever they address: design, training, or safety policies) is not infinite. Extreme protection against errors results in cognitive disinvestment of the human for the considered area. Command authority can also be degraded by overly restrictive operator policies and procedures that "hamstring" the commander's authority (Billings, 1995). Cognition is therefore less protected, and room is left for rare, but unrecovered, errors. Moreover, when errors are blocked, new errors come (Wiener, 1989).

HUMAN FACTORS SOLUTIONS

Industry Response to Identified Problems

Industry is not naive nor inactive. Manufacturers, safety officers, and regulators are aware of the incidents and accidents. They do not deny the drawbacks, but they differ from the human factors community when explaining the reasons for problems and proposing solutions.

Some decision makers consider that many accidents and incidents are due to an insufficient expertise of crews. For example, many engineers would find it surprising that a crew would disregard very low air speed and low energy on an aircraft (Bangalore A320 accident) or that a crew would not be able to counterbalance a standard bank (Tarom A310 accident). Also, there is a general consensus in the aviation community that the majority of problems arise because of crews or mechanics deviating from procedures. This picture is reinforced by several recent accident/incident analyses and by the increasing number of incident reports showing crew nonadherence to procedures.

Logically, authorities and industry have asked for a more procedure-driven approach and better training to reduce these kinds of crew nonadherence to procedure. However, the industry quickly realized that simplistic solutions had also limits. On the one hand, the number of regulations, procedures, and checklists has grown considerably in the past 10 years, encapsuling the aviation system into a very normative atmosphere with associated drawbacks of slow innovation and uncontrolled escapes. On the other hand, training issues have also been debated. There is a general agreement that cockpit resource management (CRM) concepts are a plus for training crews on automated aircraft. The debate is much more open about training duration, because it is not clear what part of training should be reinforced. Most of the industry plead for a better ab initio basic flight training, and then for short type-rating courses focusing on the aircraft type specificities and standard operating procedures (SOPs). Others, namely human factors people, plead for the addition of a continuous reinforcement of basic manual skills along with the carrier and type ratings, because these skills are poorly solicited with modern aircraft and tend to be lost. But economic pressure does not favor such a solution. Another training issue concerns the level of deep knowledge to be taught on system architecture and logic in order to guarantee an efficient crew practice and understanding.

The limited impact of all these efforts on the current (already excellent) level of safety leads industry to more and more consideration of alternative solutions, including more human factors-oriented solutions in design and certification, provided that human factors proposes viable and non-naive solutions. This is the chance for human factors. But are the knowledge and the solutions available?

Human Factors Suggestions

Human factors suggestions for improving the design and operations of automated aircraft are gathered under the banner of human-centered design (see Billings, 1997, for a revised version). The current literature related to human-centered design emphasizes five different aspects that should be included in an efficient human factors methodology: (a) respect for human authority, (b) respect for human ecology, (c) the need for a better traceability of choices, (d) the need for an early and global systemic approach during design, and (e) the need for a different perspective in interpreting and using flight experience data through aviation-reporting systems.

Respect for Human Authority. Human factors are first interested in the role and position given to the end user in the system. Human factors cannot be restricted to a list of local ergonomics principles, such as shapes, colors, or information flow; it first refers to a philosophy of human–machine coupling, giving the central role to the human. For example, imagine a system in which the crew should be a simple executor, with a perfect instrument panel allowing total comprehension, but without any possibility to deviate from orders, and with a permanent and close electronic cocoon overriding crew decisions. Such a system should no longer be considered as corresponding to a satisfactory human factors design. The crew cannot have a secondary role because the members continue to be legally and psychologically responsible for the flight. They must keep the final authority for the system.

Fortunately, crews still have the authority with glass cockpits. But the tendency of technique is clearly to cut down this privilege. Today, some limited systems already override the crew decision in certain circumstances (flight or engine protections). Tomorrow, datalink could override crews in a more effective and frequent manner. After tomorrow, what will come? The goal of the human factors community is to improve the value associated with these systems, not to enter into a debate on the safety balance resulting from these solutions (as usual, we know how many accidents/incidents these systems have caused, but we ignore how many incidents/accidents they have contributed to avoiding). But how far can we go in this direction with crews remaining efficient and responsible in the flight loop and with a satisfactory ethics of human factors? We must acknowledge that the human factors community has no solutions for an infinite growth of system complexity. We must also acknowledge that the current solutions are not applicable to future complexity. Therefore, if we want the crew to remain in command, either we limit the increasing human–machine performance until the research comes up with efficient results preserving the authority of humans, or we accept designing systems, for example, made with two envelopes. The first envelope is controlled by crews. Within this envelope, crews are fully trained to comprehend and operate the system. They share full responsibility, have full authority, and can be blamed for unacceptable consequence of their errors. Outside this first envelope, the system enters into full automatic management. Crews are passengers. Human factors are not concerned, except for the transition from one envelope to the other one. Last, of course, ergonomics and human factors only tailor the first envelope.

Respect for Human Ecology. There is extensive literature on the design of human–machine interfaces fitting the natural, ecological human behavior, for example: ecological interfaces (Flach, 1990; Rasmussen & Vicente, 1989); naturalistic decision making and its implication for design (Klein, Oranasu, Calderwood, & Zsambok, 1993);

and ecological safety (Amalberti, 1996, and this chapter). Augmented error tolerance and error visibility are also direct outcomes of such theories. All of these approaches plead for a respect for human naturalistic behavior and defenses. Any time a new system or an organization interacts negatively with these defenses, not only are the human efficiency and safety reduced to the detriment of the final human–machine performance, but also new and unpredicted errors occur. But one must acknowledge that augmented error tolerance and error visibility are already part of design processes, even though they can still improve.

The next human factors efforts should focus on fitting not only the surface of displays and commands with the human ecological needs, but the inner logic of systems. The achievement of this phase asks for much more educational effort by design engineers and human factors. How could a design fit the needs of a pilot's cognitive model if there is not a shared and accepted cognitive model and a shared comprehension of technical difficulties between technical and life-science actors? This is the reason why reciprocal education (technical education for human factors actors, human factors education for technical actors) is the absolute first priority for the next 10 years.

Traceability of Choices. Human factors asks for a better traceability of design choices. This ideally implies using a more human factors-oriented methodology to evaluate system design. Ideally, for each system evaluation, human factors methodology should ask for the performance measurement of a sample of pilots, test pilots, and line pilots, interacting with the system in a series of realistic operational scenarios. The performance evaluation should in this case rely on the quality of situation awareness, the workload, and the nature and cause of errors. However, this is pure theory, not reality. The reality is much more uncertain because of the lack of tools for measuring such human factors concepts, and the lack of standards for determining what is acceptable or unacceptable. For example, the certification team is greatly concerned with the consideration to be given to human errors. What does it mean when an error is made by a crew? Is the system to be changed, or the error to be considered as nonsignificant, or the crew to be better taught? What meaning do the classic inferential statistics on risk have when considering that the rate of accidents is less than one in 1 million departures, and that most of modern electronic system failures will arise only once in the life of an aircraft? Moreover, the existing human factors methodologies are often poorly compatible with the (economical and human) reality of a certification campaign. Urgent (pragmatic) human factors research is therefore required.

Need for an Early and Global Systemic Approach During Design. There is a growing need to consider as early as possible in the design process the impact of the new system on the global aviation system. The design not only results in a new tool, but it is always changing the job of operators. And due to the interaction within the aviation system, this change in the jobs of operators is not restricted to direct end users, such as pilots, but also concerns the hierarchy, mechanics, dispatchers, air traffic controllers, and the authorities (regulations).

Such a global approach has been underconsidered with the introduction of glass cockpits and has resulted, for example, in several misfits between traffic guidance strategies and the capabilities of these new aircraft (multiple radar vector guidance, late change of runway, etc.). The situation is now improving. However, some of the future changes are so important and so new, like the introduction of datalink or the emergence of airplane manufacturers in Asia, that adequate predictive models of how the aviation system will adapt are still challenging human factors and the entire aviation community.

Need for a Different Perpective in Interpreting and Using Flight Experience Through Aviation-Reporting Systems. We have seen throughout this chapter that the safety policies have long given priority to suppressing all identified human errors by all means (protections, automation, training). This attitude was of great value for enhancing safety while the global aviation system was not mature and the rate of accidents was over one accident per million departures. Nowadays the problem is different with a change of paradigm. The global aviation system has become extremely sure, and the solutions that have been efficient up to this level are losing their efficiency.

However, for the moment, the general trend is to continue to optimize the same solutions: asking for more incident reports, detecting more errors, suppressing more errors. The aviation reporting systems are exploding under the amount of information to be stored and analyzed (over 40,000 files per year in only the U.S. ASRS), and the suppression of errors often results in new errors occurring. There is an urgent need to reconsider the meaning of errors in a very safe environment and to reconsider the relationship between human error and accident. Such programs are in progress at Civil Aviation Authorities of the United Kindgom (CAA UK) and the French Direction Générale de l'Aviation Civile (DGAC) and could result in different data exploiting and preventative actions.

Barriers to Implementing Solutions. The FAA human factors report (Abbott et al., 1996) has listed several generic barriers to implementing new human factors approaches in industry; among them were the cost-effectiveness of these solutions, the maturity of human factors solutions, the turf protection, the lack of education, and the industry difficulty with human factors. These reasons are effective barriers, but there are also strong indications that human factors will be much more considered in the near future. The need on the part of industry is obvious with the growing complexity of systems and environment. Also, mentalities have changed and are much more oriented to listening to new directions.

An incredible window of opportunity is open. The duration of this window will depend on the capacity of human factors specialists to educate industry and propose viable and consistent solutions. To succeed, it is urgent to turn from a dominant critical attitude to a constructive attitude. It is also important to avoid focusing on the lessons from the last war and to anticipate future problems, such as the coming of datalink and the cultural outcomes.

CONCLUSION

Several paradoxical and chronic handicaps have slowed down the consideration of human factors in aviation's recent past. First and fortunately, the technique has proven its high efficiency, improving performance and safety of the aviation system so far that a deep human factors revisitation of the fundamentals of design was long judged as not useful. Second, human factors people themselves served the discipline poorly in the 1970s by presenting most solutions at a surface level. Whatever the value of ergonomics norms and standards, the very fundamental knowledge to be considered in modern human factors is based on the result of task analysis and cognitive modeling, and the core question to solve is the final position of human in command, not the size and color of displays. The European ergonomics school with Rasmussen, Leplat, Reason, de Keyser, and some U.S. scientists such as Woods or Hutchins have been early precursors in the 1980s to this change of human factors' focus. But the change in orientation takes much more time for industry, maybe because these modern human factors are less quantitative and more qualitative, and ask for a much deeper investment in psychology than ergonomics recipes did before. Third and last, many people in society think with a linear logic of progress and have great reluctance to consider that a successful solution could reach an apogee beyond which the optimization could lead to more drawbacks than benefits. In other words, human factors problems could be much more sensitive tomorrow than today if the system continues to optimize on the same basis. In contrast, the instrument is so powerful that it needs some taming. This phase will only take place through experience, doubtless by changes in the entire aeronautic system, in jobs, and in roles.

Let us not forget, in conclusion, that these changes are far from being the only ones likely to occur in the next 20 years. Another revolution, as important as that brought about by automation, may well take place when datalink systems supporting communications between onboard and ground computers control the aircraft's flight path automatically. But this is another story . . . in a few years from now, which will certainly require that a lot of studies be carried out in the field of human factors.

One could say also that the problems are largely exaggerated. It is simply true that performance has been improved with automation, and that safety is remarkable, even though better safety is always desirable.

ACKNOWLEDGMENTS

The ideas expressed in this chapter only engage the author and must not be considered as official views from any national or international authorities or official bodies to which the author belongs. Part of the work presented in this chapter (human error and human recovery) was supported by a grant from the French Ministry of Defense, DRET Gr9, Contract 94.013; Dr. Guy Veron was the technical monitor. The author thanks Robert Helmreich for his helpful suggestions and text revision.

REFERENCES

Abbott, K., Slotte, S., Stimson, D., Amalberti, R., Bollin, G., Fabre, F., Hecht, S., Imrich T., Lalley, R., Lyddane, G., Newman, T., & Thiel, G. (1996, June). *The interfaces between flightcrews and modern flight deck systems* (Report of the FAA HF Team). Washington, DC: Federal Aviation Administration.

Amalberti, R., & Deblon, F. (1992). Cognitive modelling of fighter aircraft's control process: A step towards intelligent onboard assistance system. *International Journal of Man–Machine Studies*, *36*, 639–671.

Amalberti, R. (1993, October). *Cockpit automation: Maintaining manual skills and cognitive skills* (pp. 110–118). Paper presented at IATA, Montreal.

Amalberti, R. (Ed.). (1994). *Briefings, a human factors course for professional pilots.* Paris: IFSA-DEDALE (French, English and Spanish versions).

Amalberti, R. (1996). *La conduite de systèmes à risques.* Paris: Presses Universitaires de France (PUF).

Amalberti, R., Pariès, J., Valot C., & Wibaux, F. (1997). Human factors in aviation: An introductory course. In K. M. Goeters (Ed.), *Human factors for the next millenarium* (pp. 1–32). Aldershot, Hampshire: Avebury Technical, Ashgate Publishing Limited.

Anderson, J. (1985). *Cognitive psychology and its implications*. New York: Freeman.

Bainbridge L. (1987). Ironies of automation. In J. Rasmussen, J. Duncan, & J. Leplat (Eds.), *New technology and human errors* (pp. 271–286). New York: Wiley.

Billings, C. (1997). *Human centered aviation automation*. Mahwah, NJ: Lawrence Erlbaum Associates.

Bowers, C., Deaton, J., Oser, R., Prince, C., & Kolb, M. (1994). Impact of automation on aircrew communication and decision-making performance. *International Journal of Aviation Psychology, 5*(2), 145–167.

Cacciabue, P. C. (1992). Cognitive modeling: A fundamental issue for human reliability assessment methodology? *Reliability Engineering and System Safety, 38*, 91–97.

Endsley, M. (1996). Automation and situation awareness. In R. Parasuraman & M. Mouloua (Eds.), *Automation and human performance: Theory and applications* (pp. 163–180). Mahwah, NJ: Lawrence Erlbaum Associates.

Flach, J. (1990). The ecology of human–machine systems: 1. Introduction. *Ecological Psychology*, 2, 191–205.

Flach, J., & Dominguez, C. (1995). User-centered design: Integrating the user, instrument and goal. *Ergonomics in Design, July*, 19–24.

Funk, K., Lyall, B., & Riley, V. (1995, April). *Flight desk automation problems.* Paper presented at the Eighth International Symposium on Aviation Psychology, Columbus, OH.

Gibson, J., & Crooks, L. (1938). A theoretical field analysis of automobile-driving. *American Journal of Psychology, 51*, 453–471.

Gras, A., Moricot, C., Poirot-Delpech, S., & Scardigli, V. (1994). *Faced with automation: The pilot, the controller and the engineer.* Paris: Presse de la Sorbonne. (condensed version of a book: *Face à l'automate: le pilote, le contrôleur et l'ingénieur*, 1993, Paris: Presse de la Sorbonne)

Hoc, J. M. (1996). *Supervision et contrôle de processus: La cognition en situation dynamique.* Grenoble: Presses Universitaires de Grenoble (PUG).

Hollnagel, E. (1993). *Human reliability analysis, context and control.* London: Academic Press.

Klein, G., Oranasu, J., Calderwood, R., & Zsambok, C. (1993). *Decision making in action: Models and methods.* Norwood, NJ: Ablex.

Lee, J., & Moray, N. (1994). Trust, self-confidence, and operator's adaptation to automation. *International Journal of Human–Computer Studies, 40*, 153–184.

Parasuraman, R., & Mouloua, M. (Eds). (1996). *Automation and human performance.* Mahwah, NJ: Lawrence Erlbaum Associates.

Pelegrin, C., & Amalberti, R. (1993, April). Pilot's strategies of crew coordination in advanced glass-cockpits: A matter of expertise and culture. *Proceedings of the ICAO Human Factors Seminar, Washington, DC* (Circular OACI 217-AN/12, pp. 12–15).

Plat, M., & Amalberti, R. (in press). Experimental crew training to surprises. In N. Sarter & R. Amalberti (Eds.), *Cognitive engineering in the aviation domain.* Mahwah, NJ: Lawrence Erlbaum Associates.

Rasmussen, J. (1986). *Information processing and human–machine interaction.* Amsterdam: Elsevier North Holland.

Rasmussen, J., & Vicente, K. (1989). Coping with human errors through system design implications for ecological interface design. *International Journal of Man–Machine Studies, 31*, 517–534.

Rasmussen, J. (1996, August). *Risk management in a dynamic society: A modeling problem.* Keynote address, presented at the Conference on Human Interaction with Complex Systems, Dayton, OH.

Reason, J. (1990). *Human error.* Cambridge: Cambridge University Press.

Rizzo, A., Ferrente, D., & Bagnara, S. (1994). Handling human error. In J. M. Hoc, P. Cacciabue, & E. Hollnagel (Eds.), *Expertise and technology* (pp. 99–114). Mahwah, NJ: Lawrence Erlbaum Associates.

Sarter, N., & Woods, D. (1991). Situation awareness: A critical but ill-defined phenomenon. *International Journal of Aviation Psychology, 1*(1), 45–57.

Sarter, N., & Woods, D. (1992). Pilot interaction with cockpit automation: Operational experiences with the flight management system. *International Journal of Aviation Psychology*, 2(4), 303–321.

Sarter, N., & Woods, D. (1995). *Strong, silent and "out-of-the-loop": Properties of advanced automation and their impact on human automation interaction* (CSEL Rep. No. 95-TR-01). Columbus, OH: Ohio State University Press.

Valot, C., & Amalberti, R. (1992). Metaknowledge for time and reliability. *Reliability Engineering and Systems Safety, 36*, 199–206.

Wiener, E. (1988). Cockpit automation. In E. Wiener & D. Nagel (Eds.), *Human factors in aviation* (pp. 433–461). New York: Academic Press.

Wiener, E. (1989, May). *Human factors of advanced technology ("glass cockpit") transport aircraft* (NASA Contractor Rep. No. 177528, Contract NCC2-377). Moffet Field, CA: NASA.

Wiener, E., Kanki, B., & Helmreich, R. (Eds.). (1993). *Cockpit resource management*. New York: Academic Press.

Wioland, L., & Amalberti, R. (1996, November). When errors serve safety: Towards a model of ecological safety. *Cognitive Systems Engineering in Process Control* (pp. 184–191). Kyoto, Japan.

Wioland, L., & Doireau, P. (1995, August), Detection of human error by an outside observer. *Proceedings of the fifth European Conference on Cognitive Science Approaches to Process Control* (pp. 54–65). Espoo, Finland.

Wise, J., Abbott, D., Tilden, D., Dick, J., Guide, P., & Ryan, L. (1993). *Automation in corporate aviation: Human factors issues* (Final Rep. No. ERAU, CAAR-15405-93).

Woods, D., Johannesen, D., Cook, R., & Sarter, N. (1994). *Behind human error*. CSERIAC, Wright Patterson Air Force Base, OH.

8

Team Processes and Their Training in Aviation

Carolyn Prince
NAWCTSD
FAA
UCF Partnership for Aviation Team Training

Eduardo Salas
Naval Air Warfare Center Training Systems Division

> *Group process is the intragroup and intergroup actions that transform resources into a product; these processes serve to maintain the group and to help the group directly in achieving their goals.*
>
> —Gladstein, 1984, p. 500

In a short story, the Argentinean writer Borges (1962) described three men who lived in an imaginary country. The first man lost nine coins, the second man, looking in the area where the coins had been lost, found four of the coins, and a third man found two more coins because he knew they had been lost. When the man who had lost the coins found the last three, he knew that all the coins had been recovered. This whole sequence was possible, explained Borges, because all three men were different manifestations of the same person, and each of the three had the same knowledge of the coins' status. Although teams must function, at times, like Borges' fictional man, team members do not have a single mind and memory and must compensate for this by using teamwork to achieve coordination.

With the recognition that technically competent aircrews were experiencing performance problems due to failures in crew interactions (Ruffel Smith, 1979), teamwork became an issue for cockpit crew member training. Soon afterward, it was recognized that other teams working in aviation (e.g., cabin crews, maintenance teams, controllers) also needed improved teamwork. Accordingly, a concern for all aviation training now focuses on improving the ways individuals interact.

Attempts to improve team interactions for cockpit crews resulted in the introduction of crew resource management (CRM) training in the late 1970s. They were also responsible for the establishment of research programs on crew interactions, or team processes, by the Federal Aviation Administration (FAA), the military services, and the National Aeronautics and Space Administration Ames Laboratory (NASA-Ames; see Wiener, Kanki, & Helmreich, 1993).

Even though there was a long history of research on teamwork and team processes prior to their investigation in aviation teams (see Salas, Dickinson, Converse, & Tannenbaum, 1992; for a review), there has been noticeable progress since the aviation research on teamwork began. In part, this is due to a concentration of research efforts on those who work in dynamic environments and on the need for research results in these environments. In addition, aviation researchers are from a variety of backgrounds (i.e., social psychology, industrial psychology, cognitive psychology, aviation, and human factors) who bring different approaches and specialized knowledge to the work.

In 1993, an overview of the research and training of the management of crew resources was published (Wiener et al., 1993). Because both research and training program modifications have continued since that time, we present some recent research findings on the aviation team processes and how they are trained. Although our subject is aviation team process research, we focus on cockpit crews for two reasons: (a) most of the teamwork research has been done with cockpit crews; and (b) much of what is being learned with cockpit crews is being adapted to other teams.

This chapter is divided into four parts. In the first part, theoretical developments are addressed, with a background review of team process research, a rationale for the importance of this research to aviation teams, and an overview of some research for identifying team processes in aviation environments. This is followed by a section where the measurement of those processes and the progress made in measurement tool development are discussed. Next, are some strategies for training teamwork. The chapter concludes with a short discussion of some major trends in aviation team training research.

THEORETICAL DEVELOPMENTS IN AVIATION TEAM PERFORMANCE

The basis for knowledge about aviation team performance was originally derived from general team/group research. In this research, many traditional models of teamwork were presented as system models with an input-throughput-output design (Ginnett, 1993; Salas et al., 1992). Generally, the models' input and output variables were clearly defined. Inputs included items such as the composition of the team (both their individual characteristics and the team characteristics), the group structure, the resources available, the characteristics of the task, and the organizational structure. Outputs were likely to be classified as the results of the team's work. The throughput variables were less clear.

Throughput variables for teams are often considered to be team processes. For example, Nieva, Fleishman, and Reick (1978) proposed a taxonomy of team performance functions and referred to each as a process. The functions were categorized as orientational (processes used by team members to acquire and distribute the information needed for task accomplishment); organizational (processes used to coordinate team member activities); adaptational (processes for monitoring and compensation that members use to maintain normative strategies); and motivational (processes used to define objectives and motivate team members to accomplish those objectives).

Two other theorists, Hackman and Oldham (1980), defined effectiveness criteria for teams as the level of effort brought to bear on the group task, the amount of knowledge and skill applied to task work, and the appropriateness of performance strategies used by the group. These criteria were offered to evaluate team functioning in the throughput phase.

Based on the work of humanistic theorists and decision-making and boundary-management theorists, Gladstein (1984) developed a model of group effectiveness. The elements of group process included open communication, low conflict, supportiveness, weighing individual inputs, boundary management, and strategy discussion. In testing the model, Gladstein found that the process variables represented two distinct components, intragroup processes and boundary management. Unfortunately, she was unable to find a correlation between the process variables and her measures of team effectiveness.

Tannenbaum, Beard, and Salas (1992), in reviewing the team building and team effectiveness literature, provided still another model. They identified process variables as communication, coordination, conflict resolution, decision making, problem solving, and boundary spanning. Tannenbaum et al. found research that reported that communication, coordination, and decision making variables were related to some measure of team effectiveness. However, they also found research calling into question the relationship between their team process variables and measures of effectiveness.

Establishing a relationship between proposed process variables and team outcome is crucial. The inability to find consistent relationships between team process variables and measures of team effectiveness in the general team research suggested a number of possible problems with the proposed variables. The problems include:

1. The proposed team process variables were insufficient.
2. The variables were too generally defined.
3. The team process variables are more complex in their interrelationships with other variables than they were operationalized as being.
4. The effectiveness measures selected are not under the team's control (e.g., for sales teams, the volume of sales may be more affected by the economy than by team activity; Gladstein, 1984).

A failure to identify and understand the team process variables and to establish the relationships among these variables clearly creates a problem for team training design. It is the process (e.g., the actions taken, strategies performed, and use of cues by team members) that is central to designing effective training. Process variables may be described in terms such as "teamwork, coordination, and communication" (Salas et al., 1992, p. 16), but without understanding and agreement about the composition of these concepts, the terms contribute little to explaining how individuals achieve performance as a team. For example, one proposed concept, coordination, is a complex variable that is made up of other processes. It was defined by Zalesny, Salas, and Prince (1995) as involving the structuring of both interactions and responses of team members. Its demands are determined by the task and it is a team or group construct that is dependent on individual performance. In other words, coordination requires the purposeful actions of multiple individuals and is determined by task design elements (an input variable) such as interdependency.

Although team processes had not been clearly defined when teamwork became a recognized need for aviation, general team research nonetheless provided a valuable base for aviation specific team research. Aviation researchers were able to draw from the accumulated results of existing team research and apply these lessons to their research.

Identification of Team Processes in Aviation

When research in aviation team functioning began in support of CRM training, the team variable considered most important for training was the attitude of the team members (Helmreich, 1987). Attitudes were relatively easy to measure and were affected by training programs. More important, there was evidence that certain attitudes were related to team process (Cooper, White, & Lauber, 1980). Although training for attitude change was somewhat successful, it soon became apparent that it was not sufficient for addressing team performance problems (Prince, Chidester, Bowers, & Cannon-Bowers, 1992). There was a need for training that could directly affect the actions of crews, but before such training could be developed, aviation crew behaviors important to team performance had to be identified.

In the 1990s, research reports in aviation presented three candidate sets of team process variables. Each set of variables was defined by behaviors and each resulted from research in different aviation areas, civil air transport, helicopter operations, and a variety of military communities, from transport to air-to-air combat (Helmreich & Foushee, 1993; Prince & Salas, 1993; U.S. Army Aviation Center, 1992).

Research from the air transport community was analyzed by Helmreich and Foushee (1993). Similar to researchers in basic team training who divided process into *teamwork* and *taskwork* (see McIntyre & Salas,1995; for an overview), they characterized group process as being composed of two categories (i.e., *interpersonal and cognitive functions* and *machine interface tasks*, p. 19) that must be integrated in order to achieve the desired team outcomes. Machine interface tasks, including aircraft control and the accomplishment of procedures, referred to the technical competence required to fly the aircraft. The other category, interpersonal and cognitive functions, Helmreich and Foushee subdivided into three clusters: (a) communication and decision tasks, (b) team formation and management tasks, and (c) situation awareness and workload management tasks. They further partitioned each cluster into areas. The communication and decision task cluster included three critical areas: briefings; advocacy and assertions; and self-critique. The team formation and management task cluster was separated into two areas: (a) leadership, followership, and task concern; and (b) interpersonal relationships and group climate. The situation awareness and workload management task cluster was made up of preparation/planning/vigilance, workload distribution, and distraction avoidance.

For Army helicopter operations, five objectives were identified (i.e., team relationships, mission planning and rehearsal, workload level, exchange of information, and cross-monitoring of performance; U.S. Army Aviation Center, 1992). In addition, 13 areas of importance, or "basic qualities," were named. They included: (a) flight team leadership and crew climate; (b) premission planning and rehearsal; (c) decision-making techniques; (d) action prioritization and workload distribution; (e) management of unexpected events; (f) statements and directives; (g) mission situational awareness; (h) communication and acknowledgment of decisions and actions; (i) elicitation of crew information and actions; (j) mutual cross-monitoring; (k) offering of inputs by crew; (l) advocacy and assertion; and (m) crew-level after-action reviews (U.S. Army Aviation Center, 1992). Each basic quality was assigned to one of the five objectives (e.g., flight team leadership and crew climate belong to the team relationship objective) and used for training development.

Critical incident interviews with over 200 aviation crew members were used to identify team processes in a third organizational setting (Prince & Salas, 1989). This setting, naval aviation, is composed of air transport and helicopter operations as well as basic flight training, search and rescue, mine-sweeping, and surveillance squadrons, and other communities committed to achieving military goals. Interviews were conducted with crew members from six of these communities (e.g., transport, training, fighters) and responses were recorded. Behaviors that related to good or poor team performance were extracted from the recorded interviews. These behaviors were listed on a survey form that was sent to a second large group of aviators. Respondents rated each behavior on its importance, frequency, need to be trained, and difficulty to perform. Experienced aviators and human factors specialists sorted the behaviors that survey respondents had given the highest ratings into categories based on their similarities. Seven categories identified by the sorting represented seven skill areas. These skills included: (a) leadership, (b) communication, (c) assertiveness, (d) situation awareness, (e) mission analysis (planning), (f) adaptability, and (g) decision making. Subsequently, each of these skills was confirmed through observation of crews in realistic scenarios, a mishap analysis of 225 human-factor caused accidents, and empirical work (see Prince & Salas, 1993).

Six of the seven skill areas defined by Prince and Salas (i.e., situation awareness, communication, mission analysis or planning, assertiveness, decision making, and leadership; 1993) are included in the other skill sets (Helmreich & Foushee, 1993; U.S. Army Aviation Center, 1992). The clusters or areas proposed by Helmreich and Foushee include situation awareness, leadership, assertions, planning, communication and decision tasks (Helmreich & Foushee, 1993; see Table 8.1). The basic qualities defined for Army rotary wing aircrews specifically mention flight team leadership, premission planning, decision-making techniques, mission situational awareness, communication, and assertion (U.S. Army Aviation Center, 1992) .

The agreement that exists among the three sets of team process elements is important. These variables came from three separate research efforts, with different research objectives, and each focused on dissimilar aviation communities. Nonetheless, these efforts identified at least six process variables (i.e., leadership, communication, assertion, situation awareness, decision making, and planning) with some similarity in the behaviors that are used to define them. More important, with these variables, a team process behavior/performance link was established. For example, in military research, the Army identified its process behaviors through a review of aviation mishaps (U.S. Army Aviation Center, 1992). In the Navy, the behaviors found through critical incident interviews were confirmed as important by an analysis of behaviors that contributed to naval aviation accidents (Hartel, Smith, & Prince, 1991).

Naming the process skills and the behaviors that define them was important for training development and performance evaluation. But, this was not sufficient for building training programs and measurement tools. More knowledge about these processes and their interactions was needed so effective training programs could be built.

Process Skill Research. To learn more about team processes, several experiments on the process skills (identified by the research described by Prince & Salas, 1993) were conducted. In one, Prince, Brannick, Prince, and Salas (1997) explored development of a measurement instrument for team coordination based on the process skills. They conducted an experiment in which 102 pilots (51 crews) flew two simulator scenarios. For the research, approximately 60% of the crews were inexperienced pilots (students);

TABLE 8.1
Common Terms for Aviation Team Process

	Leadership	*Decision Making*	*Situation Awareness*	*Adaptability*	*Communication*	*Assertiveness*	*Mission Analysis (Planning)*
Helmreich and Foushee (1993)	Leadership/followership/task concerns[b]	Communication & decision tasks[a]	Situation awareness and workload management[a]		Communication & decision tasks	Advocacy & assertion[b]	Preparation/planning/vigilance[b]
U.S. Army Aviation , Center (1992)	Flight team leadership[c]	Application of appropriate decision making[c]	Maintenance of mission situation awareness[c]		Decision & actions, communicate & acknowledge[c]	Advocacy & assertion practiced[c]	Mission planning & rehearsal[d]

[a]Task cluster.
[b]Task area.
[c]Basic quality.
[d]Objective.

the remainder were experienced military pilots. Observers watched the videotapes of the crews in the scenarios and used rating forms to document specific crew process behaviors. The observers rated the crews on each skill, using a 5-point scale. Crew technical performance was rated on a separate scale by an independent observer. The results showed clear differences that distinguished teams by their performance. As expected, the experienced pilots' team process performance was rated higher than student pilots' performance and the process behaviors were found to be related to the independently rated technical performance. Finally, results verified that observers could agree on the process skill classification for observed behaviors and, using these skill categories, could discriminate differences in the skills for individual crews.

Another experiment exploring the same set of process skills (i.e., communication, leadership, decision making, situation awareness, adaptability, mission analysis or planning, and assertion) was conducted to determine if the quality of team process interactions had an influence on team performance beyond the individual task proficiency of individuals (Stout, Salas, & Carson, 1994). For this experiment, 110 undergraduate students were trained to fly a simple flight trainer and then, in crews of two, flew a scenario using the trainer. Task proficiency and team process behaviors were measured. Data analysis revealed that the ratings on the process skills correlated with performance scores of the team when individual task proficiency was held constant. The "bottom line" of this research shows that team processes, as defined by the seven skill areas (Prince & Salas, 1993), are important to team performance in this simulated aviation task.

An investigation of the interaction of two team process skills was demonstrated in an experiment on planning behaviors of crews (Stout, 1994). The quality of planning behaviors of 20 crews who flew a simulator scenario was assessed by raters who observed the crews on videotape. A second set of raters observed all the crews and counted the frequency of communications in which crew members provided needed information to one another before being asked for it. It was found that teams rated high in planning quality were also those who provided more information to one another before the information was requested. Because this provision of unsolicited information by one crew member to another is considered to be one indication of crew situation awareness, the experiment's results demonstrate a relationship between planning behaviors and situation awareness of the crew.

Two of the individual team processes, decision making and situation awareness, have been recognized as especially important in aviation. These two areas, identified by Helmreich and Foushee (1993) under their interpersonal and cognitive function category, by Army researchers as 2 of the 13 basic quality areas (U.S. Army Aviation Center, 1992), and by Prince and Salas (1993) as 2 of 7 crucial skill areas, are arguably the most cognitively complex of all the team process variables. They both have strong, important components for understanding performance from the standpoint of the individual crew member, as well as for an entire crew. Research on these two areas for individual performance has been conducted for a number of years (Diehl, 1991; Endsley, 1989; Fracker, 1989; Jensen, 1982; Shrestha, Prince, Salas, & Baker, 1995). Now, there is research for the FAA, NASA-Ames, and the Navy on measurement and training of decision making and situation awareness for the whole crew. This research represents the growth of existing aviation decision making and situation awareness research on individuals into important team issues. Also, with a focus on these two complex skills, it demonstrates an expansion of aviation crew training interests from attitudes, to behaviors, to cognition.

Team Situation Awareness Research. Situation awareness is a topic that has generated considerable attention, particularly as it is applied to pilots in the military fighter communities. According to Sarter and Woods (1991), "situation awareness is based on the integration of knowledge resulting from recurrent situation assessments" (p. 50). Endsley (1989) suggested that there are three levels of perception and pattern matching that are the processes associated with situation awareness. The first level is perception of elements in the pilots' environment, the second is integration of information (or the comprehension of the elements' meaning), and the third level is projection of the status of these elements into the future.

Most of the research on situation awareness has centered on the individual pilot's assessment of situations and integration of that information to achieve situation awareness (Endsley, 1989; Fracker, 1989; Sarter & Woods, 1991; Tenney, Adams, Pew, Huggins, & Rogers, 1992). Even though it has been recognized that situation awareness within a crew is different from individual situation awareness, there has been little research on how crews develop and maintain situation awareness (Salas, Prince, Baker, & Shrestha, 1995). Beyond the need for information about the individual crew member's situation awareness, little is known about how individual team members' actions may help or hinder the development of situation awareness in one another.

To begin to understand situation awareness from the perspective of experienced crew members, a series of interviews was conducted with aviators from general aviation, military units, and the air carrier industry (Prince & Salas, 1998). A group of 50 pilots, representing a range of experience (from 400 to 20,000 flight hours) and a variety of backgrounds, were interviewed. Because we expected situation awareness to be somewhat dependent on experience, responses were examined for similarities and differences based on flight hours. For example, when crew members were asked what they felt was necessary for team situation awareness, their responses varied, based on the hours of flight experience. For those crew members with fewer than 1,000 hours, the behaviors they mentioned fit into two of the process categories: communications and preparation for the flight (or planning). The majority of those crew members who had more than 1,000 hours of flying experience confirmed the importance of planning and communications, but their responses showed that leadership and adaptability behaviors also were necessary to team situation awareness (Prince & Salas, 1997).

These interviews showed that according to crew members (and depending on their level of experience) team situation awareness is achieved and maintained through individual situation awareness and four of the process skills (i.e., communication, planning, leadership, and adaptability). This finding has been confirmed by research evidence that found situation awareness of crew members, as measured by their knowledge of ongoing flight elements, strongly related to the other team process skills, particularly those identified by the interviews (Brannick, Ellis, Prince, & Salas, in preparation).

Decision Making. Research conducted at NASA-Ames has had the goal of identifying the crew interactions involved in decision making. This included the development of a taxonomy of decisions made in the cockpit (Orasanu, 1993). In presenting the taxonomy, Orasanu suggested that it may be appropriate to alter decision-making strategies for different types of decisions rather than to use a single, optimal, decision strategy. For example, a "go/no-go" decision (a decision with preestablished parameters that determine its solution) requires quick action on the part of the controlling pilot with, perhaps, little skill in what is commonly considered to be decision making and little input from others. Orasanu also suggested that skill in situation assessment may be the major requirement for this type of decision, because once the situation is known, the decision choice is clear. Other circumstances, where crews are faced with ambiguity, multiple options, and no clear superior choice of action, may require considerable skill in decision making with the entire crew participating in the process.

In research on the appropriateness of decision strategies, Prince, Hartel, and Salas (1993) looked at the decision strategies used by crews as they flew two realistic scenarios. These crews had been evaluated on both technical and process skills in a previous experiment (Prince et al., 1997). There were three major decision points in the scenarios and each differed in type, based on Orasanu's (1993) taxonomy. One decision point allowed the crew to rely on established procedures to handle the problem they faced and a second point was one where a better decision could be made if the entire crew made inputs. The third decision point had no established procedures and the crew needed to creatively use a number of resources to arrive at the best decision. Prince et al. found that there were three decision strategies used by the crews who took part in the research. The strategies varied in the amount of input sought and the time taken to reach a decision. Crews who used the same strategy for each decision they faced, despite the type of decision, were less effective than those who varied their strategies to fit the decision (e.g., limiting information search to fit the time constraints of the situation). This finding confirmed Orasanu's (1993) observation that different strategies may be required by different decisions. These research results reenforced the need for exploring similarities and differences in decision types and determining requirements for their solutions.

Summary of Process Variable Research

The progress made from aviation research in determining variables that can be assigned to the "throughput" area of a team performance model has been considerable (e.g., Helmreich & Foushee, 1993; Prince & Salas, 1989; U.S. Army Aviation Center, 1992). Research has identified attitudes that affect crews' performance as well as categories of skills with related behaviors that are important to the crews' interactions. Similarities in behaviors identified by several researchers is evidence that these investigations are yielding information that is relevant to a number of kinds of cockpit crews. Research into particular skill areas has revealed that these skills are both complex and highly interdependent. This has meant that, in addition to attitudes and actions, cognitive elements are included in the research on team process, particularly in situation awareness and decision making.

TEAM PROCESS MEASUREMENT

One basic concern for training is measurement. Measurement is crucial for continued growth in understanding and training team processes. We cannot know what needs to be trained or if we are training successfully without a good system of measurement. Whether it is the subjective collection of information and comparison of one entity with another, or a more objective development and application of a sophisticated measuring instrument, some form of measurement needs to occur. In the absence of an objective measurement tool, even a simple description of crew activity will be used for evaluation, training, or evaluating the crew training program by comparing it with some standard (e.g., "Is the performance observed unique? Is it satisfactory? Is it better?"; see Carnap, 1995, for a discussion of classificatory, comparative and quantitative concepts). However, measuring the outcome of the crew's performance (e.g., a safe landing) is necessary but not sufficient for training. That is, it does not supply the information that is needed on how the crew accomplished that outcome.

Denson (1981) called attention to the essential role of measurement for team training and the need for systematic investigation. Despite the basic need for measurement in the sphere of teamwork, it is difficult to achieve. The challenges are many and include the difficulty presented by mixed levels of analysis, the dynamic nature of the task and its environment, and determining what needs to be measured. Teamwork is the product of individuals, but it can only occur in the context of interacting individuals. In a two-person cockpit, the pilot may be flying the plane, and the copilot may be navigating and backing up the pilot on the instruments; they can be described as two individuals doing two individual tasks. Yet, successful completion of the flight is likely to be a result of both individuals' actions and, even more likely, it is also due to the coordination that occurs between the two individuals and their independent activities. Their interactions are, at times. dictated by the situation. Therefore, measurement of individual activities is insufficient to describe the functioning of a team.

Another challenge to measurement is the dynamic nature and complexity of crew performance. Diverse tasks within the context of a single flight require different levels of coordination and may place constraints on the crew members for their interactions. These characteristics require that measurement tools capture performance at several points within the tasking in order to ensure that it is representative of the whole performance of the crew.

For CRM's first 10 years, measurement of the training was accomplished primarily by self report (i.e., measuring the change in items selected on the Cockpit Management Attitudes Questionnaire, CMAQ; Helmreich, Wilhelm, Gregorich, & Chidester, 1990) and by the Line/LOFT worksheet (Helmreich & Wilhelm, 1987). The CMAQ was widely used because it was most relevant to the content of attitude-change programs. The early Line/LOFT worksheet contained short, summary phrases that referred to team processes that could be applied to a crew flying a scenario. Raters would observe crews in a realistic simulation, check statements they felt applied to the crew members and then give them a number from 1 to 5 (poor to excellent) on each of the summary phrases (e.g., "Inquiry/questioning practiced").

Since 1990, measurement instruments reflect the progress that has been made in defining the components of team process more clearly. University of Texas researchers have updated the Line/LOFT Worksheet (Helmreich & Wilhelm, 1987) to a rating instrument with more specifically defined behaviors, the Line/LOS Worksheet (NASA/UT/FAA, 1994). The new worksheet addresses many of the same items of information about team process, but statements are more descriptive (e.g., "Crew members ask questions regarding crew actions and decisions, e.g., effective inquiry about uncertainty of clearance limits, clarification of confusing/unclear ATC instructions"; p. 2) so they may capture more specific crew behaviors.

Military researchers have been working on measurement instruments as well (Baker & Salas, 1992; Prince et al., 1997). A rating form has been developed for the Army's 13 areas of importance to team process (i.e., flight team leadership and crew climate, premission planning and rehearsal, decision-making techniques, action prioritization and workload distribution, management of unexpected events, statements and directives, mission situational awareness, communicated and acknowledged decisions, supporting information and actions sought from crew, mutually cross-monitored crew member actions, supporting information and actions offered by crew, advocacy and assertion, and crew-level after-action reviews; U.S. Army Aviation Center, 1992). Raters are provided with behavioral anchors, or descriptions, of how a crew would perform if they were at the level of 1, 4, and 7 (on a 7-point scale, ranging from "Very poor" to "Superior") for each of the 13 basic quality areas. For example, in the area of cross-monitoring, a very poor rating is described as, "Crew members seldom, if ever, check each other's task execution. Crew members are insulted if they are corrected by another crew member" (U.S. Army Aviation Center, 1992, p. E-29).

Researchers working with naval flight crews have developed and tested two different instruments and are working on a new version that combines elements of both. Based on the research that had identified the behaviors associated with effective aviation team process and the categories that those behaviors described, an initial rating instrument was developed (Prince & Salas, 1993). This instrument contained each skill dimension (i.e., leadership, decision making, communication, adaptability, assertiveness, situation awareness, and mission analysis) and its associated behaviors (e.g., "Provided information in advance," under situation awareness). Observers were encouraged to watch aircrew performance in a realistic simulation, check off examples of the behaviors for each crew member, make notes, and then rate the crew on each of the dimensions. This rating was to take into account the frequencies of behaviors noted but was not to be solely influenced by frequency. The form was simple, with all observations on one page. It contained behaviors that could be confirmed by other observers and were applicable to any scenario or any flight. Unfortunately, for effective use, it required training and full understanding of the generic behaviors and how they related to specific actions.

The second measurement instrument, the Targeted Acceptable Responses to Generated Events (TARGET; Fowlkes, Lane, Salas, Oser, & Prince, 1992), was developed next. This instrument is a checklist that is individually tailored to each scenario. Construction of the checklist requires efforts from both researchers and operational experts. Expected behaviors of the crews that are required by procedures or by the scenario's events are listed for each event included in the scenario within each phase of flight. That is, the checklist contains the flight phase (e.g., prior to takeoff), the event (e.g., takeoff clearance given by air traffic control) and the expected behavior (e.g., pilot acknowledges takeoff clearance). Raters simply check each behavior if they see it and make no judgment about the behaviors. This checklist documents a great deal of the crew's performance based on a priori decisions as to what needs to be recorded. It requires little training to use, but is specific to each scenario and takes time and effort to construct. A third rating instrument is being developed that combines the approach of the TARGET checklist with the original measuring tool.

An inspection of three different rating instruments (Helmreich et al., 1990; Prince & Salas, 1993; U.S. Army Aviation Center, 1992) shows some similarities on the level of specific behaviors (see Table 8.2). As an example, in the area of situation awareness, the description of superior situation awareness performance provided in the U.S. Army materials includes:

1. Routinely updates one another on mission and situation awareness elements' status.
2. Anticipates the situation awareness needs of others.
3. Verbalizes and acknowledges changes in the elements of situation awareness.
4. Is aware of the physical and mental state of others.
5. Alerts others to personal problems.
6. Alerts one another to the presence of obstacles.
7. Requests needed information. (U.S. Army Aviation Center, 1992)

TABLE 8.2
Specific Behaviors for Rating Team Situation Awareness

Mission Situational Awareness (U.S. Army Aviation Center, 1992)	*Situation Awareness (Prince & Salas, 1993)*
1. Routinely updated one another on mission status and situation awareness elements' status	1. Demonstrates ongoing awareness of mission status
2. Anticipates the situation awareness needs of others	2. Provides information in advance
3. Verbalizes and acknowledges changes in the elements of situation awareness	3. Comments on deviations
4. Is aware of the physical and mental state of others	4. Demonstrates awareness of task performance of self and others
5. Alerts others to personal problems	5. Identifies problems/potential problems
6. Alerts one another to the presence of obstacles	6. Verbalizes a course of action
7. Requests needed information	

These relate directly to descriptions of behaviors from Prince and Salas (1993) for situation awareness that are used on a rating checklist:

1. Demonstrates ongoing awareness of mission status.
2. Provides information in advance.
3. Comments on deviations.
4. Demonstrates awareness of task performance of self and others.
5. Identifies problems/potential problems.
6. Verbalizes a course of action.

Except for the last behaviors on both lists (requesting information and verbalizing plans), the close correspondence of the two sets of behaviors is striking. The behavioral markers that correspond to the process skills described by Helmreich and Foushee (1993) are more general but are related to the military lists: for situation awareness, "demonstrate high levels of vigilance" (monitor, scan, cross-check, attend) is closely related to many of the individual behaviors on the two rating instruments.

Measurement of individual process skills, where detailed knowledge is important, requires measurement instruments designed for that purpose. There are several measurement techniques that have been suggested for capturing different facets of situation awareness (see Fracker, 1991; for discussion). These include varieties of self-report methods and explicit testing of subject's knowledge about aspects of the situation, both during a scenario and afterward.

In the air traffic control community, a system with the ability to graphically recreate recorded data (Situation Assessment Through the Re-Creation of Incidents, SATORI; Rodgers & Duke, 1994) has been suggested as a possible contributor to helping identify those tasks for which the maintenance of controller situation awareness is problematic. It provides detailed information that can be used to analyze consistent problem areas for controllers. A measurement technique related to cockpit performance that has been advanced by Endsley (1994) is the Situation Awareness Global Assessment Technique (SAGAT). This technique is used with a high fidelity simulator and a realistic scenario. Pilots fly a scenario; periodically the simulation is stopped and the pilot is asked questions relating to information that is available within the scenario. This questioning provides specific information about each crew member's situation awareness that may not be captured by other means. However, it is an intrusive technique that may change the pilot's awareness and behavior during the scenario. It also requires considerable work to ensure the questions used are indeed tapping the situation awareness required.

Another technique for measuring situation awareness that was suggested by a number of researchers (Adams, Tenney, & Pew, 1991; Mosier & Chidester, 1991; Sarter & Woods, 1991) is the analysis of crew actions and communications in scenarios. This analysis cannot measure precisely what each crew member may know about a situation at certain times during the flight, but the measurement itself is not intrusive and therefore will not have an effect on subsequent crew performance. The design of the scenario and the method for capturing relevant communications and actions are necessary to successfully use this method. Analysis of communications in scenarios constructed to elicit situation awareness can be done with crews, although it is difficult to do "real time." As the crew increases in size, the time required for analysis is also likely to increase.

Summary of Process Measurement Research

Measurement instruments that attempt to describe or evaluate team processes have progressed from those that were designed to capture the team's attitudes, to those with more general statements to guide the observer, and on to those that include very specific behaviors expected from crews in a particular scenario. These forms impose different demands on the instrument designer and on the observer, with the general forms requiring more rater training and more specific forms demanding considerable development time. An instrument that is well designed, easy to use, that provides reliable information on crew interactions, and that is useful as a guide for specific, useful feedback to the teams is still needed. Devising instruments to look at specific skills rather than all of the team processes may be a realistic approach to this goal. By concentrating on one process skill and the contribution of other skills to that process, the task of rating becomes more contained and may be done with greater ease. At the present stage of knowledge, it is recommended that more than one measurement technique be used and results compared for the information that each provides. No matter what the training tool or method, there should be close links between the measurement tool and the objectives of the training (reflected in the training content) that are employed. Together, they form an essential part of the training strategy.

TEAM TRAINING STRATEGIES

Recently, Salas and Cannon-Bowers (1997) argued that team training strategies result from combining tools, methods, and content. They proposed that the tools of team training include: performance measures, team task analysis, feedback, task simulations and exercises, and principles of learning, training, and team performance. The methods include information-based methods (e.g., lectures), demonstration-based methods (e.g., videotapes), and practice-based methods (e.g., role plays). Tools and methods, combined with the content (i.e., the knowledge, skills, and abilities) and guided by the training objectives, shape the strategies. The strategies may include guided practice (i.e., practice with specific feedback), team coordination training, and team leader training.

There has been little published since 1993 that indicates wide research activity in exploring some of the more traditional methods (e.g., guided practice) for training team processes in aviation. Training in the 1980s for aviation teamwork emphasized awareness of safety and the importance of the entire team for effective performance. Generally, training was conducted in a seminar with a mix of lecture, discussion, some role play, exercises, and videos of accident recreations (Prince & Salas, 1993). There was some practice in the simulator, but this was not necessarily closely linked in time or content with the CRM classroom training. Because the behaviors for team process had not been identified, training had attitude change as its goal (Helmreich, 1987). The strategies used in training showed some effectiveness for changing responses on an attitude measure (Helmreich et al., 1990), but there was little evidence that the training had an effect on crew member performance in the cockpit (Prince & Salas, 1993). In some courses, emphasis was placed on determining the behavioral style of crew members, making them aware of their own style, and encouraging them to develop more effective styles in the cockpit (Hackman, 1993). This training had some potential problems. As Hackman pointed out, training crews to have a preferred style has two major drawbacks. First, there is no empirical evidence that this will create a more effective, safer crew, and second, it is unlikely that a style change will persist under conditions of stress (when it may be most needed). With the identification of behaviors that could be important both to team functioning and to explaining team processes (Helmreich & Foushee, 1993; Prince & Salas, 1993; U.S. Army Aviation Center, 1992), it became possible for training to emphasize what crews must actually do to achieve effective performance. Training in specific team skills is a way to assist the aviators to further develop their existing strengths and learn ways to compensate for their weaknesses in teamwork. This suggested that strategies for training would require practice methods and feedback tools (Hackman, 1993; Prince & Salas, 1993).

Tools for Aviation Team Training

Scenarios. As a result of the interest in practice for training, new attention was focused on the scenarios used in training. This has resulted in improving scenario construction. Previously, scenario "success" as a training tool was primarily a result of chance, because little was known about what was required to make a worthwhile training scenario. Guidelines resulting from the NASA/Industry Workshop and two FAA Advisory Circulars (FAA Advisory Circular 120-35B and 120-51A; FAA, 1989, 1990) reflected the best of the accumulated knowledge at that time (e.g., stressing realism), but did not give specific guidance on how scenario events should be chosen so that the crew's performance could be fully and fairly evaluated. Hamman, Seamster, Smith, and Lofaro (1993) addressed this problem by developing the event set framework as scenario design guidance. They started with the premise that scenarios should provide an environment where both technical and process skills can be used. In order to design scenarios for Line Oriented Flight Training (LOFT), they proposed using the event set (i.e., "a group of related events which are part of the scenario and are inserted into a LOFT session for specified training objectives"; Hamman et al., 1993, p. 590) Each set contains a condition that initiates the event, irrelevant occurrences or distractors, and supporting events. Hamman et al. asserted that using event sets allows designers to create more realistic and more complex problems within the scenario than is possible with the traditional scenario design, where a number of simple, unrelated problems are introduced.

Scenario-based research in the Navy also has explored the design of scenarios for specific training purposes. Paralleling the work of Hamman et al. (1993), Bowers, Baker, and Salas (1994) and Fowlkes, Lane, Salas, Franz, and Oser (1994) recommended that scenarios for evaluation be developed with event sets. In addition, scenarios that are used for research into specific process skills (i.e., decision making and situation awareness) have been constructed to give crews opportunities to exercise those specific skills. For this purpose, scenario design begins with the skill of interest and behaviors that define it. Development of the scenario includes providing a realistic environment where crews must use the skills to successfully complete the mission.

Simulation. Simulators may be an ideal environment for implementing scenarios for training, but their cost often limits their availability for other than technical training. Role-play scenarios can be designed where no equipment is used, but they lack many necessary task cues and can not require the crew to perform additional duties and tasks that are part of an aviation team's job. Low-fidelity trainers can provide an environment where some of the task cues can be included, and this can be done economically. A number of experiments have been conducted with aviation team members (cockpit crews and maintenance team members) that have used a low-fidelity system for the practice of teamwork (Bowers, Salas, Prince, & Brannick, 1992). Low-fidelity flight trainers can be built with off-the-shelf aviation programs, a personal computer, and simple peripherals. They are enhanced by realistic scenarios designed with the physical limitations of the systems in mind. The secret to the acceptance of these systems is carefully constructed scenarios and realistic application (Baker, Prince, Shrestha, Oser, & Salas, 1993).

In an experiment to test the use of a low-fidelity system, 50 crews of two pilots each, an instructor and a student pilot, flew two realistic scenarios using a table-top trainer (Prince et al., 1997). In this research, student pilots had less than 70 hours experience. Specific feedback to the student on the process skills, particularly communications, assertiveness, and situation awareness, was provided by the instructor during the scenario. To control for scenario difficulty, half the student/instructor crews flew one scenario first, and the other half flew the other scenario first. Independent observers rated all students higher on the process skills in the second scenario than in the first. This indicated that the practice and feedback, using the low-fidelity system, could be transferred to a new scenario on the same system. A second experiment, with a separate set of aviators, evaluated the ability of pilots to transfer the training received on the low-fidelity system to a full mission simulator. It was found that practice with a scenario and specific feedback resulted in significant transfer to the more sophisticated system (Brannick, Prince, Salas, & Stout, in preparation).

Aviators' reactions to the table-top training system have been positive. A clear majority of those who have tested the system gave it a high rating as a trainer for team process skills. They considered the scenario that is developed for the training and the way the training is implemented to be key to the system's value. Their reactions are similar to the reactions of aviators to high-fidelity simulator scenarios (Prince & Salas, 1993).

Methods for Aviation Team Training

Because skill training is likely to demand more resources than less individualized training (e.g., lecture and class exercises), there has been an interest in determining if skill training is indeed superior to other, less expensive forms of training. As a result, several experiments have been conducted that explored training methods for different team process skills. In one experiment, three different methods were compared for training crew member assertiveness (Smith & Salas, 1991). Four groups of participants were formed, a no-training control group, a group that received lecture only, a group that received lecture and modeling, and a group that received lecture, modeling, practice, and feedback. All groups, except the control group, received training of equal length. Subsequent to the training, participants flew a scenario with a research confederate who created conflict situations that required assertiveness from the participants. Those participants who received training with practice and feedback used significantly more assertive behaviors than those who had undergone the other training methods. In other words, practice and feedback as part of the training was more effective for training than lecture alone or lecture with demonstration.

A follow-up experiment to the one on assertiveness training looked at two additional variables in the transfer situation, the awareness of the participant that his or her training was being tested, and the leadership behaviors that the participant encountered (Smith, 1994). Participants were all general aviation pilots. They each flew a simulator scenario that included a captain and another crew member who were research confederates. Assertiveness training, awareness of being rated, and leader behaviors that either encouraged or discouraged inputs from the crew were manipulated. Results of the experiment showed that pilots who received assertiveness training were more assertive than untrained pilots. Among the trained group, those who knew their assertiveness was being evaluated were more assertive when the captain played an authoritarian role than those who were unaware they were being evaluated. However, when the captain played a teamwork-oriented role, all subjects were equally assertive. This research has two important implications: (a) the context in which the trained team process behaviors must be used is very important for training transfer; and (b) team leadership behaviors have a significant effect on the demonstration of assertive behaviors.

Strategies for Teamwork Training in Aviation

Another experiment was conducted to explore a technique that might be useful in helping crews translate knowledge about a system into use for decision making (Hartel, 1991). Two-person teams (102 teams in all) participated in research to determine if teams that were given training in a specific troubleshooting method would perform more effectively in a stressful scenario. Prior to participating in the experiment, all teams were trained to the same level of technical competence in the systems of the training device (i.e., a simple helicopter simulation implemented on a table-top computer) they would be using in the experiment. Half of the teams were then given training in how to identify a problem and verify it before attempting a solution; the other teams were not trained. All teams flew a scenario in which a system was failed. It was their job to diagnose and fix the problem so they could continue with the flight. Although all teams had equal technical knowledge of the system, those teams who had received the additional training in a troubleshooting technique solved the problem in less time and with fewer errors than untrained teams. This experiment demonstrated the importance of crew process behaviors, over and above technical knowledge, for solving problems.

For communication skill training, 90 undergraduate students, assigned to two-person crews, received either no communications training, knowledge training (lecture only), or skills training that included a demonstration of the communication behaviors important to the crew's task and an explanation of their relevance to the crew's success (Lassiter, Vaughn, Smaltz, Morgan, & Salas, 1990). All groups flew a simulated mission on a low-fidelity-flight simulator. Analysis of the performance effectiveness of the crews revealed that the skill training groups demonstrated a higher level of communication skill than the other two groups. Independently rated performance scores were higher for this group also.

Training strategies have progressed in team training as the goals have changed. Early training, with the goal of attitude change, could be accomplished using a team-building strategy. Present training, with changes in behaviors as the goal, is accomplished best with combining practice methods and performance-level feedback tools. Cockpit crews have been well served by simulators, where realistic practice of team skills can be easily accomplished. In some cases, innovative methods have been used to provide practice opportunities for aviation teams where a high-fidelity simulator is not available. These methods have relied on low-fidelity training systems (Bowers et al., 1992).

As knowledge about team processes progresses and the cognitive components that contribute to team performance are identified, training strategies will likely change once again. New training strategies will address the aviation teamwork attitudes and behaviors, as well as teamwork's cognitive components.

FUTURE TRENDS IN TEAMWORK TRAINING FOR AVIATION

We see at least six trends in aviation teamwork training that will have an impact in the future. These trends are the result of increasing information about team interactions, growing knowledge about the importance of teamwork, and technology advances both in the team's work environment and in training.

The first is toward using the increased knowledge about specific team processes to improve teamwork training in aviation. As we noted in the beginning of this chapter, advances in information about team process already has changed training from more general attitude-based programs to those where particular behaviors are trained. The knowledge has altered scenario design and performance measurement so that scenarios can elicit selected behaviors and measurement tools can help document them. By defining and clarifying each concept, the nature of the training content can be altered. For example, in decision making, the typology of aviation decisions (Orasanu, 1993) is leading to training crew members to adapt their decision strategies to the amount of risk and time available, rather than memorizing a single formula to be applied in all situations.

A second trend in aviation training for teamwork is the expansion of the team concept beyond the team as traditionally defined. Crews of individuals working in similar job categories (e.g., cockpit crews, air traffic controllers, and maintenance crews) have been those most often recognized as requiring teamwork and have been targeted for training. The importance of communication and coordination of activities between individuals with vastly different tasks and technical training is now being recognized. For example, the cockpit crew and the air traffic controller must interact for some portions of a flight, and, in some situations, may need to employ teamwork skills. Equally, the ground handlers, the cockpit crew, and the flight attendants often must function as a large team in preparation for the flight. The concept of an enlarged team is more advanced within military aviation where crews from various aviation communities (helicopter, fighter, attack) must function as a team so they may accomplish a task beyond the capabilities of a single community.

Another trend has been introduced by automation, which has had two major impacts on cockpit crews. The first has been an alteration in the composition of the team. This can be illustrated in advanced technology cockpits where a third crew member has been replaced by an automated system. The second has been a change in both the tasking and the interactions of the remaining crew members. The effects of automation are still being discovered and although there is much to be learned about the effects of automation on team process, training is seen as a way to help crew members overcome any process loss that may be created (see Morgan, Herschler, Wiener, & Salas, 1993). Training crews to work with automated systems is moving beyond purely technical training to include process considerations such as communicating the systems' status to other crew members.

The Federal Aviation Administration's interest in Advanced Qualification Programs (AQP) for cockpit crews has highlighted the need for better measurement of team performance. Thus, improving measurement techniques is the fourth trend that we have found in recent research. Accurate measurement tools are needed to determine the level of performance and to provide feedback information. To this end, researchers and organizations are developing and modifying measurement instruments for their specific interests (e.g., decision making in scenarios, situation awareness of crews) as well as for all the team process skills. The trend in the measurement instruments is advancing from the more general, generic type of measurement to more specific behaviors related to particular situations.

A fifth trend is the use of advanced and innovative technologies for training. In some cases, a technology is developed and research is needed to determine the most effective way to train with that technology, in other cases, innovative technology needs to be developed to deliver training that already has been identified by researchers. An example of the first case is evident in military organizations where distributed interactive simulation (DIS) is being refined. Presently, DIS is primarily an engineering development of hardware and software systems. This particular tool allows for the creation of large, heterogeneous teams that may be composed of smaller teams, each with members who have a variety of duties. For DIS, usually simulators from different locations are linked together. Because team members may be geographically dispersed, DIS provides a significant challenge for methods to be developed so that observers can view and document process behaviors. There is also a need to determine how feedback can be given to be most effective. In the second case, advances in training technologies are needed to supply practice and feedback opportunities. Practice and feedback have already been identified as important to process skill training, but existing high fidelity simulators are too expensive for many organizations to use in non-technical training. Because many aviation teams do not have simulators for their tasks, technology needs to provide alternative means for these teams to practice their team processes.

The final trend that we have noted is an increased exchange of research findings between aviation and other areas. Originally, all teamwork research was accomplished outside of aviation. Then, much of the team research was done within aviation. Now, progress is beginning to be made in general team research (Cannon-Bowers, Tannenbaum, Salas, & Volpe, 1995), particularly in determining training strategies for group processes, at the same time that researchers are active in the aviation area. In addition, researchers are coming to aviation with more diverse backgrounds. Earlier contributions to aviation were primarily from the area of social psychology; the contributing areas have been expanded to include both industrial/organizational psychology and cognitive psychology. Industrial psychologists can add a wealth of knowledge about training, task analysis, and organizational effects to the aviation team research. As to the inputs of cognitive psychology, Steiner (1972) commented that although process research focuses on the "relationships among observable behaviors of group members," it reveals only the "tip of the iceberg" (p. 184). He contended that research will be complete only by including the internal processes of individuals interacting in teams.

Research in aviation team processes for training is beginning to reflect more depth in defining processes, more breadth in defining teams, greater interest in the team's environment, a need for quantifying performance, and an openness to new training technologies. Aviation team training research is also expanding its search for useful knowledge and concepts to other professions (i.e., medicine). Finally, not only has aviation research profited from experience gained from general team research, but it has begun to contribute to that body of knowledge (Cannon-Bowers et al., 1995). This signals an exchange of information that can be beneficial to both.

REFERENCES

Adams, M. J., Tenney, Y. J., & Pew, R. W. (1991). *State-of-the-art report: Strategic workload and the cognitive management of advanced multi-task systems* (Rep. No. CSERIAC 91-6). Wright-Patterson Air Force Base, OH: Crew System Ergonomics Information Analysis Center.

Baker, D. P., Prince, C., Shrestha, L., Oser, R., & Salas, E. (1993). Aviation computer games for crew resource management training. *International Journal of Aviation Psychology, 3,* 143–156.

Baker, D. P., & Salas, E. (1992). Principles for measuring teamwork skills. *Human Factors, 34,* 469–475.

Borges, J. L. (1962). *Labyrinths: Selected stories and other writings.* New York: New Directions Books.

Bowers, C. A., Baker, D. P., & Salas, E. (1994). Measuring the importance of teamwork: The reliability and validity of job/task analysis indices for team-training design. *Military Psychology, 6,* 205–214.

Bowers, C. A., Salas, E., Prince, C., & Brannick, M. (1992). Games teams play: A method for investigating team coordination and performance. *Behavior Research Methods, Instruments, & Computers, 24,* 503–506.

Brannick, M. T., Prince, A., Prince, C., & Salas, E. (1995). The measurement of team process. *Human Factors, 37*(3), 641–651.

Brannick, M. T., & Prince, C. (1995, April). Reliability of measures of aircrew skills across events and scenarios. In R. S. Jensen & L. A. Rankovan (Eds.), *Eighth International Symposium on Aviation Psychology* (pp. 603–612). Columbus, OH: Ohio State University.

Brannick, M. T., Ellis, E., Prince, C., & Salas, E. (in preparation). *A comparison of two measurement tools for situation awareness.*

Brannick, M. T., Prince, C., Salas, E., & Stout, R. (in review). *Development and evaluation of a team training tool.*

Carnap, R. (1995). *An introduction to the philosophy of science.* New York: Dover Publications.

Cannon-Bowers, J. A., Tannenbaum, S. I., Salas, E., & Volpe, C. E. (1995). Defining competencies and establishing team training requirements. In R. A. Guzzo & E. Salas (Eds.), *Team effectiveness and decision making in organizations* (pp. 333–380). San Francisco, CA: Jossey-Bass.

Cooper, G. E., White, M. D., & Lauber, J. K. (Eds.). (1980). *Resource management on the flightdeck: Proceedings of a NASA/Industry workshop* (Rep. No. NASA CP 2120). Moffett Field, CA: NASA-Ames Research Center.

Denson, R. W. (1981). *Team training: Literature review and annotated bibliography* (Tech. Rep. No. 62205F-17100347). Brooks Air Force Base, TX: Air Force Systems Command.

Diehl, A. E. (1991). The effectiveness of training programs for preventing aircrew error. In *Proceedings of the Sixth International Symposium on Aviation Psychology* (pp. 640–655). Columbus, OH: The Ohio State University.

Endsley, M. R. (1989). *Final report: Situation awareness in advanced strategic mission* (Rep. No. NOR DOC 89-32). Hawthorne, CA: Northrop Corporation.

Endsley, M. R. (1994). Situation awareness in dynamic human decision making: Measurement. In R. D. Gilson, D. J. Garland, & J. M. Koonce (Eds.), *Situational awareness in complex systems* (pp. 79–100). Daytona Beach, FL: Embry-Riddle Aeronautical University Press.

Federal Aviation Administration. (1989). *Cockpit resource management training* (Advisory Circular 120-51A). Washington, DC: Department of Transportation.

Federal Aviation Administration. (1990). *Line operational simulations: Line oriented flight training, special purpose operational training, line operational evaluation* (Advisory Circular 120-35B). Washington, DC: Department of Transportation.

Fowlkes, J. E., Lane, N. E., Salas, E., Franz, T., & Oser, R. (1994). Improving the measurement of team performance: The TARGETs methodology. *Military Psychology, 6*, 47–61.

Fowlkes, J. E., Lane, N. E., Salas, E., Oser, R. L., & Prince, C. (1992). Targets for aircrew coordination training. In *Proceedings of the 14th Interservice/Industry Training Systems Conference, San Antonio*, TX (pp. 342–352). Arlington, VA: National Defense Industrial Association.

Fracker, M. L. (1989). Attention allocation in situational assessment. In *Proceedings of the Human Factors Society 33rd Annual Meeting* (pp. 1396–1400). Santa Monica, CA: Human Factors Society.

Fracker, M. L. (1991). Measures of situational awareness. In *Proceedings of the Human Factors Society 35th Annual Meeting* (pp. 1396–1399). Santa Monica, CA: Human Factors Society.

Ginnett, R. C. (1993). Crews as groups: Their formation and their leadership. In E. L. Wiener, B. G. Kanki, & R. C. Helmreich (Eds.), *Cockpit resource management* (pp. 71–98). San Diego, CA: Academic Press.

Gladstein, D. L. (1984). Groups in context: A model of task group effectiveness. *Administrative Science Quarterly, 29*, 499–517.

Hackman, J. R. (1993). Teams, leaders, and organizations: New directions for crew-oriented flight training. In E. L. Wiener, B. G. Kanki, & R. C. Helmreich (Eds.), *Cockpit resource management* (pp. 47–69). San Diego, CA: Academic Press.

Hackman, J. R., & Oldman, G. R. (1980). *Work redesign*. Reading, MA: Addison-Wesley.

Hamman, W. R., Seamster, T. L., Smith, K. M., & Lofaro, R. J. (1993). The future of LOFT scenario design and validation. In *Proceedings of the Seventh International Symposium on Aviation Psychology* (pp. 589–594). Columbus, OH: Ohio State University.

Hartel, C. E. J. (1991). *Improving team-assisted diagnostic decision making: Some training propositions and an empirical test.* Unpublished doctoral dissertation, Colorado State University, Fort Collins, CO.

Hartel, C. E. J., Smith, K., & Prince, C. (1991). *Defining aircrew coordination: Searching mishaps for meaning.* Paper presented at the Sixth International Symposium on Aviation Psychology, Columbus, OH.

Helmreich, R. (1987). Theory underlying CRM training: Psychological issues in flightcrew performance and crew coordination. In H. W. Orlady & H. C. Foushee (Eds.), *Cockpit resource management training: Proceedings of the NASA/MAC workshop* (NASA Rep. No. CP 2455; pp. 15–22). Moffett Field, CA: NASA-Ames Research Center.

Helmreich, R. L., & Foushee, H. C. (1993). Why crew resource management? Empirical and theoretical bases of human factors training in aviation. In E. L. Wiener, B. G. Kanki, & R. L. Helmreich (Eds.), *Cockpit resource management* (pp. 3–45). San Diego, CA: Academic Press.

Helmreich, R., & Wilhelm, J. (1987). *Reinforcing and measuring flightcrew resource management: Training captain/check airmen/instructor reference manual* (NASA/University of Texas Technical Manual 87-1). Moffett Field, CA: NASA-Ames Research Center.

Helmreich, R., Wilhelm, J., Gregorich, S. W., & Chidester, T. R. (1990). Preliminary results from the evaluations of cockpit resource management training: Performance rating of flightcrews. *Aviation, Space, and Environmental Medicine, 61*, 576–579.

Jensen, R. S. (1982). Pilot judgment: Training and evaluation. *Human Factors, 24*, 61–73.

Lassiter, D. L., Vaughn, J. S., Smaltz, V. E., Morgan, B. B., Jr., & Salas, E. (1990). A comparison of two types of training interventions on team communication performance. In *Proceedings of the Human Factors Society 34th Annual Meeting* (pp. 1372–1376). Santa Monica, CA: Human Factors Society.

McIntyre, R. M., & Salas, E. (1995). Measuring and managing for team performance: Lessons from complex environments. In R. Guzzo & E. Salas (Eds.), *Team effectiveness and decision making in organizations* (pp. 9–45). San Francisco, CA: Jossey-Bass.

Morgan, B. B., Jr., Herschler, D. A., Wiener, E. L., & Salas, E. (1993). Implications of automation technology for aircrew coordination performance. In W. B. Rouse (Ed.), *Human/technology interaction in complex systems* (Vol. 6, pp. 105–136). Greenwich, CT: JAI Press.

Mosier, K. L., & Chidester, T. R. (1991). Situation assessment and situation awareness in a team setting. *Proceedings of the 11th Congress of the International Ergonomics Association*, 798–800.

National Aeronautics and Space Administration/University of Texas/Federal Aviation Administration. (1994). *Line/LOS Checklist, version 4.* NASA/Ames Research Center, Moffett Field, CA.

Nieva, V. F., Fleishman, E. A., & Reick, A. (1978). *Team dimensions: Their identity, their measurement and their relationships* (Final Tech. Rep., Contract No. DAHC19-78-C-0001). Alexandria, VA: U.S. Army Research Institute for the Behavioral & Social Sciences.

Orasanu, J. M. (1993). Decision-making in the cockpit. In E. L. Wiener, B. G. Kanki, & R. L. Helmreich (Eds.), *Cockpit resource management* (pp. 137–168). San Diego, CA: Academic Press.

Prince, A., Brannick, M. T., Prince, C., & Salas, E. (1997). The measurement of team process behaviors in the cockpit: Lessons learned. In M. T. Brannick, E. Salas, & C. Prince (Eds.), *Team performance assessment and measurement* (pp. 289–310). Mahwah, NJ: Lawrence Erlbaum Associates.

Prince, C., Chidester, T. R., Bowers, C. A., & Cannon-Bowers, J. A. (1992). Aircrew coordination: Achieving teamwork in the cockpit. In R. W. Swezey & E. Salas (Eds.), *Teams: Their training and performance* (pp. 329–353). Norwood, NJ: Ablex.

Prince, C., Hartel, C., & Salas, E. (1993). Aeronautical decision making and consistency of crew behaviors: Implications for training. In *Proceedings of the Seventh International Symposium on Aviation Psychology* (pp. 248–251). Columbus, OH: Ohio State University.

Prince, C., & Salas, E. (1989). Aircrew performance: Coordination and skill development. In D. E. Daniel, E. Salas, & D. M. Kotick (Eds.), *Independent research and independent exploratory development (IR/IED) programs: Annual report for fiscal year 1988* (NTSC Tech. Rep. No 89-009; pp. 35–40). Orlando, FL: Naval Training Systems Center.

Prince, C., & Salas, E. (1993). Training and research for teamwork in the military aircrew. In E. L. Wiener, B. G. Kanki, & R. L. Helmreich (Eds.), *Cockpit resource management* (pp. 337–366). San Diego, CA: Academic Press.

Prince, C., & Salas, E. (1998). Situation assessment for routine flight and decision making. *International Journal of Cognitive Ergonomics.*

Rodgers, M. D., & Duke, D. A. (1994). SATORI: Situation assessment through the re-creation of incidents. In R. D. Gilson, D. J. Garland, & J. M. Koonce (Eds.), *Situational awareness in complex systems* (pp. 217–225). Daytona Beach, FL: Embry-Riddle Aeronautical University Press.

Ruffell Smith, H. P. (1979). *A simulator study of the interaction of pilot workload with errors, vigilance, and decisions* (NASA Rep. No. TM-78482). Moffett Field, CA: NASA-Ames Research Center.

Salas, E., & Cannon-Bowers, J. A. (1997). Methods, tools, and strategies for team training. In M. A. Quiñones & A. Ehrenstein (Eds.), *Training for a rapidly changing work place: Applications of psychological research* (pp. 249–279). Washington, DC: American Psychological Association.

Salas, E., Dickinson, T. L., Converse, S. A., & Tannenbaum, S. I. (1992). Toward an understanding of team performance and training. In R. W. Swezey & E. Salas (Eds.), *Teams: Their training and performance* (pp. 3–29). San Francisco: Jossey-Bass.

Salas, E., Prince, C., Baker, D. P., & Shrestha, L. (1995). Situation awareness in team performance: Implications for measurement and training. *Human Factors, 37*, 123–136.

Sarter, N. B., & Woods, D. D. (1991). Situation awareness: A critical but ill-defined phenomenon. *International Journal of Aviation Psychology, 1*, 45–57.

Shrestha, L. B., Prince, C., Salas, E., & Baker, D. P. (1995). Understanding situation awareness: Concepts, methods, and training. In W. B. Rouse (Ed.), *Human/technology interaction in complex systems* (Vol. 7, pp. 45–83). Greenwich, CT: JAI Press.

Smith, K. (1994). *Narrowing the gap between performance and potential: The effects of team climate on the transfer of assertiveness training.* Unpublished doctoral dissertation, University of South Florida, Tampa, FL.

Smith, K., & Salas, E. (1991, March). *Training assertiveness: The importance of active participation.* Paper presented at the 37th Annual Meeting of the Southeastern Psychological Association, New Orleans, LA.

Steiner, I. D. (1972). *Group process and productivity.* New York: Academic Press.

Stout, R. J. (1994). *Effects of planning on the shared understanding of team member information requirements and efficient communication strategies.* Unpublished doctoral dissertation, University of Central Florida, Orlando, FL.

Stout, R. J., Salas, E., & Carson, R. (1994). Individual task proficiency and team process: What's important for team functioning. *Military Psychology, 6*, 177–192.

Tannenbaum, S. I., Beard, R. L., & Salas, E. (1992). Team building and its influence on team effectiveness: An examination of conceptual and empirical developments. In K. Kelley (Ed.), *Issue, theory, and research in industrial/organizational psychology* (pp. 117–153). Amsterdam: Elsevier.

Tenney, Y. J., Adams, M. J., Pew, R. W., Huggins, W. F., & Rogers, W. H. (1992). *A principled approach to the measurement of situation awareness in commercial aviation* (Contractor Rep. No. NASA-18788). National Aeronautics and Space Administration, Scientific and Technical Information Program.

U.S. Army Aviation Center. (1992, December). *Aircrew coordination exportable training package.* Fort Rucker, AL: Author.

Weiner, E. L., Kanki, B. G., & Helmreich, R. L. (1993). *Cockpit resource management.* San Diego, CA: Academic Press.

Zalesny, M. D., Salas, E., & Prince, C. (1995). Conceptual and measurement issues in coordination: Implications for team behavior and performance. In G. R. Ferris (Ed.), *Research in personnel and human resources management* (Vol. 13, pp. 81–115). Greenwich, CT: JAI Press.

9

Crew Resource Management: A Time for Reflection

Daniel E. Maurino
International Civil Aviation Organization (ICAO)

> *Although there is no real possibility of a quantitative evaluation of the benefits, no airline having set up a CRM program would now consider to kill it.*
>
> —Pariès and Amalberti

Despite their rather gloom and ominous remark, it would be quite mistaken to consider Pariès and Amalberti prophets of doom. The remark reflects the concerns of many within the international aviation human factors community who believe that Crew Resource Management (CRM) is an essential prevention tool in the contemporary aviation system, and who haven taken to critically review CRM and its history to ensure that there is a meaningful future for this training in aviation. The history of CRM appears to be one of extreme success: With barely enough age to vote, CRM has already been assigned a significant role as contributor to the safety and efficiency of the aviation system. It is perceived by user population and regulatory community alike as a sound way to proceed. In fact, nobody would dare say that CRM does not work.

There is, however, more than meets the eye in the successful history of the development, implementation, and operational practice of CRM. Without casting doubts about its value, there are certain quarters that suggest caution about what the future might hold, because they perceive that the relationship between CRM and improved safety is still tenuous. In these quarters, the prevailing attitude is one of critical vigil. Neither endorsing optimists nor sceptics, nor denying eventual merits in each relative position, it is contended that there are present-day issues which, in the best interests of CRM itself, must not be ignored.

The literature in regards to CRM is abundant (Cooper, White, & Lauber, 1979; Hayward & Lowe, 1993; Orlady & Foushee, 1986; Wiener, Kanki, & Helmreich, 1993). Readers interested in CRM development, implementation, and current practices may refer to these and many other existing publications. This chapter does not discuss CRM in itself; rather, it assesses some of the issues that might affect its future. Such assessment is conducted within the framework provided by a historical review of the evolution of CRM and its associated safety paradigms. The position upon which this chapter builds is quite simple and should not be mistaken: CRM is too valuable a tool to be squandered by misunderstandings, misperceptions, incompetence, or plain ignorance.

WHY CRM TRAINING?

To clarify misunderstandings and misperceptions, the essential question that demands unambiguous answer is *why* it is desirable—or necessary—to provide CRM training to operational personnel. CRM training is not an end in itself, a means of living for consultants, a power structure for management, nor an opportunity to generate research data and statistics. Nor is CRM the last frontier of aviation safety or organizational efficiency—or a frontier of any kind. We provide CRM training to operational personnel so that they are better equipped to contribute to the aviation system production's goals: the safe and efficient transportation of people and goods. CRM training is a safety and production tool. In sociotechnical systems, operational personnel are expected to act as the last line of defense against systemic flaws and failures. Their training must then build upon an understanding of systemic safety and a safety paradigm which are relevant to contemporary civil aviation.

CRM training for operational personnel must evolve from a realistic understanding of how today's underspecified and unpredictable aviation system can fail. At the individual level, we must seek failures in the cognitive dimensions of human performance. At the collective level, we must seek failures in the social and organizational dimensions of the system. Once potential failures have been identified, the blueprint of a healthy system built upon proactive rather than reactive measures has been identified, and data regarding the user organization and its population has been collected, *then and only then* should the process of writing the CRM training curriculum start. These fundamental steps have not always been observed, and the design of CRM training has been undertaken lightly—and even unscrupulously—on occasion.

CRM training for operational personnel must be relevant to their professional needs. Evidence from accident reports suggests that this has not always been the case. It is a matter of record that CRM has saved the day in many instances and averted more serious outcomes in others. It is equally a matter of record that the lack of relevance of CRM training—or its absence thereof—has oftentimes been an absent defense that could have contained the consequences of flaws in human or system performance. Furthermore, flaws in equipment design and in the aviation system itself are contributing realities to safety breakdowns. Both are here to stay, at least for a long while. Changes in equipment design are very expensive, and we will only see them (we hope) with the next generation of flight decks. Changes in the aviation system itself are very slow. We can only work at the interface between pilots and equipment and pilots and the system through CRM training, buying time while change takes place. The relevance of CRM training to the operational context is therefore essential.

FOUR GENERATIONS OF CRM TRAINING: TWO PERSPECTIVES IN HARMONY

CRM training was introduced during the 1970s because of the "70% factor," and also because there was growing concern that ritualistic training—ticking boxes in a proficiency form—did not address the operational issues that eventually led to the "70%

factor" (Cooper et al., 1979). Today we know that it was not an idle concern. Over its almost 20 years of existence, CRM has experienced considerable evolution and change. This section provides an overview of this evolution, based on two perspectives: one European, the other North American.

Pariès and Amalberti (1995) asserted, "No training is suspended in a vacuum: any training policy is built on a specific understanding of the system's potential failures, and follows a particular safety paradigm, including an accident causation model and some remedial action principles" (p. 217). Therefore, they seek to establish a connection between changes in the understanding of aviation safety, their allied prevention strategies and the evolution of CRM training. They suggested that what started as a "cockpit crisis" prevention and management program is gradually shifting toward a macro-system education: "CRM corresponds to a revolution in accident causation models. It was a shift from 'training for the abnormal' to 'training for the normal.' It was a dramatic acceptance that the prevailing bipolar equation—pilot proficiency plus aircraft reliability equals safe flights—had been proven wrong by hard real life reasons." Pariès and Amalberti distinguished four generations of CRM training, summarized as follows.

> First generation: aimed at individual attitudes, leadership and communication. The objective was to prevent accidents due to flawed flight crew team performance. The safety paradigm was that safety was a function of flight crew performance exclusively, and that there were individuals with either the "right" or the "wrong stuff."

The resistance to initial CRM training by segments of the pilot community (Helmreich, 1992) led to a revision of the original approach, thus giving birth to the second generation of CRM training. In an attempt to overcome resistance, this second generation essentially distanced itself from the notion of the "right/wrong stuff."

> Second generation: aimed at individual attitudes, leadership and communication, but expanded to include situation awareness, the error chain, stress management and decision making. Like the first, the second generation of CRM training aimed at preventing accidents through improved crew performance, and its underlying safety paradigm was that safety was a consequence of improved crew synergy.

Both first and second generation of CRM training relied intensively in role-playing and non-aviation-related games, and they resorted to repetitive accident case studies. A distinct characteristic of first and second generation CRM training programs is that they consciously and purposefully introduced and maintained a clear separation between technical and CRM training.

The introduction of "glass cockpits" led to the development of a third generation of CRM training, with a broadened human factors knowledge base, and with particular attention to the cognitive dimensions of small teams acting in dynamic environments and to the importance of shared mental models (Orasanu, 1994). The third generation of CRM programs also revisited human–machine interface issues, in the relationship between pilots and computers. It was during this third generation that Cockpit Resource Management became Crew Resource Management.

> Third generation: aimed at improving overall system performance through improved performance of the system's basic flight operational units (aircraft/crew system; flight/cabin crew system). It added the concepts of mental models, stress and fatigue management, automation management, vigilance, and human reliability to the basic issues included in

the two first generations of CRM programmes. It further included discussions intended to develop not only *skills*, but also *understanding* and *knowledge*.

The major step forward in the third generation of CRM was a change in its underlying safety paradigm: Safety was now considered to be a proactive rather than reactive endeavor, and the consequence of a healthy system and its effective performance. Two consequences of adopting a proactive stance also represented major change: the integration of human factors knowledge into the training, thus resolving the prevailing dichotomy of "technical" versus "non-technical" training; and the gradual shift from non-aviation games and role-playing toward the realities of aviation, such as the justification of operational doctrines.

The recognition that safety as an outcome is the consequence of the global health of the system, and that training is a tool to help the process and therefore to influence the outcome, led to the development of the fourth generation of CRM training.

Fourth generation: aims at improving overall system performance through improved performance of as many as possible of the system's components. It includes topics such as interaction among teams, shared mental models, role and status, and organizational synergy. The safety paradigm corresponds to the shift in safety thinking observed since the beginning of the 1990s: safety is one positive outcome of the system's health.

Fourth generation CRM training includes maintenance (Robertson, Taylor, Stelly, & Wagner, 1995; Taggart, 1995), air traffic control (Baker et al., 1995) and flight dispatch (Chidester, 1993; Federal Aviation Administration, 1995). Furthermore, it recognizes that management actions or inactions influence the outcome of aviation operations (Maurino, 1992). In this aspect, Pariès and Amalberti advocate the term cross-corporate or Company Resource Management to reflect that the benefits of CRM extend beyond safety to include cost efficiency, service quality and job satisfaction. The term *organizational resource management* (ORM) reflects the same line of thinking (Heinzer, 1993), fully developed by Smith (1992), who views CRM as an organizational development.

On the North American side of the Atlantic, Helmreich (1994) agreed with Pariès and Amalberti and acknowledged the existence of four generations of CRM training. With slight differences in regards to milestones and emphasis—probably reflecting American empiricism vis-à-vis French encyclopedism—Helmreich drew a "road map" of the evolution of CRM which is consistent with that of his colleagues across the Atlantic. According to Helmreich, the distinctive features of each of the four generations of CRM are:

First generation: derived from classical management development; focused on management styles and interpersonal skills; aimed at fixing the "wrong stuff."

Second generation: focused on concepts (situational awareness, stress management) and modular in conception (error chain, decision-making models).

Third generation: observes a systems approach, with a focus on specific skills and behaviors. It places emphasis on team building and in the integration of CRM with technical performance. This generation includes the first attempts to assess CRM training; therefore, to allow such assessment, special training is designed for check airmen and instructors. Lastly, the training transcends beyond the cockpit door to include flight attendants, maintenance personnel, dispatchers and air traffic controllers.

Fourth generation: addresses specialized curriculum topics, including automation and fatigue, joint training between flight and cabin crew, crew performance training directly derived from incident data and, very importantly, it includes an added focus on cultural issues, including national and organizational culture and the particular issues of multinational crews.

There exists consensus of opinion among Pariès and Amalberti and Helmreich as to the evolution and raison d'être of CRM. Each "road map" would fit into the blueprint proposed by the other without major difficulties. There is coincidence in regards to the two fundamental changes experienced by CRM: the blending of CRM within technical training, and the expansion of CRM beyond the cockpit to become a systems concept. In terms of conceptual preferences, while Pariès and Amalberti made it a clear point to present CRM espoused to prevailing safety paradigms, Helmreich seemed less concerned about this relationship (while not ignoring it). Instead, and by virtue of the globalization of CRM, Helmreich placed greater emphasis on cross-cultural issues as they affect CRM training.

WHERE CRM AND CULTURE MEET

The International Civil Aviation Organization (ICAO) is the specialized agency of the United Nations tasked with establishing standards to ensure uniformity of procedures and practices in international civil aviation. ICAO started its Flight Safety and Human Factors Program in 1989 (Maurino, 1991) and, given its international nature, obtained sensible knowledge about the influence of cross-cultural issues in aviation safety and efficiency. This information suggests the need to properly take cross-cultural issues into account when developing CRM training.

The most important lesson ICAO learned is that any kind of human endeavor has strong cultural components (Maurino, 1995a, 1995b). An important and practical consequence of the cross-cultural issues is that there are no solutions valid "across the board." Problems in different context might *seem* similar on the surface, but they encode important cultural biases. Therefore, they require different, culturally calibrated solutions. It is naive to assume that simply exporting a solution that worked in one context will bring the same benefits to another. Indeed, such attempt might only set a collison course between the respective biases of the different cultures involved, making the problem worse. Another important lesson ICAO learned is that there is a tendency to deal with cultural issues from an "us/right" versus "them/wrong" position. The perceived "qualities" or "defects" involved in rating different cultures are not only without foundation, but they create barriers in understanding the implications of the cross-cultural issue (Phillips, 1994). Difficult as they may be, the implications of cultural differences in cross-border endeavors are worth considering in depth (Phelan, 1994).

Culture can be defined as "the values, beliefs and behaviours we share with other people and which help define us as a group, especially in relation to other groups. Culture gives us cues and clues on how to behave in normal and novel situations" (Merritt & Helmreich, 1995). CRM was born in North America as an American solution to the intricacies of human inter-relationship in flight decks. It is based in social psychology, a practice scarcely known in large geographical areas including Africa, Latin America, the Middle and Far East, Asia, and largely across the Pacific, except for

English-speaking enclaves. In its beginnings and for some time it was considered that "any culture, whether it is Japanese, American, or any other—fits in with the cockpit environment. And in this sense, CRM is culture free" (Yamamori, 1986). This opinion was shared by many—if not most—proponents of CRM during the 1980s. In fact, the shared perception was that with few minor cosmetic changes—or even without them—CRM could fit within the operational practices of any airline around the world. This view is still held by some presently (Blixth, 1995).

We now know that, well-intentioned and honest as it might have been, the myth of culture-free CRM is a fallacy. Not only does culture influence CRM, but available evidence suggests that CRM is not a good traveler unless culturally calibrated. In fact, implementing off-the-shelf CRM training in an organization without due consideration to national, corporate and even pilot subgroup cultures may generate aberrant situations (Maurino, 1994; Merritt, 1993, 1995).

For example, Johnston (1993a) expressed concern about the universal understanding of the concepts of CRM. He questioned the effectiveness of simple translation, arguing that meaning is also provided by the cultural and environmental context. Examples of cognitive incompatibilities and of the limitations of translation are everyday occurrences in organizations which work in several languages—such as ICAO—where interpretation and translation are frequent victims of cross-cultural perceptions. It is a logical corollary to this state of affairs that led Johnston to question "In this general context, consider how the specialist vocabulary of CRM might be received in different cultures; indeed, might it not be the case that existing CRM training may be unsuitable, or misdirected, in some cultures?"

Helmreich and Merritt are leading research to provide solutions to the challenge of making CRM a universal, albeit culturally calibrated, concept. Helmreich and Merritt build their research on culture in the cockpit upon two basic tools: the research in culture by the Dutch anthropologist Gert Hofstede, and the Flight Management Attitudes Questionnaire.

Hofstede (1980) developed a survey measure to profile the dimensions of national cultures. He defined four dimensions of culture:

> Power Distance (PD): the extent to which people accept that power is distributed unequally. In high PD cultures, subordinates do not question superiors. Those in authority are seldom accountable, since it is assumed that authority "belongs" to them.
>
> Individualism–Collectivism (IC): the extent to which individual costs and rewards or the group is the dominant factor which motivates behavior.
>
> Uncertainty Avoidance (UA): the extent to which people feel threatened by uncertain or unknown situations or conditions; the need for defined structures and highly proceduralized protocols.
>
> Masculinity: the extent to which people are concerned with achievement rather than quality of life and interpersonal relationships. (p. 43)

Of these four dimensions, Power Distance and Individualism–Collectivism have been found to be highly relevant to cross-cultural research in aviation, while Uncertainty Avoidance and Masculinity have been found to be of lesser relevance.

The Flight Management Attitudes Questionnaire (FMAQ; Helmreich, Merritt, Sherman, Gregorich, & Wiener, 1993) is an extension of a previous tool, the Cockpit Attitudes Management Questionnaire (CMAQ; Helmreich, Wilhelm, & Gregorich, 1988), which had been developed to assess crewmembers attitudes about group performance, leadership, and susceptibility of stress among flight crews. The FMAQ is an iteration which combines the CMAQ with Hoftede's measures to capture cross-cultural differences in flight deck management attitudes and to provide a measure of organizational climate.

This research has produced evidence that CRM practices which were held to be of universal application are indeed culturally influenced. It would not then seem reasonable to apply training designed for a group in one country to other groups of different nationalities without first understanding the receiver group's attitudes toward leadership and followership, communication styles, work values, team concept and the use and flexibility of regulations (Merritt, 1993). CRM is a *process*, the output of which is improved safety and efficiency of aviation operations. Culture is among the many input factors, and cultural preferences influence CRM practices.

Merritt (1993, 1994) explained that whereas Anglo-speaking countries rank high on Individualism, Asian and Latin countries are at the opposite side of the dimension. Individualism is inversely correlated to Power Distance: Individualist cultures tend to be egalitarian, while collectivist cultures tend to be hierarchical. Merritt summed up the implications of these cultural preferences in communication styles, leadership, coordination, conflict resolution, and role expectations as follows:

> People in individualist cultures consider the implications of their behaviors within narrowly defined areas of personal costs and benefits; people in collectivist cultures consider the implications of their behavior in a context that extends beyond their immediate family.
>
> Communication in individualist cultures is direct, succinct, personal, and instrumental; communication in collectivist cultures is indirect, elaborate, contextual, and affective.
>
> Differences in communication styles affect feedback and monitoring: People in individualist cultures provide precise, immediate, and verbal feedback; feedback in collectivist cultures tends to be vague, delayed, nonpersonal, and nonverbal.
>
> People in individualist cultures place more emphasis on resolving conflicts than on being agreeable; in collectivist cultures it is more important to be agreeable than to be right.
>
> Further implications in conflict resolution arise from the fact that whereas collectivist cultures address the broad context and avoid focus on specific detail, individualist cultures concentrate on the key issues and play down the broad context.
>
> Conflict resolution in collectivist cultures follows strategies based on compromise, avoidance, collaboration, and, lastly, competition. The order is almost inverse in individualist cultures: collaboration, competition, compromise, and avoidance.
>
> In collectivist cultures superiors have considerable authority and discretion. In individualist cultures leaders are bound by rules that have been previously negotiated in detail by all involved. Whereas leaders in individualist cultures "hold office" and are therefore accountable, superiors in collectivist cultures "hold the power" and are not questioned.

Failure to consider cultural issues results in what Merritt (1993) called "cultural imperialism," which in turn produces "cultural mutiny": "First the consultants promote a model of optimal CRM behaviors. Masquerading as universally applicable, the behaviors are in reality only 'optimal' in their host culture. Tolerance and polite deference for the trainers' odd ideas give way finally to open disagreement in the face of being asked to provide direct specific feedback (in essence, criticism) to members of their own in-group. In what will be a face-saving strategy for both senior management and the consultant, I suspect the remainder of the training will be provided as specified in the contract."

CRM training that is the product of "cultural imperialism" may be at odds not only with the national culture but also with the organizational culture. The research by Merritt and Helmreich (1995) also explores the relationship between organizational culture and organizational climate, which is the pilots' appraisal of the organizational culture. A positive organizational climate reflects an organizational culture that is in harmony with pilots' values; a negative organizational climate indicates conflict between organizational culture and pilots' values. Merritt and Helmreich drew an interesting correlation between pilots' attitudes and negative organizational climate: Pilots who endorse organizational climates considered negative by the majority of the peer group demonstrate less positive CRM attitudes ("macho" attitudes, hierarchical, less interactive cockpits). This is the case when the airline's management is hierarchical, uncommunicative, and unrealistic in its appraisal of stressors. In such cases, pilots with less positive CRM attitudes do not perceive a conflict between their own personal style and the way the airline is run. But pilots with positive CRM attitudes will experience a conflict between their professional style and the airline's management style.

Having drawn this connection, Merritt and Helmreich explained that, nonetheless, a positive organizational climate is positively correlated to CRM attitudes, and that airlines with poor morale have weaker attitudes toward CRM while airlines with high morale support positive CRM attitudes. They concluded: "managers can play a strong and influential role in shaping CRM within their company, simply by the way they choose to manage." This conclusion supports with data a contention first expressed by Bruggink (1990): Pilots will model their own behavior after the behavior they observe in the organization, because they believe it is the behavior management expects from them.

Cross-cultural issues in aviation are here to stay. As aviation becomes a global village, and as CRM expands beyond the flight deck and across national boundaries, intracultural and cross-cultural issues will position themselves at the leading edge of research, design, prevention, and training endeavors. There is too much at stake to engage in simplistic reasoning and cosmetic solutions or, still worse, in denial. The proclaimed objectives of the global aviation system are the same across the community: the safe and efficient transportation of people and goods. There are, however, as many different approaches to implement these objectives as cultures and subcultures exist.

THE LINK BETWEEN CRM AND ACCIDENTS

The relationship between improper CRM behaviors vis-à-vis accident causation seems to have persuasively been established in the United States as far back as 1979 (Cooper et al., 1979). On the other hand, regional safety statistics provided by Boeing (Russell, 1993) suggest differences that might encourage generalizations between CRM and safety. Johnston (1993a) suggested caution when qualitatively linking aviation accidents and CRM: Care must be exercised when drawing fast conclusions and extending generalizations across different aviation contexts. Statistical analyses show a sequence of cause–effect relationships, which reflect agreed categorizations largely determined by prevailing beliefs. For all the scientific rigor and impartiality that statistics usually reflect, the benefit of hindsight plays a role in defining the numbers. As those involved in statistical analysis know, the "slings and arrows" of temptation are real: An analyst's major and deliberate effort is to stick to absolute impartiality and to resist "seeing what one wants to see" in the data evaluated.

Moreover, while revealing relationships perceived to be prevalent in accident causation, statistics do not reveal the processes involved in such relationships. These processes—as well as their underlying and supporting beliefs—are influenced by cultural factors and bounded by environmental and contextual constraints. It is contended that the answers to the safety questions lie not in the numbers but in the understanding of the processes. Asserting that 70% of aviation accidents in a particular context—the United States, for example—are due to human error might involve processes that relate to the complexities of human interaction in small teams in dynamic environments. But the North American aviation system is arguably fit and healthy, and supports flight crews in discharging their duties. In this case, where the system is beyond serious challenge, CRM training is an answer to operational human error. However, in a developing aviation system, the "70% argument" might simply attest to the impossibility of humans to achieve the system's production goals with existing tools, and to the fact that the system generates risk-taking behaviors as normal practice. In this case, CRM, although a palliative, would be far from a solution.

Commenting with insider's knowledge about aviation safety performance of third-world countries, Faizi (1996) reflected: "Safety culture in a society is the end product of many ingredients: literacy rate, socio-economic conditions, level of available technical education, natural resources, industrial base and last, but not least, political stability" (p. 6). He prefaced a candid yet thorough analysis explaining the poor safety record of developing regions of the world by warning "what follows cannot be considered the story of any one single country. The majority of Third World countries are tormented by similar problems of varying intensity or gravity" (p. 6). After advocating for the need to collect data exclusive from third-world countries to be able to address safety deficiencies by prescribing the medicine appropriate to the patient's and not somebody else's symptoms, Faizi summed up the ailings of developing regions of the world:

- Slackness of regulatory functions.
- Inadequate professional training.
- Non-professional safety management.
- Funds: Mismanagement and scarcity.
- Aging fleet.

What Faizi enumerated are flawed organizational processes (Maurino, Reason, Johnston, & Lee, 1995; Reason, 1990) that contribute to the differences in safety statistics among different regions. Interestingly, CRM training only deserves a passing remark in Faizi's analysis: "Words like Crew Resource Management (CRM) and Line-Oriented Flight Training (LOFT), which have become household words in the 'NORTH,' are alien to the crews of the 'SOUTH.' In these countries, it is still 'stick and rudder' skills which determine the efficiency and merit of an airman . . ." (p. 6).

The foregoing would seem to validate Johnston's concerns and suggests caution when proceeding forward. While the symptoms of certain safety deficiencies may appear similar on the surface, underlying cultural and contextual factors may dictate radically different solutions. In fact, safety deficiencies that could be addressed by CRM training in North America might not be effectively addressed *at all* by training in other regions of the world (Maurino, 1994).

LATEST DEVELOPMENTS

There is growing concern about the reactive application of human factors knowledge largely favored in the past; most often after the investigation of an accident uncovered flaws in human performance due to inherent human limitations and/or fostered by deficient human–technology interfaces. Reactive approaches tend to focus on immediate rather than on root causes of problems because of the emotional contexts within which they take place. Improving safety through the application of human factors knowledge requires the proactive application of the knowledge to anticipate and minimize the consequences of human error and deficient human–technology interfaces.

A proactive stance must start from the foundations. Human factors must progress beyond the "knobs and dials" of long ergonomic standing. It is essential to incorporate existing human factors knowledge at the stage of system design, *before* the system is operational—that is, during the certification process of equipment, procedures, and personnel. Likewise, it is essential to provide end-users (flight crews, air traffic controllers, mechanics, and dispatchers) with relevant knowledge and skills related to human capabilities and limitations.

This guiding philosophy led ICAO to develop human factors-related Standards and Recommended Practices (SARPs) for inclusion in the Annexes to the Chicago Convention and associated documentation. Annex 1 (Personnel Licensing) and Annex 6 (Operation of Aircraft) have been amended to include human factors training standards. The human factors training requirements for flight crews in Annexes 1 and 6 carry important consequences for trainers and training developers, regulators and human factors researchers. A considerable number of organizations have yet to implement such training. The onus is on trainers and training developers to see that initial CRM training is optimized and that recurrent CRM training is operationally relevant and observes the philosophy underlying the requirements in Annexes 1 and 6 and discussed in this chapter.

The regulatory community will have the responsibility of developing an appropriate regulatory framework for a field in which, notwithstanding a major educational campaign by ICAO, there are still misconceptions about the aim and objective behind human factors regulations. But the real onus falls on the research community. The evaluation of CRM has been the focus of research in different countries. However, a universally accepted tool for CRM student evaluation does not yet exist. Although some progress has been made, the new ICAO requirement dictates the need for accelerated research. Until an objective evaluation tool is designed and accepted by consensus of opinion, we can evaluate human factors *knowledge*, but we must be very cautious when it comes to the evaluation of human factors *skills*.

A TALE OF TWO CONTINENTS

The response of the aviation community to ICAO's requirements and to the challenge of an integrated approach to aviation safety is biased—it could not be otherwise—by cultural preferences. In Europe, the European Joint Aviation Authorities (JAA) responded promptly to ICAO's requirements to include human factors training within operational personnel training curricula. The JAA are an associated body representing the civil aviation regulatory authorities of 23 European countries who have agreed to cooperate in developing and implementing common safety and regulatory standards, the Joint Aviation Regulations (Pearson, 1995). All JARs are based on ICAO standards, and the primary aim is to reduce national variance. Flight crew licensing is governed by the JAA Flight Crew Licensing (FCL) provisions, heretofore existing only as draft proposals.

In terms of human factors training aimed at developing knowledge, JAR-FCL drafts are consistent with recent developments in aviation human factors training. Applicants for pilot's licences are furthermore required to pass a written examination in basic human performance and limitations (HPL) knowledge. A number of books and educational materials have been published and are easily accessible (Campbell & Bagshaw, 1991; Green, Muir, James, Gradwell, & Green, 1991; Hawkins, 1993; Trollip & Jensen, 1991). The United Kingdom has been the first authority among JAA member states to make HPL training and examination mandatory, and as of 1993 some 10,000 licence applicants had taken the examination, with an initial pass rate that rose from 30% to 70%–80% between 1991 and 1993 (Barnes, 1993).

JAR-FCL also require that any applicant to a commercial pilot licence and instrument rating (CPL/IR) who intends to operate in multicrew aircraft types must satisfactorily complete a training course aimed at developing skills, referred to as an MCC (Multi-Crew Cooperation) course. The objective of the MCC course is "to enable pilots to become proficient in instrument flying and multicrew cooperation. . . ." Technical exercises are mandated in detail, including type-specific training exercises such as engine failure and fire and emergency descents. JAR-FCL also include the requirement for skills evaluation after completion of MCC training.

According to Johnston (1993b), the MCC course represents an innovative departure of existing licensing and training practices. Available JAR-FCL guidance material on human factors issues within the MCC course implies a basic CRM course. MCC includes 30 hours of simulator training, however, it does not include mandatory groundschool training nor any guidance as to the possible content of theoretical training. The human factors skills evaluation embedded in the MCC course is based on information in JAR-FCL that relates to the general conduct of checks and to the assessment of acceptable pilot performance. Tests should be conducted in a multicrew environment, and there is the requirement to assess the "management of crew cooperation" and "maintaining a general survey of the aeroplane operation by appropriate supervision" on a pass–fail basis. Interested observers from across the international community are waiting with considerable expectation to find out how the JAA will define parameters to conduct this assessment and how the pass–fail criteria will be developed.

Another interesting departure from existing requirements and practices included in JAR-FCL, which many observers perceive as a misinterpretation of the ICAO's Human Factors training requirements and the widespread implementation of CRM training, is the requirement for psychological testing for pilots. In order to fully appreciate the magnitude of this requirement it must be clearly understood that it relates specifically to psychiatrically fit pilots. The evaluation will be initiated by a physician when the physician receives information that evokes doubts concerning the aptitudes or personality of a particular individual. In such circumstances, a psychological evaluation will be initiated utilizing the expertise of an approved psychologist (Goeters, 1995).

Psychometric testing for periodic licence renewal is neither an ICAO requirement nor a foreseeable alternative to existing Standards and Recommended Practices (SARPs) in Annex 1 (Personnel Licensing and Training). Although initially considered by a number of countries, the periodic psychological evaluation will not be mandatory in the upcoming JAR-FCL, either, since it was not deemed appropriate in a routine basis (Pearson, 1995). It can nonetheless be required if the physician conducting a routine examination believes it to be necessary. Understandably, the requirement has set the European professional pilot community up in arms, who perceive that such vague requirement constitutes breeding grounds for abuse, and who also challenge (a) that physicians are suitably equipped to make psychological/operational assessments on pilots, and (b) that testing should be necessary for pilots who suffer from no psychiatric symptomatology (Murphy, 1995). When considering that Murphy speaks on behalf of the International Federation of Airlines' Pilots Associations (IFALPA), a casus belli becomes loud and clear for those who can read the telltale indicators of a gathering storm: If matters which up to now have been considered technical or professional drift towards political or industrial grounds, we will face the single greatest threat to the future of CRM training. Individual casualties aside, CRM will be the real loser should a European confrontation materialize.

Beyond emotions, the fundamental question for which available evidence does not provide a convincing answer is *why* psychological testing is deemed necessary in Europe. What exactly is the problem being addressed by this new requirement? Is the problem so important that it deserves to be addressed by a medical licensing standard? Goeters (1995) explained that "the main reason for the implementation of the psychological requirements into the licensing regulations is to improve the reliability of flight operations. The psychological evaluation should identify those subjects who are expected to develop problems in aircraft operations such as failing checks, unsafe aircraft handling, incidents, accidents, and other negative behavioral events" (p. 150). However laudable the objective, the justification for added licensing requirements and the costs they incur necessitate more than open-ended justifications in this time of dollar-sense. Furthermore, training departments—into which airline organizations invest considerable amount of money—*should* have all the specialist knowledge and be in a better position to deal with the "negative behavioral events" alluded to by Goeters than psychologists without specialist knowledge of flight operations.

To further compound the picture, the scientific community is far from a consensus position about the predictive validity of psychological testing. Whereas Goeters argued that "the application of psychological evaluation is justified by the fact that sufficient empirical data exist to indicate that the problem cases in pilot training and flight operations are predictable by psychological assessment methods" (p. 150), Damos (1995) opined that "despite the amount of effort and money that have been spent on developing pilot selection tests, there are some troubling aspects of these batteries that need to be examined" (p. 165). She further argued, "anyone who has dealt with pilot selection knows that the area is not mature and does not have a strong theoretical foundation," and concluded "one fundamental cause of the low predictive validities [of the selection tests] stems from the fact that from a rigorous, scientific viewpoint, the pilot's task has never been defined" (p. 167).

Another very relevant issue—given that the proposed JAR-FCL would apply to 23 countries extending from the Barents Sea to the Mediterranean and from the Atlantic coast of Ireland to the Urals—refers to the impact of cultural variations (Johnston, 1994). Different social and cultural responses can be expected from 23 countries with significant differences in Power Distance and Individualism measures: Whereas some countries might accept psychological testing, others might express considerable scepticism. Social and cultural understandings will undoubtedly conspire against uniformed psychological assessment inasmuch as there will not be a unified standard against which to measure "negative behavioral events," "undesirable personality," or "undesirable attitudes."

In terms of the international aviation community, the important conclusion to draw from this European contest is that it has nothing to do with human factors or with CRM. The fact that most safety breakdowns in aviation have human performance issues as underlying cause—the "70% factor"—might generate a very thin line of reasoning to connect psychological testing with the "70% factor" and therefore with human factors and CRM. Stretching the imagination, misinformed individuals might think that by conducting psychological testing they are "doing" human factors or CRM. Such reasoning holds absolutely no truth. Psychological testing has a significant role in modern technological societies and, if properly administered and interpreted, it may be successfully applied in a variety of roles. CRM is part of a strategy aimed at fighting systemic deficiencies at their roots, and CRM training evaluation has nothing to do with increased medical checks or psychological assessment of pilots.

In North America, the response to the challenge of the integration of human factors knowledge into operational personnel training took the form the Advanced Qualification Program (AQP; Federal Aviation Administration, 1991). Rather than establishing a programmed number of hours for a training course, AQP is based on the concept of training to proficiency. The AQP applicant, rather than the regulatory authority, develops a set of proficiency objectives that substitute the traditional training and checking requirements, thus assuring that valuable training resources are not expended in training and checking activities of marginal relevance.

AQP is an attempt to download the complex regulatory maze of Federal Aviation Regulations (FARs). Regulatory standards have traditionally addressed technical performance in individual positions. This approach was effective for the generation of aircraft to which it was aimed. But as technology was introduced, the relevance of existing regulatory standards became under scrutiny. Moreover, existing standards were silent regarding proficiency in crew functions. After the National Transportation Safety Board (NTSB) recommended in 1988 that CRM training should be included in simulator or aircraft training exercises, airlines gradually embraced the recommendation. However, given the absence of a regulatory requirement, widely different CRM training programs could be observed. Another deciding factor in developing AQP was that existing regulatory requirements did not reflect advances in aircraft technology or the changing function of the pilot from a controller to a systems manager. All of these developments carry significant training implications, and AQP provides a systematic methodology to accommodate them.

AQP is type-specific. The first step in AQP is to conduct an aircraft type-specific task analysis, which includes the identification of CRM behaviors pertinent to the execution of piloting tasks, within the context in which the task will be developed. Proficiency objectives thus determined are supported by enabling objectives, which prepare crews for further training in operational flightdeck environments. For recurrent training, certain proficiency objectives may be categorized as currency items, that is, activities on which proficiency is maintained by virtue of frequent exposure during routine line operations. Although verification that proficiency in such items is maintained remains a requirement, they need not be addressed for training or evaluation during periodic training.

Proficiency objectives can be further categorized as critical or noncritical, based on the significance of the consequences of operational errors, which in turn determines the interval within which training and evaluation of these items must take place. Noncritical items can be distributed over a longer period of time than critical objectives. In this way, training resources may be devoted with greater emphasis on abnormal conditions, emergencies, CRM, and other skills that line experience may identify as important and relevant to pilots' operational needs (Birnbach & Longridge, 1993).

AQP mandates the evaluation of proficiency objectives that reflect both technical and CRM performance in a crew-oriented environment. In regards to the evaluation of CRM, AC 120-54 warns: "CRM issues and measures are not completely developed at this writing. AQP is expected to support further development of CRM. Collection and analysis of anonymous data (not identified with a named individual) will validate the CRM factors as well as overall crew performance. *Until CRM performance factors can be validated, data should be collected without fail/pass consideration* (emphasis added). However, correction of below standard performance to standard performance is expected" (AC 120-54, chap. 2, p. 2-1).

Not without its problems, the implementation of AQP has been considered by some almost an obsession in the United States. Concern has been expressed that AQP may in some cases result in "no formal CRM training at all in the classroom; it is all done via the sim ride" (Komich, personal communication, April 5, 1995). Whether simulator training can effectively address all CRM issues that have contributed to aircraft accidents in the past remains an issue open to debate. Real line operations are situations of unmonitored human performance, whereas, for all its realism, flight crews undergoing simulator training are under monitored conditions: Crews might arguably act differently because they know they are being scrutinized. Although simulation can be an ideal tool to combat flight crew errors, it might not be equally efficient in dealing with violations, which take place within the social, regulated environment and which include a motivational component (Maurino et al., 1995; Reason, Parker, & Lawton, 1995). Personal and organizational factors influence motivation and therefore foster or discourage violations, and such factors are hardly replicable through simulation. While the simulator remains a vital tool, it might not be sufficient to address certain CRM issues that might more effectively be addressed in a classroom.

Practitioners involved in AQP development have also expressed concern that the continuing development and validation of CRM issues and measures might become an academic exercise and therefore lose track of the real issues at hand. Accounts abound of lengthy discussions regarding issues such as, for example, whether a briefing should be evaluated as standard or substandard if the individual involved looks away and breaks eye contact. The immediate question would be: How does this help reduce human error in the cockpit?

Lastly, AC 120-54 conveys the notion of a degree of complexity commensurate with the significant departure from traditional training practices AQP represents. Such complexity has made it necessary for many airlines to designate a manager exclusively dedicated to dealing with AQP. While the savings AQP might bring to large organizations certainly justify such deployment of resources, it remains questionable whether smaller organizations will be able to take advantage of the safety and economic benefits of AQP; such resources as needed to implement AQP might not be available to these organizations. Developing proficiency objectives following the guidelines in AC 120-54 seems a daunting task beyond the professional expertise of the average instructor training pilot, still the cornerstone of most aviation training departments. This is particularly true outside the United States. It would indeed be regrettable if only major operators can benefit from AQP-like approaches because of lack of professional expertise to follow the guidelines of AC 120-54.

APPLES AND ORANGES: AN INTERPRETATION

Readers might perceive the previous section as a comparison between the MCC and the AQP; if so, it would be the proverbial comparing of apples and oranges, because of differences in cultural perceptions and understandings between Europe and North America. The previous section should simply be considered a presentation of facts and differences. But why such differences when pursuing identical goals? This section is an attempt to briefly answer this question.

The differences in European and North American approaches to implement CRM training as a system safety tool are a case of cultural issues involved in the transfer of technology. CRM was conceived in a small power distance society, with low degree of uncertainty avoidance. These cultural dimensions are reflected both in CRM and in the caution with which the United States pursued its implementation. Large power distance societies, or societies with a high degree of uncertainty avoidance, will instead try to "regulate" themselves out of problems. It is a way of building the authority system, of evading accountability and of passing the "buck" to front-line operators in case of safety breakdowns or system failures, so that authority remains intact. Cultural dimensions are clearly reflected in the issue of CRM evaluation: AQP, observing collective-based thinking, warns about the dangers involved in evaluating individual CRM behaviors, advocates for nonjeopardy training for the time being and imposes the requirement for feedback but only to improve the system. MCC, on the other hand, following individual-based thinking, regulates student evaluation on a pass–fail basis. Neither approach is "better" than the other; both approaches are as good as their respective effectiveness within the systems in which they are implemented. We must not miss, however, the crucial point in terms of understanding the differences: If we import bits and pieces—usually those which we like—of a solution that worked in one context to another without due consideration to the social givens and without an understanding of the original context, we may generate aberrant results within the receiving context (Maurino, 1995a).

The JAA favors a highly regulated, inflexible approach. Such an approach builds almost exclusively on assuring high levels of individual competence and optimum operational behaviors, with scant consideration for the system within which humans discharge their responsibilities. Consistent with this approach, physicians play a central role in keeping humans fit, supported in the initial screening process by psychologists. Because of the reliance on individual rather than system performance, individuals deemed unfit by prevailing standards must immediately be removed, thence the emphasis in evaluation, including psychological testing. The relevance of the knowledge to operational realities is of no great consequence, because knowledge is only a tool for the authority to dictate what operational personnel must and must not do. Across the Atlantic, the FAA favors also a highly regulated yet flexible approach. In this case, system performance is the objective, although substandard individual performance is expected to be addressed. Because system operation is paramount, operational personnel must be brought into the decision-making loop to support regulators and academicians in defining the blueprint of a healthy system. Knowledge belongs to the community, because the community plays a role in defining desired standards. Because system performance is the objective, individual evaluation is a lesser concern. Likewise, because knowledge is an essential tool for all involved, its relevance to operational realities is of prime importance.

As long as both approaches remain independent, there is no conflict. However, if we try to embed unmodified CRM, which builds upon a social/collective, nonjeopardy, flexible, agreement-by-consensus approach into individualistic, accountable, inflexible, top-down contexts, we will be laying grounds for a clash of objectives. Proper consideration of differences and cultural calibration of ideas, no matter how good they may be, are essential.

CONCLUSIONS

On balance, it might be argued that these are the best of the times and the worst of the times for CRM. CRM has established itself as a central protagonist of the aviation system, and it has become an international training requirement mandated by ICAO. However, as the numerous warnings flags raised in this chapter attest, there are very good reasons to plan the future of CRM with critical intelligence. Effective CRM training requires dropping the piecemeal strategies largely favored in the past in favor of a system approach, because implementing CRM does not only mean training pilots, controllers, or mechanics, but developing the organization. The notion that CRM is another tool to improve organizational effectiveness has transcended "traditional" operational boundaries, and has gradually started to be acknowledged by those in "high places" (Harper, 1995). This encouraging development must continue to gain momentum.

The imbalance of priorities and possibilities among different contexts remains a serious obstacle to the globalization of CRM as it exists today. This situation is closely related to the cross-cultural issues discussed in this chapter, and it will acquire particular relevance as airlines start implementing CRM training in response to the ICAO requirement in Annex 6. Nothing could be more distanced from reality than assuming that because CRM—as we know it—has worked in the United States, it will work anywhere. This does not imply that the *basic principles* underlying CRM are not of universal value—CRM is indeed a global concept. It is merely a reminder that the contents of the package with the basic principles might need to be substantially different.

Airlines within the "industrial belt" of the aviation community are dealing with "rocket science" CRM, while others not fortunate enough to have their headquarters located within this belt are still struggling with "CRM 101." The integration of flight and cabin crew CRM training is an example. As a consequence of safety studies and recommendations in accident reports, it has become the focus of attention of several major airlines (Vandermark, 1991). On the other hand, Dahlburg (1995) reported that pilots in India have refused to fly with senior flight attendants who receive larger salaries than some junior co-pilots. The pilots' view expresses indignation that "they compare a co-pilot to someone who only serves coffee and tea and keeps the passengers comfortable." The dispute has reached a point where flight attendants have been put off of a number of flights. Would "American" CRM find breeding grounds within this organizational context? Faizi's plea for consideration of contexts, Johnston's doubts about the suitability of existing CRM to some cultures, and Merritt's concerns about cultural imperialism are clear reminders that frontiers exist not only to provide a means of living for customs and immigration officers.

Any discussion on cultural issues associated to the transfer of technology has sensitive overtones, some of which have been discussed in this chapter. One major airline continues to challenge the Boeing's statistics on regional accident rates mentioned earlier, arguing that although an airline may have a clean record, because the airline is based in a poor-record region such statistics may damage its corporate image. While opinions as to the fairness of this complaint (or the fairness of the Boeing statistics for that matter) will surely differ, this airline's position reflects beyond doubts the perceived legitimacy of regional comparisons originated "at the top." It is only logical that less-developed regions might perceive cultural research on the transfer of technology as an indictment against their cultures. However, as the discussion on the implementation of CRM training in Europe reminds us, cultural issues are *also* involved in the transfer of technology between industrialized societies! If one lesson is to be drawn from these skirmishes, that would be that cultural research can become extremely vulnerable, that there are potential dangers in loose interpretations of cross-cultural research data, and that there are definite dangers in going *beyond* the data assuming evidence in attempting to support such data.

The European situation also provides food for further thought. As of lately, it is frequent to hear airline pilots lamenting that in every situation in which the professional group was asked to make concessions in the name of safety—concessions which other groups would have thought unacceptable, they add—the end result was a slap on their faces. Such was, pilots contend, the case with periodic reading of the Flight Data Recorder (FDR) for quality assurance purposes. On file at IFALPA—they argue—there are cases where FDR exceedances were a pretext to get rid of "undesirable" pilots. The frightening perspective that CRM might be used for similar purposes is very real, according to pilots' sources. They regret that despite the fact that human factors experts asked for a bona fide approach to develop CRM, now that the aberration of psychological testing is in the regulatory mill, now that individuals have been disqualified (i.e., sacked) through CRM seminars, and now that research and personal data have been unethically used, the human factors experts keep a conspicuous silence. ICAO (Maurino, 1995a, 1995c) has repeatedly warned against the considerable damage potentially involved in misunderstandings and misperceptions about CRM. Helmreich (1995) was the first researcher to take a position in denouncing what he has dubbed "The European Crisis." In the best interests of the future of CRM, it remains a matter of hope that more human factors experts will join Helmreich in proving the pilots wrong in their fears about further slappings.

But the real issue is that we must not forget the past. Since the early days of CRM at the NASA Workshop at San Francisco in 1979 all the way to the present day, the "founding fathers" in North America have gone to considerable efforts to present CRM as fact and not ideal. The involvement of operational personnel, the use of operational events, establishing and maintaining a healthy distance from the notion of cheap group therapy, the clear differentiation between changing attitudes while never addressing personality, were all attempts to convey the notion of a training tool without mystical connotations. Those were also the times of nonjeopardy and nonevaluation. CRM is a process, not an outcome, and research efforts in North America to assess CRM have been aimed toward the process rather than the outcome (individual performance). Although times have changed, all these givens remain essentially the same. It would nonetheless seem that, simple and sensible as they are, these givens never obtained an outbound clearance from U.S. immigrations. If we forget that CRM is a practical tool and not a creed, if we forget that the assessment of social activities involves much more than assessing 100-feet/10-knot tolerances, CRM will become extremely vulnerable. Once that point is reached, there will hardly be room to turn back. Almost 20 years of CRM wars deserve a happier ending than that.

ACKNOWLEDGMENTS

I would like to express my profound appreciation to my colleagues, Professor Robert L. Helmreich, Captain A. Neil Johnston, and Mr. Jean Pariès, for their review of this chapter.

REFERENCES

Baker, S., Barbarino, M., Baumgartner, M., Brown, L., Deuchert, I., Dow, N., Keller, M., Kocher, B., Skoniezki, A., & Wilson, P. (1995). *Team resource management in ATC—Interim report.* Brussels, Belgium: EUROCONTROL.

Barnes, R. M. (1993). Human performance and limitations requirements: The United Kingdom experience. In International Civil Aviation Organization (Ed.), *Proceedings of the Second ICAO Global Flight Safety and Human Factors Symposium, Washington DC, April 12–15, 1993* (pp. 23–29). Montreal, Canada: ICAO.

Birnbach, R. A., & Longridge, T. M. (1993). The regulatory perspective. In E. L. Wiener, B. G. Kanki, & R. L. Helmreich (Eds.), *Cockpit resource management* (pp. 263–281). San Diego, CA: Academic.

Blixth, J. (1995). Description of the SAS Flight Academy Crew Resource Management Program. In Ministere de l'Equipment, des Transports et du Tourisme (Ed.), *Proceedings of an International Seminar on Human Factors Training for Pilots.* Paris, France: Editor.

Bruggink, G. M. (1990). Reflections on air carrier safety. *The Log, British Airline Pilot Association (BALPA), July,* 11–15. London, England: BALPA.

Campbell, R. D., & Bagshaw, M. (1991). *Human performance and limitations in aviation.* London, England: BSP Professional Books.

Chidester, T. (1993). Role of dispatchers in CRM training. In R. S. Jensen & D. Neumeister (Eds.), *Proceedings of the Seventh International Symposium on Aviation Psychology* (pp. 182–185). Columbus, OH: The Ohio State University.

Cooper, G. E., White, M. D., & Lauber, J. K. (Eds.). (1979). *Resource management on the flight deck* (NASA Conference Publication No. 2120). Moffett Field, CA: NASA Ames Research Center.

Dahlburg, J. T. (1995, June 30). Battle of the sexes taken to the skies over India. *The Gazette, Montreal* (B1, p. 8).

Damos, D. (1995). Pilot selection batteries: a critical examination. In R. Fuller, N. Johnston, & N. McDonald (Eds.), *Applications of psychology to the aviation system* (pp. 165–169). Hants, England: Averbury Technical.

Faizi, G. H. (1996). Aviation safety performance of third world countries. In *InterPilot: Quarterly Review of the International Federation of Airline Pilots' Association, March,* Surrey, England (pp. 6–15).

Federal Aviation Administration. (1991). *Advanced qualification program* (Advisory Circular AC 120-54). Washington, DC: U.S. Department of Transportation.

Federal Aviation Administration. (1995). *Dispatch resource management* (Advisory Circular AC 121-32). Washington, DC: U.S. Department of Transportation.

Goeters, K. M. (1995). Psychological evaluation of pilots: The present regulations and arguments for their application. In R. Fuller, N. Johnston, & N. McDonald (Eds.), *Applications of psychology to the aviation system* (pp. 149–156). Hants, England: Averbury Technical.

Green, R. G., Muir, H., James, M.,Gradwell, D., & Green, R. L. (1991). *Human factors for pilots.* Hants, England: Averbury Technical.

Harper, K. (1995, June 30). Pilots given too much blame for crashes: IATA chief Jeanniot. *The Gazette, Montreal* (B1, p. 9).

Hawkins, F. H. (1993). *Human factors in flight* (2nd ed.). Hants, England: Ashgate Publishing Limited.

Hayward, B. J., & Lowe, A. R. (Eds.). (1993). *Towards 2000—Future directions and new solutions.* Albert Park, Victoria, Australia: The Australian Aviation Psychology Association.

Heinzer, T. H. (1993). Enhancing the impact of human factors training. In International Civil Aviation Organization (Ed.), *Proceedings of the Second ICAO Global Flight Safety and Human Factors Symposium, Washington DC, April 12–15, 1993* (pp. 190–196). Montreal, Canada: ICAO.

Helmreich, R. L. (1992). Fifteen years of CRM wars: A report from the trenches. In B. J. Hayward & A. R. Lowe (Eds.), *Towards 2000—Future directions and new solutions* (pp. 73–88). Albert Park, Victoria, Australia: The Australian Aviation Psychology Association.

Helmreich, R. L. (1994). New developments in CRM and LOFT training. In International Civil Aviation Organization (Ed.), *Report of the Seventh ICAO Flight Safety and Human Factors Seminar, Addis Ababa, Ethiopia, October 18–21, 1994* (pp. 209–225). Montreal, Canada: ICAO.

Helmreich, R. L. (1995). Critical issues in developing and evaluating CRM. In International Civil Aviation Organization (Ed.), *Seminar manual—First regional human factors training seminar, Hong Kong, September 4–6, 1995* (pp. 112–117). Montreal, Canada: Author.

Helmreich, R. L., Merritt, A., Sherman, P., Gregorich, S., & Wiener, E. (1993). *The Flight Management Attitude Questionnaire* (NASA/UT/FAA Technical Rep. No. 93-4). Austin, TX: NASA.

Helmreich, R. L., Wilhelm, J. A., & Gregorich, S. E. (1988). *Revised versions of the Cockpit Management Attitudes Questionnaire (CMAQ) and CRM seminar evaluation form* (NASA/UT Technical Rep. No. 88-3). Moffett Field, CA: NASA Ames Research Center.

Hofstede, G. (1980). Motivation, leadership and organizations: Do American theories apply abroad? *Organizational Dynamics, Summer*, 42–63.

Johnston, A. N. (1993a). CRM: Cross-cultural perspectives. In E. L. Wiener, B. G. Kanki, & R. L. Helmreich (Eds.), *Cockpit resource management* (pp. 367–398). San Diego, CA: Academic.

Johnston, A. N. (1993b). Human factors training for the new Joint European (JAA) pilot licences. In R. S. Jensen (Ed.), *Proceedings of the Seventh International Symposium on Aviation Psychology* (pp. 736–742). Columbus, OH: The Ohio State University.

Johnston, A. N. (1994). Human factors perspectives: The training community. In International Civil Aviation Organization (Ed.), *Report of the Sixth ICAO Flight Safety and Human Factors Seminar, Amsterdam, Kingdom of The Netherlands, May 16–19, 1994* (pp. 199–208). Montreal, Canada: Author.

Maurino, D. E. (1991). The ICAO Flight Safety and Human Factors Programme. In R. S. Jensen (Ed.), *Proceedings of the Sixth International Symposium on Aviation Psychology* (pp. 1014–1019). Columbus, OH: The Ohio State University.

Maurino, D. E. (1992). Shall we add one more defense? In R. Heron (Ed.), *Third Seminar in Transportation Ergonomics, October 7, 1992*. Montreal, Canada: Transport Canada Development Centre.

Maurino, D. E. (1994). Crosscultural perspectives in human factors training: Lessons from the ICAO human factors programme. *The International Journal of Aviation Psychology, 4*(2), 173–181.

Maurino, D. E. (1995a). The future of human factors and psychology in aviation from the ICAO's perspective. In R. Fuller, N. Johnston, & N. McDonald (Eds.), *Applications of psychology to the aviation system* (pp. 9–15). Hants, England: Averbury Technical.

Maurino, D. E. (1995b). Cultural issues and safety: Observations of the ICAO Human Factors Programme. *Earth Space Review, 4*(1), 10–11.

Maurino, D. E. (1995c). Human factors training for pilots: A status report. In Ministere de l'Equipment, des Transports et du Tourisme (Ed.), *Proceedings of an International Seminar on Human Factors Training for Pilots*. Paris, France: Editor.

Maurino, D. E., Reason, J., Johnston, A. N., & Lee, R. (1995). *Beyond aviation human factors*. Hants, England; Averbury Technical.

Merritt, A. C. (1993). The influence of national and organizational culture on human performance. *Invited paper at an Australian Aviation Psychology Association Industry Seminar, October 25, 1993*. Sydney, Australia: The Australian Aviation Psychology Association.

Merritt, A. C. (1994). Cultural issues in CRM training. In International Civil Aviation Organization (Ed.), *Report of the Sixth ICAO Flight Safety and Human Factors Seminar, Amsterdam, Kingdom of The Netherlands, May 16–19, 1994* (pp. 326–343). Montreal, Canada: Editor.

Merritt, A. C. (1995, May). Designing culturally sensitive CRM training: CRM in China. *Invited paper at the Training and Safety Symposium, Guangzhou, People's Republic of China*. Guangzhou, People's Republic of China: Civil Aviation Authority of China.

Merritt, A. C., & Helmreich, R. L. (1995). Culture in the cockpit: A multi-airline study of pilot attitudes and values. In R. S. Jensen (Ed.), *Proceedings of the Eighth International Symposium on Aviation Psychology* (p. 676). Columbus, OH: The Ohio State University.

Murphy, E. (1995). JAA psychological testing for pilots: Objection and alarms. In R. Fuller, N. Johnston, & N. McDonald (Eds.), *Applications of psychology to the aviation system* (pp. 157–164). Hants, England: Averbury Technical.

Orasanu, J. M. (1994). Shared problem models and flight crew performance. In A. N. Johnston, N. McDonald, & R. Fuller (Eds.), *Aviation psychology in practice* (pp. 255–285). Hants, England: Averbury Technical.

Orlady, H. W., & Foushee, H. C. (Eds.). (1986). *Cockpit resource management training* (NASA Conference Pub. No. 2455). Moffett Field, CA: NASA Ames Research Center.

Pariès, J., & Amalberti, R. (1995). *Recent trends in aviation safety: From individuals to organizational resources management training.* In Risøe National Laboratory Systems Analysis Dept. (Eds.), *Technical Report* (Risøe Series 1; pp. 216–228). Roskilde, Denmark: Risøe National Laboratory.

Pearson, R. A. (1995). JAA psychometric testing: The reasons. In R. Fuller, N. Johnston, & N. McDonald (Eds.), *Applications of psychology to the aviation system* (pp. 139–140). Hants, England: Averbury Technical.

Phelan, P. (1994). Cultivating safety. *Flight International,* August 24–30, 1994, 22–24.

Phillips, D. (1994, August 23). Aviation safety vs. national culture? Boeing takes a flyer. *Paris Herald Tribune,* p. 27.

Reason, J. (1990). *Human error.* Cambridge, England: Cambridge University Press.

Reason, J., Parker, D., & Lawton, R. (1995). Organisational controls and the varieties of rule-related behaviour. In Economic and Social Research Council (Eds.), *Proceedings of a Conference on Risk in Organisational Settings, London, May 16–17, 1995* (pp. 67–95). York, England: University of York, Heslington.

Robertson, M. M., Taylor, J. C., Stelly, J. W., & Wagner, R. H. (1995). Maintenance CRM training: Assertiveness attitudes effect on maintenance performance in a matched sample. In R. Fuller, N. Johnston, & N. McDonald (Eds.), *Human factors in aviation operations* (pp. 215–222). Hants, England: Averbury Technical.

Russell, P. D. (1993) Crew factors accidents: Regional perspective. In International Air Transport Association (Ed.), *Proceedings of the 22nd Technical Conference: Human Factors in Aviation* (pp. 45–61). Montreal, Canada: International Air Transport Association.

Smith, P. M. (1992, Autumn). Some implications of CRM/FDM for flight crew management. *Flight Deck, 5,* 38–42. Heathrow, England: British Airways Safety Services.

Taggart, W. R. (1995). Implementing human factors training in technical operations and maintenance. *Earth Space Review, 4*(1), 15–18.

Trollip, S. R., & Jensen, R. S. (1991). *Human factors for general aviation.* Englewood, Colorado: Jeppesen Sanderson, Inc.

Vandermark, M. J. (1991). Should flight attendants be included in CRM training? A discussion of a major air carrier's approach to total crew training. *The International Journal of Aviation Psychology, 1*(1), 87–94.

Wiener, E. L., Kanki, B. G., & Helmreich, R. L. (Eds.). (1993). *Cockpit resource management.* San Diego, CA: Academic.

Yamamori, H. (1986). Optimum culture in the cockpit. In H. W. Orlady & H. C. Foushee (Eds.), *Cockpit resource management training* (NASA Conference Pub. No. 2455, pp. 75–87). Moffett Field, CA: NASA Ames Research Center.

10

Fatigue and Biological Rhythms

Giovanni Costa
Università di Verona, Italy

This chapter is concerned with temporal factors affecting human performance and work efficiency. The aim is to emphasize that a careful consideration of the temporal structure of body functions and, consequently, a proper timing of work activities can be of paramount importance to ensure high levels of performance efficiency, decreasing fatigue and enhancing health and safety.

BIOLOGICAL RHYTHMS OF BODY FUNCTIONS

The temporal organization of the biological systems is one of the most remarkable characteristics of the living organisms. In the last few decades chronobiology has highlighted the importance of this aspect for the human life, revealing the complex mechanisms underlying the temporal interactions among the various components of the body (systems, organs, tissues, cells, subcellular structures).

These are characterized by a large spectrum of rhythms having different frequencies and amplitudes. According to their periodicity (τ), three main groups of biological rhythms have been defined: (a) ultradian rhythms ($\tau \leq 20$ hr), such as heart rate, respiration, and electric brain waves; (b) circadian rhythms (20 hr $\leq \tau \leq$ 28 hr), such as the sleep/wake cycle or temperature; (c) infradian rhythms ($\tau \geq 28$ hr), among which circaseptan (weekly), circatrigintan (monthly), and circannual rhythms can be found, like immunological response, the menstrual cycle, and seasonal mood/hormonal changes, respectively.

Circadian (Latin: *circa diem* = about a day) rhythms are the most extensively studied due to their great influence on everyday life.

Circadian Rhythms and Their Mechanism

The human is a daylight creature; in the course of evolutionary adaptation, the human species has associated its own state of wakefulness and activity (ergotropic phase) with the day/light period and its sleep and rest state (trophotropic phase) with the night/dark period.

Although in modern society artificial lighting makes it possible to have light for the whole 24-hr span, the body functions (hormonal, metabolic, digestive, cardiovascular, mental, etc.) are still influenced mainly by the natural light/dark cycle, showing periodic oscillations that have, in general, peaks (acrophases) during the daytime and troughs at night.

For example, body temperature, the main integrated index of body functions, decreases during the night sleep, reaching a minimum of 35.5–36°C at about 4 a.m., and increases during the day up to a maximun of 37–37.3°C around 5 p.m.. This reflects the increased basal level of body arousal during the day in order to be fit for activity, whereas it decreases at night in order to restore the body and recover from fatigue.

After experiments with subjects living in isolation chambers or caves without any reference to external time cues ("free-running"), circadian rhythms have been proved to be sustained by an endogenous timing system, located in the suprachiasmatic nucleus of the hypothalamus in the brain, acting as an autonomous oscillator or "body clock." Its inherent oscillation shows an effective period of 25 hr that can be shortened up to 22.5 hr or lengthened up to 27 hr (*entrainment*) by varying the light/dark cycle.

In normal living conditions, this endogenous clock is entrained to the 24-hr period ("masking effect") by the rhythmic changes of external socioenvironmental synchronizers or *zeitgebers*, such as the light period, habitual sleep and meal times, the timing of work and leisure activities (Minors & Waterhouse, 1986; Wever, 1985).

As a result, a multitude of circadian rhythms of psychological and physiological variables, having different phases and amplitudes, interact with each other and harmonize on the 24-hr period to sustain normal body functioning. For example, during the day, increasing temperature levels are associated with high sympathetic nervous system activity, higher metabolic rate, increased alertness, better vigilance performance, and physical fitness; during the night, lower temperature levels are associated with increased parasympathetic activity, low metabolic rate, increased sleepiness, and poorer work efficiency. As concerns hormones, cortisol shows its acrophase in the early morning, adrenaline around noon, and melatonin after midnight.

In subjects synchronized to the normal living routine, the loss of this harmonization, or the disruption of some circadian rhythms, can be a premonitory symptom of health impairment as well as one of the clinical manifestations of disease (Reinberg & Smolenski, 1983).

Two work-related conditions are the main causes of disruption of this circadian rhythmicity: (a) shift and night work, which requires a periodic rotation of the duty period around the 24-hr span; and (b) long transmeridian flights, which impose rapid time-zone transitions and irregular changes in the light/dark regimen.

Therefore, respect for the rhythmic organization of body functions is of paramount importance for people engaged in aviation jobs. In fact, most of them have to face continuous interference with biological and social rhythms in relation to their specific work activity: Flying personnel have to cope with time-zone transitions ("jet lag") and, like ground personnel (e.g., air traffic control, maintenance, services), do shift and night work.

Circadian Rhythms of Vigilance and Performance

The circadian oscillation in the physiological state plays an important role in influencing human performance and fatigue. It is common knowledge that work efficiency during the night is not the same as during the day. There is in fact clear evidence that it can be significantly influenced by the time of day in which the task is performed.

In normal conditions, alertness and efficiency of many psychomotor functions show in general a progressive increase after waking with peaks 8–10 hr later, in the afternoon; after that they progressively worsen, with troughs at night. They appear roughly influenced by the wake–sleep cycle and parallel the body temperature rhythm (Åkerstedt, 1996; Colquhoun, 1971).

Moreover, after well-controlled laboratory experiments, these fluctuations have been shown to vary according to task demands, suggesting a different weight of endogenous (multioscillators?) and exogenous components on mental activities. For example, performance in simple perceptual motor tasks shows an improvement over the day with higher levels in the late afternoon, whereas in tasks with high short-term memory load performance decreases from morning to evening. On the other, hand, performance having high cognitive or "working memory" component (e.g., verbal reasoning and information processing) shows an alternate trend with better levels around midday (Colquhoun, 1971, 1987; Folkard, 1990; Monk, 1990).

These fluctuations can vary up to 30%, and also reflect the interactions of many other factors, in which a time of day effect can be seen as well, both in terms of phase and amplitude.

The relationship between mental efficiency and basal arousal (the inverse of sleepiness) is described by an inverted U-shaped curve (Blake, 1971); furthermore, the optimal arousal level for a task depends on its structure, with more complex tasks having lower optimal arousal levels (Folkard & Monk, 1985). Increased arousal levels, due to higher motivation or work demand, decrease the circadian oscillation of performance efficiency, counteracting by an extra effort the reduction of alertness and increasing fatigue. Lack of interest and boredom have the opposite effect, as fluctuations increase when motivation is low (Alluisi & Chiles, 1967).

A "post-lunch dip" in performance has been documented in both laboratory and field conditions (Monk & Folkard, 1985). Although it can be related to the exogenous masking effect of food ingestion (with consequent "blood steal" from the brain and increased sleepiness), there are some suggestions that it may be also due to an endogenous temporal component. In fact, after experiments with no meals and in temporal isolation conditions, it appears that at least part of it is due to a decrease of the arousal level that does not parallel the underlying increase of body temperature. According to some authors, the "post-lunch dip" reflects a bimodality of circadian rhythms due to a 12-hr harmonic of the circadian system (Hildebrandt, Rohmert, & Rutenfranz, 1974; Richardson, Carskadon, Orav, & Dement, 1982). On the other hand, others support the assumption that the sleepiness/alertness cycle has a 4-hr rhythm (Zulley & Bailer, 1988), so that a high level of performance cannot be maintained for more than 2 consecutive hours. Besides, Lavie (1985) indicated the existence of 90-min cycles (ultradian rhythms) in alertness and, consequently, in perceptual-motor performance ("biological working units").

Moreover, performance impairment and negative oscillations increase with prolonging working hours and/or sleep deprivation, particularly in more complex tasks (Alluisi, 1972). The maximum decrement during the same extended duty period can be twice as severe when work starts at midnight rather than at midday (Klein & Wegmann, 1979a). Naps inserted during work periods, especially at night, reduce the negative fluctuations, although a temporary grogginess ("sleep inertia") might appear if the person is suddenly awakened during the deep sleep phase (Gillberg, 1985).

As concerns individual factors, the circadian phase position of biological rhythms seems to have a relevant influence on performance efficiency and fatigue.

The morning active types (or "larks") appear to have more difficulties in coping with night work compared to evening active types (or "owls"), probably because of their earlier psychophysical activation. In fact, they show an advanced phase position of alertness and body temperature, with acrophase 2 hr earlier during the day, and a consequent quicker decrease during the night hours. They have less difficulties in coping with early morning activities, as their temporal structure facilitates an early awakening and higher vigilance in the first part of the day (Clodorè et al., 1987; Foret, Benoit, & Royant-Parola, 1982; Kerkhof, 1985; Ostberg, 1973).

Introverted people appear to act mainly as morning types, whereas extroverts tend to behave as evening types; the differences in operating behavior become more evident when these characteristics are associated with neurotic instability, with the neurotic introverts showing a worse phase adjustment of circadian rhythms (Colquhoun & Folkard, 1978).

Age also appears to correlate with morningness (Åkerstedt & Torsvall, 1981); moreover, aging shows a greater susceptibility to the occurrence of rhythm disturbances, sleep disorders and psychic depression, which can in turn contribute to causing performance impairment.

PROBLEMS CONNECTED WITH SHIFT WORK AND TRANSMERIDIAN FLIGHTS

Shift and Night Work

Shift work interferes with biological and social rhythms, forcing people to adapt them to unusual work schedules. On night work in particular, they must change their normal sleep–wake pattern and adjust their body functions to inverted rest–activity periods, having to work when they should sleep and sleep when they should stay awake.

Such adjustment entails a progressive phase shift of body rhythms, which increases with the number of successive night shifts, oriented forward or backward respectively according to the advanced (afternoon–morning–night) or delayed (morning–afternoon–night) rotation of the duty periods. The circadian system is exposed to a continuous stress in the attempt to adjust as quickly as possible to the new working hours, while at the same time being invariably frustrated by the continuous "changeover" imposed by the alternating shift schedules. Therefore, people seldom or never adjust completely or reach a total inversion, even in case of permanent night work or slowly rotating shift systems (7–15 consecutive nights shifts); this is also because family and social cues are diurnal and workers immediately go back to their normal habits during rest days.

In many cases, a flattening of the amplitude or a delinking ("desynchronization") among the different rhythmic functions, having different speed and direction of readjustment, can be observed (Åkerstedt, 1985a; Knauth, Rutenfranz, Herrmann, & Poppel, 1978; Knauth, Emde, Rutenfranz, Kiesswetter, & Smith, 1981; Reinberg et al., 1988).

Such perturbation of the rhythmic structure plays an important role in influencing health and work capacity. People can suffer to a greater or lesser extent from a series of symptoms ("shift-lag" syndrome), characterized by feelings of fatigue, sleepiness, insomnia, digestive problems, poorer mental agility, and impaired performance (Comperatore & Krueger, 1990; Kogi, 1985; Tilley, Wilkinson, Watson, Warren, & Drud, 1982).

Sleep is the main function altered, being decreased both in quantity and in quality. A significant reduction of hours of sleep occurs during the morning shift period in relation to the early wakeup, which also causes a reduction of REM (rapid eye movement) sleep. On night shifts, diurnal sleep is perturbed both for "circadian reasons," due to difficulty in falling asleep during the rising phase of body temperature, and for the interferences connected with unfavorable environmental conditions (light and noises in particular). Consequently, it is more fragmented and perturbed in its ultradian components (less REM sleep and phase 2), thus losing part of its restorative properties (Åkerstedt, 1985b, 1990, 1996; Knauth et al., 1980; Knauth & Rutenfranz, 1975; Kogi, 1982; Tepas & Carvalhais, 1990).

Such conditions in the long run not only can give rise to permanent and severe disturbances of sleep, but also can cause chronic fatigue and changes in behavior patterns, characterized by persistent anxiety or depression, which often requires medical treatment with administration of psychotropic drugs (Gordon, Cleary, Parker, & Czeisler, 1986; Koller, 1983; Tepas, 1989).

Shiftworkers' health impairment can include other psychosomatic disorders of the gastrointestinal tract (colitis, gastroduodenitis, and peptic ulcer) and cardiovascular system (hypertension, ischemic heart diseases) that are influenced by other time- and work-related factors and behaviors (Åkerstedt, Knutsson, Alfredsson & Theorell, 1984; Costa, 1996; International Labour Office [ILO], 1988; Waterhouse, Folkard, & Minors, 1992). In fact, eating habits are often disrupted in shift workers, who are forced to change their timetables (according to the shift schedules) and food quality. Moreover, they tend to increase the consumption of stimulating substances (coffee, tea, alcohol, tobacco smoke) that have negative effects on digestive and cardiovascular functions (Lennernas, 1993). Furthermore, shiftworkers may well experience more difficulties in keeping normal relationships at family and social level, with negative impact on marital relations, care of children, and social contacts, which can further contribute to the development of the already mentioned psychosomatic disorders.

Besides, we have to consider the multifactorial characteristic of such disorders, and their chronic-degenerative trend. In fact, they are quite common among the general population and show the influence of several factors concerning genetic and family heritage, personality, life styles, and social and working conditions. Therefore, shift work has to be seen as one of the many risk factors that favor their development, which is more likely to become apparent after a long-term exposure.

Consequently, the process of maladaptation to shift work, which according to Haider, Cervinka, Koller, and Kundi (1988) may develop through four consequent phases of adaptation, sensitization, accumulation, and manifestation, can have a different speed and intensity among shiftworkers according to different personal and environmental factors.

Jet Lag

Air crews operating on long transmeridian flights have to cope with a shift of external time besides the shift of the working period. Therefore the individual biological rhythms have to adjust to abnormal working hours in a changed environmental context, in which a shift of time has occurred as well. The short-term problems arising from these conflicts are similar to those of normal shift work, but are often aggravated by the fatigue due to the extended duty periods and by loss of the usual external time cues.

After a long transmeridian flight, the circadian system does not adjust immediately to the new local time, but requires several days in relation to the number of time zones crossed; the greater the number, the longer is the time required, considering that human circadian system can adjust to no more than 60–90 min per day (Wegmann & Klein, 1985).

The speed of resynchronization can differ among individuals (e.g., aged people adjust more slowly than youths) and between variables, leading to an "internal dissociation" (i.e., heart rate and catecholamines adjust more quickly than body temperature and cortisol; the same is for simple mental tasks in comparison with complex psychomotor activities). The adjustment is generally more rapid in westbound than eastbound flights (Gander, Myhre, Graeber, Andersen, & Lauber, 1989; Härmä, Laitinen, Partinen, Ilmarinen, & Suvanto, 1989; Suvanto, Partinen, Härmä, & Ilmarinen, 1990; Wegmann, Klein, Conrad, & Esser, 1983; Wegmann et al., 1986); in the first case there is a progressive phase delay of the circadian rhythms in relation to the extended personal day, whereas in the latter there is a phase advance due to the compressed day (*directional asymmetry*). A complete readjustment after transition of six time zones was found to take 13 days in eastward and 10 days in westward flights (Wegmann & Klein, 1985). The reason for the quicker adjustment of body rhythms to a phase delay seems to be related to the natural lengthening of the "biological" day, which arises when people are kept in complete isolation ("free running"), showing a period of 25 hr.

However, the recovery pattern is not necessarily linear, but can present a "zigzag" trend, with some of the postshift days showing worse levels of adjustment than those on the day immediately preceding (Monk, Moline, & Graeber, 1988). Furthermore, in several cases, particularly after more extended time zone transitions, a splitting of rhythms may occur; that is, some adjust by phase advance whereas others adjust by phase delay ("resynchronization by partition") (Gander et al., 1989; Klein & Wegmann, 1979b).

During this period, the person suffers from the so called "jet-lag" syndrome, characterized by a general feeling of malaise with sleepiness, fatigue, insomnia, hunger, and digestive disturbances with constipation or diarrhea. Sleep times and patterns appear much more variable and disrupted after eastward than westward flights across an equivalent number of time zones, in particular as concerns a reduction of REM sleep and increase of SWS (slow wave sleep) phases (see Graeber, 1986).

Consequent to the desynchronization (and in association with sleep deficit and fatigue), performance on many psychomotor activities (e.g., reaction time, hand-eye coordination, logical reasoning, vigilance) also shows an acute 8%–10% decrement that can last for 3–5 days (in the case of six or more time zones crossed). This appears more pronounced in the afternoon and early night hours after a westbound flight, and during the morning and early afternoon after an eastbound flight (Klein & Wegmann, 1979b; Wegmann & Klein, 1985), and corresponds to the effect of a moderate alcohol consumption (Klein, Wegmann & Hunt, 1972).

This decrement depends also on the nature of the task and can lead to a lower work efficiency, which must be compensated by higher motivation and extra effort. Moreover, in air crews operating on long distance flights, jet lag is one of the components of impaired well-being and fitness that is also considerably affected by the great irregularity of the rest/activity patterns and by prolonged duty periods. Therefore, the resulting fatigue is due to two components: (a) desynchronization of the circadian rhythms, and (b) prolonged physical and mental effort; the latter is usually compensated for by an adequate night's rest, as occurs after long north–south flights without time-zone transition.

As concerns long-term effects on health, it is still questionable whether people engaged for many years on transmeridian routes (and therefore subject to frequent and irregular perturbations of the body temporal structure) have more negative consequences. The complaints mostly reported concern sleep and nervous disturbances, chronic fatigue, and digestive problems, like normal shiftworkers; however, clinical findings do not report an incidence significantly different from the general population. However, this comparison can be affected by some confounding factors. One is that air crews are a highly selected and supervised population; a second is due to possible masking behaviors, as a negative assessment of medical fitness for work may have important economic consequences; a third one is the high interindividual variability in terms of years spent on long-haul flights during working life (Haugli, Skogtad, & Hellesøy, 1994; Wegmann & Klein, 1985).

Errors and Accidents Due to Performance Decrement and Fatigue in Workers on Irregular Work Schedules

The circadian fall in psychophysical activation at certain hours of the day, often aggravated by disruption of the biological rhythms, sleep deficit, and fatigue, certainly decreases work efficiency and increases the possibility of errors and accidents.

Significant decrements in work performance at night (up to 100%) with consequent errors or accidents have been reported in many groups of workers engaged on irregular work schedules, such as switchboard operators, gas-work log-keepers, car drivers, spinners, ship workers, nurses, and train and truck drivers (see Monk & Folkard, 1985, for a review). In all reports, a less pronounced "post-lunch dip" has been also documented, the possible causes of which have already been mentioned.

Electroencephalographic recordings (Torsvall & Åkerstedt, 1987) have clearly demonstrated the occurrence of dramatic episodes of sleepiness on the job in train drivers during night work, particularly in the second half, leading to potentially hazardous errors (failure to respond to signals). Moreover, the too early starting hours of morning shifts have been documented to influence higher frequencies of errors and accidents in train and bus drivers (Hildebrandt, Rohmert, & Rutenfranz, 1975; Pokorny, Blom, & Van Leeuwen, 1981).

Lapses in performance have been described also among air traffic controllers (Folkard & Condon, 1987) concerning the so-called "night-shift paralysis," such as a sudden immobility of the voluntary muscles during consciousness, which can last from a few seconds to a few minutes, occurring in about 6% of the subjects with peaks around 5 a.m.; it seems related to the number of successive night shifts worked and sleep deprivation.

These are important aspects to consider, particularly when high and sustained levels of performance efficiency are required as public health is at stake, and its failure may be very costly both from the social and economic point of view; such is the case with aviation activities. Besides, new technologies, which increase cognitive tasks and require more alertness and vigilance, are often more vulnerable to errors than manual work activities. On the other hand, automated systems may increase monotony and boredom, thus decreasing vigilance and safety, particularly in case of emergency. It is worth mentioning that both the two main nuclear-reactor accidents at Three Mile Island (1979) and Chernobyl (1986) and the Bophal chemical plant disaster (1984) started during the night hours (at 0400, 0125, and 0057 hr, respectively), and in all situations "human error" has been claimed as an important factor. Besides, Kelly and Schneider (1982) assessed the risk of accident in a nuclear power plant working on 12-hr shifts to be 70% higher than on 8-hr shifts.

On long-haul flights, cockpit crews periods of decrement in vigilance are more likely to occur during the monotonous cruise, and occur more frequently during eastbound than westbound flights (Cabon, Coblentz, Mollard, & Fouillot, 1993; Graeber, 1982).

Reviewing some of the main airline accidents that occurred in the period 1967–1988, Price and Holley (1990) underlined that chronic fatigue, sleep loss, and desynchronosis were three "human factors" that contributed significantly to the unfavorable events. In most cases they were the consequence of improper work scheduling, which imposed prolonged duty periods and irregular wake times in the previous hours, not allowing sufficient time to rest and sleep. In other cases the influence of circadian desynchronization due to time-zone changes or to night work (in one case concerning maintenance personnel) appeared evident. The negative effect of sleep loss and shift work on mental performance has been claimed also for the Space Shuttle Challenger accident.

However, the epidemiological studies concerning work accidents among normal shiftworkers are quite controversial; some investigations reported more accidents on night shifts, others found more on day shifts, and still others reported less frequent but more serious events on night shifts (Carter & Corlett, 1978). This can be explained considering the different tasks and work sectors examined (at major or minor risk of accident), and the daily fluctuations in work demands, usually with lower levels at night (i.e., night interruption of higher risk jobs, slowing down of work pace, added automation), which can compensate for the reduction in psychophysical performance.

Specific Problems for Women

It is legitimate to presume that irregular working hours and altered sleep/wake cycles may have more or more specific adverse effects on women's health, above all in relation to their peculiar periodical hormonal activity (the menstrual cycle) and the reproductive function.

Moreover, female shiftworkers may have to face more stressful living conditions in relation to higher work loads and time pressure connected with their additional domestic duties. For example, women with children show greater sleep problems with consequent higher cumulative fatigue (Dekker & Tepas, 1990; Estryn-Behar, Gadbois, Peigne, Masson, & Le Gall, et al., 1990; Gadbois, 1981; Gersten, Duchon, & Tepas, 1986; Uehata & Sasakawa, 1982). Therefore, these biological and sociocultural factors may affect women and men differently, not only in terms of social life, but also with respect to adaptation and tolerance to irregular working hours.

As concerns sexual differences in circadian rhythms, studies carried out in well-controlled experimental conditions did not report significant differences between males and females, although women seem more likely to desynchronize by a shortening of their sleep/wake cycle (Wever, 1985). Also in real working conditions many studies, carried out prevalently on nurses, did not report differences in adjustment of biochemical parameters and psychophysical functions, including vigilance and performance (Folkard, Monk, & Lobban, 1978, 1979; Hildebrandt & Stratman, 1979; Minors & Waterhouse, 1985; Smith, 1979; Suvanto et al., 1990).

In relation to the menstrual cycle, it has been frequently reported that arousal and mood tend to worsen in the premenstrual and menstrual phases, whereas mental tension, anxiety, and depression increase; this can determine changes in arousal and negatively affect work performance. It is still an open question whether this is related to hormonal mechanisms or more to psychological and social factors (Patkai, 1985; Redgrove, 1971).

However, women engaged on irregular shift schedules and night work show a higher incidence of menstrual disorders (irregular cycles and menstrual pains), and problems of fertility (lower rates of pregnancies, more abortions/miscarriages or preterm deliveries, low-birth-weight infants; Axelsson et al., 1989; McDonald et al., 1988; Nurminen, 1989; Uehata & Sasakawa, 1982).

Also, among airline flight attendants several studies reported high frequencies of irregular menstrual cycles and dysmenorrhea (Cameron, 1969; Preston, Bateman, Short, & Wilkinson, 1973), and possible increased risk of spontaneous fetal loss, but no other adverse outcomes on pregnancy (Daniell, Vaughan, & Millies, 1990; Lyons, 1992).

Interindividual Differences

It is generally recognized that about 20% of workers cannot stand irregular working schedules, including night work, because of manifested intolerance, particularly dealing with difficult rhythm adjustment and sleep deprivation. On the other hand, only 10% of them do not complain at all, whereas the remaining withstand shift and night work with different levels of discomforts and health disorders. In fact, the effects of such stress condition can vary widely among people in relation to many intervening variables concerning individual factors as well as working situations and social conditions.

Aging is generally associated with a progressive intolerance due to several reasons: tendency to flattening and instability of circadian rhythms; deterioration in sleep restorative efficiency; decreasing in overall psychophysical fitness and capacity to cope with stressors (Åkerstedt, 1985a). Gander, De Nguyen, Rosekind, and Connell (1993) documented a significant increase of daily sleep loss, and a decline in the amplitude of baseline temperature rhythm, with increasing age of air crew members, particularly those involved on long-haul flights: subjects aged 50–60 years showed on average 3.5 times more sleep loss per day than subjects aged 20–30 years. Also, Suvanto et al. (1990) found a significant correlation between age and desynchronosis in flight attendants.

Subjects having a greater amplitude of the oral temperature circadian rhythm, who show a more stable circadian structure and a slow adjustment after a phase change, seem to have a better long-term tolerance to shift and night work (Reinberg et al., 1980). It is the opposite for those who exhibit low amplitudes and are more prone to a persisting internal desynchronization (Reinberg et al., 1984).

Moreover, the already mentioned "evening types," who show delayed circadian peaks of temperature and alertness, have fewer sleeping problems and a better adaptation to night work than "morning types" (Breithaupt et al., 1978; Folkard & Monk, 1985; Hildebrandt & Stratmann, 1979). The same appears as concerns the adjustment to jet lag after transmeridian flights (Colquhoun, 1979).

Other authors have stressed the influence of some personality and behavioral aspects, such as the characteristics of "rigidity of sleeping habits" and "languidity to overcome drowsiness" (Costa, Lievore, Casaletti, Gaffuri, & Folkard, 1989; Folkard et al., 1979), as well as extroversion and neuroticism (Colquhoun, 1981; Gander et al., 1989; Nachreiner, 1975), in negatively conditioning both short-term adjustment and long-term tolerance to irregular work schedules and jet lag. However, it is also possible that increased neuroticism could be more a consequence than a cause of long-term shiftwork intolerance (Meers, Maasen, & Verhaegen, 1978).

On the other hand, good physical fitness, consequent to physical training interventions, is able to lessen fatigue and increase performance on night shifts (Härmä, Ilmarinen, & Knauth, 1988).

Moreover, the level of commitment to (shift) work, that is, the degree to which the person is able to structure his or her life around it (e.g., avoiding moonlighting or other stressful extrajob activities, adopting more strict sleep/wake regimens) can favor better adjustments and tolerance (Adams, Folkard, & Young, 1986; Folkard et al., 1978). This can be enhanced by a vigorous personality or high motivation (Verhaegen et al., 1987).

PREVENTIVE MEASURES

Personal Coping Strategies

As a countermeasure against jet lag, air crews are sometimes advised to try to maintain their home-base time as much as possible. This can be possible in case of a short stay and quick return home (in 2–3 days), if provided an adequate sleep length in proper bedrooms shielded from light and noise. Otherwise, increased drowsiness is likely to occur during the subsequent long-haul flight.

In case of a prolonged work period in a different time zone, it is advisable to force the speed of adjustment by immediate immersion in the local time and social activities. This can be facilitated by a proper use of exposure to bright light, both through outdoor physical activities and increasing indoor artificial lighting (>1,000 lux). Bright light, in fact, besides having a direct stimulating effect on mental activity, influences the pineal gland and suppresses the secretion of melatonin, an hormone that plays an important role in the entreinment of the circadian system. Therefore, proper timing of light exposure can affect the direction and the magnitude of the entrainment of circadian rhythms: For example, light exposure in the morning causes a phase advance, whereas light exposure in the evening causes a phase delay (Czeisler et al., 1989; Eastman, 1990; Lewy & Sack, 1989; Wever, 1989).

These effects have useful implications also for shift work, provided that bright light could be used during the night shift, as concerns not only short-term adjustments but also long-term tolerance (Costa, Ghirlanda, Minors, & Waterhouse, 1993; Eastman, 1990). In fact, bright light can reduce the symptoms of seasonal affective disorders, and some of the negative effects of night work can be likened to a mild form of endogenous depression (Healy & Waterhouse, 1991).

In recent years the oral administration of melatonin has also been tested to counteract both shift lag (Folkard, Arendt, & Clark, 1993) and jet lag (Arendt & Aldhous, 1988). It has been proved to be useful in inducing sleep and fastening the resetting of circadian rhythms, reducing feeling of fatigue and sleepiness, and increasing sleep quality and duration, without impairing performance and causing negative effects on health.

Moreover, proper timing and composition of meals can help the adaptation. In principle, people should try to maintain stable meal times, which can act as cosynchronizers of body functions and social activities. In cases when full resynchronization of circadian rhythm is required, some authors propose special diet regimens, assuming that meals with high carbohydrate content facilitate sleep by stimulating serotonin synthesis, whereas meals with high protein content, which stimulates catecholamines secretion, favor wakefulness and work activity (Erhet, 1981; Romon-Rousseaux, Lancry, Poulet, Frimat, & Furon, 1987). During night work in particular, it would be preferable that shiftworkers have the meal before 0100 hr (also to avoid the coincidence of the postmeal dip with the alertness trough), then take only light snacks with carbohydrates and soft drinks, and not later than 2 hr before going to sleep.

At present these strategies appear very promising, including the consideration of the fact that they can reduce or avoid the use of many drugs currently taken to alleviate the jet-lag symptoms. In fact, the assumption of hypnotics to induce sleep (usually benzodiazepines) has no effect on the process of resynchronization, or may even retard it by interaction with neurotransmitters and receptors; moreover, they can cause a transient (up to 12 hr) impairment in psychomotor performance (e.g., visuomotor coordination). Furthermore, in the case of a prolonged stay in different time zones, forcing the sleep recovery can also disturb the slow physiological realignment of the other circadian functions, taking into consideration the "zigzag" pattern of the readjustment process (Monk et al., 1988; Walsh, 1990).

On the other hand, the use of stimulating substances (such as xanthines contained in coffee, tea, and cola drinks) for fight drowsiness and to delay the onset of sleep, besides having a potential influence on the adjustment of the circadian system at high doses only, may disrupt sleep patterns and have negative effects on the digestive system.

Good sleep strategies and relaxation techniques should also be adopted to help alleviate desynchronosis and fatigue (Peen & Bootzin, 1990). People should try to keep a tight sleeping schedule while on shift work and avoid disturbances (e.g., by arranging silent and dark bedrooms, using ear plugs; making arrangements with family members and neighbors). Timing of diurnal sleep after a night duty should also be scheduled taking into consideration that sleep onset latency and length can be influenced more by the phase of the temperature rhythm than by prior wakefulness, so that sleeps starting in the early morning, during the rising phase of the temperature rhythm, show longer latency and shorter duration than those commencing in the early afternoon (Åkerstedt & Gillberg, 1981; Czeisler, Weitzman, Moore-Ede, & Zimmerman, 1980).

Further, the proper use of naps can be very effective in compensating for sleep loss. Useful naps can be taken before night shift or extended operations ("prophylactic naps"), during night as "anchor sleep" (Minors & Waterhouse, 1981), to alleviate fatigue ("maintenance naps"), or after early morning and night shifts to integrate normal sleep ("replacement naps") (Åkerstedt & Torsvall, 1985; Bonnet, 1990; Naitoh, Englund, & Ryman, 1982; Rosa et al., 1990).

Compensatory Measures

Many kinds of interventions, aimed at compensating for shift and night work inconveniences, have been introduced in recent years, usually in a very empirical way according to different work conditions and specific problems arising in different companies, work sectors, and countries. Such interventions can act as *counterweights*, aimed only at compensating for the inconveniences, or as *countervalues*, aimed at reducing or eliminating the inconveniences (Thierry, 1980; Wedderburn, 1991).

The main counterweight is monetary compensation, adopted as a worldwide basic reward for irregular work schedules and prolonged duty periods. It is a simple monetary translation of the multidimensional aspects of the problem, and can have a dangerous masking function. Other counterweights may be represented by interventions aimed at improving work organization and environmental conditions.

As concerns countervalues, most are aimed at limiting the consequences of the inconveniences, for example, medical and psychological health checks; the possibility of early retirement or transfer from night work to day work; availability of extra time off and/or more rest periods at work; canteen facilities; and social support (transports, housing, children care). One important preventive measure can be the exemption from shiftwork for transient periods during particular life phases, due to health impairments or significant difficulties in family or social life. In some German factories shiftworkers over 50 may be admitted into specialized hospitals for 2 or 3 weeks, at intervals of 2–3 years, in order to normalize their circadian rhythmicity and have psychotherapeutic support and a general health check (Rutenfranz, Haider, & Koller, 1985). Andlauer et al. (1982) pointed out that "6 weeks of unbroken rest per year is a minimum requirement to compensate the adverse effects of shift work," thus allowing an effective recovery of biological functions.

The possibility, or the priority, for transfer to day work after a certain number of years on night shifts (generally 20), or over 55 years of age, has been granted by collective agreements in some countries (International Labour Office, 1988). Passing from shift work that includes night work to schedules without night work brought with it an improvement in physical, mental, and social well-being (Åkerstedt & Torsvall, 1981). Moreover, some national legislation and collective agreements enable the night workers having a certain amount of night work to their credit (at least 20 years) to retire some years before (from 1 to 5) the normal age of retirement (International Labour Office, 1988).

Some countervalues are aimed at reducing the causes of inconveniences, that is, reduction of working hours, night work in particular; adoption of shift schedules based on physiological criteria (see later discussion); reduced work load at night; and sleep strategies and facilities. For example, the introduction of supplement crews is a positive measure that constitutes reduction of the amount of night work of the individual worker by sharing it with a larger number of workers. This makes it possible also to reduce the number of hours on night shift to 7 or 6 or even less, particularly when there are other stress factors, such as heavy work, heat, noise, or high demands on attention. As already mentioned, Andlauer et al. (1982) proposed the doubling up the night shifts in factories where there may be a "public risk."

Some Guidelines for the Arrangement of Shift Work Schedules According to Chronobiological Criteria

Designing shift systems also taking into consideration psychophysiological and social criteria has a positive effect on shiftworkers' performance efficiency and well-being. In recent years many authors addressed this topic and gave some recommendations aimed at making shift schedules more respectful of human characteristics, in particular the biological circadian system (Knauth, 1993; Knauth & Rutenfranz, 1982; Monk, 1988; Rosa et al., 1990; Wedderburn, 1991).

They deal with the following points in particular: number of consecutive night duties, speed and direction of shift rotation, timing and length of each shift, regularity and flexibility of shift systems, and distribution of rest and leisure times. The most relevant can be summarized as follows.

The number of consecutive night shifts should be reduced as much as possible (preferably one or two at most); this prevents accumulation of sleep deficit and fatigue, and minimize the disruption of the circadian rhythms.

Consequently, rapidly rotating shift systems are preferable to slowly rotating shifts (weekly or forthnightly) or permanent night work. This helps also to avoid prolonged interferences with social relations, which can be further improved by keeping the shift rotation as regular as possible and inserting some free weekends. Moreover, at least one rest day should be scheduled after the night-shift duty.

The forward or "clockwise" rotation of the duty periods (morning–afternoon–night) is to be preferred to the backward one (afternoon–morning–night), because it allows a longer rest interval between shifts, and parallels the "natural" tendency of phase delay of circadian rhythms over 24 hr, as appears in "free-running" conditions.

Therefore, shift systems including fast changeovers or doublebacks (e.g., morning and night shifts in the same day), which are very attractive for the long blocks of time off, should be avoided as they do not leave sufficient time for sleeping between the duty shifts.

Morning shift should not start too early, in order to allow a normal sleep length (as people go to bed at the usual time) and to save the REM sleep, which is more concentrated in the second part of the night sleep. This can decrease fatigue and risk of accident on the morning shift, which often has the highest workload. A delayed start of all the shifts (e.g., 0700–1500–2300 or 0800–1600–2400 hr) could favor a better exploitation of leisure time in the evening also for those working on night shift.

The length of the shifts should be arranged according to the physical and mental load of the task. Therefore a reduction of the duty hours can become a necessity in job activities requiring high levels of vigilance and performance for their complexity or safety reasons (e.g., firefighters, train and aircraft drivers, pilots and air traffic controllers, workers in nuclear and petrochemical plants). For examples, Andlauer et al. (1982), after the Three Mile Island accident, proposed doubling up the night shift with two teams and providing satisfactory rest facilities for the off-duty team so that no operator should work more than 4.5 hr in the night shift.

On the other hand, extended work shifts of 9–12 hr, which are generally associated to compressed working weeks, should only be contemplated if the nature of work and the workload is suitable for prolonged duty hours, the shift system is designed to minimize accumulation of fatigue and desynchronization, and when there are favorable environmental conditions (e.g., climate, housing, commuting time).

Besides, in case of prolonged or highly demanding tasks, it may be useful to insert short nap periods, particularly during the night shift. This has been found to have favorable effects on performance (Gillberg, 1985; Rogers, Spencer, Stone, & Nicholson, 1989), physiological adjustment (Matsumoto, Matsui, Kawamori, & Kogi, 1982; Minors & Waterhouse, 1981) and tolerance of night work (Kogi, 1982). Costa (1993) showed that the strain connected with night duty among air traffic controllers is considerably reduced, with consequent maintenance of a satisfactory performance level, when a short sleep is allowed. Therefore, the use of naps during the night shift should be promoted and negotiated officially, taking into consideration that night workers in many factories or hospitals very often take naps or "unofficial" rest periods during the night shifts through informal arrangements among colleagues and under the tacit agreement of the management. According to Gillberg (1985), even a nap with a duration of 1 hr appears to have positive effects in normal working conditions.

Furthermore, it is important to give the opportunity to maintain usual meal times as fixed as possible, by scheduling sufficiently long breaks and providing hot meals.

Anyway, it is quite clear that there is no "optimal shift system" in principle, as each shift system has advantages and drawbacks, or in practice, as different work sectors and places have different demands. Therefore there may be several "best solutions" for the same work situation, and flexible working time arrangements appear to be very useful strategies in favoring adaptation to shift work, as already evidenced with the so-called "time autonomous working groups" (De Haan, 1990; Knauth, 1993).

Some Suggestions for Air Crew Scheduling and Crew Behavior

A proper strategy in flight schedules arrangement as well as in timing rest and sleep periods can be of paramount importance in counteracting performance impairment and fatigue due to desynchronosis and prolonged duty period. This can be achieved by restricting flight duty periods of excessive length and/or reducing maximum flight time at night and/or extending the rest periods prior to or after long-haul flights.

In reviewing the regulations adopted on this respect in nine industrialized countries, Wegmann, Conrad, and Klein (1983) identified 12 factors that played a major role in the formulation of rules and criteria. Among these, 2 concerned the biological rhythms, in particular, time of day, that is, irregular flight schedules within the 24-hr activity/rest cycle, and time zones, including transmeridian routes. Surprisingly, they found that, besides the general agreement on the stress connected to flying at unusual hours and at night, only two countries provided specific regulations on time-of-day scheduling, and three on night flights and perturbation of circadian rhythms. On the other hand, several models have been proposed to predict the operational load on air crews, taking into consideration departure and arrival time, night duty hours, and time zones crossed (Wegmann, Hasenclever, Christoph, & Trumbach, 1985).

It is obviously impossible to fix rules able to deal with all the possible flight schedules and routes all over the world, but it seems right and proper to consider these aspects and try to incorporate some indications coming from chronobiological studies on transmeridian flights in flight scheduling (Graeber, 1986; Klein & Wegmann, 1979b; Wegmann, Klein, Conrad, and Esser, 1983) such as the following.

In general, nighttime between 2200 and 0600 hr is the least efficacious time to start a flight, as it coincides with the lowest levels of psychophysical activation.

The resynchronization on a new time zone should not be forced, but crew should return as soon as possible to their home base and be provided with a sufficient rest time to prevent sleep deficits (e.g., 14 hr of rest is considered the minimum after crossing four or more time zones).

After the return home from transmeridian flights, the length of the postflight rest period should be directly related to the number of time zones crossed. According to Wegmann, Klein, Conrad, and Esser (1983), the minimum rest period should as long as the number of time zones crossed multiplied by 8, in order to avoid a residual de-synchronization of no more than 3 hr (that seems to have no operational significance) before beginning a new duty period.

The final section of long transmeridian flights should be scheduled to avoid its coincidence with the nocturnal trough of alertness and performance efficiency (Klein & Wegmann, 1979a). For example, the most advantageous time for departure of eastward flights would be in the early evening, as this allows a nap beforehand, which can counteract sleepiness during the first part of the flight; moreover, the circadian rising phase of psychophysiological functions, occurring in correspondence to the second part of the flight, may support a better performance for approach and landing.

Preadjustment of the duty periods in the 2–3 days preceding long and complex transmeridian flights, in order to start work either progressively earlier or later according to the direction of the flight, can avoid abrupt phase shifts and increase performance efficiency.

Rest and sleep schedules should be carefully disciplined to help compensating for fatigue and desynchronosis. For example, in case of prolonged layover after eastward flights, it would be advisable to limit sleep immediately after arrival and prolong the subsequent wake according to the local time. This would increase the likelihood of an adequate duration of sleep immediately preceding the next duty period.

In the case of flights involving multiple segments and layovers in different times zones, sleep periods should be scheduled taking into consideration the two troughs of the biphasic (12-hr) alertness cycle, such as a nap of 1–2 hr plus a sleep of 4–6 hr. This would allow better maintainance of performance levels during the subsequent periods of higher alertness, in which work schedules might be optimally adjusted (Dement et al., 1986).

To post entire crews overseas for prolonged periods of time would be the best for chronobiological adjustment, but not for family and social relationships.

Naps may be very helpful; they pay an essential role in improving alertness (Nicholson et al., 1985). They can be added at certain hours of the rest days to integrate sleep periods, and can be inserted during flight duty. After several studies on long-haul and complex flights, showing that circadian rhythms remain close to home time for about the first 2 days, Sasaki, Kurosaki, Spinweber, Graeber, and Takahashi (1993) suggested that crew members should schedule their sleep or naps to correspond to early morning and afternoon of home time in order to reduce sleepiness and minimize the accumulation of sleep deficit. On the other hand, it could be preferable to permit and schedule flight deck napping for single crew members, if operationally feasible, instead of letting it happen unpredictably.

For air crews not involved on long transmeridian flights, the general guidelines suggested for rapid rotating shiftworkers may be followed, but they should be further adapted in relation to the more irregular patterns of duty sections during the working day.

Finally, it may be advisable to try to take advantage from some individual chronobiological characteristics. It could be useful to consider the different activation curve between morning and evening types, as already mentioned, when scheduling flight timetables, in order to allow people to work in periods when they are at their best levels. For example, morning type crew members would certainly be fitter on flights scheduled on the first part of the day, whereas evening types would show a lower sleepiness on evening and night flights. Some suggestions on this arise from the study of Sasaki, Kurosaki, Mori, and Endo (1986).

Medical Surveillance

Good medical surveillance is essential to ensure that operators are in good health and able to carry out their job without excessive stress and performance impairment. Besides the careful application of ICAO precise norms and recommendations for the medical certification of licence holders, medical checks should be oriented toward preserving physical and mental health with regard to the temporal organization of body functions.

In the light of the possible negative consequences connected to desynchronization of the biological rhythms, both selection and periodical checks of workers engaged on irregular work schedules should take into consideration some criteria and suggestions proposed by several authors and institutions (International Labour Office, 1988; Koller, 1989; Rutenfranz, 1982; Rutenfranz et al., 1985; Scott & LaDou, 1990).

Work at night and on irregular shift schedules should be restricted for people suffering from severe disorders that are associated with or can be aggravated by shift lag and jet lag, in particular, important gastrointestinal diseases (e.g., peptic ulcer, chronic hepatitis, and pancreatitis); insulin-dependent diabetes, as regular and proper food intake and correct therapeutic timing are required; hormonal pathologies (e.g., thyroid and suprarenal gland), because they demand regular drug assumption strictly connected to the activity/rest periods; epilepsy, as the seizures can be favored by sleep deprivation and the efficacy of treatment can be hampered by irregular wake–rest schedules; chronic psychiatric disorders, depression in particular, as they are often associated with a disruption of the sleep/wakefulness cycle and can be influenced by the light/dark periods; chronic sleep disturbances; and coronary heart diseases, severe hypertension, and asthma, as exacerbations are more likely to occur at night and treatment is less effective at certain hours of the day.

Moreover, occupational health doctors should consider very carefully those who may be expected to encounter more difficulty in coping with night work and jet lag on the basis of their psychophysiological characteristics, health, and living conditions, such as age over 50; low amplitude and stability of circadian rhythms; rigidity of sleeping habits and low ability to overcome drowsiness; extreme morningness; high neuroticism; unsatisfactory housing conditions; and women with small children but lacking social support.

Therefore, medical checks have to be focused mainly on sleeping habits and troubles, eating and digestive problems, psychosomatic complaints, drug consumption, housing conditions, transport facilities, work loads, and off-job activities, preferably using standardized questionnaires, checklists, and rating scales, in order to monitor the worker's behavior throughout the years.

Besides this, permanent education and counseling should be provided for improving self-care strategies for coping, in particular as concerns sleep, smoking, diet, drugs, stress management, and physical fitness.

The adoption of these criteria could also improve the efficacy of preemployment screenings, in order to avoid allocating some people, who are more vulnerable in circadian rhythmic structure and psychophysical homeostasis to jobs that require shift and night work.

REFERENCES

Adams, J., Folkard, S., & Young, M. (1986). Coping strategies used by nurses on night duty. *Ergonomics, 29*, 185–196.

Åkerstedt, T. (1985a). Adjustment of physiological circadian rhythms and the sleep-wake cycle to shiftwork. In S. Folkard & T. H. Monk (Eds.), *Hours of work: Temporal factors in work scheduling* (pp. 185–197). Chichester, England: John Wiley & Sons.

Åkerstedt, T. (1985b). Shifted sleep hours. *Annals of Clinical Research, 17*, 273–279.

Åkerstedt, T. (1990a). Psychological and psychophysiological effects of shiftwork. *Scandinavian Journal of Work Environment and Health, 16*, 67–73.

Åkerstedt, T. (1990b). *Wide awake at odd hours. Shift work, time zones and burning the midnight oil* (pp. 1–116). Stockholm: Swedish Council for Work Life Research.

Åkerstedt, T., & Gillberg, M. (1981). The circadian variation of experimentally displaced sleep. *Sleep, 4*, 159–169.

Åkerstedt, T., Knuttson, A., Alfredsson, L., & Theorell, T. (1984). Shift work and cardiovascular disease. *Scandinavian Journal of Work Environment and Health, 10,* 409–414.

Åkerstedt, T., & Torsvall, L. (1981). Age, sleep and adjustment to shift work. In W. P. Koella (Ed.), *Sleep 80* (pp. 190–194). Basle: Karger.

Åkerstedt, T., & Torsvall, L. (1985). Napping in shift work. *Sleep, 8,* 105–109.

Alluisi, E. A. (1972). Influence of work–rest scheduling and sleep loss on sustained performance. In W. P. Colquhoun (Ed.), *Aspects of human efficiency* (pp. 199–215). London: The English Universities Press.

Alluisi, E. A., & Chiles, W. D. (1967). Sustained performance, work–rest scheduling and diurnal rhythms in man. *Acta Psychologica, 27,* 436–442.

Andlauer, P., Rutenfranz, J., Kogi, K., Thierry, H., Vieux, N., & Duverneuil, G. (1982). Organization of night shifts in industries where public safety is at stake. *International Archives of Occupational and Environmental Health, 49,* 353–355.

Arendt, J., & Aldhous, M. (1988). Further evaluation of the treatment of jet-lag by melatonin: A double-blind crossover study. *Annual Review of Chronopharmacology, 5,* 53–55.

Axelsson, G., Rylander, R., & Molin, I. (1989). Outcome of pregnancy in relation to irregular and inconvenient work schedules. *British Journal of Industrial Medicine, 46,* 393–398.

Blake, M. J. F. (1971). Temperament and time of day. In W. P. Colquhoun (Ed.), *Biological rhythms and human performance* (pp. 109–148). London: Academic Press.

Bonnet, M. H. (1990). Dealing with shift work: Physical fitness, temperature, and napping. *Work & Stress, 4,* 261–274.

Breithaupt, H., Hildebrandt, G., Dohre, D., Josch, R., Sieber, U., & Werner, M. (1978). Tolerance to shift of sleep as related to the individual's circadian phase position. *Ergonomics, 21,* 767–774.

Cabon, P. H., Coblentz, A., Mollard, R. P., & Fouillot, J. P. (1993). Human vigilance in railway and long-haul flight operation. *Ergonomics, 36,* 1019–1033.

Cameron, R. G. (1969). Effect of flying on ther menstrual function of air hostesses. *Aerospace Medicine, 40,* 1020–1023.

Carter, F. A., & Corlett, E.N. (1978). *Shift work and accidents.* Dublin: European Foundation for the Improvement of Living and Working Conditions.

Clodorè, M., Benoit, O., Foret, J., Touitou, Y., Touron, N., Bouard, G., & Auzeby, A. (1987). Early rising or delayed bedtime: Which is better for a short night's sleep? *European Journal of Applied Physiology, 56,* 403–411.

Colquhoun, W. P. (1971). Circadian variations in mental efficiency. In W. P. Colqhuoun (Ed.), *Biological rhythms and human performance* (pp. 39–107). London: Academic Press.

Colquhoun, W. P. (1979). Phase shift in temperature rhythm after trasmeridian flights, as related to pre-shift phase angle. *International Archives of Occupational and Environmental Health, 42,* 149–157.

Colquhoun, W. P. (1981). Rhythms in performance. In J. Aschoff (Ed.), *Biological rhythms: Handbook of behavioural neurobiology* (pp. 333–348). New York: Plenum Press.

Colquhoun, W. P., & Folkard, S. (1978). Personality differences in body temperature rhythm, and their relation to its adjustment to night work. *Ergonomics, 21,* 811–817.

Comperatore, C. A., & Krueger G. P. (1990). Circadian rhythm desynchronosis, jet-lag, shift lag, and coping strategies. *Occupational Medicine: State of Art Reviews, 5,* 323–341.

Costa, G. (1993). Evaluation of work load in air traffic controllers. *Ergonomics, 36,* 1111–1120.

Costa, G. (1996). The impact of shift and night work on health. *Applied Ergonomics, 27,* 9–16.

Costa, G., Apostoli, P., D'Andrea, F., & Gaffuri, E. (1981). Gastrointestinal and neurotic disorders in textile shift workers. In A. Reinberg, A. Vieux, & P. Andlauer (Eds.), *Night and shift work: Biological and social aspects* (pp. 215–221). Oxford: Pergamon Press.

Costa, G., Ghirlanda, G., Minors, D. S., & Waterhouse, J. (1993). Effect of bright light on tolerance to night work. *Scandinavian Journal of Work Environment and Health, 19,* 414–420.

Costa, G., Lievore, F., Casaletti, G., Gaffuri, E., & Folkard, S. (1989). Circadian characteristics influencing interindividual differences in tolerance and adjustment to shiftwork. *Ergonomics, 32,* 373–385.

Czeisler, C. A., Kronauer, R. E., Allan, J. S., Duffy, J. F., Jewett, M. E., Brown, E. N., & Ronda, J. M. (1989). Bright light induction of strong (type O) resetting of the human circadian pacemaker. *Science, 244,* 1328–1333.

Czeisler, C. A., Weitzman, E. D., Moore-Ede, M. C., & Zimmerman, J. C. (1980). Human sleep: Its duration and organization depend on its circadian phase. *Science, 210,* 1264–1267.

Daniell, W. E., Vaughan, T. L., & Millies, B. A. (1990). Pregnancy outcomes among female flight attendants. *Aviation Space and Environment Medicine, 61,* 840–844.

De Haan, E. G. (1990). Improving shift work schedules in a bus company: Towards more autonomy. In G. Costa, G. C. Cesana, K. Kogi, & A. Wedderburn (Eds.), *Shiftwork: Health, sleep and performance* (pp. 448–454). Frankfurt: Verlag Peter Lang.

Dekker, D. K., & Tepas, D. I. (1990). Gender differences in permanent shiftworker sleep behaviour. In G. Costa, G. C. Cesana, K. Kogi, & A. Wedderburn (Eds.), *Shiftwork: Health, sleep and performance* (pp. 77–82). Frankfurt: Verlag Peter Lang.

Dement, W. C., Seidel, W. F., Cohen, S. A., Bliwise, N. G., & Carskadon, M. A. (1986). Sleep and wakefulness in aircrew before and after transoceanic flights. In R. C. Graeber (Ed.), *Crew factors in flight operations: IV. Sleep and wakefulness in international aircrews* (pp. 23–47). [Technical Memorandum 88231]. Moffett Field, CA: NASA Ames Research Center.

Eastman, C. I. (1990). Circadian rhythms and bright light: recommendations for shiftwork. *Work & Stress, 4*, 245–260.

Ehret, C. F. (1981). New approaches to chronohygiene for the shift worker in the nuclear power industry. In A. Reinberg, A. Vieux, & P. Andlauer (Eds.), *Night and shift work: Biological and social aspects* (pp. 263–270). Oxford: Pergamon Press.

Estryn-Behar, M., Gadbois, C., Peigne, E., Masson, A., & Le Gall, V. (1990). Impact of night shifts on male and female hospital staff. In G. Costa, G. C. Cesana, K. Kogi, & A. Wedderburn (Eds.), *Shiftwork: Health, sleep and performance* (pp. 89–94). Frankfurt: Verlag Peter Lang.

Folkard, S. (1990). Circadian performance rhythms: some practical and theoretical implications. *Philosophical Transactions of the Royal Society of London, B327*, 543–553.

Folkard, S., Arendt, J., & Clark, M. (1993). Can Melatonin improve shift workers' tolerance of the night shift? Some preliminary findings. *Chronobiology International*, 10, 315–320.

Folkard, S., & Condon, R. (1987). Night shift paralysis in air traffic control officers. *Ergonomics, 30*, 1353–1363.

Folkard, S., & Monk, T. H. (1985). Circadian performance rhythms. In S. Folkard & T. Monk (Eds.), *Hours of work: Temporal factors in work scheduling* (pp. 37–52). Chichester, England: John Wiley & Sons.

Folkard, S., Monk, T. H., & Lobban, M. C. (1978). Short and long-term adjustment of circadian rhythms in "permanent" night nurses. *Ergonomics, 21*, 785–799.

Folkard, S., Monk, T. H., & Lobban, M. C. (1979). Towards a predictive test of adjustment to shift work. *Ergonomics, 22*, 79–91.

Foret, J., Benoit, O., & Royant-Parola, S. (1982). Sleep schedules and peak times of oral temperature and alertness in morning and evening "types." *Ergonomics, 25*, 821–827.

Gadbois, C. (1981). Women on night shift: Interdependence of sleep and off-job activities. In A. Reinberg, A. Vieux, & P. Andlauer (Eds.), *Night and shift work: Biological and social aspects* (pp. 223–227). Oxford: Pergamon Press.

Gander, P. H., De Nguyen, B. E., Rosekind, M. R., & Connell, L. J. (1993). Age, circadian rhythms, and sleep loss in flight crews. *Aviation Space and Environmental Medicine, 64*, 189–195.

Gander, P. H., Myhre, G., Graeber, R. C., Andersen, H. T., & Lauber, J. K. (1989). Adjustment of sleep and circadian temperature rhythm after flights across nine time zones. *Aviation Space and Environmental Medicine, 60*, 733–743.

Gersten, A. H., Duchon, J. C., & Tepas, D. I. (1986). Age and gender differences in night worker's sleep lengths. In G. Costa, G. C. Cesana, K. Kogi, & A. Wedderburn (Eds.), *Shiftwork: Health, sleep and performance* (pp. 467–670). Frankfurt: Verlag Peter Lang.

Gillberg, M. (1985). Effects of naps on performance. In S. Folkard & T. Monk (Eds.), *Hours of work: Temporal factors in work scheduling* (pp. 77–86). Chichester, England: John Wiley & Sons.

Gordon, N. P., Cleary, P. D., Parker, C. E., & Czeisler, C. A. (1986). The prevalence and health impact of shiftwork. *American Journal of Public Health, 76*, 1225–1228.

Graeber, R. C. (1982). Alterations in performance following transmeridian flight. In F. M. Brown & R. C. Graber (Eds.), *Rhythmic aspects of behavior* (pp. 173–212). Hillsdale, NJ: Lawrence Erlbaum Associates.

Graeber, R. C. (Ed.) (1986). *Crew factors in flight operations: IV. Sleep and wakefulness in international aircrews* [Technical Memorandum 88231]. Moffett Field, CA: NASA Ames Research Center.

Haider, M., Cervinka, R., Koller, M., & Kundi, M. (1988). A destabilization theory of shiftworkers effects. In J. M. Hekkens, G. A. Kerkhof, & W. J. Rietveld (Eds.), *Trends in chronobiology* (pp. 209–217). Oxford: Pergamon Press.

Härmä, M., Ilmarinen, J., & Knauth, P. (1988). Physical fitness and other individual factors relating to the shiftwork tolerance of women. *Chronobiology International, 5*, 417–424.

Härmä, M., Laitinen, J., Partinen, M., Ilmarinen, J., & Suvanto, S. (1989). The effect of light on the adaptation of the circadian rhythms among flight attendants. In G. Costa, G. C. Cesana, K. Kogi, & A. Wedderburn (Eds.), *Shiftwork: Health, sleep and performance* (pp. 254–259). Frankfurt: Verlag Peter Lang.

Haugli, L., Skogtad, A., & Hellesøy, O. H. (1994). Health, sleep, and mood perceptions reported by airline crews flying short and long hauls. *Aviation Space and Environmental Medicine, 65*, 27–34.

Healy, D., & Waterhouse, J. M. (1991). Reactive rhythms and endogenous clocks [Editorial]. *Psychologie Medicale, 21*, 557–564.

Hildebrandt, G., Rohmert, W., & Rutenfranz, J. (1974). 12 & 24h rhythms in error frequency of locomotive drivers and the influence of tiredness. *International Journal of Chronobiology, 2*, 175–180.

Hildebrandt, G., Rohmert, W., & Rutenfranz, J. (1975). The influence of fatigue and rest period on the circadian variation of error frequency of shift workers (engine drivers). In W. P. Colquhoun, S. Folkard, P. Knauth, & J. Rutenfranz (Eds.), *Experimental studies of shiftwork* (pp. 174–187). Opladen: Westdeutscher Verlag.

Hildebrandt, G., & Stratmann, I. (1979). Circadian system response to nightwork in relation to the circadian phase position. *International Archives of Occupational and Environmental Health*, 43, 73–83.

International Labour Office. (1988). *Night work.* Geneva.

Kelly, R. J., & Schneider, M. F. (1982). The twelve-hour shift revisited. Recent trends in the electric power industry, *Journal of Human Ergology, 11* (Suppl.), 369–384.

Kerkhof, G. (1985). Individual differences in circadian rhythms. In S. Folkard & T. Monk (Eds.), *Hours of work: Temporal factors in work scheduling* (pp. 29–35). Chichester, England: John Wiley & Sons.

Klein, E. K., & Wegmann, H. M. (1979a). Circadian rhythms of human performance and resistance: operational aspects. In *Sleep, wakefulness and circadian rhythm* (pp. 2.1–2.17). London: AGARD Lectures Series No. 105.

Klein, E. K., & Wegmann, H. M. (1979b). Circadian rhythms in air operations. In *Sleep, wakefulness and circadian rhythm* (pp. 10.1–10.25). London: AGARD Lectures Series No. 105.

Klein, E. K., Wegmann, H. M., & Hunt, B. I. (1972). Desynchronization of body temperature and performance circadian rhythm as a result of outgoing and homegoing transmeridian flights. *Aerospace Medicine, 43*, 119–132.

Knauth, P. (1993). The design of shift systems. *Ergonomics, 36*, 15–28.

Knauth, P., Emde, E., Rutenfranz, J., Kiesswetter, E., & Smith, P. (1981). Re-entrainment of body temperature in field studies of shiftwork. *International Archives of Occupational and Environmental Health, 49*, 137–149.

Knauth, P., Landau, K., Droge, C., Schwitteck, M., Widynski, M., & Rutenfranz, J. (1980). Duration of sleep depending on the type of shift work. *International Archives of Occupational and Environmental Health*, 46, 167–177.

Knauth, P., & Rutenfranz, J. (1975). The effects of noise on the sleep of night-workers. In W. P. Colquhoun, S. Folkard, P. Knauth, & J. Rutenfranz (Eds.), *Experimental studies of shiftwork* (pp. 57–65). Opladen: Westdeutscher Verlag.

Knauth, P., Rutenfranz, J., Herrmann, G., & Poppel, S. J. (1978). Reentrainment of body temperature in experimental shift work studies. *Ergonomics, 21*, 775–783.

Knauth, P., & Rutenfranz, J. (1982). Development of criteria for the design of shiftwork systems. *Journal of Human Ergology, 11* (Suppl.), 337–367.

Kogi, K. (1982). Sleep problems in night and shift work. *Journal of Human Ergology, 11* (Suppl.), 217–231

Kogi, K. (1985). Introduction to the problems of shiftwork. In S. Folkard & T. Monk (Eds), *Hours of work: Temporal factors in work scheduling* (pp. 165–184). Chichester, England: John Wiley & Sons.

Koller, M. (1983). Health risk related to shift work. *International Archives of Occupational and Environmental Health, 53*, 59–75.

Koller, M. (1989). Preventive health measures for shiftworkers. In M. Wallace (Ed.), *Managing shiftwork* (pp. 17–24). Bundoora: Brain-Behaviour Research Institute, Department of Psychology, La Trobe University.

Lavie, P. (1985). Ultradian cycles in wakefulness. Possible implications for work-rest schedules. In S. Folkard & T. Monk (Eds.), *Hours of work: Temporal factors in work scheduling* (pp. 97–106). Chichester, England: John Wiley & Sons.

Lennernas, M. A. (1993). *Nutrition and shift work.* Uppsala: Acta Universitatis Upsaliensis.

Lewy, A. J., & Sack, R. L. (1989). The dim light melatonin onset as a marker for circadian phase position. *Chronobiology International, 6*, 93–102.

Lyons, T. J. (1992). Women in the fast jet cockpit: Aeromedical considerations. *Aviation Space Environmental Medicine, 63*, 809–818.

Matsumoto, K., Matsui, T., Kawamori, M., & Kogi, K. (1982). Effects of nighttime naps on sleep patterns of shiftworkers. *Journal of Human Ergology, 11* (Suppl.), 279–289.

McDonald, A. D., McDonald, J. C., Armstrong, B., Cherry, N. M., Nolin, A., & Robert, D. (1988) Fetal death and work in pregnancy. *British Journal of Industrial Medicine, 45*, 148–157.

Meers, A., Maasen, A., & Verhaegen, P. (1978). Subjective health after six months and after four years of shift work. *Ergonomics, 21*, 857–859.

Minors, D. S., & Waterhouse, J. M. (1981). Anchor sleep as a synchronizer of rhythms on abnormal routines. In L. C. Johnson, D. I. Tepas, W. P. Colquhoun, & M. J. Colligan (Eds.), *Advances in sleep research: Vol. 7. Biological rhythms, sleep and shift work* (pp. 399–414). New York: Spectrum.

Minors, D. S., & Waterhouse, J. M. (1986). Circadian rhythms and their mechanisms. *Experientia, 42*, 1–13.

Monk, T. (1988). *How to make shift work safe and productive.* Pittsburgh: University of Pittsburgh School of Medicine.

Monk, T. (1990). Shiftworker performance. *Occupational Medicine: State of Art Reviews, 5*, 183–198.

Monk, T. H., & Folkard, S. (1985). Shiftwork and performance. In S. Folkard & T. Monk (Eds.), *Hours of work: Temporal factors in work scheduling* (pp. 239–252). Chichester, England: John Wiley & Sons.

Monk, T. H., Moline, M. L., & Graeber R. C. (1988). Inducing jet-lag in the laboratory: patterns of adjustment to an acute shift in routine. *Aviation Space and Environmental Medicine, 59*, 703–710.

Nachreiner, F. (1975). Role perceptions, job satisfaction and attitudes towards shiftwork of workers in different shift systems as related to situational and personal factors. In W. P. Colquhoun, S. Folkard, P. Knauth, & J. Rutenfranz (Eds.), *Experimental studies of shiftwork* (pp. 232–243). Opladen: Westdeutscher Verlag.

Naitoh, P., Englund, C. E., & Ryman D. (1982). Restorative power of naps in designing continuous work schedules. *Journal of Human Ergology, 11* (Suppl.), 259–278.

Nicholson, A. N., Pascoe, P. A., Roehrs, T., Roth, T., Spencer, M. B., Stone, B. M., & Zorik, F. (1985). Sustained performance with short evening and morning sleep. *Aviation Space and Environmental Medicine, 56*, 105–114.

Nurminen, T. (1989). Shift work, fetal development and course of pregnancy. *Scandinavian Journal of Work Environment and Health, 15*, 395–403.

Ostberg, O. (1973). Circadian rhythms of food intake and oral temperature in "morning" and "evening" groups of individuals. *Ergonomics, 16*, 203–209.

Patkai, P. (1985). The menstrual cycle. In S. Folkard & T. Monk (Eds.), *Hours of work: Temporal factors in work scheduling* (pp. 87–96). Chichester, England: John Wiley & Sons.

Peen, P. E., & Bootzin, R. R. (1990). Behavioural techniques for enhancing alertness and performance in shift work. *Work & Stress, 4*, 213–226.

Pokorny, M., Blom, D., & Van Leeuwen, P. (1981). Analysis of traffic accident data (from bus drivers): An alternative approach (I). In A. Reinberg, A. Vieux, & P. Andlauer (Eds), *Night and shift work: Biological and social aspects* (pp. 271–278). Oxford: Pergamon Press.

Preston, F. S., Bateman, S. C., Short, R. V., & Wilkinson, R. T. (1973). Effects of time changes on the menstrual cycle length and on performance in airline stewardnesses. *Aerospace Medicine, 44*, 438–443.

Price, W. J., & Holley, D. C. (1990). Shiftwork and safety in aviation. *Occupational Medicine: State of Art Reviews, 5*, 343–377.

Redgrove, J. A. (1971). Menstrual cycles. In W. P. Colquhoun (Ed.), *Biological rhythms and human performance* (pp. 211–240). London: Academic Press.

Reinberg, A., Andlauer, P., De Prins, J., Malbecq, W., Vieux, N., & Bourdeleau, P. (1984). Desynchronisation of the oral temperature circadian rhythm and intolerance to shift work. *Nature, 308*, 272–274.

Reinberg, A., Andlauer, P., Guillet, P., Nicolai, A., Vieux, N., & Laporte, A. (1980). Oral temperature, circadian rhythm amplitude, ageing and tolerance to shiftwork. *Ergonomics, 23*, 55–64.

Reinberg, A., Motohashi, Y., Bourdeleau, P., Andlauer, P., Levi, F., & Bicakova-Rocher, A. (1988). Alteration of period and amplitude of circadian rhythms in shift workers. *European Journal of Applied Physiology, 57*, 15–25.

Reinberg, A., & Smolenski, M. H. (1983). *Biological rhythms and medicine.* New York: Springer Verlag.

Richardson, G. S., Carskadon, M. A., Orav, E. J., & Dement, W. C. (1982). Circadian variation in sleep tendency in elderly and young subjects. *Sleep, 5* (Suppl. 2), 82–94.

Romon-Rousseaux, M., Lancry, A., Poulet, I., Frimat, P., Furon, D. (1987). Effects of protein and carbohydrate snacks on alertness during the night. In A. Oginski, J. Pokorski, & J. Rutenfranz (Eds.), *Contemporary advances in shiftwork research* (pp. 133–141). Krakow: Medical Academy.

Rogers, A. S., Spencer, M. B., Stone, B. M., & Nicholson, A. N. (1989). The influence of a 1h nap on performance overnight. *Ergonomics, 32*, 1193–1205.

Rosa, R. R., Bonnet, M. H., Bootzin, R. R., Eastman, C. I., Monk, T., Penn, P. E., Tepas, D. I., & Walsh, J. K. (1990). Intervention factors for promoting adjustment to nightwork and shiftwork. *Occupational Medicine: State of Art Reviews, 5*, 391–414.

Rutenfranz, J. (1982). Occupational health measures for night and shiftworkers. *Journal of Human Ergology, 11*(Suppl.), 67–86.

Rutenfranz, J., Haider, M., & Koller, M. (1985). Occupational health measures for nightworkers and shiftworkers. In S. Folkard & T. Monk (Eds.), *Hours of work: Temporal factors in work scheduling* (pp. 199–210). Chichester, England: John Wiley & Sons.

Sasaki, M., Kurosaki, Y. S., Mori, A., & Endo, S. (1986). Patterns od sleep-wakefulness before and after transmeridian flight in commercial airline pilots. In R. C. Graeber (Ed.), *Crew factors in flight operations: IV. Sleep and wakefulness in international aircrews* (pp. 49–68). [Technical Memorandum 88231]. Moffett Field, CA: NASA Ames Research Center.

Sasaki, M., Kurosaki, Y. S., Spinweber, C. L., Graeber, R. C., & Takahashi, T. (1993). Flight crew sleep during multiple layover polar flights. *Aviation Space and Environmental Medicine, 64,* 641–647.

Scott, A. J., & LaDou, J. (1990). Shiftwork: Effects on sleep and health with recommendations for medical surveillance and screening. *Occupational Medicine: State of Art Reviews, 5,* 273–299.

Smith, P. (1979). A study of weekly and rapidly rotating shift workers. *International Archives of Occupational and Environmental Health, 46,* 111–125.

Suvanto, S., Partinen, M., Härmä, M., & Ilmarinen, J. (1990). Flight attendant's desynchronosis after rapid time zone changes. *Aviation Space and Environmental Medicine, 61,* 543–547.

Tepas, D. I., & Carvalhais, A. B. (1990). Sleep patterns of shiftworkers. *Occupational Medicine: State of Art Reviews, 5,* 199–208.

Thierry, H. K. (1980). Compensation for shiftwork: A model and some results. In W. P. Colquhoun & J. Rutenfranz (Eds.), *Studies of shiftwork* (pp. 449–462). London: Taylor & Francis.

Tilley, A. J., Wilkinson, R. T., Warren, P. S. G., Watson, B., & Drud, M. (1982). The sleep and performance of shift workers. *Human Factors, 24,* 629–641.

Torsvall, L., & Åkerstedt, T. (1987). Sleepiness on the job: Continuously measured EEG changes in train drivers. *Electroencephalography and Clinical Neurophysiology, 66,* 502–511.

Uehata, T., & Sasakawa, N. (1982). The fatigue and maternity disturbances of night work women. *Journal of Human Ergology, 11*(Suppl.), 465–474.

Verhaegen, P., Cober, R., De Smedt, M., Dirkx, J., Kernstens, J., Ryvers, D., & Van Daele, P. (1987). The adaptation of night nurses to different work schedules. *Ergonomics, 30,* 1301–1309.

Walsh, J. K. (1990). Using pharmachological aids to improve waking function and sleep while working at night. *Work & Stress, 4,* 237–243.

Waterhouse, J. M., Folkard, S., & Minors D. S. (1992). *Shiftwork, health and safety. An overview of the scientific literature 1978–1990.* London: Her Majesty's Stationery Office.

Wedderburn, A. (1991). Compensation for shiftwork. *Bulletin of European Shiftwork Topics* (No. 4). Dublin: European Foundation for the Improvement of Living and Working Conditions.

Wegmann, H. M., Conrad, B., & Klein, K. E. (1983). Flight, flight duty, and rest times: A comparison between the regulations of different countries. *Aviation Space and Environmental Medicine, 54,* 212–217.

Wegmann, H. M., Gundel, A., Naumann, M., Samel, A., Schwartz, E., & Vejvoda, M. (1986). Sleep, sleepiness, and circadian rhythmicity in aircrews operating on transatlantic routes. In R. C. Graeber (Ed.), *Crew factors in flight operations: IV. Sleep and wakefulness in international aircrews* (pp. 85–104). [Technical Memorandum 88231]. Moffett Field, CA: NASA Ames Research Center.

Wegmann, H. M., Hasenclever, S., Christoph, M., & Trumbach, S. (1985). Models to predict operational loads of flight schedules. *Aviation Space and Environmental Medicine, 56,* 27–32.

Wegmann, H. M., & Klein, K. E. (1985). Jet-lag and aircrew scheduling. In S. Folkard & T. Monk (Eds.), *Hours of work: Temporal factors in work scheduling* (pp. 263–276). Chichester, England: John Wiley & Sons.

Wegmann, H. M., Klein, K. E., Conrad, B., & Esser, P. (1983). A model of prediction of resynchronization after time-zone flights. *Aviation Space and Environmental Medicine, 54,* 524–527.

Wever, R. A. (1985). Man in temporal isolation: Basic priciples of the circadian system. In S. Folkard & T. Monk (Eds.), *Hours of work: Temporal factors in work scheduling* (pp. 15–28). Chichester, England: John Wiley & Sons.

Wever, R. A. (1989). Light effects on human circadian rhythms: A review of recent Andechs experiments. *Journal of Biological Rhythms, 4,* 161–185.

Zulley, J., & Bailer, J. (1988). Polyphasic sleep/wake patterns and their significance to vigilance. In J. P. Leonard (Ed.), *Vigilance: Methods, models, and regulations* (pp. 167–180). Frankfurt: Verlag Peter Lang.

11

Situation Awareness in Aviation Systems

Mica R. Endsley
SA Technologies

In the aviation domain, maintaining a high level of situation awareness is one of the most critical and challenging features of an aircrew's job. Situation awareness (SA) can be thought of as an internalized mental model of the current state of the flight environment. This integrated picture forms the central organizing feature from which all decision making and action takes place. A vast portion of the aircrew's job is involved in developing SA and keeping it up to date in a rapidly changing environment. Consider this excerpt demonstrating the criticality of situation awareness for the pilot and its frequent elusiveness:

> Ground control cleared us to taxi to Runway 14 with instructions to give way to two single-engine Cessnas that were enroute to Runway 5. With our checklists completed and the Before Takeoff PA [public announcement] accomplished, we called the tower for a takeoff clearance. As we called, we noticed one of the Cessnas depart on Runway 5. Tower responded to our call with a "position and hold" clearance, and then cleared the second Cessna for a takeoff on Runway 5. As the second Cessna climbed out, the tower cleared us for takeoff on Runway 5.
>
> Takeoff roll was uneventful, but as we raised the gear we remembered the Cessnas again and looked to our left to see if they were still in the area. One of them was not just in the area, he was on a downwind to Runway 5 and about to cross directly in front of us. Our response was to immediately increase our rate of climb and to turn away from the traffic. . . . If any condition had prevented us from making an expeditious climb immediately after liftoff, we would have been directly in each other's flight path. (Kraby, 1995, p. 4)

The problem can be even more difficult for the military pilot who must also maintain a keen awareness of many factors pertaining to enemy and friendly aircraft in relation to a prescribed mission in addition to the normal issues of flight and navigation, as illustrated by this account:

> We were running silent now with all emitters either off or in standby. . . . We picked up a small boat visually off the nose, and made an easy ten degree turn to avoid him without making any wing flashes. . . .
>
> Our RWR [radar warning receiver] and ECM [electronic counter measures] equipment were cross checked as we prepared to cross the worst of the mobile defenses. I could see a pair of A-10's strafing what appeared to be a column of tanks. I was really working my head back and forth trying to pick up any missiles or AAA [anti-aircraft artillery] activity and not hit the ground as it raced underneath the nose. I could see Steve's head scanning outside with only quick glances inside at the RWR scope. Just when I thought we might make it through unscathed, I picked up a SAM [surface to air missile] launch at my left nine o'clock heading for my wingman! . . . It passed harmlessly high and behind my wingman and I made a missile no-guide call on the radio. . . .
>
> Before my heart had a chance to slow down from the last engagement, I picked up another SAM launch at one o'clock headed right at me! It was fired at short range and I barely had time to squeeze off some chaff and light the burners when I had to pull on the pole and perform a last ditch maneuver. . . . I tried to keep my composure as we headed down towards the ground. I squeezed off a couple more bundles of chaff when I realized I should be dropping flares as well! As I leveled off at about 100 feet, Jerry told me there was a second launch at my five o'clock. (Isaacson, 1985, pp. 24–25)

In order to perform in the dynamic flight environment, aircrew must not only know how to operate the aircraft and the proper tactics, procedures, and rules for flight, but they must also have an accurate, up-to-date picture of the state of the environment. This is a task that is not simple in light of the complexity and sheer number of factors that must be taken into account in order to make effective decisions. Situation awareness (SA) does not end with the simple perception of data, but also depends on a deeper comprehension of the significance of that data based on an understanding of how the components of the environment interact and function, and a subsequent ability to predict future states of the system.

Having a high level of SA can be seen as perhaps the most critical aspect for achieving successful performance in aviation. Problems with SA were found to be the leading causal factor in a review of military aviation mishaps (Hartel, Smith, & Prince, 1991), and in a study of accidents among major aircarriers, 88% of those involving human error could be attributed to problems with situation awareness (Endsley, 1995a). Due to its importance and the significant challenge it poses, finding new ways of improving SA has become one of the major design drivers for the development of new aircraft systems. Interest has also increased within the operational community, which is interested in finding ways to improve SA through training programs. The successful improvement of SA through aircraft design or training programs requires the guidance of a clear understanding of SA requirements in the flight domain, the individual, system, and environmental factors that affect SA, and a design process that specifically addresses SA in a systematic fashion.

SITUATION AWARENESS DEFINITION

Situation awareness is formally defined as "the perception of the elements in the environment within a volume of time and space, the comprehension of their meaning and the projection of their status in the near future" (Endsley, 1988). Situation awareness therefore involves perceiving critical factors in the environment (Level 1 SA), understanding what those factors mean, particularly when integrated together in relation to the aircrew's goals (Level 2), and at the highest level, understanding what will happen with the system in the near future (Level 3). These higher levels of SA allow pilots to function in a timely and effective manner.

Level 1 SA—Perception of the Elements in the Environment

The first step in achieving SA is to perceive the status, attributes, and dynamics of relevant elements in the environment. A pilot needs to perceive important elements such as other aircraft, terrain, system status, and warning lights, along with their relevant characteristics. In the cockpit, just keeping up with all of the relevant system and flight data, other aircraft, and navigational data can be quite taxing.

Level 2 SA—Comprehension of the Current Situation

Comprehension of the situation is based on a synthesis of disjointed Level 1 elements. Level 2 SA goes beyond simply being aware of the elements that are present, to include an understanding of the significance of those elements in light of one's goals. The aircrew puts together Level 1 data to form a holistic picture of the environment, including a comprehension of the significance of objects and events. For example, upon seeing warning lights indicating a problem during takeoff, the pilot must quickly determine the seriousness of the problem in terms of the immediate air worthiness of the aircraft and combine this with knowledge on the amount of runway remaining in order to know whether it is an abort situation or not. A novice pilot may be capable of achieving the same Level 1 SA as more experienced pilots, but may fall far short of being able to integrate various data elements along with pertinent goals in order to comprehend the situation as well.

Level 3 SA—Projection of Future Status

It is the ability to project the future actions of the elements in the environment, at least in the very near term, that forms the third and highest level of situation awareness. This is achieved through knowledge of the status and dynamics of the elements and a comprehension of the situation (both Level 1 and Level 2 SA). Amalberti and Deblon (1992) found that a significant portion of experienced pilots' time was spent in anticipating possible future occurrences. This gives them the knowledge (and time) necessary to decide on the most favorable course of action to meet their objectives.

SITUATION AWARENESS REQUIREMENTS

Fully understanding SA in the aviation environment rests on a clear elucidation of its elements (at each of the three levels of SA), identifying which things the aircrew needs to perceive, understand, and project. These are specific to individual systems and contexts, and, as such, must be determined for a particular class of aircraft and missions (e.g., commercial flight deck, civil aviation, strategic or tactical military aircraft, etc.). In general, however, across many types of aircraft systems certain classes of elements are needed for situation awareness that can be described.

Geographical SA. Location of own aircraft, other aircraft, terrain features, airports, cities, waypoints, and navigation fixes; position relative to designated features; runway and taxiway assignments; path to desired locations; climb/descent points.

Spatial/Temporal SA. Attitude, altitude, heading, velocity, vertical velocity, G's, flight path; deviation from flight plan and clearances; aircraft capabilities; projected flight path; projected landing time.

System SA. System status, functioning, and settings; settings of radio, altimeter, and transponder equipment; air traffic control (ATC) communications present; deviations from correct settings; flight modes and automation entries and settings; impact of malfunctions/system degrades and settings on system performance and flight safety; fuel; time and distance available on fuel.

Environmental SA. Weather formations (area and altitudes affected and movement; temperature, icing, ceilings, clouds, fog, sun, visibility, turbulence, winds, microbursts); instrument flight rules (IFR) versus visual flight rules (VFR) conditions; areas and altitudes to avoid; flight safety; projected weather conditions.

In addition, for military aircraft, elements relative to the military mission will also be important.

Tactical SA. Identification, tactical status, type, capabilities, location and flight dynamics of other aircraft; own capabilities in relation to other aircraft; aircraft detections, launch capabilities and targeting; threat prioritization, imminence and assignments; current and projected threat intentions, tactics, firing and maneuvering; mission timing and status.

Determining specific SA requirements for a particular class of aircraft is dependent on the goals of the aircrew in that particular role. A methodology for determining SA requirements has been developed and applied to fighter aircraft (Endsley, 1993), bomber aircraft (Endsley, 1989), and air traffic controllers (Endsley & Rodgers, 1994).

INDIVIDUAL FACTORS INFLUENCING SITUATION AWARENESS

In order to provide an understanding of the processes and factors that influence the development of SA in complex settings such as aviation, a theoretical model describing factors underlying situation awareness has been developed (Endsley, 1988, 1994, 1995c). Key features of the model are summarized here and are shown in Fig. 11.1. (The reader is referred to Endsley, 1995c, for a full explanation of the model and supporting research.) In general, SA in the aviation setting is challenged by the limitations of human attention and working memory. The development of relevant long-term memory stores, goal-directed processing, and automaticity of actions through experience and training are seen as the primary mechanisms used for overcoming these limitations to achieve high levels of SA and successful performance.

Processing Limitations

Attention. In aviation settings, the development of situation awareness and the decision process are restricted by limited attention and working memory capacity for novice aircrew and those in novel situations. Direct attention is needed for perceiving

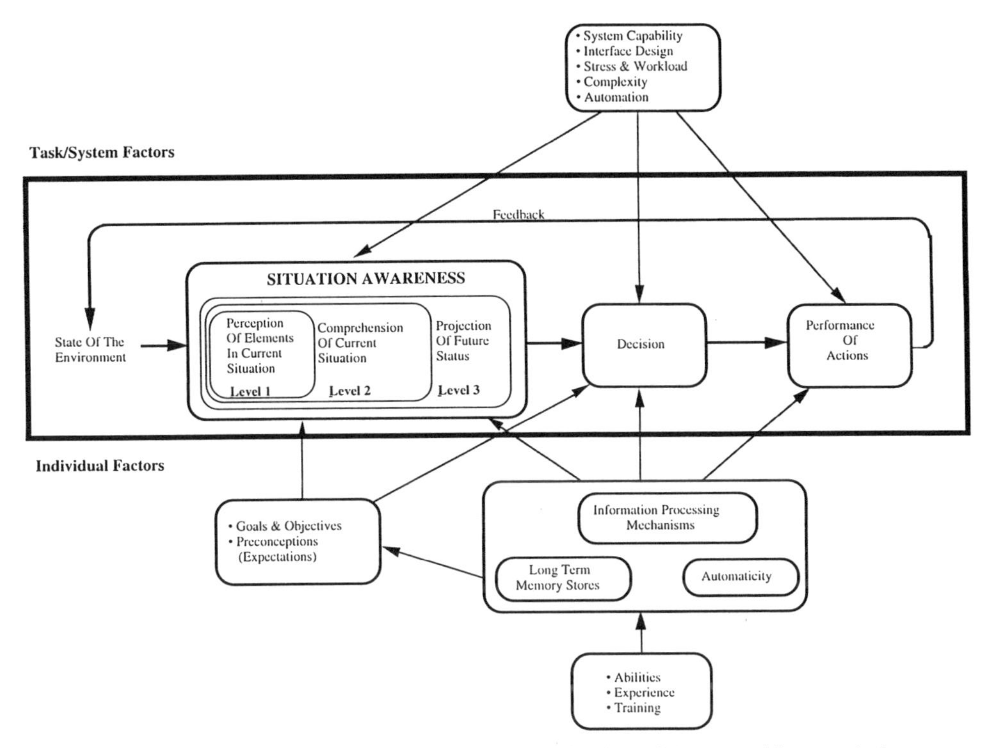

FIG. 11.1. Model of situation awareness. From Endsley (1995c). Reprinted by permission.

and processing the environment to form SA, for selecting actions and executing responses. In the complex and dynamic aviation environment, information overload, task complexity, and multiple tasks can quickly exceed the aircrew's limited attention capacity. Because the supply of attention is limited, more attention to some information may mean a loss of SA on other elements. The resulting lack of SA can result in poor decisions leading to human error. In a review of NTSB aircraft accident reports, poor SA resulting from attention problems in acquiring data accounted for 31% of accidents involving human error (Endsley, 1995a).

Pilots typically employ a process of information sampling to circumvent attention limits, attending to information in rapid sequence following a pattern dictated by long-term memory concerning the relative priorities and the frequency with which information changes. Working memory also plays an important role in this process, allowing the pilot to modify attention deployment on the basis of other information perceived or active goals. For example, in a study of pilot SA, Fracker (1990) showed that a limited supply of attention was allocated to environmental elements on the basis of their ability to contribute to task success.

Unfortunately, people do not always sample information optimally. Typical failings include: (a) forming non-optimal strategies based on a misperception of the statistical properties of elements in the environment, (b) visual dominance—attending more to visual elements than information coming through competing aural channels, and (c) limitations of human memory, leading to inaccuracy in remembering statistical properties to guide sampling (Wickens, 1984). In addition, in the presence of information overload, a frequent occurrence, pilots may feel that the process of information sampling is either not sufficient or not efficient, in which case the pilot may choose to attend to certain information, to the neglect of other information. If the pilot is correct in this selection, all is well. However, in many instances this is not the case.

As a highly visible example, reports of controlled descent into the terrain by high-performance fighter aircraft are numerous (McCarthy, 1988). Although various factors can be implicated in these incidents, channelized attention (31%), distraction by irrelevant stimuli (22%), task saturation (18%), and preoccupation with one task (17%) have all been indicated as significant causal factors (Kuipers, Kappers, van Holten, van Bergen, & Oosterveld, 1990). Some 56% of respondents in the same study indicated a lack of attention for primary flight instruments (the single highest factor) and having too much attention directed toward the target plane during combat (28%) as major causes. Clearly, this demonstrates the negative consequences of both intentional and unintentional disruptions of scan patterns. In the case of intentional attention shifts, it is assumed that attention was probably directed to other factors that the pilots erroneously felt to be more important, because their SA was either outdated or incorrectly perceived in the first place. This leads to a very important point. In order to know which information to focus attention on and which information can be temporarily ignored, the pilot must have at some level an understanding about all of it—that is, "the big picture."

The way in which information is perceived (Level 1 SA) is affected by the contents of both working memory and long-term memory. Advanced knowledge of the characteristics, form, and location of information, for instance, can significantly facilitate the perception of information (Barber & Folkard, 1972; Biederman, Mezzanotte, Rabinowitz, Francolin, & Plude, 1981; Davis, Kramer, & Graham, 1983; Humphreys, 1981; Palmer, 1975; Posner, Nissen, & Ogden, 1978). This type of knowledge is typically gained through experience, training, or preflight planning and analysis. One's preconceptions or expectations about information can effect the speed and accuracy of the perception of information. Repeated experience in an environment allows people to develop expectations about future events that predisposes them to perceive the information accordingly. They will process information faster if it is in agreement with those expectations and will be more likely to make an error if it is not (Jones, 1977). As a classic example, readback errors, repeating an expected clearance instead of the actual clearance to the air traffic controller, are common (Monan, 1986).

Working Memory. Working memory capacity can also act as a limit on SA. In the absence of other mechanisms, most of a person's active processing of information must occur in working memory. The second level of SA involves comprehending the meaning of the data that is perceived. New information must be combined with existing knowledge and a composite picture of the situation developed. Achieving the desired integration and comprehension in this fashion is a very taxing proposition that can seriously overload the pilot's limited working memory and will draw even further on limited attention, leaving even less capacity to direct toward the process of acquiring new information.

Similarly, projections of future status (Level 3 SA) and subsequent decisions as to appropriate courses of action will draw on working memory as well. Wickens (1984) stated that the prediction of future states imposes a strong load on working memory by requiring the maintenance of present conditions, future conditions, rules used to generate the latter from the former, and actions that are appropriate to the future conditions. A heavy load will be imposed on working memory if it is taxed with achieving the higher levels of situation awareness in addition to formulating and selecting responses and carrying out subsequent actions.

Coping Mechanisms

Mental Models. In practice, however, experienced aircrew may use long-term memory stores, most likely in the form of schemata and mental models, to circumvent these limits for learned classes of situations and environments. These mechanisms provide for the integration and comprehension of information and the projection of future events. They also allow for decision making on the basis of incomplete information and under uncertainty.

Experienced aircrew often have internal representations of the system they are dealing with—a mental model. A well-developed mental model provides (a) knowledge of the relevant "elements" of the system that can be used in directing attention and classifying information in the perception process, (b) a means of integrating elements to form an understanding of their meaning (Level 2 SA), and (c) a mechanism for projecting future states of the system based on its current state and an understanding of its dynamics (Level 3 SA). During active decision making, a pilot's perceptions of the current state of the system may be matched to related schemata in memory that depict prototypical situations or states of the system model. These prototypical situations provide situation classification and understanding and a projection of what is likely to happen in the future (Level 3 SA).

A major advantage of these mechanisms is that the current situation does not need to be exactly like one encountered before due to the use of categorization mapping (a best fit between the characteristics of the situation and the characteristics of known categories or prototypes). The matching process can be almost instantaneous due to the superior abilities of human pattern matching mechanisms. When an individual has a well-developed mental model for the behavior of particular systems or domains, it will provide (a) for the dynamic direction of attention to critical environmental cues, (b) expectations regarding future states of the environment (including what to expect as well as what not to expect) based on the projection mechanisms of the model, and (c) a direct, single-step link between recognized situation classifications and typical actions, providing very rapid decision making.

The use of mental models also provides useful default information. These default values (expected characteristics of elements based on their classification) may be used by aircrew to predict system performance with incomplete or uncertain information, providing more effective decisions than for novices who will be more hampered by missing data. For example, experienced pilots are able to predict within a reasonable range how fast a particular aircraft is traveling just by knowing what type of aircraft it is. Default information may furnish an important coping mechanism for experienced aircrew in forming SA in many situations where information is missing or overload prevents them from acquiring all the information they need.

Well-developed mental models and schemata to can provide the comprehension and future projection required for the higher levels of SA almost automatically, thus greatly off-loading working memory and attention requirements. A major advantage of these long-term stores is that a great deal of information can be called on very rapidly, using only a very small amount of attention (Logan, 1988). When scripts have been developed, tied to these schema, the entire decision-making process can be greatly simplified, and working memory will be off-loaded even further.

Goal-Driven Processing. In the processing of dynamic and complex information, people may switch between data-driven and goal-driven processing. In a data-driven process, various environmental features are detected whose inherent properties determine which information will receive further focalized attention and processing. In this mode, cue salience will have a large impact on which portions of the environment are attended to and thus SA. People can also operate in a goal-driven fashion. In this mode, situation awareness is affected by the aircrew's goals and expectations, which influence how attention is directed, how information is perceived, and how it is interpreted. The person's goals and plans direct which aspects of the environment are attended to; that information is then integrated and interpreted in light of these goals to form level 2 SA. On an ongoing basis, one can observe trade-offs between top-down and bottom-up processing, allowing the aircrew to process information effectively in a dynamic environment.

With experience, aircrew will develop a better understanding of their goals, which goals should be active in which circumstances, and how to acquire information to support these goals. The increased reliance on goal-directed processing allows the environment to be processed more efficiently than with purely data-driven processing. An important issue for achieving successful performance in the aviation domain lies in the ability of the aircrew to dynamically juggle multiple competing goals effectively. They need to rapidly switch between pursuing information in support of a particular goal to responding to perceived data activating a new goal, and back again. The ability to hold multiple goals has been associated with distributed attention, which is important for performance in the aviation domain (Martin & Jones, 1984).

Automaticity. SA can also be effected by the use of automaticity in processing information. Automaticity may be useful in overcoming attention limits, but may also leave the pilot susceptible to missing novel stimuli. Over time, it is easy for actions to become habitual and routine, requiring a very low level of attention. When something is slightly different, however—for example, a different clearance than usual—the pilots may miss it and carry out the habitual action. Developed through experience and a high level of learning, automatic processing tends to be fast, autonomous, effortless, and unavailable to conscious awareness in that it can occur without attention (Logan, 1988). Automatic processing is advantageous in that it provides good performance with minimal attention allocation. Although automaticity may provide an important mechanism for overcoming processing limitations, thus allowing people to achieve SA and make decisions in complex, dynamic environments like aviation, it also creates an increased risk of being less responsive to new stimuli as automatic processes operate with limited use of feedback. When using automatic processing, a lower level of SA can result in nontypical situations, decreasing decision timeliness and effectiveness.

Summary. In summary, situation awareness can be achieved by drawing on a number of internal mechanisms. Due to limitations of attention and working memory, long-term memory may be heavily relied on to achieve SA in the highly demanding aviation environment. The degree to which these structures can be developed and effectively used in the flight environment, the degree to which aircrew can effectively deploy goal driven processing in conjunction with data driven processing, and the degree to which aircrew can avoid the hazards of automaticity will ultimately determine the quality of their SA.

CHALLENGES TO SITUATION AWARENESS

In addition to SA being affected by the characteristics and processing mechanisms of the individual, many environmental and system factors will have a large impact on SA. Each of these factors can act to seriously challenge the ability of the aircrew to maintain a high level of SA in many situations.

Stress

Several types of stress factors exist in the aviation environment that may affect SA, including:

1. Physical stressors—noise, vibration, heat/cold, lighting, atmospheric conditions, boredom, fatigue, cyclical changes, G's.
2. Social/psychological stressors—fear or anxiety, uncertainty, importance or consequences of events, self-esteem, career advancement, mental load, and time pressure (Hockey, 1986; Sharit & Salvendy, 1982).

A certain amount of stress may actually improve performance by increasing attention to important aspects of the situation. A higher amount of stress can have extremely negative consequences, however, as accompanying increases in autonomic functioning and aspects of the stressors can act to demand a portion of a person's limited attentional capacity (Hockey, 1986).

Stressors can affect SA in a number of different ways, including attentional narrowing, reductions in information intake, and reductions in working memory capacity. Under stress a decrease in attention has been observed for peripheral information, those aspects that attract less attentional focus (Bacon, 1974; Weltman, Smith, & Egstrom, 1971), and there is an increased tendency to sample dominant or probable sources of information (Broadbent, 1971). This is a critical problem for SA, leading to the neglect of certain elements in favor of others. In many cases, such as in emergency conditions, it is those factors outside the person's perceived central task that prove to be lethal. An L-1011 crashed in the Florida Everglades killing 99 people when the crew became focused on a problem with a nose gear indicator and failed to monitor the altitude and attitude of the aircraft (National Transportation Safety Board, 1973). In military aviation, many lives are lost due to controlled flight into the terrain accidents, with attentional narrowing serving as a primary culprit (Kuipers et al., 1990).

Premature closure, arriving at a decision without exploring all information available, has also been found to be more likely under stress (Janis, 1982; Keinan, 1987; Keinan & Friedland, 1987). This includes considering less information and attending more to negative information (Janis, 1982; Wright, 1974). Several authors have also found that scanning of information under stress is scattered and poorly organized (Keinan, 1987; Keinan & Friedland, 1987; Wachtel, 1967). A lowering of attention capacity, attentional narrowing, disruptions of scan patterns, and premature closure may all negatively effect Level 1 SA under various forms of stress.

A second way in which stress may negatively effect SA is by decreasing working memory capacity and hindering information retrieval (Hockey, 1986; Mandler, 1982). The degree to which working memory decrements will impact SA depends on the resources available to the individual. In tasks where achieving SA involves a high working memory load, a significant impact on SA Levels 2 and 3 (given the same Level 1 SA) would be expected. If long-term memory stores are available to support SA, however, as in more learned situations, less effect will be expected.

Overload/Underload

High mental workload is a stressor of particular importance in aviation that can negatively affect SA. If the volume of information and number of tasks are too great, SA may suffer as only a subset of information can be attended to, or the pilot may be actively working to achieve SA, yet suffer from erroneous or incomplete perception and integration of information. In some cases, SA problems may occur from an overall high level of workload, or, in many cases, due to a momentary overload in the tasks to be performed or in information being presented.

Poor SA can also occur under low workload. In this case the pilot may have little idea of what is going on and may not be actively working to find out due to inattentiveness, vigilance problems, or low motivation. Relatively little attention has been paid to the effects of low workload (particularly on long-haul flights, for instance) on SA; however, this condition can pose a significant challenge for SA in many areas of aviation and deserves further study.

System Design

The capabilities of the aircraft for acquiring needed information and the way in which it presents that information will have a large impact on aircrew SA. Although a lack of information can certainly be seen as a problem for SA, too much information poses an equal problem. Associated with improvements in the avionics capabilities of aircraft in the past few decades has been a dramatic increase in the sheer quantity of information available. Sorting through this data to derive the desired information and achieve a good picture of the overall situation is no small challenge. Overcoming this problem through better system designs that present integrated data is currently a major design goal aimed at alleviating this problem.

Complexity

A major factor creating a challenge for SA is the complexity of the many systems that must be operated. There has been an explosion of avionics systems, flight management systems, and other technologies on the flight deck that have greatly increased the complexity of the systems aircrew must operate. System complexity can negatively effect both pilot workload and SA through an increase in the number of system components to be managed, a high degree of interaction between these components, and an increase in the dynamics or rate of change of the components. In addition, the complexity of the pilot's tasks may increase through an increase in the number of goals, tasks, and decisions to be made in regard to the aircraft systems. The more complex the systems are to operate, the greater is the increase the mental workload required to achieve a given level of SA. When that demand exceeds human capabilities, SA will suffer.

System complexity may be somewhat moderated by the degree to which the person has a well-developed internal representation of the system to aid in directing attention, integrating data, and developing the higher levels of SA. These mechanisms may be effective for coping with complexity; however, developing those internal models may require a considerable amount of training. Pilots have reported significant difficulties in understanding what their automated flight management systems are doing and why (Sarter & Woods, 1992; Wiener, 1989). McClumpha and James (1994) conducted an extensive study of nearly 1,000 pilots across varying nationalities and aircraft types. They found that the primary factor explaining variance in pilots' attitudes toward advanced technology aircraft was their self-reported understanding of the system. Although pilots are eventually developing a better understanding of automated aircraft with experience, many of these systems do not appear to be well designed to meet their SA needs.

Automation

SA may also be negatively impacted by the automation of tasks as it is frequently designed to put the aircrew "out of the loop." System operators working with automation have been found to have a diminished ability to detect system errors and perform tasks manually in the face of automation failures as compared to purely manual performance on the same tasks (Billings, 1991; Moray, 1986; Wickens, 1992; Wiener & Curry, 1980). In 1987, a Northwest Airlines MD-80 crashed on takeoff at Detroit Airport due to an improper configuration of the flaps and slats, killing all but one passenger (National Transportation Safety Board, 1988). A major factor in the crash was the failure of an automated takeoff configuration warning system that the crew had become reliant on. They did not realize the aircraft was improperly configured for takeoff and had neglected to check manually (due to other contributing factors). When the automation failed, they were not aware of the state of the automated system or the critical flight parameters they counted on the automation to monitor. Although some of the out-of-the-loop performance problem may be due to a loss of manual skills under automation, loss of SA is also a critical component leading to this accident and many similar ones.

Pilots who have lost SA through being out of the loop may be slower to detect problems and additionally will require extra time to reorient themselves to relevant system parameters in order to proceed with problem diagnosis and assumption of manual performance when automation fails. This has been hypothesized to occur for a number of reasons, including (a) a loss of vigilance and increase in complacency associated becoming a monitor with the implementation of automation, (b) being a passive recipient of information rather than an active processor of information, and (c) a loss of or change in the type of feedback provided to the aircrew concerning the state of the system being automated (Endsley & Kiris, 1995). In their study, Endsley and Kiris found evidence for an SA decrement accompanying automation of a cognitive task that was greater under full automation than under partial automation. Lower SA in the automated conditions corresponded to a demonstrated out-of-the-loop performance decrement, supporting the hypothesized relationship between SA and automation.

SA may not suffer under all forms of automation, however. Wiener (1993) and Billings (1991) stated that SA may be improved by systems that provide integrated information through automation. In commercial cockpits, Hansman et al. (1992) found automated flight management system input was superior to manual data entry, producing better error detection of clearance updates. Automation that reduces unnecessary manual work and data integration required to achieve SA may provide benefits to both workload and SA. The exact conditions under which SA will be positively or negatively affected by automation need to be determined.

ERRORS IN SITUATION AWARENESS

Based on this model of SA, a taxonomy for classifying and describing errors in SA was created (Endsley, 1994, 1995c). The taxonomy, presented in Table 11.1, incorporates factors affecting SA at each of its three levels. Endsley (1995a) applied this taxonomy to an investigation of causal factors underlying aircraft accidents involving major air carriers in the United States based on National Transportation Safety Board (NTSB) accident investigation reports over a 4-year period. Of the 71% of the accidents that could be classified as having a substantial human error component, 88% involved problems with SA. Of 32 SA errors identified in these accident descriptions, 23 (72%) were attributed to problems with Level 1 SA, a failure to correctly perceive some pieces of information in the situation. Seven (22%) involved a Level 2 error in which the data was perceived but not integrated or comprehended correctly, and two (6%) involved a Level 3 error in which there was a failure to properly project the near future based on the aircrew's understanding of the situation.

More recently, Jones and Endsley (1995) applied this taxonomy to a more extensive study of SA errors based on voluntary reports in NASA's Aviation Safety Reporting System (ASRS) database. This provides some indication of the types of problems and relative contribution of causal factors leading to SA errors in the cockpit, as shown in Fig. 11.2.

Level 1—Failure to Correctly Perceive the Situation

At the most basic level, important information may not be correctly perceived. In some cases, the data may not be available to the person, due to a failure of the system design to present it or a failure in the communications process. This factor accounted for 11.6%

TABLE 11.1
SA Error Taxonomy

Level 1: Failure to correctly perceive information
Data not available
Data hard to discriminate or detect
Failure to monitor or observe data
Misperception of data
Memory loss
Level 2: Failure to correctly integrate or comprehend information
Lack of or poor mental model
Use of incorrect mental model
Overreliance on default values
Other
Level 3: Failure to project future actions or state of the system
Lack of or poor mental model
Overprojection of current trends
Other
General
Failure to maintain multiple goals
Habitual schema

Note. From Endsley (1995a). Adapted with permission.

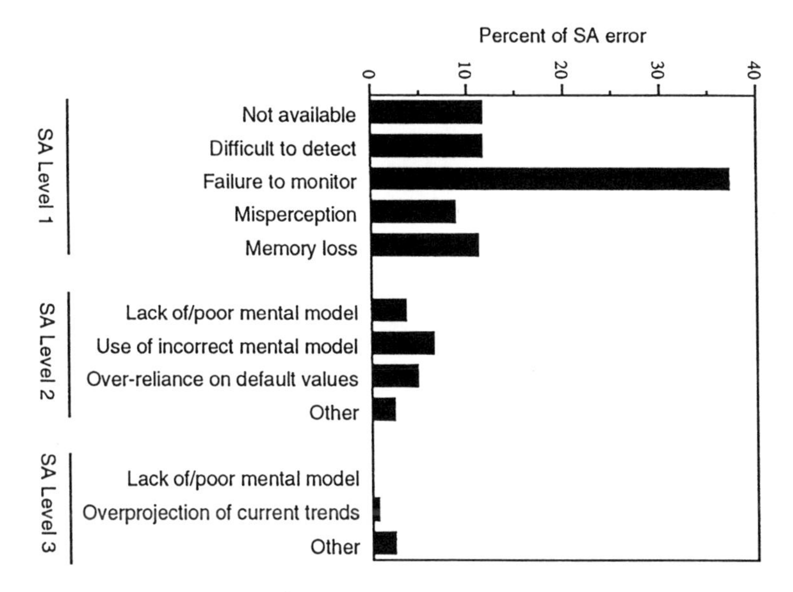

FIG. 11.2. SA error causal factors. From Jones and Endsley (1995). Reprinted by permission.

of SA errors, most frequently occurring due to a failure of the crew to perform some necessary task (such as resetting the altimeter) to obtain the correct information. In other cases, the data is available, but is difficult to detect or perceive, accounting for another 11.6% of SA errors in this study. This included problems due to poor runway markings and lighting and problems due to noise in the cockpit.

Many times, the information is directly available, but for various reasons is not observed or included in the scan pattern, forming the largest single causal factor for SA errors (37.2%). This is due to several factors, including simple omission—not looking at a piece of information, attentional narrowing, and external distractions that prevent the crew from attending to important information. High taskload, even momentary, is another a major factor that prevents information from being attended to.

In other cases, information is attended to, but is misperceived (8.7% of SA errors), frequently due to the influence of prior expectations. Finally, in some cases it appears that a person initially perceives some piece of information but then forgets about it (11.1% of SA errors), which negatively effects SA as it relies on keeping information about a large number of factors in memory. Forgetting was found to be frequently associated with disruptions in normal routine, high workload, and distractions.

Level 2 SA—Failure to Comprehend the Situation

In other cases, information is correctly perceived, but its significance or meaning is not comprehended. This may be due to the lack of a good mental model for combining information in association with pertinent goals. Of SA errors, 3.5% were attributed to the lack of a good mental model, most frequently associated with an automated system.

In other cases, the wrong mental model may be used to interpret information, leading to 6.4% of the SA errors in this study. In this case, the mental model of a similar system may be used to interpret information, leading to an incorrect diagnosis or understanding of the situation in areas where that system is different. A frequent problem is where aircrew have a model of what is expected and then interpret all perceived cues into that model, leading to a completely incorrect interpretation of the situation.

In addition, there may also be problems with overreliance on defaults in the mental model used, as was found for 4.7% of the SA errors. These defaults can be thought of as general expectations about how parts of the system function that may be used in the absence of real-time data. In other cases, the significance of perceived information relative to operational goals is simply not comprehended or several pieces of information are not properly integrated. This may be due to working memory limitations or other unknown cognitive lapses. 2.3% of the SA errors were attributed to miscellaneous factors such as these.

Level 3 SA—Failure to Project Situation into the Future

Finally, in some cases, individuals may be fully aware of what is going on, but be unable to correctly project what that means for the future, accounting for 2.9% of the SA errors. In some cases this may be due to a poor mental model or due to over projecting current trends. In other cases, the reason for not correctly projecting the situation is less apparent. Mental projection is a very demanding task at which people are generally poor.

General

In addition to these main categories, two general categories of causal factors are included in the taxonomy. First, some people have been found to be poor at maintaining multiple goals in memory, which could impact SA across all three levels. Second, there is evidence that people can fall into a trap of executing habitual schemata, doing tasks automatically, which renders them less receptive to important environmental cues. Evidence for these causal factors was not apparent in the retrospective reports analyzed in the ASRS or NTSB databases.

SA IN MULTICREW AIRCRAFT

Although SA has been discussed mainly at the level of the individual, it is also relevant for the aircrew as a team. The crew may be constructed of two or three members in a commercial aircraft and as many as five to seven members in some military aircraft. In some military settings, several aircraft may also be deployed as a flight, forming a more loosely coupled team in which several aircraft must work together to accomplish a joint goal.

Team SA has been defined as "the degree to which every team member possesses the SA required for his or her responsibilities" (Endsley, 1989). If one crew member has a certain piece of information, but another who needs it does not, the SA of the team has suffered and its performance may suffer as well unless the discrepancy is corrected. In this light, a major portion of intercrew coordination can be seen as the transfer of information from one crew member to another, as required for developing SA across the team. This coordination involves more than just sharing data. It also includes sharing the higher levels of SA (comprehension and projection), which may vary widely between individuals depending on their experiences and goals.

The process of providing shared SA can be greatly enhanced by shared mental models, which provide a common frame of reference for crew-member actions and allow team members to predict each other's behaviors (Cannon-Bowers, Salas, & Converse, 1993; Orasanu, 1990). A shared mental model may provide more efficient communications by providing a common means of interpreting and predicting actions based on limited information, and therefore may be important for SA. For instance, Mosier and Chidester (1991) found that better performing teams actually communicated less than poorer performing teams.

Impact of CRM on SA

Crew resource management (CRM) programs have received a great deal of attention and focus in aviation in recent years, as a means of promoting better teamwork and use of crew resources. Robertson and Endsley (1995) investigated the link between SA and CRM programs and found that CRM can have an effect on crew SA by directly improving individual SA, or indirectly through the development of shared mental models and by providing efficient distribution of attention across the crew. They hypothesized that CRM could be used to improve team SA through various behaviors measured by the Line/LOS Checklist (LLC), as shown in Fig. 11.3, which have been shown to be positively impacted by CRM (Butler, 1991; Clothier, 1991).

Individual SA. First, improved communication between crew members can obviously facilitate effective sharing of needed information. In particular, improved inquiry and assertion behaviors by crew members help to insure needed communication. In addition, an understanding of the state of the human elements in the system (intercrew SA) also forms a part of SA. The development of good self-critique skills can be used to provide an up-to-date assessment of one's own and other team member's abilities and performance, which may be impacted by factors such as fatigue or stress. This knowledge allows team members to recognize the need for providing more information

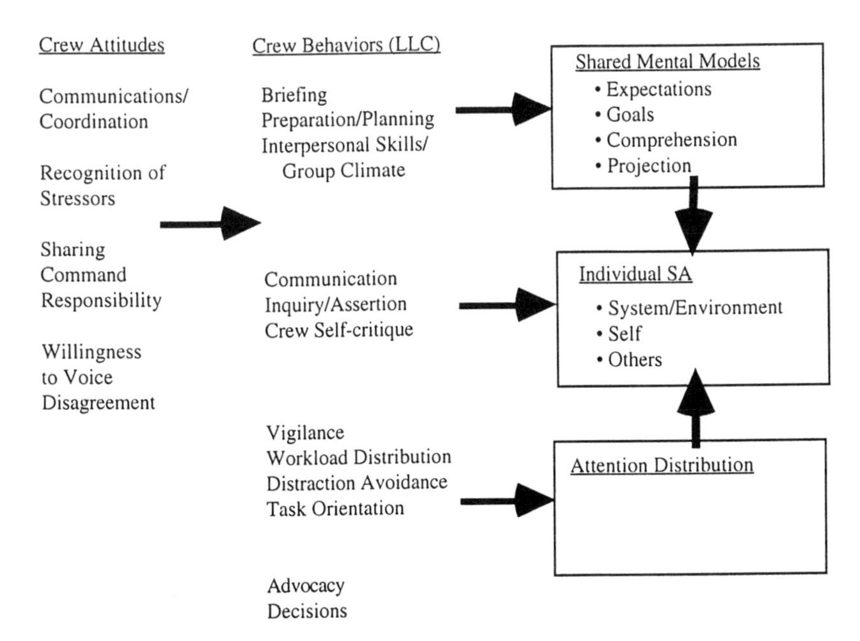

FIG. 11.3. CRM factors affecting SA. From Robertson and Endsley (1995). Reprinted by permission.

and taking over functions in critical situations, an important part of effective team performance.

Shared Mental Models. Several factors can help to develop shared mental models between crew members. The crew briefing establishes the initial basis for a shared mental model between crew members, providing shared goals and expectations. This can increase the likelihood that two crew members will form the same higher levels of SA from low-level information, improving the effectiveness of communications. Prior preparation and planning similarly can help establish a shared mental model. Effective crews tend to "think ahead" of the aircraft, allowing them to be ready for a wide variety of events. This is closely linked to Level 3 SA—projection of the future. The development of interpersonal relationships and group climate can also be used to facilitate the development of a good model of other crew members. This allows individuals to predict how others will act, forming the basis for Level 3 SA and efficiently functioning teams.

Attention Distribution. The effective management of the crew's resources is extremely critical, particularly in high taskload situations. A major factor in effectively managing these resources is ensuring that all aspects of the situation are being attended to—avoiding attentional narrowing and neglect of important information and tasks. CRM programs that improve task orientation and the distribution of tasks under workload can directly impact how crew members are directing their attention, and thus their SA. In addition, improvements in vigilance and the avoidance of distractions can be seen to directly impact SA.

Thus, there are a number of ways that existing CRM programs can effect SA at the crew level, as well as within individuals. Further development of CRM and other training programs to focus on the problems of SA is a current need.

FUTURE DIRECTIONS

Design

Cockpit design efforts can be directed toward several avenues for improving SA, including searching for (a) ways to determine and effectively deliver critical cues, (b) ways to ensure accurate expectations, (c) methods for assisting pilots in deploying attention effectively, (d) methods for preventing the disruption of attention, particularly under stress and high workload, and (e) ways to develop systems that are compatible with pilot goals. Many ongoing design efforts are aimed at enhancing SA in the cockpit by taking advantage of new technologies such as advanced avionics and sensors, datalink, global positioning systems (GPS), three-dimensional visual and auditory displays, voice control, expert systems, helmet mounted displays, virtual reality, sensor fusion, automation, and expert systems. The glass cockpit, advanced automation techniques, and new technologies such as the Traffic Collision Avoidance System (TCAS) have become a reality in today's aviation systems.

Each of these technologies provides a potential advantage: new information, more accurate information, new ways of providing information, or a reduction in crew workload. Each can also effect SA in unpredicted ways, however. For instance, recent evidence shows that automation, which is often cited as a potential benefit to SA through the reduction of workload, can actually reduce SA, thus contributing to the out-of-the-loop performance problem (Carmody & Gluckman, 1993; Endsley & Kiris, 1995). Three-dimensional displays, also touted as beneficial for SA, have been found to have quite negative effects on pilots' ability to accurately localize other aircraft and objects (Endsley, 1995b; Prevett & Wickens, 1994).

As many factors surrounding the use of new technologies and design concepts may act to both enhance and degrade SA, significant care should be taken to evaluate the impact of proposed concepts on SA. Only by testing new design concepts in carefully controlled studies can the actual impact on these factors be identified. This testing needs to include not only an examination of how the technologies affect basic human processes, such as accuracy of perception, but also how they affect the pilot's global state of knowledge when used in a dynamic and complex aviation scenario where multiple sources of information compete for attention and must be selected, processed, and integrated in light of dynamic goal changes. Real-time simulations employing the technologies can be used to assess the impact of the system by carefully measuring aircrew performance, workload and situation awareness. Direct measurement of situation awareness during design testing is recommended for providing sufficient insight into the potential costs and benefits of design concepts for aircrew SA, allowing a determination of the degree to which the design successfully addresses the issues put forth. Techniques for measuring SA within the aviation system design process are covered in more detail in Endsley (1996).

Training

In addition to improving SA through better cockpit designs, it may also be possible to find new ways of training aircrew to achieve better SA with a given aircraft design. The potential role of CRM programs in this process has already been discussed. It may also be possible to create "SA oriented" training programs that seek to improve SA directly. This may include programs that provide aircrew with better information needed to develop mental models, including information on their components, the dynamics and functioning of the components, and projection of future actions based on these dynamics. The focus should be on training aircrew to identify prototypical situations of concern associated with these models by recognizing critical cues and what they mean in terms of relevant goals.

The skills required for achieving and maintaining good SA also need to be identified and formally taught in training programs. Factors such as how to employ a system to best achieve SA (when to look for what where), appropriate scan patterns, or techniques for making the most of limited information need to be determined and explicitly taught in the training process. A focus on aircrew SA would greatly supplement traditional technology-oriented training that concentrates mainly on the mechanics of how a system operates.

In addition, the role of feedback as an important component of the learning process may be more fully exploited. It may be possible to provide feedback on the accuracy and completeness of pilot SA as a part of training programs. This would allow aircrew to understand their mistakes and better assess and interpret the environment, leading to the development of more effective sampling strategies and better schema for integrating information. Training techniques such as these need to be explored and tested to determine methods for improving SA with existing systems.

Conclusion

Maintaining situation awareness is a critical and challenging part of an aircrew's job. Without good SA, even the best trained crews can make poor decisions. Numerous factors that are a constant part of the aviation environment make the goal of achieving a high level of SA at all times quite challenging. Enhancing SA through better cockpit design and training programs remains the major challenge for aviation research through the next decade.

REFERENCES

Amalberti, R., & Deblon, F. (1992). Cognitive modeling of fighter aircraft process control: A step towards an intelligent on-board assistance system. *International Journal of Man–Machine Systems, 36*, 639–671.

Bacon, S. J. (1974). Arousal and the range of cue utilization. *Journal of Experimental Psychology, 102*, 81–87.

Barber, P. J., & Folkard, S. (1972). Reaction time under stimulus uncertainty with response certainty. *Journal of Experimental Psychology, 93*, 138–142.

Biederman, I., Mezzanotte, R. J., Rabinowitz, J. C., Francolin, C. M., & Plude, D. (1981). Detecting the unexpected in photo interpretation. *Human Factors, 23*, 153–163.

Billings, C. E. (1991). *Human-centered aircraft automation: A concept and guidelines* (NASA Technical Memorandum 103885). Moffet Field, CA: NASA Ames Research Center.

Broadbent, D. E. (1971). *Decision and stress*. London: Academic Press.

Butler, R. E. (1991). Lessons from cross-fleet/cross airline observations: Evaluating the impact of CRM/LOS training. In R. S. Jensen (Ed.), *Proceedings of the Sixth International Symposium on Aviation Psychology* (pp. 326–331). Columbus, OH: The Ohio State University Press.

Cannon-Bowers, J. A., Salas, E., & Converse, S. (1993). Shared mental models in expert team decision making. In N. J. Castellan (Ed.), *Current issues in individual and group decision making* (pp. 221–247). Hillsdale, NJ: Lawrence Erlbaum Associates.

Carmody, M. A., & Gluckman, J. P. (1993). Task specific effects of automation and automation failure on performance, workload and situational awareness. In R. S. Jensen & D. Neumeister (Eds.), *Proceedings of the Seventh International Symposium on Aviation Psychology* (pp. 167–171). Columbus, OH: The Ohio State University Press.

Clothier, C. (1991). Behavioral interactions in various aircraft types: Results of systematic observation of line operations and simulations. In R. S. Jensen (Ed.) *Proceedings of the Sixth International Conference on Aviation Psychology* (pp. 332–337). Columbus, OH: The Ohio State University Press.

Davis, E. T., Kramer, P., & Graham, N. (1983). Uncertainty about spatial frequency, spatial position, or contrast of visual patterns. *Perception and Psychophysics, 5*, 341–346.

Endsley, M. R. (1988). Design and evaluation for situation awareness enhancement. In *Proceedings of the Human Factors Society 32nd Annual Meeting* (pp. 97–101). Santa Monica, CA: Human Factors Society.

Endsley, M. R. (1989). *Final report: Situation awareness in an advanced strategic mission* (NOR DOC 89-32). Hawthorne, CA: Northrop Corporation.

Endsley, M. R. (1993). A survey of situation awareness requirements in air-to-air combat fighters. *International Journal of Aviation Psychology, 3*(2), 157–168.

Endsley, M. R. (1994). Situation awareness in dynamic human decision making: Theory. In R. D. Gilson, D. J. Garland, & J. M. Koonce (Eds.), *Situational awareness in complex systems* (pp. 27–58). Daytona Beach, FL: Embry-Riddle Aeronautical University Press.

Endsley, M. R. (1995a). A taxonomy of situation awareness errors. In R. Fuller, N. Johnston, & N. McDonald (Eds.), *Human factors in aviation operations* (pp. 287–292). Aldershot, England: Avebury Aviation, Ashgate Publishing.

Endsley, M. R. (1995b). Measurement of situation awareness in dynamic systems. *Human Factors, 37*(1), 65–84.

Endsley, M. R. (1995c). Toward a theory of situation awareness. *Human Factors, 37*(1), 32–64.

Endsley, M. R. (1996). Situation awareness measurement in test and evaluation. In T. O'Brien & S. Charlton (Eds.), *Human factors testing and evaluation* (pp. 159–180). Hillsdale, NJ: Lawrence Erlbaum Associates.

Endsley, M. R., & Kiris, E. O. (1995). The out-of-the-loop performance problem and level of control in automation. *Human Factors, 37*(2), 381–394.

Endsley, M. R., & Rodgers, M. D. (1994). *Situation awareness information requirements for en route air traffic control* (DOT/FAA/AM-94/27). Washington, DC: Federal Aviation Administration Office of Aviation Medicine.

Fracker, M. L. (1990). Attention gradients in situation awareness. In *Situational awareness in aerospace operations (AGARD-CP-478,* Conference Proceedings Nos. 478, pp. 6/1–6/10). Neuilly Sur Seine, France: NATO-AGARD.

Hansman, R. J., Wanke, C., Kuchar, J., Mykityshyn, M., Hahn, E., & Midkiff, A. (1992, September). *Hazard alerting and situational awareness in advanced air transport cockpits.* Paper presented at the 18th ICAS Congress, Beijing, China.

Hartel, C. E., Smith, K., & Prince, C. (1991, April). *Defining aircrew coordination: Searching mishaps for meaning.* Paper presented at the Sixth International Symposium on Aviation Psychology, Columbus, OH.

Hockey, G. R. J. (1986). Changes in operator efficiency as a function of environmental stress, fatigue and circadian rhythms. In K. Boff, L. Kaufman, & J. Thomas (Eds.), *Handbook of perception and performance* (2, pp. 44/1–44/49). New York: John Wiley.

Humphreys, G. W. (1981). Flexibility of attention between stimulus dimensions. *Perception and Psychophysics, 30,* 291–302.

Isaacson, B. (1985). A lost friend. *USAF Fighter Weapons Review, 4*(33), 23–27.

Janis, I. L. (1982). Decision making under stress. In L. Goldberger & S. Breznitz (Eds.), *Handbook of stress: Theoretical and clinical aspects* (pp. 69–87). New York: The Free Press.

Jones, D. G., & Endsley, M. R. (1995). Investigation of situation awareness errors. In *Proceedings of the 8th International Symposium on Aviation Psychology* (pp. 746–751). Columbus, OH: The Ohio State University Press.

Jones, R. A. (1977). *Self-fulfilling prophecies: Social, psychological and physiological effects of expectancies.* Hillsdale, NJ: Lawrence Erlbaum Associates.

Keinan, G. (1987). Decision making under stress: Scanning of alternatives under controllable and uncontrollable threats. *Journal of Personality and Social Psychology, 52*(3), 639–644.

Keinan, G., & Friedland, N. (1987). Decision making under stress: Scanning of alternatives under physical threat. *Acta Psychologica, 64,* 219–228.

Kraby, A. W. (1995). A close encounter on the Gulf Coast. *Up front: The flight safety and operations publication of Delta Airlines, 2nd Quarter,* 4. Atlanta: Delta Airlines.

Kuipers, A., Kappers, A., van Holten, C. R., van Bergen, J. H. W., & Oosterveld, W. J. (1990). Spatial disorientation incidents in the R.N.L.A.F. F16 and F5 aircraft and suggestions for prevention. In *Situational awareness in aerospace operations (AGARD-CP-478,* pp. OV/E/1–OV/E/16). Neuilly Sur Seine, France: NATO-AGARD.

Logan, G. D. (1988). Automaticity, resources and memory: Theoretical controversies and practical implications. *Human Factors, 30*(5), 583–598.

Mandler, G. (1982). Stress and thought processes. In L. Goldberger & S. Breznitz (Eds.), *Handbook of stress: Theoretical and clinical aspects* (pp. 88–104). New York: The Free Press.

Martin, M., & Jones, G. V. (1984). Cognitive failures in everyday life. In J. E. Harris & P. E. Morris (Eds.), *Everyday memory, actions and absent-mindedness* (pp. 173–190). London: Academic Press.

McCarthy, G. W. (1988, May). Human factors in F16 mishaps. *Flying Safety,* pp. 17–21.

McClumpha, A., & James, M. (1994). Understanding automated aircraft. In M. Mouloua & R. Parasuraman (Eds.), *Human performance in automated systems: Current research and trends* (pp. 183–190). Hillsdale, NJ: Lawrence Erlbaum Associates.

Monan, W. P. (1986). *Human factors in aviation operations: The hearback problem* (NASA Contractor Rep. No. 177398). Moffett Field, CA: NASA Ames Research Center.

Moray, N. (1986). Monitoring behavior and supervisory control. In K. Boff (Ed.), *Handbook of perception and human performance* (Vol. 2, pp. 40/1–40/51). New York: Wiley.

Mosier, K. L., & Chidester, T. R. (1991). Situation assessment and situation awareness in a team setting. In Y. Queinnec & F. Daniellou (Eds.), *Designing for everyone* (pp. 798–800). London: Taylor & Francis.

National Transportation Safety Board. (1973). *Aircraft accidents report: Eastern Airlines 401/L-1011, Miami, Florida, December 29, 1972.* Washington, DC: Author.

National Transportation Safety Board. (1988). *Aircraft accidents report: Northwest Airlines, Inc., McDonnell-Douglas DC-9-82, N312RC, Detroit Metropolitan Wayne County Airport, August, 16, 1987* (NTSB/AAR-99-05). Washington, DC: Author.

Orasanu, J. (1990, July). *Shared mental models and crew decision making.* Paper presented at the 12th Annual Conference of the Cognitive Science Society, Cambridge, MA.

Palmer, S. E. (1975). The effects of contextual scenes on the identification of objects. *Memory & Cognition, 3*, 519–526.

Posner, M. I., Nissen, J. M., & Ogden, W. C. (1978). Attended and unattended processing modes: The role of set for spatial location. In H. L. Pick & E. J. Saltzman (Eds.), *Modes of perceiving and processing* (pp. 137–157). Hillsdale, NJ: Lawrence Erlbaum Associates.

Prevett, T. T., & Wickens, C. D. (1994). *Perspective displays and frame of reference: Their interdependence to realize performance advantages over planar displays in a terminal area navigation task* (ARL-94-8/NASA-94-3). Savoy, IL: University of Illinois at Urbana-Champaign.

Robertson, M. M., & Endsley, M. R. (1995). The role of crew resource management (CRM) in achieving situation awareness in aviation settings. In R. Fuller, N. Johnston, & N. McDonald (Eds.), *Human factors in aviation operations* (pp. 281–286). Aldershot, England: Avebury Aviation, Ashgate Publishing.

Sarter, N. B., & Woods, D. D. (1992). Pilot interaction with cockpit automation: Operational experiences with the flight management system. *International Journal of Aviation Psychology, 2*(4), 303–321.

Sharit, J., & Salvendy, G. (1982). Occupational stress: Review and reappraisal. *Human Factors, 24*(2), 129–162.

Wachtel, P. L. (1967). Conceptions of broad and narrow attention. *Psychological Bulletin, 68*, 417–429.

Weltman, G., Smith, J. E., & Egstrom, G. H. (1971). Perceptual narrowing during simulated pressure-chamber exposure. *Human Factors, 13*, 99–107.

Wickens, C. D. (1984). *Engineering psychology and human performance.* Columbus, OH: Charles E. Merrill.

Wickens, C. D. (1992). *Engineering psychology and human performance* (2nd ed.). New York: HarperCollins.

Wiener, E. L. (1989). *Human factors of advanced technology ("glass cockpit") transport aircraft* (NASA Contractor Report No. 177528). Moffett Field, CA: NASA-Ames Research Center.

Wiener, E. L. (1993). Life in the second decade of the glass cockpit. In R. S. Jensen & D. Neumeister (Eds.), *Proceedings of the Seventh International Symposium on Aviation Psychology* (pp. 1–11). Columbus, OH: The Ohio State University Press.

Wiener, E. L., & Curry, R. E. (1980). Flight deck automation: Promises and problems. *Ergonomics, 23*(10), 995–1011.

Wright, P. (1974). The harassed decision maker: Time pressures, distractions, and the use of evidence. *Journal of Applied Psychology, 59*(5), 555–561.

12

Aviation Personnel Selection and Training

David L. Pohlman
J. D. Fletcher
Institute for Defense Analyses

This chapter discusses the selection and training of people for work in aviation. This work encompasses a full spectrum of activity from operators of aircraft (i.e., pilots), to flight attendants, dispatchers, flight controllers, mechanics, engineers, baggage handlers, ticket agents, airport managers, and accountants. The topic covers a lot of territory. For manageability, we concentrate on three categories of aviation personnel: pilots and aircrew, maintenance support staff, and flight controllers.

Work load within most categories of aviation work has been increasing since aviation began. In the earliest days, aircraft were pushed to their limits and aviation operations were limited. Pilots flew the airplane from one place to another. Maintainers serviced the airframe and engine, both of which were adapted from relatively familiar, nonaviation technologies and materials. Flight controllers, if they were present at all, were found standing on the airfield waving red and green flags. Since those days, aviation operations have progressed. The aircraft is no longer a limiting factor. Pilots, maintainers, and controllers are all pushed to the edge of the human performance envelope by the aircraft they operate, maintain, and control. To give an idea of the work for which we are selecting and training people, it may help to discuss the work loads that different specialties impose on aviation personnel.

Pilots

Control of aircraft in flight has been viewed as a challenge from the beginning of aviation—if not before. McRuer and Graham (1981) reported that in 1901 Wilbur Wright addressed the Western Society of Engineers as follows:

> Men already know how to construct wings or aeroplanes, which when driven through the air at sufficient speed, will not only sustain the weight of the wings themselves, but also that of the engine, and of the engineer as well. Men also know how to build screws of sufficient lightness and power to drive these planes at sustaining speed. . . . Inability to balance and steer still confronts students of the flying problem. . . . When this one

feature has been worked out, the age of flying machines will have arrived, for all other difficulties are of minor importance (p. 353).

The "age of flying machines" is now approaching the century mark. Many problems of aircraft balance and steering—of operating aircraft—have been solved, but, as McRuer and Graham concluded, many remain. A pilot flying an approach in bad weather with most instruments nonfunctional or a combat pilot popping up from a high speed ingress to roll over and deliver ordnance on a target, is working at the limits of human ability. Control of aircraft in flight still "confronts students of the flying problem."

To examine the selection and training of pilots, it is best, as with all such issues, to begin with the requirements. What are pilots required to know and do? The U.S. Federal Aviation Administration (FAA) tests for commercial pilots reflect the growth and current maturity of our age of flying machines. They cover the following areas of knowledge (U.S. Department of Transportation, 1995b):

1. FAA regulations that apply to commercial pilot privileges, limitations, and flight operations.
2. Accident reporting requirements of the National Transportation Safety Board (NTSB).
3. Basic aerodynamics and the principles of flight.
4. Meteorology to include recognition of critical weather situations, wind sheer recognition and avoidance, and the use of aeronautical weather reports and forecasts.
5. Safe and efficient operation of aircraft.
6. Weight and balance computation.
7. Use of performance charts.
8. Significance and effects of exceeding aircraft performance limitations.
9. Use of aeronautical charts and magnetic compass for pilotage and dead reckoning.
10. Use of air navigation facilities.
11. Aeronautical decision making and judgment.
12. Principles and functions of aircraft systems.
13. Maneuvers, procedures, and emergency operations appropriate to the aircraft.
14. Night and high altitude operations.
15. Descriptions of and procedures for operating within the National Airspace System.

Amid the concern for rules and regulations reflected by these knowledge areas, it is still possible to discern a requirement to fly the airplane. All pilots must master basic airmanship, operation of aircraft systems, and navigation. Military pilots must add to these basic skills the operation of weapons systems while meeting the considerable work load requirements imposed by combat environments.

Basic Airmanship. There are four basic dimensions to flight: altitude (height above a point), attitude (position in the air), position (relative to a point in space), and time (normally a function of airspeed). A pilot must control these four dimensions simultaneously. Doing so allows the aircraft to take off, remain in flight, travel from Point A to Point B, approach, and land.

Basic aircraft control is a psychomotor task. Most student pilots need 10 to 30 hours of flying time to attain minimum standards in a slow-moving single engine aircraft. Experience in flying schools suggests that of the four basic dimensions, time is the most difficult to master. A good example might be the touchdown portion of a landing pattern. Assuming that the landing target is 500 feet beyond the runway threshold, and the aircraft is in the appropriate dimensional position as it crosses the threshold at about 55 miles per hour, a student in a single engine, propeller aircraft has about 6.2 seconds to make and implement the necessary decisions to touch the aircraft down. A student in a military trainer making a no flap, heavy weight landing at about 230 miles per hour has approximately 1.5 seconds to make and implement the same decisions. This time compression prevents many student pilots from graduating to more advanced aircraft. As flight instructors put it: "This student is never going to catch up with the airplane. If it crashes on the runway, he'll still be on final approach."

Aircraft Systems Operation. Pilots must also operate the aircraft systems. These systems include engine controls, navigation, fuel controls, communications, airframe controls, and environmental controls, among others. Some aircraft have on-board systems that can be run by other crew members, but the pilot remains responsible for them and must be aware of the status of each system at all times. For instance, the communications system can be operated by other crew members, but the pilot must quickly recognize from incessant radio chatter the unique call sign in use that day and respond appropriately. Increases in the number of aircraft systems, time compression, and aircraft system density have all combined to increase the complexity of operating aircraft.

Navigation. Once pilots master basic airmanship and the use of basic aircraft systems, they must learn to navigate. Navigating in four dimensions is markedly different from navigating in two dimensions. Pilots must know and remember all the types of airspace while maintaining the aircraft in an assigned position, on an assigned course, and on an assigned heading. Pilots must also maintain altitude, or modify it at an assigned rate and airspeed. They must accomplish all of the above while acknowledging and implementing new instructions over the radio. They may be required to do all these tasks in difficult weather (clouds, fog, rain, or snow) and turbulence.

Combat Weapons Systems. Combat aircraft confront pilots with all the usual problems of "balance and steering," along with many additional systems to contend with. Each weapon the aircraft carries affects flight parameters in different ways. Combat pilots must understand how each weapon affects the aircraft when it is aboard and when it is launched. They must understand the launch parameters of the weapons, their flight characteristics, and the additional system controls they require. These controls consist of buttons, switches, rockers, and sliders located on the throttles and stick grip. Some of these switch control between different weapons, others change the mode of the selected weapons, while others may manipulate systems such as radar and radios. The pilot must understand, monitor, and operate properly (while wearing flight gloves) all controls belonging to each weapon system. It is not surprising to find that the capabilities of fighter aircraft exceed pilots' capabilities to use them. But we have yet to get our overloaded pilot into combat.

Combat Work Load. The task of flying fighter aircraft in combat is one of the most complex cognitive and psychomotor tasks imaginable. "Fifty feet and the speed of heat" is an expression military fighter pilots use to describe the most effective way to ingress a hostile target area. A fighter pilot in combat must be so versed in the flying and operation of the aircraft that nearly all the tasks just described are assigned to background, or "automatic," psychomotor and cognitive processing. The ability to operate an aircraft in this manner is described as *strapping the aircraft on.* A combat pilot must:

- Plan the route through space in relation to the intended target, suspected threats, actual threats, other known aircraft, wingmen, and weapons.
- Monitor the aircraft displays for electronic notification of threats.
- Differentiate among threat displays (they can portray 15 or more different threats).
- Plan ingress to and egress from the target.
- Set switches for specific missions during specific periods of the flight.
- Monitor radio chatter on multiple frequencies for new orders and threat notification.
- Monitor progress along the planned route.
- Calculate course, altitude, and airspeed corrections.
- Plan evasive maneuvers for each type of threat and position during the mission.
- Plan weapons delivery.

This work load approaches the realm of the impossible. However, other aviation specialties also present impressive work loads. One of the most highly publicized of these is that of the flight controller.

Flight Controllers

In semiformal terms, flight controllers are responsible for the safe, orderly, and expeditious flow of air traffic on the ground at airports and in the air where service is provided using instrument flight rules (IFR) and visual flight rules (VFR) depending on the airspace classification. In less formal terms, they are responsible for dampening the potential for chaos around airports where as many as 2,000 flights a day may require their attention.

In good conditions, all airborne and ground-based equipment is operational and VFR rules prevail. However, as weather deteriorates and night approaches, pilots increasingly depend on radar flight controllers to guide them, keeping them a safe distance from obstacles and other aircraft. The radar images used by controllers are enhanced by computers that add to each aircraft's image such information as the call sign, aircraft type, airspeed, altitude, clearance limit, and course. If the ground radar becomes unreliable or otherwise fails, controllers must rely on pilot reports and "raw" displays, which consist of small dots, (*blips*) with none of the additional information provided by computer-enhanced displays. During a radar failure, controllers typically calculate time and distance mechanically, drawing pictures on the radar scope with a grease pencil. The most intense condition for flight controllers occurs when all ground equipment is lost save radio contact with the aircraft which may be declaring an emergency during IFR conditions. This condition is rare, but not unknown in modern aircraft control.

Using whatever information is available to them, flight controllers must attend to the patterns of all aircraft (often as many as 15) in their three-dimensional airspace. They must build a mental, moving image of the current situation and project it into the near future. Normally, controllers will sequence aircraft in first-in, first-out order so that the closest aircraft begins the approach first. The controller changes courses, altitudes, aircraft speeds, and routing to achieve "safe, orderly, and expeditious flow of aircraft." During all this, the controller must prevent the aircraft from flying closer to each other than 1 mile horizontally and 1,000 feet of altitude vertically.

The orderly flow of aircraft may be disrupted by emergencies. An emergency aircraft is given priority over all normally operating aircraft. The controller must place a bubble of safety around the emergency aircraft by directing other aircraft to clear the airspace around it and the path of its final approach. The controller must also determine the nature of the emergency so that appropriate information can be relayed to emergency agencies on the ground. If the ground equipment fails, the only separation available for control may be altitude with no feedback to verify that the reported altitude is correct. The controller must expedite the approach of the emergency aircraft while mentally reordering the arriving stack of aircraft.

Knowledge and skill requirements for aircraft controller certification include the following (U.S. Department of Transportation, 1995c):

1. Flight rules.
2. Airport traffic control procedures.
3. En-route traffic control procedures.
4. Communications procedures.
5. Flight assistance services.
6. Air navigation and aids to air navigation.
7. Aviation weather and weather reporting procedures.
8. Operation of control tower equipment.
9. Use of operational forms.
10. Knowledge of the specific airport, including rules, runways, taxiways, and obstructions.
11. Knowledge of control zones, including terrain features, visual checkpoints, and obstructions.
12. Traffic patterns, including use of preferential runways, alternate routes and airports, holding patterns, reporting points, and noise abatement procedures.
13. Search and rescue procedures.
14. Radar alignment and technical operation.

Aircraft Maintenance Technicians

A typical shift for an aircraft maintenance technician (AMT) may consist of several calls to troubleshoot and repair problems ranging from burnt out landing lights to finding a short in a cannon plug that provides sensor information to an inertial navigation system. To complicate matters, some problems may only be present when the aircraft is airborne—there may be no way to duplicate an airborne problem on the ground. The problem may be that one of many switches indicates that the aircraft is not airborne when it actually is or it may arise from changes in the aircraft frame and skin due to temperature variations and condensation, intermittent electrical shorts due to vibration, or interference from systems used only in flight. Despite these complications, the AMT is usually under pressure to solve problems quickly because many aircraft are scheduled to fly within hours and sometimes minutes after landing. Additionally, an AMT may have to contend with inadequate descriptions of the problem(s), unintelligible handwriting by the person reporting the problem, and weather conditions from 140° F in bright sun to –60° F with 30 knots of wind blended with snow. All these factors combine to increase the challenge of maintaining modern aircraft.

Although some research on maintenance issues had been performed for the U.S. military, until about 10 years ago most human factors research in aviation, including research on selection and training, concerned cockpit and air traffic control issues. Concern with maintenance as a human factors issue was almost nonexistent. This point of view has been changing rapidly in recent years (Jordan, 1996). Although the selection, training, and certification of maintenance technicians have lagged behind recent increases in the complexity and technological sophistication of modern aircraft, they also have been evolving. Appreciation of aviation maintenance as a highly skilled, often specialized profession requiring training in institutions of higher learning has been developing, albeit slowly (Goldsby, 1996).

Current FAA certification of aviation maintenance technicians still centers on mechanical procedures involving the airframes and power plants. Technicians are required to possess knowledge and skills concerning (U.S. Department of Transportation, 1995a):

1. Basic electricity.
2. Aircraft drawings.
3. Weight and balance in aircraft.
4. Aviation materials and processes.
5. Ground operations, servicing, cleaning, and corrosion control.
6. Maintenance publications, forms, and records.
7. Airframe wood structures, coverings, and finishes.
8. Sheet metal and nonmetallic structures.
9. Welding.
10. Assembly and rigging.
11. Airframe inspection.
12. Hydraulic and pneumatic power systems.
13. Cabin atmosphere control systems.
14. Aircraft instrument systems.
15. Communication and navigation systems.
16. Aircraft fuel systems.
17. Aircraft electrical systems.
18. Position and warning systems.
19. Ice and rain systems.
20. Fire protection systems.

21. Reciprocating engines.
22. Turbine engines.
23. Engine inspection.
24. Engine instrument systems.
25. Lubrication systems.
26. Ignition and starting systems.
27. Fuel and fuel metering systems.
28. Induction and engine airflow systems.
29. Engine cooling systems.
30. Engine exhaust and reverser systems.
31. Propellers.

This is a long list, but still more areas of knowledge need to be covered if maintenance training and certification are to keep pace with developments in the design and production of modern aircraft. The list needs to include specialization in such areas as (a) aircraft electronics to cover the extensive infusion of digital electronics, computers, and fly-by-wire technology in modern aircraft; (b) composite structures, which require special equipment, special working environments, and special precautions to protect the structures themselves and the technicians' own health and safety; and (c) nondestructive inspection technology, which involves sophisticated techniques using technologies such as magnetic particle and dye penetrants, x-rays, ultrasound, and eddy currents. Even within more traditional areas of airframe and powerplant maintenance, current business practices and trends are creating pressures for more extensive and specialized training and certification. Goldsby (1996) suggested that these pressures arise from increasing use of (a) third-party providers to provide increasing amounts of modification and repair work, (b) aging aircraft, (c) leased aircraft requiring greater maintenance standardization and inspection techniques, (d) noncertified airframe specialists, and (e) second- and third-party providers of noncertified technicians.

The most important problem-solving skills for AMTs may be those of logical interpretation and diagnostic proficiency. These higher order cognitive skills can only be developed by solving many problems provided either by extensive and broad experience working on actual aircraft or by long hours spent with appropriately designed and employed maintenance simulators. Talent and logical thinking help, which is to say that personnel selection and classification remain relevant, but there appears to be no real substitute for experience in developing troubleshooting proficiency. Selection, training, and certification increasingly need to emphasize problem-solving skills in addition to the usual capacities for learning and systematically employing complex procedural skills.

PERSONNEL SELECTION AND CLASSIFICATION FOR AVIATION

How people are selected for employment and classified for occupational specialties affects the performance and capabilities of every organization. Effective selection and classification procedures save time, material, and funding in training, and they improve the quality and productivity of job performance. They help ensure worker satisfaction, organizational competence, productivity, and, in military circles, operational readiness.

Of personnel selection and classification, selection is viewed as the earlier and perhaps more general process—people are first selected for employment and then classified into specific jobs or career paths. In much civilian practice, personnel selection and classification are indistinguishable, individuals with relevant pretraining are recruited to perform specific jobs, and selection is tantamount to classification. In large organizations such as the military services, which provide appreciable amounts of training to their employees, the processes of selection and classification are more separate. For instance, people are selected for military service based on general, but well-observed standards. Those selected are then classified and assigned for training to one of many career fields with which they may have had little or no experience. These personnel management efforts pay off. Zeidner and Johnson (1989) determined that the U.S. Army's selection and classification procedures save the Army about $263 million per year.

It should be noted that classification may matter as much as selection. Researchers have found that how well people are classified for specific jobs or career paths has a major impact on job performance, job satisfaction, and attrition, regardless of how carefully they are initially selected for employment. One study found that personnel retention rates over a 5-year period differed by 50% for well-classified versus poorly classified individuals (Stamp, 1988).

Because of the expense, complexity, and limited tolerance for error in aviation work, more precise selection and classification have been sought almost from the beginning of the age of flying machines (at the very beginning the Wright bothers just flipped a coin). Hunter (1989) wrote that "almost every test in the psychological arsenal has been evaluated at one time or another to determine its applicability for aircrew selection" (p. 129). Hilton and Dolgin (1991) wrote that there may be no other "occupation in the world that benefits more from personnel selection technology than that of military pilot" (p. 81).[1]

A Brief Historical Perspective

Aviation and many personnel management procedures began their systematic development at about the same time. This fact is not entirely coincidental. The development of each increased the requirement for the other.

In the 1990s, selection and classification procedures are applied across the full range of aviation personnel, but the development of systematic personnel management procedures in aviation initially focused on selection of pilots, rather than on aviation support personnel. These procedures grew to include physical, psychomotor, mental ability, and psychological (personality) requirements, but they began with self-selection.

Self-Selection. Perhaps from the time of Daedelus and certainly from the time of the Wright brothers, people have been drawn to aviation. In the early days of World War I, many pilots were volunteers who came from countries other than the one providing the training (Biddle, 1968). Some of these early pilots could not even speak the language of the country for which they flew, but they wanted to fly, and the

[1]The history of selection in aviation has not always been honorable. The term *fly-by-night* comes from early aviators who would descend on a town, "select" (through not entirely scientific means) individuals who were proclaimed to have a talent for flying, collect a fee for training these individuals, and fly out at night before the lessons were to begin (Roscoe, Jensen, & Gawron, 1980).

Americans among them established the base of America's early capabilities in aviation during and after that war.

Self-selection continues to be a prominent factor in pilot and aircrew selection in both military and civilian aviation. Only people with a strong desire to fly civil aircraft are likely to try and obtain a license to fly. Advancement past the private pilot stage and acquiring the additional ratings required of commercial pilots is demanding, time-consuming, and expensive. The persistence of a prospective pilot in finishing training and pursuing an aviation career beyond a private pilot license constitutes a form of natural selection. That aviation continues to attract and hold so many able people who select themselves for careers in aviation attests to its strong and continuing appeal.

Early on it was observed that training pilots was an expensive undertaking, and selection for aircrew personnel soon evolved from self-selection alone to more systematic and formal procedures. The arguments for this evolution frequently cite the costs of attrition from flight training. These costs have always been high, and they have risen steadily with the cost and complexity of aircraft. Today, it costs more than $1 million to train a jet pilot, and the current cost to the Air Force for each failed aviation student is estimated to be $59,000 (Duke & Ree, 1996). This latter expense excludes the very high cost of aircraft whose loss might be prevented by improved selection and classification procedures.

In the case of private, civilian instruction, attrition costs are usually borne by individuals, but in the military Services the costs are borne by organizations with both the motivation and resources to seek means for their reduction. Research, development, implementation, and evaluation of procedures to select and classify individuals for aviation training have been a major investment and contribution of the world's military services. These procedures began with those used in general for selecting and classifying military personnel—physical qualifications.

Physical Qualification Selection. With World War I, the demand for flyers grew, and the number of applicants for flying training increased. Military organizations reasonably assumed that physical attributes play a significant role in a person's ability to successfully undertake flight training and to later assume the role of pilot. Flight physicals became a primary selection tool.

At first, these physicals differed little from the standard examinations of physical well-being used to select all individuals for military service (Brown, 1989; Hilton & Dolgin, 1991).[2] Soon, however, research aimed specifically to improve selection of good pilot candidates began in Italy and France (Dockeray & Isaacs, 1942). Need for balance in air, psychomotor reaction, appropriate concentration and distribution of attention, emotional stability, and rapid decision making were assumed to be greater than those for nonaviation personnel, and more stringent procedures were established for selecting aviation personnel.

Italian researchers, who may have initiated this line of research, developed measures of reaction time, emotional reaction, equilibrium, attention, and perception of muscular effort and added them to the standard military physical examinations specifically used

[2]Vestiges of early physical standards for military service held on long after the need for them was gone. As late as the Korean War, fighter pilots were required to have opposing molars. This requirement was eventually traced to the Civil War era need to bite cartridges before they could be fired. Only when fighter pilots became scarce in the early 1950s did anyone question its enforcement.

to select pilots. Similar research and development efforts were soon undertaken by other countries, including the United States.

Rigorous flight physicals continue to be used today to qualify and retain individuals in flight status, for both military and civilian pilots. The FAA defines standards for first-, second-, and third-class medical certificates covering eyesight, hearing, mental health, neurologic conditions (epilepsy and diabetes are cause for disqualification), cardiovascular history (annual electrocardiographic examinations are required of people over age 40 with first-class certificates), and general health as judged by a certified federal air surgeon (U.S. Department of Transportation, 1996).

Mental Ability Selection. During World War I, the military Services also determined that rigorous flight physicals for selecting pilots were not sufficient. Other methods were needed to reduce the costs and time expended on candidates who were washing out of training despite being physically qualified. A consensus developed that pilots need to make quick mental adjustments using good judgment in response to intense, rapidly changing situations. It was then assumed that pilot selection would benefit from methods that would measure mental ability. These methods centered on use of newly developed paper-and-pencil tests of mental ability. What was new about these tests was that they could be inexpensively administered to many applicants all at the same time.

Assessment procedures administered singly to individuals by especially trained examiners had been used in the United States at least as early as 1814 when both the Army and the Navy used examinations to select individuals for special appointments (Zeidner & Drucker, 1988). In 1883, the Civil Service Commission initiated the wide use of open, competitive examinations for appointment to government positions. Corporations such as General Electric and Westinghouse developed and implemented employment testing programs in the early 1900s. However, it took the efforts of the Vineland Committee working under the supervision of Robert Yerkes in 1917 to develop reliable, parallel paper-and-pencil tests that could be administered by a few individuals to large groups of people using simple, standardized procedures (Yerkes, 1921).

The Vineland Committee developed a plan for the psychological examination of the entire U.S. Army. It produced the Group Examination Alpha (the Army Alpha) which was "an intelligence scale for group examining . . . [making] possible the examination of hundreds of men in a single day by a single psychologist" (Yerkes, 1921, p. 310). The Army Alpha provided the basis for many paper-and-pencil psychological assessments that were developed for group administration in the succeeding years. It was used by the U.S. Committee on Psychological Problems of Aviation to devise a standard set of tests and procedures that were adopted in 1918 and used to select World War I pilots (Hilton & Dolgin, 1991).

The Army Alpha test, then, laid the foundation for psychological assessment of pilots performed by the U.S. Army in World War I, by the Civil Aeronautics Authority in 1939, and after that by the U.S. Army and Navy for the selection of aircrew personnel in World War II. Reducing the number of aircrew student washouts throughout this period saved millions of dollars that were thereby freed to support other areas of the war effort (U.S. Dept. of the Air Force, 1996). It is also likely that the aircrews selected and produced by these procedures were of higher quality than they might have been without them, thereby significantly enhancing military effectiveness. However, the impact of personnel selection and classification on the ultimate goal of military effectiveness—or on productivity in nonmilitary organizations—then and now has received infrequent and limited attention from researchers (Kirkpatrick, 1976; Zeidner & Johnson, 1989).

After World War I, there was a flurry of activity concerning psychological testing and pilot selection. It differed from country to country (Dockeray & Isaacs, 1942; Hilton & Dolgin, 1991). Italy emphasized psychomotor coordination, quick reaction time, and constant attention. France used vasomotor reactions during apparatus testing as indicators of emotional stability. Germany concentrated on apparatus tests to measure resistance to disorientation. Great Britain emphasized physiological signs as indicators of resistance to altitude effects. Germany led in development of personality measures for pilot selection. The United States, Japan, and Germany all used general intelligence as an indicator of aptitude for aviation.

In the United States, this activity was short lived. The civil aircraft industry was embryonic, and there was a surplus of aviators available to fly the few existing civil aircraft. Only in the mid-1920s, when monoplanes started to replace postwar military aircraft, did civil air development gain momentum and establish a growing need for aviation personnel. Hilton and Dolgin reported that this pattern was found in many countries, consisting of a rigorous physical examination, a brief background questionnaire, perhaps a written essay, and an interview.

In the 1920s and 1930s, as aircraft became more sophisticated and expensive, the selection of civilian pilots became more critical. The development of a United States civilian aviation infrastructure was based on the U.S. Post Office airmail service and first codified through the Contract Mail Act (the Kelly Act) of 1925 (Hansen & Oster, 1997). This infrastructure brought with it requirements for certification and standardized management of aviation and aviation personnel. It culminated in the Civil Aeronautics Act in 1938, which established the Civil Aeronautics Authority, later reorganized as the Civil Aeronautics Board in 1940.

Another world war and an increased demand for aviation personnel both appeared likely in 1939. For these reasons, the Civil Aeronautics Authority created a Committee on Selection and Training of Aircraft Pilots, which immediately began to develop qualification tests for screening civilian aircrew personnel for combat duty (Hilton & Dolgin, 1991). This work formed the basis for selection and classification procedures developed by the Army Air Force Aviation Psychology Program Authority. A comprehensive summary description of this program and its accomplishments was published at the end of World War II by Viteles (1945).

The procedures initially developed by the Aviation Psychology Program were a composite of paper-and-pencil intelligence and flight aptitude tests. They were implemented in 1942 as the Aviation Cadet Qualifying Examination (ACQE) and used thereafter by the U.S. Army Air Force to select aircrew personnel for service in World War II (Flanagan, 1942; Hilton & Dolgin, 1991; Hunter, 1989; Vitales, 1945). These procedures used paper-and-pencil tests, motion picture tests, and apparatus tests. The Army's procedures were designed to assess five factors that had been found to account for washouts in training: intelligence and judgment, alertness and observation including speed of decision and reaction, psychomotor coordination and technique, emotional control and motivation, and ability to divide attention. The motion picture and apparatus tests were used to assess hand and foot coordination, judgment of target speed and direction, pattern memory, spatial transposition, and skills requiring timed exposures to visual stimuli.

Flanagan (1942) discussed issues in classifying personnel after they had been selected for military aviation service. Basically, he noted that pilots need to exhibit superior reaction speed and the ability to make decisions quickly and accurately, that bombardiers need superior fine motor steadiness under stress (for manipulating bomb sights), concentration and ability to make mental calculations rapidly under distracting conditions, and that navigators need superior ability to grasp abstractions such as those associated with celestial geometry and required to maintain spatial orientation, but not the high level of psychomotor coordination needed by pilots and bombardiers.

In contrast, the U.S. Navy relied primarily on physical screening, paper-and-pencil tests of intelligence and aptitude (primarily mechanical comprehension), the Purdue Biographical Inventory, and line officer interviews to select pilots throughout World War II (Fiske, 1947; Jenkins, 1946). The big differences between the two Services were that the Army used apparatus (we might even call them simulators today), whereas the Navy did not and that the Navy used formal biographical interviews, whereas the Army did not. The Army studied the use of interviews and concluded that even those that were reliable contributed little in terms of time, effort, and costs (Viteles, 1945).

Today, the military Services depend on a progressive series of selection instruments. These include academic performance records, medical fitness, a variety of paper-and-pencil tests of general intelligence and aptitude, possibly a psychomotor test such as the Air Force's Basic Abilities Test (BAT), and flight screening (flying lessons) programs. Commercial airlines rarely hire a pilot who has no experience. They use flight hours to determine if candidates will be able to acclimate to the life of airline pilots. They capitalize on aviation personnel procedures developed by the military by hiring large numbers of pilots, maintainers, controllers, and others who have been selected, classified, and trained by the military Services.

Three conclusions may be drawn from the history of selection for aircrew personnel. First, most research in this area has focused on the selection of individuals for success in training, not performance in the field, in operational units, or on the job. Nearly all validation studies of aircrew selection measurements concern their ability to predict performance in training.[3] This practice makes good monetary sense—the attrition of physically capable flight candidates is very costly. We certainly want to maximize the probability that individuals selected for aircrew training will successfully complete it. Also, it is not unreasonable to expect some correlation between success of individuals in training and their later performance as aircrew members. However, almost 100 years into the age of flying machines, information relating selection measures to performance on the job remains scarce.[4] We would still like to identify those individuals who, despite their successes in training, are unlikely to become good aviators on the job. And we would like to identify, earlier than we can now, those exceptional individuals who are likely to become highly competent performers, if not aces, in our military forces and master pilots in our civilian aircraft industry.

The second and third conclusions were both suggested by Hunter (1989). His review of aviator selection concludes that there seems to be little relationship between general intelligence and pilot performance. It is certainly true that our tests of intelligence do not predict very well either performance in aircrew training or on the job. These tests are largely measures of verbal intelligence intended to predict success in academic institutions—as these institutions are currently organized and operated. Newer multi-

[3]Notably, they concern the prediction of success in training, given our current training procedures. Different training procedures could yield different "validities."

[4]There are exceptions. See for example the efforts discussed by Carretta and Ree (1996) to include supervisory performance ratings in assessment of selection and classification validities.

faceted measures of mental ability (e.g., Gardner, Kornhaber, & Wake, 1996), may more successfully identify aspects of general intelligence that predict aviator ability and performance. Also, by limiting variability in the population of pilots, our selection and classification procedures may have made associations between measures of intelligence and the performance of pilots difficult to detect. In any case, our current measures of intelligence find limited success in accounting for pilot performance.

A third conclusion is also suggested by Hunter. After a review of 36 studies performed between 1947 and 1978 to assess various measures used to select candidates for pilot training, Hunter found that only those concerned with instrument comprehension and mechanical comprehension were consistent predictors of success—validity coefficients for these measures ranged from .20 to .40. Other selectors, concerned with such matters as physical fitness, stress reactivity, evoked cortical potentials, age, and education were less successful. A follow-up study by Hunter and Burke (1995) found similar results. The best correlates of success in pilot training were job samples, gross dexterity, mechanical understanding, and reaction time. General ability, quantitative ability, and education were again found to be poor correlates of success.

In brief, selection for aircrew members in the 1990s centers on predicting success in training, and includes measures of physical well-being, general mental ability, instrument and mechanical comprehension, and psychomotor coordination, followed by a brief exposure to flying an inexpensive, light airplane or a simulator. Attrition rates for training by the military Services now range around 22% (Duke & Ree, 1996).

The best hope for reducing these rates further and for increasing the precision of our selection and classification procedures in general may be the use of computer-based testing. These tests go well beyond the computerized, adaptive testing approaches explored earlier using items adopted from paper-and-pencil formats. Adaptive techniques use the correct and incorrect responses made by individuals to branch them rapidly through pools of items with known psychometric characteristics until the computer program settles on an appropriate level of ability within an appropriate band of confidence. Newer tests may still use branching techniques, but they go beyond the use of items originally developed for paper-and-pencil testing (Kyllonen, 1995). These tests capitalize on multimedia, timing, and response-capturing capabilities that are only available through the use of computers. These computerized tests and test items have required and engendered new theoretical bases for ability assessment. Most of the theoretical bases that are emerging are founded on information-processing models of human cognition. These are discussed, briefly and generically, in the next section.

A Brief Theoretical Perspective

Over the years, work in aviation has changed. The leather-helmeted, white-scarfed daredevil fighting a lone battle against the demons of the sky, overcoming the limited mechanical capabilities of his aircraft, and evading the hostile intent of an enemy at war is gone. The problems remain: The sky must, as always, be treated with respect, maintenance will never reach perfection, and war is still with us, but the nature of aviation work and the requisite qualities of people who perform it have evolved with the evolution of aviation technology.

Today, in place of mechanical devices yoked together for the purposes of flight and requiring mostly psychomotor reflexes and responses, we have computer-controlled, highly specialized, integrated aviation systems requiring judgment, abstract thinking, abstract problem solving, teamwork, and a comprehensive grasp of crowded and complex airspaces along with the rules and regulations that govern them (Driskell & Olmstead, 1989). Aviation work has evolved from realms of the psychomotor to those of information processing and from individual dash and élan to leadership, teamwork, and managerial judgment. With this evolution of demands on both the qualitative and quantitative aspects of human performance in aviation, it is not surprising to find information-processing models increasingly sought and applied in the selection, classification, assignment, training, and assessment of aviation personnel.

The complexity of human performance in aviation has always inspired similarly complex models of human cognition. Primary among the models to grow out of aviation psychology in World War II was Guilford's (1967) well-known and wonderfully heuristic Structure of the Intellect which posited 120 different ability factors based on all combinations of five mental operations (memory, cognition, convergent thinking, divergent thinking, and evaluation), six types of products (information, classes of units, relations between units, systems of information, transformations, and implications), and four classes of content (figural, symbolic, semantic, and behavioral). An appropriate combination and weighting using "factor-pure" measures of these abilities would significantly improve the selection and classification of individuals for work in aviation. Despite the significant research and substantial progress they engendered in understanding human abilities, Guilford's ability factors—or perhaps our ability to assess them—failed to prove as independent and factor pure as hoped, and the psychological research community moved on to other, more dynamic models. These models center on notions of human information processing and cognition and are characterized by Kyllonen's (1995) cognitive abilities measurement (CAM) approach.

Information processing encompasses a set of notions, or a method, intended to describe how people think, learn, and respond. Most human information-processing models use stimulus–thought–response as a theoretical basis (Bailey, 1989; Wickens & Flach, 1988). The information-processing model depicted in Fig. 12.1 differs from that originally developed by Wickens and Flach (1988), but it is derived from and based on their model. Figure 12.1 covers four major activities in information processing: short-term sensory store, pattern recognition, decision making, and response execution.

Short-Term Sensory Store

The model presented here is an extension of the Wickens and Flack model. It assigns stimuli input received by the short-term sensory store into separate buffers, or registers, for the five senses. Input from internal sensors for such factors as body temperature, heart and respiration rates, blood chemistry, limb position and rates of movement, and

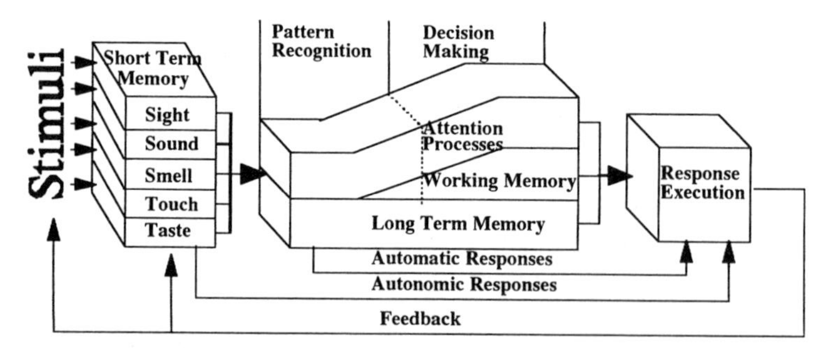

FIG. 12.1. Generic information-processing model (adapted from Wickens & Flach, 1988).

other internal functions could be added (Bailey, 1989), but are not needed in this summary discussion.

Visual and auditory sensory registers have been fairly well supported as helpful constructs that account for research findings. Evidence to support the other sensory short-term memory registers is more limited, but it is not unreasonable to posit these as constructs in a human information processing model. They have been added and included here.

Pattern Recognition

Over the past 30 years general theories of perception and learning have changed. They have evolved from the fairly strict logical positivism of behavioral psychology, which emphasizes the study of directly observable and directly measurable actions, to what may be called *cognitive psychology*. Cognitive psychology gives more consideration to internal, less observable processes that are assumed to mediate and enable human learning—and to produce the directly observable behavior that is the subject of behaviorist approaches.

The keynote of these notions, which currently underlies our understanding of human perception, memory, and learning, may have been struck by Neisser (1967), who stated, "The central assertion is that seeing, hearing, and remembering are all acts of *construction*, which may make more or less use of stimulus information depending on circumstances" (p. 10).[5] Neisser was led to this point of view by a large body of empirical evidence showing that many aspects of human behavior, such as seeing and hearing, simply could not be accounted for by external physical cues reaching human perceptors, such as eyes and ears. Additional processes, including an internally, one might say cognitively, generated analysis by synthesis process, had to be posited to account for well-established and observable human abilities to detect, identify , and process physical stimuli.

Human cognition, then, came to be viewed as an overwhelmingly constructive process. Perceivers and learners are not viewed as blank slates, passively recording bits of information transmitted to them over sensory channels, but as active participants who use the fragmentary cues permitted them by their sensory receptors to construct, verify, and modify their own cognitive simulations of the outside world. Human perception (and learning) are understood to be enabled through the use of simulations of the world that the perceiver constructs and modifies based on sensory cues received from the outside world. Even memory has come to be viewed as constructive with recollections assumed to be reconstructed in response to stimuli rather than retrieved whole cloth from long-term storage.

Attention Processes. For a stimulus to be processed, it must be detected by the information-processing system. Stimulus detection and processing distribute human ability to attend to stimuli. When there is little or no work load, attention resources are distributed in an unfocused random pattern (Huey & Wickens, 1993). As more sensory input becomes available, the individual must begin to prioritize what stimuli are going

[5]These ideas were, of course, around long before Neisser published his book. But after many years of wrestling with strictly beharviorist models which only reluctantly considered internal processes such as cognition, this book seems to have freed the psychological research community to pursue new, more "constructivist" approaches to perception, memory, learning, and cognition.

to be selected for interpretation. The attention process decides, based on pattern recognition from both long-term and working memory resources, which to process further. Which signals receive attention may be guided by the following (Wickens & Flach, 1988):

- Knowledge: Knowing how often a stimulus is likely to be presented, and if that stimulus is likely to change enough to affect a desired outcome, will influence the attention it receives.
- Forgetting: Human memory will focus attention on stimuli that have already been adequately sampled, but lost to memory.
- Planning: A plan of action that is reviewed before an activity is to take place will focus attention on some stimuli at the expense of others.
- Stress: Stress reduces the number of stimuli that can receive attention. Stress can also focus attention on stimuli that are of little consequence. For instance, a pilot may focus on a burned out gear down indicator light and fly into the ground.

Stimuli attended to may not be the brightest, loudest, or most painful, but they will be those deemed most relevant to the situation (Gopher, Weil, & Siegel, 1989). As suggested by the rationale just cited, the likelihood that a stimulus will be detected depends on the perceived penalty for missing it.

Working Memory. In an unpublished study, Pohlman and Tafoya (1979) investigated the fix-to-fix navigation problem in a T-38 instrument simulator. They found two primary differences between student pilots and instructor pilots. First, the accuracy of students in solving a fix-to-fix problem was inconsistent, whereas the instructor pilots were consistently accurate. Second, student pilots used a classic geometric approach to solve the problem in contrast to the instructors who used a rate-of-change comparison approach. Notably, almost every instructor denied using rate-of-change comparison until it was demonstrated they were in fact doing that, showing once again that experts may be unaware of the techniques they use (Gilbert, 1992).

Although students were working geometry problems in the cockpit, instructors were merely comparing the rates that the distance and bearing were changing, and flew the aircraft so that the desired range and desired bearing were arrived at simultaneously. A real bonus was that the rate-of-change comparison method automatically accounted for wind. Because current rate-of-change information is kept in working memory rather than in long-term memory (Wickens & Flach, 1988), the use of current rate-of-change information by these experts indicates that working memory is an integral portion of the process of attention distribution. Observations such as this support the inclusion of a working memory interface between the attention process and the long-term memory used primarily for pattern matching.

Long-Term Memory. Long-term memory becomes relevant in pattern matching and perception when the signal attended to requires interpretation. Long-term memory is the primary repository of patterns and episodic information. Patterns of dark and light can be converted into words on a page or pictures remembered and linked to names, addresses, and events. Memory that is linked to the meaning of the patterns is usually called *semantic memory*. Memory relating to events and the people, places, things, and emotions involved in them is usually called *episodic memory*.

Automaticity. Humans are capable of different types of learning. One of these involves choosing responses at successively higher levels of abstraction. For instance, in learning to read one may first attend to individual letters, then, with increased practice and proficiency, one may attend to individual words, then to phases, and finally perhaps to whole ideas. There are different levels of automaticity imposed by individual talents and abilities. As a boy, Oscar Wilde often demonstrated (for wagers) his ability to read both facing pages of a book at the same time and to complete entire, three-volume novels in 30 minutes or less (Ellmann, 1988). Clearly, there are levels of automaticity to which most of us can only aspire. In general, automaticity is more likely to be attained in situations where there are strict rules governing the relationship between stimuli and responses (Huey & Wickens, 1993). The key for aviation tasks, with all their time pressures and demands for attention, is that automatic processing frees up attention resources for allocation to other matters (Bailey, 1989; Shiffrin & Schneider, 1977).

Decision Making

Once stimuli have been detected, selected, and pattern matched, a decision must be made. As the process proceeds, cues are sought to assist the decision maker in gathering information that will help with the decision. These cues are used to construct and verify the simulation, or runnable model, of the world that an individual constructs, verifies, and modifies in order to perceive and learn. As each situation is assessed, the individual chooses among possible responses by first "running" them in the simulation.

One difficulty in aviation is the lack of time to make decisions. Unlike other vehicles, an aircraft cannot stop in mid-air and shut down its systems to diagnose a problem. Decision making is often stressed by this lack of time combined with the inevitable uncertainly and incompleteness of relevant sensory input. On the other hand, stress may be increased when sensory input is increased because of the greater work load placed on pattern recognition to filter out what is relevant and what is not. A pilot, controller, or maintenance technician may have to ignore so much of the sensory input as to miss cues needed for problem solution or to adapt to new situations.

An individual may also miss relevant cues because they do not support the individual's analysis by synthesis, or simulation, of the situation. If the cues do not fit, an individual can either modify the underlying model or ignore the cue, and the decision made may not always be correct. All these factors influence what cues are available to long-term and working memory for situation assessment. Tversky and Kahneman (e.g., 1974) discussed these and other factors as biases and heuristics in the decision-making processes.

Determining the ways that prospective aviation personnel process information and their capacities for doing so should considerably strengthen our procedures for selecting, classifying, and training them. For instance, the ability to filter sensory cues quickly and accurately may be critical for aircrew personnel, especially combat pilots, and flight controllers who must frequently perform under conditions of sensory overload. Creative, accurate, and comprehensive decision making that takes account of all salient cues and filters out the irrelevant ones may be critical for AMTs. Rapid decision making that quickly adjusts situation assessment used to select among different decision choices may be at a premium for pilots and controllers. A large working memory capacity may be especially important for combat pilots whose lives often depend on the number of cues they process rapidly and accurately.

Emerging models of human information processing are, in any case, likely to find increasing application in the selection, classification, and training of aviation personnel. The dynamic nature of these models requires similarly dynamic measurement capabilities. These capabilities are now coming inexpensively and readily available through the use of computer-based assessment which can measure aspects of human cognitive processes that heretofore were inaccessible given the military's need for inexpensive, standard, procedures to assess hundreds of people in a single day by a single examiner. Development of these capabilities may represent as important a milestone in selection and classification as did the work of the Vineland Committee to produce the Army Alpha Test. These are currently being pursued by U.S. Air Force laboratory personnel who are performing leading research in this area (Carretta, 1996; Kyllonen, 1995; Ree & Carretta, 1998).

Finally, it should be noted that improvements in selection and classification procedures are needed for many aviation personnel functions, not just for potential aircrew members. Among U.S. scheduled airlines, domestic passenger traffic (revenue passenger enplanements) increased by 83% over the years from 1980 to 1995, and international passenger traffic doubled in the same period (*Aviation and Aerospace Almanac*, 1997). Combined domestic and international commercial passenger traffic for U.S. scheduled airlines is projected to increase another 42% from 1996 to 2005. Thousands of new aviation mechanics and flight controllers are needed to meet this demand, to operate and maintain the new digital equipment and technologies being introduced into modern aircraft and aviation work, and to satisfy the expansion of safety inspection requirements brought about by recent policies of deregulation.

The FAA has stated that there is an unacceptably high attrition rate in air traffic controller training, costing the FAA about $9,000 per washout. It therefore called for both modernized training and more precise selection and classification (U.S. Department of Transportation, 1989). The plan is to introduce more simulation into the processes of selection and classification. It raises significant questions about the psychometric properties—the reliability, validity, and precision—of simulation used to measure human capabilities and performance. These questions are by no means new, but they remain inadequately addressed by the psychometric research community.

Today's personnel selection and classification procedures contribute much to the efficiency with which we prepare people for work in aviation. Although these procedures fall short of perfection, they provide significant savings in funding, resources, and personnel safety over less systematic approaches. Still, our current selection and classification procedures rarely account for more than 25% of the variance in human performance observed in training and on the job (e.g., U.S. Dept. of the Air Force, 1996). There remains plenty of leverage to be gained by improving the effectiveness and efficiency of other means for securing the human competencies we need for aviation. Prominent among these means is training. As the age of flying machines has developed and grown, so too has our reliance on training.

TRAINING FOR AVIATION

A Little Background

Training and education may be viewed as opposite ends of a common dimension we might call *instruction*. Training may be viewed as a means to an end—as preparation to perform a specific job. Education, on the other hand, may be viewed as an end in its own right and as preparation for all life experiences—including training. The contrast matters because it affects the way we develop, implement, and assess instruction—especially with regard to trade-offs between costs and effectiveness. In education, the emphasis is on maximizing the achievement—the improvements in human knowledge, skills, and performance—returned from whatever resources can be brought to bear on it. In training, the emphasis is on the other side of the cost-effectiveness coin—on preparing people to perform specific, identifiable jobs. Rather than maximize learning of a general sort, in training we seek to minimize the resources that must be allocated to produce a specified level of learning—a specifiable set of knowledge, skills, and attitudes determined by the job to be done.

These distinctions between education and training are (of course) not hard and fast. In military training, as we pass from combat systems support (e.g., depot maintenance, hospital care, finance and accounting), to combat support (e.g., field maintenance, field logistics, medical evacuation), to combat (i.e., warfighting) the emphasis in training shifts from a concern with minimizing costs toward one of maximizing capability and effectiveness. In education, as we pass from general cultural transmission to programs of professional preparation and certification, the emphasis shifts from maximizing achievement within given cost constraints toward minimizing the costs to produce specifiable thresholds of instructional accomplishment.

These considerations suggest that no assessment of an instructional technique for application in either education or training is complete without some consideration of both effectiveness and costs. During early stages of research, studies may honestly be performed to assess separately the cost or effectiveness of an instructional technique. However, once the underlying research is sufficiently complete to allow implementation, evaluations to effect change and inform decision makers will be incomplete unless both costs and effectiveness considerations are included in the data collection and analysis.

Those familiar with assessments of training programs will note that the inclusion of cost and effectiveness considerations together in the same evaluation study occurs less frequently than desirable. Assessments of instruction for aviation are not innocent of this neglect. However, perhaps because of the pragmatic culture of aviation and the high stakes involved, aviation training assessments have been more likely than others to consider both cost and effectiveness. Even though more could and should be done, aviation assessments have helped devise techniques and set standards for cost-effectiveness analyses in many forms of training.

It may also be worth noting that selection, classification, assignment, training, human factoring, and job and career design are all components of systems designed to produce needed levels of human performance. As in any system, all these components interact. More precise selection and classification reduce requirements for training. Better designed equipment will reduce the need for training and either ease or change standards for selection and classification. Addition of job performance aids will do the same, and so on. Any change in the amount and quality of resources invested in any single component of the system is likely to affect the need for resources invested in other components—as well as the return to be expected from these investments.

Comprehensive, cost-effectiveness consideration of this complex decision space, in which all components interact, poses a sizable problem in optimal control. It has yet to be successfully articulated, let alone solved. What is the return to training from investments in recruiting or selection? What is the return to training or selection from investment in ergonomic design? What is the impact on training and selection from investment in electronic performance support systems? What, even, is the impact on training, selection, and job design from investments in spare parts? More questions could be added to this list. These comments are just to note the context within which training in general and aviation training in particular operate to produce human competence. Properly considered, training in aviation and elsewhere does not occur in a vacuum separate from other means used to produce requisite levels of human competence.

Learning and Training

At the most general level, training is intended to bring about human learning. Learning is said to take place when an individual alters his or her knowledge and skills through interaction with the environment. Instruction is characterized by the purposeful design and construction of that environment to produce learning. Theories of learning, which are mostly descriptive, and theories of instruction, which are mostly prescriptive, help inform the many decisions that must be made to design, develop, and implement training environments and the training programs that use them.

Every instructional program represents a view of how people perceive, think, and learn. As discussed earlier, these views have evolved over the past 30 years to include more consideration of the internal processes that are assumed to mediate and enable human learning. These "cognitive," "constructive" notions of human learning are reflected in our current systems of instruction. They call into question the view of instruction as straightforward information transmission.

They suggest instead that the role of instruction is to supply appropriate cues for learners to use in constructing, verifying, and modifying their cognitive simulations—or runnable models—of the subject matter being presented. The task of instruction design is not so much to transmit information from teacher to student as to create environments in which students are enabled and encouraged to construct, verify, and correct these simulations. A learning environment will be successful to the extent that it too is individualized, constructive, and active. Systems intended to bring about learning, systems of instruction, differ in the extent to which they assist learning by assuming some of the burdens of this individualized, constructive, and active process for the student.

Training Program Design and Development

These considerations do not, however, lead to the conclusion that all instruction, especially training, is hopelessly idiosyncratic and thereby beyond all structure and control. There is still much that can and should be done to design, develop, and implement instructional programs beyond simply providing opportunities for trial and error with feedback. Systematic development of instruction is especially important for programs intended to produce a steady stream of competent individuals, an intention that is most characteristic of training programs. All aspects of the systematic development of training are concerns of what is often called Instructional System Design (ISD; Logan, 1979) or the Systems Approach to Training (SAT; Guptill, Ross, & Sorensen, 1995). ISD/SAT approaches apply standard systems engineering to the development of instructional programs. They begin with the basic elements of systems engineering, which are shown in Fig. 12.2. These are the generic steps of analysis, design, production, implementation, and evaluation. ISD/SAT combines these steps with theories of learning and instruction to produce systematically designed and effective training programs.

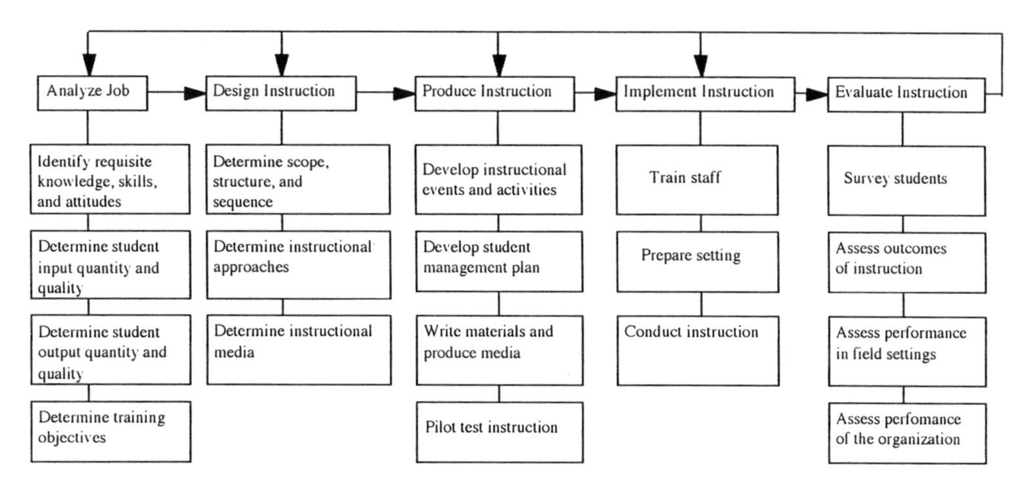

FIG. 12.2. Example procedures for instructional system development.

Training *analysis* is based on systematic study of the job and the task(s) to be performed. It identifies training inputs and establishes training objectives to be accomplished in the form of student flow and the knowledge, skill, and attitude outcomes to be produced by the training. Training *design* devises the instructional interactions needed to accomplish the training objectives identified by training analysis. It is also used to select the instructional approaches and media used to present these interactions. Training *production* involves the development and preparation of instructional materials, which may include hardware such as simulators, software such as computer programs and audiovisual productions, and databases for holding information such as subject matter content and the performance capabilities of weapon systems. Training *implementation* concerns the appropriate installation of training systems and materials in their settings and attempts to ensure that they will perform as designed. Training *evaluation*, determines if the training does things right (verification), and if it does the right things (validation). As discussed by Kirkpatrick (1976), it provides verification that the training system meets its objectives (Kirkpatrick's Level II) and the validation that meeting these objectives prepares individuals to better perform the targeted tasks or jobs (Kirkpatrick's Level III) and improves the operation of the organization overall (Kirkpatrick's Level IV). Notably, evaluation provides formative feedback to the training system for improving and developing it further.

Many ISD/SAT systems for instructional design have been devised—Montemerlo and Tennyson (1976) found manuals for more than 100 such systems had been written as of 1976, more doubtless exist now—but all these systems have some version of the basic steps for systems engineering in common. An ISD/SAT approach seeks to spend enough time on the front end of the system life cycle to reduce its costs later on. It is a basic principle of systems development that down line modifications are several magnitudes more expensive than designing and building something properly the first time. The same is true for training systems. It is more efficient to develop and field a properly designed training system than simply to build the system and spend the rest of its life fixing it. But the latter approach is pursued far more frequently than the former. For that matter, many training systems are in use that have never been evaluated, let alone subjected to Kirkpatrick's four levels of assessment. To some extent, training for aviation is an exception to these very common and haphazard approaches.

Training in Aviation

An aircraft pilot performs a continuous process of what A. Williams (1980) described as discrimination and manipulation. A pilot must process a flood of stimuli arriving from separate sources, identify which among them to attend to, generate from a repertoire of discrete procedures an integrated plan for responding to the relevant stimuli, and perform a series of discrete acts, such as positioning levers, switches, and controls, and continuous manual control movements requiring small forces and adjustments based on counter pressures exerted in response to the control movements. Williams suggested that the heart of these actions is decision making and that it concerns (a) when to move the controls, (b) which controls to move, (c) which direction to move the controls, (d) how much to move the controls, and (e) how long to continue the movement. It is both straightforward and complicated.

The task of flight controllers might be described the same way. Both pilots and controllers must contend with significant time pressures and with the possibilities of severe consequences for error. Both require psychomotor responses and both properly involve some degree of artistry and personal expression. No two people will perform these activities in precisely the same way, and they may be most effectively accomplished in ways that are consonant with other aspects of personal style (A. Williams, 1980).

The responses involve performance of pretrained procedures, but the procedures must be assembled into an integrated, often unique, response. As described by Roscoe et al. (1980), the performance of aviation personnel concern procedural, decisional, and perceptual–motor responses. Responses chosen are generative and created to meet the demands of the moment. They involve sensing, transforming, recollecting, recognizing, and manipulating of concepts, procedures, and devices. These responses are controlled by decision making that is basically cognitive, but with emotional overtones. Responses made by pilots and controllers key on this decision making, but the decision making is more tactical than strategic. The decisions may be guided by general principles, but they are made under significant time pressures and resemble those of a job shop or a military command post more than those of an executive suite.

How do we prepare people for jobs of this sort? What training works in these situations? Aviation has developed so rapidly that little, probably insufficient time has been devoted to systematic review of the requirements it makes of individuals and the training needed to satisfy these requirements.

Aviation training is just now beginning to evolve from the World War I days of the Lafayette Escadrille as described by Charles Biddle, an American who enlisted in the French Foreign Legion Aviation Section in 1917. Biddle was later commissioned in the U.S. Army Air Force where he performed with distinction as a fighter pilot[6] and a squadron commander. He was also a prolific letter writer. His letters, which were collected and published, provide a grass-roots description of training for pilots in World War I (Biddle, 1968).

This early training consisted mostly of an accomplished (hence, instructor) pilot teaching each student one-on-one in the aircraft. Ground training consisted of academic classes and some small group sessions with an instructor pilot. Each individual was briefed on what to do and then allowed to practice the action under the guidance of a monitor. Flying began, as it does today, with students taxiing the aircraft around on

[6]He attributed much of his success in air combat to his earlier experience with duck hunting—learning how to track and lead moving targets in three-dimensional space.

the ground, learning to balance and steer.[7] As subsequent steps were mastered, and certified by the instructor, the student proceeded to actual flight, and new, more difficult, and often more specialized stages of learning with more capable aircraft to fly and more complex maneuvers to complete.[8] Today's flight instruction follows the same basic pattern—probably because it works. It leads trainees reliably to progressively higher levels of learning and performance.

This approach has led to a robust set of assumptions concerning how aircrew training must be done. It emphasizes one-on-one student instruction for both teaching and certification, a focus on the individual, the use of actual aircraft to provide the training, and hours of experience to certify proficiency. Each of these assumptions deserves some discussion.

One-on-One Instruction. One-on-one instruction receives somewhat more emphasis in aviation training than elsewhere. For an activity as complex and varied as piloting an airplane, it is difficult to imagine an alternative to this approach. One-on-one instructor to student ratios have long been recognized as effective—perhaps the most effective format for instruction. Bloom (1984) found that the difference between students taught in classroom groups of 30 and those taught one-on-one by an individual instructor providing individualized instruction was as large as two standard deviations in achievement. Unfortunately, one-on-one instruction is also very expensive. We cannot afford an Aristotle for every Alexander, or even a Mark Hopkins for the rest of us.

This form of teaching has been described as an instructional imperative and an economic impossibility (Scriven, 1975). Databased arguments have been made (e.g., Fletcher, 1992) that technology, such as computer-based instruction that tailors the pace, content, sequence, difficulty, and style of presentations to the needs of individual students, can help fill this gap between what is needed and what is affordable. Technology can be used more in aviation training,[9] and FAA efforts have been made to encourage and increase not just the use of technology but the use of relatively inexpensive personal computers in aviation training.

For instance, a successful line of research was undertaken at Embry Riddle University to develop PC-based training that emphasizes number of flight hours in aircraft less, knowledge and competencies of the trainees more, and improved validity for FAA certification (e.g., K. Williams, 1994). Hampton, Moroney, Kirton, and Biers (1993) found that students trained using PC-based training devices needed fewer trials and less time to reach pilot test standards for eight maneuvers performed in an aircraft. They also found that the per-hour operating costs of the PC-based devices were about 35% less than those of an FAA-approved generic training device costing about $60,000 to buy.

The Air Force Human Resources Laboratory early pursued some of this work and found that PC-based approaches produced superior achievement compared to paper-based approaches (programmed instruction) used in F-16 weapons control training (Pohlman & Edwards, 1983). The same laboratory, now the Air Force Armstrong Laboratory, has developed a Basic Flight Instruction Tutoring System (BFITS) using a

---

[7]This is the so-called "penguin system" in which a landborne airplane, in Biddle's case a Bleriot monoplane with reduced wingspan, is used to give students a feel for its controls.

[8]As early as 1915 in World War I, these maneuvers included aerobatics, which Biddle credited with saving the lives of many French-trained aviators—some of whom were, of course, American.

[9]One of the first applications of speech recognition technology in technology-based instruction was for training naval flight controllers (Breaux, 1980).

PC-equipped with a joystick and rudder petals and intended for *ab initio* (from the beginning) flight training (Benton, Corriveau, Koonce, & Tirre, 1992). Koonce, Moore, and Benton (1995) reported positive transfer of BFITS training to subsequent flight instruction.

It may be worth noting that many benefits of one-on-one instruction can be lost through improper implementation—with no reductions in their relatively high cost. Instructors who have not themselves received instruction in how to teach and assess student progress may do both poorly despite their own high levels of proficiency and best intentions (Semb, Ellis, Fitch, & Matheson, 1995).

Roscoe et al. (1980) stated that "there is probably more literal truth than hyperbole in the frequent assertion that the flight instructor is the greatest single source of variability in the pilot training equation" (p. 173). Instructors must both create an environment in which students learn and be able to assess and certify students' learning progress. Much can be done to simplify and standardize the subjective assessment of student achievement accomplished during flight checks. Early on, Koonce (1974) found that it is possible to achieve interrater reliabilities exceeding .80 in flight checks, but these are not typical. In practice, instructors still, as reported by Roscoe and Childs (1980), vary widely in their own performance of flight maneuvers and the indicators of competence they consider in assessing the performance of their students.

Despite the expense and difficulty of one-on-one instruction and despite technology-based opportunities for providing means that are both more effective and less costly for achieving many aviation training objectives, the use of individual instructors is likely to remain a key component of aviation training for some time to come.

Focus on the Individual. The barnstorming, ruggedly individualistic pilots of the past had a exhilarating run in the sky. Their day, however, is mostly gone. Even combat pilots fly under the tightening control of attack coordinators and radar operators, and they must coordinate their actions with wingmen. Commercial airline pilots must deal with an entire crew of people who are specialists in their fields and whose knowledge of specifics aspects of aviation may well exceed that of the aircraft captain. The culture of the individual master of the craft, however, remains. This cultural bias may be less than ideal in an age of aircrews and teams. It represents a challenge for training.

Foushee and Helmreich (1988), among others, pointed out that group performance has received little attention from the aviation training community and the attention it has received has been stimulated by unnecessary and tragic accidents. Generally, these accidents seem to occur because of failures to delegate tasks (attention being focused on a relatively minor problem leaving no one to mind the store) or an unwillingness to override the authority and perceived majesty of the aircraft captain. Still, it is interesting to note that the 1995 areas of knowledge listed earlier and required by the FAA for pilot certification are silent with regard to crew, team, and group abilities.

Communication skills are particularly important in this regard. Roby and Lanzetta (1958) and Olmstead (1992) reported empirical studies in which about 50% of team performance was accounted for by the presence and timing of particular kinds of communications. These were problem-solving teams placed under the sort of time pressures that are likely to occur in aviation. An interesting study reported by Foushee and Helmreich compared the performance of preduty (rested) with postduty (fatigued) crews. The study is notable because the postduty crews performed better than the preduty crews on operationally significant measures—and others—despite their fatigue.

This relative superiority may be attributed to learning by the postduty crews to perform as a team, something the preduty crews had yet to accomplish. Communication patterns were key to these differences.

In brief, communications and other crew skills can and probably should be both taught and certified in aviation training programs. These issues are currently addressed under the heading of cockpit resource management (Wiener, Kanki, & Helmreich, 1993). They deserve the attention of the military and civilian aviation communities. This is not to suggest that a focus on individuals is undesirable in aviation training. Rather, it suggests that crew and team communication, management, and behavior should be added to current aviation training and certification requirements.

Aircraft Versus Simulators. To a significant extent, the study of aviation training is the study of training simulators. This is true of training for aircrew members, flight controllers, and aviation maintenance technicians. Simulation is a sufficiently important topic on its own to deserve a separate chapter in this book. Comments here are of a general nature and focused on the use of simulation in training.

Rolfe and Staples (1986), Caro (1988), and others have provided useful and brief histories of flight simulators. The first flight simulators were developed early in the age of flying machines and were often aircraft tethered to the ground, but capable of responding to aerodynamic forces. The Sanders Teacher, one of the first of these, was introduced in 1910. Some of these devices depended on natural forces to provide the wind needed to give the student experience in learning to balance and steer, and some, like the Walters trainer, also introduced in 1910, used wires and pulleys manipulated by flight instructors to give students this experience. Motion was made possible through the use of compressed air actuators developed for aviation simulators by Lender and Heidelberg in 1917 and 1918—although the use and value (cost-effectiveness) of motion in flight simulation was as much a matter of discussion then as it is now (e.g., Hays, Jacobs, Prince, & Salas, 1992; Koonce, 1979; Pfeiffer & Horey, 1987; Waag, 1981).

As instrumentaton for aircraft improved, the need to include instruments coupled to the simulation of flight increased. The Link Trainers succeeded in doing this. By the late 1930s, they were able to present both the instrument layout and performance of specific aircraft to students. Simulators using electrical components to model characteristics of flight were increasingly used as World War II progressed. In 1943, Bell Telephone Laboratories produced an operational flight trainer/simulator for the U.S. Navy's PBM-3 aircraft using electrical circuitry to solve flight equations in real time and display their results, realistically, using the complete system of controls and instruments available in the aircraft. Modern simulators evolved further with the incorporation of computers that could not only respond to controls in simulators and display the results of flight equations on aircraft instruments, but could provide motion simulation and generate out the window visual displays as well.

Rolfe and Staples (1986) pointed out that a faithful simulation requires: (a) a complete model of the response of the aircraft to all inputs, (b) a means of animating the model (rendering it runnable in real time), and (c) a means of presenting this animation to the student using mechanical, visual, and aural responses. They noted that the degree to which all this is necessary is another question. The realism, or "fidelity," needed by simulation to perform successful training of all sorts is a perennial topic of discussion. Much of this discussion is based either in actuality or in effect on the intuitive appeal of E. L. Thorndike's (1903) early argument for the presence and necessity of identical elements in training to ensure successful transfer of what is learned in training to what is needed on the job. Thorndike suggested that such transfer is always specific, never general and keyed to either substance or procedure.

In dynamic pursuits such as aviation where unique situations are frequent and expected, this point of view often leads to an insistence in training on maximizing fidelity. We do not know precisely what will happen on the job and therefore assume we must provide as many identical elements in training as we can. Unfortunately, fidelity does not come free. As fidelity rises so do costs, reducing the number, availability, and/or accessibility of training environments that can be provided to students. If the issue ended here, we might solve the problem by throwing more money at it—or not as policy dictated.

However, there is another issue involving fidelity, simulation, and training. Simulated environments permit the attainment of training objectives that cannot or should not be attempted without simulation. As discussed by Orlansky et al. (1994) among many others, aircraft can be crashed, expensive equipment ruined, and lives hazarded in simulated environments in ways that range from impractical to unthinkable without them. Simulated environments provide other benefits for training. They can make the invisible visible, compress or expand time, and repeatedly reproduce events, situations, and decision points. They can compress years of some experience into hours. Training using simulation is not just a degraded, less expensive reflection of the realism we would like to provide. It enables the attainment of training objectives that are otherwise inaccessible.

Training using simulation adds value and reduces cost. Evidence of this utility comes from many sources. In aircrew training the issue keys on transfer—are the skills and knowledge acquired in simulation of value in flying actual aircraft, do they transfer from one situation to the other? Many attempts to answer this question rely on transfer effectiveness ratios (TER; e.g., Roscoe & Williges, 1980). These ratios may be defined for pilot training in the following way:

$$\text{TER} = \frac{A_c - A_s}{S}$$

Where:

TER = Transfer Effectiveness Ratio

A_c = Aircraft time required to reach criterion performance, without access to simulation

A_s = Aircraft time required to reach criterion performance, with access to simulation

S = Simulator time.

Roughly, this TER is the ratio of aircraft time savings to the expenditure of simulator time—it reveals how much aircraft time is saved for every unit of simulator time invested. If the TER is small, a cost-effectiveness argument may still be made for simulation because simulator time is likely to cost much less than aircraft time. Orlansky and String (1977) investigated precisely this issue in a now classic and often cited study. They found (or calculated, as needed) 34 TERs from assessments of transfer performed from 1967 to 1977 by military, commercial, and academic organizations. The

TERs[10] ranged from –0.4 to 1.9, with a median value of .45. Orlansky, Knapp, and String (1984) also compared the cost to fly actual aircraft with the cost to "fly" simulators. Very generally they found that (a) the cost to operate a flight simulator is about one tenth the cost to operate a military aircraft, (b) 1 hour in a simulator saves about 30 minutes in an aircraft, so that (c) use of flight simulators is cost-effective if the TER is .20 or greater.

At a high level of abstraction, this finding is extremely useful and significant. Because nothing is simple, however, a few caveats may be in order. First, as Provenmire and Roscoe (1973) pointed out, not all simulator hours are equal—early hours in the simulator appear to save more aircraft time than later ones. This consideration leads to learning curve differences between cumulative TERs and incremental TERs with diminishing returns best captured by the latter. Second, transfer is not a characteristic of the simulator alone. Estimates of transfer from a simulator or simulated environment must also consider what the training is trying to accomplish—the training objectives. This issue is well illustrated in a study by Holman (1979), who found 24 TERs for a CH-47 helicopter simulator ranging from 2.8 to .0, depending on which training objective was under consideration. Third, there is an interaction between knowledge of the subject matter and the value of simulation alone. Gay (1986) and Fletcher (1990) found that the less the student knows about the subject matter, the greater the need for tutorial guidance in simulation. The strategy of throwing a naive student into a simulator with the expectation that learning will occur does not appear to be viable. Fourth, the operating costs of aircraft differ markedly and will create quite different trade-offs between the cost-effectiveness of training with simulators and without them. In contrast to the military aircraft considered by Orlansky, Knapp, and String where the cost ratio was about .10, Provenmire and Roscoe were concerned with flight simulation for the Piper Cherokee, where the cost ratio was .73. Other caveats may well occur to the reader. Specific applications deserve specific attention.

Nonetheless, many data-based studies have demonstrated the ability of simulation to both increase effectiveness and lower costs for many aspects of flight training. Hays et al. (1992) reviewed 26 studies of transfer from training with flight simulators to operational equipment. They found that there was significant positive transfer from the simulators to the aircraft, that training using a simulator and an aircraft was almost always superior to training with a simulator alone, and that self-paced simulator training was more effective than lock-step training. The usual ambiguities about the value of including motion systems in flight simulators emerged. Beyond this, the findings of Orlansky, Knapp, and String (1979) remain the best evidence of lowered costs in flight training obtained through the use of simulators.

The value of simulation is not limited to flight. From a broad review of interactive multimedia capabilities used for simulation, Fletcher (1997) extracted 11 studies in which simulated equipment was used to train maintenance technicians. These studies compared instruction with simulators to instruction with actual equipment, held overall training time roughly equal, and assessed final performance using actual (not simulated) equipment. Over the 11 studies, the use of simulation yielded an effect size (which is the measure of merit in such meta-analyses) of .40 standard deviations suggesting an improvement from 50th percentile to about 66th percentile achievement among students using simulation. Operating costs using simulation were about .40 of those without it because the equipment being simulated did not break and could be presented and

[10]They found only one negative TER.

manipulated on devices costing one to two orders of magnitude less than the actual equipment that was the target of the training.

Process Versus Performance Measurement. Experience is a thorough teacher and especially valuable when the nuances and enablers of human proficiency in the targeted area of performance are ill defined and poorly captured by instructional objectives. However, 1 hour of experience will produce different results in different people. An instructional strategy based solely on the assumption that time in the aircraft, or working with actual equipment in the case of flight controllers and maintenance technicians, equates to learning is sorely limited. Training and the certification that it bestows may be better served by increased emphasis on performance assessment in place of such process measurement as hours of experience.

These comments are not to suggest that one-on-one instruction, use of actual aircraft, and hours of experience should be eliminated from training programs. They do suggest that simply doing things the way they have always been done sooner or later leads to inefficiency, ineffectiveness, and stagnation. These assumptions, and others that are emphasized in aviation training, should be routinely subjected to analytical review and the possibility of change.

Pathways to Aviation Training

As of 1993 there were about 737,000 people working in the (civilian) air transport industry (Hansen & Oster, 1997). There are five types of pilot certificates: student, private, commercial, airline transport, and instructor. Except for student pilot, ratings for aircraft category (airplane, rotorcraft, glider, and lighter-than-air), aircraft class (single-engine land, multiengine land, single-engine sea, and multiengine sea), aircraft type (large aircraft, small turbojet, small helicopters, and other aircraft), and aircraft instruments (airplanes and helicopters) are placed on each certificate to indicate the qualification and limitations of the holder. AMTs are certified for two possible ratings (airframe and power plant) and for repairman. As discussed earlier, the number of maintenance certifications may be increased to meet the requirements posed by modern aircraft design. Separate certification requirements also exist for air traffic controllers, aircraft dispatchers, and parachute riggers. The aviation workforce is large and both technically and administratively complex.

In response to Congressional concerns, the National Research Council (NRC) undertook a study (Hansen & Oster, 1997) to assess our ability to train the quantity and quality of people needed to sustain this workforce. The NRC identified the following five "pathways" to aviation careers:

1. Military training has been a major source of aviation personnel in the past and its diminution provided a major impetus for the NRC study. The military is likely to become much less prominent and civilian sources are likely to become substantially more important as the military Services continue to downsize and the air transport industry continues to expand and replace its aging workers.

2. Foreign hiring has been used little by U.S. airlines and is not expected to increase in the future. In fact, many U.S.-trained pilots are expected to seek employment in other countries when U.S. openings are scarce.

3. On-the-job training allows individuals to earn FAA licenses and certificates by passing specific tests and without attending formal training programs. U.S. airlines prefer to hire people who have completed FAA certificated programs, and on-the-job training is not likely to grow as a source of training in the future.

4. Collegiate training is offered by about 280 postsecondary institutions tracked by the University Aviation Association currently located at Auburn University. Collegiate training is already the major source for AMTs, and the NRC report suggested that it will become significantly more important as a source of aircrew personnel. The report also points out, however, that pilots, even after they complete an undergraduate degree in aviation, must still work their way up through nonairline flying jobs before accumulating the hours and ratings certifications currently expected and required by the airlines for placement.

5. Ab initio ("from the beginning") training is offered by some foreign airlines to selected individuals with no prior flying experience. As yet, U.S. airlines have not seen it necessary to provide this form of training.

The NRC study concluded that civilian sources will be able to meet market demand, despite the downsizing of the military. However, they stressed the need to sustain and develop the professionalization and standardization of collegiate aviation programs—most probably by establishing an accreditation system similar to that in engineering and business and supported by the commercial aviation industry and the FAA. As described earlier in this chapter, the U.S. aviation industry continues to grow, as it does worldwide. The next 5 to 10 years will be both interesting and challenging to those concerned with support and growth of the aviation workforce. The NRC study suggests means for accomplishing these ends successfully. The community concerned with human competence in aviation has been given a significant opportunity to rise to the challenge.

REFERENCES

Aviation and Aerospace Almanac 1997. (1997). New York: Aviation Week Group, McGraw-Hill.

Bailey, R. W. (1989). *Human performance engineering.* Englewood Cliffs, NJ: Prentice-Hall.

Benton, C., Corriveau, P., Koonce, J. M., & Tirre, W. C. (1992). *Development of the basic flight instruction tutoring system* (BFITS) (AL-TP-1991-0060). Brooks Air Force Base, TX: Armstrong Laboratory Human Resources Directorate (ADA 246 458).

Biddle, C. J. (1968). *Fighting airman: The way of the eagle.* Garden City, NY: Doubleday & Company.

Bloom, B. S. (1984). The 2 sigma problem: The search for methods of group instruction as effective as one-to-one tutoring. *Educational Researcher, 13,* 4–16.

Breaux, R. (1980). Voice technology in military training. *Defense Management Journal, 16,* 44–47.

Brown, D. C. (1989). Officer aptitude selection measures. In M. F. Wiskoff & G. M. Rampton (Eds.), *Military personnel measurement: Testing, assignment, evaluation* (pp. 97–127). New York: Praeger.

Caro, P. W. (1988). Flight training and simulation. In E. L. Wiener & D. C. Nagel (Eds.), *Human factors in aviation* (pp. 229–261). New York: Academic Press.

Carretta, T. R. (1996). *Preliminary validation of several US Air Force computer-based cognitive pilot selection tests* (AL/HR-TP-1996-0008). Brooks Air Force Base, TX: Armstrong Laboratory Human Resources Directorate.

Carretta, T. R., & Ree, M. J. (1996). Factor structure of the Air Force Officer Qualifying Test: Analysis and comparison. *Military Psychology, 8,* 29–43.

Dockeray, F. C., & Isaacs, S. (1921). Psychological research in aviation in Italy, France, England, and the American Expeditionary Forces. *Comparative Psychology, 1,* 115–148.

Driskell, J. E., & Olmstead, B. (1989) Psychology and the military: Research applications and trends. *American Psychologist, 44,* 43–54.

Duke, A. P., & Ree, M. J. (1996). Better candidates fly fewer training hours: Another time testing pays off. *International Journal of Selection and Assessment, 4,* 115–121.

Ellmann, R. (1988). *Oscar Wilde.* New York: Vintage Books.

Fiske, D. W. (1947). Validation of naval aviation cadet selection tests against training criteria. *Journal of Applied Psychology, 5,* 601–614.

Flanagan, J. C. (1942). The selection and classification program for aviation cadets (aircrew—bombardiers, pilots, and navigators). *Journal of Consulting Psychology, 6,* 229–239.

Fletcher, J. D. (1990). *The effectiveness and cost of interactive videodisc instruction in Defense training and education* (IDA Paper P-2372). Alexandria, VA: Institute for Defense Analyses (ADA 228 387).

Fletcher, J. D. (1992). Individualized systems of instruction. In M. C. Alkin (Ed.) *Encyclopedia of educational research* (6th ed., pp. 613–620). New York: Macmillan.

Fletcher, J. D. (1997). What have we learned about computer based instruction in military training? In R. J. Seidel & P. R. Chatelier (Eds.), *Virtual reality, training's future?* (pp. 169–177). New York: Plenum.

Foushee, H. C., & Helmreich, R. L. (1988). Group interaction and flight crew performance. In E. L. Wiener & D. C. Nagel (Eds.), *Human factors in aviation* (pp. 189–227). New York: Academic Press.

Gardner, H., Kornhaber, M., & Wake, W. (1996) *Intelligence: Multiple perspectives.* Fort Worth, TX: Harcourt Brace.

Gay, G. (1986). Interaction of learner control and prior understanding in computer-assisted video instruction. *Journal of Educational Psychology, 78,* 225–227.

Gilbert, T. F. (1992). Foreword. In H. D. Stolovitch & E. J. Keeps (Eds.), *Handbook of human performance technology* (pp. xiii–xviii). San Francisco, CA: Jossey-Bass.

Goldsby, R. (1996). Training and certification in the aircraft maintenance industry: Technician resources for the twenty-first century. In W. T. Shepherd (Ed.), *Human factors in aviation maintenance—Phase Five progress report* (DOT/FAA/AM-96/2) (pp. 229–244). Washington, DC: Department of Transportation, Federal Aviation Administration (ADA 304 262).

Gopher, D., Weil, M., & Siegel, D. (1989). Practice under changing priorities: An approach to the training of complex skills. *Acta Psychologica, 71,* 147–177.

Guilford, J. P. (1967). *The nature of human intelligence.* New York: McGraw-Hill.

Guptill, R. V., Ross, J. M., & Sorenson, H. B. (1995). A comparative analysis of ISD/SAT process models. In *Proceedings of the 17th Interservice/Industry Training System and Education Conference (I/ITSEC). November 1995, Albuquerque, NM* (pp. 20–30). Arlington, VA: National Security Industrial Association.

Hampton, S., Moroney, W., Kirton, T., & Biers, W. (1993). *An experiment to determine the transfer effectiveness of PC-based training devices for teaching instrument flying* (CAAR-15471-93-1). Daytona Beach, FL: Center for Aviation/Aerospace Research, Embry-Riddle Aeronautical University.

Hansen, J. S., & Oster, C. V. (Eds.). (1997). *Taking flight: Education and training for aviation careers.* Washington, DC: National Research Council, National Academy Press.

Hays, R. T., Jacobs, J. W., Prince, C., & Salas, E. (1992). Flight simulator training effectiveness: A meta-analysis. *Military Psychology, 4,* 63–74.

Hilton, T. F., & Dolgin, D. L. (1991) Pilot selection in the military of the free world. In R. Gal & A. D. Mangelsdorff (Eds.), *Handbook of military psychology* (pp. 81–101). New York: Wiley.

Holman, G. J. (1979). *Training effectiveness of the CH-47 flight simulator* (ARI-RR-1209). Alexandria, VA: Army Research Institute for the Behavioral and Social Sciences (ADA 072 317).

Hunter, D. R. (1989). Aviator selection. In M. F. Wiskoff & G. M. Rampton (Eds.), *Military personnel measurement: Testing, assignment, evaluation* (pp. 129–167). New York: Praeger.

Hunter, D. R., & Burke, E. F. (1995). Predicting aircraft pilot training success: A meta-analysis of published research. *International Journal of Aviation Psychology, 4,* 297–313.

Huey, B. M., & Wickens, C. D. (1993). *Workload transition.* Washington, DC: National Academy Press.

Jenkins, J. G. (1946). Naval aviation psychology (II): The procurement and selection organization. *American Psychologist, 1,* 45–49.

Jordan, J. L. (1996). Human factors in aviation maintenance. In W. T. Shepherd (Ed.), *Human factors in aviation maintenance—Phase Five progress report* (DOT/FAA/AM-96/2) (pp. 251–253). Washington, DC: U.S. Department of Transportation, Federal Aviation Administration (ADA 304 262).

Kirkpatrick, D. L. (1976). Evaluation of training. In R. L. Craig (Ed.), *Training and development handbook* (2nd ed., pp. 18.1–18.27). New York: McGraw-Hill.

Koonce, J. M. (1974). *Effects of ground-based aircraft simulator motion conditions upon prediction of pilot proficiency* (AFOSR-74-1292). Savoy, IL: Aviation Research Laboratory, University of Illinois (AD A783 256/257).

Koonce, J. M. (1979). Predictive validity of flight simulators as a function of simulation motion. *Human Factors, 21,* 215–223.

Koonce, J. M., Moore, S. L., & Benton, C. J. (1995). *Initial validation of a Basic Flight Instruction tutoring system (BFITS).* Columbus, OH: Eighth International Symposium on Aviation Psychology.

Kyllonen, P. C. (1995). CAM: A theoretical framework for cognitive abilities measurement. In D. Detterman (Ed.), *Current topics in human intelligence, Vol. IV: Theories of intelligence.* Norwood, NJ: Ablex.

Logan, R. S. (1979). A state-of-the-art assessment of instructional systems development. In H. F. O'Neil, Jr. (Ed.), *Issues in instructional systems development* (pp. 1–20). New York: Academic Press.

McRuer, D., & Graham, D. (1981). Eighty years of flight control: Triumphs and pitfalls of the systems approach. *Journal of Guidance and Control, 4*(4), 353–362.

Montemerlo, M. D., & Tennyson, M. E. (1976). *Instructional systems development: Conceptual analysis and comprehensive bibliography* (NAVTRAEQUIPCENIH 257). Orlando, FL: Naval Training Equipment Center.

Neisser, U. (1967). *Cognitive psychology.* New York: Appleton-Century-Crofts.

Olmstead, J. A. (1992). *Battle staff integration* (IDA Paper P-2560). Alexandria, VA: Institute for Defense Analyses (ADA 248 941).

Orlansky, J., Dahlman, C. J., Hammon, C. P., Metzko, J., Taylor, H. L., & Youngblut, C. (1994). *The value of simulation for training* (IDA Paper P-2982). Alexandria, VA: Institute for Defense Analyses (ADA 289 174).

Orlanksy, J., Knapp, M. I., & String, J. (1977). *Operating costs of military aircraft and flight simulators* (IDA Paper P-1733). Alexandria, VA: Institute for Defense Analyses (ADA 144 241).

Orlansky, J. & String, J. (1977). *Cost-effectiveness of flight simulators for military training* (IDA Paper P-1275). Alexandria, VA: Institute for Defense Analyses (ADA 052 801).

Pfeiffer, M. G., & Horey, J. D. (1987). *Training effectiveness of aviation motion simulation: A review and analyses of the literature* (Special Report No. 87-007). Orlando, FL: Naval Training Systems Center (ADB 120 134).

Pohlman, D. L., & Edwards, D. J. (1983). Desk top trainer: Transfer of training of an aircrew procedural task. *Journal of Computer-Based Instruction, 10,* 62–65.

Pohlman, D. L., & Tafoya, A. F. (1979). *Percieved rates of motion in cockpit instruments as a method for solving the fix to fix navigation problem.* Unpublished Technical Paper, Williams Air Force Base, AZ: Air Force Human Resources Laboratory.

Provenmire, H. K., & Roscoe, S. N. (1973). Incremental transfer effectiveness of a ground-based aviation trainer. *Human Factors, 15,* 534–542.

Ree, M. J., & Carretta, T. R. (1998). Computerized testing in the U.S. Air Force. *International Journal of Selection and Assessment, 6,* 82–89.

Roby, T. L., & Lanzetta, J. T. (1958). Considerations in the analysis of group tasks. *Psychological Bulletin, 55,* 88–101.

Rolfe, J. M., & Staples, K. J. (1986). *Flight simulation.* Cambridge, England: Cambridge University Press.

Roscoe, S. N., & Childs, J. M. (1980). Reliable, objective flight checks. In S. N. Roscoe (Ed.), *Aviation psychology* (pp. 145–158). Ames, IA: Iowa State University Press.

Roscoe, S. N., Jensen, R. S., & Gawron, V. J. (1980). Introduction to training systems. In S. N. Roscoe (Ed.), *Aviation psychology* (pp. 173–181). Ames, IA: Iowa State University Press.

Roscoe, S. N., & Williges, B. H. (1980). Measurement of transfer of training. In S. N. Roscoe (Ed.), *Aviation psychology* (pp. 182–193). Ames: Iowa State University Press.

Scriven, M. (1975). Problems and prospects for individualization. In H. Talmage (Ed.), *Systems of individualized education* (pp. 199–210). Berkeley, CA: McCutchan.

Semb, G. B., Ellis, J. A., Fitch, M. A., & Matheson, C. (1995). On-the-job training: Prescriptions and practice. *Performance Improvement Quarterly, 8,* 19.

Shiffrin, R. M., & Schneider, W. (1977). Controlled and automatic human information processing II: Perceptual learning, automatic attending, and a general theory. *Psychological Review, 84,* 127–190.

Stamp, G. P. (1988). *Longitudinal research into methods of assessing managerial potential* (Tech. Rep. No. 819). Alexandria, VA: U.S. Army Research Institute for the Behavioral and Social Sciences (ADA 204 878).

Thorndike, E. L. (1903). *Educational Psychology.* New York: Lemcke & Buechner.

Tversky, A., & Kahneman, D. (1974). Judgment under uncertainity: Heuristics and biases. *Science, 185,* 1124–1131.

U.S. Department of Transportation. (1989). *Flight plan for training: FAA training initiatives management plan.* Washington, DC: U.S. Department of Transportation, Federal Aviation Administration, Office of Training and Higher Education.

U.S. Department of Transportation. (1995a). *Aviation mechanic general, airframe, and powerplant knowledge and test guide* (AC 61-28). Washington, DC: U.S. Department of Transportation, Federal Aviation Administration.

U.S. Department of Transportation. (1995b). *Commercial pilot knowledge and test guide* (AC 61-114). Washington, DC: U.S. Department of Transportation, Federal Aviation Administration.

U.S. Department of Transportation. (1995c). *Control tower operator (CTO) study guide* (TS-14-1). Washington, DC: U.S. Department of Transportation, Federal Aviation Administration.

U.S. Department of Transportation. (1996). *Commercial flight regulations chapter 1, Part 67* (14 CFR). Washington, DC: U.S. Department of Transportation, Federal Aviation Administration.

Viteles, M. S. (1945). The aircraft pilot: Five years of research, a summary of outcomes. *Psychological Bulletin, 42*, 489–521.

Waag, W. L. (1981). *Training effectiveness of visual and motion simulation* (AFHRL-TR-79-72). Brooks Air Force Base, TX: Air Force Human Resources Laboratory (ADA 094 530).

Wickens, C. D., & Flach, J. M. (1988). Information Processing. In E. L. Wiener & D. C. Nagel (Eds.), *Human factors in aviation* (pp. 111–155). New York: Academic Press.

Wiener, E. L., Kanki, B. J., & Helmreich, R. L. (Eds.). (1993). *Cockpit resource management*. San Diego, CA: Academic Press.

Williams, A. C. (1980). Discrimination and manipulation in flight. In S. N. Roscoe (Ed.), *Aviation psychology* (pp. 11–30). Ames, IA: Iowa State University Press.

Williams, K. W. (Ed.). (1994). *Summary proceedings of the Joint Industry-FAA Conference on the development and use of PC-based aviation training devices* (DOT/FAA/AM-94/25). Washington, DC: Office of Aviation Medicine, Federal Aviation Administration, U.S. Department of Transportation (ADA 286-584).

Yerkes, R. L. (1921). *Memoirs of the National Academy of Sciences* (Vol. 15). Washington, DC: National Academy of Sciences.

Zeidner, J., & Drucker, A. J. (1988). *Behavioral science in the Army: A corporate history of the Army Research Institute*. Alexandria, VA: U.S. Army Research Institute for the Behavioral and Social Sciences.

Zeidner, J., & Johnson, C. (1989). *The economic benefits of predicting job performance* (IDA Paper P-2241). Alexandria, VA: Institute for Defense Analyses (ADA 216 744).

III

AIRCRAFT

13

Pilot Performance

Lloyd Hitchcock
Hitchcock and Associates, Inc.

The determination of pilot performance and the efforts to maximize it are central to aviation safety. It is generally conceded that two out of three aviation accidents are attributable to inappropriate responses of the pilot or crew. Although the catch phrase "pilot error" is all too often laid on the pilot who is guilty only of making a predictable response to "mistakes waiting to happen" that are intrinsic to the design of his cockpit controls or displays or to the work environment surrounding him (or her), there is no question that the greatest improvement in flight safety can be achieved by eliminating the adverse elements of the human component in the aircraft system. Although the most important contributor to aviation safety, the pilot is also the most complicated, variable, and least understood of the aviation "subsystems." Pilot performance has been shown to be affected by everything from eating habits to emotional stress, both past and present. Scheduling decisions can disrupt the pilots' sleep and rest cycles and impose the requirement for pilots to execute the most demanding phase of flight at the point of their maximum fatigue. Illness and medication can degrade performance markedly, as can the use of alcohol and tobacco. Although a complete exposition of all the factors that serve to determine or delimit pilot performance is all but impossible within the constraints of a single text, it is hoped that the following will at least sensitize the reader to many of the variables that have impact on the skill and ability of the commercial and general aviation pilot.

PERFORMANCE MEASUREMENT

Before the role played by any factor in determining pilot behavior can be objectively assessed, we must first be able to quantitatively measure the performance within the cockpit environment. In aviation's infancy, the determination of pilot performance was simple and direct: Those who flew and survived were considered adequate aviators. Since that time, the increased complexity and demands of the airborne environment have continued to confound the process of evaluating the performance of those who fly. The earliest measures of pilot skill were the subjective ratings of the pilot's instructors. The "up-check" was the primary method of evaluation used by the military flight training programs through World War II and, to a great extent, remains the dominant method of pilot assessment today. The general aviation pilot receives his or her license based on the subjective decision of a Federal Aviation Administration (FAA) certified flight examiner.

The appearance of flight simulators not only has enhanced the training of aviators but has made possible a level of quantitative assessment of pilot performance that was not possible before the age of the simulator. An early basic text on flight contains a photograph of a student pilot manipulating a scrap lumber stick and rudder bar, anchored by bungee cords, while seated on an upturned bucket. As cited by Kelly (1970) in his book *The Pilot Maker*, the late 1920s saw the emergence of the most famous flight simulator of them all, the Link Trainer. This simple, tiltable cockpit, actuated by pneumatic bellows, was connected to a chart table course tracking device and was the primary instrument flight trainer for over 500,000 pilots through the end of World War II. The development of more and more sophisticated technology has made possible the level of realism enjoyed by today's simulation efforts. The simulator came into its own in its support of the nation's space program. Because there was no way that such activities as the docking of two space craft or a lunar landing could be practiced in real-world conditions, the simulator was the only way for the astronauts to develop the precise skills demanded by these mission elements. The appearance of computer-automated acquisition and analysis of simulator-based pilot performance data has given rise to a new problem facing those who attempt to understand and interpret such findings. As pointed out by De Maio et al. (1983), automated performance measurement systems (APMS) are generally keyed to quantitative descriptions of aircraft state (e.g., altitude, airspeed, bank angle, etc.), which are usually plotted as a function of elapsed flight time. This time-referenced methodology can ignore the variable of pilot intention and can result in the averaging of performance inputs that may well have been made to accomplish totally different objectives but were grouped together solely because they occurred at the same temporal point in the task sequence. Some widely divergent measures of pilot performance in the course of simulations are found in the literature. It is not surprising that the measures used are generally dictated by the specific issues that the simulations are intended to study. In the joint Navy/FAA study of penetration of severe turbulence by sweptwing transport aircraft, Hitchcock and Morway (1968) developed a statistical methodology allowing them to place probability values on the occurrence of given magnitudes of variation in airspeed, angle-of-attack, roll angle, altitude, and G-load as a function of aircraft weight, penetration altitude, storm severity, and the use of a penetration programmed flight director. This technique permitted the combination of several variables (e.g., G-loading, angle-of-attack variation, and airspeed deviation) into a multidimensional probability surface that described the statistical boundaries of the sampled population of simulated turbulence penetrations. At its Technical Center in Atlantic City, NJ, the FAA has conducted a series of man-in-the-loop simulations to determine the safety implications of proposed changes in the separation between parallel runways at several major airports (Department of Transportation, 1989a, 1989b). Experienced air traffic controllers were tasked with maintaining separation between inbound aircraft even when challenged with a potential collision resulting from a converging deviation by an adjacent aircraft that had "blundered" away from its assigned approach path. The performance of the currently certified airline pilots who were flying the simulators integrated into this study was measured in terms of the likelihood that they would be able, and willing, to take the evasive actions needed to avoid contact with a 500-ft-radius safety "bubble" that was computationally imposed around the center of the straying aircraft. Simulated air-to-ground weapon delivery success is traditionally measured by the dispersion of bomb/rocket impacts around the assigned target. Simulated formation flight is often quantitatively measured by the mean and variance of the distance between the aircraft involved. Carrier landing performance has been quantified through such measurements as the inbound aircraft's deviations about the optimum approach path, the likelihood that the aircraft will trap the target of the third retaining wire, or the computation of probability "footprints" that describe the distribution of the simulated aircraft's expected touchdown points on the flight deck. In summary, there are virtually as many schemes for the measurement of pilot performance as there are questions to be asked.

WORKLOAD

Although the cockpit does impose a certain level of physical demand, with few exceptions, the muscular loadings and stamina demands on the flight crew are well within the range of the reasonably fit pilot. When the term *cockpit workload* is used it almost invariably refers to the combined mental and perceptual demands imposed by the time critical pressures of the flight environment. Within the extensive literature describing the efforts performed to date in an attempt to determine the cognitive workloading imposed on a pilot during flight, according to Crabtree, Bateman, and Acton (1984) the preponderance of these studies were concerned with gaining insight into four principal issues:

1. Will the pilot's current workload permit him or her to take on any additional tasks?
2. Is the pilot too overworked to properly handle an emergency should one arise?
3. Can the crew station and/or the piloting task be modified in such a way that the level of loading will be reduced?
4. Will a proposed new system reduce or add to the pilot's workload, either real or perceived?

Most empirical workload assessment procedures produce dependent measures that can be subsumed under one of three general headings: (a) subjective assessments, (b) performance measures, or (c) physiological measures. Among the most popular techniques for the subjective assessment of workloading within the piloting environment are the Subjective Workload Assessment Technique or SWAT (Reid & Nygren, 1988), the Subjective Workload Dominance Technique (SWORD), described by Vidulich (1989), and NASA's Task Load Index, the NASA-TLX (Hart & Staveland, 1988). In an attempt to better define the elements of pilot workload, Hart and Staveland performed a study that assessed a number of separate measures of pilot workload—(a) a communications analysis, (b) subjective ratings of workload, (c) subjective ratings of other factors (stress, fatigue, mental effort, time pressure, and performance), and (d) heart rate—and showed that generic workload was approximately the same for both pilot and copilot during seven flight segments encompassing all activities from preflight to landing rollout. Although the subjective rating of fatigue showed a significant increase across the flight duration, this factor was not found to correlate with the other measures of loading. The correlations between the subjective workload ratings and heart rate were significant for both pilot and copilot but were higher for the left seat. The authors suggested that this difference reflects the fact that "the responsibility for piloting the aircraft affected the pilot's unconscious response to stress" (p. 152), because these correlations were more pronounced than were those between the purely subjective components. By assuming a direct relationship between the complexity of the displays and controls associated with a driving task, Atsumi, Sigura, and Kimura (1993) also showed a relationship between what they classed as "mental work load" and both heart rate variability and respiratory sinus arrhythmia. Within the aviation context, Doherty (1991) has demonstrated that there are identifiable changes in voice pattern as a function of task-induced stress, although the direct relationship between these changes and workload is far more tenuous and rests only on the finding that speech "jitter" shows a weak negative correlation with a measure of workload. Selcon, Taylor, and Koritsas (1991) had pilots use both the NASA-TLX workload scale and the Royal Air Force (RAF) Situational Awareness Rating Technique (SART) to rate videotapes of aircombat simulations. Although both techniques were sensitive to task difficulty, only the SART reflected difference in pilot experience. Boyer, Pollack, and Eggemaier (1992) found that the subjective workload ratings of memory demands increased as a positive correlate of age, whereas the ratings of psychophysical tasks showed no differences between age groups. Although intuitively satisfying and potentially useful in equipment design and flight procedure evaluations, these measurement methodologies and their findings, unfortunately, provide little new insight into the basic nature of nonphysical workload. At the very least, such findings as these indicate the need to be constantly alert to interactive uncertainties and limitations associated with the measurement of workload. The topic of pilot workload (and, indeed, of air traffic control [ATC] and maintenance personnel as well) is an extremely complex area and has been the subject of a significant number of book-length treatments on its own. Anyone interested in exploring this issue further could profitably start with Jahns's (1973) monograph, *Operator Workload: What Is It and How Should It Be Measured?*, and Hancock and Meshkati's (1988) excellent text, *Human Mental Workload.*

REST AND FATIGUE

Hurst and Hurst (1982), in their book *Pilot Error: The Human Factors*, made a strong case for the fact that *fatigue* is a term that has yet to be defined in a way that is satisfactory to all parties. The failure of aviation medicine to provide an objective, quantifiable, physiological definition of fatigue has allowed those investigating aircraft accidents and incidents to elect to disregard fatigue as a contributing factor in those cases where there is less than overwhelming evidence of its role as a cause. Nance (1986) defined this reluctance as "the traditional view of aviation engineers, an axiom as they saw it, that what cannot be measured with precision cannot be controlled" (p. 75). McFarland (1953) concurred that fatigue's "symptoms . . . are manifested in so many different ways that, thus far, no precise definition of fatigue has been formulated and no very satisfactory method of studying it has been developed" (p. 326). On the other hand, McFarland did concede that "In the experience of every airline pilot, there are certain objective and subjective symptoms that he interprets as "fatigue" and that the operations manager often perceives as evidence of poor performance" (p. 326). The potential danger posed by fatigue is heightened by the fact that the very nature of the pilot's task combines the opportunity for fatigue, arising from a sustained period of flight, with the maximum demand for piloting skill and alertness posed by a landing that is often made at night and frequently in bad weather. The pressures on aircrews to fly as long and as often as possible are the product of a number of factors intrinsic within the airline industry. As both the cost of aircraft and their rate of depreciation continue to grow, the need to maximize aircraft utilization time increases proportionately. The fact that current aircraft are so well designed that they can be flown for many hours with only a minimum of maintenance shifts the burden of being the limiting factor to the flight crew. The large number of different aircraft that many airlines now operate tends to limit the degree of interchangeability within their pilot pool. On those occasions when pilots do switch between aircraft types, Lyall (1992) showed that the change in aircraft is often associated with an increase in both communication frequency and hand-flying—changes that carry a potential for increased fatigue. In addition, the need to minimize personnel costs by pilot reduction has further constrained the operations manager's crew scheduling options. Indeed, the current trend to the use of two-person flight crews, as opposed to the three- and sometimes four-person crews of the past, has removed the option of carrying a "rested" pilot along in the cockpit in case one were needed. However, more recent work by Feyer and Williamson (1992) revealed that the use of a relief driver does not significantly reduce the sense of fatigue in the case of truck drivers, although this may have been due to the tendency to schedule longer routes for the team drivers than for the lone driver. The need to balance schedule compliance with flight time regulations and crew preferences often leads to some degree of controversy. As early as 1949, the scheduled time for Pan American Airways (PAA) flight between Miami, Florida, and Belem, Brazil, exceeded the 8-hr flight time limitation. Establishing a relief crew layover in Port-of-Spain resulted in a need for the replaced crews to wait for more than a day for a return flight. The crews did not appreciate an extra day away from their home base and, because the total flight time for the route was just over 11 hours, the pilots, with the support of the Airline Pilots Association (ALPA), petitioned the Civil Aeronautics Board (CAB) for permission to exceed the 8-hr flight time limitation. The CAB ruled in their favor (Regulation Serial No. SR-345, May 10, 1950) thus transforming the flight time regulation from that of a mandate into the much more fuzzy realm of a negotiated agreement.

Over 50 years ago, Drew (1940) published a seminal study showing that such measured aspects of precision pilotage as deviations in airspeed, sideslip, course heading, and altitude holding were all markedly affected by flight duration. In his book *Fatal Words*, Cushing (1994) cited the role that fatigue can play in missed or misunderstood communications. However, the specific operational and safety implications of such findings have been far less clearly demonstrated. A summary of 59 fatal domestic airline accidents by McFarland failed to show any correlation between these accidents and the duration of the pilot's off-duty period prior to the fatal flight, time on duty, or duration of flight time during the preceding 24-hr period. The introduction of long-range, high-speed commercial jet aircraft has introduced the issue of the *circadian rhythm phenomenon*, or "jet lag," into concerns about pilot fatigue. The work of Gander et al. (Gander, Connell, & Graeber, 1986; Gander & Graeber, 1987; Gander, Kronauer, & Graeber, 1985; Gander, McDonald, Montgomery, & Paulin, 1991; Gander, Myrhe, Graeber, Anderson, & Lauber, 1989) and Foushee, Lauber, Baetge, and

Acombe (1986) described the negative impact of changes in the pilot's day–night cycles on their sleep and rest patterns. Although there is still considerable discussion about the true extent of the role that changes in circadian rhythm play in the determination of sleep disruption fatigue, there was sufficient concern for the NASA Ames Research Center to hold a workshop on the subject and launch the Fatigue Countermeasures Program (NASA, 1980). This program makes use of such measurement devices as questionnaires, flight log analysis, physiological recordings (including electroencephalographic, electro-oculographic, and electromyographic), and a psychomotor vigilance task (PVT) to assess the impact of changes in sleep patterns on pilot performance. The findings of this study effort are slated to become the basis for regulatory action by the government and operational planning by the airline industry. Investigations by Neri, Shappell, and DeJohn (1992) and Neville, Bisson, French, Boll, and Storm (1994) have shown that fatigue and sleep disruption remain current challenges to effective military operations. Studies are currently underway to evaluate the feasibility and effectiveness of such fatigue countermeasures as in-cockpit naps, although such periods of in-flight rest are not currently sanctioned by federal regulation. In addition, the development of an expert scheduling system that incorporates the current physiological and pilot performance data has been proposed as an aid to commercial flight operations managers.

STRESS EFFECTS

Acceleration

The dominant impact of linear acceleration on the pilot is a reduction in peripheral vision and ultimate loss of consciousness associated with sustained high levels of positive G-loadings (+Gz).[1] Such effects are of great importance to the military combat aviator pilot and the aerobatic pilot but are far less of a challenge for the commercial or general aviation pilots, who, hopefully, will never experience the acceleration levels necessary to bring about such physical consequences. These acceleration effects are the result of two factors, the pooling of blood in the lower extremities and the increase in the effective vertical distance (hemodynamic column height) that the heart must overcome to pump blood to the brain. Chambers and Hitchcock (1963) showed that highly motivated pilots would voluntarily sustain up to 550 sec of +Gx (eyeballs in), and even the most determined would tolerate exposures of approximately 160 sec of +Gz (eyeballs down). The seminal work on acceleration-induced loss of vision (grayout) was done by Alice Stoll in 1956. She demonstrated that grayout, blackout, and subsequent unconsciousness are determined not only by the magnitude of the acceleration level but also by the rate of onset (the time required to reach the programmed G-level). More rapid rates of onset apparently do not allow the body time to adapt to the acceleration imposed changes in blood flow. A great deal of effort has been expended in the

[1]Traditionally, the direction in which acceleration is imposed on the body is defined in terms of the change in weight felt by the subject's eyeballs. Thus, positive acceleration (+Gz) such as that felt in tight high-speed turns or in the pullout from a dive is known as "eyeballs down." The forward acceleration (+Gx) associated with a dragster or an astronaut positioned on his or her back during launch would be "eyeballs in." Accelerations associated with sharp level turns (+ or –Gy) would result in an "eyeballs right" during a left turn and "eyeballs left" while in a flat right turn. The negative loading (–Gx) associated with a panic stop in an automobile would be "eyeballs out" and the loading (–Gz) associated with an outside loop would be "eyeballs up."

development of special suiting to constrain blood pooling and the use of greater reclining angles as ways in which the pilot's tolerance to acceleration can be enhanced. In addition, the work of Chambers and Hitchcock (1963) demonstrated the roles that such variables as control damping, cross-coupling, balancing, and number of axes being controlled have in the impact of acceleration on a pilot's tracking control precision, with well-damped, balanced, and moderately cross-coupled controls achieving the best performance. A general review of the effects of sustained acceleration is available in Fraser's chapter on sustained linear acceleration in the NASA *Bioastronautics Data Book* (NASA, 1973). More recent work has focused not just on the physical effects of acceleration but also on its impairment of a pilot's cognitive capabilities. Research by Deaton, Holmes, Warner, and Hitchcock (1990) and Deaton and Hitchcock (1991) has shown that the seatback angle of centrifuge subjects has a significant impact on their ability to interpret the meaning of four geometric shapes even though the variable of back angle did not affect the subjects' physical ability to perform a psychomotor tracking task. A much earlier unpublished study by Hitchcock, Morway, and Nelson (1966) showed a strong negative correlation between acceleration level and centrifuge subjects' performance on a televised version of the Otis Test of Mental Abilities. Such findings are consistent with the pilot adage that states that "All men are morons at 9G."

Vibration

The boundaries of acceptable human body vibration are established by the International Standards Organization *Guide for the Evaluation of Human Exposure to Whole Body Vibration* (1985) and the Society of Automotive Engineers *Measurement of Whole Body Vibration of the Seated Operator of Off-Highway Work Machines* (1980). The dynamic vibration environment experienced by the pilot is the product of many factors including maneuver loads, wing loading, gust sensitivity, atmospheric conditions, turbulence, aircraft size, structural bending moments, airframe resonant frequency, and the aircraft's true airspeed. A clear picture of the impact of vibration on pilot performance is not easily obtained. Investigations of vibration stress have used so many diverse tasks involving such a variety of control systems and system dynamics that it is difficult to integrate their findings. Ayoub (1969) found significant (40%) reduction in a single-axis side-arm controller compensatory tracking task during a 1-hr exposure to a ±.2g sinusoidal vibration at 5 Hz (hertz) or cycles per second.[2] Recovery had not been completed for at least 15 min after exposure. Hornick and Lefritz (1966) exposed subject pilots to 4-hr simulation of three levels of a terrain following task using a two-axis side-stick controller. The vibration spectrum used ranged from 1 to 12 Hz with the peak energy falling between 1 and 7 Hz and with g loadings of .10, .15, and .20g. There was no tendency for tracking error to increase as a function of exposure time for the two easier task levels, although performance degraded after 2.5 hr of exposure to the heaviest loading. Further, these researchers found that reaction time to a thrust change command was almost four times longer during vibration exposure than during the nonvibratory control period. In general the effects of vibration on pilot performance, as measured by tracking performance during simulation, can be summarized as:

[2]Although the uppercase G is used to denote steady-state acceleration, convention dictates that the lowercase g should be used to designate the level of vibration exposure.

- Low-frequency (5 Hz) sinusoidal vibrations from .2 to .8g can reduce tracking proficiency up to 40%.
- When vibration-induced performance decrement is experienced, the effect can persist for up to $\frac{1}{2}$ hr after exposure.
- Higher levels of random vibration exposure are required to affect performance than are required for sinusoidal exposure.
- For gz exposure, vertical tracking performance is more strongly affected than is horizontal.

Under sufficiently high levels of vibration exposure, visual capabilities and even vestibular functioning can be impaired. Although the role of vibration exposure in determining pilot performance should not be ignored, the level of exposure routinely experienced in the commercial aviation environment would not generally be expected to introduce any significant challenge to pilot proficiency.

Combined Stresses

The appearance of other stressors in the flight environment raises the possibility of interactive effects between the individual variables. For example, heat tends to lower acceleration tolerance, whereas cold, probably due to its associated vascular constriction, tends to raise G tolerance. In the same vein, preexisting hypoxia reduces the duration and magnitude of acceleration exposure required to induce peripheral light loss (Burgess, 1958). The nature of stress interactions is determined by (a) their order of occurrence, (b) the duration of their exposure, (c) the severity of exposure, and (d) the innate character of their specific interaction. Any analysis of the flight environment should include a consideration of the potential for synergy between any stressors present. An excellent tabulation of the known interactions between environmental stresses is contained in Murray and McCalley's chapter on combined environmental stresses in the NASA *Bioastronautics Data Book* (NASA, 1973).

PHYSICAL FITNESS

Aging

The interactive role of the potentially negative impact of the aging process and the safety enhancements that are assumed to accompany the gaining of additional operational experience have been assessed in a comprehensive overview of the subject by Guide and Gibson (1991). These authors cite the studies of Shriver (1953), who found that the physical abilities, motivation, skill enhancement, and piloting performance declined with age, and Gerathowohl (1978), who found that both the psychological (cognitive) and physical capabilities of pilots deteriorated with age. In large part, the FAA imposition of the so-called "Age 60 Rule," which prohibits anyone from serving as pilot or copilot of an aircraft heavier than 7,500 pounds after their 60th birthday, is based on a concern for the potential for "sudden incapacitation" by the older pilot (General Accounting Office, 1989). However, a number of studies have shown that this concern is most probably misplaced. Buley (1969) found that the average pilot experiencing sudden inflight incapacitation resulting in an accident was 46 years old. This finding was subsequently confirmed by Bennett (1972), who found that most incapacitation accidents were not related to age. However, age does have an observable impact on aviation safety in that the accident rate for private pilots aged 55–59 (4.97/1,000) is almost twice that for the 20–24 (2.63/1,000) age group (Guide & Gibson, 1991). On the other hand, the accident rate for airline transport rated (ATR) pilots aged 55–59 (3.78/1,000) is approximately one-third that of pilots with the same rating who are aged 20–24 (11.71/1,000). This difference between the age effects for the private and ATR pilot population is most likely the result of two factors. The first is the far more stringent physical and check ride screening given to the airline pilots. Downey and Dark (1990) found that the first-class medical certificate failure rate of ATR pilots went from 4.3/1,000 for the 25–29 age group to 16.2/1,000 for pilots in the 55–59 age group. Thus, many of those age-related disabilities that are seen in the private pilot population appear to have been successfully eliminated from the airline pilot group before they have had a chance to impact safety. The second factor is proposed by Kay et al. (1994), who found that the number of recent flight hours logged by a pilot is a far more important determinant of flight safety than is the age of the pilot. The Kay study authors concluded that their "analyses provided no support for the hypothesis that the pilots of scheduled carriers had increased accident rates as they neared the age of 60" (p. 42). To the contrary, pilots with more than 2,000 hr total time and at least 700 hr of recent flight time showed a significant reduction in accident rate with increasing age. These findings replicate and confirm the conclusions of Guide and Gibson (1991), who also found that the recent experience gained by the aviator was, at least for the mature ATR-rated pilot population, a major determinant of flight safety. According to the comprehensive analyses of flight safety records performed by these researchers, pilots flying more than 400 hr per year have fewer than a third of the accidents per hours flown than do those with less than 400 hr annually. In addition, though the senior pilots would appear to be slightly less safe than those in their forties, they are "safer" than the younger (25–34) pilots who would be most apt to replace them when they are forcibly retired by the Age 60 Rule. Hultsch, Hertzog, and Dixon (1990) and Hunt and Hertzog (1981) also point out that extensive recent experience enables many individuals to develop compensatory mechanisms and thus significantly reduce the negative effects of many of the more general aspects of aging. Stereotyping may play a part in the perception of the aging pilot. Hyland et al. (Hyland, Kay, & Deimler, 1994), in an experimental simulation study of the role of aging in pilot performance, found that the subjective ratings given to the subject pilots by the evaluating check pilots declined as a function of the age of the pilots being graded. However, the objective measures of piloting performance, taken at the same time, showed no corresponding decrement. This would indicate the possibility of an "age bias" influence on the more subjective measures to which pilots are routinely subjected. Tsang (1992), in her extensive review of the literature on the impact of age on pilot performance, pointed out that much of the information on the impact of aging comes from the general psychological literature due to the "sparcity of systematic studies with pilots." She cautioned against the uncritical transfer of findings from the general literature to the tasks of the pilot because most laboratory studies on the effects of aging on cognitive and perceptual processes tend to concentrate on a single isolated function, but the act of flying involves an integration of interactive mental and physical functions.

A corollary of aging that is critical to flight safety is the degradation in vision that all too often afflicts the mature aviator. Whether the problem is an impairment of the ability to focus on near objects (presbyopia) or on far objects (myopia), the result is a need for the pilot to rely on some form of corrective lenses for at least some portion of his or her visual information acquisition. Using a hand to remove and replace glasses as the pilot switches back and forth between the view out of the cockpit to the instrument panel is less than desirable, to say the least. The use of bifocal or trifocal glasses imposes a potentially annoying requirement for the wearer to tilt the head forward and backward to focus through the proper lens. In addition, a representative study by Birren and Shock (1950) determined that the aviator's dark adaptation ability can be expected to degrade progressively from about the age of 50.

The older pilot (40 and over) also shows a marked degradation in auditory sensitivity. The older pilot can show a decline of 15 decibels or more compared to that of the typical 25-year-old. In earlier days, Graebner (1947) reported that the age-related decline of auditory sensitivity, particularly at the higher frequencies (200 cps [cycles per second] and above), was more pronounced for pilots than for the general population. This was attributed to the high cockpit noise levels associated with the reciprocal engines in use at the time. It is reasonable to assume that the transition to the jet engine will have significantly reduced this effect.

Those who are interested in a more comprehensive study and detailed evaluation of the role of age in determining flight safety are referred to two recent studies supported by the FAA Office of Aviation Medicine. The first is an annotated bibliography of age-related literature performed by Hilton Systems, Inc. (1994), under contract to the Civil Aeromedical Institute in Oklahoma. The second is an analytic review of the scientific literature, compiled by Hyland, Kay, Deimler, and Gurman (1994), relative to aging and airline pilot performance.

Effects of Alcohol

A number of general reviews of the impact of alcohol on both psychological and physiological performance are available (Carpenter, 1962; McFarland, 1953; Ross & Ross, 1995). In general, the documented effects of alcohol are all deleterious, with alcohol consumption adversely affecting a wide range of sensory, motor, and mental functions. The drinker's visual field is constricted, which could affect both instrument scan and the detection of other aircraft (Moskowitz & Sharma, 1974). Alcohol reduces a pilot's ability to see at night or at low levels of illumination, with the eye of one who has consumed alcohol requiring up to twice the intensity to see a target light as needed before the ingestion of the alcohol. In addition, the intensity of light required to resolve flicker has been found to be a direct function of the observer's blood alcohol concentration. Alcohol consumption has also been found to reduce the sense of touch. The effects of alcohol ingestion on motor behavior are considered to be the result of its impairment of nervous functions rather than as direct degradation of muscle action. Such activities as reflex actions, steadiness, and visual fixation speed and accuracy are adversely affected by the consumption of even a small amount of alcohol. The consumption of sufficient quantities of alcohol can result in dizziness, disorientation, delirium, or even loss of consciousness. However, at the levels that would most often be encountered in the cockpit, the most significant effects would most likely be in the impairment of mental behavior rather than a degradation of motor responses. A detailed review of the literature by Levine, Kramer, and Levine (1975) confirmed the alcohol-induced performance deterioration in the area of cognitive domain, perceptual-motor processes, and psychomotor ability, with the psychomotor domain showing the greatest tolerance for alcohol effects. Alcohol also has been found to degrade memory, judgment, and reasoning. More recent work by Barbre and Price (1983) showed that alcohol intake not only increased search time in a target detection task but also degraded touch accuracy and hand travel speed. In addition, alcohol was found to reduce the subject's motivation to complete a difficult task. Both Aksnes (1954) and Henry, Davis, Engelken, Triebwasser, and Lancaster (1974) demonstrated the negative effects of alcohol on Link Trainer performance. Billings, Wick, Gerke, and Chase (1973) showed similar alcohol-induced performance decrements in light aircraft pilots. Studies by Davenport and Harris (1992) showed the impact of alcohol on pilot performance in a landing simulation. Taylor, Dellinger, Schilling, and Richardson (1983) found similar degradation of both holding pattern performance and instrument landing system (ILS) approaches as a function of alcohol intake. Ross and Mundt (1988) evaluated the performance of pilots challenged with simulated VOR tracking, vectoring, traffic avoidance, and descent tasks. Using a multiattribute modeling analysis, pilot performance was evaluated by flight instructor judgments under 0.0% and 0.04% blood alcohol concentrations (BACs). The multiattribute approach was sufficiently sensitive to reveal "a significant deleterious effect on overall pilot performance" associated with alcohol consumption of even this rather low level which is the maximum allowable by FAA regulation in 1985 and 1986. Ross, Yeazel, and Chau (1992) using light aircraft simulation studies of pilots under BACs ranging from 0.028 to 0.037% challenged pilots with the demands of simulated complicated departures, holding patterns, and approaches under simulated instrument meteorological conditions (IMC) or instrument landing approaches involving turbulence, cross winds, and wind shear. Significant alcohol-related effects were found at the higher levels of workload. Of particular significance for those interested in the effects of alcohol on pilots is the synergistic relationship between alcohol and the oxygen lack associated with altitude. Early studies by McFarland and Forbes (1936), McFarland and Barach (1936), and Newman (1949) established the facts that, even at altitudes as low as 8,000 ft, the ingestion of a given amount of alcohol results in a greater absorption of alcohol into the blood than at sea level and that, at altitude, it takes the body significantly longer to metabolize the alcohol out of the blood and spinal fluid. More recent studies by Collins et al. (Collins & Mertens, 1988; Collins, Mertens, & Higgins, 1987) confirmed the interaction of alcohol and altitude in the degradation in the performance of complex tasks. A recent survey by Ross and Ross (1992) assessed the perception of professional pilots of the seriousness of the alcohol usage problem. The average overall level of concern over pilot drinking was found to be just below 3 on a scale of 0 (no problem) to 10 (a very serious problem). Noncarrier pilots rated alcohol usage as a more serious problem for the scheduled airline pilot than did the major carrier pilots themselves. The majority of commercial pilots approved of the proposal to enact laws making drinking and flying a felony and also approved of random blood alcohol concentration testing, although they were almost evenly divided on the potential effectiveness of such testing and expressed significant concern about the possibility that such a testing program could violate pilots' rights.

Drug Effects

In 1953, McFarland published one of the first and most comprehensive descriptions of the potential negative effects of commonly used pharmaceuticals on flight safety. Some of the more common antibiotic compounds have been found to adversely affect the aviator's tolerance to altitude-induced hypoxia and therefore psychomotor performance. Of course, those antihistamines that advise against the operation of machinery after use should be avoided by the pilot, as should any use of sedatives prior to or during flight operations. The use of hyoscine (scopolamine) as a treatment of motion sickness was found to reduce visual efficiency in a significant number of users. In general, the use of common analgesics, such as aspirin, at the recommended dosage levels, does not appear to be a matter of concern. However, because any medication has the potential for adverse side affects in the sensitized user, the prudent pilot would be well advised to use no drug except under the direction of his flight surgeon.

Tobacco

The introduction of nicotine into the system is known to have significant physiological effects. Heart rate is increased by as much as 20 beats per minute, systolic blood pressure goes up by 10–20 mm Hg, and the amount of blood flowing to the extremities is reduced. Although these effects have clear significance for the pilot's potential risk of in-flight cardiac distress, perhaps the most significant impact of smoking on flight safety lies in the concomitant introduction of carbon monoxide into the pilot's bloodstream. Human hemoglobin has an affinity for carbon monoxide that is over 200 times as strong as its attraction to oxygen (O_2). Hemoglobin cannot carry both oxygen and carbon dioxide molecules. Therefore, the presence of carbon monoxide will degrade the body's capability to transport oxygen, essentially producing a temporary state of induced anemia. McFarland, Roughton, Halperin, and Niven (1944) and Sheard (1946) demonstrated that the smoking-induced level of carboxyhemoglobin (COHb) of 5–10%, the level generally induced by smoking a single cigarette, can have a significant negative effect on visual sensitivity although this CO content is well below the 20% or more COHb considered necessary to induce general physiological discomfort. Trouton and Eysenck (1960) reported some degradation of limb coordination at 2–5% COHb levels. Schulte (1963) found consistent impairment of cognitive and psychomotor performance at this same COHb level. Putz (1979) found that CO inhalation also adversely affected dual-task performance. These findings are not unanimously accepted. Hanks (1970) and Stewart et al. (1970) found no central nervous system functions at COHb levels below 15%. The carbon monoxide anemia induced by smoking synergizes with the oxygen deficits imposed by altitude. According to McFarland et al. (1944), by both decreasing the effectiveness of the oxygen transport system and increasing the metabolic rate, and thus the need for oxygen, smoking can raise the effective altitude experienced by the pilot by as much as 50%, making the physiological effects of 10,000 ft on the smoker equivalent to those felt by the nonsmoker at 15,000 ft. Although most commercial flights now restrict the occurrence of smoking in flight, the uncertainties about the rate with which the effects of smoking prior to flight are dissipated will cause the issue of smoking to continue to be of concern for those interested in optimizing pilot performance. The in-flight use of tobacco by the general aviation pilot will remain as a potential concern. To date, no studies defining the role of second-hand smoke inhalation on pilot performance were located.

Nutrition

Perhaps the earliest impact of nutrition on pilot performance was reported by McFarland, Graybiel, Liljencranz, and Tuttle (1939) in their description of the improvement in vision brought about by vitamin A supplementation of the diet of night-vision-deficient airmen. Hecht and Mendlebaum (1940) subsequently confirmed this effect by experimentally inducing marked degradation in the dark-adaptation capability of test subjects fed a vitamin A-restricted diet. Currently, the ready availability of daily vitamin supplements and the general level of nutrition of the population as a whole have tended to virtually eliminate any concern about a lack of vitamin C on the health of skin, gums, and capillary system or a degradation in the pilot's nervous system, appetite, or carbohydrate metabolism due to a deficiency in the B vitamin complex. However, the intrinsic nature of airline operations inevitably results in some irregularity in the eating habits of the commercial pilot. Extended periods without eating can result in low blood sugar (hypoglycemia). Although the effects of long-term diet deficiency are generally agreed on (marked reduction in endurance and a correspondingly smaller degradation of physical strength), the exact relationship between immediate blood sugar level and performance is less well established. Keys (1946) demonstrated that reaction time was degraded at blood sugar levels below 65–70 mg%.

SUMMARY

The importance of each and every one of the variables described in this section is sufficient that all are the subjects of book chapters and, in many cases, entire texts in their own right. The best that can be hoped is that the foregoing will create a sensitivity to the complexity of the topic field of pilot performance. There is much work that remains to be done in developing more objective methods for measuring the essential components of piloting skill. Even more challenging is the pressing need to define and quantify the cognitive components of the concept of pilot workload. Because of the economic and safety implications of aging on both the airline industry and the pilot ranks, the issue of aging will remain a major topic of interest and concern. Because age does not seem to be a prime determinant of sudden in-flight incapacitation, additional effort is clearly needed to determine the physical factors that can be effective in predicting such occurrences. We already know enough to be certain of the negative impacts of alcohol, smoking, and controlled substances on pilot performance. In short, it is unfortunately clear that, although pilot performance is unquestionably the most critical element in flight safety, it is the aircraft system area about which we know far less than we should.

REFERENCES

Aksnes, E. G. (1954). Effects of small doses of alcohol upon performance in a link trainer. *Journal of Aviation Medicine, 25*, 680–688.

Atsumi, B., Sigura, S., & Kimura, K. (1993). Evaluation of mental workload in vehicle driving by analysis of heart rate variability. *Proceedings of the Human Factors and Ergonomics Society 37th Annual Meeting, 1*, 574–578.

Ayoub, M. M. (1969). *Performance and recovery under prolonged vibration*. Unpublished manuscript, School of Engineering, Texas Technological College, Lubbock, TX.

Barbre, W. E., & Price, D. L. (1983). Effects of alcohol and error criticality on alphanumeric target acquisition. *Proceedings of the Human Factors Society 27th Annual Meeting, 1*, 468–471.

Bennett, G. (1972, October). Pilot incapacitation. *Flight International*, pp. 569–571.

Billings, C. E., Wick, R. L., Gerke, R. J., & Chase, R. C. (1973). Effects of ethyl alcohol on pilot performance. *Aerospace Medicine, 44*, 379–382.

Birren, J. E., & Shock, N. W. (1950). Age changes in rate and level of visual dark adaptation. *Applied Physiology, 2*(7), 407–411.

Boyer, D. L., Pollack, J. G., & Eggemeier, F. T. (1992). Effects of aging on subjective workload and performance. *Proceedings of the Human Factors Society 36th Annual Meeting, 1,* 156–160.

Buley, L. E. (1969). Incidence, causes, and results of airline pilot incapacitation while on duty. *Aerospace Medicine, 40*(1), 64–70.

Burgess, B. F. (1958). The effect of hypoxia on tolerance to positive acceleration. *Journal of Aviation Medicine, 29,* 754–757.

Carpenter, J. A. (1962). Effects of alcohol on some psychological processes. *Quarterly Journal of Studies on Alcohol, 24,* 274–314.

Chambers, R. M., & Hitchcock, L. (1963). *The effects of acceleration on pilot performance* (Tech. Rep. No. NADC-MA-6219). Warminster, PA: Naval Air Development Center.

Collins, W. E., & Mertens, H. W. (1988). Age, alcohol, and simulated altitude: Effects on performance and breathalyser scores. *Aviation, Space, and Environmental Medicine, 59,* 1026–1033.

Collins, W. E., Mertens, H. W., & Higgins, E. A. (1987). Some effects of alcohol and simulated altitude on complex performance scores and breathalyser readings. *Aviation, Space, and Environmental Medicine, 58,* 328–332.

Crabtree, M. S., Bateman, R. P., & Acton, W. H. (1984). Benefits of using objective and subjective workload measures. *Proceedings of the Human Factors Society 28th Annual Meeting, 2,* 950–953.

Cushing, S. (1994). *Fatal words* (p. 71). Chicago: University of Chicago Press.

Davenport, M., & Harris, D. (1992). The effect of low blood alcohol levels on pilot performance in a series of simulated approach and landing trials. *International Journal of Aviation Psychology, 2*(4), 271–280.

Deaton, J. E., & Hitchcock, E. (1991). Reclined seating in advanced crew stations: Human performance considerations. *Proceedings of the Human Factors Society 35th Annual Meeting,* 1, 132–136.

Deaton, J. E., Holmes, M., Warner, N., & Hitchcock, E. (1990). *The development of perceptual/motor and cognitive performance measures under a high G environment* (Tech. Rep. No. NADC-90065-60). Warminster, PA: Naval Air Development Center.

De Maio, J., Bell, H. H., & Brunderman, J. (1983). Pilot oriented performance measurement. *Proceedings of the Human Factors Society 27th Annual Meeting, 1,* 463–467.

Department of Transportation. (1989a). *Atlanta tower simulation* (Report No. DOT/FAA/CT-TN89/27). Atlantic City, NJ: FAA Technical Center.

Department of Transportation. (1989b). *Dallas/Fort Worth simulation* (Report No. DOT/FAA/CT-TN89/28). Atlantic City, NJ: FAA Technical Center.

Doherty, E. T. (1991). Speech analysis techniques for detecting stress. *Proceedings of the Human Factors Society 35th Annual Meeting, 1,* 689–693.

Downey, L. E., & Dark, S. J. (1990). *Medically disqualified airline pilots in calendar years 1987 and 1988* (Report No. DOT-FAA-AM-90-5). Oklahoma City, OK: Office of Aviation Medicine.

Drew, G. C. (1940). *Mental fatigue* (Report 227). London: Air Ministry, Flying Personnel Research Committee.

Feyer, A., & Williamson, A. M. (1992). Work practices in the long distance road transport industry in Australia. *Proceedings of the Human Factors Society 36th Annual Meeting, 2,* 975–979.

Foushee, H. C., Lauber, J. K., Baetge, M. M., & Acombe, D. B. (1986). *Crew factors in flight operations: III, The operational significance of exposure to short-haul air transport operations* (Technical Memorandum 88322). Moffett Field, CA: National Aeronautics and Space Administration.

Gander, P. H., Connell, L. J., & Graeber, R. C. (1986). Masking of the circadian rhythms of body temperature by the rest-activity cycle in man. *Journal of Biological Rhythms Research, 1,* 119–135.

Gander, P. H., & Graeber, R. C. (1987). Sleep in pilots flying short-haul commercial schedules. *Ergonomics, 30,* 1365–1377.

Gander, P. H., Kronauer, R., & Graeber, R. C. (1985). Phase-shifting two coupled circadian pacemakers: Implications for jetlag. *American Journal of Physiology, 249,* 704–719.

Gander, P. H., McDonald, J. A., Montgomery, J. C., & Paulin, M. G. (1991). Adaptation of sleep and circadian rhythm to the Antarctic summer: A question of zeit-geber strength. *Aviation, Space, and Environmental Medicine, 62,* 1019–1025.

Gander, P. H., Myrhe, G., Graeber, R. C., Anderson, H. T., & Lauber, J. K. (1989). Adjustment of sleep and the circadian temperature rhythm after flights across nine time zones. *Aviation, Space, and Environmental Medicine, 60,* 733–743.

Gander, P. H., Nguyen, D., Rosekind, M. R., & Connell, L. J. (1993). Age, circadian rhythm and sleep loss in flight crews. *Aviation, Space, and Environmental Medicine, 64,* 189–195.

General Accounting Office. (1989). *Aviation safety: Information on FAA's Age 60 Rule for pilots* (GAO-RCED-90-45FS). Washington, DC: Author.

Graebner, H. (1947). Auditory deterioration in airline pilots. *Journal of Aviation Medicine, 18*(1), 39–47.

Guide, P. C., & Gibson, R. S. (1991). An analytical study of the effects of age and experience on flight safety. *Proceedings of the Human Factors Society 35th Annual Meeting, 1*, 180–184.

Hancock, P. A., & Meshkati, N. (Eds.). (1988). *Human mental workload*. Amsterdam: North-Holland.

Hanks, T. H. (1970, February). *Analysis of human performance capabilities as a function of exposure to carbon monoxide*. Paper presented at Conference on the Biological Effects of Carbon Monoxide, New York Academy of Sciences.

Hart, S. G., & Staveland, L. (1988). Development of NASA-TLX (Task Load Index): Results of empirical and theoretical research. In P. A. Hancock & N. Meshkati (Eds.), *Human mental workload* (pp. 139–183). Amsterdam: North-Holland.

Hecht, S., & Mendlebaum, J. (1940). Dark adaptation and experimental human vitamin A deficiency. *Journal of General Physiology, 130*, 651–664.

Henry, P. H., Davis, T. Q., Engelken, E. J., Triebwasser, J. A., & Lancaster, M. C. (1974). Alcohol-induced performance decrements assessed by two Link Trainer tasks using experienced pilots. *Aerospace Medicine, 45*, 1180–1189.

Hilton Systems, Inc. (1994). *Age 60 Rule research, Part I: Bibliographic database* (Report No. DOT/FAA/AM-94/20). Oklahoma City, OK: Civil Aeromedical Institute, FAA.

Hitchcock, L., & Morway, D. A. (1968). *A dynamic simulation of the sweptwing transport aircraft in severe turbulance* (Tech. Rep. No. NADC-MR-6807, FAA Report No. FAA-DS-68-12). Warminster, PA: Naval Air Development Center.

Hitchcock, L., Morway, D. A., & Nelson, J. (1966). *The effect of positive acceleration on a standard measure of intelligence*. Unpublished study, Aerospace Medical Acceleration Laboratory, Naval Air Development Center, Warminster, PA.

Hornick, R. J., & Lefritz, N. M. (1966). A study and review of human response to prolonged random vibration. *Human Factors, 8*(6), 481–492.

Hultsch, D. F., Hertzog, C., & Dixon, R. A. (1990). Ability correlates of memory performance in adulthood and aging. *Psychology and Aging, 5*, 356–358.

Hunt, E., & Hertzog, C. (1981). *Age related changes in cognition during the working years* (final report). Department of Psychology, University of Washington, Seattle.

Hurst, R., & Hurst, L. (1982). *Pilot error: The human factors*. London: Granada Press.

Hyland, D. T., Kay, E. J., & Deimler, J. D. (1994). *Age 60 study, Part IV: Experimental evaluation of pilot performance* (Office of Aviation Medicine Report No. DOT/FAA/AM-94/23). Oklahoma City, OK: Civil Aeromedical Institute, FAA.

Hyland, D. T., Kay, E. J., Deimler, J. D., & Gurman, E. B. (1994). *Age 60 study, Part II: Airline pilot age and performance—A review of the scientific literature* (Office of Aviation Medicine Report No. DOT/FAA/AM-94/21). Washington, DC: FAA.

International Standards Organization. (1985). *Guide for the evaluation of human exposure to whole body vibration* (ISO 2631). Geneva: Author.

Jahns, D. W. (1973). Operator workload: What is it and how should it be measured? In K. D. Cross and J. J. McGrath (Eds.), *Crew system design* (pp. 281–287). Santa Barbara, CA: Anacapa Sciences.

Kay, E. J., Hillman, D. J., Hyland, D. T., Voros, R. S., Harris, R. M., & Deimler, J. D. (1994). *Age 60 study, Part III: Consolidated database experiments final report* (Office of Aviation Medicine, Report No. DOT/FAA/AM-94/22). Washington, DC: FAA.

Kelly, L. L. (1970). *The pilot maker*. New York: Grosset & Dunlap.

Keys, A. (1946). Nutrition and capacity for work. *Occupational Medicine, 2*(6), 536–545.

Levine, J. M., Kramer, G. G., & Levine, E. N. (1975). Effects of alcohol on human performance: An integration of research findings based on an abilities classification. *Journal of Applied Psychology, 60*, 285–293.

Lyall, E. A. (1992). The effects of mixed-fleet flying of the Boeing 737-200 and -300. *Proceedings of the Human Factors Society 36th Annual Meeting*, 35–39.

McFarland, R. A. (1953). *Human factors in air transportation*. New York: McGraw Hill.

McFarland, R. A., & Barach, A. L. (1936). The relationship between alcoholic intoxication and anoxemia. *American Journal of Medical Science, 192*(2), 186–198.

McFarland, R. A., & Forbes, W. H. (1936). The metabolism of alcohol in man at high altitudes. *Human Biology, 8*(3), 387–398.

McFarland, R. A., Graybiel, A., Liljencranz, E., & Tuttle, A. D. (1939). An analysis of the physiological characteristics of two hundred civil airline pilots. *Journal of Aviation Medicine, 10*(4), 160–210.

McFarland, R. A., Roughton, F. J. W., Halperin, M. H., & Niven, J. I. (1944). The effect of carbon monoxide and altitude on visual thresholds. *Journal of Aviation Medicine, 15*, 382–394.

Moskowitz, H., & Sharma, S. (1974). Effects of alcohol on peripheral vision as a function of attention. *Human Factors, 16,* 174–180.
Nance, J. J. (1986). *Blind trust.* New York: William Morrow.
National Aeronautics and Space Administration. (1973). *Bioastronautics data book.* Washington, DC: Scientific and Technical Information Office, NASA.
National Aeronautics and Space Administration. (1980). *Pilot fatigue and circadian desynchronis* (Technical Memorandum 81275). Moffett Field, CA: Author.
Neri, D. F., Shappell, S. H., & DeJohn, C. A. (1992). Simulated and sustained flight operations and performance: Part I. Effects of fatigue. *Military Psychology, 4,* 137–155.
Neville, K. J., Bisson, R. U., French, J., Boll, P. A., & Storm, W. F. (1994). Subjective fatigue of C-141 aircrews during Operation Desert Storm. *Human Factors, 36*(2), 339–349.
Newman, H. W. (1949). The effect of altitude on alcohol tolerance. *Quarterly Journal of Studies on Alcohol, 10*(3), 398–404.
Putz, V. R. (1979). The effects of carbon monoxide on dual-task performance. *Human Factors, 21,* 13–24.
Reid, G., & Nygren, T. E. (1988). The subjective workload assessment technique: A scaling procedure for measuring mental workload. In P. A. Hancock & N. Meshkati (Eds.), *Human mental workload* (pp. 139–183). Amsterdam: North-Holland.
Ross, L. E., & Mundt, J. C. (1988). Multiattribute modeling analysis of the effects of low blood alcohol level on pilot performance. *Human Factors, 30*(3), 293–304.
Ross, L. E., & Ross, S. M. (1992). Professional pilots' evaluation of the extent, causes, and reduction of alcohol use in aviation. *Aviation and Space Environment Medicine, 63,* 805–808.
Ross, L. E., & Ross, S. M. (1995). Alcohol and aviation safety. In R. R. Watson (Ed.), *Drug and alcohol abuse reviews, Vol. 7: Alcohol, cocaine, and accidents.* Totowa, NJ: Humana.
Ross, L. E., Yeazel, L. M., & Chau, A. W. (1992). Pilot performance with blood alcohol concentrations below 0.04%. *Aviation Space and Environmental Medicine, 63,* 951–956.
Schulte, J. H. (1963). Effects of mild carbon monoxide intoxication. *AMA Archives of Environmental Medicine, 7,* 524.
Selcon, S. J., Taylor, R. M., & Koritsas, E. (1991). Workload or situational awareness?: NASA TLX versus SART for aerospace systems design. In *Proceedings of the Human Factors Society 35th Annual Meeting* (pp. 62–66). Santa Monica, CA: The Human Factors Society.
Sheard, C. (1946). The effect of smoking on the dark adaptation of rods and cones. *Federation Proceedings, 5*(1,2), 94.
Society of Automotive Engineers. (1980). *Measurement of whole body vibration of the seated operator of off-highway work machines* (Recommended Practice J1013). Detroit: Author.
Stewart, R. D., Peterson, J. E., Baretta, E. D., Blanchard, R. T., Hasko, M. J., & Herrmann, A. A. (1970). *AMA Archives of Environmental Medicine, 21,* 154.
Stoll, A. M. (1956). Human tolerance to positive G as determined by the physiological endpoints. *Journal of Aviation Medicine, 27,* 356–359.
Taylor, H. L., Dellinger, J. A., Schilling, R. F., & Richardson, B. C. (1983). Pilot performance measurement methodology for determining the effects of alcohol and other toxic substances. *Proceedings of the Human Factors Society 27th Annual Meeting, 1,* 334–338.
Trouton, D., & Eysenck, H. J. (1960). The effects of drugs on behavior. In H. J. Eysenck (Ed.), *Handbook of abnormal psychology.* London: Pitman Medical.
Tsang, P. S. (1992). A reappraisal of aging and pilot performance. *International Journal of Aviation Psychology, 2*(3), 193–212.
Vidulich, M. A. (1989). The use of judgement matrices in subjective workload assessment: The Subjective WORkload Dominance (SWORD) technique. *Proceedings of the Human Factors Society 33rd Annual Meeting, 2,* 1406–1410.

14

Controls, Displays, and Workplace Design

John M. Reising
Kristen K. Liggett
U.S. Air Force Research Laboratory

Robert C. Munns
Royal Air Force

Aircraft control and display (C/D) technology has changed dramatically over the past 30 years. The advent of compact, high-power, rugged digital devices has allowed the on-board, real-time processing of data electronically. The digital impact has allowed a major shift from electromechanical to electro-optical devices and has also had a far-reaching effect on the way in which C/D research is being conducted. Because electro-optical C/Ds are computer controlled and therefore multifunctional, there has been a shift away from experiments concerned with optimal arrangement of physical instruments within the cockpits, and an added emphasis placed on the packaging of the information that appears on the display surface. The reason for this shift is that multifunction displays can show many formats on the same display surface and portray the same piece of information in many different ways. Also, with the advent of such technologies as touch-sensitive overlays and eye control, the same physical devices serve both as control and display, blurring the previously held careful distinction between the two. The first section of this chapter discusses the history of cockpit technology from the mechanical era through the electro-optical era. Subsequent sections in this chapter discuss the impact and application of the new technology in the military environment.

TRANSITION OF COCKPITS WITH TIME AND TECHNOLOGY

The history of cockpit technology is divided into a number of different eras. For this chapter we chose three mechanization eras—the mechanical, electromechanical (E-M), and electro-optical (E-O), because they have a meaningful relationship to instrument design changes. Although we can, and do, discuss these as separate periods, the time boundaries are very vague, even though design boundaries are clear (Nicklas, 1958). Mechanical instruments, of course, were used first. Nevertheless, the use of electromechanical instruments can be traced to the very early days of flight, around 1920.

Electro-optical instruments were investigated in the 1930s. For example, in 1937, a cathode ray tube (CRT) based electro-optical display called the Sperry Flightray was evaluated on a United Air Lines "Flight Research Boeing" (Bassett & Lyman, 1940). The fact that all operators, private, commercial, and military, have flown with instruments incorporating all three designs also makes the era's boundaries fuzzy.

For the purpose of this section, we consider the mechanical era as that time from the beginning of flight until the introduction of the integrated instrument system (IIS) by the U.S. Air Force in the late 1950s (Klass, 1956). The electromechanical era extends from that point until the introduction of the U.S. Navy's F-18 aircraft, which makes extensive use of multipurpose CRT displays. The issues of the electro-optical era, and beyond, comprise the primary subject matter of this chapter.

The Mechanical Era

The importance of instrumenting the information needed to fly an airplane was recognized by the Wright brothers very early in their flying adventures. The limitations of measuring airspeed by the force of the wind on one's face were not very subtle. From the time these famous brothers first installed an anemometer, a mechanical device used to measure wind velocity, and a weather vane, to measure angle of incidence, aviators and designers have been concerned about cockpit instrument issues such as weight, size, shape, accuracy, reliability, and environmental effects (Nicklas, 1958). As aviators gained more flying experience, they recognized the need for additional pieces of information in the cockpit, which in turn meant that there was a need for some kind of instrument. It did not take many engine failures before the need for data that would warn of an impending failure became obvious.

The requirement for displaying most pieces of information in a cockpit can be traced to the need to identify or solve a problem. So the "research" process during most of the mechanical era was to invent a device or improvise from something that already existed in the nonaviation world. Any testing was generally done in flight. Simulators, as we have come to know them over the past 25 years, were virtually nonexistent during the mechanical era. The first simulators were modified or upgraded flight trainers, and were not generally regarded as an adequate substitute for flight trials. During this era it was not unusual for a potential solution to progress from conception to a flight trial in a matter of weeks, as opposed to the years it currently takes.

It would certainly be wrong to leave one with the impression that the mechanical era was one of only simple-minded evolutionary changes in the cockpit. On the contrary, the history of instrument flying, even as we know it today, can be traced back to the early flying days of Lt. James Doolittle of the Army Air Corps (Glines, 1989). In 1922 he performed the first crossing of the United States accomplished in less than 24 hr. Hampered by darkness and considerable weather, he claimed that the trip would have been impossible without the "blessed bank and turn indicator," an instrument invented in 1917 by Elmer Sperry. In his Gardner Lecture, Doolittle claimed that it was the "blind flying" pioneering exploits of a number of other aviators that provided the "fortitude, persistence and brains" behind the blind flying experiments of the 1920s and early 1930s (Doolittle, 1961). In 1929, Doolittle accomplished the first flight that was performed entirely on instruments. It was obvious to these pioneers that instrument flying, as we know it today, was going to become a pacing factor in the future of all aviation.

Although many milestones in the development of instrument flying technology took place in the mechanical era, technology had advanced sufficiently by 1950 to begin to shift the emphasis from wind-driven mechanical instruments to instruments powered by electricity.

The Electromechanical Era

As mentioned earlier, this era began when the U.S. Air Force introduced the integrated instrument system (IIS), often simplistically referred to as the "T-line" concept, for high-performance jet aircraft. This was the first time that the U.S. Air Force had formed an internal team of engineers, pilots, and human factors specialists to produce a complete instrument panel. The result was a revolutionary change in how flight parameters were displayed to pilots. These changes were necessitated because aircraft were flying more rapidly and weapons systems were becoming more complex. This complexity reduced the time available for the pilot to perform an instrument cross check, and the fact that each parameter was displayed on a dedicated 3- or 4-inch round dial compounded the problem. The solution was to display all air data, that is, angle of attack, mach, airspeed, altitude, and rate of climb, on vertical moving tapes that were read on a fixed horizontal lubber line that extended continuously across all of the tape displays and the attitude director indicator (ADI). Lateral information was read on a vertical reference line that traversed through the center of the ADI and the horizontal situation indicator (HSI). The two reference lines thus formed the "T" (Fig. 14.1). Programmable or manually selectable command markers were added to the tape displays to provide easily noticeable deviations from a desired position.

Again, flight trials provided the "proof of the pudding" as well as being critical to the design and development process. Ideas were incorporated, flown, changed, flown, changed again, and flown until all of the design team members were satisfied. Seem-

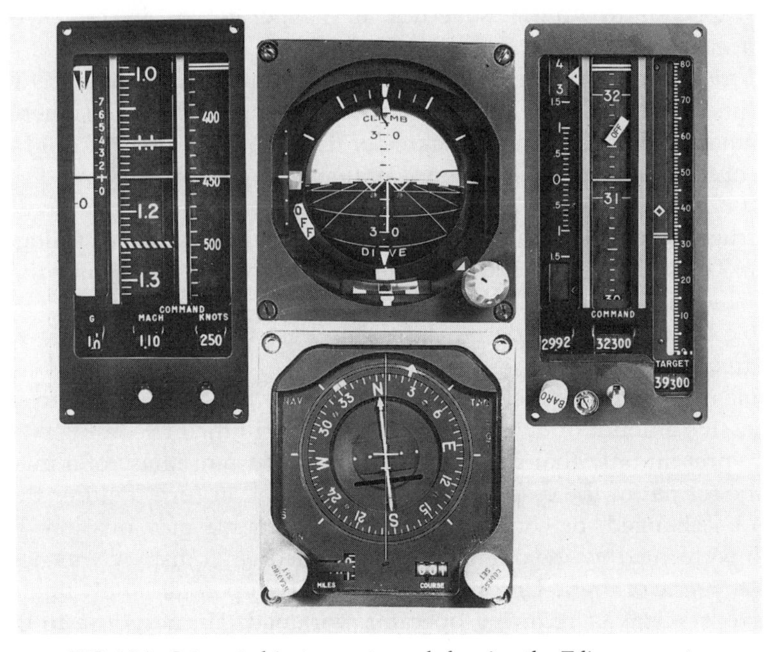

FIG. 14.1. Integrated instrument panel showing the T-line concept.

ingly simple questions, such as which direction the individual tapes should move, and how they should move in relation to each other, were answered through many flying hours. In the end, a system emerged that was easier to read and cross check than the old mechanical round dials. Although the displays were simpler, the electromechanization was orders of magnitude more complex. The servomechanisms, with their tremendously complex mechanical gearing, were a watchmaker's nightmare, but, even so, the data was processed in a relatively simple fashion within the constraints imposed by analog processing of electrical signals and mechanical gear mechanisms. The concept, although mechanically complex, has stood the test of time and can be seen on many of this era's aircraft. However, the pure economics of maintaining this type of instrumentation is fueling the transition to solid-state displays. For example, both the new Boeing 777 on the commercial side and the F-22 on the military side are intended to have multifunction displays that cover the vast majority of the front instrument panel. A major reason for this trend is the increasing cost to maintain and support E-M instruments (Galatowitsch, 1993).

The Electro-Optical Era

The advent of the F-18 is generally regarded as a watershed in cockpit display design, and can be considered as the beginning of the electro-optical era. The cockpit displays of this era are composed largely of CRTs presenting data that had been digitally processed by the aircraft's on-board systems. The real impact of this digital processing was the design flexibility of the displays, and the ability to vary the display according to the information required by the user. Because of this characteristic, the displays are generally known as multifunction displays (MFDs). The ability to show large amounts of information on a limited display surface shifted the emphasis of cockpit research from packaging of instrumentation to packaging of information. Specifically, the concern was how best to format the displays and how to structure the control menus so that the user did not drown in an overflow of data, or get lost in the bowels of a very complicated menu structure.

The F-18 cockpit truly broke new ground, but its introduction represented only the tip of a technological iceberg in terms of the challenge for the designer's electronic cockpits. Although the MFD gave a degree of freedom over what it could display, the technology of the CRT (size, power consumption, and weight) still posed some serious limitations on the positioning of the display unit itself. Since then, there has been a continual struggle to reduce the bulk of the display devices while increasing the display surface area. The ultimate aim is to provide the operator with a display that covers all of the available viewing area with one contiguous, controllable display surface. This would enable the ultimate in "designability," but are we in a position to adapt to this amount of freedom?

The problem given to the crew station designer by the MFD is how to show the air crew the massive amount of data now available without their becoming swamped. The answer is to present only that data required for the current phase of a mission and to configure the format of the display accordingly, which in turn requires the ability for displays to be changed, or controlled, during the course of a mission. Initially this change was performed by the operator, who decided what display was needed to suit the particular phase of flight. Unfortunately, extensive operator involvement was counterproductive in terms of reducing operator workload. The response to this problem is to develop continually more sophisticated decision aids to predict the requirements of the user and then display recommendations (Reising, Emerson, & Munns, 1993). This subject is addressed later in the chapter.

The current generation of display devices is typically 6 × 8 or 8 × 8 inches. This is a halfway house to our ultimate goal, but already we are confronting some of the problems associated with freedom of design. There is a continual struggle between the mission planners who wish to use the now flexible displays for the portrayal of tactical, mission-oriented data and those designers concerned with the safe operation of the aircraft from a fundamental instrument flying point of view. The latter see the real estate previously dedicated to primary flight instrumentation now being shared with, or usurped by, "secondary" displays. There are still many questions to be answered concerning the successful integration of the various display types. It is essential that the operator maintains situational awareness both from a battle management perspective and from a basic flight control standpoint.

Further freedom is offered by the advent of E-O technology in that the cockpit need no longer be positioned at the front of the aircraft offering a direct view forward. These displays are capable of taking real-world images from a variety of sources and assimilating them with data symbology. In the military sphere, there are distinct advantages to be gained from blocking the outside view. These are mainly concerned with the vulnerability to laser attack weapons. Indeed, if we can supply the operator with an enhanced, or even virtual, view of the world, do we need instruments in the conventional sense?

It is clear that there are a great many paradigms to be broken. To a large extent we have followed the design precedents set when displays were constrained by mechanical limitations. This will change as a greater body of research is developed to indicate the way in which the human will respond to the E-O technology. Indeed, in the same way that the advent of faster aircraft forced the display designer's hand at the start of the electromechanical era, it could well be the introduction of the new generation of high-agility fighters, capable of sustained angles of attack in excess of 70°, that will force the full exploitation of electronic media. Time will also see the growth of a population of operators not steeped in the traditional designs, thus allowing a more flexible approach and less of a penalty in terms of retraining.

As always, the role of the designer is to provide the operator with the information needed in the most intuitive and efficient manner. The difference now is that the interface can be designed to meet the requirements of the human, without the human having to be redesigned to meet the engineering constraints of the system to be controlled.

DISPLAYS AND CONTROLS

As the E-O era unfolds, active matrix liquid crystal displays (AMLCD) are the leading multifunction contender, but other flat panel display technologies, including a "flat" cathode ray tube (field emitter display), have the possibility of taking a significant portion of a future market (Gray, 1993). Coupled with advances in displays are advances in control technologies, such as touch and voice control. A third component, the "behind the panel" advances in graphic generators and aircraft digital avionics systems, is also proceeding apace, so that all three components can give the pilot an effective means of interacting with the aircraft through its pilot–computer interface. Projections are for this trend of advances in all three areas to continue until synthetic cockpit environments are created by virtual reality techniques, although this will not occur until well beyond the year 2000 (Grossman, 1991; Oliveri, 1994).

Even before the virtual cockpit becomes a reality, more near-term display and control technology has the potential of providing a substantial increase to the pilot's efficiency. Translating that potential into actuality is, however, another matter. The very nature of the interaction with the controls and displays has changed. New control modes, coupled with the intuitive nature of the displays' graphic formats, have the possibility of making the pilot–computer interface, for all practical purposes, transparent. In this section are examples of research studies that address a major issue in the crew station design world, that is, how the pilot might take advantage of the unique opportunities offered by glass cockpits and helmet-mounted displays with computer-generated display formats and new subsystem control modes.

Head Up Displays and Standardization

As designers become more and more challenged by the advent of new control and display technologies, display designs have become abundant. Every air-framer has his or her own version of tactics displays, situational awareness displays, map displays, and head up display (HUD) symbology. On the one hand, the copious formats allow for creativity and invention of new ways to display important information; on the other hand, pilots lack transfer of training from one aircraft to the next. Each new cockpit suite poses new display formats for the pilot to learn and become proficient with in a short period of time. Because of this dilemma, there has recently been an emphasis on standardizing certain display formats—especially the HUD format, because it is capable of being a primary flight instrument. The standardization of the HUD symbology will allow pilots to maintain familiarity with the symbology regardless of the aircraft they fly.

The HUD evolved out of a need for a display that referenced the outside world and could be used for weapon aiming purposes. At first this consisted of a simple weapon aiming reticule, but it quickly developed into a more sophisticated projection device through which the user could correlate the position or vector of the airframe or weapon with the outside world. Although the HUD started its evolution as a weapon aiming device, it did not take long for the community to realize that a great deal of information could also be displayed to aid with the basic control of the aircraft. As the displays matured over the years, data was added to the HUD in a piecemeal fashion without any central coordination or philosophy. This haphazard growth resulted in a great deal of diversity in the design. In 1991 the U.S. Air Force (USAF) started a program to develop and test baseline formats for its electronic displays. The first phase of work led to a published design of HUD symbology for fighter-type aircraft (Mil-Std 1787B; Department of Defense, in press), and work is currently ongoing for head-down displays in transport aircraft. The aim is to define tested designs for all electronic media in USAF aircraft to form the basis for any future development work.

Transport Aircraft HUDs

Although developed originally for use in fighter-type aircraft, HUDs have recently been incorporated into transport aircraft, both military and civilian. In the civilian transport arena, the primary reason for including a HUD is to enable takeoffs and landings in low visibility conditions. Alaska Airlines has led the way with the incorporation of HUDs into their 727s. "With the HUDs, Alaska can go down to Cat IIIa landing minima on a Cat II ILS beam" (Adams, 1993, p. 27). Because of this advantage in operating under adverse weather conditions, other airlines are also considering incorporating HUDs into their fleets.

As far as traditional military transports are concerned, the C-17 is the only current transport to incorporate a HUD. The primary use of a HUD is to aid in visual approaches to austere fields that possess little or no landing guidance. An additional use is to aid the pilot in low-altitude parachute extraction maneuvers that require steep angles of descent.

Helmet-Mounted Displays

The advantage of the HUD is that it does not require users to bring their eyes into the cockpit. It also provides the real-world correlation mentioned earlier. However, one of the limitations of the HUD is its field of view. Pilots can only benefit from the HUD's information when looking straight ahead. Because of this limitation, there has been a push in the helmet-mounted display (HMD) symbology arena so pilots can constantly benefit from the primary flight information superimposed on the real world—regardless of where they are looking. The HUD's field of view is typically 30° in the horizontal. It is thus not possible for information (or weapon aiming reticules) to be presented to the operator outside this limited field of view. Helmet- or head-mounted displays projected onto the visor or onto a combining glass attached to the helmet have been developed to overcome this problem. By the use of miniature display technology producing a display for each eye, combined with accurate head and, in some cases, eye-pupil tracking, it is theoretically possible to present a stereoscopic, full-color image to the user in any direction (Adam, 1994). This could be anything from a simple overlay of information on the outside scene to a totally artificial virtual image. Although the technology of displays and processing speed of graphics generators has not yet matured to the point where the virtual world image is possible in an aircraft environment, it will undoubtedly materialize in the near future.

The first military group to embrace HMD technology has been the rotor-craft community—of the approximately 500 aircraft utilizing HMDs, most are helicopters (Lucas, 1993). Clearly the field of view limitations of a conventional HUD are raised to a new level of significance where the aircraft is capable of moving sideways and even engaging in reverse! Similarly, it is likely that the new generation of high-agility, fixed-wing fighters, capable of sustaining angles of attack in excess of 70°, will present challenges that may only be satisfactorily addressed by the adoption of HMD technology.

Besides the use of HUDs and HMDs, the E-O era has also created a head-down electronic blackboard on which almost any display format can be drawn. Because of their versatility, the head-down displays can be configured in nontraditional ways. The trend of duplicating the E-M instrumentation with an E-O display format may finally disappear as a paradigm shift takes place and the full potential of E-O displays becomes apparent. The research described next gives an example of an investigation aimed at taking advantage of the digitally based displays.

Background Attitude Indicator

This study deals with one of the basic aspects of flying—maintaining flight safety when there is no dedicated head-down primary attitude indicator. If one grants the premise that the more mission-related information the better, the logical conclusion is that all the glass displays in a modern cockpit should contain this type of information, with the baseline MIL-STD HUD (Department of Defense, in press) used as the primary flight display. Because of this idea, the elimination of a dedicated head-down primary attitude indicator would free up head-down real estate for mission-related glass displays. Loss of attitude awareness (a potential flight safety problem) could result when the pilot is focusing head down to do mission-related tasks.

This problem was investigated by researchers at Lockheed–Fort Worth (Spengler, 1988), who created a background attitude indicator (BAI) using only a ¾-inch "electronic border" around the outer edge of the display (Fig. 14.2). The three displays on the front instrument panel presented mission-related information on the central rectangular portion of each, and presented, on background border, a single attitude display format that extended across *all three* displays. The attitude information, in essence, framed the mission essential display information and acted as one large attitude indicator (Fig. 14.3). The BAI consisted of a white horizon line with blue above it to represent positive pitch, and brown below it to represent negative pitch. This display worked very well for detecting deviations in roll, but was less successful in showing deviations in pitch because, once the horizon line left the pilot's field of view, the only attitude information present in the BAI was solid blue (sky) or brown (ground). Because the concept was effective in showing roll deviations but was lacking in the pitch axis, enhancing the pitch axis became the focus of work done at Wright Laboratory's Cockpit Integration Division, Wright Patterson Air Force Base, Ohio.

The lab's initial work began by enhancing the pitch cues for a BAI that framed *one* display format only (as opposed to framing three display formats as in the original Lockheed work) (Liggett, Reising, & Hartsock, 1992). The lab's BAI contained wing reference lines, digital readouts, and a ghost horizon (a dashed horizon line that appeared when the true horizon left the pilot's field of view, and that indicated the direction of the true horizon) (Fig. 14.4). The BAIs also contained variations of color shading, color patterns, and pitch lines with numbers.

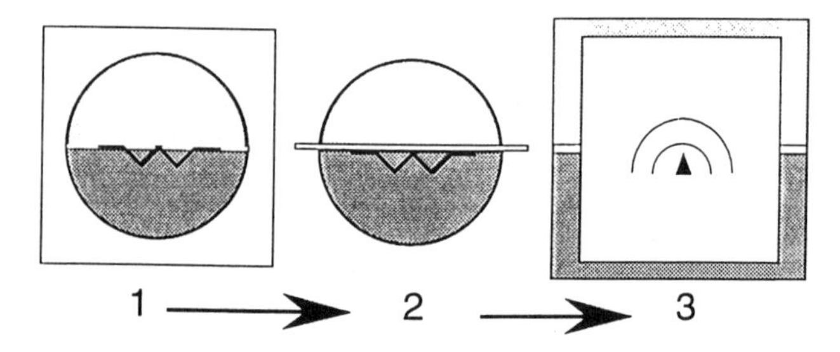

FIG. 14.2. Evolution from attitude director indicator to background attitude indicator.

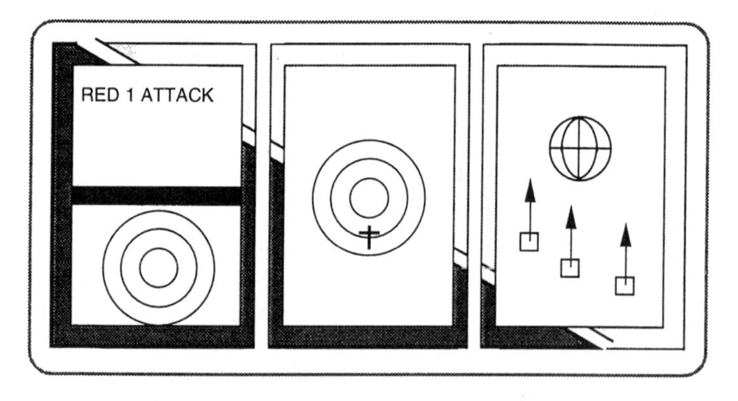

FIG. 14.3. Spengler background attitude indicator.

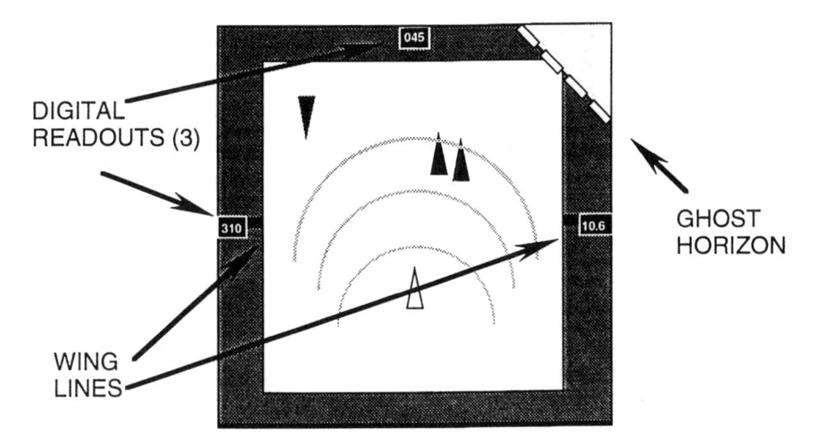

FIG. 14.4. Wright Laboratory's background attitude indicator.

Experimental results revealed that the combination of color shading and color patterns (Fig. 14.5) was the format that provided the pilot with the best performance when recovering from unusual attitudes. When using this format, the pilots moved the control stick to begin their successful recoveries more quickly than when using any other format. This measure of initial stick input time relates to the interpretability of the format because the pilots looked at the format, determined their attitude via the cues on the BAI, and began their recovery as quickly as possible.

The design ideas from the initial Wright Laboratory study were transferred to framing three displays as in the original Lockheed work to provide the pilot with one large attitude indicator, which pilots highly favored. This display provided effective peripheral bank cues, as well as two types of pitch cues—the shaded patterns supplied qualitative cues, whereas the pitch lines with numbers gave quantitative indications of both the degree of pitch and pitch rate information. Based on the results of these simulation studies, background attitude indicators appear to be a viable means of enabling the pilot to recover from unusual attitudes.

This research does indeed proclaim a paradigm shift from the old way of displaying attitude information head down on a dedicated piece of real estate for an ADI, to an

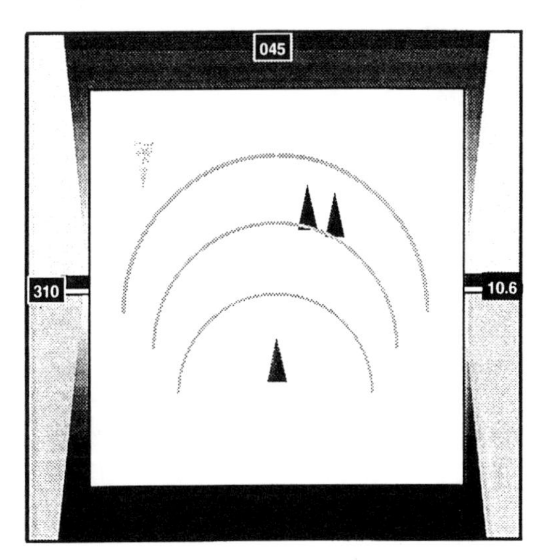

FIG. 14.5. Background attitude indicator with color shading and patterns.

innovative new way of displaying the same information. Another prime example of a paradigm shift is the use of three-dimensional (3-D) stereo display formats. MFD displays with 3-D computer graphics have the potential of creating map formats that closely match the essential three-dimensional aspects of the real world. The next study deals with how pilots would control objects within a 3-D map.

Cursor Control Within 3-D Display Formats

Mental models play an important role in the efficient operation of systems (Wickens, 1992). A mental model is the picture operators have in their heads of the way a system operates. Because direct views of the inner workings of a system are often not possible (e.g., the flow of electrons inside the avionics system), displays are a major means of conveying the operation of a system. The closer the display formats conform to the user's mental model, the more beneficial they are. In the airborne arena, the pilot is operating in a three-dimensional world; consequently, the more accurately a display can portray this three-dimensional aspect, the more accurately it can conform with the pilot's mental model.

A perspective view of terrain features for low-altitude missions will aid pilots because this view conforms very well to their three-dimensional mental model of the world. Perspective views, however, only contain monocular depth cues. Adding 3-D stereo cues can enhance agreement between pilots' mental model and the actual display by making it more representative of the real world.

Once a three-dimensional perspective map is created, an obvious question is, how does the operator manipulate a cursor in the 3-D map world? Moving a cursor to mark items is one of the most important tasks involved in using map displays. The operator may be required to mark geographic features such as hilltops or river bends, as well as man-made features such as dams or bridges. The 3-D perspective view can be interpreted as x, y, and z coordinates. The problem now arises as to how to move a cursor to areas of interest in these displays.

The lab's research in this area has focused on two types of continuous cursor controllers (a joystick and a hand tracker) and one discrete controller (a voice control system) to manipulate a cursor in 3-D space so as to designate targets on a map. The joystick and hand tracker had been used in previous 3-D research (Ware & Slipp, 1991), and the voice was chosen based on researchers' experience with it in the two-dimensional (2-D) arena.

Based on previous research in the cursor control area (Reising, Liggett, Rate, & Hartsock, 1992) it was determined that using aiding techniques with continuous controllers could enhance the pilot's performance when designating targets. This study investigated two types of aiding. *Contact aiding* provided subjects with position feedback information via a color change in the target once the cursor came in contact with it (Fig. 14.6). This aiding eliminated some of the precise positioning necessary when using a cursor to designate targets. *Proximity aiding* (Osga, 1991) used the Pythagorean Theorem to calculate the distance between the cursor and all other targets on the screen. The target in closest proximity to the cursor was automatically selected; therefore, the requirement for precise positioning was completely eliminated.

The display formats consisted of a perspective view map containing typical features, targets, and terrain. The targets could be presented in different depth volumes within the 3-D scene (Fig. 14.7).

Subjects designated targets significantly more rapidly with proximity aiding (with the hand tracker or joystick) than when using either voice or contact aiding (with the

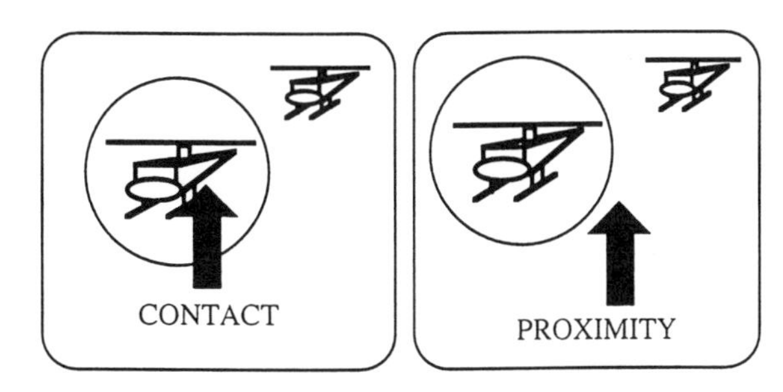

FIG. 14.6. Types of aiding. Solid circle indicates selected target.

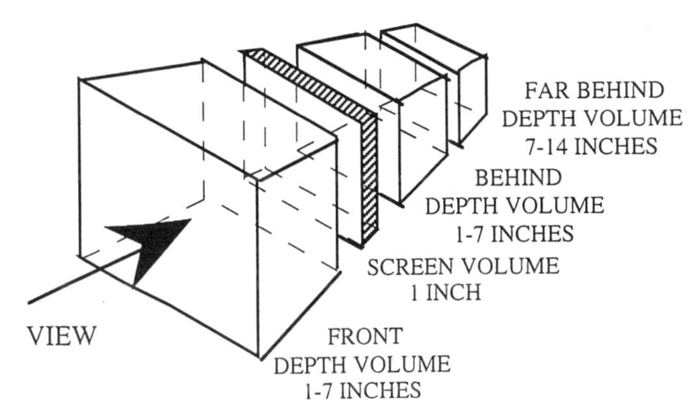

FIG. 14.7. Depth volumes within the 3-D scene.

hand tracker or joystick) (Fig. 14.8). When using a continuous controller, there are two components to positioning: gross and precise movements. The addition of proximity aiding to both continuous controllers greatly reduced gross positioning and eliminated precise positioning. Contact aiding, on the other hand, did not affect gross positioning but decreased the amount of precise positioning.

Another interesting finding was that the voice control system performed significantly better than either of the continuous controllers with contact aiding. The reason for superior performance of the voice control system relates to the components of the positioning task. The continuous controllers with contact aiding both had gross and

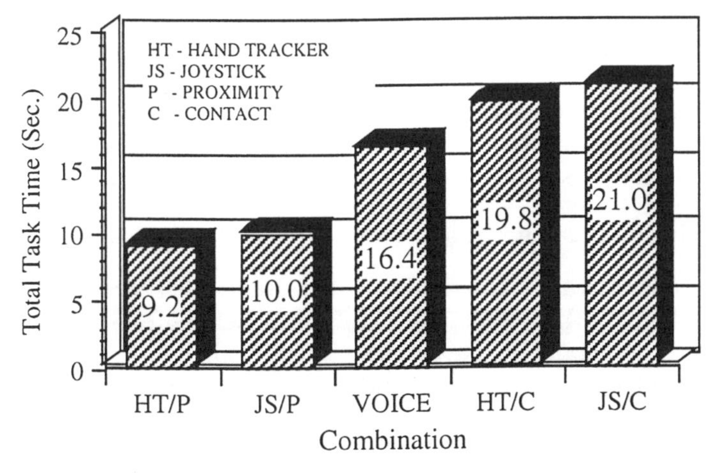

FIG. 14.8. Effect of proximity and contact aiding on target designation times.

fine positioning to deal with. The voice control system and the controllers with proximity aiding, however, eliminated the fine positioning factor to a large extent. Because the target was large enough to visually identify in all cases, the movement to the target was basically reduced to a gross positioning task, and fine adjustment was eliminated.

Because of this strong potential payoff for voice, voice systems were examined more closely, especially with regards to their recognition accuracy. The next section describes a study designed to test the International Telephone & Telegraph (ITT) system's connected speech capabilities.

More Natural Language for Voice Command

Unlike discrete speech, in which a pause must occur between each word, connected speech recognizers do not require pauses. The result is a more natural speech for pilots—it's the way they normally speak while operating in the airborne environment. In this study, test phrases were divided into three groups: *complex* phrases, which contained more than five words, for example, "North three four seven six point one two," and two groups of *simple* phrases, which contained five words or less, for example, "Configure for landing." The simple phrases were subdivided into *alternate* and *no-alternate* phrases, depending on whether or not the particular phrase was the only phrase in the entire set of phrases with a particular meaning. For example, "Select Maverick" is a simple no-alternate phrase because there is no other phrase that results in a selection of the Maverick missile. However, "Set up landing" is a simple alternate phrase because "Configure for landing" means the same thing.

Three types of accuracy were analyzed. Word accuracy consisted of the ratio of total words correctly recognized divided by the total words presented. Phrase accuracy consisted of the ratio of total phrases correctly recognized divided by total phrases presented. Intent accuracy consisted of the ratio of total correct actions divided by the total actions presented. For example, if the phrase presented was "Configure for takeoff" and the phrase recognized was "Set up takeoff," the word accuracy would be 33%, the phrase accuracy would be 0%, but the intent accuracy would be 100% because the action carried out by the phrase "Set up takeoff" is the same as the action carried out by the phrase "Configure for takeoff."

Results showed that when phrase accuracy was considered, simple no-alternate phrases had significantly higher recognition rates than either the simple alternate or the complex phrases (Fig. 14.9). However, simple alternate phrases had the highest recognition accuracy when the intention of the phrase was the measure of success (Fig. 14.10). For a more detailed discussion of the results of this study, see Barry, Solz, Reising, and Williamson (1994).

Simple alternate phrases were created to provide pilots with multiple speech options in accomplishing the tasks, thus providing a more natural interface to the cockpit.

Overall Thoughts on the Benefits of New Cockpit Technologies

The research studies discussed thus far show that HMDs, HUDs, and head-down MFDs, both 2-D and 3-D, have the potential of significantly helping pilots efficiently maintain adequate attitude and situational awareness. These display technologies, coupled with innovative control technology such as hand trackers and voice, allow pilots to continue to utilize these displays for traditional purposes, such as target designation. However, this technology by itself is no panacea; in fact, if not implemented in an intelligent manner, it could become a detriment to the pilot. The designers still need to spend the majority of

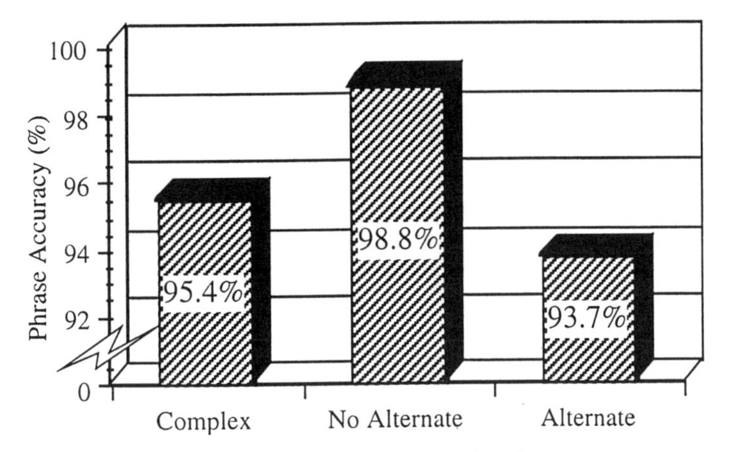

FIG. 14.9. Phrase accuracy comparison for phrase types.

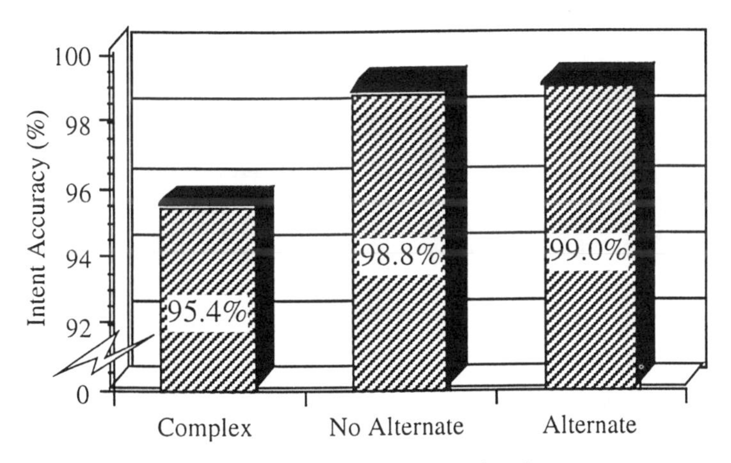

FIG. 14.10. Intent accuracy comparison for phrase types.

their time figuring out how the subcontrol modes, coupled with the myriad of possible formats, "play" together to present pilots with a clear picture of what the aircraft is doing and how to change its subsystems, if required. This new technology is a two-edged sword—it offers the designers virtually unlimited freedom to present information to pilots; on the other hand, it also gives them the opportunity to swamp pilots in data. The clever application of the glass cockpit will be the key to ensure that technology helps, rather than hinders, pilots.

The intelligent design of these controls and displays, and their integration into cockpits, can be facilitated by using a structured design process and taking advantage of the computer aided design tools that complement the process. The next section covers the design process and its supporting design tools.

CURRENT CREW STATION DESIGN

The overall design process invoked in human–machine systems is well documented (Gagne, 1962). A paradigm specifically related to the crew station design process for aircraft is shown in Fig. 14.11. It consists of five steps: Mission Analysis, Preliminary

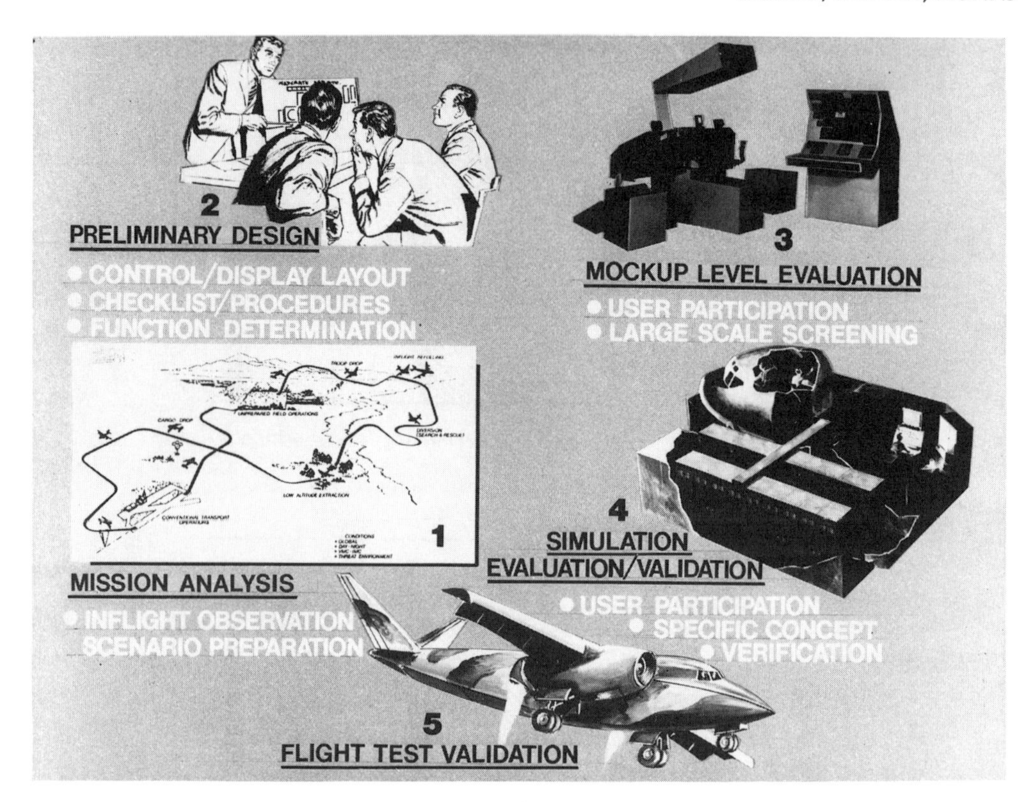

FIG. 14.11. Crew system design process.

Design, Mockup Level Evaluation, Simulation Evaluation/Validation, and Flight Test Validation (Kearns, 1982). The steps in the figure are numbered numerically to show the order in which they should be addressed. The order should be followed to ensure a good design. Before the process is described in detail, the design team, or players who participate in the design process, are discussed.

The Process and the Team

The Design Team. To be successful, each step in the process needs strong user involvement. A multidisciplined design team is formed to follow the design from birth to implementation. Certain players take the lead during different steps of the process. The team should include, as a minimum, pilots, design engineers, avionics specialists, human factors engineers, computer engineers, hardware specialists, and software specialists. Participation from each of the players throughout the process will allow for a more thorough design of the system. The ultimate goal of the design team is to "get it right the first time."

Mission Analysis. The first step, Mission Analysis, is often referred to as problem definition because it specifies a problem with the current system that needs to be solved, or it identifies deficiencies in the crew station where a problem may occur without the incorporation of a new system. This step is initiated with a thorough examination of the intended operational use of the system to be designed. This examination is followed by a derivation and documentation of total system and individual component requirements. The Statement of Need or requirements data published by the future user of the system provide important baseline material for this step. Typically, the documentation produced during this step includes a mission profile describing a sequential listing of all the operations the system must perform in order to be effective in the flight environment. This profile is decomposed from a very generic state of aircraft operations to a very detailed state that includes all of the specific tasks performed by the aircraft, its systems, and each of the crew members during the mission profile (ORLOC, 1981). To augment this data, the design team may also perform an analysis of the decisions that have to be made by crew members as the mission progresses. An essential output of this step is the identification of the information that the crew needs to perform its mission.

Preliminary Design. The second step in the cockpit design process, as depicted in Fig. 14.11, is Preliminary Design. This step is often referred to as "develop a solution." During this part of the process, most of the activity is devoted to generating a design. The requirements generated in the first step are reviewed, and decisions are made regarding how the functions necessary to fly the mission profile will be performed. The functions can be allocated to the pilot, the computer, or a combination of both. A series of trade studies is often performed to determine who will do what. These studies will determine the level of automation within the cockpit and will play a major role in the cockpit design. The cockpit design will also be driven by the information requirements determined from step 1. A key element in the evolving design is operator and user involvement. The sustained participation of operators with relevant experience results in fewer false starts, better insight in how and why the mission is performed, and a great savings in time, as well as money, in the latter steps of the process. By getting the operator involved from the beginning, the costly problem of making design changes further down the road is avoided.

The dividing line between problem definition and solution development is often vague. Specific designs may affect task sequencing during the mission profile. This change in sequencing can reveal workload problems within the cockpit. Because of this overtasking, the pilot may shed tasks, which in turn alter the mission profile. Once the profile has changed, the designs may affect the tasks in a different way, and thus the cycle begins. The design process is indeed an iterative process.

Mockup Evaluation, Simulation Evaluation/Validation, Flight Test. The last three steps are interdependent but are all very critical to the successful completion of an effective and proven cockpit design. These three steps all work synergistically to "prove the solution." Mockup Evaluation marks the initial introduction of the implemented design concepts to the user. Although the users should be involved for the preliminary design step, the actual implementation into a mockup will show the design in a whole new light. The design concepts are evaluated in a limited context, and suggestions are made by the user as to which designs should move forward to simulation. This step weeds out unfeasible design concepts. Simulation Evaluation provides a more realistic and robust testing of the design concepts. It is common in simulation evaluation to compare the design concept to an existing design in order to measure the "goodness" of the design concept. This step provides the final recommendation of a design concept for flight test. Flight testing typically involves only one design to be tested in operational use. For the purpose of this discussion, these final steps are combined to provide "Solution Evaluation." Once again, there may not be a clear break between the solution evaluation and the solution definition step. It has been observed that most designers design, evaluate, redesign, and so on as they go. The transition from solution definition to solution evaluation occurs when formal, total-mission, total-system, pilot-in-the-loop evaluations begin. But even then, decisions made during the problem and solution definition steps are often revisited, changes made, and simulation sessions (or even flight tests) rescheduled—all resulting in, as previously suggested, a very iterative or cyclic process.

Traceability. As the process evolves, it is important that the design team maintain an accurate record of the changes that have taken place along the way, the decisions that were made that influenced the design, and the rationale behind their decisions. This information provides traceability of the design from requirements to final product. Traceability is important because the design process can take a long time, and it is helpful to know why things were done the way they were. The traceability document provides a record of past decisions, which may be reviewed periodically, so the design flows in an evolutionary manner, as opposed to a revolutionary manner, and thus, avoids regression. Also, the design of a new product can benefit from the traceability information of previous products, thus saving time and effort. This discipline of documenting the design is (or should be) a MUST feature of the design process, not a "nice to have" feature.

Facilitating the Crew Station Design Process With Computer Support

The preceding discussion serves as a guideline for crew station designers. The crew station design process, in some form, has been employed for many years. With the advent of computer-aided design tools, this process has been complemented. These new tools allow designers to visualize and modify their design ideas much easier than the traditional way of hand-drawing design concepts.

Design Tools. One popular design tool, Computer-Aided Three-Dimensional Interactive Application (CATIA), can assist with the stages of product design, development, manufacture, and delivery, while improving product quality and saving money. Dassault Systèmes, Paris, France, designed and developed CATIA, and the system is marketed and supported worldwide by IBM. The latest, Version 4.0, supports the entire life cycle of a product. CATIA has an integrated approach to the entire product design, and because of this it is internationally recognized as an industry leader (EDGE, 1993). A key aspect of this tool is that it allows everyone on the design team access to the same data in a common format with all updates. This facilitates concurrent activity among the design team, which speeds up the entire process.

CATIA has played a major part in the design of the Boeing 777 (Hughes, 1994). Not only does it check the physical layout of parts of the aircraft, but it is also used by pilots, safety experts, and the airline advisory committee to perform normal and emergency procedures. CATIA uses its three-dimensional human models to test and evaluate these procedures. Additionally, CATIA facilitates the use of digital mockups that can eliminate the need for physical mockups on sections of the aircraft, which results in a significant cost savings (Rich, 1989).

Along with CATIA, JACK is another computer package that contains an interactive 3-D human model that can interface with computer-aided design/computer-aided engineering (CAD/CAE) processes. JACK can be used to simulate human performance in a crew station. It incorporates strength modeling, human kinematics, reach envelopes, visibility envelopes, and anthropometric data (Maida, 1993). JACK currently runs on Silicon Graphics IRIS 4D and INDIGO workstations. A PC-based tool similar to JACK is Mannequin. Mannequin can model body positioning, articulation, visual range of motion, and do a torque assessment. It runs without a CAD package, but data files exchange can occur between the more popular CAD packages and the PC-designed Mannequin graphics program (Gross, 1991). For more in-depth model-based design of the crew station, Man–Machine Integration Design and Analysis System (MIDAS) is available: "MIDAS contains tools to describe the operating environment, equipment, and mission of manned systems, with embedded models of human performance/behavior to support static and dynamic 'what-if' evaluations of the crew station design and operator task performance" (Smith & Hartzell, 1993, p. 13).

The tools described, as well as others available, all have the same goal—to assist the designers during the crew system design process. This section was meant to introduce the reader to some available products. Obviously, the list of tools described in this section is not all-inclusive.

Automating the Crew Station Design Process

Although CATIA and similar computer-aided tools are helpful in *assisting* the designer with the crew system design process, there has been a program underway since 1986 that was developed to *automate* the entire crew system design process.

The Crew-Centered Cockpit Design (CCCD) Program (formerly called Cockpit Automation Technology) is developing both a detailed, step-by-step methodology that mirrors the five-step design process described in Kearns (1982), and a set of computer-hosted tools to assist the design team in conducting the steps within the design process. The key to this new automated design technique is a tool called the Design Traceability Manager (DTM). As the name implies, this tool is critical to maintaining a record of the design process as it evolves. The tool, in essence, gathers the methodology in electronic format, and a series of support tools in a central location, and allows the design team to maintain a record of design decisions. Specifically, the tool helps the manager keep track of the crew system design process as it evolves, documenting what activities are in progress, completed, or deleted. The tool also suggests and then accesses appropriate tools to facilitate activities needed to accomplish a section within the crew station design process. Most importantly, the tool facilitates the tracking of all design decisions and the rationale behind those decisions (Garcia, 1994).

The Crew Station Design Process (CSDP). The CSDP is depicted in Fig. 14.12. The major difference between this process and the process in Kearns is the Planning Step. Although this step is inferred in Kearns, it is more important in the use of the automated process because it involves tailoring the entire process to perform only the tasks that are necessary for a specific design. The automated process includes possible activity that may be performed to produce a design. Often, parts of the process have been completed or are not applicable to a specific design. Tailoring the process beforehand will make the design process more manageable and keep the design team focused once the process begins.

The Support Tools. The computerized tools that support the methodology provide assistance in the analysis, design, and test of components with the cockpit system, as well as the cockpit system itself. The current tools employed include a mission

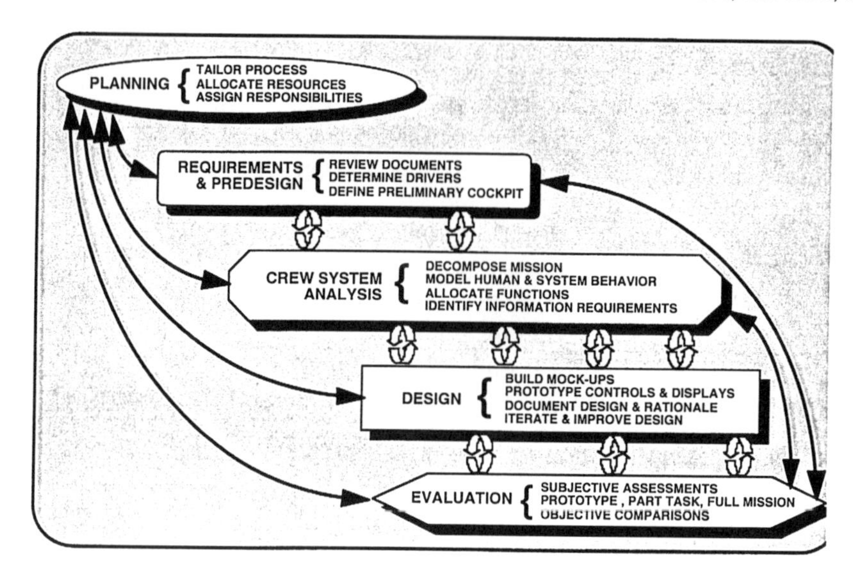

FIG. 14.12. Five steps in the crew station design process.

decomposition tool (MDTOOL), a timeline management tool (TMT), a workload analysis tool, the graphical modeling system (GMS) for rapid prototyping, an automated quality functional deployment (QFD) process for trade-off studies, and an engineering design simulator (EDSIM) (Martin, 1994).

The MDTOOL assists the operational expert in structuring the mission scenario. What was once done on paper charts is now coordinated by a tool that combines aircraft states, location/navigation information, and threat activity. The mission timeline can be decomposed using TMT into phases, functions, subfunctions, and tasks. Once broken down into tasks, a workload analysis can be conducted based on a time-required versus time-available digital workload model. The use of GMS allows the designers to quickly view their design ideas by combining common aircraft components from a database of such contents. This viewing capability facilitates quick changes and component critiques. Once a number of alternatives have been developed and designed to support the given mission, the QFD process is used to determine the most likely candidate for success-based system requirements, equipment constraints, and timing requirements of critical mission phases. The chosen design can be rapidly generated on EDSIM, which is a physical cockpit simulator in which the pilot can interactively test the design concepts in a semirealistic environment. This simulator can be quickly configured to represent numerous design options. Using GMS, the formats can be displayed on CRTs in the cockpit, functioning dynamically as the pilot flies the scenario generated and piped into the EDSIM from MDTOOL. Again, this iterative process continues until one best design is determined and set forward for evaluation/validation.

The use of such tools to support the CSDP illustrate how technological advances have facilitated crew station design. What was once done with paper and pencil by pilots and designers sitting around a table is now done by designers sitting at a computer. These tools allow the designer, with inputs from pilot experts, to design and redesign the specifics of a display, the location of the display, and how the display will accomplish mission objectives. The tools allow real-time evaluation of the design concepts by a graphical interface either on a computer monitor, at a desk, or a monitor in a breadboard simulator.

With the advent of these tools to support crew station design, it may be possible once again to take a design from original idea to testing, in simulation if not in flight, in a matter of weeks or less.

Summary of CCCD. The process, supported by the computer tools, works together to provide a coherent and traceable design. According to Storey, Roundtree, Kulwicki, and Cohen (1994, p. 690), "the result of a good cockpit design should include 1) a comprehensive set of understandable activities that are completed to produce the design, 2) a fully explained tradeoff rationale that lead to the design, and 3) an understanding of why the design is good." The CCCD Program is trying to provide designers a method to accomplish this goal.

The design of the crew station is intimately tied with its control and display technology. In the crew stations of today's highly complex airborne systems, the design is also inextricably interwoven within the automation philosophy used in the creation of the system. Although flightdeck automation is the subject of another chapter, it is also discussed in the next section because of the unique requirements of the military environment to which it is applied.

AUTOMATION IN MILITARY AIRCRAFT

As the design of crew stations evolves, the tasks necessary to successfully complete the mission are being identified. To relieve workload problems, these tasks must be divided between the pilot and computer, and an optimum balance should exist between the two. As shown by research within the artificial intelligence community, the on-board computer, or electronic crewmember (EC), is becoming more capable of accomplishing tasks for the pilot. The issue now centers on how much automation is optimum. This section discusses issues of automation, the teaming of the pilot with the EC, and team trust in the military flight environment.

The military flight environment faces some unique circumstances not encountered in the civilian flight arena. The fighter/attack aircraft, with one pilot, flying very low, and attempting to locate a target at night, in adverse weather conditions, is one of the most difficult cases. Under these circumstances, automation becomes a key enabling technology, but questions such as, "What functions should be automated?" and once decided, "Can they be automated?" must be answered before automation can be successfully employed.

Much has been written about the effects of automation on crew workload and flight safety in the commercial airline world. Although statistics provide strong evidence that civil air transportation is very safe, the two most publicized automation issues, or problem areas, have not been solved. These are the role automation should play during heavy-workload, high-density operations in airport terminal areas, and on the opposite end of the workload spectrum, how automation should be employed to help in the often very boring long-haul situation. The best example of the latter is a long oceanic flight at night (Last, 1988).

The Air Force, more or less, shares the conventional terminal area and long-haul problems of the civil fleet, as well as the automation problems or issues related to its mission: to fly, fight, and win, in a hostile combat environment.

The Single-Seat Fighter/Attack Aircraft

We are generally familiar with the common tasks and priorities in the commercial cockpit. The overriding requirement is to maintain safe and controlled flight, despite the vagaries of weather, poor visibility, and high volumes of local traffic. Even the demands of systems intended to assist the pilot, such as the air traffic control (ATC) system, can provide annoying distractions and additional workload at critical times. The military environment presents a whole series of additional elements in the equation, and under certain circumstances even changes the basic premise. That is, the overriding aim is not necessarily to maintain safe and controlled flight, but to maintain a balance between survival and the ability to successfully complete the mission (Price, 1992). This places on military pilots unique psychological pressures that are simply not part of the commercial pilot's considerations.

The Mission. The workload demands on military pilots are extremely variable. To illustrate this point, let us consider a typical single-seat, ground-attack mission against enemy high-value targets. The mission actually starts about 3 hr before takeoff time with mission briefing and planning. At this stage crucial routing decisions will be made depending on topography over the route, other friendly forces that have been committed, and intelligence on the position of enemy formations. It is now that many of the preconceptions of the mission that will later color decisions are formed, and yet it is at this stage that automation has been incorporated to a very large extent with the introduction of computer-driven flight planning systems. It is important that the teamwork and trust between operator and machine begin at this point.

The first phase of flight itself consists of loading the aircraft with the necessary preflight data prepared during mission planning. This is followed by takeoff and then a transit toward the target area, usually at high to medium level, in preparation for descent to low altitude, prior to entering enemy territory. This phase may include several rendezvous requirements for force gathering, air-to-air refueling, and so forth. Generally, however, it is a time of relatively low workload, but very high anticipatory stress. This forms the ideal scenario for neglect to perform routine actions.

Ingress to the Target. In order to reach the target the aircraft must cross enemy territory. This typically necessitates crossing the region where the rival land forces are in conflict. This is generally known as the forward edge of the battle area (FEBA). With descent to low altitude and penetration of the FEBA comes the first peak of workload. Upon entering the FEBA, the aircraft may be anywhere between 50 and 500 ft above the ground, probably at night and in less than optimal weather conditions. The pilot is relying on a number of complementary systems working in the visible, infrared, and radar frequencies, all of which have unique properties in terms of the information provided, but none of which is probably capable of providing sufficient information in isolation. In addition to the basic flying task, the pilot is passing over territory with a densely packed assortment of threats arrayed in a confusing and constantly changing tactical situation.

In order to make adequate sense of the environment it is necessary to compare data being received from on-board sensors with datalink information being received from other aircraft and control centers. This combined information must then be compared with on-board databases and mission planning requirements in order to identify any possible conflicts and to develop a coherent picture of the world and its threats, as perceived by the aircraft. The results of comparing and combining disparate types of information must be collated into a coherent strategy for the successful completion of the mission in terms of an optimum route and countermeasures deployment. The data from this myriad of sources are fused into a minimum data set necessary for the pilot to perform the task at hand (Rowntree, 1993).

Having successfully negotiated the FEBA, the next phase is a low-level transit to the target site. It is during this phase that the workload varies the most. In the ideal scenario, the pilot avoids all enemy defense activity and the challenge comes from physically flying the machine at dangerous altitudes in appalling weather at night. In the real world, the pilot will likely be presented with a series of unexpected situations arising from enemy action (unexpected threats, air-defense fighter activity) or possibly his or her own system malfunctions that will require reevaluation of the flight plan and mission capability.

Weapon Delivery and Egress. The mission workload rises to a crescendo over the target area: Aircraft system activity for weapon deployment, a concentration of enemy defense activity, coordinated attack timing considerations, and precision flying requirements for weapon targeting all compound the operator's task.

Egress from the target site, through enemy territory, through the FEBA, and, ultimately, recovery to base, presents essentially the same set of problems and fluctuating workloads to the pilot. However, they are compounded by fatigue and possible battle damage sustained by aircraft during the sortie.

Philosophy of Automation

Early Automation Philosophy. "It appears that the best arrangement is one in which inanimate components work as a team with human operators to provide safe and accurate control and guidance" (Draper, Whitaker, & Young, 1964, p. 5). The key concept in this philosophy is that the operator and machine form a team. The active participation of the operator, a key component of the teaming arrangement, was reinforced in 1966 paper by Knemeyer and Yingling of the Air Force Flight Dynamics Laboratory. They expressed that we cannot expect pilots, in an emergency, to cope with a problem that they have not been following. By monitoring only and not performing, they will fail to notice important information and, consequently, will not react properly. Knemeyer and Yingling went on to assert that automation is desirable, and should exist not as a conflict but as an aid, an adjunct to the pilot.

For several years (early to mid 1960s) the Air Force research and development community pursued a technically sound and logical solution to the pilot/autopilot integration problem. The concept was first evaluated by the Germans during World War II and then introduced in this country with names such as "pilot control force steering" and "force wheel steering." The system linked pilots to the control surfaces through the autopilot by placing electronic force sensors in the control column and rudder pedals of the aircraft. The control pressures applied by pilots were converted to electronic signals, which were sent to the autopilot computer, where they were summed with the commands being provided by the basic flight director system to, in turn and accordingly, move the control surfaces. This system provided the means for the pilot, in an emergency or otherwise, to assume control of the aircraft smoothly and in a conventional fashion without having to uncouple or disengage the autopilot. To a certain extent, this concept has been overcome by the introduction of computer controlled "fly-by-wire" flight control systems in which the pilot provides inputs to the flight control computer, which sums them with inputs from the aircraft's attitude sensors and manipulates the control surfaces to provide the required flight vector. However, this does not provide the pilot with the degree of autopilot control visualized in the original "force wheel steering" concept.

Today's Automation Philosophy

Despite the work in the early 1960s, the team arrangement design philosophy has rarely, if ever, been carried out in implementation of automation in today's aircraft systems. In order to create teamwork, the designer must examine the roles of human and machine in the system (Gagne, 1962). One of the key components in this process is functional allocation between the human and the machine. Ideally this division of responsibilities between the two "team members" occurs by taking into account the strengths and weaknesses, workload limitations, and so forth of each and then assigning roles accordingly. However, in actual system design that is not how the process usually occurs; functional allocation is largely a myth and rarely applied in system design and development (Fuld, 1993). What basically happens is that everything possible is automated, and the human gets left with doing what the machine cannot do, or fails to do, because of a malfunction.

"Somewhat paradoxically, machines that can do more, and do it faster, provide the basis for systems that are increasingly demanding of the human operator, particularly in terms of cognitive requirements" (Howell, 1993, p. 235). The demand comes about because the operator is not "in the loop," but rather is a bystander—as long as the system functions normally. When emergencies occur, the operator is expected to take control of the system, diagnose the problem, and bring the system back to its nominal state. However, as was discussed in the introduction to this chapter, a design driver should be to make the operator an integral part of an automated system and not just an observer.

Future Automation Philosophy. It is interesting to note that nearly 30 years after the teamwork automation philosophy was espoused, it has once again come to the forefront. The current term used is *human-centered automation* (Billings, 1991), which starts with the operator as the heart of the system and then incorporates the automation. From the operator's point of view automation is designed as it should be—to augment or assist operators in areas where they show limitations (Wickens, 1992). Although this automation philosophy is consistent with earlier years, the implementation is much more difficult because many more avionics systems are contained aboard modern combat aircraft. On the other hand, because of present-day capabilities in computers and software, the resulting product can be much closer to a true team. Operator–machine relationships are being created that emulate those occurring between two human crewmembers—mutual support and assistance. As the sophistication of decision aids increases, the result will be an electronic crewmember (EC).

Human and Electronic Crewmembers Are a Team

In order to function effectively, the pilot and the EC must work together as a close-knit team. The ideal relationship between pilot and machine can be likened to that of good managers and their staff. The manager must be sufficiently aware of the work of the staff to be able to predict problems, but not so involved that their work is hindered.

The manager must be involved enough to be able to offer assistance when called on, and yet must not micromanage and risk becoming overloaded and prevented from making strategic decisions. The good manager will know which staff members can be relied on to act without supervision, as pilots will form opinions of which of the aircraft systems do not require frequent attention. As in the conventional management situation, the aircraft system must maintain a knife-edge balance of providing sufficient data exchange without swamping the pilot system manager, and achieving sufficient autonomy without alienating the manager.

It must be remembered that the EC will be performing many of the functions now allocated to the second human crewmember in today's military attack aircraft, such as the Royal Air Force's GR-1 Tornado and the U.S. Air Force's F-15 Strike Eagle. The effectiveness of these aircraft is significantly affected by the degree to which the two human crewmembers become an integrated team. Southeast Asia experience with F-4 Phantoms demonstrates this statement. With constantly changing crewmembers, a conscious effort to build up the team relationships was critical. A problem arose because crew rest, illness, sortie generation requirements, and individual rotation schedules precluded the same team from flying together on each mission. The result was that often a pilot and a weapons system officer, who had not worked together before, were paired for a particular mission. Time and again it was proven that learning the individual aspects of each team member became an important factor in a successful mission.

Team Trust

One essential feature of a successful team is trust in the other partner. This, in turn, implies that the partner behaves in a rational and reliable manner. One partner cannot initiate actions that, even though they are logical to it, appear to be illogical to the other. In order to avoid arbitrary actions, there must be some overall governing rules that provide the logical structure under which both members operate. As examples of explicitly stated governing rules, consider the three laws of robotics (Asimov, 1950):

1. A robot may not injure a human being, or through inaction, allow a human being to come to harm.
2. A robot must obey the orders given to it by human beings except where such orders would conflict with the first law.
3. A robot must protect its own existence as long as such protection does not conflict with the first or second law.

These rules provide the guidance required to allow the robot to perform its job in a reasonable and consistent manner. If the word "pilot" is substituted for the word "human" in the preceding example, a possible basis for governing the behavior of the EC exists. The three laws stated here are only examples of governing rules, and they would require major changes to be applicable in a military setting. For instance, without modification, the ideal robot would not allow pilots to take off, knowing that they were deliberately going in harm's way. The point is that rules of this type provide the basis for consistent behavior for the EC and thereby provide a basis of trust for the pilot. It is through this trust that an effective team can be built. Trust, however, is not acquired instantaneously; it must be built up gradually. Trust can be envisioned to develop in three stages. At first, trust is based on the predictability of individual behaviors. In the second stage, trust is based on dependability. "Dependability may be thought of as a summary statistic of an accumulation of behavioral evidence, which expressed the extent to which a person can be relied upon" (Muir, 1987, p. 532). In the third stage of trust, faith is the major component because one team member is willing to bet that the other member will be dependable in the future.

Once trust is built between the crewmember and the EC, the continued overall efficiency of the system depends on such factors as machine accuracy, compliance with the suggestions of the EC, and degree of faith in the continued accuracy of the decision aid (Riley, 1989).

Summary of Automating the Cockpit

The need for teamwork or a teaming arrangement between the crew and the automatic control system spans the time from when automatics were first introduced into aviation until the present introduction of modern computational capabilities including artificial intelligence in the form of decision support systems. Today's avionics designers are replacing their concept of an "autopilot" with high-technology support concepts such as the pilot's associate or the electronic copilot or, as we have referred to it, the electronic crewmember. International conferences, sponsored jointly by the air forces of Germany, the United Kingdom, and the United States were held in Germany in 1988 and in 1990 to address important issues relating to the human–electronic crew (Emerson, Reinecke, Reising, & Taylor, 1989, 1992). The first meeting dealt heavily in defining the state-of-the-art of artificially intelligent systems for airborne use and defining the areas that required concentration of research and development resources in order to bring reality to the concept. The second conference assessed the progress made during the previous 2 years, relative to applying artificial intelligence into the cockpit. The papers presented indicated that significant progress had indeed been made. Also examined was the teamwork issue and the question, "Was the team maturing?"

One of the conclusions of these conferences was that, although some progress has been made in creating teamwork, a fully functioning EC probably will not reach maturity until well into the next century. Current research is being oriented toward smaller, achievable parts of an EC. It may take quite a while to build the team, but the performance capability of the team has the potential of being orders of magnitude beyond current systems.

The EC is but one aspect of future crew stations. Progress in control and display technology will keep apace of automation, and the crew stations of tomorrow may offers some radical departures from those of today. The next section makes some educated guesses at what these crew stations may contain.

WHAT WILL THE FUTURE HOLD?

Embedded Cockpit

In cockpits of many future aircraft, the pilot's forward view may be blocked, at least during some parts of the mission. In the commercial arena the high-speed civil transport (HSCT) is being planned to land without lowering the aircraft's nose, which is currently done in the Concorde (Hatfield & Parrish, 1990; Saville, 1993). The reasons for eliminating the "droop snoot" relate to cost savings achieved by eliminating the weight and complexity associated with lowering the nose. The cockpit impact of not lowering the nose is that the pilots will not be able to see the runway by looking out the front windows; therefore, current designs have only side windows.

In the military arena, the closed cockpit could come about for another reason—laser protection. It has been predicted that lasers of the near future could be used to dazzle, flash blind, or even injure pilots to prevent them from carrying out their missions (Anderberg & Wolbarsht, 1992). In order to protect against lasers, pilots may be required to wear a visor that effectively blocks out the frequency of the laser (Thomas, Cartledge, Grahm, Patterson, & Poe, 1993). Frequency-agile lasers, which cover a wide range of the visual spectrum, are even more difficult to protect against. The result is that measures used to protect the pilots' eyes may cause occlusion of vision, which effectively prevents seeing the outside environment.

Although the reasons are different in the civilian and military arenas, the challenge for the cockpit designer is the same—build a cockpit so that the pilots can safely perform their mission during crucial phases when they cannot see outside. The term *embedded cockpit* implies more than just occluding the pilots' outside vision. Embedded means burying the cockpit within the airframe structure so that the canopy "bulge" disappears. There are aerodynamic advantages to this cockpit design, and, in the case of the military, the additional advantage of laser protection. Therefore, the cockpits of both the HSCT and possible future military aircraft can be described as embedded; however, the HSCT cockpit, even though embedded, would not be fully closed because of the side windows, whereas the military cockpit would be both.

One of the key aspects of safe mission performance for both types of cockpits is obtaining adequate situational awareness. Although there are a number of definitions of this concept, Endsley (1994, p. 31) captured the essence of the idea in her definition: "Situation[al] awareness is the perception of the elements in the environment within a volume of time and space, the comprehension of their meaning, and the projection of their status in the near future."

In the commercial arena, one of the most difficult missions is landing in low-visibility conditions. For military aircraft, locating and destroying a target under adverse weather conditions is one of the most difficult missions. Both missions require synthetically presenting the essential elements of situational awareness, and in the HSCT, this synthetic generation of the outside world will be required all the time because the frontal pilot vision is permanently blocked. The military requirement is not quite as severe because once out of the laser threat environment, protective equipment can be removed (except if the cockpit is embedded). Nevertheless, both military and civilian aircraft will require the creation of situational awareness within constrained environments.

There are a number of new technologies that the crew station designer can use to provide pilots a view of the outside world from the enclosed cockpit. Among these technologies are synthetic vision using 3-D stereo, enhanced vision utilizing various video sensors, and a combination of the two. What impact will they have on future crew stations, and what are the human factors implications? This next section speculates on these, because very little concrete research has been performed in this area.

Synthetic Vision and Enhanced Vision

If the pilot cannot see the outside world, the essential elements of the outside world must be created within the cockpit. The two approaches currently being examined are synthesized vision and enhanced vision. With the synthesized vision approach, the outside world is largely created through computer graphic techniques; that is, information from databases is used to portray the world by rendering (drawing) a representation of the outside world similar to that shown in a very sophisticated, interactive video game. In the enhanced vision approach, various sensors (television, radar, forward-looking infrared radar, etc.) are used to extend the pilot's vision by allowing the pilot to "see" at night, in fog, or during heavy rain. Thus the pilot's vision is extended or enhanced, and becomes effective in regions pilots would not normally be capable of seeing.

Cockpit Configuration

Because the HSCT has no front windows, the entire front "wall" of the crew station can be used as a display. Large-area flat-panel displays, such as AMLCDs, would be the display medium to create these large viewing areas for pilots. However, the exact configuration of the display device depends not only on the field of view provided by large displays, but also on such factors as display resolution and update rate, as well as magnification level of the sensor image (Harris & Parrish, 1992).

Research has been conducted (Haralson, Reising, & Gharyeb, 1989) comparing the effectiveness of large-area display formats. The panoramic cockpit control and display system (PCCADS) was developed to compare a traditional F-15 cockpit suite with a cockpit suite that contained one 10 × 10 display with two 5 × 5 inch displays. Results showed that pilots performed better using the large-area display and also liked the large display formats. PCCADS 2000 compared the two previously mentioned cockpit suites with one large display format. Again, the pilots preferred and performed better with one large display format. The pilots were able to tailor the format by putting information where they wanted it. In this way, they knew exactly where to look to find critical information. This, is turn, enhanced performance.

Another aspect of the large display surface is that it could be three-dimensional (3-D). Current 3-D techniques require some sort of eyeware in order to present the third dimension, but commercial airline pilots would find this headgear objectionable, thus necessitating another way of presenting 3-D information. The answer lies in an autostereo display, which does not require glasses (Eichenlaub & Martens, 1992).

In military cockpits pilots do wear helmets, so presenting information on their visors is a natural means of display presentation. The creation of a "virtual cockpit" on the visor, in 3-D, is a natural follow-on to the fact that pilots cannot see the outside world because laser protection devices prevent outside light from reaching their eyes.

CONCLUSIONS

Aircraft crew stations have progressed from those of Doolittle's day containing myriad mechanical devices to those of today based almost entirely on electro-optical devices where the distinction between controls and displays continues to blur. In addition, automation has progressed from simple autopilots to flight management systems with numerous software decision aids. With the increasing emphasis on unmanned aerial vehicles, there is even some discussion in the military environment as to how many future systems will possess human crewmembers (Wilson, 1995). No matter how this issue is resolved, as long as there are manned aircraft, airborne crew stations will offer one of the most interesting and challenging areas of work for the human factors professional.

REFERENCES

Adam, E. C. (1994). Head-up displays vs. helmet-mounted displays: The issues. *Proceedings of SPIE—The International Society for Optical Engineering: Cockpit Displays*, 13–21.

Adams, C. (1993, November). HUDs in commercial aviation. *Avionics*, pp. 22–28.

Anderberg, B., & Wolbarsht, M. L. (1992). *Laser weapons: The dawn of a new military age.* New York: Plenum.

Asimov, I. (1950). *I, robot.* New York: Fawcett.

Barry, T., Solz, T., Reising, J., & Williamson, D. (1994). The use of word, phrase, and intent accuracy as measures of connected speech recognition performance. In *Proceedings of the Human Factors Society 38th Annual Meeting* (pp. 325–329). Santa Monica, CA: Human Factors and Ergonomics Society.

Bassett, P., & Lyman, J. (1940, July). The flightray, a multiple indicator. *Sperryscope*, p. 10.

Billings, C. (1991). *Human centered automation* (NASA Technical Memorandum 103885). Moffett Field, CA: Ames Research Center.

Department of Defense (in press). *Military Standard Aircraft Display Symbology* (MIL-STD-1787B Draft). Wright-Patterson AFB, OH: Author.

Doolittle, J. (1961). Blind flying Gardner Lecture. *Aerospace Engineering*, p. 14.

Draper, C. S., Whitaker, H. P., & Young, L. R. (1964, September). The roles of men and instruments in control and guidance systems for spacecraft. *Proceedings of the XVth International Astronautical Congress*, Warsaw, Poland.

EDGE. (1993). CAD/CAM/CAE solutions: New IBM offerings for industry-leading CATIA software further speed product design, manufacture, and delivery. *Work-Group Computing Report*, *4*(178), 28.

Eichenlaub, J., & Martens, A. (1992, March). 3D without glasses. *Information Display*, pp. 9–12.

Emerson, T., Reinecke, M., Reising, J., & Taylor, R. (Eds.). (1989). *The human–electronic crew: Can they work together?* (WL-TR-89-7008). Dayton, OH: Wright Laboratory, Advanced Cockpit Branch.

Emerson, T., Reinecke, M., Reising, J., & Taylor, R. (Eds.). (1992, July) *The human-electronic crew: Is the team maturing* (WL-TR-92-3078). Dayton, OH: Wright Laboratory, Advanced Cockpit Branch.

Endsley, M. (1994). Situational awareness in dynamic human decision making: theory. In D. Gilson, D. Garland, & J. Koonce (Eds.), *Situational awareness in complex systems* (pp. 27–58). Daytona, FL: Embry-Riddle Aeronautical University Press.

Fuld, R. B. (1993, January). The fiction of function allocation. *Ergonomics in Design*, pp. 20–24.

Gagne, R. M. (1962). *Psychological principles in systems development.* New York: Holt, Rinehart, & Winston.

Galatowitsch, S. (1993, May). Liquid crystals vs. cathode ray tubes. *Defense Electronics*, p. 26.

Garcia, L. (1994, May). A tool for design traceability management. *Proceedings of the National Aerospace and Electronics Conference*, Dayton, OH, pp. 807–813.

Gray, H. (1993). The field-emitter display. *Information Display*, *3*, 9–14.

Glines, C. V. (1989, September). Flying blind. *Air Force Magazine*, pp. 138–141.

Gross, C. M. (1991, Fall). Mannequin™: Human computer-aided design on a pc. *CSERIAC Gateway*, pp. 11–14.

Grossman, L. (1991, November). Fighter 2020. *Air Force Magazine*, pp. 30–35.

Haralson, D. G., Reising, J. M., & Gharyeb, J. (1989, May). Toward the panoramic cockpit, and 3-d displays. *Proceedings of the National Aerospace and Electronics Conference*, Dayton, OH.

Harris, R. L., & Parrish, R. V. (1992, October). Piloted studies of enhanced and synthetic vision display parameters. *Proceedings of SAE Aerotech '92*, Anaheim, CA, pp. 1–9.

Hatfield, J., & Parrish, R. (1980, December). Advanced cockpit technology for future civil transport aircraft. *Proceedings of Symposium on The Human-Machine Interface in Airborne Systems. AESS-IEEE Conference*, Dayton, OH.

Howell, W. C. (1993). Engineering psychology in a changing world. *Annual Review of Psychology*, *44*, 231–263.

Hughes, D. (1994, January). Aerospace Sector Exploits CAD/CAM/CAE. *Aviation Week & Space Technology*, pp. 56–58.

Kearns, J. H. (1982). *A systems approach for crew station design and evaluation.* (Tech. Report AFWAL-TR-81-3175). Wright-Patterson Air Force Base, OH: Flight Dynamics Laboratory.

Klass, P. (1956, July 23). USAF reveals new instrument concept. *Aviation Week*, p. 62.

Knemeyer, S., & Yingling, G. L. (1966, November). Results of the USAF pilot factors program as they apply to flight safety. *Proceedings of the 19th Annual International Air Safety Seminar*, Madrid, Spain.

Last, S. R. (1988). Should technology assist or replace the pilot? In *Sixth Aerospace Behavioral Engineering Technology Conference Proceedings—Human Computer Technology: Who's in Control* (pp. 151–155). Warrendale, PA: Society of Automotive Engineers.

Liggett, K. K., Reising, J. M., & Hartsock, D. C. (1992). The use of a background attitude indicator to recover from unusual attitudes. In *Proceedings of the 36th Annual Meeting of the Human Factors Society* (pp. 43–46). Santa Monica, CA: Human Factors and Ergonomics Society.

Lucas, T. (1993, November). HUDs and HMDs in military aircraft. *Avionics*, pp. 30–37.

Maida, J. C. (1993, July). Software: An evaluation of JACK. *Ergonomics in Design*, pp. 35–36.

Martin, C. D. (1994). Application of a crew-centered cockpit design process and toolset. In *Proceedings of the National Aerospace and Electronics Conference* (pp. 701–708). Dayton, OH: IEEE.

Muir, B. M. (1987). Trust between humans and machines, and the design of decision aids. *International Journal of Man-Machine Studies, 27*, 527–539.

Nicklas, D. (1958). *A history of aircraft cockpit instrumentation 1903–1946* (WRDC Technical Report No. 57-301). Wright-Patterson Air Force Base, OH: Wright Air Development Center.

Oliveri, F. (1994, January). Virtual warriors. *Air Force Magazine*, pp. 30–34.

ORLOC. (1981). *KC-135 crew system criteria* (Tech. Rep. No. AFWAL-TR-81-3010). Wright-Patterson Air Force Base, OH: Flight Dynamics Laboratory.

Osga, G. A. (1991). Using enlarged target area and constant visual feedback to aid cursor pointing tasks. *Proceedings of the Human Factors Society 35th Annual Meeting* (pp. 369–373). Santa Monica, CA: Human Factors Society.

Price, A. (1992, December). Tornado in the desert. *Air Force Magazine*, pp. 42–46.

Reising, J. M., Emerson, T. J., & Munns, R. C. (1993). Automation in military aircraft. In *Proceedings of the HCI International '93: 5th International Conference of Human–Computer Interaction*, Orlando, FL (pp. 283–288). New York: Elsevier.

Reising, J. M., Liggett, K. K., Rate, C., & Hartsock, D. C. (1992). 3-D target designation using two control devices and an aiding technique. In *Proceedings of the SPIE/SPSE Symposium on Electronic Imaging Science and Technology*, San Jose, CA.

Rich, M. A. (1989). Digital mockup (airplane design and production using computer techniques). In *Proceedings for the AIAA, AHS, and ASEE, Aircraft Design, Systems and Operations Conference*. Seattle, WA: AIAA.

Riley, V. (1989). A general model of mixed-initiative human–machine systems. In *Proceedings of the 31st Annual Meeting of the Human Factors Society* (pp. 124–128). Santa Monica, CA: Human Factors Society.

Rowntree, T. M. (1993, January). The intelligent aircraft. *IEE Review*, pp. 23–27.

Saville, K. (1993, May 24). Supersonic research takes off. *Hampton Daily Press* p. A1.

Smith, B. R., & Hartzell, E. J. (1993). A^3I: Building the MIDAS touch for model-based crew station design. *CSERIAC Gateway, IV*(3), 13–14.

Spengler, R. P. (1988). *Advanced fighter cockpit* (Report No. ERR-FW-2936). Fort Worth, TX: General Dynamics.

Storey, B. A., Roundtree, M. E., Kulwicki, P. V., & Cohen, J. B. (1994, May). Development of a process for cockpit design. In *Proceedings of the National Aerospace and Electronics Conference*. Dayton, OH (pp. 688–695).

Thomas, S. R., Cartledge, R. M., Grahm, M. R., Patterson, J. A., & Poe, D. (1993). Field investigations of laser eye protection F-15E aircraft lighting compatibility. In *Proceedings of the 37th Annual Meeting of the Human Factors Society* (pp. 55–59). Santa Monica, CA: Human Factors and Ergonomics Society.

Ware, C., & Slipp, L. (1991). Using velocity control to navigate 3-D graphical environments: A comparison of three interfaces. In *Proceedings of the 35th Annual Meeting of the Human Factors Society* (pp. 300–304). Santa Monica, CA: Human Factors Society.

Wickens, C. (1992). *Engineering psychology and human performance* (2nd ed.). New York: HarperCollins.

Wilson, J. R. (1995, July). Finding a niche: US unmanned aerial vehicles finally get some respect. *Armed Forces Journal*, pp. 34–39.

15

Flight Simulation

William F. Moroney
University of Dayton

Brian W. Moroney
University of Cincinnati

The U.S. Army Signal Corps' Specification Number 486 (1907) for the first "air flying machine" has a very straightforward "human factor" requirement: "It should be sufficiently simple in its construction and operation to permit an intelligent man to become proficient in its use within a reasonable period of time." Less than 3 years later, Haward (1910, as quoted in Rolfe & Staples, 1986) described an early flight simulator as "a device which will enable the novice to obtain a clear conception of the workings of the control of an aeroplane, and of the conditions existant in the air, without any risk personally or otherwise" (p. 15).

The capabilities of both aircraft and flight simulators have evolved considerably since that time. Modern flight simulators have the same purpose except they are used not only by novices but by fully qualified aviators seeking a proficiency rating in a particular type of aircraft. Indeed, after qualifying on a simulator, pilots may proceed directly from a simulator to a revenue-producing flight.

Flight simulation is a worldwide industry (Sparaco, 1994), with sales of $3 billion/year for commercial airlines and $2.15 billion/year for the U.S. Department of Defense. Individual simulators range in price from $3,000 for a PC-based simulation with basic controls up to an average of $10–$13 million for a motion-based simulator (down from $15–$17 million in the early 1990s).

Flight simulation is essentially the representation of aircraft flight and system characteristics with varying degrees of realism for research, design, or training purposes (Cardullo, 1994a). Cardullo listed three categories of training simulators: (a) the Operational Flight Trainer (OFT), used to train individual pilots or crews in all aspects of flight and the use of flight, navigation, and communication systems; (b) the Weapons Systems Trainer (WST), used to train in the use of offensive and defensive systems; and (c) the Part Task Trainer (PTT), used to train flight crews for specific tasks (e.g., in-flight refueling).

Most flight simulators have the following features:

1. Visual displays: Most simulators provide an external view of the world along with flight, navigation, and communication instruments. In addition, depending on its mission, some simulators display radar and infrared data.

2. Control/input devices: Usually a yoke or a stick combined with a control loader is used to mimic the "feel" of the real aircraft. The original control loaders were mechanical devices, but today most are hydraulic, electronic, or electro-hydraulic. In high-fidelity simulators, switches and knobs identical to those in the aircraft are used, whereas in lower fidelity devices a mouse or a keyboard may be used to input changes in switch position.

3. An auditory display: These may include a synthetically generated voice, warning and advisory tones, and/or "intercommunication systems."

4. Computational systems: These units may include the flight dynamics model, image generation, control, and data collection software.

In addition, some simulators (usually OFTs) have a motion base that provides rotation and translation motion cues to the crewmember(s), whereas others may use G-seats or anti-G suits to simulate motion and G cues.

The more sophisticated simulators are used typically in the commercial and military aviation communities, while less sophisticated simulators are used by general aviation communities. Some military simulators are full mission simulators and may include enemy threats (e.g., surface-to-air missiles, communications jamming, etc.) as well as other simulated aircraft with aircrew, simulating wingmen or enemy aircraft, and other "players" simulating air traffic controllers, airborne command posts, and so on.

This chapter is intended to provide a broad overview of flight simulation with an emphasis on emerging areas. The chapter begins with a brief history of flight simulators and a discussion of the advantages and disadvantages of flight simulators. Following this, the topics of simulator effectiveness, including cost and transfer measurement strategies, and the issue of fidelity are examined. Next is a description of different types of visual and motion systems, as well as a discussion of the debate surrounding the use of motion in flight simulators. Considerable discussion is then devoted to the issue of simulator sickness and strategies to minimize its deleterious effects. To broaden the reader's appreciation of the wide variety of simulators, a brief overview of five unique simulators is presented. Next, the often ignored, but critical area of instructional features is explored, followed by an overview of an area of tremendous potential growth—personal computer (PC) based simulation. Finally, the technical differences between simulators and training devices are delineated. The chapter closes with a listing of the authors' perceptions of the future and opportunities.

HISTORY OF FLIGHT SIMULATORS

Adorian, Staynes, and Bolton (1979) described one of the earliest simulators, an Antoinette trainer (circa 1910), in which a student was expected to maintain balanced flight while seated in a barrel (split the long way) equipped with short "wings." The barrel, with a universal joint at its base, was mounted on a platform slightly above shoulder height so that instructors could push or pull on these "wings" to simulate "disturbance" forces. The student's task was to counter the instructors' inputs and align a reference bar with the horizon by applying appropriate control inputs through a series of pulleys.

In an attempt to introduce student pilots to the world of flying prior to actual liftoff, the French Foreign Legionnaires realized that an airframe with minimal fabric on its wings would provide trainees with insight into the flight characteristics of the aircraft while limiting damage to the real aircraft and the student (Caro, 1988). Winslow (1917, as reported in Rolfe & Staples, 1986), described this device as a "penguin" capable of hopping at about 40 miles per hour. Although this may seem of limited use, it was a considerable improvement from the earlier flight training method of self-instruction in which trainees practiced solo until basic flight maneuvers had been learned. Instructors would participate in the in-flight training only after the trainees had, through trial and error, learned the relationship between input and system response (Caro, 1988). Apparently, the French Foreign Legionnaires understood the value of a skilled flight instructor.

The origins of modern flight simulation can be traced to 1929, when Edward A. Link received a patent for his generic ground-based flight simulator. His initial trainer was designed to demonstrate simple control surface movements and was later upgraded for instrument flight instruction. Link based his design on the belief that the trainer should be as analogous to the operational setting as possible. Through the use of compressed air which actuated bellows (adapted by Link from his father's pipe-organ factory), the trainer had motion capabilities of pitch, yaw, and roll that enabled student pilots to gain insight into the relationship between stick inputs and movement in three flight dimensions. Originally marketed as a coin-operated amusement device (Fischetti & Truxal, 1985), the value of Link's simulator was recognized when the Navy and Army Air Corps began purchasing trainers in 1934. Flight instructors, watching from outside the "Blue Box," would monitor the movements of the ailerons, elevator, and rudder to assess the student's ability to make the correct stick movements necessary for various flight maneuvers.

When the United States entered World War II there were over 1,600 trainers in use throughout the world. The necessity for the trainers increased as the Allied forces rushed to recruit and train pilots. As part of this massive training effort, 10,000 Link trainers were used by the United States military during the war years (Caro, 1988; Stark, 1994). After the war, simulations developed for military use were adapted by commercial aviation. Loesch and Waddell (1979) reported that by 1949, the use of simulation had reduced airline transition flight training time by half. Readers interested in details on the intriguing history of simulation would do well to consult the excellent three-volume history entitled *Fifty Years of Flight Simulation* (Royal Aeronautical Society, 1979). Also, Jones, Hennessy, and Deutsch (1985), in *Human Factors Aspects of Simulation*, provided an excellent overview of the state of the art in simulation and training through the early 1980s.

Following the war and throughout the 1950s, increases in aircraft diversity and complexity resulted in the need for aircraft-specific simulators, that is, simulators that represent a specific aircraft in instrument layout, performance characteristics, and flight handling qualities. Successful representation of instrument layout and performance characteristics was readily accomplished; however, the accurate reproduction of flight handling qualities was a more challenging task (Loesch & Waddell, 1979). Exact replication of flight is based on the unsupported belief that higher fidelity simulation would result in greater transfer of training from the simulator to the actual aircraft. This belief has prevailed for many years and continues today. However, even 40 years ago, researchers were questioning the need for duplicating every aspect of flight in the simulator (Miller, 1954; Stark, 1994).

Caro (1979) described the purpose of a flight training simulator as "to permit required instructional activities to take place" (p. 84). However, from his examination of existing simulators, simulator design procedures, and the relevant literature Caro concluded that "designers typically are given little information about the instructional activities intended to be used with the device they are to design and the functional purpose of those activities" (p. 84). Fortunately, some progress has been made in this area. Today, as part of the system development process, designers (knowledgeable about hardware and software), users/instructors (knowledgeable about the tasks to be learned), and trainers/psychologists (knowledgeable about skill acquisition and evaluation) interact as a team in the development of training systems (Stark, 1994). The objective of this development process is to maximize training effectiveness while minimizing the cost and time required to reach the training objective (Stark, 1994).

WHY USE SIMULATION?

Simulation is effective and efficient. It provides a means for experiencing critical conditions that may never be encountered in flight, as well as an opportunity for initial qualification or requalification in type. In the context of a broader training program, simulation provides an excellent training environment and is generally well accepted by the aviation community.[1]

Advantages

Part of the efficiency of simulators may be attributed to their almost 24-hr/day availability, and their ability to provide immediate access to the operating area. For example, simulators allow a student to complete an Instrument Landing System (ILS) approach and return immediately to the Final Approach Fix (FAF) for the next ILS approach, without consuming time and fuel. Indeed, because simulators are not realistic, conflicting traffic in the landing approach can be eliminated to further increase the number of approaches flown per training session. In short, simulators provide more training opportunities than could be provided by an actual aircraft in the same time. As noted by Jones (1967), simulators can provide training time in nonexistent aircraft or in aircraft where an individuals first performance in a new system is critical (consider the first Space Shuttle landings or single-seat aircraft).

Due to safety concerns, simulators may be the only way to teach some flight maneuvers or to expose aircrew to conditions that they are unlikely to experience under actual flight conditions (e.g., wind shear, loss of hydraulic systems, engine loss, engine fire, exposure to wake turbulence and clear air turbulence). Additionally, automation has increased the need for simulators, as Wiener and Nagal (1988, p. 453) commented, "It appears that automation tunes out small errors and creates the opportunities for larger ones." In automated glass (cathode ray tube–laden) cockpits, improvements in system reliability have reduced the probability and frequency of system problems, thus inducing a sense of complacency among the aircrew. However, when an unanticipated

[1]Simulator effectiveness is discussed in a separate section of this chapter.

event occurs, the crew must be trained to respond rapidly and correctly. Simulators provide an opportunity for that type of training.

Simulator usage also reduces the number of flight hours on the actual aircraft, which in turn reduces mechanical wear and tear, associated maintenance costs, and the load on the National Airspace System. Additionally, airlines do not incur the loss of revenue associated with using an aircraft for in-flight training. Simulator usage also reduces environmental problems, not only pollution and noise but the damage to land and property associated with military training.

Simulators also provide an improved training environment by incorporating instructional features that enhance student learning, and facilitate instructor intervention. Such features are described later in this chapter. Additionally, simulators provide standardized training environments with identical flight dynamics and environmental conditions. Thus, the same task can be repeated until the required criteria are attained, and, indeed, until the task is overlearned (automated). Unlike the airborne instructor, the Simulator Instructor (SI) can focus on the teaching task without safety of flight responsibilities, or concerns about violations of regulations. Thus, he or she may deliberately allow a student to make mistakes such as illegally entering a terminal control area.

Simulators allow performance data to be collected, which according to (Stark, 1994) permits:

1. Performance comparison: As part of the diagnosis process, the Instructor Pilot (IP) can compare the student's performance with the performance criteria, and the performance of students at the same stage of training.
2. Performance and learning diagnosis: Having evaluated the student's performance, the IP can gain some insight into the students learning process and suggest new approaches in problem areas.
3. Performance evaluation: Performance measurement can be used to evaluate the efficacy of different approaches to training a particular task.

Despite the emphasis on high fidelity and "realism," *simulators are not realistic*. In a sense, the lack of realism may contribute to their effectiveness. Indeed, Lintern (1991) believed that transfer can be enhanced by "carefully planned distortions of the criterion task" (p. 251). Additionally, most instructional features found in simulators do not exist in the cockpit being simulated. Indeed, if real cockpits had the same features as simulators, the "PAUSE" button would be used routinely.

Disadvantages

Let us now examine some of the alleged "disadvantages" of simulators. We must recognize that performance in a simulator does not necessarily reflect how an individual will react in flight. Because there is no potential for an actual accident, the trainee's stress level may be lower in a simulator. However, the stress level can be high when an individual's performance is being evaluated or when he or she is competing for a position or a promotion.

To the extent that aircrew being evaluated or seeking qualification-in-type expect an emergency or unscheduled event to occur during their time in the simulator, their performance in a simulator may not reflect in-flight performance. Since the aircrew would, in all probability, have reviewed operating procedures prior to their turn in the simulator. Nonetheless, it should be recognized that a review of procedures even in preparation for a check ride is of value.

Performance in simulators rarely reflects the fatigue and/or boredom common to many cockpits. Therefore, performance in a simulator may be better than actually expected in flight. In addition, simulators reduce the utilization of actual aircraft, which leads to fewer maintenance personnel and reduced supply requirements. These apparent savings may create personnel shortages and logistic problems when the operational tempo rises beyond the training level.

Simulators, particularly dome and motion-based simulators, usually require unique air-conditioned facilities, and maintenance personnel, which reduces the assets available to operational personnel. In addition, when used excessively, simulators may have a negative effect on morale and retention. This attitude is usually reflected as, "I joined to fly airplanes, not simulators." Finally, the acceptance and use of simulators are subject to the attitudes of simulator operators, instructors, aircrew, and management.

Overall, the advantages significantly outweigh any real or perceived disadvantages as evidenced by the general acceptance of simulators by the aviation community and regulatory agencies.

SIMULATOR EFFECTIVENESS

One objective of effective training systems is to provide the required training at the lowest possible cost. Simulation is a means for achieving that objective. Baudhuin (1987) stated, "the degree of transfer from the simulator to the system often equates to dollars saved in the operation of the real system and in material and lives saved" (p. 217). Within the aviation community, the effectiveness of simulators is accepted as an article of faith. Indeed, the aviation industry could not function without simulators and Flight Training Devices (FTDs), whose existence is mandated by Federal Aviation Administration (FAA) regulations (1991, 1992).

In a very detailed analysis of cost effectiveness, Orlansky and String (1977) reported that flight simulators for military training can be operated at between 5% and 20% of the cost of operating the aircraft being simulated; median cost is approximately 12%. They also reported that commercial airlines can amortize the cost of a simulator in less than 9 months and the cost of an entire training facility in less than 2 years.

Roscoe (1980) provided sufficient data illustrating the effectiveness of fixed-base simulators for teaching the skills needed in benign flight environments. Spears, Sheppard, Roush, and Richetti (1981a, 1981b) provided detailed summaries and evaluations of 196 research and development reports related to simulator requirements and effectiveness. Pfeiffer, Horey, and Butrimas (1991) supplied additional support in their report of positive transfer of instrument training to instrument and contact flight in a operational flight training aircraft (U.S. Navy T-2C).

Often because of the high cost of true transfer of training experiments, quasi-experiments are performed to determine the transfer between a FTD or part-task trainer and a representative high-fidelity simulator that serves as a surrogate for the real aircraft. However, Jacobs, Prince, Hays, and Salas (1990) in a meta-analysis of data culled from 247 sources identified 19 experiments in which training transfer between the simulator and the actual jet aircraft was evaluated. They concluded that simulators reliably produced superior training relative to aircraft-only training. They also reported that for jet aircraft, takeoffs, landings, and approaches benefited from the use of a simulator, with the landing approach showing the greatest benefit. However, similar conclusions regarding the effectiveness of helicopter simulators could not be drawn because only seven experiments involving helicopters meet the criterion for inclusion in the meta-analysis.

Today's effectiveness questions are focused on how the required skills can be taught rapidly and inexpensively. Thus, we have seen an emphasis on the systems approach to training (Department of the Army, 1990), which like the Instructional Systems Development (ISD) approach, emphasizes requirement definition and Front End Analysis (FEA) early in the system development process with an evaluation at the end. Roscoe and Williges (1980), Roscoe (1980), and Baudhuin (1987) provided excellent descriptions of strategies for evaluating transfer of training, including the development of Transfer Effectiveness Ratios (TERs), and of Incremental Transfer Effectiveness Functions (ITEF), and Cumulative Transfer Effectiveness Functions (CTEF). All of these approaches attempt to measure the degree to which performing the desired task in the actual aircraft is facilitated by learning an intervening task on a training device or simulator. The resulting measure is usually expressed in terms of time saved. The critical concern as emphasized by Roscoe (1980) was not simply measuring training effectiveness but determining cost-effectiveness. Specifically, Roscoe was concerned with identifying the region in which increasing the investment in the training device (by improving fidelity, adding additional instructional features, etc.) did not result in a significant increase in transfer. However, as noted by Beringer (1994), because the cost of simulation has decreased as the capabilities of simulators have increased, today's question is more often phrased as, "If we can get more simulation for the same investment, what is the 'more' that we should ask for?" Thus, according to Beringer, cost is seen as a facilitating, rather than a prohibitive, factor.

Measuring effectiveness is a fairly complicated process that has performance measurement at its core. Lane's (1986) report is "must reading" for individuals interested in measuring performance in both simulators and the real world. Mixon and Moroney (1982) provided an annotated bibliography of objective pilot performance measures in both aircraft and simulators. Readers interested in measuring transfer effectiveness are referred to Boldovici's (1987) chapter on sources of error and inappropriate analysis for estimating transfer effectiveness.

FIDELITY

Hays and Singer, in their book *Simulation Fidelity in Training System Design* (1989), provided an excellent, comprehensive examination of the complex issue of fidelity. They defined simulation fidelity as:

> the degree of similarity between the training situation and the operational situation which is simulated. It is a two dimensional measurement of this similarity in terms of: (1) the physical characteristics, for example, visual, spatial, kinesthetic, etc.; and (2) the functional characteristics (for example, the informational, and stimulus response options of the training situation). (p. 50)

Interestingly, the simulation community also appears to be divided into two camps on the issue of fidelity. One group (usually the simulator developers and regulatory agencies) believes that the simulation should be as realistic as technically possible. Thus, they emphasize high fidelity of both the simulator cockpit and the environment. They are concerned that failure to properly represent the cockpit of the environment may increase the probability of a critical error, which could result in the loss of life. The other group (usually the trainers and researchers) emphasizes the functional characteristics. They contend, as Bunker (1978) stated, that "instead of pondering how to achieve realism, we should ask how to achieve training" (p. 291). Lintern (1991) noted that "similarity, as it is normally viewed is not a sufficient element of a conceptual approach to skill transfer" (p. 253). He argued further that simulator designers must distinguish between the informational "invariants" critical for skill acquisition and the irrelevant elements (i.e., the "extras" often included in the name of fidelity). The "invariants," according to Lintern, are the properties of the events that remain unchanged as other properties change. Such a property remains constant across events that are perceived as similar but differs between events that are perceived as different.

Recently, however, due in part to economic considerations, the two camps are interacting more often, which should lead to improved training systems. Increased computing capability, increased memory capacity and rapid access, multiple-processor architecture, improved image generation capability, and so on have lead to new simulation technologies ranging from PC-based flight simulation (discussed later in this chapter), to virtual environments (Garcia, Gocke, & Johnson, 1994), to distributed simulations such as SIMNET (Alluisi, 1991), and to even more complex real-time virtual-world interactions (Seidensticker, 1994). Because there are such widely varied technologies available in very disparate cost ranges, the focus appears to be gradually evolving from developing new technology to deciding what needs to be simulated. Clearly, the Department of the Army's (1990) Systems Approach to Training (SAT), emphasizes the use of task analysis to identify training requirements and has tasked the training community with defining the simulator requirements for the developers. Thus, today's simulators are purchased as part of a training system, which may include a variety of training devices, media, and educational strategies to achieve the desired outcome.

Hays and Singer (1989) advised that the effectiveness of a simulator is not only a function of the characteristics and capabilities of the simulator but how those features support the total training system. They indicate that simulator fidelity should vary as a function of stage of learning, type of task, and type of task analysis. Each of these factors is described independently below.

Stage of Learning. Fitts (1962) developed a tripartite model of skill development consisting of a cognitive phase, an associative phase, and an autonomous phase, that has served the aviation community well. Although the boundaries between the phases are not clearly delineated, the skills needed in aviation progress in this sequence. During the cognitive phase, the novice attempts to understand the task, the expected behavior, the sequence of required procedures, and the identification of relevant cues. Instructions and demonstrations are most effective during this phase. During the associative phase, the student integrates skills learned during the cognitive phase and new patterns emerge, errors are gradually eliminated, and common features among different situations begin to be recognized. Hands-on practice is most appropriate during this phase. Finally, during the autonomous phase, the learner's performance becomes more automatic, integrated, and efficient, thus requiring less effort. Individuals at this level of skill development are more resistive to overload because they have well-developed "subroutines." At this stage, learners can perform previously learned tasks while a new skill is being acquired. Whole task and mission simulations are most appropriate for individuals at this skill level.

During the early phase of learning, less expensive, lower fidelity simulators will suffice. Caro (1988) provided an interesting case study in which wooden mockups with fairly simple displays were as effective as a much more expensive cockpit procedures trainer. Warren and Riccio (1985) noted that simulations providing stimuli that experienced pilots tend to ignore make learning more difficult, because the trainee has to learn how to ignore those irrelevant stimuli. More recently, Kass, Herschler, and Companion (1991) demonstrated that students trained in a "reduced stimulus environment that presented only task-relevant cues performed better in a realistic battle field test condition than did those who were trained in the battle field test condition" (p. 105). Similarly, Lintern, Roscoe, and Sivier (1990) trained two groups of flight-naive subjects in landing procedures; one group trained with crosswinds and the other group trained without crosswinds. When the performance of both groups was evaluated on a 5 knot crosswind landing task, the group trained without the crosswinds performed better. Apparently, training with the crosswinds confounded the students' understanding of the relationship of control action and system response, whereas training without the crosswinds did not interfere with the students' learning. Thus it has been demonstrated that higher fidelity does not necessarily lead to more efficient transfer of training.

Type of Task. The fidelity required to teach a cognitive task (information processing) is very different from the fidelity required to learn a psychomotor (tracking) task. For example, the type of simulation required to facilitate the development of a cognitive map of a fuel flow system is very different from the type of simulation required to demonstrate an individual's knowledge of the same fuel flow system under emergency conditions. In the former case, a model board indicating valve positions, fuel tank locations and quantities, and so forth would be appropriate, whereas in the latter a full-cockpit simulation is more appropriate. However, even in very complex flight conditions such as Red-Flag simulated air combat exercises, it has been demonstrated that individuals trained on lower fidelity simulators showed a higher level of ground attack skills than individuals who did not receive any simulator training (Hughes, Brooks, Graham, Sheen, & Dickens, 1982).

Type of Task Analysis. The type of task analysis performed will significantly influence the level of fidelity incorporated into the simulator. Baudhuin (1987) emphasized the need for comprehensive Front-End Analysis (FEA) in developing meaningful task modules that ultimately are incorporated into the specifications to which the system is developed. Warren and Riccio (1985) and Lintern (1991) argued that appropriate task analysis would help distinguish between the necessary and the irrelevant cues. Inappropriate task analysis will lead to inadequate, perhaps even inappropriate, training and low transfer to the operational setting.

Thus, decisions regarding the required level of fidelity are multifaceted. Alessi (1988) reported that the relationship between learning and simulation has not been adequately researched and therefore he provides a taxonomy of fidelity considerations. Cormier (1987) and Lintern (1991) provided insights into the role of appropriate cues in the transfer of training, whereas Baudhuin (1987) provided guidance on simulator design. The critical element to remember is this: Simulator fidelity is not the end, but rather it is a means to the end—effective, efficient training.

VISUAL AND MOTION SYSTEMS

This section introduces the reader to visual and motion-based systems, describes display strategies, discusses the motion versus no-motion controversy, and force-cuing devices.

Visual Systems

Early simulators, such as Link's "Blue Box," served primarily as Instrument Flight Rules (IFR) trainers and thus provided no information about the correspondence between control inputs (pushing the stick) and changes in the external visual scene. Gradually, simple but effective visual displays such as a representation of the horizon on a tiltable blackboard to represent varying glide slopes evolved (Flexman, Matheny, & Brown, 1950). During the 1950s, simulator designers, in their quest for realism, developed additional methods to present external visual information. For example, model boards using closed-circuit television in which a gantry-mounted video camera (steered by the pilot) moved over a prefabricated terrain model were developed. Although model board technology systems were used successfully in developing procedural skills requiring vision, the resolution and depth-of-field constraints imposed by video camera limitations reduced their ability to help develop complex perceptual and psychomotor skills (Stark, 1994). Additionally, they were expensive to construct and modify, and due to physical limits in the area that they could represent, aircrew quickly learned the terrain.

The development of the digital computer in the 1960s and improved mathematical models afforded the creation of complex external visual scenes. The use of Computer-Generated Imagery (CGI) allows for the dynamic presentation of an enormous amount of visual input.[2] However, simulator designers must distinguish between the required information (cues) and the noise content of the visually presented material in order to define the necessary image fidelity (Chambers, 1994). Armed with knowledge of the system requirements, the available technology, and the associated life-cycle cost, designers must then make trade-offs to determine whether the external view from the cockpit should be displayed as either a real or virtual image. Real-image displays project an image onto a surface 10 or 20 ft away from the pilot's eye, whereas virtual-image displays project an image at or near optical infinity.

Real-Image Displays. Real images are usually projected onto flat screens. However, to provide a large Field of View (FOV), dome-shaped screens are often used to ensure that the entire image is presented at a constant distance from the observer. Large FOV images, greater than 40–50 degrees horizontally and 30–40 degrees vertically, are generally achieved by coordinating a number of projectors (Stark, 1994). Currently however, systems that provide a large FOV with very accurate scene detail are technically difficult to build and maintain, and extremely expensive to develop and operate.

Therefore, designers developed systems that maintain a high degree of detail within the pilot's area-of-interest (see Fig. 15.1). These area-of-interest systems operate in a number of ways. Many simply provide the greatest detail off the nose of the aircraft, with less scene detail in the periphery. An alternate strategy utilizes a head-slaved area-of-interest display that creates a highly detailed scene based on the head movements of the pilot. ESPRIT (Eye-Slaved Projected Raster InseT), uses two projectors to

[2]Fortin (1994) provided an excellent technical presentation on computer image generation.

FIG. 15.1. Generic tactical aircraft simulator installed in a dome. Note high-quality imagery immediately forward of the "nose" of the simulator. Photo courtesy of McDonnell Douglas Corporation.

present the image to the pilot. The first projector displays a low-resolution background scene with information to be processed by the peripheral visual system. The second projector positions a very detailed scene along the pilot's line of sight. The positioning of the second image is controlled through the use of a servo system which utilizes an oculometer to monitor the movements of the pilot's eyes. The interval between visual fixations is sufficiently long to allow the servo system to position the highly detailed image in the pilot's line of sight (Haber, 1986; Stark, 1994). While the pilot moves his or her eyes to another fixation point no visual information is processed, thus the pilot does not see the image as it is moved to the new fixation point. ESPIRIT is currently used in some Royal Air Force Simulators.

The high cost of dome systems and their support requirements has led to the development of a relatively low-cost (approximately $1.0 million) air combat simulator (Mosher, Farmer, Cobasko, Stassen, & Rosenshein, 1992). Originally tested on a 19-inch CRT with a 30 degree vertical and 35 degree horizontal FOV, the innovative approach uses aircraft icons. When an aircraft is outside the pilot's FOV an aircraft icon is presented at the appropriate location at the edge of the display. This icon provides information to the pilot about the relative position, orientation, and closure rate of aircraft outside the displayed field of view. This approach has led to the development of the F-16 unit training device, which utilizes a rear projection display, to create a 60 degree vertical and 78 degree horizontal FOV out-of-the-cockpit "virtual dome system."

Virtual-Image Displays. Virtual-image displays present collimated images (i.e., images at or near optical infinity) to the pilot, who must be positioned at the correct focal plane to observe the image. Collimation is sometimes achieved by projecting a CRT image through a beamsplitter onto an appropriately designed spherical mirror. Head-Up Displays (HUDs) installed in aircraft also use collimating optics to project CRT-generated images onto a combiner (beamsplitter) surface mounted in the front of the cockpit. As noted by Randle and Sinacori (1994), when collimated systems are used, the optical message is that "all elements in the scene are equally distant and far away." However, the absence of parallax cues (since the entire image is at the same distance), makes it difficult for pilots to discriminate objects in the foreground from objects in the background. Distance must be inferred from perspective, occlusion, and texture cues, without the support of stereopsis, vergence, and accommodation (Stark, 1994; Randle & Sinacori, 1994). Despite the loss in perceptual fidelity, the illusion is compelling and becomes "virtual reality" when the pilot becomes involved in his or her flying tasks.

Collimation technology is also employed in helmet mounted displays (HMDs) used for simulation. Training HMDs usually consist of a helmet with two half-silvered mirrors mounted on the helmet and positioned in front of the eyes. The presentation of the optical infinity image directly in front of the subject's eye(s) eliminates many of the problems associated with domes. However, HMDs require precise alignment of the two images. Whereas the weight of early versions of HMDs limited user acceptance, recent advances in image projection (lighter weight CRTs, fiber optics, polycarbonate optics, etc.) have now decreased the weight of these systems to more acceptable levels.

Both real and virtual imagery systems work well under conditions (landing, take-off, air combat) in which most of the elements in the real-world scene are at a considerable distance from the pilot's eye position. Nevertheless, there are still many unknowns regarding visual cues and the appropriate dynamics for low-level (nap-of-the-earth) rotary-wing simulations. As will be discussed later, the highest rates of simulator sickness are reported in helicopter simulators. Part of this may perhaps be attributed to the nature of the helicopter, which Bray (1994) described as a small, very agile, low-stability, highly responsive aircraft capable of motions which are difficult to simulate. In addition, at the nap-of-the-earth levels, the visual display requirements of a helicopter simulator are demanding. Randle and Sinacori (1994) described the pilot's gaze-points as being distributed primarily in the "immediate impact field from 3 to 5 sec ahead." The requirements for "in-close" viewing need to be specified carefully and they will vary as a function of altitude and airspeed. Much work remains to be done in defining and justifying the requirements for rotary wing simulators.

Motion Systems

This section describes degrees of freedom (DOF) and synergistic platform motion simulators; the current debate over the use of motion platforms and a brief description of unique simulators follow.

Simulators have from zero (no movement) to six DOF (yaw, pitch, and roll; heave, surge, and sway). The first three—yaw, pitch, and roll—require rotation about an axis. Yaw is rotation about the aircraft's vertical axis, pitch is rotation about the lateral axis, and roll is rotation about the longitudinal axis of the aircraft. The latter three—heave, surge, and sway—require displacement. Heave refers to up and down displacement, surge refers to forward and backward displacement, and sway refers to lateral displacement.

Synergistic platform motion simulators are the most common type of motion simulators. The hexapod platform uses six hydraulic posts, while other platforms use combinations of lifting devices and posts (see Fig. 15.2). Although hexapod platforms have several DOF, their nominal excursion ranges are perhaps 40–50 degrees for yaw, pitch, and roll with up to 6 ft in heave, surge, and sway. Typical yaw, pitch, and roll velocities may range from 5–20 deg/sec with displacement rates of 1–2 ft/sec. The motion capabilities of a system are based on both the characteristics of the simulated aircraft and the physical limitations of the individual components of the motion platform. In commercial and transport systems a "cockpit" (see Fig. 15.3) with stations for the aircrew and sometimes a checkride pilot or instructor are mounted atop the platform. The movements of the posts are coordinated to produce the motion required for vestibular and kinesthetic input "similar" to the movements of the actual vehicle. Acceleration and displacement provide the initial sensory input, and washout techniques are then used to return the platform to its initial position. Because washout of the movement theoretically occurs below the pilot's motion detection threshold, the pilot's perception is that his or her vehicle is still moving in the direction of the initial motion (Rolf & Staples, 1986).

An additional concern with motion systems is the magnitude of the lag between an airborne pilot's input and the corresponding movement of the platform. Delays of approximately 150 msec between the aircraft and the simulator have minimal effect; however, delays of greater than 250 msec significantly reduce the quality and transferability of the simulation (McMillan, 1994; Stark, 1994).

FIG. 15.2. Motion-based platform with Boeing 737 simulator mounted on platform. Photo courtesy of Frasca Corporation.

FIG. 15.3. Boeing 737 simulator cockpit installed on motion-based platform shown in Fig. 15.2. Photo courtesy of Frasca Corporation.

Finally, although synergistic platform motion simulators do induce the feeling of motion, it is difficult to coordinate the smooth movement of the hydraulic components due to interactions between the various degrees of freedom. Additionally, these platforms are expensive to operate and maintain, and require special facilities.

Motion Versus No-Motion Controversy

The focus of most simulator research has been on measuring the training provided by a specific simulator in a particular training program. As discussed previously, there has been an untested belief, based primarily on face validity, that the more closely the simulator duplicates the aircraft, the better the transfer. In part because of the costs associated with motion systems, there has been considerable controversy about the contribution of platform motion to training. In order to examine the contribution of platform motion to simulator training effectiveness for basic contact (non-IFR) flight E. L. Martin and Wagg (1978a, 1978b) performed a series of studies. They reported (1978a) no differences between the groups trained in either the fixed or motion based simulators. However, students in both simulator groups performed better than students in the control group who received all their training in the T-37 aircraft. Later, they extended the study (1978b) to include aerobatic tasks and found that platform motion did not enhance performance in the simulator or in the aircraft. They concluded that aerobatic skills may be more cost-effectively trained in the aircraft. In her review of the six studies in the series, E. L. Martin (1981) concluded that the procurement of six-post synergistic platform motion systems was not necessary for teaching pilot contact skills.

More recently, two different groups have examined the issue of simulator motion from different perspectives. Jacobs et al. (1990) performed a meta-analysis of flight simulator training research, whereas Boldovici (1992) performed a qualitative analysis based on the opinions of 24 well-known authorities in the field of simulator motion. Jacobs et al. concluded that, for jet aircraft, motion cuing did not add to simulator effectiveness and in some cases may have provided cues that reduced the effectiveness of the simulator. However, they advised that this conclusion be accepted with caution because (a) the calibration of the motion cuing systems may not have been performed as frequently as necessary, and (b) the conclusion is based on all tasks combined not on specific tasks (thus any gain on a task that could have been attributed to motion may have been canceled by a decrement on another task). No conclusion was possible for helicopter simulators because only one study compared the transfer between the simulators and the actual aircraft. However, that one study by McDaniel, Scott, and Browning (1983) reported that certain tasks (aircraft stabilization equipment off, free-stream recovery, and coupled hover) benefited from the presence of motion, whereas takeoffs, approaches, and landings did not. Bray (1994) believes that platform motion "might offer a bit more in the helicopter simulator than it does in the transport aircraft simulator, because control sensitivities are higher and stability levels are lower in helicopters." However, with reference to motion platforms for helicopters, he comments that "if the benefits of 6-DOF cockpit motion are vague, its cost is not."

With respect to the motion simulation literature in general, Boldovici (1992) argued that finding no differences (null hypothesis) between the effect of motion and no-motion conditions does not prove that an effect does not exist, only that no effect was obtained. He also noted that the statistical power of some of the literature examined may be inadequate to detect existing differences and that most of the literature failed to adequately describe the characteristics of the motion platform. Sticha, Singer, Blacksten, Morrison, and Cross (1990) suggested that perhaps there are real differences between the effectiveness of fixed and motion-based systems but inappropriate lags in the motion systems, problems in drive algorithms, lack of synchronization of the visual and motion systems, and so on may preclude the advantage of motion-based simulation from being noted. They propose that the results "may simply show that no motion is better than bad motion" (p. 60). Lintern and McMillan (1993) supported their position and suggested that motion provides neither an advantage nor a disadvantage, since most flight transfer studies show no transfer effects attributable to motion.

Boldovici (1992), on the other hand, asked 24 well-known authorities in the field of simulator motion to provide arguments both for and against the use of motion platforms.

Their arguments for using motion platforms included: reducing the incidence of motion sickness; low cost compared to aircraft use; user's and buyer's acceptance; trainee motivation; learning to perform time-constrained, dangerous tasks; motion as a distraction to be overcome by practice; application of adaptive or augmenting techniques; and finally the inability to practice some tasks without motion. Their arguments against the use of motion platforms included: absence of supporting research results; possible learning of unsafe behavior; possible achievement of greater transfer by means other than motion cuing; undesirable effects of poor synchronization of the motion cues; direct, indirect, and hidden costs; existing alternatives to motion bases for producing motion cuing; and finally, the relatively benign force environments encountered under most flight conditions. Boldovici examined each of the sometimes conflicting positions previously listed and concluded that:

1. Results of transfer-of-training studies are insufficient to support the decisions about the need for motion systems.
2. Greater transfer can be achieved by less expensive means than using motion platforms. Therefore, if cost-effectiveness is used as a metric, motion platforms will never demonstrate an advantage.
3. From a statistical viewpoint, the research results concluding no differences in transfer to parent vehicles do not prove that no differences exist. Boldovici recommended that researchers report the results of power tests to determine the number of subjects required to detect treatment effects.
4. Because much of the transfer-of-training literature does not adequately address test reliability, we cannot adequately assess the validity of our inferences.
5. Because some of the conditions under which a simulator is "flown" cannot be repeated safely in the aircraft, some transfer of training cannot be evaluated. On the other hand, adequate training for flying an aircraft in benign environments can be provided by a fixed-base simulator.
6. Training in either motion-based or fixed-base simulators can promote learning unsafe or counterproductive behavior.
7. No evidence exists regarding the effect of motion on trainee motivation.
8. The use of motion-based platforms to reduce simulator sickness is inappropriate (see also Sharkey & McCauley, 1992).
9. User's and buyer's acceptance is not an appropriate reason for the use of motion platforms.
10. Incentives (such as job advancement for working in high-tech projects) for purchasing expensive simulators may be greater than incentives for purchasing less expensive simulators.
11. Some tasks may require force motion cuing, which can be provided by seat shakers, G-seats, and motion bases. Sticha, Buede, et al. (1990) developed a rule-based model for determining which, if any, of these force-cuing strategies is necessary. Their model for the optimization of simulation-based training systems (OBATS) requires the developer of the training system to develop specifications which identify the cues required for proper learning.

Thus, although the controversy continues, progress has been made in this area even to the extent of developing a promising decision support model. Caro (quoted in

Boldovici, 1992) asked the incisive question: "Does the motion permit the operator to discriminate between conditions that otherwise could not be distinguished?" (p. 20). Although it appears that the answer to this question will be more often negative, if the discrimination is essential and cannot be induced visually, then perhaps the use of motion should be considered seriously.

Force-Cuing Devices

Force-cuing devices have been used to simulate motion in fixed base platforms. Two devices, the G-suit and G-seat, have been used to simulate the effects of motion on the pilot's body during high G-load situations (Cardullo, 1994b). The G-suit (more properly the anti-G suit) is used in aircraft to maintain the blood level in the brain by preventing the blood from pooling in the pilot's lower extremities. The G-suit used in simulators consists of a series of air bladders, imbedded in a trouser-like assembly, which inflate as a function of the simulated G-load. Thus the pilot has some of the sensations of being exposed to G-forces. On the other hand, the G-seat consists of independently operating seat and backrest panels, and mechanisms that vary the pressure exerted on the restrained pilot. The properties of the seat (shape, angle, and hardness) are manipulated to correspond with changes in the G-load imposed by specific maneuvers. The use of a G-suit or G-seat during simulation provides the pilot with additional cues regarding the correct G-load needed for certain flight maneuvers (Stark, 1994). Some helicopter simulators use seat shakers to simulate the vibratory environment unique to rotary wing aircraft.

In addition to tactile cues, dimming the image intensity has been used to mimic the "graying" of the visual field which occurs under high G-loads (Cardullo, 1994b). Cardullo also describes other strategies such as variable transparency visors which mimic the graying of the visual field by varying the amount of light transmitted through the visor as a function of G-load. Harness loading devices, usually used in conjunction with G-seats, simulate the G-load by tightening and releasing the crewmember's restraint system as a function of G-load.

SIMULATOR SICKNESS

People and other animals show symptoms of motion sickness in land vehicles, ships, aircraft, and spacecraft (Money, 1970). Consequently, while attempting to simulate the motion and the external visual environment of these vehicles, it was reasonable to expect a form of motion sickness to occur. This form of motion sickness is referred to as simulator sickness. As noted by Kennedy and Fowlkes (1992), simulator sickness is polygenic and polysymptomatic. The symptoms of motion sickness (cold sweats, stomach awareness, emesis, etc.) are, at the very least, disruptive in the operational environment. Simulator sickness threatens and perhaps destroys the efficacy of the training session and may decrease simulator usage (Frank, Kennedy, Kellogg, & McCauley, 1983; McCauley, 1984; Kennedy, Hettinger, & Lilienthal, 1990).

During and after a simulator session, the foremost concern is the safety and health of the trainee. Secondary to safety is the value of the training session. Trainees more concerned about avoiding simulator sickness than learning the assigned task are unlikely to benefit from simulator training. Additionally, if simulators produce effects that differ from the real-world situation, then the skills learned in the simulator may be of limited value in the operational setting. Furthermore, the perceptual aftereffects of a simulator session may interfere with the pilot's flight readiness, that is, the ability to fly an aircraft safely or operate a vehicle immediately or shortly after a simulator training session (McCauley, 1984; Kennedy et al., 1990).

To determine the incidence rate of simulator sickness, Kennedy, Lilienthal, Berbaum, Baltzley, and McCauley (1989) surveyed 1,186 "flights," conducted in 10 different U.S. Navy simulators during a 30-month period. All the simulators had a wide field-of-view visual system. The reported incidence rate, based on the Motion Sickness Symptom Questionnaire (MSSQ), ranged from 10% to an astonishing 60%. The lowest incidence rates occurred in fixed-wing, fixed-base, dome-display simulators, whereas the highest reported sickness rate occurred in rotary wing (helicopter) simulators that employed six degree of freedom motion systems. It should be noted that in many instances simulator sickness was induced even in stationary simulators. The latter case may be explained by the strong correlation between simulator sickness and the perception of vection, that is, the sensation of self-motion (Hettinger, Berbaum, Kennedy, Dunlap, & Nolan, 1990). A major contributor to the perception of vection is visual flow (i.e., the movement of the surround as the observer moves past it). Sharkey and McCauley (1991) have reported that increased levels of global visual flow (GVF) are associated with an increased incidence of simulator sickness. McCauley (personal communication, October 1994) believes that, "In fixed-base and typical hexapod motion bases, sickness occurs only with a wide field-of-view representation of the outside world, which leads to vection." These higher levels of visual flow are more common in aircraft (or simulators) flying at lower altitudes than in aircraft flying at higher altitudes. Thus, the higher incidence of simulator sickness in rotary wing simulators may be attributed in part to the higher visual flow rates common at lower altitudes. More specifically, as reported by Sharkey and McCauley (1991), the increased incidence may be associated with *changes* in that visual flow.

Sensory conflict theory, the more commonly accepted explanation for simulator sickness, states that motion sickness occurs when current visual, vestibular, and other sensory inputs are discordant with expectations based on prior experience (Reason, 1978). Support for this theory is found in studies which indicate that individuals with more experience (higher flight hours) in the operational vehicle report a higher incidence of simulator sickness than less experienced individuals (Kennedy et al., 1989). The authors attributed this finding to a greater sensitivity to the disparity between the operational system and the simulator among experienced individuals (Kennedy et al., 1990).

Stoffregen and Riccio (1991) noted that the disparity between actual and expected sensory input may be impossible to measure because the baseline cannot be determined. They proposed an alternate theory, which contends that simulator sickness is produced by prolonged postural instability. This theory predicts that individuals who become sick in a simulator have not identified the appropriate constraints on bodily motion imposed by the simulator and thus have failed to implement the correct postural control strategies necessary for that situation. Irrespective of which theory is correct, the presence of simulator sickness may be detrimental to learning and performance.

If we are to improve the efficacy of simulators and training devices, we must identify the possible causal factors contributing to simulator sickness. Factors proposed include:

1. Mismatch between visual and vestibular cuing (Kennedy et al., 1990).
2. Visual and inertial lag discrepancies produced by the computational limitations of the simulator computer system (Kennedy et al., 1990).

3. Motion systems with resonant frequencies in the nausoegenic region (Frank et al., 1983).
4. Geometric distortions of the visual field that occur when the crewmember moves his or her head outside the center of projection (Rosinski, 1982).

Although this may seem to be an area amenable to additional research, both Guedry (1987) and Boldovici (1992) noted that, without incidence data obtained in the actual aircraft, objective assessments of the contribution of platform motion to simulator sickness will be difficult to obtain. Indeed, would the elimination of all simulator sickness be desirable, as that would change trainees' expectancies when they start flight training? Nonetheless, simulator sickness is a problem that does interfere with learning and even leads individuals to avoid using some simulators. Therefore, the following preventative strategies, proposed by McMillan (1994), McCauley and Sharkey (1991), Kennedy et al. (1990), and/or are contained in the *Simulator Sickness Field Manual* (Naval Training Systems Center, 1989) should be applied:

1. Monitor trainees new to the simulator more closely. Trainees with considerable flight time are especially vulnerable to simulator sickness.
2. Only use trainees, who are in their usual state of fitness. Avoid subjects with symptoms of fatigue, flu, ear infections, hangover, emotional stress, upset stomach, and so on.
3. For optimal adaptation there should be a minimum of 1 day and a maximum of 7 days between simulator sessions.
4. Simulator sessions should not exceed 2 hrs, indeed, shorter sessions are more desirable.
5. Minimize changes in orientation especially when simulating low-level flights.
6. Take steps to minimize abrupt changes in direction (e.g., altitude, roll, porpoising).
7. Use the freeze option only during straight and level flight.
8. Do not slew while the visual scene is displayed.
9. Use a reduced field of view for nauseogenic situations.
10. If the trainee shows initial signs of sickness have the trainee use flight instruments. If the symptoms increase the trainee should not return to the simulator until all symptoms have subsided (10–12 hrs).
11. Advise the trainee to minimize head movements during new situations.
12. When the trainee enters and exits the simulator, the visual display should be off and the simulation should be at zero degrees of pitch, yaw, and roll.
13. Maintain proper calibration of the visual and motion systems.

The occurrence of simulator sickness also has significant implications for "total immersion" proposed as part of virtual reality and the increased use of simulators in amusement parks. Indeed, in 1992, McCauley and Sharkey used the term "cybersickness" to describe virtual-reality-induced motion sickness. Kennedy et al. (1990), cautioned that the increased fidelity promised by virtual reality may lead to a higher incidence of simulator sickness. Clearly, further basic etiological research on the undesirable side effects of simulation and the development of appropriate countermeasures are needed.

INSTRUCTIONAL FEATURES OF SIMULATORS

Simulators incorporate many Advanced Instructional Features (AIFs), designed to enhance training. Although the list of AIFs presented in Table 15.1 is impressive, Polzella and Hubbard in 1986 reported that most AIFs are underutilized due to the minimal training provided to Simulation Instructors (SIs). Apparently the situation hasn't changed, for in 1992, Madden reported that most SI training was on-the-job, and indeed only 10% of training was classroom training or involved the use of an instructor's manual. Many manuals were described as "written for engineers," "user unfriendly," and "too technical."

Polzella and Hubbard (1986) reported that some AIFs may be more appropriate for initial-level training than for more advanced training. For example, the use of AIFs during initial-level training affords an opportunity for immediate feedback, whereas during advanced training the use of AIFs would disrupt the continuity of a realistic scenario. Jacobs et al. (1990) in their meta-analysis noted that the use of AIFs was rarely reported in the literature that they examined.

Little research has been performed on the training efficacy of AIFs in flight simulation, although most of the AIF strategies are based on the training and education literature. Hughes, Hannon, and Jones (1979) reported that playback was more effective in reducing errors during subsequent performance than demonstration. However, record/playback was no more effective than simple practice. Moreover, inappropriate use of the AIFs can contribute to problems. For example, use of the rewind and slew features while the scene is being observed by the trainee, or freezing the simulator in an unusual attitude, can contribute to simulator sickness (Kennedy et al., 1990).

The research specific to the use of AIFs in flight simulation indicates that appropriate use of AIFs can greatly facilitate learning. Backward chaining, a teaching strategy in which a procedure is decomposed into a chain of smaller elements and the student's training starts at the endpoint and proceeds back along the chain, appears to have considerable promise. For example, using backward chaining, a student would learn touchdown procedures first and gradually be repositioned further back on the glideslope. Backward chaining has been utilized successfully to train 30-degree dive-bombing maneuvers (Bailey, Hughes, & Jones, 1980) and simulated carrier landings (Wightman & Sistrunk, 1987).

Recently, under laboratory conditions, the time manipulation capability of simulators has produced some promising results. Using Above Real-Time Training (ARTT), in an F-16 part-task flight simulator, Guckenberger, Uliano, and Lane (1993) evaluated performance by F-16 pilots trained under varying rates of time compression (1.0×, 1.5×, 2.0×, and random order of time compression). When tested under real-time conditions and required to perform an emergency procedure in a simulated air combat task, the following differences were noted. Groups trained under ARTT conditions performed the emergency procedures tasks significantly more accurately than the group trained under the real-time condition. In addition, the ARTT groups "killed" six times more MIGs than the 1.0× group. Thus, it appears that ARTT can be used to train individuals to perform procedural tasks more accurately and in less time than traditional techniques.

Although AIFs have considerable promise, their use must be justified in terms of savings and transfer to the real world. A theory or model that estimates the amount of transfer and savings resulting from the use of particular AIFs is needed.

The authors believe that, as in most training, the skills, knowledge, and enthusiasm of the instructor as well as the management policy (and level of enforcement) greatly determine how the simulator is used and its ultimate effectiveness. Unfortunately, the

TABLE 15.1
Advanced Instructional Features

	Simulator Instructor (SI) Options
Preset/reset:	Starts/restarts the task at preset coordinates, with a predetermined aircraft configuration and environmental conditions.
Demonstration:	Demonstrates desired performance to the student.
Briefing:	Provides student with an overview of the planned training.
Slew/reposition:*	Moves the aircraft to a particular latitude, longitude, and altitude.
Repeat/Fly Out:	Allows the student to return to an earlier point, usually to where a problem has occurred in the training exercise, and "fly the aircraft out" from those conditions.
Crash/kill override:	Allows simulated flight to continue after a crash or kill.
Playback, replay:	Replays a selected portion of the flight. The playback may be time in real time, compressed time (higher rate) or expanded time (slower rate).
System Freeze:*	Temporarily stops the simulation while maintaining the visual scene and other data.
Automated/ adaptive training:	Computer algorithms vary level of task difficulty based on student performance. Under predetermined conditions augmented feedback may be provided.
Record:	Records student performance usually for either a set period of time or a portion of the training.
Motion:*	Turns motion parameters on or off.
Sound:	Turns sound on or off.
Partial Panel:	Selects which instrument to blank, thus simulating instrument failure.
Reliability:	Assigns a probability of failure to a particular system or display.
Scenery Select:	Selects terrain over which aircraft is to travel, defines level of detail, and amount/type of airborne traffic.
	Task Features
Malfunction:	Simulates sensor or instrument malfunction or failure.
Time Compression:	Reduces time available to perform the required tasks.
Time Expansion:	Increases the time available to accomplish the required tasks.
Scene Magnification:	Increases/decreases the magnification of the visual scene.
Environmental:	Manipulates time of day, seasons, weather, visibility, wind direction and velocity, etc.
Flight Dynamics:	Manipulates flight dynamic characteristics such as stability, realism, gain, etc.
Parameter Freeze:	"Locks in" a parameter such as altitude or heading; used to reduce task difficulty.
	Performance Analysis/Monitoring Features
Automated performance measurement and storage:	Collects data on student's performance during training and is used by the SI to evaluate student performance. On some systems data on the student's prior performance may be recovered and used for comparison purposes. These data can also become part of the normative database for the system.
Repeaters:	Displays cockpit instruments and switch status at the SI's console.
Closed Circuit:	Allows SI to visually monitor student's performance.
SI displays:	Presents student performance in an integrated or pictorial format such as a map overlay or sideview of an instrument approach.
Warnings:	Advises SI that student has exceeded a preset parameter (e.g., lowering gear above approved airspeed). Sometimes alerts, advising the student that a performance parameter is about to be exceeded or has been exceeded, are also presented in the cockpit during the training.
Debriefing Aids:	Presents student performance in a pictorial format; on some simulators selected parameters (airspeed, range to target) are also displayed or may be called up. Measures of effectiveness may also be provided.
Automated checkride:	An evaluation on a predetermined series of maneuvers for which performance test standards have been specified.

Note. Not all AIFs will be utilized during a given training period and not all simulators have all features. Material integrated from: Caro (1979), Polzella, Hubbard, Brown, and McLean (1987), Hays and Singer (1989), and Sticha, Singer, Blacksten, Morrison, and Cross (1990).

*See Simulator Sickness section of this chapter for cautions regarding use of these features.

SI is the forgotten component in the simulator system. As Hays, Jacobs, Prince, and Salas (1992) note, much simulator research is dependent on the subjective judgement of the SI. This is also true for pilot performance evaluations. In both the research and the operational world, strategies for improving the reliability and validity of subjective ratings need to be developed and evaluated. Greater emphasis on instructor training in the proper use of AIFs, and improved evaluation procedures, possibly combined with the development of expert system "trainers" as part of the software package promises considerable payoff.

PC-BASED FLIGHT SIMULATIONS

The increased capability of PC-based flight simulation has benefited from advances in computer technology (increased memory capability and processing speed) and reducing hardware and software costs (Sinnett, Oeting, & Selberg, 1989). The increased use of PC-based flight simulation had been documented at the American Society of Mechanical Engineers' Symposium (Sadlowe, 1991), and by Peterson (1992) and Williams (1994). Not only have computer hardware improvements resulted in near real-time flight simulation characteristics, but sophisticated interface media to represent yoke, throttle, and cockpit controls have been developed to better emulate the psychomotor aspects of aircraft control. Of particular significance are the capabilities today's software provides: (a) more realistic characterization of instrument navigation aids, (b) more realistic presentations of aircraft handling characteristics and instrumentation, and (c) a wide range of instructional features. As would be expected, most PC-based simulations have considerably less fidelity and lower cost than full-scale simulations. However, the lower levels of fidelity may be adequate in many research and training situations. Lower fidelity simulations have proven effective in evaluating the effects of: (a) automation (Thornton, Braun, Bowers, & Morgan, 1992; Bowers, Deaton, Oser, Prince, & Kolb, 1993), (b) scene detail and field of view during the introductory phases of flight training (Lintern, Taylor, Koonce, & Talleur, 1993), and (c) the development of aircrew coordination behavior (Bowers, Braun, Holmes, Morgan, & Salas, 1993). Beringer (1994) networked five PCs and combined two commercially available flight simulation packages to develop the simulator presented in Fig. 15.4. This approximately $25,000 apparatus is currently being used at the FAA's Civil Aeromedical Institute to compare two levels of navigational displays.

Another PC-based system, BFITS (Basic Flight Instruction Tutoring System), was developed as a research tool by the U.S. Air Force to determine learning rates of basic flying skills, ranging from climbs and descents to traffic pattern procedures (Benton, Corriveau, Koonce, & Tirre, 1992). The BFITS syllabus develops and evaluates both knowledge and flight skills. It contains an instructor module, a simulator module, and a performance evaluator module. Although BFITS is currently being validated, it contains the basic information that could be used in the development of an expert system, which would significantly reduce the workload imposed on the instructor by each student. However, as alluded to by Benton, Corriveau, Koonce, and Tirre, development of a system that could replace the on-board instructor in all circumstances would be difficult.

Because of the availability, flexibility, and low costs of PC-based simulations, efforts to determine the effectiveness of their transfer of training for general aviation (private,

FIG. 15.4. FAA's PC-based simulation facility. Photo courtesy of FAA Civil Aeromedical Institute.

noncommercial) have been undertaken. Taylor (1991) described a series of studies which utilized the ILLIMAC (University of Illinois Micro Aviation Computer) flight simulation system. The system utilizes an 8086, 16-bit microprocessor to control a fixed-base, general aviation trainer with the flight characteristics of the Piper Lance. Taylor reported that providing students, who have completed their private pilot certification program, with a concentration of instrument procedures on the ILLIMAC prepared them well for their commercial training. Based on the findings of these studies, an accelerated training program was developed and approved by the FAA. Under this program, students saved a full semester of flight training.

In a study at Embry-Riddle Aeronautical University, Moroney, Hampton, Biers, and Kirton (1994) compared the in-flight performance of 79 aviation students trained on one of two PC-Based Aircraft Training Devices (PCATDs) or a FAA-approved generic Training Device (TD). Student performance on six maneuvers and two categories of general flight skills was evaluated, based on criteria specified in the FAA's Performance Test Standards (PTS) for an instrument rating (FAA, 1989). For those factors evaluated, no significant difference in either the number of trials or hours to instrument flight proficiency in the aircraft was noted among those students taught in any of the three training devices. However, differences in student performance were noted in the number of trials/hours to proficiency in the TDs. Compared to students trained in the approved generic training device, students trained in the PCATDs required: (a) significantly fewer total trials, trials per task, and hours to reach the overall PTS, and (b) significantly fewer trials to reach proficiency in the following maneuvers: precision approach, nonprecision approach, timed turn to magnetic compass heading, and general flight skills (partial panel). Relative to cost, the training received in the PCATDs cost 46% less than the training received in the approved generic training device (mean savings of $463). Finally, the initial cost of the PCATDs, associated hardware, and software was approximately 8% of that of the approved TD ($4,600 and $60,000, respectively). Based on these findings, the authors recommended: (a) the use of PCATDs by general aviation, and (b) that steps be initiated to PCATDs as flight training devices, which could be used to accrue instrument rating credit.

More recently, Taylor et al. (1997) compared the performance of a group trained on a PCATD with the performance of a group trained entirely in an aircraft. Students trained on the PCATD completed the course in significantly less time than students trained in the aircraft. Taylor et al. (1997) reported substantial transfer from the PCATD to the aircraft for tasks such as ILS approaches, localizer back-course, and Non-Directional Beacon approaches. However, they reported lower transfer when the PCATD was used for reviewing tasks learned earlier in the course. They recommended that PCATD training be focused on those areas in which substantial transfer to the aircraft has been documented.

The research efforts just described contributed to the FAA's release of Advisory Circular 61-126 (1997), *Qualification and Approval of Personal Computer-Based Aviation Training Devices (AC No: 61-126)*. This document permits approved PCATDs, meeting the qualification criteria, to be used in lieu of up to 10 hours of time that ordinarily may be acquired in a flight simulator or flight training device authorized for use under Part 61 or Part 141 regulations. In a related effort, Moroney, Hampton, and Biers (1997) surveyed flight instructors, and used instructor focus groups to describe how the instructional features of PCATDs could be best utilized, how the instructor-software interface could be improved, and strategies for presenting and evaluating student performance.

SIMULATOR OR TRAINING DEVICE?

Throughout this chapter we have referred to all devices that use simulation as simulators. However, the FAA does not classify all such devices as "simulators." The FAA's Advisory Circular, *Airplane Simulator Qualification* (AC120-40B; FAA, 1991), defined an airplane simulator as:

> a full size replica of a specific type or make, model and series airplane cockpit, including the assemblage of equipment and computer programs necessary to represent the airplane in ground and flight operations, a visual system providing an out-of-the-cockpit view, and a force cuing system. (pp. 3, 4)

The FAA specifies four levels of simulators, ordered in increasing complexity from Level A through Level D. For example, the optical system for Levels A and B must have a minimum field of view of 45 degrees horizontal and 30 degrees vertical, and Levels C and D must provide an FOV of at least 150 degrees horizontal and 75 degrees vertical. Thus, many of the simulators discussed in this chapter do not meet the FAA's definition of a simulator but rather are classified as airplane training devices, which are defined in the *Airplane Flight Training Device Circular* (AC 120-45A; FAA, 1992) as:

> a full scale replica of an airplane's instruments, equipment, panels, and controls in an open flight deck area or an enclosed aircraft cockpit, including the assemblage of equipment and computer software programs necessary to represent the airplane in ground and flight conditions to the extent of the systems installed in the device; does not require force (motion) cuing or visual system. (p. 3)

There are seven levels of Flight Training Devices (FTDs). Level 1, the lowest level, is deliberately ambiguous and perhaps PC-based systems may qualify for this level. However, Level 7 FTDs must have the same lighting as the aircraft; use aircraft seats that can be positioned at the design-eye position; simulate all applicable flight, navigation, and systems operation; and provide significant aircraft noises (precipitation, windshield wipers); and so on.

Boothe (1994) commented that in simulators and FTDs, the emphasis is not just on accomplishing the required task, but on obtaining maximum "transfer of behavior" the task must be performed exactly as it would be in the aircraft. Thus the same control strategies and control inputs must be made in both the aircraft and the simulator. He believed that the emphasis should be on appropriate cues, as identified by pilots, who are the subject-matter experts. To achieve this end, Boothe argued for replication of form and function, flight and operational performance, and perceived flying (handling) qualities. He noted that these advisory circulars are developed by government and industry working groups, which utilize realism as their reference and safety as their justification.

Roscoe (1991) offered a counterposition. He argued that "qualification of ground-based training devices for training needs to be based on their effectiveness for that purpose and not solely on their verisimilitude to an airplane" (p. 870). Roscoe concluded that pilot certification should be based on demonstrated competence, not hours of flight experience. Lintern et al. (1990) argued further that for "effective and economical training, absolute fidelity is not needed nor always desirable, and some unreal-worldly training features can produce higher transfer than literal fidelity can" (p. 870). Caro (1988) added: "The cue information available in a particular simulator, rather than stimulus realism per se, should be the criterion for deciding what skills are to be taught in that simulator" (p. 239). Thus there are significant differences of opinion regarding both the definition and the requirements for the qualification of simulators.[3]

UNIQUE SIMULATORS

There are a number of unique simulators that merit discussion; however, only five of them are described briefly here. For an excellent review of the various unique simulators currently in use see Martin (1994). First is the LAMARS (Large Amplitude Multimode Aerospace Research Simulator) located at Wright-Patterson Air Force Base. The LAMARS (Martin, 1994) has a flight cab located at the end of a 20-ft movable arm. The cab can heave, sway, yaw, pitch, and roll, but cannot surge.

Second, the Dynamic Flight Simulator (DFS, see Fig. 15.5), located at the Naval Air Warfare Center (NAWC), has a cockpit in a two-axis gimbaled gondola at the end of a 50-ft arm in a centrifuge. The DFS can generate 40 g and has an onset rate of 13 g/sec

[3]Readers interested in U.S. Air Force requirements for flight simulators are referred to AFGS-87241 Guide Specification Simulators, Flight (U.S. Air Force, 1990).

FIG. 15.5. Dynamic Flight Simulator and centrifuge arm. Photo courtesy of Naval Air Warfare Center, Aircraft Division.

(Eyth & Heffner, 1992; Kiefer & Calvert, 1992). The pilot views three cathode ray tubes (see Fig. 15.6) which present the outside scene while the enclosed gondola responds with yaw, pitch and roll appropriate to the pilot's input (Cammarota, 1990). The DFS has been used to simulate G-forces sustained in air combat maneuvering, recoveries from flat spins, and high angle of attack flight.

A third type of motion simulator uses a cascading motion platform. Cascading refers to the approach of stacking one movable platform (or degree of freedom) on another so that although each platform in the stack has only one DOF, because it is mounted on other platforms additional DOF can be achieved without interactions between the platforms. The Vertical Motion Simulator (VMS) located at NASA Ames is used to simulate handling qualities of vertical take-off and landing (VTOL) aircraft. The VMS (Martin, 1994) has a 50-ft heave capability with a 16 ft/sec velocity. Limitations of cascading platforms include the size and cost of the facility.

The fourth simulator, TIFS (Total-In-Flight Simulation) is owned by the U.S. Air Force but operated by Calspan (see Fig. 15.7). TIFS is a simulator installed in an turboprop aircraft, and can be adapted to provide a variety of handling characteristics. The aircraft being "simulated" is flown from the simulator located in the nose of the aircraft, while a "safety" crew located in the aft cockpit is ready to take control of the aircraft if a hazardous condition arose. The TIFS has been used to simulate the handling qualities of aircraft as diverse as the Concorde, C-5, B-2, X-29, YF-23, and the Space Shuttle (V. J. Gawron, personal communication, September 1994).

The final simulator is the recently developed SIRE (Synthesized Immersion Research Environment) facility located within the Air Force Research Laboratory at Wright-Pat-

FIG. 15.6. Cockpit installed in gondola of Dynamic Flight Simulator (the top and bottom portions of the gondola have been removed). Photo courtesy of Naval Air Warfare Center, Aircraft Division.

FIG. 15.7. Total-in-Flight Simulation aircraft. Photo courtesy of Calspan Corporation.

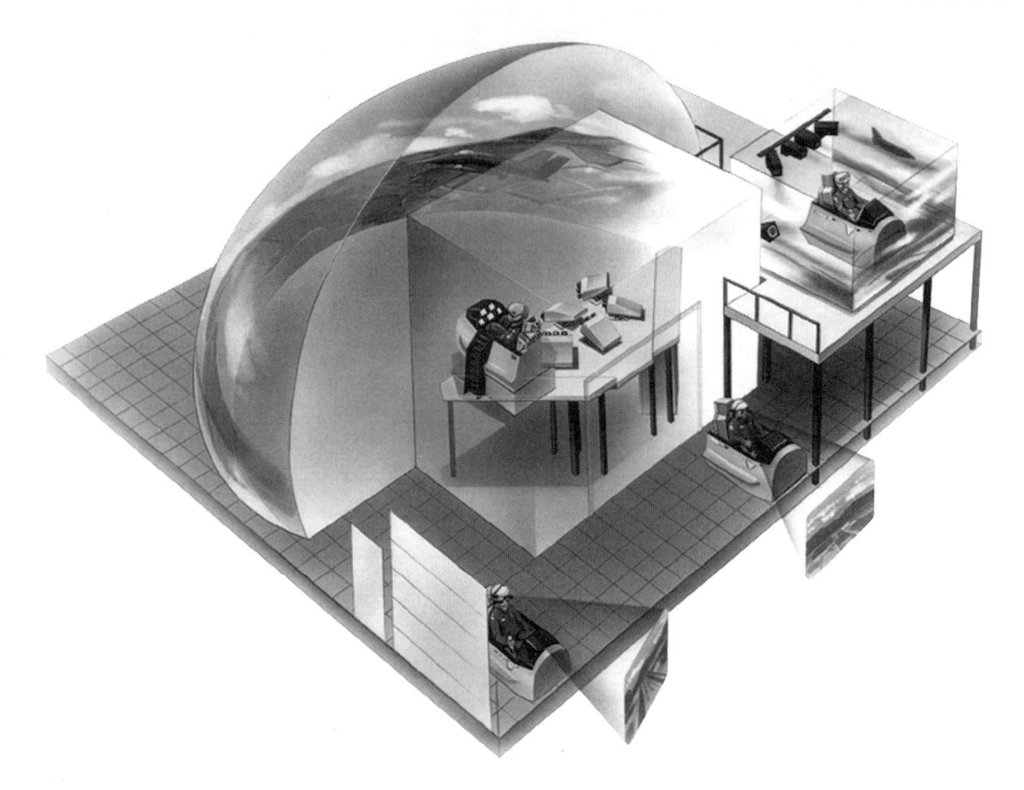

FIG. 15.8. The Synthesized Immersion Research Environment (SIRE) facility. Photo courtesy of the Crew Systems Interface Division of the Air Force Research Laboratory.

terson Air Force Base. SIRE (see Fig. 15.8) is a virtual environment research facility designed to develop and evaluate advanced, multi-sensory virtual interfaces for future U.S. Air Force crewstations. The main station of the SIRE facility is a 40 foot diameter dome, with a high-resolution, large FOV (70 degree vertical by 150 degree horizontal) visual imaging system. The station can be used to present three-dimensional sound information and has a electro-hydraulic control loader system. Several smaller independent cockpit simulators are tied into the main station, thus providing the capability for wingmen and adversary aircraft.

THE FUTURE OF FLIGHT SIMULATION

In lieu of the traditional summary, the authors feel that, at a minimum, a brief listing of their expectancies and research opportunities regarding flight simulation would be appropriate.

Expectancies

1. The use of simulation as a cost-effective alternate to operational training will increase. Although face validity (the look and feel of the vehicle being simulated) will remain a predominant factor in system design, the advantages of less costly, lower

fidelity simulation will reduce the emphasis on face validity. Thus the use of lower fidelity systems will increase.

2. Knowledge from the field of flight simulation will be transferred to fields as diverse as elementary schools, entertainment, and nuclear power plants.

3. Simulators will be linked into even larger, real-time interactive networks.

4. Simulators will continue to be used as procedure trainers and their role in evaluating and providing training in decision-making skills and cockpit/crew resource management will increase.

5. A number of active, operational, large-dome type simulators will be "retired" because of operating costs and changes in technology. However, dome type simulators will still serve as useful research tools.

6. The use of PC-based simulations and helmet-mounted display simulations will increase.

7. The Systems Approach to Training (SAT) or its analogues will encourage the use of: (a) Front End Analysis, (b) PC-based training systems, and (c) artificial intelligence and expert systems in training.

8. With the increasing use of simulators, greater emphasis should be placed on the role of the instructor. In time, training systems will incorporate instructor associates (i.e., interactive expert systems that will describe the goal of the training, demonstrate what is expected, and provide diagnostic feedback). The development of expert-system-based instructor associates promises considerable payoff.

Research Opportunities

1. Studies are needed that differentiate between tasks which can be learned most effectively and efficiently in training devices, simulators, and aircraft. Once the common characteristics of selected tasks have been identified, it should be possible to generalize these findings to other tasks.

2. Studies are needed which identify the cues necessary for effective and efficient transfer of training. In addition to these studies, task analytical techniques that can identify the essential cues need to be developed and validated. To maximize the return on investment, we need to identify the critical visual and motion cues and communicate that knowledge, in an appropriate form, to the system developer.

3. The role of the instructor is pivotal to flight simulation. The evaluation function performed by the instructor is primarily subjective, and the reliability and validity of instructor ratings could be improved. Additionally, the use of objective performance measures, and more formal strategies for displaying and evaluating student performance would greatly improve the contribution of the instructors.

4. Presently a variety of Advanced Instructional Features (AIFs) are available. To learn when and how to use AIFs, we need a theoretical base and a model that would provide valid and reliable estimates of the amount of transfer and savings resulting from the use of a particular AIF.

5. Developments in the area of virtual reality and virtual environments will require increased knowledge regarding cue presentation and human perception. Higher fidelity isn't necessarily better; indeed it may lead to increased simulator sickness.

6. As simulation expands into the vertical flight environment, we need to increase our knowledge of the control, display, and cue requirements unique to that environment.

7. Studies demonstrating the cost effectiveness of lower fidelity simulations are needed. These simulations could then be utilized by general, military, and commercial aviation.

CONCLUSION

This chapter began with the specification of the "human factor requirements" for the U.S. Army's first heavier-than-air flying machine. If we were to rewrite this statement as a requirement for today's flight simulator, perhaps it might read: "The flight simulator's cost-effective design should incorporate only those cues (at the appropriate level of fidelity) and instructional features necessary to permit an intelligent person to effectively learn and demonstrate the required skills at an appropriate level of proficiency within a reasonable period of time." Wilbur and Orville Wright delivered their heavier-than-air flying machine within 7 months after contract award. However, responding to the specification just given will take considerably longer and require more assets. Nonetheless, this specification is presented as a challenge to individuals involved in flight simulation. Indeed, if flight simulation is to advance, we must respond to the elements of this revised specification.

ACKNOWLEDGMENTS

The authors acknowledge the assistance of their colleagues and students at the University of Dayton and the University of Cincinnati. The assistance of friends in the aviation community, who not only shared material from their archives with us, but reviewed portions of the chapter, is also recognized. Finally, the authors thank their wives, Kathy and Hope, for their patience and support.

REFERENCES

Adorian, P., Staynes, W. N., & Bolton, M. (1979). The evolution of the flight simulator. *Proceedings of conference, 50 years of Flight Simulation* (Vol. 1, pp. 1–23). London: Royal Aeronautical Society.

Alessi, S. M. (1988). Fidelity in the design of instructional simulations. *Journal of Computer-Based Instruction, 15*(2), 40–47.

Alluisi, E. A. (1991). The development of technology for collective training: SIMNET, a case history. *Human Factors, 33*, 343–362.

Bailey, J., Hughes, R., & Jones, W. (1980). *Application of backward chaining to air to surface weapons delivery training* (AFHRL Tech. Rep. No. 79-63). Brooks Air Force Base, TX: Air Force Human Resources Laboratory.

Baudhuin, E. S. (1987). The design of industrial and flight simulators. In S. M. Cormier & J. D. Hagman (Eds.), *Transfer of learning* (pp. 217–237). San Diego, CA: Academic Press.

Benton, C. J., Corriveau, P., Koonce, J. M., & Tirre, W. C. (1992). *Development of the basic flight instruction tutoring system (BFITS)* (Vol. AL-TP-1991-0060). Brooks Air Force Base, TX: Air Force Systems Command.

Beringer, D. B. (1994). Issues in using off-the-shelf PC-based flight simulation for research and training: Historical perspective, current solutions and emerging technologies. *Proceedings of the Human Factors and Ergonomics Society 38th Annual Meeting, 1994* (pp. 90–94). Santa Monica, CA: Human Factors and Ergonomics Society.

Boldovici, J. A. (1987). Measuring transfer in military settings. In S. M. Cormier & J. D. Hagman (Eds.), *Transfer of learning* (pp. 239–260). San Diego, CA: Academic Press.

Boldovici, J. A. (1992). *Simulator motion* (TR 961, AD A257 683). Alexandria, VA: U.S. Army Research Institute.

Boothe, E. M. (1994). A regulatory view of flight simulator qualification. *Flight simulation update, 1994* (10th ed.). Binghamton, NY: SUNY Watson School of Engineering.

Bowers, C. A., Braun, C. C., Holmes, B. E., Morgan, B. B. J., & Salas, E. (1993). The development of aircrew coordination behaviors. In R. S. Jensen & D. Neumeister (Eds.), *Proceedings of the Seventh International Symposium on Aviation Psychology* (pp. 758–761). Columbus, OH: Aviation Psychology Laboratory, Ohio State University.

Bowers, C., Deaton, J., Oser, R., Prince, C., & Kolb, M. (1993). The impact of automation on crew communication and performance. In R. S. Jensen & D. Neumeister (Eds.), *Proceedings of the Seventh International Symposium on Aviation Psychology* (pp. 573–577). Columbus, OH: Aviation Psychology Laboratory, Ohio State University.

Bray, R. S. (1994). Cockpit motion in helicopter simulation. In W. E. Larsen, R. J. Randle, & L. N. Popiah (Eds.), *Vertical flight training: An overview of training and flight simulator technology with reference to rotary wing requirements* (NASA 1373, DOT/FAA/CT-94/83). Moffett Field, CA: NASA Ames.

Bunker, W. M. (1978). Training effectiveness versus simulation realism. *The Eleventh NTEC/Industry Conference Proceedings*. Orlando, FL: Naval Training Equipment Center. As cited in Hays, R. T., & Singer, M. J. (1989), *Simulation fidelity in training system design* (p. 53). New York, NY: Springer-Verlag.

Cammarota, J. P. (1990). Evaluation of full-sortie closed-loop simulated aerial combat maneuvering on the human centrifuge. *Proceedings on the National Aerospace and Electronics Conference* (pp. 838–842). Piscataway, NJ: IEEE.

Cardullo, F. M. (1994a). Motion and force cuing. *Flight simulation update, 1994* (10th ed.). Binghamton, NY: SUNY Watson School of Engineering.

Cardullo, F. (1994b). Simulation purpose and architecture. *Flight simulation update, 1994* (10th ed.). Binghamton, NY: SUNY Watson School of Engineering.

Caro, P. W. (1979). Development of simulator instructional feature design guides. *Proceedings of Conference: 50 Years of Flight Simulation* (pp. 75–89). London: Royal Aeronautical Society.

Caro, P. W. (1988). Flight training and simulation. In E. L. Weiner & D. C. Nagel (Eds.), *Human factors in aviation* (pp. 229–261). New York, NY: Academic Press.

Chambers, W. (1994). Visual simulation overview. In *Flight simulation update, 1994* (10th ed.). Binghamton, NY: SUNY Watson School of Engineering.

Cormier, S. M. (1987). The structural process underlying transfer of training. In S. M. Cormier & J. D. Hagman (Eds.), *Transfer of learning* (pp. 151–181). San Diego, CA: Academic Press.

Department of the Army. (1990). *Systems approach to training analysis* (TRADOC Pamphlet 351-4). Fort Monroe, VA: U.S. Army Training and Doctrine Command.

Eyth, J., & Heffner, P. (1992). *Design and performance of the centrifuge-based dynamic flight simulator* (AIAA Paper No. 92-4156, Flight Simulation Technologies Conference, Hilton Head, SC). Washington, DC: American Institute of Aeronautics and Astronautics.

Federal Aviation Administration. (1989). *Instrument rating: Practical test standards*. Washington, DC: Author.

Federal Aviation Administration. (1991). *Airplane simulator qualification*. Washington, DC: Author.

Federal Aviation Administration. (1992). *AC 120-45A: Airplane flight training device qualification*. Washington, DC: Author.

Federal Aviation Administration. (1997). *AC No: 61-126: Qualification and approval of personal computer-based aviation training devices*. Washington, DC: Author.

Fischetti, M. A., & Truxal, C. (1985, March). Simulating "The right stuff." *IEEE Spectrum*, 38–47.

Fitts, P. M. (1962). Factors in complex skill training. In R. Glaser (Ed.), *Training research and education* (pp. 77–197). Pittsburgh, PA: University of Pittsburgh Press.

Flexman, R. E., Matheny, W. G., & Brown, E. L. (1950). *Evaluation of the school link and special method of instruction in a 10 hour private pilot flight training program* (Bull. 47, No. 80). Champaign-Urbana, IL: University of Illinois.

Fortin, M. (1994). Image generation. In *Flight Simulation Update, 1994* (Vol. 10). Binghamton, NY: Watson School of Engineering, SUNY.

Frank, L. H., Kennedy, R. S., Kellogg, R. S., & McCauley, M. E. (1983). *Simulator sickness: Reaction to a transformed perceptual world. I. Scope of the problem* (Rep. No. NAVTRAEQUIPCEN TN-65). Orlando, FL: Naval Training Equipment Center.

Garcia, A. B., Gocke, R. P. J., & Johnson, N. P. (1994). *Virtual prototyping: Concept to production* (Rep. No. DSMC 1992-93). Fort Belvoir, VA: Defense Systems Management College Press.

Guckenberger, D., Uliano, K. C., & Lane, N. E. (1993). *Teaching high-performance skills using above-real-time training* (NASA Contractor Rep. No. 4528). Edwards AFB, CA: NASA Dryden.

Guedry, F. E. (1987). *Motion cues in flight simulation and simulator induced sickness* (Advisory Group for Aerospace Research-CP-433). Neuilly Sur Seine, France: NATO.

Haber, R. N. (1986). Flight simulation. *Scientific American, 255*(1), 96–103.

Hays, R. T., & Singer, M. J. (1989). *Simulation fidelity in training system design.* New York: Springer-Verlag.

Hays, R. T., Jacobs, J. W., Prince, C., & Salas, E. (1992). Requirements for future research in flight simulation training: Guidance based on a meta-analytic review. *International Journal of Aviation Psychology, 2*(2), 143–158.

Hettinger, L. J., Berbaum, K. S., Kennedy, R. S., Dunlap, W. P., & Nolan, M. D. (1990). Vection and simulator sickness. *Military Psychology, 2*, 171–181.

Hughes, R. G., Hannon, S. T., & Jones.W. E. (1979). *Application of flight simulator record/playback feature* (Rep. No. AFHRL-TR-79-52, AD A081 752). Williams AFB, AZ: Air Force Human Resources Laboratory.

Hughes, R., Brooks, R., Graham, D. Sheen, R., & Dickens, T. (1982). Tactical ground attack: On the transfer of training from flight simulator to operational Red Flag range exercise. *Proceedings of the Human Factors and Ergonomics Society 26th Annual Meeting, 1982* (pp. 596–600). Santa Monica, CA: Human Factors and Ergonomics Society.

Jacobs, J. W., Prince, C., Hays, R. T., & Salas, E. (1990). *A meta-analysis of the flight simulator research* (Tech. Rep. No. 89-006). Orlando, FL: Naval Training Systems Center.

Jones, E. R. (1967). *Simulation applied to education.* Unpublished manuscript, McDonnell-Douglas Corporation, St. Louis, MO.

Jones, E., Hennessy, R., & Deutsch, S. (Eds.). (1985). *Human factors aspects of simulation.* Washington, DC: National Academy Press.

Kass, S. J., Herschler, D. A., & Companion, M. A. (1991). Training situational awareness through pattern recognition in a battlefield environment. *Military Psychology, 3*(2), 105–112.

Kennedy, R. S., & Fowlkes, J. E. (1992). Simulator sickness is polygenic and polysymptomatic: Implications for research. *International Journal of Aviation Psychology, 2*(1) , 23–38.

Kennedy, R. S., Hettinger, L. J., & Lilienthal, M. G. (1990). Simulator sickness. In G. H. Crampton (Ed.), *Motion and space sickness* (pp. 317–341). Boca Raton, FL: CRC Press.

Kennedy, R. S., Lilienthal, M. G., Berbaum, K. S., Baltzley, D. R., & McCauley, M. E. (1989). Simulator Sickness in U.S. Navy flight simulators. *Aviation, Space and Environmental Medicine, 60*, 10–16.

Kiefer, D. A., & Calvert, J. F. (1992). *Developmental evaluation of a centrifuge flight simulator as an enhanced maneuverability flying qualities tool* (Paper No. 92-4157). Washington, DC: American Institute of Aeronautics and Astronautics.

Lane, N. E. (1986). *Issues in performance measurement for military aviation with applications to air combat maneuvering.* Orlando, FL: Naval Training Systems Center.

Lintern, G. (1991). An informational perspective on skill transfer in human–machine systems. *Human Factors, 33*(3), 251–266.

Lintern, G., & McMillan, G. (1993). Transfer for flight simulation. In R. A. Telfer (Ed.), *Aviation instruction and training* (pp. 130–162). Brookfield, VT: Ashgate Publishing Co.

Lintern, G., Roscoe, S. N., & Sivier, J. E. (1990). Display principles, control dynamics, and environmental factors in augmentation of simulated visual scenes for teaching air-to-ground attack. *Human Factors, 32*, 299–371.

Lintern, G., Taylor, H. L., Koonce, J. M., & Talleur, D. A. (1993). An incremental transfer study if scene detail and field of view effects on beginning flight training. *Proceedings of Seventh International Symposium on Aviation Psychology* (pp. 737–742). Columbus, OH: Ohio State University.

Loesch, R. L., & Waddell, J. (1979). The importance of stability and control fidelity in simulation. *Proceedings of Conference: 50 Years of Flight Simulation* (pp. 90–94). London: Royal Aeronautical Society.

Madden, J. J. (1992). Improved instructor station design. *Proceedings of the 15th Interservice/Industry Training Systems Conference* (pp. 72–79). Orlando, FL: Naval Training Systems Center.

Martin, E. A. (1994). Motion and force simulation systems I. *Flight simulation update-1994* (10th ed.). Binghamton, NY: SUNY Watson School of Engineering.

Martin, E. L. (1981). *Training effectiveness of platform motion: Review of motion research involving the advanced simulator for pilot training and the simulator for air-to-air combat* (Rep. No. AFHRL-TR-79-51). Williams Air Force Base, AZ: Air Force Human Resources Laboratory.

Martin, E. L., & Wagg, W. L. (1978a). *Contributions of platform motion to simulator training effectiveness: Study 1-Basic Contact* (Tech. Rep. No. AFHRL TR-78-15, AD A058 416). Williams Air Force Base, AZ: Air Force Human Resources Laboratory.

Martin, E. L., & Waag, W. L. (1978b). *Contributions of platform motion to simulator training effectiveness: Study II-aerobatics* (Tech. Rep. No. AFHRL TR-78-52. AD A064 305). Williams Air Force Base, AZ: Air Force Human Resources Laboratory.

McCauley, M. E. (Ed.). (1984). *Research issues in simulator sickness*. Washington, DC: National Academy Press.

McCauley, M. E., & Sharkey, T. J. (1991). Spatial orientation and dynamics in virtual reality systems: Lessons from flight simulation. *Proceedings of the Human Factors and Ergonomics Society 35th Annual Meeting, 1991* (pp. 1348–1352). Santa Monica, CA: Human Factors and Ergonomics Society.

McCauley, M. E., & Sharkey, T. J. (1992). Cyberspace: Perception of self-motion in virtual environments. *Presence, 1*(2), 311–318.

McDaniel, W. C., Scott, P. G., & Browning, R. F. (1983). *Contribution of platform motion simulation in SH-3 helicopter pilot training* (TAEG Rep. No. 153). Orlando, FL: Training Analysis and Evaluation Group, Naval Training Systems Center.

McMillan, G. (1994). System integration. *Flight simulation update, 1994* (10th ed.). Binghamton, NY: SUNY Watson School of Engineering.

Miller, R. B. (1954). *Psychological considerations in the design of training equipment* (Tech. Rep. No. 54-563). Wright Patterson AFB, OH: Wright Air Development Center.

Mixon, T. R., & Moroney, W. F. (1982). *An annotated bibliography of objective pilot performance measures* (Rep. No. NAVTRAEQUIPCEN IH 330). Orlando, FL: Naval Training Equipment Center.

Money, K. E. (1970). Motion sickness. *Physiological Reviews, 50*, 1–39.

Moroney, W. F., Hampton, S., & Beirs, D. W. (1997). Considerations in the design and use of personal computer-based aircraft training devices (PCATDs) for instrument flight training: A survey of instructors. *Proceedings of Ninth International Symposium on Aviation Psychology*. Columbus, OH: Ohio State University.

Moroney, W. F., Hampton, S., Beirs, D. W., & Kirton, T. (1994). The use of personal computer-based training devices in teaching instrument flying: A comparative study. *Proceedings of the Human Factors and Ergonomics Society 38th Annual Meeting, 1994* (pp. 95–99). Santa Monica, CA: Human Factors and Ergonomics Society.

Mosher, S., Farmer, D., Cobasko, J., Stassen, M., & Rosenshein, L. (1992). *Innovative display concepts for field-of-view expansion in air combat simulation* (WLTR-92-3091). Wright Patterson AFB, OH: Flight Dynamics Directorate, Wright Laboratory.

Naval Training Systems Center. (1989). *Simulator sickness*. Orlando, FL: Naval Training Systems Center.

Orlansky, J., & String, J. (1977). *Cost-effectiveness of flight simulator for military training* (Rep. No. IDA NO. HQ 77-19470). Arlington, VA: Institute for Defense Analysis.

Peterson, C. (1992, June). Simulation symposium: Are PC-based flight simulators coming of age? *Private Pilot*, pp. 75–79.

Pfeiffer, M. G., Horey, J. D., & Butrimas, S. K. (1991). Transfer of simulated instrument training to instrument and contact flight. *International Journal of Aviation Psychology, 1*(3), 219–229.

Polzella, D. J., & Hubbard, D. C. (1986). Utility and utilization of aircrew training device advanced instructional features. *Proceedings of the Human Factors Society 30th Annual Meeting, 1986* (pp. 139–143). Santa Monica, CA: Human Factors Society.

Polzella, D. J., Hubbard, D. C., Brown, J. E., & McLean, H. C. (1987). *Aircrew training devices: Utility and utilization of advanced instructional features* (Tech. Rep. No. AFHRL-TR-87-21). Williams Air Force Base, AZ: Operation Training Division.

Randle, R. J., & Sinacori, J. (1994). Visual space perception in flight simulators. In W. E. Larsen, R. J. Randle, & L. N. Popiah (Eds.), *Vertical flight training: An overview of training and flight simulator technology with reference to rotary wing requirements* (NASA 1373, DOT/FAA/CT-94/83). Moffett Field, CA: NASA Ames.

Reason, J. T. (1978). Motion sickness adaptation: A neural mismatch model. *Journal of the Royal Society of Medicine, 71*, 819–829.

Rolfe, J. M., & Staples, K. J. (Eds.). (1986). *Flight simulation*. Avon: Cambridge University Press.

Roscoe, S. N. (1991). Simulator qualification: Just as phony as it can be. *International Journal of Aviation Psychology, 1*(4), 335–339.

Roscoe, S. N. (1980). Transfer and cost effectiveness of ground-based flight trainers. In S. N. Roscoe (Ed.), *Aviation psychology* (pp. 194–203). Ames, IA: Iowa State University Press.

Roscoe, S. N., & Williges, B. H. (1980). Measurement of transfer of training. In S. N. Roscoe (Ed.), *Aviation psychology* (pp. 182–193). Ames, IA: Iowa State University Press.

Rosinski, R. R. (1982). *Effect of projective distortions on perception of graphic displays* (Rep. No. 82-1). Washington, DC: Office of Naval Research.

Royal Aeronautical Society. (1979). *Fifty years of flight simulation* (Vols. I, II, and III). London, England: Royal Aeronautical Society.

Sadlowe, A. R. (Ed.). (1991). *PC-based instrument flight simulation: A first collection of papers*. New York: The American Society of Mechanical Engineers.

Seidensticker, S. (1994). Distributed interactive simulation (DIS). In *Flight simulation update, 1994* (10th ed.). Binghamton, NY: SUNY Watson School of Engineering.

Sharkey, T. J., & McCauley, M. E. (1991). The effect of global visual flow on simulator sickness. *Proceedings of the AIAA Flight Simulation Technologies Conference* (Rep. No. AIAA-91-2975-CP) (pp. 496–504). Washington, DC: American Institute of Aeronautics and Astronautics.

Sharkey, T. J., & McCauley, M. E. (1992). Does a motion base prevent simulator sickness? *Proceedings of the AIAA Flight Simulation Technologies Conference* (pp. 21–28). Washington, DC: American Institute of Aeronautics and Astronautics.

Sinnett, M. K., Oetting, R. B., & Selberg, B. P. (1989). Improving computer technologies for real-time digital flight simulation. *SAE Aerospace Technology Conference and Exposition* (pp. 1826–1829). Long Beach, CA: SAE.

Sparaco, P. (1994, June). Simulation acquisition nears completion. *Aviation Week & Space Technology*, pp. 71–72.

Spears, W. D., Sheppard, H. J., Roush, M. D., & Richetti, C. L. (1981a). *Simulator training requirements and effectiveness study (STRES)* (Tech. Rep. No. AFHRL-TR-80-38, Part I). Dayton, OH: Logistics and Technical Training Division.

Spears, W. D., Sheppard, H. J., Roush, M. D. I., & Richetti, C. L. (1981b). *Simulator training requirements and effectiveness study (STRES)* (Tech. Rep. No. AFHRL-TR-80-38, Part II). Dayton, OH: Logistics and Technical Training Division.

Stark, E. A. (1994). Training and human factors in flight simulation. *Flight simulation update* (10th ed.). Binghamton, NY: SUNY Watson School of Engineering.

Sticha, P. J., Buede, D. M., Singer, M. J., Gilligan, E. L., Mumaw, R. J., & Morrison, J. E. (1990). *Optimization of simulation-based training systems: Model description, implementation, and evaluation* (Tech. Rep. No. 896). Alexandria, VA: U.S. Army Research Institute.

Sticha, P. J., Singer, M. J., Blacksten, H. R., Morrison, J. E., & Cross, K. D. (1990). *Research and methods for simulation design: State of the art* (Rep. No. ARI-TR-914). Alexandria, VA: U.S. Army Research Institute for the Behavioral and Social Sciences.

Stoffregen, T. A., & Riccio, G. E. (1991). An ecological critique of the sensory conflict theory of motion sickness. *Ecological Psychology, 3*(3), 159–194.

Taylor, H. L. (1991). ILLIMAC: A microprocessor based instrument flight trainer. In A. R. Sadlowe (Ed.), *PC-based instrument flight simulation: A first collection of papers* (pp. 1–11). New York: The American Society of Mechanical Engineers.

Taylor, H. L., Lintern, G., Hulin, C. L., Talleur, D. A., Emanuel, T. W., & Phillips, S. I. (1997). Effectiveness of personal computers for instrument training. *Proceedings of Ninth International Symposium on Aviation Psychology*. Columbus, OH: Ohio State University.

Thornton, C., Braun, C., Bowers, C., & Morgan, B. B. J. (1992). Automation effects in the cockpit: A low fidelity investigation. *Proceedings of the Human Factors Society 36th Annual Meeting, 1992* (pp. 30–34). Santa Monica, CA: Human Factors and Ergonomics Society.

U.S. Air Force (1990). USAF Guide Specification: Simulators (AFSGS-8724A ed.). Wright Patterson AFB, OH: Aeronautical Systems Directorate (Code ENES).

U.S. Army Signal Corps Specification No. 486. (1907). *Advertisement and specification for a heavier than air flying machine.* Washington, DC: Department of the Army. (On display at USAF Museum, Wright Patterson AFB, Dayton, OH)

Warren, R., & Riccio, G. E. (1985). Visual cue dominance hierarchies: Implications for simulator design. *Proceedings of the Aerospace Technology Conference* (pp. 61–74). Washington, DC: AIAA.

Wiener E. L., & Nagel, D. C. (1988). *Human factors in aviation.* New York: Academic Press.

Williams, K. W. (1994). *Proceedings of the joint industry–FAA conference on the development and use of PC-based aviation training devices* (Rep. No. DOT/FAA/AM-94.25). Oklahoma City, OK: Civil Aeromedical Institute.

Wightman, D. C., & Sistrunk, F. (1987). Part-task training strategies in simulated carrier landing final approach training. *Human Factors, 29*(3), 245–254.

16

Human Factors Considerations in Aircraft Cabin Design

Lori Emenaker Kovarik
R. Curtis Graeber
Peter R. Mitchell
The Boeing Company, Seattle, WA

"Why are airplane seats so uncomfortable?" Aircraft cabin designers, especially human factors specialists, hear this question all too frequently. The short answer is, of course, that comfort is relative. The long answer is a good example of how human factors considerations are applied to the design of the entire aircraft cabin.

Good seat design, as well as good aircraft cabin design, depends on the effective management of design requirements. These design requirements usually include a wide range of frequently competing aspects, only some of which involve human factors considerations. Aircraft manufacturing costs, operating economics, safety, available technology, producibility, regulations, and maintainability are just a few of the key considerations in the design process. Specialists in each of these fields are equally passionate about the importance of their particular requirements in the final design. It is the job of human factors specialists to serve as advocates for all users—from the passenger to the crew, the installation mechanics, and the aircraft cleaning and maintenance personnel—to ensure that all their needs are met.

As technology moves forward, airplanes can fly farther, faster, and higher and operate more efficiently than ever before. Throughout aviation history, these technological advances have consistently placed new and different demands on human physiology, ergonomics, and performance, thus providing a continual challenge for human factors engineers.

EVOLUTION OF AIRCRAFT CABIN HUMAN FACTORS

It all began when the first paying passenger boarded an airmail carrier for a cross-country flight. Because airmail carriers were already making money, any additional revenue they could generate from carrying passengers was pure profit. The earliest revenue passengers were seated with the pilot in open cockpits. The only human factors goals then were to keep passengers alive and unharmed. Passengers were given items, such as leather overcoats, helmets, hot water bottles, and cotton balls (to place in their ears), to help protect them from the elements.

Competition between air carriers for revenue passengers quickly drove the evolution of cabin human factors from merely keeping passengers alive to keeping them comfortable, and, ultimately, happy. Open cockpits gave way to larger, enclosed passenger cabins, but comfort was still not a main design consideration. Not all enclosed cabins had windows, and passengers often experienced cold temperatures, loud noises, noxious fumes, and excessive vibration during their flight. Airsickness was common, and buckets were often provided for passenger convenience (Serling, 1992).

As air travel became more commonplace in the 1920s, airlines realized they garnered more repeat business by keeping passengers more comfortable. The interior of the aircraft cabin began to resemble railroad train interiors, with fixed seat rows and mesh overhead luggage racks. Seats became adjustable and windows were fitted with roller blinds to give passengers some control over their environment. The Fokker Company introduced the first onboard lavatory in 1924. The same aircraft offered sliding windows to provide fresh air, finally offering some relief from airsickness (Edwards & Edwards, 1990).

Such amenities as leather-covered seats, heated cabins, individual reading lights, and hot and cold running water appeared. In 1925, one airline tried to distract passengers from air travel discomforts by showing an in-flight film. One year later, a European carrier began serving lunch and drinks on flights between London and Paris, and soon after, Boeing Air Transport hired registered nurses to attend to passengers' safety and comfort needs (Serling, 1992).

During the 1930s, cabin luxury seemed effective in taking passengers' minds off flying, as well as luring them to travel by air instead of by land or sea. As airplanes flew increasingly longer routes, more amenities were necessary to make the flight comfortable. Fortunately, technology enabled airplanes to become larger, and more of these luxurious amenities could be offered. Sleeping berths; honeymoon suites; promenade decks; carpets; wood-paneled walls and ceilings; padded, upholstered chairs; and separate lavatories for men and women were available in soundproofed, ventilated cabins. Hot meals, prepared in the airplane, were served in dining rooms at tables laid with linen and heavy silverware. In 1940, Boeing introduced the first pressurized cabin into airline service (Edwards & Edwards, 1990). The ability to fly above the weather went a long way toward eliminating the problem of airsickness.

Appealing to passengers' sense of luxury was gradually replaced by an appeal to their sense of economy. Airlines recognized that they could entice more people with cheaper fares than with seven-course gourmet meals. Passenger airplanes were becoming bigger, and jet engines were replacing propellers. The Boeing 707 that made its inaugural transatlantic flight in 1958 could seat 84 passengers. The Douglas DC-8 made its debut in 1960 and carried up to 179 passengers. In 1970, Pan Am inaugurated Boeing 747 service between London and New York with a capacity of 490 passengers. As airlines attempted to maximize the number of passengers per flight, the human factors emphasis was refocused from surrounding a few passengers in luxury to keeping as many passengers as safe and comfortable as possible.

Airplane cabins of the past were typically designed to accommodate the needs of just the passengers and crew. Today, the definition of the cabin *end user* has expanded. A variety of factors, including airline operating economics, new regulations, and the need to avoid flight delays, has prompted airlines and manufacturers to consider maintenance and manufacturing to be important end-user tasks also.

As for passengers and crew, more people are flying, and they want to fly longer distances in less time at reduced costs. Aircraft design technology is attempting to meet these needs. Bigger airplanes, faster speeds, and longer flights continue to provide new issues for human factors specialists to consider.

AIRCRAFT CABIN DESIGN TODAY

After more than 70 years of evolution and technological advances, some airplane seats are still considered uncomfortable. Again, this is because comfort is only one of many requirements that are considered in the seat design process. To begin narrowing down the huge array of design options and requirements, a designer first needs to know the specifics of the overall aircraft design. Defining the boundaries, or envelope, of a new or derivative aircraft is known as the *aircraft configuration process*. Once the configuration is established, the design is completed during the detail design process.

Aircraft Configuration

Before any configuration design can begin, the aircraft mission requirements must be defined. Flight range, passenger and cargo capacity, airport infrastructure requirements, and various regulatory requirements are all addressed and defined based on current and predicted market needs. Once these are established, the aircraft configuration designers have some guidelines within which to work.

Overall configuration is a *macrolevel* design process conducted mainly by engineering and marketing organizations. This is where the aircraft cabin concept, typically in the form of a floor plan and cross section, is created. The goal is to develop an aircraft configuration that is certifiable, economically viable, and competitive in current and future markets, with future derivative growth potential. The aircraft configuration that is defined at this stage will drive the detail design of the entire aircraft.

Before a detail design can begin, a scheduled, disciplined process for establishing cross-functional responsibilities is necessary to ensure that requirements are evaluated in the context of the overall design. Even at this early stage, engineering and marketing groups are compromising and conducting *trade-offs* among various factors as these groups work together to develop a safe design that can successfully meet a wide variety of customer needs.

At this larger, macrolevel, human factors considerations are usually limited to passengers and crew. One of the human factors elements that designers must consider in evaluating design trade-offs is passenger comfort. When designing long-range airplanes, more consideration is given to passenger comfort because this is an important factor for passengers making long-distance travel plans (Brauer, 1996).

Cross-Section Development

Many physical factors affect passenger comfort: sidewall clearances, ceiling and stowbin clearances, cabin altitude pressurization, temperature, decor, lighting, noise, humidity, ozone levels, total number of passengers, and seating configuration are just a sample.

During the aircraft configuration process, the architectural aspects of these factors are explored by developing conceptual cross sections of the airplane. The following cross sections show how head, shoulder, armrest, feet clearances, and stowage bin

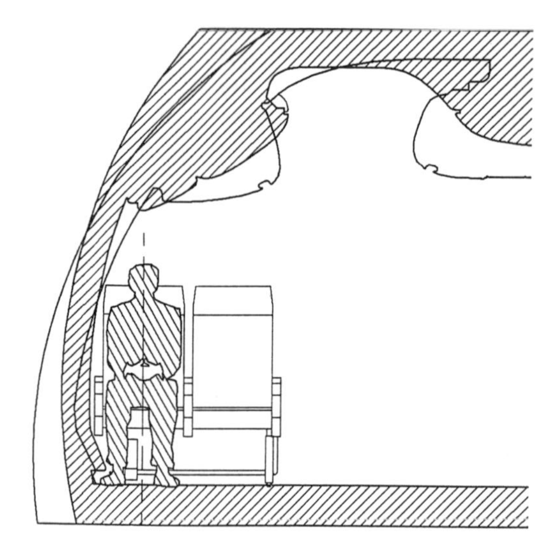

FIG. 16.1. Cross sections showing different sidewall and stowbin clearances.

capabilities are affected by the radius of the fuselage and the location of the floor (see Fig. 16.1). In addition to considering the physical aspects of comfort, designers also consider the perceptual aspects of comfort. Cabin architecture and lighting play important roles in these perceptions. The Boeing 777 cross section was designed to maximize stowage bin volume while maintaining an open, roomy feeling in the cabin. Designers accomplished this by integrating the architectural requirements with the functional needs of passengers, attendants, and maintenance personnel (see Fig. 16.2).

Seat spacing and configuration have a great influence on the perception of comfort. Economy-class passengers feel more comfortable when seated next to a window, aisle, or empty seat than they do when seated between two passengers (Brauer, 1996). With this in mind, a cross section can be developed that places the maximum number of passengers next to windows, or aisles, or even an empty seat when load factors allow. Computing tools enable engineers and designers to calculate a passenger's total personal space in various seating configurations, with the goal of maximizing that passenger's perception of comfort (Brauer, 1996; see Fig. 16.3).

Aircraft Floor Plan

Once the cross section has been developed in conjunction with the overall aircraft configuration, the possible variations of the airplane interior are laid out. Key considerations are architectural integration, flexibility to adapt to unique customer requirements, and manufacturability. Activities in this part of the design process include identification of the relationships among major passenger cabin components and determining the potential locations of seats, galleys, lavatories, closets, attendant seats, entertainment equipment, passenger and cargo doors, emergency exits, system locations, evacuation exits, and the like. Many design trade-offs (also known as *trades*) are evaluated, and the results depend on aircraft mission requirements, regulatory requirements, and the type or combination of service—economy, business, or first class—the airline intends to offer.

One such trade-off involves the development of customer-variable (or flexibility) zones. These are areas within the aircraft cabin that can be easily converted into different seating or cargo configurations after delivery. (Remember, an airplane may be in service

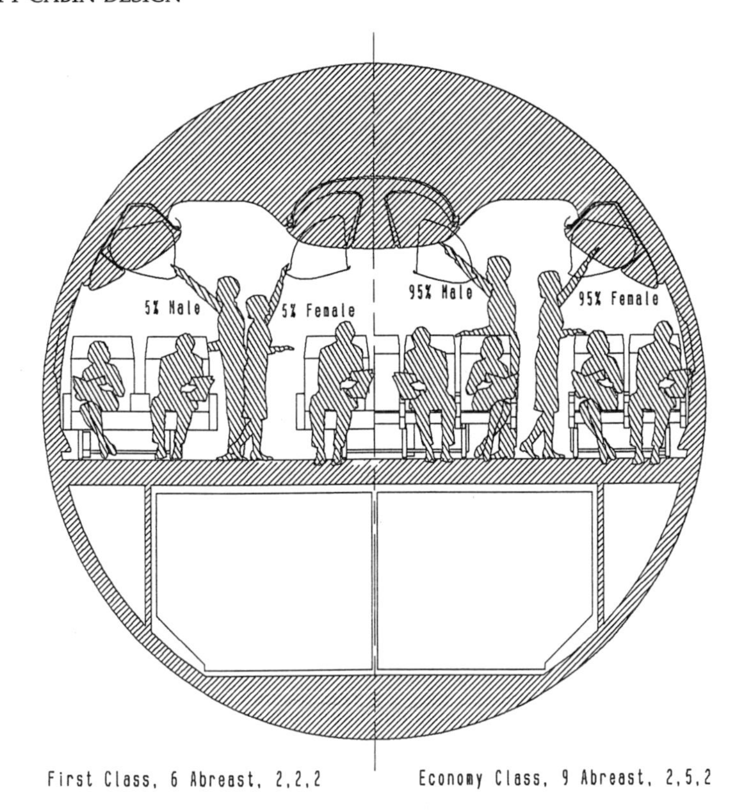

FIG. 16.2. Overhead cabin architecture.

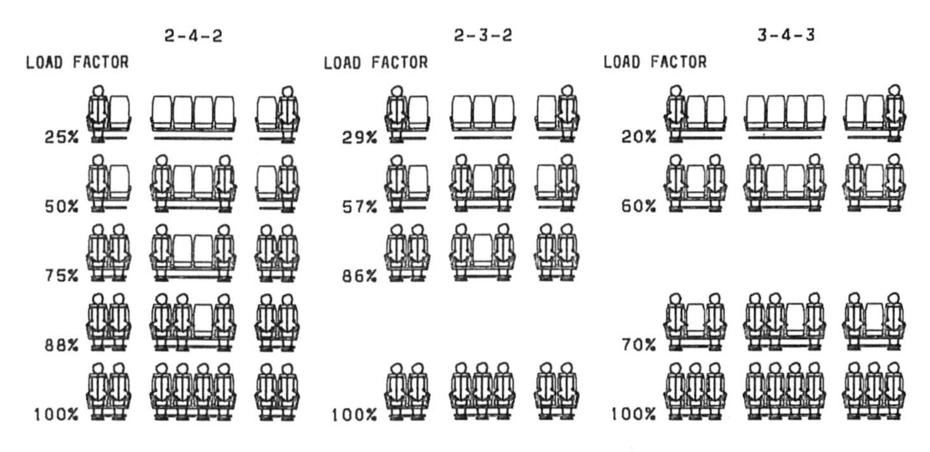

FIG. 16.3. Comparison of seating configurations for maximum/minimum passenger comfort.

for more than 30 years.) To analyze this trade-off, many design and support organizations, airline customers, manufacturers, and maintenance engineering groups identify the benefits and costs of making these areas fixed or flexible. The trade-off variables may include weight, potential mission, system complexity, cost, manufacturability, engineering design hours, and customer needs. Technical issues involve manufacturing materials (weight vs. increased corrosion resistance), geometry changes (creation of areas that isolate systems and architecture that do not change with seating, partition, lavatory, and galley changes), and provisions for systems.

Another example of this type of trade-off is the determination of the locations of the airplane doors and hatches. Technical issues for door and hatch location include body loads; wing and engine locations; Federal Aviation Administration (FAA) and Joint Aviation Authorities (JAA) regulations; technical limitations on escape slide design (also strongly influenced by FAA and JAA regulations); anticipated customer requirements for shifting zones (size and isolation); and door use for boarding, deplaning, and evacuation.

In addition to the mission requirements that must be satisfied by conducting design trade-offs, certification authorities, such as the FAA, have established safety and operational certification regulations relating to passengers and cabin crew. These regulations ensure that at least minimum safety equipment and features are designed into the cabin. They offer design requirements and usually represent the starting point of a design effort. Designing efficiently to these regulatory requirements can be challenging because, although they are explicit in some areas, they are not all-encompassing. Table 16.1 lists a sample of some of these requirements under Federal Air Regulations (FAR) Part 25. Designers must also consider the airlines' use of an aircraft under FAR Part 121.

Supplementing mission and regulatory requirements, various competitive and airline standards offer minimum ratios for such items as galley volume per passenger by class and number of passengers per lavatory by class. Figure 16.4 illustrates how much these items can vary. When the mission, certification, and standard requirements are met, the designers can then offer cabin amenities to make the aircraft more competitive on the current market. Enhanced facilities for disabled passengers, increased galley and

TABLE 16.1
Federal Air Regulations Applicable to Aircraft Cabin Design

- FAR 25.561, Emergency Landing Conditions—General
- FAR 25.562, Emergency Landing Dynamic Conditions
- Advisory Circular 25.562-1, Dynamic Evaluation of Seat Restraint Systems and Occupant Protection on Transport Airplanes
- FAR 25.783, Doors
- Advisory Circular 25.783-1, Fuselage Doors, Hatches, and Exits
- FAR 25.785, Seats, Berths, Safety Belts, and Harnesses
- Advisory Circular 25.785-1, Flight Attendants Seat Requirements
- FAR 25.787, Stowage Compartments
- FAR 25.789, Retention of Items of Mass
- FAR 25.791, Passenger Information Signs
- FAR 25.803, Emergency Evacuation
- Advisory Circular 25.803-1, Emergency Evacuation Demonstrations
- FAR 25.807, Passenger Emergency Exits
- FAR 25.809, Emergency Exit Arrangement
- FAR 25.811, Emergency Exit Marking
- FAR 25.812, Emergency Lighting
- Advisory Circular 25.812-1A, Floor Proximity Emergency Escape Path Marking
- Advisory Circular 20-38A, Measure of Cabin Interior Emergency Illumination in Transport Airplanes
- FAR 25.813, Emergency Exit Access
- FAR 25.815, Width of Main Aisle
- FAR 25.819, Lower Deck Service Compartments
- FAR 25.851, Fire Extinguishers
- Advisory Circular 20-42C, Hand Fire Extinguishers for Use in Aircraft
- FAR 25.853, Compartment Interiors
- Advisory Circular 25.853-1, Flammability Requirements for Aircraft Seat Cushions
- FAR 25.1411, Safety Equipment—General
- FAR 25.1447, Equipment for Oxygen Dispensing Units
- FAR 25.1450, Chemical Oxygen Generators
- FAR 25.1541, Markings and Placards—General
- FAR 25.1557, Miscellaneous Markings and Placards
- FAR 25.1561, Safety Equipment
- Advisory Circular 25-9, Smoke Detection, Penetration, and Evacuation Tests, and Related Flight Manual Emergency Procedures
- Advisory Circular 25-17, Transport Airplane Cabin Interiors Crashworthiness Handbook
- Advisory Circular 120-38, Transport Category Airplanes Cabin Ozone Concentrations
- Advisory Circular 121-24A, Passenger Safety Information Briefing and Briefing Cards

Interior Arrangement

777-200

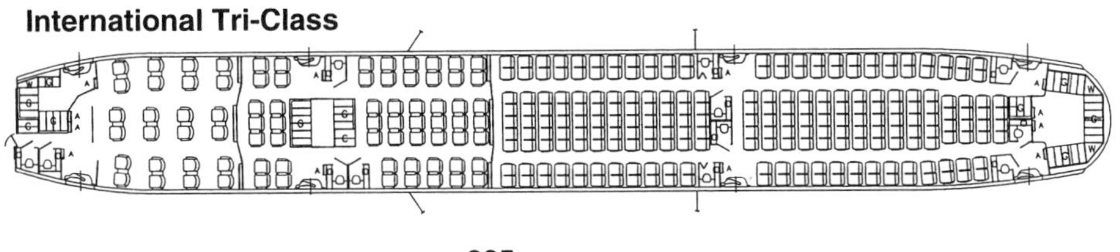

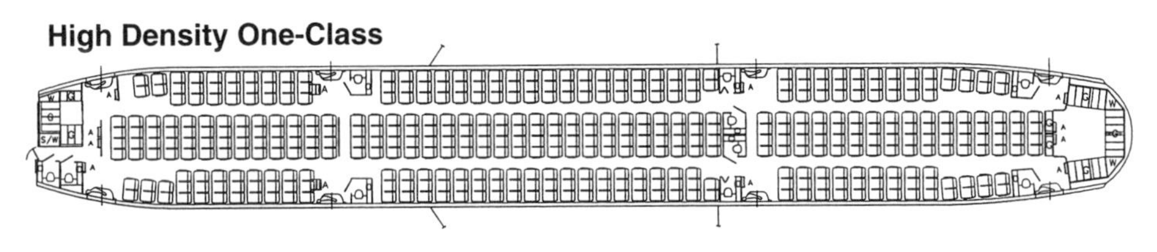

FIG. 16.4. Comparative layout of passenger accommodations (LOPA).

lavatory size and distribution, and accommodations for more flight attendants per passenger all help to foster user acceptance.

By the time overall configuration for a modern passenger airplane is reached, there have been more than 500 design trade studies conducted and more than 1 million alternatives considered. The designers and engineers have created an aircraft cross section and alternative floor plans that will be used as a map by the detail component design teams. This map also helps direct the architectural style, or look, of the aircraft interior.

Flexibility is a valuable design feature of the interior systems. The cross section and floor plans do not strictly dictate the location of every major feature, but rather they provide options on where features can go—when the aircraft is first manufactured and when it is reconfigured by the airline.

Aircraft Detail Design

Design of the aircraft cabin is a matter of integrating components, assemblies, and installations that are usually under the control of various engineering groups. In a large airplane, the number of these components can easily be in the hundreds of thousands. Table 16.2 contains typical human factors items that are addressed in the detail design phase. Timely communications and collaboration among all the organizations responsible for every aspect of each component are key elements in effective design. Communication and collaboration are accomplished using such devices as *integrated product teams* (IPT)—groups of specialists representing different engineering design functions—and various digital and physical mockup tools.

TABLE 16.2
Typical Human Factors Items in the Detail Design Phase

Ambient Environment	*Cabin Accommodations*	*Additional Topics*
Noise and vibration • Intensity • Duration Air quality • Exchange rate • Temperature • Velocity • Humidity • Filtering Air pressurization • Changes Lighting Aesthetics • Architecture • Interior decor	Basic anthropometrics Personal space • Seats • Aisle • Storage Safety • Evacuation • Crash survivability • Fire and smoke • Injuries Passenger services • Serving carts • Lavatories • Galleys Cultural and regional requirements	Special-needs passengers • Limited strength • Range of motion • Sensory impaired Medical needs • Therapeutic oxygen • Litters • Wheelchairs Communication systems • Cabin crew to flight deck • Cabin crew to passenger • Cabin crew to cabin crew Other cabin systems • Cabin management • Entertainment/business systems Long-range requirements/amenities

Integrated Product Team

The fundamental concept of the IPT design organization is the provision for a method and forum for professionals in various specialties to work together—at the same time—to create the best design. One of the first steps for the IPT members is determining which disciplines need to be involved with the component being designed. For example, a seat would have to be defined, integrated, planned, released to manufacturing, manufactured, installed, delivered, and supported. From the product definition area alone, an IPT would have representatives from seat design, materials, structures, weights, human factors, tooling, certification, stress analysis, electrical, passenger entertainment, and manufacturing (see Fig. 16.5). It is not necessary for all disciplines to be represented at all times; instead, organizational structure, discipline, commitment, and communication need to be strong enough within the team for members to know when their participation is required.

One of the most important tools for enhancing communication is an integrated, agreed-on schedule. With a good product development schedule, team members can take responsibility for making contributions on time. No one is surprised when their support is needed. The common schedule ties together the many team members working on varying parts of a product.

The human factors specialist's purpose on an IPT is to anticipate all the users that the component will have; understand their needs, expectations, and requirements; and to be an advocate for their needs early and throughout the design process. Early human factors design input is critical if the team is to effectively translate all end-user needs into system design elements. If these human factors are not considered early on, the team members will invariably end up creating costly workarounds for the usability problems they encounter.

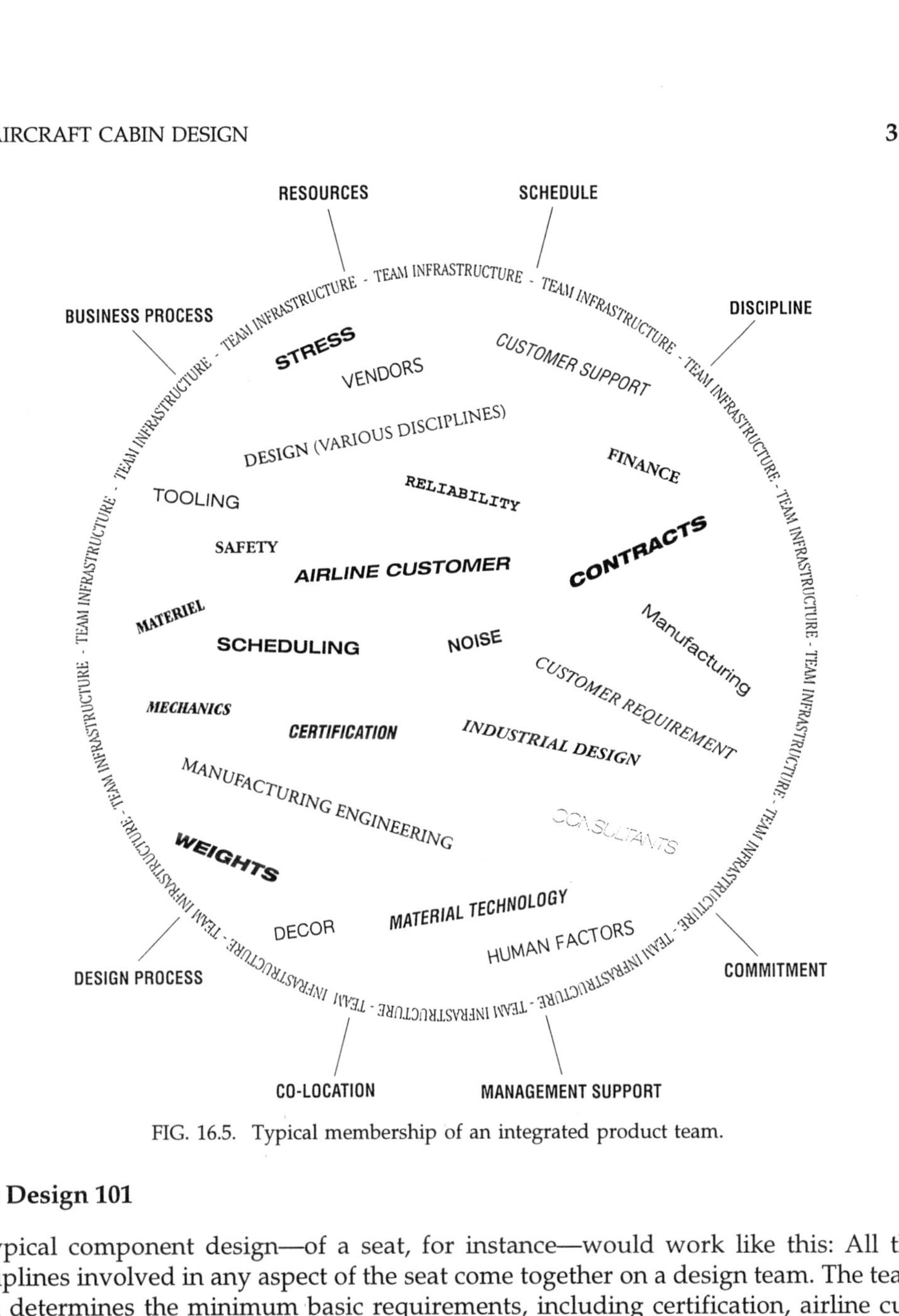

FIG. 16.5. Typical membership of an integrated product team.

Seat Design 101

A typical component design—of a seat, for instance—would work like this: All the disciplines involved in any aspect of the seat come together on a design team. The team then determines the minimum basic requirements, including certification, airline customer, and other hard requirements, for the new seat. Each team member goes back to his or her respective discipline and determines how he or she can best meet or exceed the team's requirements. In addition, each specialist establishes their own discipline's needs and requirements. The team meets periodically to review each discipline's design and developmental work, identify required trade studies, and ensure integration. Team members compromise a bit, redesign a bit, and work together to develop the very best solutions to their design challenges.

The Human Factors Specialist's Role

At the configuration level, the human factors issues primarily concern the passengers and cabin crew. At the detail design level, the human factors specialist's duties still include understanding these users and their task requirements but also expand to include understanding the users and task requirements of maintenance and manufacturing. This broad user population often means that human factors specialists are not only conducting design trade-offs with other disciplines (e.g., should the seat cushion be comfortably thick and heavy, or uncomfortably thin and lightweight?) but are also conducting design trade-offs within aspects of their own discipline (should a seat cushion be made of foam that is particularly comfortable but also has a particularly short life?).

The challenge for the human factors specialist today is to design products and systems in a way that understands the user as a system component (Rouse, 1991). There are three primary objectives in trying to achieve this goal. These objectives are especially important in the early stages of design. The first objective is that the product or system should enhance the user's abilities (Rouse, 1991). This means that designers and engineers should identify and take advantage of human abilities in the roles of interest: passenger, crew, maintenance person, or mechanic.

The second objective is that the product or system should help overcome human limitations (Rouse, 1991). Again, this requires that the limitations of the various user groups be identified and addressed in design. A good example of this is that a human head is not designed to withstand substantial impacts. Moving at high speeds in close quarters increases the likelihood that a head strike may occur. In an airplane, this limitation can be compensated for with restraints, placing seats farther apart, ensuring that anything within striking distance of a head is soft and round, installing airbags, and so on.

The third objective is that the product or system should foster user acceptance (Rouse, 1991). This requires that the user's preferences and concerns be identified and incorporated in the design. The product or system should hold appeal for all users. For example, the typical airplane passenger would like to think the designers had him or her in mind when they designed the seat in which he or she sits. If an airplane is to be used for long-distance routes, the seat base might be softer and wider. If the airplane is to be flown mainly between business centers, the seat backs should be equipped for phoning, faxing, and computing. If the aircraft is to be routed mainly to tourist destinations, video entertainment, gaming, and shopping should be available for entertainment and revenue enhancement.

Understand the Users

So, who is the user? This is the question that should initiate a human-centered design effort. The key issue is identifying a set of people whose tasks, abilities, limitations, attitudes, and values are representative of the user population of interest. Because any single user or group of users, in other words, passengers from a particular culture, reflect their own particular characteristics, it is usually necessary to sample multiple groups—that is, flight passengers from many different cultures.

A danger here is that rather than specifically identifying and researching users, designers may think they understand or have enough knowledge of a particular user group to act as user surrogates. For example, an engineer who frequently travels by air or a designer who was once a flight attendant can both argue that they know the capabilities and limitations of an airplane passenger or crew member. To an extent, this is true. However, it does not capture the abilities, attitudes, and aspirations of all current or potential passengers or crew. Flight attendant procedures at another airline might vary greatly from the procedures learned by the flight attendant or engineer. The passenger capabilities, limitations, and preferences of a 35-year-old U.S. engineer can be quite different from those of a 65-year-old Asian farmer. When designers make the effort to research user groups, they often discover how little they really know about these groups. This information is often invaluable in conducting educated design trade-off activity and preventing some potentially unfortunate results.

Sometimes it is unclear who will be the users of a product or system. For example, when designing airplanes for the year 2010, we can only project the characteristics of the passenger population. Will they be older, younger, more or less able-bodied, with long or short attention spans, singles, families, or retirees? What cultural aspects will predominate? This user group currently does not exist, yet we must still try to design for it. These questions become critically important because the service life of a modern airplane can exceed 30 years.

In addition, particularly in the case of application of new technologies, a user group exists, but it is very difficult to determine how the new technology will interact with that group's capabilities, limitations, and preferences. Manufacturers are developing technology that will allow airplanes to fly almost 20 hours. What effect will there be on such a diverse user population, both physically and mentally? It is extremely difficult and costly to gather a group of people with enough variation to represent the entire flying population; load them into an airplane fuselage, and pressurize it to typical cabin altitude pressure for 20 hours at a time. For situations like this, a designer still must identify future users and attempt to characterize their likely physical and mental abilities, limitations, and preferences (Rouse, 1991).

Again, the goal is understanding users as system or product components with a resulting requirements hierarchy, rather than convincing them of the general benefits of a particular idea or new technology. At this point, design success depends on listening to users (Rouse, 1991). Designers get plenty of time in later phases of the design process to discuss the relative merits of detail design features.

So, who is the user of an aircraft seat? Table 16.3 lists four categories of seat users and only some of the various capabilities, limitations, and preferences of these users.

Product Definition

Once the capabilities, limitations, and preferences of the user groups are understood, they need to be integrated with the design requirements, product functional requirements, and product objectives. The previously established mission, regulatory, and aircraft standard requirements—and resulting cabin cross section and floor plans—are particularly useful for providing design guidance. These requirements are considered guidelines because there are usually multiple solutions. Minimizing seat weight is an example of an engineering design requirement that can be satisfied in many ways.

Product functional requirements define what the seat should do but not necessarily how it should be done. Consequently, there are alternative ways to provide for each function. These type of requirements are typically covered in a specifications control document (SCD), which specifies physical interfaces, envelope requirements, weight targets or maximums, required functionality, testing, environmental impact issues, and related regulatory, subsystem, and other airframe manufacturer documents. The SCD may specify both design and manufacturing processes. A good example of a particular design specification for an airplane seat is the airplane-model-specific sidewall contour into which a seat must be installed and in which it must function, in both upright and reclining positions, along the length of the fuselage, including tapered sections of the airplane body. (Few airplanes have a constant cross section throughout their entire length.)

TABLE 16.3
Four Main Seat User Groups With Their Physical and Cognitive Needs and Expectations

Seat User	*Physical Needs and Expectations*	*Cognitive Needs and Expectations*
• Passengers	• Safely and comfortably operate seat features, such as seat adjustments, dining tray, lighting, air vent, and communication devices • Seating safe and comfortable for the duration of their flight • Seats will protect passenger during and after a survivable crash	• Can locate seat in aircraft? • Understand how to operate seat features and safety equipment, such as seat belt and life vest retrieval? • Understand aircraft rules about when not to smoke or leave seat?
• Crew	• Seat will protect them during and after a survivable crash • Can reach and operate all communication features while seated • Must be able to view passengers and manage passenger access while seated • Must be able to determine potentially dangerous passenger activities while seated	• Understand how to operate and when to use cabin-to-cockpit communication equipment? • Know how to make minor in-flight repairs to seats and seating area? • Know how to safely operate harnesses, evacuation alarms, and stow seat?
• Manufacturing	• Easily reach and perform all required assembly procedures • Not be exposed to common assembly hazards (noise, repetitive motions, fumes, etc.)	• Easily understand seat assembly methods?
• Maintenance	• Safely and easily clean and maintain seats and seat area • Scheduled maintenance and repairs can be performed safely and efficiently	• Maintenance and repair procedures easily understandable and concise? Maintenance procedures use standard assembly and tools?

Seat *objectives* are those things that must be accomplished by the seat in order to provide the required functions. Seat objectives are determined by the user, airline, seat vendor, and airframe manufacturer. In the currently very competitive environment, the integration into a seat design of today's ever-expanding list of necessary functions requires human factors contributions in the areas of packaging, accessibility, multiuse interfaces, miniaturization, safety, and aesthetics. The concrete manifestations of this integration of functions include telephones, video displays, video controls, game controls, financial transactions, meal services, environmental control, seat configuration control, mass retention, easy access, and, finally, the ability of the seat to adequately support a human body for up to 20 hours.

Design Development

As explained previously, the detail design of the aircraft cabin involves the integration of components, assemblies, and installations that are under the control of various functional disciplines in the engineering organization. The members of the IPT must thoroughly understand and agree on the product functional requirements and design objectives.

Timely communication among the specialists in these disciplines is a key element in effective integration. Organizational arrangements, such as the IPT structure, develop and maintain concurrent schedules, identify resource requirements, and establish a report-out and action-item structure. Powerful computer-aided design tools help designers assemble their components, check for fit, and analyze the design with a human model.

Design development is further complicated by the "make/buy" decision process. Some components and assemblies are procured outside the airframe manufacturer, resulting in added definition, coordination, and business activities.

Design reviews may reveal conflicting inputs. Compromises are required in some areas to support hard requirements elsewhere. The key principle in this stage of the design is that the required inputs be available for analysis and discussion early in the design process, thereby avoiding late design changes. Unscheduled changes of a "thought-to-be-finished" design are typically very expensive and disruptive, and there is always a risk that a solution that causes minimum change will be chosen, instead of the best solution.

Human factors engineering continually helps to maintain a user-centered perspective. Throughout the design process, human factors input is balanced against the many other design trade-offs. As the design evolves from concept to finished product, trades are made for various reasons. IPT members must be knowledgeable of the importance of other team members' requirements, and they must be willing to compromise when needed (see Table 16.4).

Analyzing the Design

A design goes through many phases of analysis and user testing throughout its development process. These phases range from early informal observations of people using various versions of a product or system to formal validation testing of a finished design. One difficulty of this process is the inability to simulate realistically enough the wide variety of flight lengths and situations in order to accurately test design options or safety procedures.

Most products and systems must be analyzed for many things. In addition to being tested for comfort (which may seem relatively easy), a seat and seating configuration must be tested for crash survivability and cabin evacuation capability. This testing is crucial and difficult to simulate. Many testing methods have been developed to attempt to model the components of a panic-stricken, full-scale evacuation of a commercial airplane; however, these techniques are not universally accepted by regulatory agencies.

Full-scale evacuation is a testing method that must be conducted on all airplane models to ensure that they meet minimum regulatory requirements. This testing is usually performed in a controlled situation using a population cross section of volunteers defined by a regulatory agency. The safety of volunteers is a primary consideration; therefore, these tests use only volunteers who claim to be reasonably fit and who fall within specified age groups. The resulting test data are sufficient for certification and useful as comparative data, but they still lack the true human behavior patterns of a life-or-death evacuation situation.

Another testing method for this situation is the use of computer models that have been developed to show evacuation trends. Although they are usually devoid of human characteristics, they can fairly accurately predict evacuation bottlenecks and pinch points and even provide a ballpark estimate of evacuation times, if the evacuation goes as planned. Currently, international efforts focus on developing more accurate computer evacuation simulations to replace the full-scale evacuation tests required for certification.

TABLE 16.4
Design Trade-Off Drivers

Airframe vendor requirements:	Manufacturability
• May address areas not addressed by regulatory requirements	• Process requirements
• May exceed regulatory requirements	• Qualified materials
Availability	• Qualified procedures
Basic cabin architecture integration	• Size and sequence of assembly
Cleanability	Postdelivery requirements
Customer contractual requirements	Regulatory requirements
Cycle time	• Foreign
Flexibility	• FAA
Functional requirements	Safety
Human factors	Tooling requirements
Maintenance access and frequency	Vendor issues
	Volume integration (many systems vie for the same space)

Several researchers in the United Kingdom devised a method to motivate volunteer evacuees to behave in a more urgent manner during an evacuation study. These volunteers were instructed that their task was to evacuate the airplane as quickly as possible, once the exits had been opened by the staff. They were also told that a £5 bonus would be paid to the first half of the volunteers who passed through the opened exits (Muir, Marrison, & Evans, 1989). The study results showed that the use of incentive payments to produce a competitive evacuation did have the potential to provide both the behavioral and statistical data required to assess such products and systems as seats and seating configurations in a more realistic way. The study recommends that this motivational technique be used sparingly because it can be potentially hazardous for volunteers.

Seat safety and evacuation are two very important and difficult systems to test. Testing other components, such as seat comfort, function, and maintainability, are somewhat easier, but no less interesting. The final seat design represents the optimum balance of all these tested design factors.

WHAT DOES THE FUTURE HOLD?

Technological advancements are allowing airframe manufacturers to design and manufacture aircraft that carry more people longer distances more efficiently. Future trends will be for aircraft to connect more destinations point to point, rather than routing passengers through a variety of hubs to their final destination (Condit, 1997). This will result in flights of longer range. The human factors issues that the future will present are subtle, but important.

One issue to consider is that the physiological effects of longer range flights (20 hours or more) may be complicated by the fact that the average aircraft passenger is getting older. How will increased time at high altitude and limited inflight mobility affect this future flying population? With increased flight time comes increased risk of encountering inflight medical emergencies. Will aircraft and crew be able to accommodate these new passenger needs?

As flights become longer and technology becomes more sophisticated, the quantity and variety of passenger activities, such as eating, drinking, sleeping, exercising, movie viewing, electronic shopping and gaming, and conducting business, will increase. What will be the physical and cognitive effects of these activities on the well-being of this "captive audience"?

Future changes should be made to the aviation human factors discipline in general. There is a need for an increase in the extent and accuracy of physical and statistical computer human-modeling tools. Current methods of modeling human figures within electronic mockups of aircraft parts and interiors are still relatively crude for accurate analysis. As for user data, state-of-the-art methods do not yet bridge gaps in user data for various populations. There still may be data gaps between designers and users. There is a wealth of information available, not only from passengers but also from crew members, maintenance personnel, and assembly mechanics, that goes virtually untapped, even at world-class aircraft manufacturing companies. There is currently no method or tool for bridging language or cultural barriers. And, finally, improvements are necessary in the aviation industry to ensure that human factors engineering methods are applied to cabin interior products and systems early in the design process.

Currently, only a limited number of procurement contracts require human factors involvement and signoff in the design activity. If this practice were to be made standard, it would help to make the human factors discipline a useful resource across all areas. Increased use of human factors methods should bring increased evidence of their benefits.

CONCLUSION

The engineering profession is making exponential technology advances in many directions. Human factors specialists today must be very diligent about trying to match these new advanced technologies to users. The temptation to focus only on technology in the spirit of advancement is great. It is the responsibility of the human factors specialists to keep aircraft—or any technology—from growing beyond the capabilities and limitations of its users. The solution is for human factors professionals to work together with design engineers to develop and apply new technologies that incorporate the human as a system component. Developing measurable and defensible user requirements is the key to making this happen.

REFERENCES

Brauer, K. (1996). *Seating configuration and passenger comfort.* Seattle: The Boeing Company.

Condit, P. (1997). Broadcast interview, MacNeil/Lehrer Report.

Edwards, M., & Edwards, E. (1990). *The aircraft cabin: Managing the human factors.* Hants, England: Gower.

Muir, H., Marrison, C., & Evans, A. (1989). *Aircraft evacuations: The effect of passenger motivation and cabin configuration adjacent to the exit* (Civil Aviation Authority Paper No. 89019). London, England: Civil Aviation Authority.

Rouse, W. B. (1991). *Design for success.* New York: Wiley.

Serling, R. J. (1992). *Legend & legacy: The story of Boeing and its people.* New York: St. Martin's Press.

17

Helicopter Human Factors

Bruce E. Hamilton
Johnson Engineering Corporation

Helicopters are just like fixed-wing aircraft except that helicopters are different. The differences are not in the men and women who fly helicopters, for they can be, and sometimes are, the same men and women who fly fixed-wing aircraft. Their abilities and limitations are the same regardless of the kind of aircraft they fly. Helicopters and fixed-wing aircraft differ in how the crew makes flight control inputs, the information required to decide what control movements are necessary, and the missions assigned to the crew. There are many areas of similarity, such as in navigation, communication, subsystem management, monitoring vehicle status, coordination between crew members, and interaction between the helicopter and other aircraft. Helicopters and fixed-wing aircraft follow, for the most part, the same flight rules and procedures. Minor differences exist in the flight rules, mostly about minimum visual ranges and decision heights. Although rotary- and fixed-wing flight is mostly the same, the differences are important and often overshadow the similarities.

One difference is in how helicopters fly. Fixed- and rotary-wing aircraft all obey the same laws of physics and use the same principle of differential pressure caused by air flowing across and under a shaped surface to generate lift. The difference is that the rotary wing, as the name implies, rotates the wing about a mast to generate airflow while the fixed wing moves forward through the air. The difference in method of generating lift accounts for the helicopter's ability to hover and move at slow speeds in any direction. Figure 17.1 illustrates the method by which the helicopter balances opposing forces in order to fly. In short, the rotating blades (rotor disk) generate lift. Tilting the rotor disk provides thrust, with the resultant vector a function of how much lift (pitch of the blades) and thrust (degree of tilt) are commanded. This resultant vector counters the force of gravity acting on the mass of the helicopter and payload, and the drag of the fuselage as it moves through the air. Increasing the pitch of the blades (more lift) without tilting the rotor disk (thrust constant) causes the helicopter to rise, whereas increasing pitch and tilting the disk causes movement in the direction of the tilt. When hovering, the resulting vector is vertical (without thrust) to balance the force

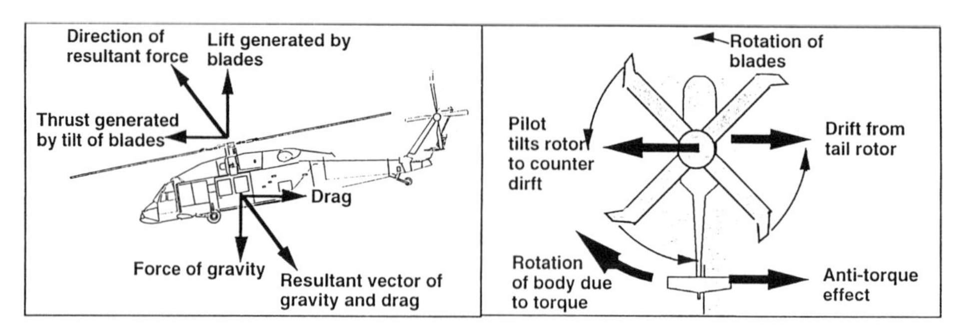

FIG. 17.1. Example of forces that must be balanced in order to fly a helicopter.

of gravity. However, because the blades are rotated by the engines, the body of the helicopter tends to rotate in the opposite direction due to torque effects. A small set of blades is mounted on a tail boom and oriented so that its lift counters the torque of the main rotor blades. However, because the torque effect is rotational and the anti-torque tail rotor applies lateral force, the tail rotor tends to push the helicopter across the ground. This is countered by tilting the main rotor disk to counter the tail thrust. Changing the amount of pitch of main or tail rotor or changing the tilt of the main rotors determines the flight of the helicopter; however, any change in a force results in imbalances, which may or may not have to be corrected for by the pilot.

The controls of a helicopter manipulate the aircraft's airfoils differently than in fixed wings but, in many respects, the result is functionally the same, especially at higher speeds. For instance, the cyclic provides pitch and roll control as does the stick or yoke in a fixed wing, the collective controls "power" as does the fixed-wing throttle, and the pedals control lateral forces about the tail just as does the rudder. However, rotary-wing flight requires more frequent adjustments, and each control interacts with the other controls as indicated earlier. As a result, special attention is paid to control placement in the helicopter cockpit.

The ability to generate and maintain lift in a variety of directions leads to a second significant difference between fixed- and rotary-wing aircraft, namely, the missions they fly. Helicopters are able to move slowly, at very low altitudes, and hover stationary over a point on the earth. This allows the helicopter to be used in a variety of unique missions. These unique missions have an impact on the design of the helicopter and the way the crew uses its ability to fly.

Besides differences in controlling flight and the flight missions, maintenance of helicopters is more demanding than for fixed-wing aircraft. The need for more frequent maintenance, more complicated control systems, and limited access within compartments makes the time required for helicopter maintenance as well as the costs high relative to fixed wing. Recognition and consideration of the human factors of maintenance early in the design process will be significantly rewarded in the cost of ownership.

ISSUES UNIQUE TO HELICOPTERS

Control of the helicopter is different from fixed-wing aircraft due to how lift is generated. In the helicopter, lift is generated by rotating blades (airfoils) and varying the angle, or pitch, of the blade as it rotates. The act of increasing pitch causes the blade to rotate about its long axis and increases the lift generated but at the cost of requiring more power. Adjusting pitch is accomplished using a control called the collective. The collective is located on the pilot's left and arranged so that pulling up on the collective increases pitch and pushing down decreases pitch.

Where the fixed wing has rudder pedals to induce lateral forces against the tail of the fuselage, the helicopter has pedals that control lateral force by varying the pitch of blades mounted at the end of the tail boom. Together, the collective, cyclic, and pedals are used to control and stabilize lift in varying directions thereby bestowing the helicopter with its freedom of movement. Controlling flight and following a flight path are a matter of balancing lift (via collective pitch), thrust of the rotor disk (via cyclic), and anti-torque (via pedals), with the balance point changing with every control input (in helicopters without engine speed governors, engine speed must also be adjusted with each control input). To fly, the pilot must make continuous control adjustments with both hands and feet. This imposes severe restrictions on nonflight tasks such as the tuning of radios or subsystem management. This must be compensated for in the design of the crew station. Advances in flight control and handling qualities promise to reduce much of the demands on the crew by automating the balancing act.

As a result of the differences in the controls, visual requirements in helicopters differ from those in fixed-wing units, especially during low speed or hover flight. In these modes, constant visual contact with the outside world is used to determine minor changes in position (fore/aft, left/right, up/down, rotation), to compensate and station keep. At hover, the pilot maintains awareness by focusing at distant visual points with quick cross-checks close by to sense small movement. A rock or bush may be used to determine if the helicopter is moving forward/backward or to a side (which is why hovering at altitude, in bad weather, or over water is so difficult). In addition, the pilot must visually check to insure that there is clearance between the main rotors and anti-torque rotors and objects such as trees, wires, and so on. During takeoffs the body of the helicopter can pitch down as much as 45° as the pilot shifts the rotor disk severely forward. Similarly, on approaches, the body may pitch up by 45°. The need for unobstructed vision determines how much and where in the cockpit glass is required. The loop between visual cues, control movements, and compensation is continuous and demanding and is one of the primary components of pilot workload.

The helicopter gains freedom of movement by adjustments of rotating blades (airfoils) overhead. This has the undesirable side effects of causing vibration and noise. As each blade moves through space its form causes differential airflow between the top and bottom surfaces, which then merges at the rear of the blade, resulting in turbulent air. This fact, coupled with normal disturbances of the air mass and the additional fact that the blade is always alternatively advancing toward and retreating (and stalling) from the flight path on each revolution, leads to vibration. Vibrations are transmitted along each blade to the mast and then into the airframe to be added to transmission and airframe vibrations. At low airspeeds the blades act as individual airfoils, whereas at higher airspeeds the blades act like a unified disk. The transition from individual to group behavior is another contributor to vibration.

All the movement, vibration, and blade stall contribute to noise in the cockpit. The vibrating environment created by the rotor dynamics also affects display readability and control (switch, knobs, and dials) movements. Noise also interferes with effective communications and contributes to fatigue. Light passing through the rotor system is intermittently blocked by the blades and causes some flicker. Certain combinations of blade size, blade number, rotation speed, and color of transparencies can cause visual disturbances and are to be avoided. All of these impacts are a result of the helicopter's method of generating lift.

The freedom with which helicopters fly leads to unique challenges in that the missions of helicopters vary widely. The same airframe, with minimal modification, may be used as an air taxi, air ambulance, search and rescue, air-to-ground attack, air-to-air attack, antisubmarine warfare, heavy lift of goods and raw materials, aerial fire fighting, police surveillance, sightseeing, and aerial film platform, to name a few. Any of these missions might be conducted during the day or at night, sometimes using night vision devices. The same helicopter can be expected to fly under visual meteorological conditions (VMC) or under instrument meteorological conditions (IMC), under either visual or instrument flight rules (VFR or IFR). These different missions involve the addition of equipment to the helicopter and some minor cockpit modifications. The cockpit usually retains its original configuration with some controls and displays moved around to make room for the control panels of the add-on equipment. The pilot–vehicle interface of the helicopter must be unusually flexible.

Another issue for helicopter pilots is the frequent operation at the power limits of the engines. For instance, desired payloads, coupled with density altitude and wind conditions, may tax the helicopter's lift capability to such an extent that a "standard" takeoff can not be accomplished. The pilot has to recognize this condition and fly forward at only a few feet of altitude to build up forward airspeed. This allows the pilot to use ground effect for additional lift. Ground effect is the benefit to lifting the helicopter that comes from the blade downwash being trapped beneath the rotors and acting as a cushion. A rough analog is the lift generated by a hovercraft sitting on trapped air. After airspeed increases, the blades transition from acting as separate airfoils to acting as a coherent disk, which increases their efficiency, and more lift for the same power is available. The margin provided by this flight technique is relatively small and difficult to judge. Failure to correctly judge the margin may mean the difference between a successful edge-of-the-envelope takeoff and possible disaster. Once aloft, environmental conditions may change (such as density altitude, due to arriving at a touchdown point at higher elevation), and adequate power margin may no longer be available. Again, the safety margin may be small and difficult to judge, and numerous accidents have occurred in which helicopters crashed after several failed attempts at landing or takeoff. The pilots were apparently convinced that the helicopter was within its operating envelope or, perhaps, outside the published envelope but within the mystical "extra design margin" that all pilots believe engineers give them.

Another major mission with unique human factors impact is the requirement to fly in close formation with the earth. Helicopters are routinely used in low-altitude missions such as aerial survey, installation, and maintenance of power lines. In a military setting, helicopters are expected to fly using trees for concealment from detection. In nap of the earth (NOE) flight, the pilot flies slowly (often at speeds at which the aircraft is unstable), and as close to the ground as conditions permit. This might be below treetop level or behind small hills. Confined-area landings and hover above short trees but next to tall trees can be expected. All of this is expected day or night. At night, limited night vision is augmented by vision aids such as night vision goggles (which amplify available star-/moonlight) or infrared sensors (which sense minute differences in temperature in a scene and encode the differences as various greyscales on a cathode ray tube [CRT]). Night vision aids all change the visual world, usually by reducing the field of view, changing the color of the world into shades of green, reducing visual acuity, and, in the case of the infrared image displayed on a helmet-mounted display, reducing vision to a single eye. If a pilot applied for a license but was colorblind, could see out of only one eye, and had only 20/60 visual acuity, he would be laughed at and denied a license. However, as a helicopter pilot, he may be reduced to that visual capability and told to ignore the fog, go hover behind a tree, and land in small, impromptu landing zones.

Other missions routinely expected of helicopters impose their own unique challenges. For example, some helicopters are expected to fly in high winds, and land on a spot on a ship's deck scarcely bigger than the rotor disk, while all the time the ship is pitching and rolling. Those helicopters have standard cockpits without specialized flight controls or displays. The message is that the designers of helicopter cockpits should expect almost any mission.

Another way in which helicopters present challenges is in the maintenance and support of the helicopter itself. The engine of the helicopter (there may be anywhere from one to three engines) drives a transmission that reduces the high speed of the engine to the relatively low speed of the blades and provides drive for the tail rotor. The engines must have fuel flow constantly adjusted to keep the blades turning at a constant speed against the forces trying to slow or speed the blades. The transmission also drives hydraulic and electrical generators for use by other systems. The flight control system must translate cyclic, collective, and pedal movements into adjustment of blade pitch while the blade rotates around the mast. This requires a fairly complex mixing of control by rotor position all superimposed upon the normal requirement to compensate for the fact that the blade generates lift during the forward portion of the rotation and stalls when retreating. In older helicopters this control system is completely mechanical, whereas in newer helicopters hydraulic systems are used to provide the required forces to move the blades. The electrical system powers the avionics suite, which can range from minimal to extremely complex systems including radar, infrared, and/or satellite communications, among other systems. All of these systems, mechanical and electronic, require maintenance. Providing easy access, with simple, quick, low-workload procedures, is a major challenge with direct impact on the cost of ownership and safety of operation.

Many human factors challenges are posed by the method of vertical flight and missions to which vertical flight is applied. Those mentioned here are important ones but not the only ones. The pilots of helicopters face major challenges in the control and monitoring of the aircraft health and systems status just as the fixed-wing pilots do. Communications between the helicopter and other aircraft and the ground represent significant workload. The unique issues derive from how vertical flight is achieved and what it is used for.

THE CHANGING NATURE OF HELICOPTER DESIGN

Human factors engineers traditionally provide information on human capabilities and limitations to the design community and serve in a "check" function once the design was completed. The role of the human factors engineer has been to provide specific details such as how tall are people when sitting, what size should the characters on the labeling be, and what color should the warning indicator be. Often the human factors engineers found themselves helping to select which vendor's display should be selected, trying to change completed designs that failed to take something into account, or answering why a design was not acceptable to the customer. These roles were generally all that was required when the issues were primarily those of pilot fit and arrangement of selected displays in general utility helicopters. The advent of the computer and its impact on aviation in general is changing the ground rules and the way in which the human factors engineer interacts with the design process.

The impact of the computer on the cockpit and the way in which cockpits are developed has been significant in many areas but has created the most change in two areas. The first is that what used to be discrete hardware functions have become integrated into a computer system. The whole infrastructure of the aircraft was assumed to be a "given" with only simple cross-checking required before actions were taken. In a computerized system, not only does consideration have to be given to status from a health monitor but also to the health of the monitoring devices. The meaning of the information provided must also be considered. For instance, a chip detector used to be a simple design in which the metallic chip physically completed an electrical circuit illuminating a light on the caution/warning panel. Now the size of chips, the number of chips, and the rate of chip accumulation can be monitored and passed to a central processor through a digital data bus. Besides a greater range of information and prognosis about the developing situation, the health of this detection and reporting network has to be considered. That is, is the detector performing its function, is it talking to the digital bus, and is the software that uses the data and alerts the crew functioning? The pilots of the emerging generation of helicopters are becoming managers of systems designed to free them from the workload of housekeeping the aircraft, but managers also have information and control needs.

The second major area of computer impact is that the displays being purchased are blank and can display almost anything desired in any format, whereas formerly displays were built function specific and limited. For example, attitude displays used to be selected by comparing the vendor's marketing material and specification sheet to select the one that fitted the available space, interfaced with the sensor system, and appeared most pleasing to the customer. Now displays are picked for size, weight, type (CRT, liquid crystal, etc.), pixel density, interface protocol, color, and so forth. What can be shown on a display is a function of the capability of the display generator, throughput of the processors, and data rates of the available sensors. Often the decision is whether to use one vendor's flight management system (including displays, controls, information, moding, capabilities, etc.) or another. The other option is to develop a purpose-built cockpit by mixing and matching or developing components.

The cost to buy a helicopter is usually equivalent to or less than a comparable fixed-wing aircraft; however, the costs of support and maintenance are significantly more for helicopters. The primary cost drivers are the person-hours required and the necessity to replace parts after a fixed number of hours of use. Human factors engineers can reduce the cost of ownership by reducing the complexity of the maintenance and support tasks. Generally speaking, the time to complete safety and maintenance inspections is short compared to the time required to open and close inspection hatches and access that which needs to be inspected. Careful attention during the design phase to how the inspection process is conducted can significantly reduce the person-hours required. Once the determination is made to conduct maintenance, how the task is completed should be engineered to reduce the number of steps, number of tools, and number of person-hours required. When conducting these analyses, the human factors engineer should consider the education level, training requirements, and job descriptions of the population of maintainers and supporters.

The next generation of helicopters will be incorporating computerized maintenance support devices in an effort to reduce the time required to determine what maintenance must be done and replace parts as a function of usage or performance rather than time.

Most avionics will be designed with built-in test (BIT) capability that will continuously report on the health of the device. The trend is toward a hand-held maintenance aid that can access a storage device located on the airframe that holds data taken during the flight. This portable device will be capable of diagnostics, cross-reference to parts lists, and display of maintenance procedures. It will also interact with ground-based systems for record keeping, updates, and prognostics. There exists an urgent need for human factors engineers to assist in the development of the electronic manuals, diagnostic procedures, and record-keeping interfaces of these electronic devices.

THE ROLE OF HUMAN FACTORS IN FUTURE HELICOPTER DESIGN

Future helicopter crewstation design will require many of the human factors subspecialties. The particular skills will depend on the phase of development. The phases of a development program are shown generically in Fig. 17.2. The airframe development starts with general requirements definition and the creation of basic data. In this phase the outer lines of the helicopter are established along with size, weight, capabilities, etc., to form the general arrangements. This phase ends with a review of the data and conceptual design. Once the requirements are defined, the preliminary design phase begins. During preliminary design, interior lines are established and the general arrangements are refined. Analysis is conducted to define technical parameters and to determine how well the design meets its goals. Prototype components are built and tested to reduce

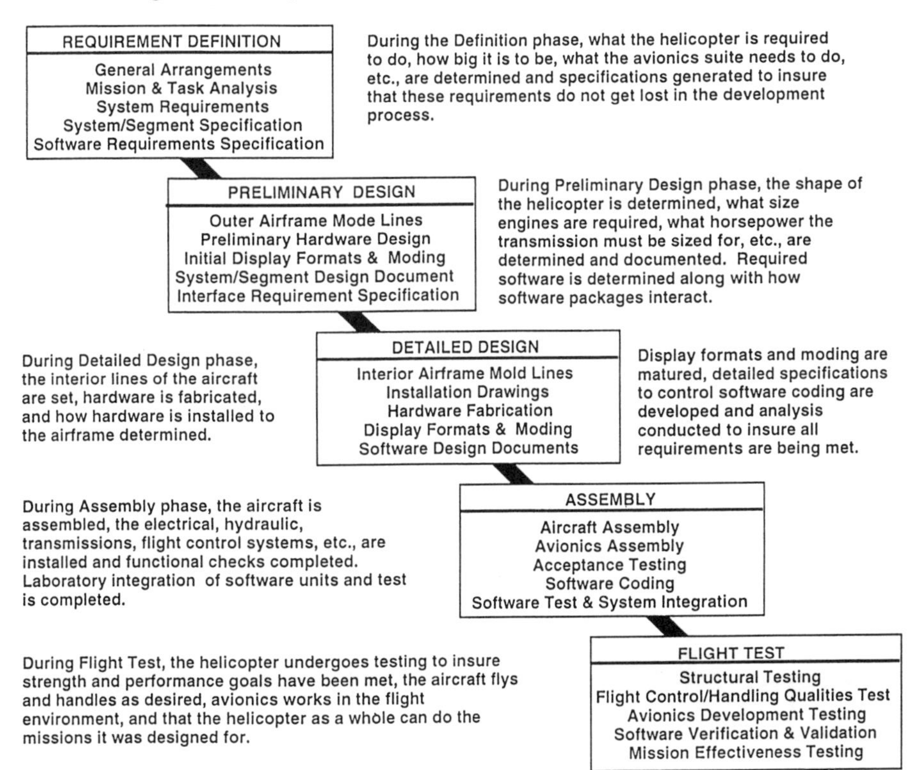

FIG. 17.2. Generic phases of a helicopter development program.

the risks associated with new designs. This phase ends with a preliminary design review and an accounting of how well the original requirements are being met, as well as progress on meeting weight and cost goals. The next phase after the preliminary design phase is the detailed design phase. In this phase, the helicopter design matures into something that can be built and assembled. The design is evaluated in a critical design review—again to verify that all systems are meeting design goals and components are ready to be manufactured. This is followed by assembly of the helicopter and acceptance testing of components. Tests of the helicopter are conducted to insure strength requirements are met and operations of the subsystems (hydraulics, engines, transmissions, etc.) are evaluated. This phase ends with an airworthiness review and safety of flight release. Flight tests are conducted following the assembly phase. The flight test program is conducted to allow a safe, orderly demonstration of performance and capabilities. The purpose of the flight test is to validate the "envelope" of performance and insure safe flight under all specified conditions.

A human factors program that supports analysis, design, integration, and test of the pilot–vehicle interface should be conducted in support of the development and validation of hardware and software requirements. This program, depicted in Fig. 17.3, should be iterative and interactive with the other disciplines involved in the helicopter development. Within this program, the human factors engineer analyzes the proposed missions to the task level to determine what the pilot and aircraft must do. This is done during the requirements definition phase and is updated during preliminary and detailed design. In other words, the mission and task analysis is first conducted independent of specific implementation to define the human information requirements and required tasks. Later in the development cycle, the early mission and task analysis will be revisited using the proposed designs and finally the actual controls and display formats. The message is that although the name is the same, the products during the three phases are different and serve different purposes.

Requirements Definition Phase

The beginning of any program, whether a new aircraft development, a variant, or a retrofit, includes analysis to document and disseminate the requirements of the program. This is done by generating a system/segment specification (system if a large

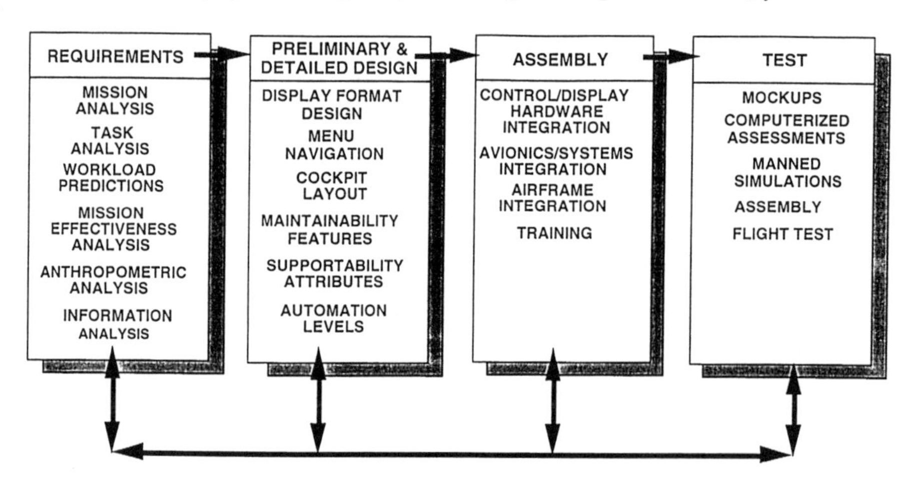

FIG. 17.3. Features of notional human factors engineering program supporting helicopter development.

program, or segment if more restricted in scope). This document informs the engineering staff of what the design is expected to do. For the cockpit, this often consists of creating a list of instruments required for flight and support of navigation, communication, and mission peculiar equipment. The kind of requirements generated during this stage of a new helicopter program might be that the aircraft is to be single engine, dual crew, side-by-side seating, instrument flight rule certified, three radios with waveforms in the regions of VHF-FM, VHF-AM, and UHF, and so on.

These requirement specifications are then provided to the avionics system and airframe developers. The airframe developers, using approximate sizes from the vendors, begin the task of arranging the cockpit. At this point, tradition is most often the moving force behind the development of the cockpit.

The impact of computers on the airframe and the cockpit has changed this portion of the development process. The list of requirements provided to the airframe engineers includes specifications for the number and sizes of displays. The airframe developers, however, cannot practically talk to display vendors about weight, space, and power because the vendors wish to known type of display, pixel density, video interfaces, and so on. The group generating the system/segment specification will still specify the requirements as before, but the requirements list must be "interpreted" into usable information for vendors prior to making the traditional information available to airframe designers. In turn, what data are needed to create the attitude display, how many displays are required, at what rate data must be displayed, what colors are used, what resolution is required, and so on, must be identified and provided to the avionics system designers so that they can determine the mix of displays and controls required. These decisions are documented in the software requirement specifications. Human factors engineers are well qualified to help the avionics system designers address these issues as well as to assist the airframe designers in arranging the cockpit, given that a mission and task analysis has been conducted. If this process is not followed, when detailed design of the software begins, what can be achieved will be limited by what hardware was selected and where it is placed.

Preliminary Design Phase

Once the helicopter's general arrangements have been created and the outer mold lines established, work begins on preliminary design of the crewstation. This typically results in generating two-dimensional drawings specifying instrument panel size, location, control placement, and so on. It is during this phase that the size and location of glass is fixed, controls positioned, and more. Human factors engineers are involved to insure that the location of the supporting structure allows adequate vision down, up, and to the sides. The problem of vision is more, however, than just placement of the glass. Designers must consider the allowable distortion of vision through the glass, the angle of viewing through the glass, transmissivity of the glass, and internal reflections, among other design issues. Other topics to be addressed by the human factors engineers working in conjunction with the airframe developers during this phase include how the crew gets in and out of the aircraft normally and during emergency exits, reach requirements for normal and emergency operations, location of display surfaces and control panels, and safety requirements.

The preliminary design of helicopter cockpits must also take into consideration the fact that helicopters crash differently than fixed wings. Helicopters tend to impact the ground with high vertical descent rates, whereas fixed-wing aircraft tend to have a lot of forward speed at impact. As a result, helicopter cockpit designers need to be concerned about instrument panel entrapment (panel tearing away and dropping onto the legs), attenuating vertical movement of the crew, and keeping the cyclic and collective from striking the crew during the crash sequence. Information about how the human body moves during the crash sequence can help guide the airframe developer during cockpit layout. A tradeoff usually results in that if controls and displays are placed out of the way for crash environments, they usually are too far away to be used during normal operations, and if they are too close they may pose a hazard during crash.

During the preliminary design phase the software system is also being designed. Software development starts with the general requirements defined in the requirements phase and derives requirements that must be satisfied to fulfill the original requirements. This process is called requirements decomposition. For instance, a general requirement might be the ability to store and use flight plans from a data transfer device. A requirement derived from this general requirement might be that a menu is to be presented to allow the pilot to select one flight plan from up to five stored plans. The decomposition process continues until all high-level requirements have been decomposed into specific discrete requirements that allow a programmer to write and test code for discrete functions. Part of this process is the assigning of responsibility for functions to specific software groups, along with how fast the task is to be completed, how often the task must be done, what information is needed, and in what format the product must be. In this way the software system is laid out analogous to the physical layout of the aircraft. These decisions are documented in the system/segment design documents and the interface requirement specification.

The human factors engineer assists this process by updating the mission and task analysis created in the first phase based on the increased understanding of implementation and probable uses of the helicopter. Specification of the display formats at this point in the program provides insight into the information required and the software processes that will be required to produce that information. Menu structure can be designed to control number of key presses, to reduce confusion in menu navigation, and to provide a low-workload interface. This information allows the software system to provide what is needed by the crew and in an optimal format while simultaneously easing the task of software decomposition.

If the proposed interface is defined in this manner during this phase, there are additional benefits besides the flow down of requirements. For instance, the proposed interface and its operation can be simulated from this information on any one of a number of devices ranging from desktop to high-fidelity motion simulators. Proposed users of the helicopter can be brought in to evaluate the software design before coding. The result is a mature design and less probability of costly recoding due to mistakes in understanding, failure to identify requirements, and misconceptions by the customer as to what is technically feasible for a given cost and schedule.

The effort of the human factors engineer to create task and workload analyses and to conduct manned simulations is costly. It is always difficult to persuade program planners to expend money on these tasks, especially during the beginning phase of the program when the demand is to transition almost ready technology from the laboratory to production. The primary selling point for the early human factors analysis and simulation is cost avoidance. Significant changes that occur after software has been validated for flight are extremely expensive in both time and money. Avoiding a single major problem will pay for all the up-front expenditures. This is the same rationale for doing wind tunnel testing on a blade design before flight test. No one would actually propose that a blade design could be drawn and would work as desired without testing (either real or virtual) and redesign. The same is true of the pilot–vehicle interface and the information content and menu structure.

Detailed Design Phase

Reality sets in during the detailed design phase. The cockpit layout during preliminary design is clean and tidy, but during detailed design a million and one little details have to be accommodated. For example, everything that floated freely in earlier drawings must now have mounting brackets with retaining bolts. Provisions must be made to allow the components to be assembled and maintained. The furnishings and equipment, like first aid kit, fire extinguisher, document holder, and seat adjust, must all be located and installed in the cockpit.

Round dials and display tapes, if used, must fit within the instrument panel space. If a dedicated caution/warning panel is used, the number of cautions/warnings must be determined, legends created, and colors assigned. If the cockpit is to have multifunction displays, then the displays must be specified as to pixel density, colors, update rates of display generators, formulas for movement of symbols on the screen generated, filter algorithms developed, bitmaps defining symbols provided, etc. The distance between switches must be large enough for fingers to use easily without becoming so great that the resulting panel does not fit into the mounting rails. Control grips for the collective and cyclic must be created and the functionality of switches determined.

The menu structure must be completed and the meaning of each pilot-selectable option defined. What had started out as a simple, clean menu usually must be modified to accommodate the emerging requirements to turn devices on and off, set up and adjust functions, enter local variations into the navigation system such as magnetic variations and coordinate system adjustments, and so forth. Allowances must be made for how the aircraft is started, checked for functionality, and shut down. Procedures for verification of functionality and means to reset/recycle balky systems must be considered. The menu now has a myriad of add-on steps and routines that obscure the simple and clear structure once envisioned. Display formats must be detailed out to the pixel level, identifying where each character or letter goes. Formats become burdened with additional information and festooned with buttons and options. Formats that once handled a few related functions now sport unrelated functions rather than add new branches to an already burdened tree. Controlling the structure of the menu as it grows from concept to practicality is a daunting task.

Timelines and throughput become major issues during detailed design. The display that was designed to have graphics updated at 30 Hz now might have to update at 15 Hz to help ease apparent processor overloads. This degrades the graphics and, coupled with lags in the speed at which the information is processed, might lead to control loop problems. Events now take seconds rather than the virtually instantaneous speed dreamed of during preliminary design, with obvious impacts on effectiveness, workload, and acceptability.

Detailed design can be summarized by stating that the devil is in the details. As hardware starts to mature, its functionality turns out to be less than envisioned, triggering a scramble to compensate. In other cases, capabilities are provided that you would really like to take advantage of but that had not previously been identified. The problem is that to meet software delivery schedules you have to stop making changes, and the earlier in the cycle you stop making changes, the smoother is the development.

Also, changes beget changes in that fixing a problem in one place in the system often forces rippling effects throughout the rest of the system.

Finally, in this phase, the users of the helicopter start to see what had previously only been talked about. They may not like what is emerging and may start providing their own helpful hints about how things should be. All of this is during a time when the system is supposed to be finished and the development money has been spent. Every change during detailed design and subsequent phases is evaluated for cost and schedule impact by people not pleased by any change, good or bad.

Assembly Phase

During the assembly phase the components of the aircraft come together and the helicopter is built. Final detailing of the design occurs in the cockpit. For instance, intercom cable lengths must be trimmed from the delivered length to that length that gives maximum freedom without excess cable draped over the crew. Decisions have to be made as to the ease with which adjustable items are allowed to move. These decisions and design adjustments come under the heading of "known unknowns," because it was known that the actual length, friction, setting, and so on were unknown until assembly. During assembly you may also encounter "unknown unknowns," problems that had not been anticipated. An example might be that a rotating control handle hits another installed component. Care had been taken to make sure the handle did not touch the airframe but another component had been moved or rotated itself, resulting in interference. These problems have to be dealt with as they arise.

An important aspect of the assembly phase is component testing. As hydraulic lines are completed, they are pressurized and checked for leaks. As electrical lines are laid in, they are tested and eventually power goes on the aircraft. The airframe side of the helicopter requires little human factors engineering support other than occasionally supplying information or details about how the crew or maintainer may use software-based interfaces to status and control systems. If the helicopter uses significant amounts of software to interface with the crew, and the human factors engineers have been involved in the design and development of the interface, then significant involvement in the testing phase can be expected. This involvement typically would be in the form of "Is this what was intended?" or, more to the point, "This can't be right!" If adequate documentation had been done during detailed design and person-in-the-loop simulation accomplished, then there should be no surprises. However, actual operation is always different from envisioned, and adjustments may be required.

Flight Test Phase

There are two major areas of involvement for the human factors engineer during the flight test phase. The first area is in obtaining the safety of flight releases for the aircraft's first flight. The second major area is in assessing workload and operational effectiveness. The specific nature and degree of effort are dependent on the kind of helicopter being built, whom it is being built for, and the aircraft's intended purpose.

The following human factors areas should be considered during a safety of flight evaluation:

- Ingress/egress—Can the crew enter and exit the cockpit both in normal and emergency situations?

- Visibility—Does adequate vision exist for flight test?
- Functional reaches—Are all flight-required controls within easy reach?
- Controls and display functional check—Do the controls and displays work as expected?
- Flight symbology—Is all the information necessary for flight available and are control lags and jitter acceptable?
- Emergency procedures—Have all conceivable failures of components in the aircraft been considered and emergency procedures created for the serious ones?

If the previous phases of development have included human factors engineers, addressing these questions should be merely a formality and a matter of providing documentation of work previously completed. If the program did not include significant human factors effort, then these questions may be difficult to answer or cause last-minute rework.

The other major area during flight test is workload and operational effectiveness. Depending on the customer and mission of the helicopter, testing may be required to demonstrate that workload levels, timelines, and situational awareness goals have been met. This may require dedicated flight time to conduct training and rehearsal of specific mission segments or tasks. It is cost-effective to conduct the majority of the workload and operational effectiveness studies in a high-fidelity full-mission simulator before flight test. During actual flight, selected tasks or segments can be evaluated to verify the simulation data. Full-mission simulation is highly recommended as a method of finding operational problems. If the full-mission simulation is done early in the development, then the cost to conduct operational effectiveness analysis in simulation during flight test phase is minimized and the impact of the discoveries to cost and schedule will also be minimal.

WORKLOAD IN THE HELICOPTER COCKPIT

Workload in the helicopter cockpit is the result of the demands of the flight tasks, the tasks required to monitor, status, and control the helicopter systems, and the demands of the mission. This is no more than saying that flying results in workload. A more useful way of looking at the genesis of workload is to regroup the demands into those from outside the cockpit and those from within the cockpit. Demands from outside the cockpit are usually environmental, flight, and mission conduct related. The within-the-cockpit demands are considered those that directly support flight demands (e.g., adjusting flight controls), mission goals, and those demands that are part of the routine housekeeping of the helicopter (startup, shutdown, navigation system alignment, etc.). This view of the sources of workload is useful because it allows the human factors engineer to recognize that the designer is trying to cope with the workload associated with an external task. The internal workload, especially the housekeeping workload, is the result of the cockpit design and is under the designer's control. It is always important when designing to recognize when the design is trying to cope with someone else's workload or is itself the source of workload. Although it is pleasant to think that the crewstation designer is always reducing the workload caused by others, it is more often the case that the crewstation designer is the source of workload.

Sources of Helicopter Cockpit Workload

The most common source of workload in the helicopter cockpit is tasks that require looking in two different places at the same time. That is, the pilot is usually trying to look out the window to attend to the demands of flight or mission tasks while simultaneously being required to look at displays within the cockpit. This results in having to switch visual attention constantly and increases workload while reducing performance. Head-up displays are recognized as effective workload reducers because they reduce the amount of time spent switching from inside to outside tasks.

Another common workload problem is having to listen to multiple radios, a copilot, and the auditory cues of the aircraft simultaneously. Auditory workload can build to the point that the pilot has no choice but to stop listening or stop looking to be able to pay more attention to the audio. This may be why alerting systems that talk to the crew often receive low marks despite the apparently intuitive "fact" that a nonvisual interface is the solution to high workload. Visual overload may have been traded for auditory overload.

Another high-workload problem is producing and maintaining situation awareness. The human factors engineer should recognize that situation awareness is a general term and that, in fact, many types of situation awareness have to exist simultaneously. For instance, pilots need to be situationally aware of the helicopter's movement through space and time. They must also be aware of where they are in the world and where they are trying to go. They must be aware of the health of the helicopter and its systems. They must be aware of the radio message traffic, who it is from, what they want, and so on. Each of these types of awareness requires its own set of information that creates and maintains the awareness.

The pilot must continually be assimilating information as a result of the need for maintaining awareness. The information may be simply confirmation that nothing has changed or that a new element has been added for integration into awareness. This information demand will result in either a degradation of awareness without readily available information or increased workload as the pilot searches for the information needed. In a glass cockpit, the searching may require menu navigation to get information followed by menu navigation to return to the original display. This is the source of the constant button pushing seen in some glass-cockpit helicopters. Menu navigation formation to support situation awareness competes directly with situation awareness.

One part of the task of creating and maintaining situation awareness is not obvious. This is the problem that the information presented may be raw data that must be processed by the pilot. Processing raw data into that required for situation awareness results in more workload. For instance, the temperature of a critical section of the engine may be continuously displayed. A red range may be marked on the display to indicate excessive temperature. However, how long the temperature can stay in this range and how long it has been in this range are awareness issues that are typically considered the responsibility of the pilot. The pilot must note when the red range is reached, track how long temperature stays red, and remember how long the engine can operate in the red. The desired awareness is built from processing the raw data of engine temperature, current time, and memorized information. A better situation would be one where these answers are displayed along with the raw data of temperature.

Another source of workload generated by cockpit design is the typical requirement for memorization of setup. This means that the pilots must know which switches, in which order, result in the desired effect. Many times an incorrectly set switch, or a switch thrown in the wrong order, precludes the desired effect but without clear indication of what the problem was. A typical result is to start the task over because the setup is remembered more as a sequence of actions than a table of positions.

This introduces another related source of workload, namely, error checking and error recovery. The pilot must recognize that goals are not being met because an action he or she thought had been initiated did not actually happen. Awareness of the lack of an action requires workload to monitor and compare the information over time to determine that the commanded action is not taking place. Determining why the action commanded is not taking place requires additional searching through information to compare the actual condition with the expected condition.

The ongoing external flight and mission demands continually conflict with internal flight, mission, and housekeeping, with the result that the pilot must constantly interrupt ongoing tasks to attend to a task that has become of higher priority. After the interruption, the pilot must return to the original task or, after assessing the situation, decide that another task is higher priority. The continued task interruptions result in workload to manage concurrent tasks and affects situation awareness.

Glass cockpits pose unique problems of their own. A multifunction display, by definition, presents different information at different times. The crew must control what is displayed by making selections from presented options. The options presented at any moment constitute a menu from which the pilot chooses. Each option chosen may change the information on the display or lead to a new menu of options. Moving through these menus to find information or controls is what is referred to as menu navigation. The structure of the menu determines how many selections must be made before accessing the information or control desired. Creating a mental map that guides the pilot to the proper display requires memorized knowledge of the shape of the menu trees and the contents of each associated display. Although the ability to customize a display with information or controls helps the problems of presentation format, allows each display to be presented in an easier to assimilate format, and allows squeezing all required functions into a small physical space, it brings the workload associated with menu navigation, the burden of memorizing the system structure, and the possibility for confusion.

As computers take over control of the helicopter and free the pilot from the routine tasks of status monitoring and function setup, and extend the number of things the crew can be expected to do, additional workload is created if something fails in the system. Should some part of the computer system or a device fail, there are many options to be examined to determine what exactly has failed and what can be done about it. As automation levels increase, the crew is less able to understand and control the automation, with the result that workload increases and awareness drops.

Engineering Solutions to Helicopter Workload

Once the sources of workload are understood, the human factors engineer can combine knowledge of required tasks (from mission and task analysis) with knowledge of sources of workload to create, during the detailed design phase of development, a cockpit that is low in workload and high in situation awareness. This is done by designing out the sources of workload and designing in the attributes of consistency, predictability, and simplicity. In early helicopters, the tasks were merely to fly and monitor the engine. Current and developing helicopters include a wide range of sensors and mission equipment and computerized control of the aircraft. Automation has been added to prevent workload overload. However, automation should not be added indiscriminately. Overautomation can lead to the pilot not being aware of what the helicopter is doing, what the limitations are, and leave him or her helpless when the automation fails. The successful cockpit design will be a subtle blending of automation, information, and control options that allow the pilot to conduct flight and mission tasks with high situation awareness without the costs of being out of control. The difficulty is in translating the goals espoused here into solutions. The following are a number of guidelines for the cockpit designer taken from Hamilton (1993). The purpose is to help the designer understand what the attributes of a "good" design are and how to achieve those attributes.

- Switching from one situation awareness to another should be recognized as an explicit task, and mechanisms should be provided to help the pilot switch and maintain situation awareness views.
- Displays should provide information in a format compatible with the information demands of the crew.
- Information needed to support decision making and situational awareness should be clustered together.
- All information required for tasks should be located together and controls for those tasks also should be located together.
- More options do not always make for a happier pilot. In fact, more choices increases pilot reaction times.
- Switch setups should be automated so that the pilot selects functions rather than settings.
- Incorrect switch choices (i.e., those whose order or setting does not apply) should not be presented.
- If the function is normally provided, but is not available because of faults, then the switch should indicate nonfunctionality and whether the function is broken or merely prevented from operation by an external state (e.g., data not available).
- Good menu structure organization is based on a human factors engineering task analysis at the mission segment level.
- Display pages should be based on a set of activities that must be performed by the pilot to complete a portion of the mission segment.
- Data that are commonly used, are frequently cross-checked, or may be time critical should be readily available.
- Information, controls, and options should be segregated into groups that functionally represent maintenance, setup, normal operations, training, and tactical applications so that the menu structure is sensitive to mission phase.
- Consistency is the biggest ally of the pilot, whereas surprise is the biggest enemy.
- Pilots will avoid automation if it requires significant setup or option selection and will avoid systems that they are not completely familiar with.
- Bezel switch labels should indicate function, current status, and indicate impact on menu navigation task (do you go to another page, just change information on the same display, turn something on or off, etc.) in order to take the guesswork out of menu navigation.
- Switch labels and status indicators should be in the language of the pilot, not the engineer who designed the system.

- Data and tasks should be arranged so that the pilot does not have to go to the bottom of a menu tree to find high-level or routine data and controls.
- All tasks will be interrupted before completion, so tasks should be designed so that interrupting tasks can be entered easily and quickly and it is simple to return to the interrupted task.
- Recognize the strengths and weaknesses of the various subsystems and integrate data from various subsystems to create better information than any one system can provide.

REQUIREMENTS DOCUMENTATION, VERIFICATION, AND FLOWDOWN

Computers have changed the way in which helicopters are designed and how they are operated. This change will be permanent in all but a few very restricted situations. This is because computers offer more capability, in less space, for less weight, and at cheaper costs than individual, purpose-built black boxes. In order to realize their potential, computers will have to have adequate throughput, buses connecting computers will have to have adequate bandwidth, and all functions of the aircraft will have to tie into the computer system. The problem is in defining what is to be built and what "adequate" means in the context of a specific program. The human factors engineer can provide significant insight into what needs to be done, what data at what rates are required, what kind of control lags can be tolerated, and how the crew interface displays information and controls the computer system that controls the aircraft. The system's designers benefit from the human factors engineer's knowledge, insight, and designs, but only if they are properly documented.

Software requirements are decomposed from general requirements to detailed, unique, and testable requirements suitable for software code development. This decomposition process results in the family of hierarchical documents as shown generically in Fig. 17.4. These documents, as tailored to a program, describe what needs to be implemented, how to implement it, and how software interacts with software. Requirements start with a system specification that contains the top-level requirements. The next level of specification is the system/segment specification, depending on program scope. In a large system, there may be more than one segment, and system-level requirements may be allocated partially to one segment and partially to another. At this point, the goal is to separate requirements by type so that flight control requirements, for instance, are not mixed in with airframe requirements. Once segments have been created and segment level requirements derived (to do task *X*, functions *A* and *B* must be done), then a system/segment design document is generated. The system/segment design document outlines how the segment is to be implemented, breaking all the requirements into functional groups (computer software configuration items) such as navigation, aircraft management, and crew interface. Software requirement specifications can then be created for these functional groups to describe the specific software routines to be developed and what the routines must do. The flow of information on the data buses must also be managed, so interface control documents are generated that define the messages and rate of transmission between the various computer software configuration items. Hardware requirements are specified in prime item development specifications and critical item development specifications. Finally,

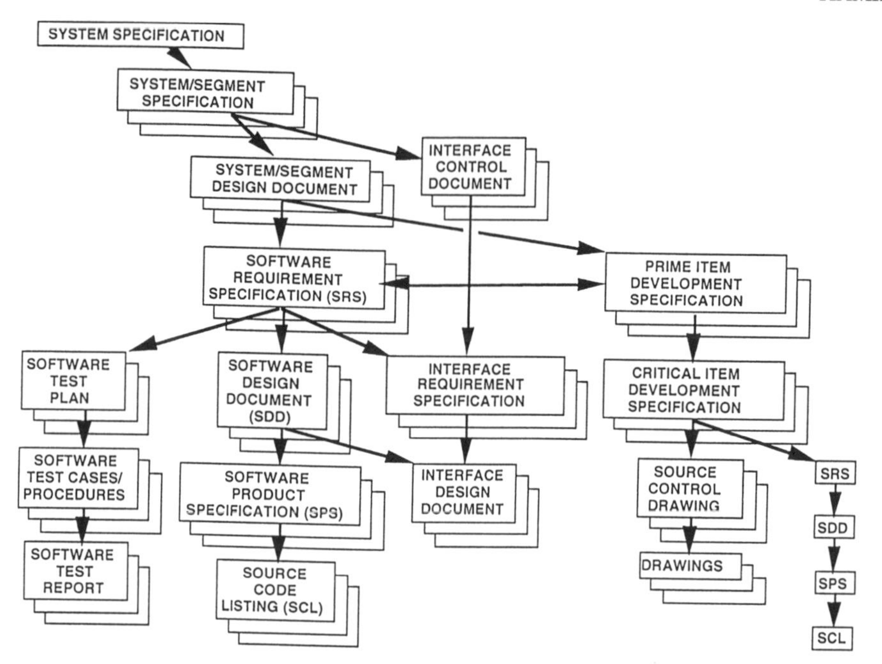

FIG. 17.4. Family of software decomposition documents showing hierarchy and interrelationships.

a host of test and qualification plans is generated to verify that the software does what the flowed-down requirements dictate.

This decomposition process governs what gets implemented. It is this process that the human factors engineers need to influence. What are finally implemented in the cockpit and the pilot–vehicle interface are only those things called for by these documents. The decomposed requirements, however, are specific statements about what software does and they do not address how the computer interacts with the pilot. For instance, a requirement in a software requirement specification might state that the routine is to present current location to the pilot and obtain a new position for purposes of updating the navigation system. The menu structure and display formats are not typically specified and are left to an informal process as to where in the system the update function is found, how the data are displayed, and how new data are entered. As a result, how the interface works is generally a fallout of design rather than a driver of the software decomposition process.

The only way to compete in the world of requirements decomposition and flowdown is for the human factors engineer to create his or her own set of requirements and flow them into the process. How to do this is described in Hamilton and Metzler (1992). The pilot–vehicle requirements should include the mission and task analysis conducted to determine information requirements and general tasks. It should include specific design goals and implementation rules. The pilot–vehicle requirements should provide the specifics of each display format, each data entry or control method, and the specifics of the menu structure. The movement of symbols, definition of symbols (bitmaps), update rates, and smoothing and filtering algorithms should also be included. This document must be generated before the software requirements review but be iteratively updated before software preliminary design review as a function of person-in-the-loop testing and hardware/software development. After software preliminary design review, the requirements should be configuration managed (changed only by program directive) in response to known unknowns and unknown unknowns. Wherever possible, the user pilots of the helicopters should be consulted early in development and kept aware of the developing interface. A representative sample of the user community, environmental conditions, and missions should be included in a simulation program.

Creating a pilot–vehicle interface specification early in the program will address many of the questions and issues of cockpit development that have been raised in this chapter. Mission and task analysis are a requirement of most major development programs, as well as generating human engineering design approaches. The human engineering approach proposed here is cost-effective because it centers on the continued iteration of analyses already conducted during the early stages of most programs. Eventually the display formats, control methods, and menu structure will have to be documented for testing, and for development of training courses and devices. Although the effort shifts from an after-the-fact documentation to a design driver, it is not a new, unfunded activity. No new tasks are being defined by this approach, although some tasks are in more detail earlier in the program than previously. The human factors engineer must broaden his or her outlook and repertoire of skills, but the benefit is that the interface has the attributes and characteristics desired by design rather than by luck.

SUMMARY

Helicopters present many of the same issues to the human factors engineer as do fixed-wing aircraft, but helicopters do have unique challenges. These issues are related mostly to how helicopters generate and control lift and to what is done with the unique flight capabilities. Human factors engineers have always had an important role in designing a low-workload, high-situation-awareness cockpit, and that role will be more important in the computer age. Mission equipment development is now as expensive as airframe development, with a large portion of that cost due to software. Human factors engineers must understand how computerized systems are developed and join in the process if acceptable cockpit workload and situation awareness are to be maintained in the face of ever-increasing capabilities and expanding missions. Just as in airframe development, oversights in requirements and confusions in meaning can have very serious impacts on cost and schedule of software-intensive systems. Like an airframe, software must have the inherent risks in the proposed design reduced by a systematic program of test and design maturation. This process of software requirement decomposition and verification will benefit from the participation of human factors engineers and will result in increased responsibilities for them. No new technological breakthroughs are required; the tools for design and test are available, but must be used in new ways.

RECOMMENDED READING

Table 17.1 provides a list of references useful in the area of helicopter human factors. The list is composed primarily of military specifications (MIL-SPECs). MIL-SPECs have been condemned in recent years as the source of unnecessary cost and waste in defense procurement, and it may well be the case that elimination of chocolate chip cookie

TABLE 17.1
Recommended Readings

Reference	*Title*	*Summary*
Military Aeronautical Design Standard ADS-36	Rotary Wing Aircraft Crash Resistance	Defines helicopter crashworthiness characteristics. Primarily applicable to design of airframe and subsystems.
Semple et al. (1971)	Analysis of Human Factors Data for Electronic Flight Display Systems	Not a specification or standard. Provides excellent background information related to human factors requirements for electronic displays. Provides historical information about a range of displays.
U.S. Army Regulation AR 602-1	Human Factors Engineering Program	Establishes structure of HFE program in response to acquisition of U.S. Army materiel.
U.S. Army Regulation AR 602-2	Manpower and Personnel Integration (MANPRINT) in Materiel Acquisition Process	Directs establishment and structure of MANPRINT programs in response to U.S. Army Acquisitions.
SAE (1996)	Pilot–System Integration	Newly released document that describes a very good approach for development of a pilot–vehicle interface.
Boff and Lincoln (1988)	Engineering Data Compendium: Human Perception and Performance	Not a specification or standard. One of the best sources of human factors, human psychophysiology, and human performance available. Standard arrangement and user's guide make the data easily accessible. Highly recommended.
Military Specification Data Item DI-HFAC-80740	Human Engineering Program Plan	Establishes structure for execution of formal human engineering program for development of vehicles, equipment, and facilities.
Military Specification Data Item DI-HFAC-80743	Human Engineering Test Plan	Directs structure and submittal of human engineering test plans.
Military Specification Data Item DI-HFAC-80744	Human Engineering Test Report	Directs reporting structure and submittal of test results.
Department of Defense (1983)	Anthropometry of Military Personnel	Documents anthropometry of various military service populations.
Military Specification MIL-STD-490	Specification Practices	Provides guidelines for types of specifications required, content, and layout of specifications.
Military Specification MIL-STD-2167	Defense System Software Development	Provides guidelines for defense-related software development programs.
Military Specification MIL-STD-2168	Defense System Software Quality Program	Provides guidelines for software development quality assurance program.
Ketchel and Jenney (1968)	Electronic and Optically Generated Aircraft Displays: A Study of Standardization Requirements	Not a specification or standard. Provides excellent summary of information requirements, display principles, and human factors principles specifically oriented toward the aviation environment.

Military Specification MIL-C-81774	Control Panel, Aircraft General Requirement for	Delineates documents for design of control panels; establishes display and control selection, utilization, and arrangement; provides for verification data of these requirements.
Military Specification MIL-D-23222	Demonstration Requirements	Defines demonstrations of the air vehicle. The only specific human engineering requirements are related to cockpit and escape system design.
Military Specification MIL-H-46855	Human Engineering Requirements of Military Systems, Equipment, and Facilities	Establishes the requirements for applying human factors engineering to the development of all vehicles, equipment, and facilities for the U.S. military.
Military Specification MIL-L-18276C	Lighting, Aircraft Interior, Installation of	Requirements for primary, secondary, instrument, and emergency lighting systems and controls. Addresses visual signals for retractable gear warning.
Military Specification MIL-L-6503H	Lighting Equipment, Aircraft, General Specification for Installation of	Requirements for installation of exterior and interior lighting, except instrument and aircrew station visual signals. Definition of emergency lighting system controls and fixtures, formation lights, position lights, landing and searchlights (and stick-mounted controls for same), etc. Specification used primarily by human factors engineering as information source on lighting intensities, cones of illumination, and orientation of lighting systems.
Military Specification MIL-L-85762	Lighting Aircraft, Interior, AN/AVS/6 Aviator's Night Vision Imaging System (ANVIS) Compatible	Provides performance requirements and testing methodology to ensure effective and standardized aircraft interior lighting for ANVIS compatibility. This specification imposes very specific design and test requirements.
Military Specification MIL-M-18012B	Markings for Aircrew Station Displays, Design and Configuration of	Design requirements and configuration of letters, numerals, and identification for aircrew displays and control panels.
Military Specification MIL-M-8650	Mockups, Aircraft, General Specification for	General requirements for construction of aircraft and related systems mockups for formal evaluation and perpetration of mockup data. Provides mockup checklist for reviewers.
Military Specification MIL-P-7788E	Panel, Information, Integrally Illuminated	Covers general requirements for integrally illuminated panels.
Military Specification MIL-STD-12D	Military Standard Abbreviations for Use on Drawings, and in Specifications, Standards, and Technical Documents	Establishes standard abbreviations and acronyms for documents and drawings submitted to government customer.

(Continued)

TABLE 17.1
(Continued)

Reference	*Title*	*Summary*
Military Specification MIL-STD-250	Aircrew Station Controls and Displays for Rotary Wing Aircraft	Establishes standardized requirements for the design, uniform assignment, arrangement, location, and actuation of controls and displays used in crew stations.
Military Specification MIL-STD-411	Aircrew Station Signals	Covers requirements for aircrew station alerting systems. Includes general functions, operational logic, information content of messages, physical characteristics of alerting system, visual, auditory, and tactual signals.
Military Specification MIL-STD-783	Legends for Use in Aircrew Stations and on Airborne Equipment	Establishes requirements for legends used for marking controls and displays in aircrew stations and on airborne equipment.
Military Specification MIL-STD-850	Aircrew Station Vision Requirements for Military Aircraft	Establishes requirements for providing adequate external vision from cockpit.
Military Specification MIL-STD-1290	Light Fixed and Rotary Wing Aircraft Crashworthiness	Establishes minimum crashworthiness design criteria. Defines the crashworthiness envelope of flight crew and passengers.
Military Specification MIL-STD-1294	Acoustical Noise Limits In Helicopters	Establishes steady-state acoustical noise limits within personnel-occupied spaces.
Military Specification MIL-STD-1295	Human Factors Engineering Design Criteria for Helicopter Cockpit, Electro-Optical Display Symbology	Establishes general information, symbology, and display format requirements for hover, position, transition, cruise, and weapon delivery modes of rotary-wing aircraft.

Military Specification MIL-STD-1333	Aircrew Station Geometry for Military Aircraft	Establishes design requirements for aircrew station geometry. Objective is to provide design that is efficient, safe, and comfortable for operation by aircrew fitting body sizes specified by the customer.
Military Specification MIL-STD-1472	Human Engineering Design Criteria for Military Systems, Equipment, and Facilities	The primary design document enforcing human engineering characteristics within the system or component. It presents quantitative and qualitative human design criteria required for achievement of mission success through integration of humans into systems, subsystems, equipment, and/or facilities. It is a design guideline that summarizes human factor applications derived through more than 50 years of laboratory and applied research.
Military Specification MIL-STD-1787	Military Standard Aircraft Display Symbology	Describes symbols, formats, and information content for electro-optical displays. Provides characteristics for takeoff, navigation, terrain following/avoidance, weapon delivery, and landing. Defines symbol geometry, fonts, recommended dimensions, and mechanization.
Military Specification MS-33575	Dimensions, Basic, Cockpit, Helicopter	Provides basic dimensional guidelines for cockpit on single sheet.
Atkins, Dauber, Karas, and Pfaff (1975)	Study to Determine Impact of Aircrew Anthropometry on Airframe Configuration	Not a specification or standard. Very broad study of cockpit geometry as related to aircrew sizing, equipment requirements, controls operation, visibility, display placement, entry/exit criteria, etc.
Zimmermann and Merrit (1989)	Aircraft Crash Survival Design Guide	Not a specification or standard. Very broad study of cockpit geometry as related to aircrew sizing, equipment requirements, controls operation, visibility, display placement, entry/exit criteria, etc.
U.S. Army (1988)	Job and Task Analysis Handbook	Not a specification or standard. Serves as a guideline in choosing a task analysis methodology and instructions for implementing the same.

specifications may reduce the cost of federal cookies without impacting taste. However, not all MIL-SPECs are therefore bad. The ones listed here are generally very good in that they define a design space or processes rather than specify a solution. Most are applicable to either fixed- or rotary-wing aircraft.

REFERENCES

Atkins, E. R., Dauber, R. L., Karas, J. N., & Pfaff, T. A. (1975). *Study to determine impact of aircrew anthropometry on airframe configuration* (Report No. TR 75-47). St. Louis, MO: U.S. Army Aviation System Command.

Boff, K. R., & Lincoln, J. E. (1988). *Engineering data compendium: Human perception and performance.* Wright-Patterson Air Force Base, OH: Harry G. Armstrong Aerospace Medical Research Laboratory.

Department of Defense. (1983). *Anthropometry of military personnel* (Handbook DOD-HDBK-743). Washington, DC: Author.

Hamilton, B. E. (1993, October). *Expanding the pilot's envelope. Technologies for highly manoeuvrable aircraft.* North Atlantic Treaty Organization Advisory Group for Aerospace Research and Development, Conference Proceedings No. 548, Annapolis, MD.

Hamilton, B. E., & Metzler, T. (1992, February 3–6). Comanche crew station design. In *Proceedings: 1992 Aerospace Design Conference,* American Institute of Aeronautics and Astronautics, AIAA-92-1049, Irvine, CA.

Ketchel, J. M., & Jenney, L. L. (1968). *Electronic and optically generated aircraft displays: A study of standardization requirements* (Report No. JANAIR 680505, AD684849). Washington, DC: Office of Naval Research.

Semple, C. A., Jr., Heapy, R. J., Conway, E. J., Jr., & Burnette, K. T. (1971). *Analysis of human factors data for electronic flight display systems* (Report No. AFFDL-TR-70-174). Wright-Patterson Air Force Base, OH: Flight Dynamics Laboratory.

Society of Automotive Engineers, Inc. (1996). *Pilot–system integration* (Aerospace Recommended Practice 4033). Warrendale, PA: Author.

U.S. Army. (1988). *Job and task analysis handbook* (Training and Doctrine Command Pamphlet 351-4(T), TRADOC PAM 351-4(T)). Washington, DC: U.S. Department of Defense.

Zimmermann, R. E., & Merrit, N. A. (1989). *Aircraft crash survival design guide Volume I—Design criteria and checklists* (Report No. TR 89-D-22). Fort Eustis, VA: Aviation Applied Technology Directorate.

IV

AIR TRAFFIC CONTROL

18

Air Traffic Control

Michael S. Nolan
Purdue University

The primary function of an air traffic control (ATC) system is to keep aircraft participating in the system separated from one another. Secondary reasons for the operation of an ATC system are to make more efficient use of airspace, and to provide additional services to pilots such as traffic information, weather avoidance, and navigational assistance.

Not every aircraft may be required to participate in an air traffic control system, however. Each nation's regulations only obligate certain aircraft to participate in the ATC system. ATC participation in each country may range from mandatory participation of all aircraft, to no ATC services offered at all.

The level of ATC services provided is usually based on each nation's priorities, technical abilities, weather conditions, and traffic complexity. To more specifically define and describe the services that can be offered by an ATC system, the International Civil Aviation Organization (ICAO) has defined different aircraft operations and classes of airspace within which aircraft may operate. Different rules and regulations apply to each type of aircraft operation, and these rules vary depending on the type of airspace within which the flight is conducted. Although ICAO publishes very specific guidelines for the classification of airspace, it is the responsibility of each country's aviation regulatory agency to categorize its national airspace.

AIRCRAFT OPERATIONS

Visual meteorological conditions (VMC) are defined as weather conditions where pilots are able to see and avoid other aircraft. In general, pilots flying in VMC conditions comply with *visual flight rules* (VFR). VFR generally require that 3 to 5 miles of flight visibility be maintained at all times, that the aircraft remain clear of clouds, and that pilots have the responsibility to see and avoid other aircraft. Pilots provide their own air traffic separation. The ATC system may assist the pilots, and may offer additional services, but the pilot has the ultimate responsibility to avoid other air traffic.

Instrument meteorological conditions (IMC) are generally defined as weather conditions where the visibility is below that required for VMC or whenever the pilot cannot remain clear of clouds. Pilots operating in IMC must comply with *instrument flight rules* (IFR), which require the filing of a flight plan, and ATC normally provides air traffic separation. Pilots may operate under IFR when flying in VMC conditions. Under these circumstances, ATC will separate only those aircraft complying with IFR. VFR aircraft provide their own separation, and IFR aircraft have the responsibility to see and avoid VFR aircraft.

AIRSPACE CLASSES

National governments define the extent to which they wish to offer ATC services to pilots. In general, ICAO recommendations suggest three general classes of airspace within which different services are provided to VFR and IFR pilots. These three general classes are uncontrolled, controlled, and positive controlled airspace.

Uncontrolled airspace is that within which absolutely no aircraft separation is provided by ATC, regardless of weather conditions. Uncontrolled airspace is normally that airspace with little commercial aviation activity.

Controlled airspace is that within which ATC separation services may be provided to certain select categories of aircraft (usually those complying with instrument flight rules). In controlled airspace, pilots flying VFR must remain in VMC, and are not normally provided ATC separation and therefore must see and avoid all other aircraft. Aircraft who wish to utilize ATC services in controlled airspace must file a flight plan and comply with instrument flight rules. IFR aircraft are permitted to operate in both VMC and IMC. When operating within controlled airspace, IFR aircraft are separated by ATC from other aircraft operating under IFR. When operating in VMC in controlled airspace, IFR pilots must see and avoid aircraft operating under VFR.

In positive controlled airspace, all aircraft, whether IFR or VFR, are separated by air traffic control. All aircraft operations require an ATC clearance. VFR pilots must remain in VMC conditions, but are separated by ATC from both VFR and IFR aircraft. IFR aircraft are also separated from both IFR and VFR aircraft. Table 18.1 describes the general rules that both IFR and VFR pilots must comply with when operating in these three classes of airspace.

AIR TRAFFIC CONTROL PROVIDERS

In most countries, a branch of the national government normally provides air traffic control services. The ATC provider may be civilian, military, or a combination of both. Some national ATC services are now being operated by private corporations funded primarily by user fees. Other governments are experimenting with ATC system privatization. Some of these initiatives propose to transfer all ATC responsibility to private agencies, whereas others propose to transfer only certain functions, such as weather dissemination and the operation of low-activity control towers, to private or semipublic entities.

Privatized air traffic control is a fairly recent historical development with roots tracing back to the 1930s. When an air traffic control system was first started in the

TABLE 18.1
Requirements for Operation and ATC Services Provided to Flight Operations Within General Airspace Categories

	Uncontrolled Airspace	*Controlled Airspace*	*Positive Controlled Airspace*
VFR flight operations	Must remain in VMC (VMC minima are generally fairly low, typically one mile flight visibility). If VMC conditions exist, VFR operations are permitted and no ATC clearance is required. If IMC conditions exist, VFR operations are not authorized. No ATC separation services provided. Pilots must see and avoid both VFR and IFR aircraft.	Must remain in VMC (VMC minima usually require 3–5 miles flight visibility). If VMC conditions exist, VFR operations are permitted and no ATC clearance is required. If IMC conditions exist, VFR operations are not authorized. No ATC separation services provided. Pilots must see and avoid both VFR and IFR aircraft. Aircraft operating in controlled airspace may be required to meet class-specific regulations.	Must remain in VMC (VMC minima usually require 3–5 miles flight visibility). If VMC conditions exist, VFR operations are permitted but an ATC time clearance is required. If IMC conditions exist, VFR operations are not authorized. ATC separation services are mandatory. ATC clearance always required. Pilots will be separated by ATC from both VFR and IFR aircraft. Aircraft operating in positive controlled airspace may be required to meet class-specific regulations.
IFR flight operations	No ATC clearance required. ATC separation services will not be provided to any aircraft. Pilots must see and avoid both VFR and IFR aircraft.	ATC clearance required. ATC separation will be provided. ATC will separate IFR aircraft from other IFR aircraft. When operating in VMC conditions, pilots must see and avoid VFR aircraft. Aircraft operating in controlled airspace may be required to meet class-specific regulations.	ATC clearance required. ATC separation will be provided. ATC will separate IFR aircraft from all other aircraft. Aircraft operating in positive controlled airspace may be required to meet class- specific regulations.

United States, control towers were operated by the municipalities that owned the airports. Enroute air traffic control was provided through a consortium of airlines. Only in the 1940s was air traffic control taken over and operated by the national government.

The concept behind privatized ATC is that if freed from cumbersome government procurement requirements, employment regulations, and legislative pressures, private corporations might provide service at less cost, be more efficient, and be more responsive to users' needs because they would be funded and controlled by the users. Possible disadvantages of such a system include lack of governmental oversight and responsibility, possible conflict of interest between system users and operators, little incentive to assist military aviation activities, and restricted access to the capital funding needed to upgrade and operate such a complex system.

ATC ASSIGNMENTS

Every nation is responsible for providing ATC services within its national borders. In order to provide for a common method of air traffic control, ICAO promulgates standardized procedures that most countries generally adhere to. These standards include universally accepted navigation systems, a common ATC language (English), and general ATC separation standards. ICAO is a voluntary organization of which most countries are members. Every ICAO signatory nation agrees to provide ATC services to all aircraft operating within its boundaries and agrees to require that their pilots abide by other national ATC systems when operating within foreign countries.

Every nation's ATC procedures can and do occasionally deviate from ICAO recommend practices. Each operational procedure that deviates from ICAO standards is published by the national ATC service provider in the Aeronautical Information Publication.

ICAO has been granted the responsibility for providing ATC services in international airspace, which is comprised mostly of oceanic and polar airspace. ICAO has assigned separation responsibility in those areas to individual states both willing and able to accept that responsibility. Some countries that have accepted this responsibility include the United States, United Kingdom, Canada, Australia, Japan, Portugal, and the Philippines.

ATC SERVICES

Airspace with little or no potential traffic conflicts requires little in the way of sophisticated ATC systems. If air traffic density increases, if aircraft operations increase in complexity, or if special, more hazardous operations are routinely conducted, additional control of aircraft is usually required to maintain an acceptable level of safety. The easiest method of defining these increasing ATC system requirements and their associated operating rules is to define different classes of airspace within which different ATC services and requirements exist.

Standard ICAO airspace classifications include classes labeled A, B, C, D, E, F, and G. In general, Class A airspace is positive controlled, where ATC services are mandatory for all aircraft. Class G is uncontrolled airspace where no ATC services are provided to either IFR or VFR aircraft. Classes B, C, D, E, and F provide declining levels of ATC services and requirements.

It is each nation's responsibility to describe, define, explain, and chart the various areas of airspace within their respective boundaries. In general, areas with either high-density traffic or a mix of different aircraft operations are classified as class A, B, or C airspace. Areas of low traffic density are usually designated as class D, E, F or G.

AIR TRAFFIC CONTROL SERVICES OFFERED WITHIN EACH TYPE OF AIRSPACE

The requirements to enter each airspace classification and the level of ATC services offered within each area are listed here with the associated United States airspace names in parentheses.

Class A Airspace (U.S. Positive Control Area, PCA). All operations must be conducted under IFR and are subject to ATC clearances and instructions. ATC separation is provided to all aircraft. Radar surveillance of aircraft is usually provided.

Class B Airspace (U.S. Terminal Control Area, TCA). Operations may be conducted under IFR or VFR. However, all aircraft are subject to ATC clearances and instructions. ATC separation is provided to all aircraft. Radar surveillance of aircraft is usually provided.

Class C Airspace (U.S. Airport Radar Service Area, ARSA). Operations may be conducted under IFR or VFR; however, all aircraft are subject to ATC clearances and instructions. ATC separation is provided to all aircraft operating under IFR and, as necessary, to any aircraft operating under VFR when any aircraft operating under IFR is involved. All VFR operations will be provided with safety alerts and, on request, conflict resolution instructions. Radar surveillance of aircraft is usually provided.

Class D Airspace (U.S. Control Zones for Airports with Operating Control Towers and Airport Traffic Areas that are not associated with a TCA or an ARSA). Operations may be conducted under IFR or VFR; however, all aircraft are subject to ATC clearances and instructions. ATC separation is provided to aircraft operating under IFR. All aircraft receive safety alerts and, on pilot request, conflict resolution instructions. Radar surveillance of aircraft is not normally provided.

Class E Airspace (U.S. General Controlled Airspace). Operations may be conducted under IFR or VFR. ATC separation is provided only to aircraft operating under IFR within a surface area. As far as practical, ATC may provide safety alerts to aircraft operating under VFR. Radar surveillance of aircraft may be provided if available.

Class F Airspace (United States has no equivalent). Operations may be conducted under IFR or VFR. ATC separation will be provided, so far as practical, to aircraft operating under IFR. Radar surveillance of aircraft is not normally provided.

Class G Airspace (U.S. Uncontrolled Airspace). Operations may be conducted under IFR or VFR. ATC separation is not provided. Radar surveillance of aircraft is not normally provided.

AERONAUTICAL NAVIGATION AIDS

Air traffic separation can only be accomplished if the location of an aircraft can be accurately determined. Therefore an air traffic control system is only as accurate as its ability to determine an aircraft's position. The navigation systems currently in use were developed in the 1950s, but are undergoing a rapid change in both technology and cost. As integrated circuitry and computer technology continue to become more robust and inexpensive, the newly certified *global navigation satellite system* (GNSS) global positioning system promises unprecedented navigational performance at a relatively low cost. ICAO has just recently affirmed its preference for GNSS as the future primary international navigation standard. Various experts predict that existing navigation systems will be either decommissioned or relegated to a GNSS backup role by the turn of the century.

In general, the accuracy of existing navigation aids is a function of system cost and/or aircraft distance from the transmitter. Relatively inexpensive navigation systems are generally fairly inaccurate. The most accurate systems tend to be the most expensive. Table 18.2 describes the type, general cost, advantages, and disadvantages of many common aeronautical navigation systems.

TABLE 18.2
Navigation System Capabilities, Advantages, and Disadvantages

System	*General Cost*	*Effective Range*	*Accuracy*	*Ease of Use*	*Advantages*	*Disadvantages*
Nondirectional Beacon	Inexpensive transmitter. Inexpensive receiver.	50 to 1,000 nautical miles.	Fairly inaccurate.	Somewhat difficult to use.	Inexpensive and easy to install transmitter.	Susceptible to static interference. Does not transmit distance information.
VORTAC	Moderately expensive transmitter. Fairly inexpensive receiver.	25 to 200 nautical miles.	Fairly accurate.	Fairly easy to use.	Current international enroute standard. Provides distance information if aircraft properly equipped.	Large number of transmitters needed to provide worldwide coverage. Pilots must fly to and from VORTAC stations.
LORAN-C	Expensive transmitter. Inexpensive receiver.	Up to 1,000 nautical miles.	Fairly accurate.	Fairly easy to use.	Only a limited number of transmitters needed. Provides direct routing between pilot-selected points.	Originally a marine navigation system, therefore commonly found near oceans.
Inertial navigation	No transmitters required. Very expensive receivers.	Independent of transmitters, therefore unlimited.	Fairly accurate.	Somewhat easy to use.	Very independent because system does not require any transmitters.	Needs to be programmed by pilot. No indication if inaccurately programmed. Accuracy degrades as time goes on.

Global navigation satellite system	Extremely expensive satellite transmitters. Very inexpensive receivers.	Worldwide.	Very accurate. Can be made extremely accurate when augmented.	Fairly easy to use.	Inexpensive, worldwide coverage. Point to point navigation.	System controlled by one agency (currently U.S. military). System accuracy intentionally degraded.
Instrument landing system	Expensive transmitter. Fairly inexpensive receiver.	10–30 nautical miles.	Fairly accurate.	Fairly easy to use.	Current world standard for precision approaches.	Limited frequencies remain available for expansion. Only one path provided to a single runway. Expensive site preparation may be required for installation.
Microwave landing system	Expensive transmitter. Somewhat expensive receiver.	10–30 nautical miles.	Fairly accurate.	Easy to use.	New world standard for precision approaches. Can serve multiple runways. Can provide multiple and curved flight paths. The U.S. has recently announced it will no longer support MLS, preferring to develop GNSS as a precision approach navigation aid.	Although accepted as world standard, may be superseded if GNSS can provide equivalent accuracy.

GLOBAL NAVIGATION SATELLITE SYSTEM

Global navigation satellite systems have just recently been adopted as the future navigation standard by ICAO. Currently, GNSS systems are as accurate as most current enroute navigation systems. Inherent inaccuracies (and some intentional signal degradation) require that GNSS be augmented if it is to replace ILS and/or MLS as a precision navigation system. Satellite accuracy augmentation (*wide-area augmentation system*, WAAS) has been proposed as one method to provide general improvements to accuracy that may permit GNSS to replace ILS as the precision approach standard. Ground-based augmentation (*local-area differential global positioning system*, LADGPS) may be required before GNSS will be sufficiently accurate for all-weather automatic landings. Which system or combination of systems will be eventually used is still undetermined.

RADAR SURVEILLANCE IN AIR TRAFFIC CONTROL

Radar is used by air traffic controllers to monitor aircraft position, detect navigational blunders, reduce separation if possible, and make more efficient use of airspace. Controllers can utilize radar to provide aircraft navigational assistance during both the enroute and approach phases of flight. If radar is able to provide more accurate aircraft positional information than existing navigation systems can provide, it may be possible to reduce the required separation between aircraft.

Three different types of radar are used in ATC systems. Primary surveillance radar was first developed during World War II, and can detect aircraft without requiring onboard aircraft equipment. *Secondary surveillance radar* (SSR) requires an interrogator on the ground and an airborne transponder in each aircraft. SSR provides more accurate aircraft identification and positioning, and can transmit aircraft altitude to the controller. Mode-S secondary radar is a recent improvement to secondary radar systems that will provide unique aircraft identification and the ability to transmit flight information to the controller, and air traffic control instructions and other information directly to the aircraft. Table 18.3 lists the functional advantages and disadvantages of each radar surveillance system.

AIRCRAFT SEPARATION IN AN ATC SYSTEM

The airspace within which ATC services are provided is normally divided into three-dimensional blocks of airspace known as sectors. Sectors have well-defined lateral and vertical limits, and normally are shaped according to traffic flow and airspace structure. Only one controller has ultimate responsibility for the separation of aircraft within a particular sector. The controller may be assisted by other controllers, but is the one person who makes the decisions (in accordance with approved procedures), concerning the separation of aircraft within that particular sector.

If pilots of participating aircraft within the sector can see other nearby aircraft, the pilots can simply "see and avoid" nearby aircraft. Or if a controller can see one or both aircraft, the controller may issue heading and/or altitude instructions that will keep the aircraft separated. This informal but effective method of aircraft separation is known as visual separation. Although a simple concept, it is very effective and efficient when

TABLE 18.3
Air Traffic Control Radar Systems

Radar System	*Operational Theory*	*Information Provided*	*Advantages*	*Disadvantage*
Primary surveillance radar	Extremely powerful electrical transmission is reflected by aircraft back to the ATC system.	Range and azimuth.	Detects all aircraft within range regardless of aircraft equipment.	Also detects other moving objects and large stationary objects. Weather and terrain can reflect and block signal. System prone to numerous false targets.
Secondary surveillance radar, also known as the air traffic control radar beacon system (ATCRBS)	Electrical signal sent from ground triggers response from every airborne transponder within range.	Range and azimuth, assigned aircraft code, and altitude.	Detects only aircraft. If ground system is properly equipped, aircraft identity and altitude can be displayed directly on controller's radar display.	Requires aircraft to be equipped with transponder. Operation restricted to single frequency pair, can be overwhelmed by too many aircraft within specific area.
Mode-S	Selective signal sent from ground triggers response from the specified transponder installed in the aircraft.	Range and azimuth, aircraft identity, and altitude. Capability exists to transmit additional flight information, pilot and ATC requests.	Detects only aircraft. If ground system is properly equipped, aircraft identity and altitude can be displayed directly on controller's radar display. Capability exists to transmit controller requests and weather information directly to aircraft.	Requires aircraft to be equipped with transponder. New system will replace secondary radar. Requires all new equipment on ground and in aircraft.

properly used. As long as aircraft can be spotted and remain identified, the use of visual separation permits aircraft to operate in much closer proximity than if the aircraft cannot be seen. Most airports utilize visual separation and visual approaches during busy traffic periods. If weather conditions permit visual separation to be applied, the capacity of most major airports can be significantly increased.

Visual separation can only be employed if one pilot sees the other aircraft, or if the controller can see both aircraft. The primary disadvantage of visual separation is that it can only be employed when aircraft are flying fairly slowly. It would be next to impossible to utilize visual separation during high-altitude, high-speed cruising conditions common to modern aircraft. Visual separation can therefore only be effectively employed within the immediate vicinity of airports. The use of visual separation near airports requires that aircraft remain continuously in sight of one another. This is a difficult proposition at best during the approach to landing or departure phase of flight because these are two of the busiest times for pilots.

NONRADAR SEPARATION

When visual separation cannot be employed, controllers must use either radar or nonradar separation techniques. Due to range and curvature of the earth limitations inherent to radar, there are many situations where radar cannot be used to identify and separate aircraft. Radar coverage exists near most medium- and high-density airports, and at altitudes of 5,000 ft or above in the continental United States and Europe. Outside of these areas, and over the ocean, radar surveillance may not exist and the controller must employ some form of nonradar separation to provide air traffic control.

Nonradar separation depends on accurate position determination and the transmittal of that information to the controller. Due to navigation and communication system limitations, ATC is unable to precisely plot the position of each aircraft in real time. Because navigation systems have inherent inaccuracies, it is impossible to know exactly where each aircraft is at any given time. Nonradar separation therefore assumes that every aircraft is located within a three-dimensional block of airspace. The dimensions of the airspace are predicated on the speed of the aircraft and the accuracy of the navigation system being used. In general, if VORs [VHF Omni-directional Range] are being utilized for aircraft navigation, the airspace assigned to each aircraft may have lateral width of about 8 nautical miles, vertical height of 1,000 ft (2,000 ft when operating above 29,000 ft), and a longitudinal length that varies depending upon the speed of the aircraft. In general, the longitudinal extent of the airspace box extends about 10 min of flight time in front of the aircraft. Depending on the speed of the aircraft, this longitudinal dimension could extend from 10 to 100 miles in front of the aircraft.

Because neither the controller nor the pilot knows exactly where within the assigned airspace box each aircraft is actually located, the controller must assume that aircraft might be located anywhere within the box. The only way to insure that aircraft do not collide is to insure that airspace boxes assigned to different aircraft never overlap. Airspace boxes are permitted to get close to one another, but as long as they never overlap, aircraft separation is assured.

Nonradar separation is accomplished by assigning aircraft either different altitudes or nonoverlapping routes. If aircraft need to operate on the same route at the same altitude, they must be spaced accordingly to prevent longitudinal overlap. Controllers may separate potentially conflicting aircraft either through the use of nonoverlapping holding patterns, or by delaying departing aircraft on the ground. If there is a sufficient speed differential between two conflicting aircraft, the controller can normally permit the faster aircraft to lead the slower aircraft using the same route and the same altitude. Depending on the speed difference between the aircraft, the longitudinal separation criteria can normally be reduced.

The controller uses flight progress strips to visualize the aircraft's position and therefore effect nonradar separation. Pertinent data are written on a flight strip as the aircraft progresses through each controller's sector. The controller may request that the pilot make various position and altitude reports, and these reports are written on the flight strip.

The primary disadvantage of nonradar separation is that its application depends on the pilot's ability to accurately determine and promptly report the aircraft's position, and the controller's ability to accurately visualize each aircraft's position. To reduce the probability of an in-flight collision occurring to an acceptably low level, the separation criteria must take into account these inherent inaccuracies and built-in communications delays. This requires that fairly large areas of airspace be assigned to each aircraft. An aircraft traveling at 500 knots might be assigned a block of airspace 1,000 feet in height, covering close to 400 square miles! This is hardly an efficient use of airspace.

RADAR SEPARATION

Radar can be utilized in air traffic control to augment nonradar separation, possibly reducing the expanse of airspace assigned to each aircraft. Radar's design history causes it to operate in ways that are not always advantageous to air traffic control, however. Primary radar was developed in World War II as a defensive, anti-aerial invasion system. It was also used to locate enemy aircraft and direct friendly aircraft on an intercept course. It was essentially designed to bring aircraft together, not keep them apart.

Primary radar is a system that transmits high-intensity electromagnetic pulses focused along a narrow path. If the pulse is reflected off of an aircraft, the position of the aircraft is displayed as a bright blip, or target, on a display screen known as a plan position indicator (PPI). This system is known as primary surveillance radar.

The radar antenna rotates slowly to scan in all directions around the radar site. Most radars require 5 to 15 sec to make one revolution. This means that once an aircraft's position has been plotted by radar, it will not be updated until the radar completes another revolution. If an aircraft is moving at 600 knots, it might move 2 to 3 miles before it is replotted on the radar display.

Primary radar is limited in range based on the curvature of the earth, the antenna rotational speed, and the power level of the radar pulse. Radars used by approach control facilities have an effective range of about 75 nautical miles. Radars utilized to separate enroute aircraft have a range of about 300 nautical miles.

Secondary surveillance radar is a direct descendent of a system also developed in World War II known as *identification friend or foe* (IFF). Secondary radar enhances the radar target and can be integrated with a ground-based computer to display the aircraft's identity, altitude, and ground speed. This alleviates the need for the controller to constantly refer to flight progress strips to correlate this information. However, flight progress strips are still used by radar controllers to maintain other information, and as a backup system utilized in case of radar system failure.

Although one might think that radar dramatically reduces aircraft separation, in fact it only normally significantly reduces the longitudinal size of the airspace box assigned to each aircraft. The vertical dimension of the airspace box remains 1,000 ft, the lateral dimension may be reduced from 8 to 5 nautical miles (sometimes 3 miles), but longitudinal separation is reduced from 10 flying minutes, to 3 to 5 nautical miles.

RADAR SYSTEM LIMITATIONS

There are various physical phenomena that hamper primary radar effectiveness. Weather and terrain can block radar waves, and natural weather conditions such as temperature inversions can cause fake or false targets to be displayed by the system. Radar also tracks all moving targets near the airport, which may include highway, train, and in some cases ship traffic. While controlling air traffic, the controller can be distracted and even momentarily confused when nonaircraft targets such as these are displayed on the radar. It is difficult for the controller to quickly determine whether a displayed target is a "false target" or an actual aircraft.

Another major limitation of radar is its positional accuracy. Because the radar beam is angular in nature (usually about ½ degree wide), the beam widens as it travels away from the transmitter. At extreme ranges, the radar beam can be miles wide. This makes it difficult to accurately position aircraft located far from the antenna, and makes it impossible to differentiate between two aircraft operating close to one another. Because radar system accuracy decreases as the aircraft distance from the radar antenna increases, aircraft close to the radar antenna (less than about 40 miles) can be laterally or longitudinally separated by 3 miles. Once the aircraft is greater than 40 miles from the radar antenna, 5 nautical miles of separation must be used. The size of the airspace box using radar is still not reduced vertically, but can now be as little as 9 square miles (compared to 600 when using nonradar separations).

ADDITIONAL RADAR SERVICES

Radar can also be used by the controller to navigate aircraft to provide a more efficient flow of traffic. During the terminal phase of flight, as the aircraft align themselves with the runway for landing, radar can be used by the controller to provide navigational commands (vectors) that position each aircraft at the optimal distance from one another, something impossible to do if radar surveillance is not available. This capability of radar is at least as important as the ability to reduce the airspace box assigned to each aircraft.

Air traffic controllers can also utilize radar to assist the pilot to avoid severe weather, although the radar used in air traffic control does not optimally display weather. The controller can also advise the pilot of nearby aircraft or terrain. In an emergency, the controller can guide an aircraft to the nearest airport, and can guide the pilot through an instrument approach. All of these services are secondary to the primary purpose of radar, which is to safely separate aircraft participating in the ATC system.

RADAR IDENTIFICATION OF AIRCRAFT

Before controllers can utilize radar for air traffic control separation, they must positively identify the target on the radar. Due to possible false target generation, unknown aircraft in the vicinity, and weather-induced false targets, it is possible for a controller to be unsure of the identity of any particular radar target. Therefore, the controller must use one or more techniques to positively verify the identity of any target before radar separation criteria can be utilized. If positive identity can not be ascertained, nonradar separation techniques must be utilized.

Controllers can verify the identify of a particular target using either primary or secondary radar. Primary methods require that the controller correlate the pilot's reported position with a target on the radar, or by asking the pilot to make a series of turns and watching for a target to make similar turns. Secondary radar identification can be established by asking the pilot to transmit an IDENT signal (which causes a distinct blossoming of the radar target), or, if the radar equipment is so equipped, asking the pilot to set the transponder to a particular code, and verifying that the radar displays that code (or the aircraft identification) next to the target symbol on the radar.

None of these methods are foolproof, and all have the potential for aircraft misidentification. During the identification process, the wrong pilot may respond to a controller's request, equipment may malfunction, or multiple aircraft may follow the controller's instruction. If an aircraft is flying too low or is outside the limits of the radar display, the target may not even show up on the radar scope. Once identified, the controller may rely completely on radar positioning information when applying separation, so multiple methods of radar identification are usually utilized to insure that a potentially disastrous misidentification does not occur and that the aircraft remains identified. If positive radar identification or detection is lost at any time, the controller must immediately revert to nonradar separation rules and procedures until aircraft identify can be reestablished.

RADAR SEPARATION CRITERIA

Radar accuracy is inversely proportional to the aircraft's distance from the radar antenna. The further away an aircraft is, the less accurate is the radar positioning of that aircraft. Radar separation criteria have been developed with this limitation in mind. One set of criteria has been developed for aircraft that are less than 40 nautical miles from the radar site. An additional set of criteria has been developed for aircraft 40 or more nautical miles from the antenna. Because the display system used in air route traffic control centers uses multiple radar sites, controllers using this equipment must always assume that aircraft might be 40 miles or farther from the radar site when applying separation criteria. Table 18.4 describes the separation criteria utilized by air traffic controllers when using radar. The controller must utilize at least one form of separation.

As stated previously, radar serves only to reduce the nonradar separation criteria previously described. It does nothing to reduce the vertical separation between aircraft. Radar primarily serves to reduce lateral and longitudinal separation. Nonradar lateral separation is normally 8 nautical miles, but the use of radar permits lateral separation to be reduced to 3 to 5 nautical miles. Radar is especially effective when reducing longitudinal separation, however. Nonradar longitudinal separation requires 5 to 100 nautical miles, whereas radar longitudinal separation is 3 to 5 nautical miles. It is this

TABLE 18.4
Radar Separation Criteria

Aircraft Distance from Radar Antenna	*Vertical Separation*	*Lateral Separation*	*Longitudinal Separation*
Less than 40 nautical miles	1,000 ft at flight level 290 and below. 2,000 ft above flight level 290.	3 nautical miles	3 nautical miles. Additional separation may be required for wake turbulence avoidance if smaller aircraft are following larger aircraft.
40 or more nautical miles	1,000 ft at flight level 290 and below. 2,000 ft above flight level 290.	5 nautical miles	5 nautical miles. Additional separation may be required for wake turbulence avoidance if smaller aircraft are following larger aircraft.

separation reduction that is most effective in maximizing the efficiency of the ATC system. Instead of lining up aircraft on airways 10 to 50 miles in trail, controllers using radar can reduce the separation to 3 to 5 miles, therefore increasing the airway capacity 200% to 500%. While under radar surveillance, pilots are relieved of the responsibility of making routine position and altitude reports. This dramatically reduces frequency congestion and pilot/controller miscommunications.

Another advantage of radar is that controllers are no longer restricted to assigning fixed, inflexible routes to aircraft. Because aircraft position can be accurately determined in near real time, controllers can assign new routes to aircraft that may shorten the pilot's flight, using the surrounding airspace more efficiently.

Radar vectors such as these are most effective in a terminal environment where aircraft are converging on one or more major airports, and are in a flight transitional mode where they are constantly changing altitude and airspeed. A controller using radar is in a position to monitor the aircraft in the terminal airspace, and can make overall adjustments to traffic flow by vectoring aircraft for better spacing, or by issuing speed instructions to pilots to close or widen gaps between aircraft. It is because of these advantages that most national air traffic control organizations first install radar in the vicinity of busy terminals. Only later (if at all) are enroute navigation routes provided radar monitoring.

CURRENT TRENDS IN AUTOMATION

Early forms of radar provided for the display of all moving targets within the radar's area of coverage. This included not only aircraft, but weather, birds, vehicular traffic, and other atmospheric anomalies. Using technology developed in World War II, air traffic controllers have been able to track and identify aircraft using the *air traffic control radar beacon system* (ATCRBS). ATCRBS, sometimes known as secondary surveillance radar, or simply secondary radar, requires a ground-based interrogator and an airborne transponder installed in each aircraft. When interrogated by the ground station, the transponder replies with a unique code that can be used to identify the aircraft, and if so equipped can also transmit the aircraft's altitude to the controller.

This system is tremendously beneficial to the controller because all aircraft can easily be identified. Nonpertinent aircraft and other phenomena observed by the radar can be ignored by the controller. If the ground-based radar is properly equipped, aircraft identity and altitude can also be constantly displayed on the radar screen, relieving the controller of mentally trying to keep each radar target properly identified.

The ground-based component of the secondary radar system has since been modified to perform additional tasks that benefit the air traffic controller. If the ground radar is properly equipped, and the computer knows the transponder code a particular aircraft is using, the aircraft can be tracked and flight information can be computer processed and disseminated. As the radar system tracks each aircraft, basic flight information can be transmitted to subsequent controllers automatically as the aircraft nears each controller's airspace boundary. Future aircraft position can also be projected based on past performance, and possible conflicts with other aircraft and with the ground can be predicted and prevented. These last two systems (known as *conflict alert* for aircraft–aircraft conflicts, and *minimum safe altitude warning* for aircraft–terrain conflicts) only provide the controller with a warning when aircraft are projected to be in danger. The system does not provide the controller with any possible remediation of the impending problem. Future enhancements to the computer system should provide the controller with options that can be selected to resolve the problem. This future system is to be known as *conflict resolution advisories.*

The ATCRBS system is limited in the amount of data it can transmit. Future system enhancements envision the replacement of the ATCRBS system with *Mode-S*, which stands for selective addressing. Mode-S will be able to interrogate and transmit data to individual aircraft and has the bandwidth to transmit more data than the limited ATCRBS. Expanded data that may be transmitted include control instructions and weather and traffic information. Pilots will also be able to use Mode-S to make silent requests for information and to acknowledge controller instructions. Eventually all routine controller–pilot communications should be able to be handled by the Mode-S system, thereby reducing frequency congestion and improving comprehension.

Possible future enhancements to the system include the ability of the air traffic control computer to make traffic decisions and transmit them directly to the aircraft control system via Mode-S. This enhancement to the system will lead to a very different air traffic control/pilot environment, with the job of each changing from that of an active participant to acting as a systems monitor. Although technically feasible at this time, the convergence of ground and airborne capabilities will most likely not occur until after the turn of the century. This future system is known as *automated enroute air traffic control* (AERA).

AIRBORNE SYSTEMS

Engineers and researchers have experimented with aircraft-based traffic avoidance systems since the 1960s. These prototype systems were not designed to replace but rather to augment and back up the current ground-based ATC system. Only recently has the Federal Aviation Administration (FAA) approved and users begun installing an airborne traffic avoidance system. This new device is known as *threat collision and avoidance system* (TCAS). TCAS was developed with three different levels of service and capabilities.

TCAS is an aircraft-based system that monitors and tracks nearby transponder-equipped aircraft. The position and relative altitude of nearby aircraft are constantly displayed on a TCAS display located in the cockpit of each aircraft. TCAS I provides proximity warning only, to assist the pilot in the visual acquisition of intruder aircraft. No recommended avoidance maneuvers are provided nor authorized as a direct result of a TCAS I warning. It is intended for use by smaller commuter aircraft holding 10 to 30 passenger seats, and general aviation aircraft. TCAS II provides traffic advisories and resolution advisories. Resolution advisories provide recommended maneuvers in a vertical direction (climb or descend only) to avoid conflicting traffic. Airline aircraft, and larger commuter and business aircraft holding 31 passenger seats or more, use TCAS II equipment. TCAS III provides all the capabilities of TCAS II but adds the capability to provide horizontal maneuver commands. All three versions of TCAS monitor the location of nearby transponder equipped aircraft. Current technology does not permit TCAS to monitor aircraft not transponder equipped.

CONFLICT ALERT/VFR INTRUDER

The ATCRBS has recently been enhanced with a new conflict alert program known as *conflict alert/VFR intruder*. The old conflict alert program only advised the controller of impending collisions between participating IFR aircraft. It did not track nonparticipating aircraft such as those operating under VFR. Conflict alert/VFR intruder tracks all IFR and VFR aircraft equipped with transponders and alerts the controller if a separation error between the VFR and a participating IFR aircraft is predicted. The controller can then advise the pilot of the IFR aircraft and suggest alternatives to reduce the risk of collision.

TRAFFIC MANAGEMENT SYSTEMS

It has become recently apparent that the current air traffic control system may not be able to handle peak traffic created in a hub-and-spoke airline system. Much of this is due to inherent limitations of the ATC system. ATC system expansion is planned in many countries, but until it is completed other methods of ensuring aircraft safety have been developed. To preserve an acceptable level of safety, special traffic management programs have been developed to assist the controllers in their primary function, the safe separation of aircraft.

Airport Capacity Restrictions

During hub-and-spoke airport operations, traffic can become intense for fairly short periods of time. During these intense traffic periods, if optimal weather and/or airport conditions do not exist, more aircraft may be scheduled to arrive than the airport and airspace can safely handle. In the past, this traffic overload would be handled through the use of airborne holding of aircraft. Controllers would try to land as many aircraft as possible, with all excess aircraft assigned to nearby holding patterns until space became available.

This method of smoothing out the traffic flow has many disadvantages. The primary disadvantage is that while holding, aircraft consume airspace and fuel. In today's highly competitive marketplace, airlines can ill afford to have aircraft circle an airport for an extended period of time.

In an attempt to reduce the amount of airborne holding, the FAA has instituted a number of new traffic management programs. One program seeks to predict near-term airport acceptance rates (AAR), and match arriving aircraft to that number. One program in use is the controlled departure program. This program predicts an airport's acceptance rate over the next 6 to 12 hr and matches the inbound flow of aircraft to that rate. Aircraft flow is adjusted through the delaying of departures at remote airports.

Overall delay factors are calculated, and every affected aircraft is issued a delayed departure time that will coordinate its arrival to the airport's acceptance rate.

The primary disadvantage of such a system is twofold. First, it is very difficult to predict 6 to 12 hr in advance conditions that will affect a particular airport's acceptance rate. These conditions include runway closures, adverse weather, and so on. As unforeseen events occur that require short-term traffic adjustments, many inbound aircraft are already airborne and therefore cannot be delayed on the ground. This means that the only aircraft that can be delayed are those that have not yet departed and are still on the ground at nearby airports. This system inadvertently penalizes airports located close to hub airports because they absorb the brunt of these unpredictable delays. In other situations, traffic managers may delay aircraft due to forecasted circumstances that do not develop. In these situations, aircraft end up being delayed unnecessarily. Unfortunately, once an aircraft has been delayed, that time can never be made up.

Once aircraft are airborne, newer traffic flow management programs attempt to match real-time airport arrivals to the AAR. These programs are known as aircraft metering. Metering is a dynamic attempt to make short-term adjustments to the inbound traffic flow to match the airport acceptance rate. In general terms, a metering program determines the number of aircraft that can land at an airport during a 5- to 10-min period, and determines and then applies a delay factor to each inbound aircraft so that they land in sequence with proper spacing. The metering program dynamically calculates the appropriate delay factor, and reports this to the controller as a specific time that each aircraft should cross a specific airway intersection. The controller monitors the progress of each flight, and issues speed restrictions to ensure that every aircraft crosses the appropriate metering fix at the computer specified time. This should, in theory, ensure that aircraft arrive at the arrival airport in proper order and sequence.

ATC SYSTEM OVERLOADS

Due to the procedural limitations placed upon aircraft participating in the ATC system, many ATC sectors far away from major airports can become temporarily overloaded with aircraft. In these situations, controllers would be required to separate more aircraft than they could mentally handle. This is one major limitation to the expansion of many ATC systems.

Various programs are being researched to counteract this problem. A prototype system has been developed in the United States known as *enroute sector loading* (ELOD). The ELOD computer program calculates every sectors current and predicted traffic load and alerts ATC personnel whenever it predicts that a particular sector may become overloaded. When this occurs, management personnel determine whether traffic should be rerouted around the affected sector. This particular program is successful at predicting both systemic overloads, and transient overloads due to adverse weather and traffic conditions.

PILOT/CONTROLLER COMMUNICATIONS-RADIO SYSTEMS

Most air traffic control instructions, pilot acknowledgments, and requests are transmitted via voice radio communications. By international agreement, voice communication in air traffic control is usually conducted in the English language using standardized phraseology. This phraseology is specified in ICAO documents and is designed to formalize phrases used by all pilots and controllers, regardless of their native language. This agreement permits pilots from the international community to be able to fly to and from virtually any airport in the world with few communication problems.

Voice communications between pilots and controllers are accomplished using two different formats and multiple frequency bands. The most common form of voice communication in ATC is simplex communications, where the controller talks to the pilot and vice versa utilizing a single radio frequency. This method makes more efficient use of the narrow radio frequency bands assigned to aviation, but has many inherent disadvantages. Because one frequency is used for both sides of the conversation, when one person is transmitting, the frequency is unavailable to others for use. To prevent radio system overload, simplex radios are designed to turn off their receiver whenever transmitting.

These conditions make it difficult for a controller to issue instructions in a timely manner when using simplex communications. If the frequency is in use, the controller must wait until a break in communications occurs. More problematic is the occasion when two or more people transmit at the same time or if someone's transmitter is inadvertently stuck on. Due to the way radios operate, if two people try to transmit at the same time, no one will be able to understand the transmission, and neither of the individuals transmitting would be aware of the problem, because their receivers are turned off when transmitting.

Duplex transmission utilizes two frequencies, one for controller-to-pilot communications, and another for pilot-to-controller communications. This communication method is similar to that utilized during telephone conversations. Both individuals can communicate simultaneously and independently, are able to interrupt one another, and can listen while talking. Duplex transmission schemes have one major disadvantage, however. To prevent signal overlap, two discrete frequencies must be assigned to every controller–pilot communication. This essentially requires that double the number of communications frequencies be made available for air traffic control. Due to the limited frequencies available for aeronautical communications, duplex transmissions can seldom be used in ATC.

Most short-range communications in ATC utilize the very-high-frequency (VHF) radio band located just above those used by commercial FM radio stations. Just as FM radio stations, aeronautical VHF is not affected by lightning and other electrical distortion, but is known as a line-of-sight frequency band. This means that the radio signal travels in a straight line and does not follow the curvature of the earth. Airborne VHF radios must be above the horizon line if they are to receive any ground-based transmissions. If an aircraft is below the horizon, it will be unable to receive transmissions from the controller and vice versa.

This problem is solved in the ATC system through the use of *remote communications outlets* (RCO). RCOs are transmitter/receivers located some distance from the ATC facility. Whenever a controller transmits, the transmission is first sent to the RCO using land-based telephone lines, and then is transmitted to the aircraft. Aircraft transmissions are relayed from the RCO to the controller in the same manner. Each RCO is assigned a separate frequency to prevent signal interference. This system permits a single controller to communicate with aircraft over a wide area, but requires the controller to monitor and operate multiple radio frequencies. The use of RCOs extends the controller's communications range, but also makes the ATC communications system vulnerable to ground-based telephone systems that may malfunction or be damaged, thereby causing serious ATC communication problems.

Most civil aircraft utilize VHF communications equipment. Military aircraft utilize ultra-high-frequency (UHF) band transmitters. UHF is located above the VHF band. UHF communications systems are preferred by most military organizations because UHF antennas and radios can be made smaller and more compact than those utilized for VHF. UHF is also a line-of-sight communications system. Most ATC facilities are equipped with both VHF and UHF radio communications systems.

Extended-range communication is not possible with VHF/UHF transmitters. RCOs can help extend the range of the controller, but need solid ground on which to be installed. VHF/UHF radios are unusable over the ocean, the poles, or in sparsely populated areas. For long-range, over-ocean radio communications, high-frequency (HF) radios are used. HF uses radio frequencies just above the medium-wave or AM radio band. HF radios can communicate without line-of-sight limitations, as far as 3,000 miles in some instances, but can be greatly affected by sunspots, atmospheric conditions, and thunderstorm activities. This interference is hard to predict and depends on the time of day, season, sunspot activity, local and distant weather, and the specific frequency in use. HF radio communication requires the use of multiple frequencies, with the hope that at least one interference-free frequency can be found for communications at any particular time. If controllers cannot directly communicate with aircraft, they may be required to use alternate means of communications, such as using the airline operations offices to act as communication intermediaries. This limitation requires that controllers who rely on HF communications not place the aircraft in a position where immediate communications may be required.

Experiments have been conducted using satellite transmitters and receivers to try to overcome the limitations of HF/VHF/UHF transmission systems. Satellite transmitters utilize frequencies located well above UHF and are also line-of-sight. But if sufficient satellites can be placed in orbit, communications anywhere in the world will be virtually assured. Satellite communications have already been successfully tested on overseas flights and should become commonplace within a few years.

VOICE COMMUNICATIONS PROCEDURES

As previously stated, virtually every ATC communication is currently conducted by voice. Initial clearances, taxi and runway instructions, pilot requests, and controller instructions are all primarily conducted utilizing voice. This type of communication is fairly unreliable due to both the previously mentioned technical complications and communications problems inherent in the use of one common language in ATC. Although every air traffic controller utilizes English, they may not be conversationally fluent in the language. In addition, different cultures pronounce words and letters in different ways. Many languages do not even use the English alphabet. And every controller has idioms and accents peculiar to their own language and culture. All these factors inhibit communications and add uncertainty to ATC communications.

When using voice radio communications, it can be very difficult for a controller to insure that correct and accurate communication with the pilot has occurred. Pilots normally read back all instructions, but this does not solve miscommunication problem. Informal and formal surveys lead experts to believe that there are literally millions of miscommunications worldwide in ATC every year. Obviously most of these are immediately identified and corrected, but some are not, leading to potential problems in the ATC system.

ELECTRONIC DATA COMMUNICATIONS

In an attempt to minimize many of these communications problems, various schemes of nonvoice data transmission have been tried in ATC. The most rudimentary method still in use is the ATCRBS transponder. If the aircraft is properly equipped, its identity and altitude will be transmitted to the ground station. Existing ATCRBS equipment is currently incapable of transmitting information from the controller to the aircraft. The new Mode-S transponder system will be able to transmit more information in both directions. This information might include aircraft heading, rate of climb/descent, airspeed, and rate of turn, for example. Mode-S should also be able to transmit pilot requests and controller instructions. Mode-S is slowly being installed on the ground and airborne equipment is gradually being upgraded. Until a sufficient number of aircraft have Mode-S capability, the ATCRBS system will still be utilized. Full Mode-S implementation will probably be completed sometime after the year 2000.

An intra-airline data communications system known as the *aircraft communications addressing and reporting system* (ACARS) has been utilized by the airlines for years to send information to and from properly equipped aircraft. ACARS essentially consists of a keyboard and printer located on the aircraft and corresponding equipment in the airline's flight operations center. ACARS is currently used by the airlines to transmit flight planning and load information. A few air traffic control facilities are now equipped to transmit initial ATC clearances to aircraft using ACARS. This limited service will probably be expanded until Mode-S becomes widespread.

CONTROLLER COORDINATION

Because controllers are responsible for the separation of aircraft within their own sector, they must coordinate the transfer of aircraft as they pass from one sector to another. In most situations, this coordination is accomplished using voice communications between controllers. In most cases, unless the controllers are sitting next to each other within the same facility, coordination is accomplished using the telephone.

Hand-offs are one form of coordination and consist of the transfer of identification, communications, and control from one controller to the next. During a hand-off, the controller with responsibility for the aircraft contacts the next controller, identifies the aircraft, and negotiates permission for the aircraft to cross the sector boundary at a specific location and altitude. This is known as the transfer of identification. Once this has been accomplished, and all traffic conflicts are resolved, the first controller advises the pilot to contact the receiving controller on a specific radio frequency. This is known as the transfer of communication. Separation responsibility still remains with the first controller until the aircraft crosses the sector boundary. Once the aircraft crosses the boundary, separation becomes the responsibility of the receiving controller. This is known as the transfer of control.

To simplify hand-offs, standardized procedures and predefined altitudes and routes are published in a document known as a *letter of agreement* (LOA). LOAs simplify the coordination process because both controllers already know what altitude and route the aircraft will be utilizing. If the controllers wish to deviate from these procedures, they must agree to an *approval request* (*appreq*).

The transferring controller usually initiates an appreq verbally, requesting a different route and/or altitude for the aircraft to cross the boundary. If the receiving controller approves the appreq, the transferring controller may deviate from the procedures outlined in the LOA. If the receiving controller does not approve the deviation, the transferring controller must amend the aircraft's route/altitude to conform to those specified in the LOA.

There are many problems inherent in this system of verbal communication/coordination. When both controllers are busy, it is very difficult to find a time when both are not communicating with aircraft. Controllers are also creatures of habit, and may sometimes "hear" things that were not said. There are many situations in ATC where aircraft are delayed or rerouted, not due to conflicting traffic, but because required coordination could not be accomplished in a timely manner.

Automated hand-offs have been developed in an attempt to reduce these communication/coordination problems. An automated hand-off can be accomplished if the two sectors are connected by computer, and the routes, altitudes, and procedures specified in the letter of agreement can be complied with. During an automated hand-off, as the aircraft nears the sector boundary, the transferring controller initiates a computer program that causes the aircraft information to be transferred and start to flash on the receiving controller's radar display. This is a request for a hand-off and implies that all LOA procedures will be complied with. If the receiving controller determines that the hand-off can be accepted, computer commands are entered that cause the radar target to flash on the transferring controller's display.

This implies that the hand-off has been accepted, and the first controller then advises the pilot to contact the next controller on the appropriate frequency. Although this procedure may seem quite complex, in reality it is very simple, efficient, and reduces voice coordination between controllers significantly. Its primary disadvantage is that the route and altitudes permissible are reduced and the ATC system becomes less flexible overall.

FLIGHT PROGRESS STRIPS

Virtually all verbal communications are written down for reference on paper flight progress strips. Flight strips contain most of the pertinent information concerning each aircraft. When a controller verbally issues or amends a clearance or appreqs a procedural change with another controller, this information is hand written on the appropriate flight progress strip. Flight progress strips are utilized so that controllers do not need to rely on their own memory for critical information. Flight strips also make it easier for other controllers to ascertain aircraft information if the working controller needs assistance or when a new controller comes on duty. Due to differences in each controller's handwriting, very specific symbology is used to delineate this information. Figure 18.1 contains examples of some common flight strip symbology.

FLIGHT INFORMATION AUTOMATION

The constant updating of flight progress strips and the manual transferring of information consume much of a controller's time, and may necessitate the addition of another controller to the sector to keep up with this essential paperwork. This process is forecast to become somewhat more automated in the future. Future ATC systems have been

Symbol	Meaning
↑	climb and maintain
↓	descend and maintain
-M>	maintain
RR	report reaching
RL	report leaving
RX	report crossing
⍏	cross at or above
⍖	cross at or below
X	cross
@	at
C	contact
⥱	join an airway
>	before
<	after
Ȼ	cancel flight plan
RV	radar vectors

FIG. 18.1. Sample flight progress strip symbology.

designed with flight strips displayed on video screens. It is theoretically possible that as controllers issue verbal commands, these commands will be automatically interpreted and the electronic flight strips will be updated. Future enhancements may make it possible for the controller to update an electronic flight strip, and that information might be automatically and electronically transmitted to the pilot or even to the aircraft's flight control system.

CONTROLLER RESPONSIBILITIES IN THE ATC SYSTEM

Controllers are responsible for the separation of participating aircraft within their own sector. They also provide additional services to aircraft, such as navigational assistance and providing weather advisories. Additional responsibilities placed on the controller include maximizing the use of the airspace and complying with air traffic management procedures.

To accomplish these tasks, the controller must constantly monitor both actual and predicted aircraft positions. Due to rapidly changing conditions, a controller's plan of action must remain flexible and subject to constant change. The controller must continuously evaluate traffic flow, plan for the future, evaluate the problems that may occur, determine appropriate corrective action, and implement this plan of action. In the recent past when traffic moved relatively slowly and the airspace was not quite as crowded, a controller might have minutes to evaluate situations and decide on a plan of action. As aircraft speeds have increased, and the airspace has become more congested, controllers must now make these decisions in seconds. As in many other career fields, experts feel that the current system may have reached its effective limit, and increased ATC system expansion will not be possible until many of the previously mentioned tasks become automated.

FUTURE ENHANCEMENTS TO ATC SYSTEMS

ICAO has recently agreed that GNSS should become the primary aircraft positioning system by the year 2000. It appears at this time that uncorrected GNSS systems should supplant VORTAC as both an enroute and nonprecision instrument approach aid. WAAS should permit GNSS to be used as a CAT I precision approach replacement for ILS. LADGPS should correct GNSS to meet CAT II and possibly CAT III ILS standards.

The GNSS system can be modified to permit the retransmission of aircraft position back to air traffic control facilities. This system, known as *automatic dependent surveillance* (ADS), should supplant radar as a primary aircraft surveillance tool. Not only should this system be more accurate than radar surveillance, but also it will not have the range and altitude limitations of radar and will be able to transmits additional data both to and from the controller. This might include pilot requests, weather information, traffic information, and more. ADS has already been demonstrated experimentally and is being tested for aircraft separation over oceanic airspace.

Many other changes are planned. ICAO has completed a future air navigation system (FANS) that defines changes to navigation, communication, and surveillance systems. FANS is a blueprint for the future of international aviation and air traffic control. Table 18.5 summarizes FANS.

Once these improvements have taken place, automated ATC systems can be introduced. Various research programs into automation have been initiated by many ATC organizations, but it is highly likely that it will be well past the year 2000 before automated systems such as AERA can be designed, constructed, installed, and made operational. In the meantime, the FAA has begun to study an air traffic management (ATM) system called "free flight."

The concept of free flight has been discussed since the early 1980s. Only since the demise of the FAA's planned advanced automation system (AAS) has it come into favor in the United States. Free flight proposes to change air traffic control separation standards from a static, fixed set of standards to dynamic separation that takes into account aircraft speed, navigational capability, and nearby traffic. Based on these

TABLE 18.5
Future ATC and Navigation System Improvements

Function	*Type*	*Current Standard*	*Future Standard*
Navigation	Enroute	VORTAC LORAN-C Inertial navigation	Global navigation satellite system
	Approach	VORTAC and NDB for nonprecision, ILS and MLS for precision.	GNSS without augmentation for nonprecision, GNSS with augmentation for precision approaches.
Communication	Short-range	VHF and UHF	VHF and UHF
	Long-range	HF	Satellite
ATC surveillance		Radar	Radar and automatic dependent surveillance
Datalink		ATCRBS	Mode-S

parameters, each aircraft will be assigned a "protected" zone that will extend ahead, to the sides and above and below the aircraft. This zone will be the only separation area protected for each aircraft. This differs from the current system that assigns fixed airway dimensions and routes for separation.

Assuming that each aircraft is equipped with an accurate flight management system (FMS), free flight proposes that each aircraft transmit to ground controllers its FMS-derived position. On the ground, computer workstations will evaluate the positional data to determine whether any aircraft conflicts are predicted to exist, and if so, offer a resolution instruction to the air traffic controller. The controller may then evaluate this information and pass along appropriate separation instructions to the aircraft involved.

The free flight concept is still being developed, but if found feasible will soon be implemented at higher altitudes within the U.S. airspace structure. As confidence in the system is gained, it will likely be extended overseas and into the low-altitude flight structure.

SUGGESTED READING

Federal Aviation Administration. (1976). *Takeoff at mid-century.* Washington, DC: Department of Transportation.

Federal Aviation Administration. (1978). *Bonfires to beacons.* Washington, DC: Department of Transportation.

Federal Aviation Administration. (1979). *Turbulence aloft.* Washington, DC: Department of Transportation.

Federal Aviation Administration. (1980). *Safe, separated and soaring.* Washington, DC: Department of Transportation.

Federal Aviation Administration. (1987). *Troubled passage.* Washington, DC: Department of Transportation.

International Civil Aviation Organization. (various). *ICAO Bulletin,* Montreal, Canada.

International Civil Aviation Organization. (various). *Annexes to the Convention of International Civil Aviation.* Montreal, Canada: Author.

Jackson, W. E. (1970). *The federal airways system.* Institute of Electrical and Electronic Engineers.

Nolan, M. S. (1994). *Fundamentals of air traffic control.* Wadsworth.

Other FAA Publications

Aeronautical Information Publication (1988)

Air Traffic Handbook (1995)

National Airspace System Plan (1988)

19

Air Traffic Controller Memory: Capabilities, Limitations, and Volatility

Daniel J. Garland
Embry-Riddle Aeronautical University

Earl S. Stein
Federal Aviation Administration, William J. Hughes Technical Center

John K. Muller
Santa Teresa Laboratory
IBM Corporation

> *Tell me and I'll forget;*
> *show me and I may remember;*
> *involve me and I'll understand.*
> —Chinese Proverb

On February 1, 1991, at 1807 Pacific Standard Time, a USAir Boeing 737-300, while landing on a runway at Los Angeles International Airport, collided with a SkyWest Fairchild Metroliner, which was positioned on the same runway awaiting clearance for takeoff. The Boeing 737 and a large portion of the Metroliner, which was crushed beneath the Boeing 737's left wing, continued down the runway, veered left, and impacted on a vacant building, resting 1,200 ft from the point of collision (NTSB, 1991). As a result of the collision and postcrash fire, both airplanes were destroyed. All 10 passengers and 2 crewmembers aboard the Metroliner and 20 passengers and 2 crewmembers aboard the Boeing 737 were fatally injured. An article, by a National Transportation Safety Board member, on air traffic control human performance issues used this accident as an example of the importance of our understanding of air traffic controller memory for aviation safety:

> On the surface, the accident was simple: the local controller forgot that she had cleared SkyWest into position on the runway, on which she subsequently cleared USAir to land. Controller error "caused" this accident. But one must ask, Was this a simple controller error accident, or do other factors need to be considered? Most importantly, what must be done to minimize the probability of another accident of this kind? . . . "Forgetting" is not necessarily the same as "not attending to duty," and all the signs are that she was attending to her duties that day. Secondly, clearing an aircraft into position on a runway and then "forgetting" about that aircraft is not a unique event. A search of NASA's Aviation Safety Reporting System data base revealed that several instances had been reported, and a very similar incident occurred not long ago at San Diego. A fundamental characteristic of human performance is that "forgetting" is all too easy. Short-term memory . . . is highly vulnerable to intervening events disrupting it. Add a distraction here and a little time

pressure there, and presto, people forget—even very important things. (Lauber, 1993, pp. 24–25)

The primary objective of this chapter is to raise an awareness. Memory requirements of air traffic control (ATC) tactical operations are vital. The chapter presents information on working memory processes in ATC tasks and shows the vulnerability of these processes to disruption. This chapter focuses on the role working memory plays in air traffic controller performance and emphasizes the mechanisms of working memory, with its limitations and constraints. It also examines how controllers might overcome or minimize memory loss of critical ATC information. Awareness of the limitations and constraints of working memory and the conditions under which they occur is critically necessary to avoid situations that can result in airspace incidents and accidents. The final section of this chapter briefly deals with some of the potential human factors consequences of new automated technologies on air traffic controller working memory.

During the last decade, the ATC system has been strained due to increases in the amount of and changes in the distribution of air traffic in the United States. The Federal Aviation Administration (FAA) anticipates that will continue to increase into the 21st century. The current system was not designed to handle this level of activity, and technology must be developed to help the control workforce do its job.

The safe, orderly, and expeditious flow of air traffic is traditionally the fundamental objective of air traffic control (Federal Aviation Administration, 1989). There were 46.7 million instrument operations logged by the Federal Aviation Administration's airport control towers in 1994 (FAA, 1995, p. VII-4). The FAA forecasts 57 million instrument operations to be handled by FAA airport control towers in 2006 (FAA, 1995, p. VII-12). Further, the FAA estimates that, in the United States alone, delays resulting in air traffic problems result in economic losses of over $5 billion per year. These losses are expected to exceed $10 billion per year by the year 2000 if no changes are made (Wise, Hopkin, & Smith, 1991).

Controllers are at the forefront of this system, which to this day remains highly person ascendant. It is their strengths that keep it going as well as it is and their qualities that may cause it to break down if they do not get the appropriate tools to keep up. One of these human qualities is the fact that we have limited working memory capacity, which is further constrained by the dynamic nature of the control process. Things keep happening, often causing interference with the coding and storage process (Stein & Bailey, 1994).

Working memory allows the controller to retain intermediate (i.e., transient) products of thinking and the representations generated by the perceptual system. Functionally, working memory is where all cognitive operations obtain their information and produce their outputs (i.e., responses). Working memory allows the controller to retain relevant information for tactical operations. Such tactically relevant information may include altitude, airspeed, heading, call/sign, type of aircraft, communications, weather data, runway conditions, current traffic "picture," projected traffic "picture," immediate and projected conflicts, and so forth. Working memory is heavily dependent on long-term memory for such cognitive tasks as information organization, decision making, and problem solving. Working memory is also heavily constrained and limited by such time-dependent processes as attention, capacity, and forgetting. Essentially, working memory permeates every aspect of the controller's ability to process air traffic information and control live traffic.

The FAA invests considerable energy in attempting to discover the causes and methods for preventing actual and potential operational errors of air traffic control

(Operational Error Analysis Work Group, 1987). The planned development and implementation of free flight procedures heightens this concern. There will be a need for effective transition training for controllers who must be able to use new technologies and procedures to control live traffic. The dramatic system changes of a free flight environment not only may replace existing ATC technology and procedures, but also will fundamentally change the way air traffic controllers conduct their job. This may change how controllers cognitively process information. Increased automation, new technologies, and procedures may impose requirements on the controller that are incompatible with the way he or she processes information and the way a controller attends, perceives, remembers, thinks, decides, and responds.

The cognitive requirements of air traffic control as it currently exists have involved the processing of a great volume of dynamically changing information (Kirchner & Laurig, 1971; Means et al., 1988). Cognitive processing of flight data (i.e., call/sign, aircraft type, sector number, planned route, assigned speed, heading, altitude, time over posted fix, etc.) is crucial to virtually every aspect of a controller's performance. It is essential for the controller to manage available information resources in such a way that accurate information is available when needed. The ease with which information (e.g., flight data) is processed and remembered depends on how it is displayed and how the operator interacts with it. As information displays change with evolving technology, controllers will process flight information in different ways, potentially affecting ATC performance and possibly influencing flight safety and efficiency.

It is important to understand these cognitive processes. Controller performance measurements have consistently involved tasks and variables derived from air traffic control and produced findings expressed in air traffic control terms (Hopkin, 1980). Another, possibly more beneficial, approach involves tracing the origins of the practical difficulties (e.g., memory lapses) that the controller encounters to fundamental limitations in human cognitive capabilities. One can then use basic psychological knowledge to explain, measure, and resolve these difficulties. Hopkin (1980) has stated that it is fundamental to consider the controller's task in psychological terms in order to provide perspectives, explanations, and insights into the cognitive processes that support air traffic control.

The air traffic control environment is characterized by a continuous sequence of ever-changing, transient information (e.g., series of aircraft being handled by an air traffic controller), which must be encoded, retained primarily for tactical use (3–5 min) and secondarily for strategic planning, and subsequently discarded. The ability to manage flight information is complicated by limitations and constraints of human memory, in particular working memory (Finkelman & Kirschner, 1980; Kirchner & Laurig, 1971; Wickens, 1992). Essentially, working memory is the part of the memory system that deals with ongoing, transient information (working memory is further defined in a subsequent section). Working memory limitations and constraints are routinely severe enough to significantly degrade performance.

HUMAN INFORMATION PROCESSING SYSTEM

Researchers have studied memory issues for a considerable period of time, as shown by a three-volume work providing an annotated compilation of 798 references dealing with short-term memory, covering the time period from 1959 to 1970 (Fisher, 1969; Fisher, 1971; Fisher & Wiggins, 1968). Unfortunately, many of the early memory studies had nothing to do with understanding. In fact, early studies often deliberately employed nonsense syllables because they were incomprehensible (Hopkin, 1982). Studies of this type did not require the participants to incorporate new material with existing knowledge, and therefore have no direct relevance to memory for complex material in applied operational settings. The popularity of memory as a research topic has not faded with time, as shown by the number of recent articles, chapters and books on memory (e.g., Baddeley, 1996; Cowan, 1995; Healy & McNamara, 1996; Jonides, 1995; Logie, 1995; Lyon & Krasnegor, 1996; Shaffer, 1993; Squire, Knowlton, & Musen, 1993).

Human information processing may provide clues to how working memory influences controller behavior. Several information-processing models have been developed (e.g., Broadbent, 1958; Card, Moran, & Newell, 1986; Smith, 1968; Sternberg, 1969; Welford, 1976; Wickens, 1984); each assumes various stages of information processing, characterized by stage-specific transformations on the data. The present approach follows a simplified description of human information processing, consisting of three interacting subsystems, similar to the Card et al. (1986) model.

The three interacting subsystems, (a) the perceptual system, (b) the cognitive system, and (c) the motor system, each have their own information-processing capabilities. The perceptual system consists of sensory systems and associated memories, responsible for translating information about external stimuli (i.e., the physical world) into internal representations. The cognitive system receives information from the perceptual system, puts it into working memory, and uses long-term memory information to make decisions about how to respond. This subsystem is the most complex and is the focus of this chapter. The motor subsystem is responsible for carrying out the response. The interaction of these three subsystems allows the controller to perform the duties of monitoring and if necessary intervention to keep the traffic flowing safely and efficiently. The interaction of the subsystems is extremely dynamic. It must adapt to the information-processing capabilities and limitations of the operator while considering task and situational demands. The three subsystems may interact in series or in parallel. For example, some tasks (e.g., marking the flight-strip in response to an altitude change) require serial processing. Other tasks (e.g., radar/flight strip scanning, flight-strip marking, ground–air–ground communications) may require integrated, parallel operation of the three subsystems.

The following brief description of information processing in the ATC system demonstrates the interaction of the three information-processing subsystems. Human information processing is a necessary component of all ATC operations. Although technical support is necessary for communication between the ATC system and the aircraft, the controller is the primary information processor. Technical equipment supports the guidance of aircraft from the ground. It provides feedback that serves to guide the execution of controller instructions and provides new information about the changed situation for guidance of future controller actions.

After receiving information about the present condition of the traffic, the controller evaluates the situation based on safety and efficiency criteria. If a potential conflict arises, demanding intervention, the controller takes the necessary control actions. The control actions, once implemented, change the situation, providing new information to the controller. The control actions require two basic information-processing steps. First, the present situational information is received, analyzed, and evaluated. The operator must have an adequate knowledge base, training, and experience. Second, the controller responds based on the available data, training, and experience. In addition to the immediate demands on information processing, the controller must process additional system information derived from coordination between different controllers. This coordination is essential to traffic planning and keeping the "picture" of the traffic under control (for more detailed information, see Ammerman et al., 1983; Ammerman & Jones, 1988; Bisseret, 1971; Kinney, Spahn, & Amato, 1977; Kirchner & Laurig, 1971; Means et al., 1988). For example, the controller detects a potential conflict between TCA483 and TWA358. The controller places these flight strips next to each other to call attention to them. TCA483 is contacted and instructed to climb to altitude 320. The controller crosses out 300 and writes 320, the new proposed altitude. Concurrently, TWA358 informs the controller it has reached the assigned cruising altitude of 300 and the controller makes a notation next to the altitude.

This illustration is an obvious simplification of the ATC system. In practice, there would be a far greater number of aircraft in the traffic pattern and the controller would potentially have to resolve a number of conflicts simultaneously. However, this illustration provides a characterization of the information-processing components in the ATC system and the basis for a closer examination of the mechanisms underlying information processing, with particular attention to cognitive research on memory and its application to the ATC system.

AIR TRAFFIC CONTROLLER MEMORY

Controllers are human, and human memory can be viewed as a continuously active system that receives, retrieves, modifies, stores, and acts on information (Baddeley, 1976, 1986; Klatzky, 1980). Researchers have referred to working memory as the "site of ongoing cognitive activities. These include the meaningful elaboration of words, symbol manipulation such as that involved in mental arithmetic, and reasoning" (Klatzky, 1980, p. 87). Further, Chase and Ericsson (1982, p. 40) defined working memory as "the part of the memory system where active information processing takes place," whereas other investigators see it as the "space in which information can be stored temporarily while it is being processed" (Klapp, Marshburn, & Lester, 1983, p. 240).

The discussion here focuses more on the transient characteristics of working memory than on long-term memory (see shaded box in Fig. 19.1). This emphasis is based on the psychological knowledge that long-term memory storage and retrieval are relatively automatic processes. They present fewer formidable disruptions to performance (Baddeley, 1976, 1986; Klatzky, 1980; Wickens, 1992). In contrast, working memory is severely affected by the limitations and constraints of limited processing resources. Wickens (1992) emphasized that occasional limitations of, and constraints on, working memory are often responsible for degraded decision making.

Working memory allows the controller to retain intermediate (i.e., transient) products of thinking and the representations generated by the perceptual system. The mechanisms of working memory and the nature of its limitations and constraints that directly and/or indirectly influence ATC are the focus of this chapter and are presented in the following sections covering memory codes, code interference, attention, capacity, chunking, organizing, and forgetting.

Memory Codes

Memory models and theory can help us understand what is going on in the everyday activities of air traffic controllers. Immediately after the presentation of an external visual stimulus such as an aircraft target with accompanying data tag on the radar

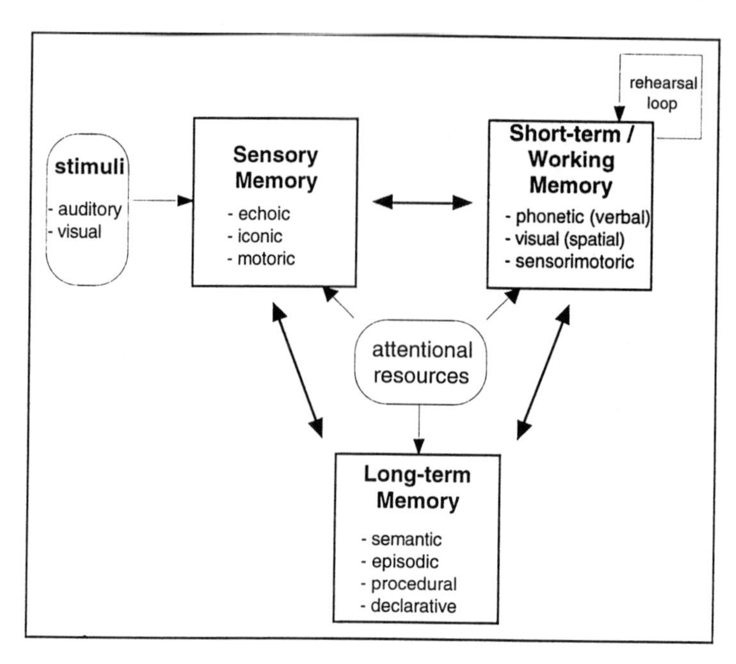

FIG. 19.1. Memory model.

display, a representation of the stimulus appears in the visual image store (i.e., iconic memory) of the perceptual system. There is also a corresponding auditory image store (i.e., echoic memory) for external auditory stimulus (e.g., ground–air–ground communications). These sensory codes or memories are representations of external physical stimuli. These sensory memories are vital to working memory, in that they prolong the external stimulus representations enough (usually measured in milliseconds) for relevant processing of the stimulus representations to take place in working memory (Card et al., 1986). The sensory memories, although not demanding the operator's limited attentional resources, are important for partial activation of the visual (i.e., iconic) and phonetic (i.e., echoic) primary codes in working memory (Baddeley, 1986; Wickens, 1984, 1992).

Although the sensory codes are generated exclusively by external physical stimuli, primary visual and phonetic codes may be activated by external stimuli via the perceptual system (i.e., sensory codes) or from inputs into working memory from long-term memory. The primary visual and phonetic codes, along with semantic and motoric codes, form the foundation of our attention demanding working memory, which is necessary for all ATC tactical operations (Baddeley & Hitch, 1974). Semantic codes are abstract representations based on the meaningfulness of the stimuli (e.g., the controller's knowledge of specifics of the sector map, the location of data on the flight strip, aircraft characteristics, etc.) and are vital for activating information in long-term memory. Motoric codes are sensory and motor representations of actions, which are involved in the encoding of past and future activities (Koriat, Ben-Zur, & Nussbaum, 1990). The encoding of future actions, which has been a neglected issue in memory research, is of critical importance to air traffic control operations. It is gaining more acceptance, with the research being conducted under the heading of situational awareness. This work was begun in the cockpit environment and has been making inroads into ATC research (Endsley, 1990; Sollenberger & Stein, 1995). For example, a controller instructs

TWA348 to climb to a new cruising altitude of 290, having forgotten to previously instruct AAL584 to descend from 290 to 270 for eventual handoff. This forgotten to-be-performed action, located in prospective memory (to use the language of situational awareness), may subsequently result in an airspace conflict.

In the following pages the visual, phonetic, semantic, and motoric codes are given further treatment. These memory codes play a significant role in the ATC process. Information is provided about the characteristics of these codes and their susceptibility to disruption and enhancement.

Visual Codes. Visual representations or images of spatial information (e.g., a controller's "pictorial" mental representation of an aircraft's location, orientation, and velocity after a brief scan of the radar display and/or the flight strip bay) are normally maintained in working memory using visual codes (Wickens, 1992). However, visual input is not necessary or sufficient for the generation of visual representations. External visual stimuli do not automatically produce visual or spatial images. That is, simply looking at something will not ensure its processing in working memory. In addition, Kosslyn (1981) reported evidence indicating visual images can be generated by nonvisual sources, such as information that has been experienced and subsequently stored in long-term memory (e.g., sector map, previous conflict situations), and by verbal (i.e., phonetic) stimulus material.

Primary visual codes are highly transient in nature, requiring a great deal of effortful attention. They demand processing (i.e., maintenance rehearsal) in order to persist in working memory (Goettl, 1985; Posner, 1973, 1978; Posner & Mitchell, 1967). Research conducted by Bencomo and Daniel (1975), using a same–different recognition task, suggests visual codes (i.e., visual representations or images) are more likely to persist when processing involves more natural visual/spatial materials (e.g., sector map, radar display), than verbal or auditory materials.

Phonetic Codes. Verbal information (e.g., the controller at Chicago Center instructs TWA484 to "descend and maintain one thousand, report leaving one two thousand") is normally maintained in working memory by phonetic or auditory rehearsal (Wickens & Flach, 1988). This process in working memory is known as "maintenance rehearsal" (also called Type I, primary, or rote rehearsal) and is used only to maintain information in working memory, presumably by renewing the information before it is subject to time-dependent loss (Bjork, 1972; Craik & Watkins, 1973; Klatzky, 1980). The phonetic primary code is automatically generated from an echoic sensory code and represents continued processing at a shallow, acoustic level (Wickens, 1992). In addition, Conrad (1964) demonstrated that phonetic codes can be automatically generated from visual stimuli (i.e., iconic codes). Conrad's (1964) results indicated that when subjects were to recall visually presented letters, recall intrusion errors tended to be acoustic rather than visual. For example, an air traffic controller may have a tendency to write, by mistake, letters such as Z instead of T. There is more potential intrusion or opportunity for error based on the associated sounds rather than on visual appearance. Further, Conrad and Hull (1964) demonstrated that recall information that was phonetically similar created greater recall confusion than information that was phonetically dissimilar.

A series of laboratory studies on phonetic codes and information presentation has concluded that verbal working memory can be enhanced by employing speech (i.e., verbal information) as an information display mode (Murdock, 1968; Nilsson, Ohlsson, & Ronnberg, 1977; Wickens, Sandry, & Vidulich, 1983). This conclusion is based on the facts that echoic (i.e., auditory) memory is retained longer that iconic (i.e., visual) memory, and that auditory displays are more compatible with the auditory nature of maintenance rehearsal in working memory (Wickens, 1992).

There are also significant human factors implications of using an auditory information display for the presentation of transient information to be used in working memory. Such information will be less susceptible to loss when presented via auditory channels, such as natural or synthetic speech. For example, Wickens et al. (1983) demonstrated that pilots can retain navigational information better with auditory display in comparison to visual display, and this finding was enhanced under high-workload conditions. These findings suggest that auditory display of information may be advantageous when rapid information presentation is necessary; the information is of a transient nature; the information is not overly complex, and visual display space cannot afford further cluttering (Wickens, 1992). However, auditory displays present formidable challenges to the human factors specialist. They cannot be easily monitored on a time-sharing basis, and once the information is gone from working memory it cannot be returned to as can visual displays.

Semantic Codes. Semantic codes are responsible for representing information in working memory in terms of meaning rather than physical (i.e., auditory, visual) attributes. Semantic codes provide the critical link between working memory and the permanent long-term memories. Card et al. (1986), when noting the intimate association between working memory and long-term memory, suggested that "structurally, working memory consists of a subset of the elements in long-term memory that have become activated" (p. 45-7). Semantic codes are primarily responsible for information storage and organization in working memory, and subsequently in long-term memory. The creation and use of semantic codes involves a process that is substantively different from maintenance rehearsal. This is elaborative rehearsal.

Elaborative rehearsal involves deep, meaningful processing in which new information is associated with existing meaningful knowledge in long-term memory. This processing, in contrast to the previously cited maintenance rehearsal, facilitates the retention of information in working memory and enhances information transfer to long-term memory by way of semantic codes. Elaborative rehearsal (i.e., semantic encoding) in working memory requires thinking about information, interpreting the information, and relating the information to other information in long-term memory. These processes enhance retrieval of information from long-term memory and facilitate planning future actions (Klatzky, 1980).

Semantic codes afford working memory the ability to actively retain and analyze information. Wingfield and Butterworth (1984) suggested that, rather than passively retaining auditory and visual information in working memory, we are "continuously forming hypotheses about the structure of what they are hearing and forming predictions about what they have yet to hear. These are working hypotheses, either confirmed or modified with the arrival of new information" (p. 352). Klapp (1987) noted that working memory actively formulates and stores hypotheses, resulting in abstract representations (i.e., semantic codes) in working memory in addition to auditory or visual codes.

Semantic codes are vital for the organization, analyses, and storage of ATC tactical information in working memory and long-term memory. They are the invaluable link between working memory and long-term memory, providing and facilitating the ability to actively manipulate and analyze data and to generate decision-making and problem-resolution alternatives. For example, in order for an air traffic controller to make an informed and accurate assessment of a potential conflict between two aircraft, a great deal of flight information is required about the two aircraft (e.g., altitude, heading, airspeed, type of aircraft, current traffic "picture," projected traffic "picture," etc.). These flight data must in turn be analyzed and interpreted against a knowledge and experience database in long-term memory to accurately construct and assess a "pictorial" mental representation of the current and projected airspace. Alternative hypotheses about the traffic situation can be generated from long-term memory and retained in working memory to be analytically integrated with the flight data. This process of hypothesis formulation and evaluation is complicated by the limitations and constraints of working memory and long-term memory decision biases (Wickens & Flach, 1988). Researchers have been searching for new tools to help controllers use their memory more effectively. Stein and Sollenberger (1995) summarized the FAA's program on controller memory at the eighth Biannual Symposium on Aviation Psychology. The program has focused on the system as it exists today and how controllers can use tools available to them in order to avoid systems errors. Most complex command and control systems involve memory processing using multiple levels of coding. Motoric codes take in another dimension.

Motoric Codes. Motoric codes are integrated sensory and motor (i.e., sensorimotor) representations of actions retained in working memory, which are involved in the encoding of past and future activities (Koriat et al., 1990). Controllers need to encode action events that surround them. For example, a controller, in examining the flight-strip bay, detects a potential conflict between TCA483, AAL284, and TWA343 before TCA483 is displayed on the radar display. The controller could, in order to remember to take an appropriate control action or to inform the radar-side controller of the conflict, cock the flight strip for TCA483 or write something down.

Recent research on memory for action events has focused on memory for past activities (e.g., Anderson, 1984; Backman, Nilsson, & Chalom, 1986; Cohen, 1981, 1983; Johnson, 1988; Kausler & Hakami, 1983; Koriat & Ben-Zur, 1988; Koriat, Ben-Zur, & Sheffer, 1988). A consistent and general finding of these studies is that memory for performing a task is superior to memory for verbal materials, due to the beneficial effects of *motoric enactment*. That is, the process of physically performing a task seems to enhance the encoding of and subsequent memory for the task. The superior memory for performing tasks "has been generally attributed to their multimodal, rich properties, assumed to result in richer memorial representations than those formed for the verbal instructions alone" (Koriat et al., 1990, p. 570).

These results are particularly relevant when discussing the impact of automation on ATC systems and the potential human factors consequences. Several researchers (e.g., Hopkin, 1988, 1989, 1991b; Narborough-Hall, 1987; Wise & Debons, 1987; Wise et al., 1991) suggested that routine task performance facilitates controller tactical operations (e.g., the understanding of and the memory for traffic situations). Hopkin (1991b) asserted that physical interaction with the flight progress strip is critical to support a controller's memory for immediate and future traffic situations.

There have been a series of studies conducted by the University of Oklahoma and the Civil Aeromedical Institute that have called this belief into question. They have demonstrated under very limited simulated conditions that controllers can work without strips without significant loss to their working memory (Vortac, Edwards, Fuller, & Manning, 1993, 1994, 1995; Vortac, Edwards, & Manning, 1995). However, these studies were conducted primarily with FAA academy instructors and using a low- to moderate-fidelity simulator. Zingale, Gromelski, and Stein (1992) attempted to study the use of flight strips using aviation students at a community college. The study demonstrated the importance of using actual controllers, because the students could not keep up with or without strips. Zingale, Gromelski, Ahmed, and Stein (1993), in a follow-on study using controllers and the same low-fidelity simulator, found that controllers did find the strips useful but were uncomfortable using a PC-based simulation that required them to key in their responses. The FAA has decided to maintain paper flight strips as operational tools for the foreseeable future.

The air traffic control environment is characterized not only by the necessity to remember past activities to support ongoing tactical operations but also activities to be performed in the future. Memory for future activities is known as prospective memory (Harris, 1984; Wilkins & Baddeley, 1978; Winograd, 1988). In some cases, information for future control actions need only be retained for a short period of time. A recent study investigating the nature of the representations underlying memory for future actions (i.e., prospective memory) found a significant beneficial effect of *imaginal-motoric enactment* of the future activity (Koriat et al., 1990). This imaginal enactment of the future activity is consistent with the research on memory for past activities. This beneficial effect can also be attributed to the multimodal and contextual properties of having actually performed the task. It is also seen with the intentional (or unintentional) visualization of the task, which promotes visual and motor encoding (Backman & Nilsson, 1984; Koriat et al., 1990).

Koriat et al. (1990) suggested that the process of encoding future activities involves an internal, symbolic enactment of the tasks, which enhances memory. This implies that rehearsal (i.e., maintenance and/or elaborative) or repeated internal simulation of the procedure to be performed will enhance memory at the time of testing, in much the same manner that maintenance rehearsal retains verbal material in working memory. Koriat et al. (1990) also suggested that if rehearsal takes advantage of the modality-specific properties of the future task, not only will memory for content be enhanced, but memory retrieval cues will be enhanced under proper conditions.

Given the previous example of a potential conflict between TCA483, AAL284, and TWA343 before TCA483 is displayed on the radar display, the controller is responsible for retaining and eventually conveying this information to the relief controller on the next shift, along with additional information concerning the status of other aircraft under control. In order to remember this potential crisis situation, the controller encodes the future task (i.e., briefing or execution of control actions needed to avoid a pending crisis situation) in terms of the sensorimotor properties (e.g., internal visual representation of the projected traffic picture and/or physical performance requirements of the control action) of the task that will enhance the actual performance at the time of task. This type of encoding will facilitate the activation of memory with the appropriate external retrieval cues (e.g., the flight strips for TCA483, AAL284, and TWA343 being placed adjacent to each other, with the TCA483 flight strip cocked; Koriat et al., 1990).

The previous example is indicative of the significant role flight strips play in facilitating motoric encoding and planning future actions. Several researchers have identified the significant cognitive value of flight strips in preparing for future actions (Hopkin, 1989, 1991b; Vortac, 1991). One reason for the cognitive value of flight strips is they represent the history of actions, goals, intentions, and plans of pilots and controllers.

These functions are elaborated in the following controller interview extract (Harper, Hughes, & Shapiro, 1989):

> It's a question of how you read those strips. . . . An aircraft has called and wants to descend, now what the hell has he got in his way? and you've got ping, ping, ping, those three, where are those three, there they are on the radar. Rather than looking at the radar, one of the aircraft on there has called, now what has he got in his way? Well, there's aircraft going all over the place, now some of them may not be anything to do with you, . . . your strips will show you whether the aircraft are above or below them, . . . or what aircraft are below you if you want to descend an aircraft, and which will become a confliction. You go to those strips and you pick out the ones that are going to be in conflict if you descend an aircraft, and you look for those on the radar and you put them on headings of whatever, you find out whether those, what those two are—which conflict with your third one. It might be all sorts of conflicts all over the place on the radar, but only two of them are going to be a problem, and they should show up on my strips. (p. 9)

This interview extract provides a good example of the role flight strips may play in assisting information processing and its significance in planning future actions. Harper et al. (1989) pointed out that "paradoxically, the 'moving' radar screen is from an interpretative point of view relatively static, while the 'fixed,' 'hard copy' strip is interpretatively relatively dynamic" (p. 5). For ATC tactical operations, planned actions are the purview of flight progress strips, and past actions are reflected in feedback on the radar and flight strip markings (Vortac, 1991).

The "generation effect" is directly related to memory codes, particularly motoric encoding (Dosher & Russo, 1976; Erdelyi, Buschke, & Finkelstein, 1977; Johnson, Taylor, & Raye, 1977; Slamecka & Graf, 1978). Simply stated, the generation effect refers to the fact that information actively and effortfully generated (or information which you are actively involved) is more memorable than passively perceived information. The essence of this memory phenomenon is expressed in the "sentiment that there is an especial advantage to learning by doing, or that some kind of active or effortful involvement of the person in the learning process is more beneficial than merely passive reception of the same information" (Slamecka & Graf, 1978, p. 592).

The generation effect has direct relevance to ATC tactical operations, where the active integration of the controller's information processing capabilities with the relevant support systems (e.g., flight progress strips, radar, etc.) is a critical component of how controllers work traffic. Means et al. (1988), using a "blank flight strip recall task," demonstrated that controllers' memory for flight data is a function of the level of control exercised. Their data indicated that memory for flight information of "hot" aircraft, which required extensive control instructions, was significantly better than memory for flight information for "cold" aircraft, which required little controller intervention (e.g., overflight).

The foregoing discussion suggests the importance of a direct manipulation environment (Hutchins, 1986; Jackson, 1989; Jacob, 1986; Schneiderman, 1983) for ATC. Such an environment seems essential to maintain and potentially enhance the integrity of ATC tactical operations. In a most insightful analysis of flight progress strips, Hopkin (1991b) indicated the cognitive significance of flight strip manipulation:

> Strips help the controller to organize work and resolve problems, to plan future work, and to adjust current work in accordance with future plans. The physical act of transferring the strip from the pending to the active bay or assuming control responsibility for an

> aircraft involves a recapitulation and review of knowledge and previous decisions. This process reinforces the picture of the traffic as a whole, and the details recalled about each aircraft. The physical action in moving a strip aids memory of its contents, of its location on the board, and of why it is there. Writing on flight strips seems more memorable than watching the automatic updating of information [on electronic flight strips]. (p. 63)

Hopkin (1991b) further commented:

> Whatever form electronic flight strips take, it is essential to define beforehand all the functions of paper flight strips, in order to discard any unneeded functions deliberately and not inadvertently, to confirm that familiar essential functions can still be fulfilled electronically, and to appreciate the functional and cognitive complexity of paper flight strips. Electronic flight strips have major advantages in compatibility with computer-based air traffic control systems, but their compatibility with human roles is less obvious, requires positive planning, and depends on matching functions correctly with human capabilities. (p. 64)

Manipulative control actions, both routine and strategic, required by the controller appear to be important. Although not everyone in or working for the FAA agrees with this and although some TRACON facilities have actually gone to a stripless environment, the controversy about flight strips as memory tool will continue for the foreseeable future. An obvious concern for current and future ATC systems is optimizing controllers' direct manipulation of the system. This optimal manipulation has been vital for ATC system performance in the past. The advent of new techology and procedures will have to include ways and means for keeping controllers in the loop or a new generation of systems errors may result.

Code Interference in Working Memory

The primary phonetic (i.e., acoustic, verbal) and visual (i.e., spatial) codes essentially form two independent systems of working memory, one for processing phonetic information and the other for processing visual information (Baddeley, Grant, Wight, & Thompson, 1975; Baddeley & Hitch, 1974; Baddeley & Lieberman, 1980; Brooks, 1968; Crowder, 1978; Healy, 1975). Different concurrent tasks can cause interference in these two systems (Baddeley et al., 1975). Essentially, recall declines as items become more similar in memory. This similarity refers to the mental representation (e.g., phonetic, visual) of the item retained in working memory (Card et al., 1986). Given phonetic or verbal rehearsal (i.e., maintenance rehearsal) as the primary maintenance technique for retaining information in working memory, items in working memory will be more susceptible to phonetic interference. For example, intrusion errors are more likely to occur between items that sound similar (e.g., B for P, K for J).

The practical human factors implication of the distinction between the two primary codes or systems of working memory is tasks should be designed to minimize code interference and to take advantage of the cooperative nature of the two primary codes (Posner, 1978). For example, air traffic controllers must create and maintain a transient, dynamic "pictorial" representation or mental model of the airspace traffic under control (Schlager, Means, & Roth, 1990; Sperandio, 1974; Whitfield, 1979; Whitfield & Jackson, 1982). The construction (and/or reconstruction) of this airspace traffic "picture" requires a great deal of spatial working memory. In order to minimize visual code interference and maintain the integrity of spatial working memory, this primary task should not be performed concurrently with tasks that require similar spatial demands in working memory. Rather, concurrent tasks will be better served if they take advantage of phonetic (i.e., verbal, acoustic) representations in working memory (Wickens, 1992).

Questions still remain as to whether the codes just described are an exhaustive representation of those present in working memory. For example, if there are auditory–verbal and visual–spatial codes or systems, perhaps there are also olfactory or kinesthetic codes (Klapp, 1987). It is also not clear whether separate systems exist within working memory each with specific processing codes or different codes within the same working memory system (Klapp, 1987; Phillips & Christie, 1977). Several memory loading studies have concluded that a single-system view of working memory is tenuous at best and largely unsubstantiated (e.g., Hellige & Wong, 1983; Klapp et al., 1983; Klapp & Philipoff, 1983; Roediger, Knight, & Kantowitz, 1977).

A general implication of these studies is that tasks using systems with different codes (e.g., visual vs. auditory) will not result in performance degradation due to interference as readily as tasks using similar system codes. These studies are consistent with the multiple-resource view of information processing (Monsell, 1984; Navon & Gopher, 1979; Wickens et al., 1983), which essentially predicts that if two tasks use the same resources (e.g., auditory-verbal), interference will be reliably greater than if the two tasks use different resources (e.g., auditory-verbal vs. visual-spatial). This means that better system designs take advantage of the operators' abilities to parallel process more effectively if the demands made on them are using more than one processing modality. Too much in the visual or acoustic store and the system overloads resulting in coding and/or retrieval errors.

Attention and Working Memory

The volatility of information in working memory is potentially the greatest contributor to operational errors in ATC tactical operations. A series of experiments in the late 1950s demonstrated that in the absence of sustained attention, information is forgotten from working memory in approximately 15 sec (Brown, 1958; Peterson & Peterson, 1959). Over the past 30 years, hundreds of experiments have confirmed this finding.

Working memory information loss is particularly profound when distracting or concurrent events demand an attention shift. Controllers, for example, frequently find themselves in situations where they must perform some kind of distracting activity (e.g., notations on flight strips, cocking a flight strip, consulting a chart, adjusting their eyeglasses) between the time primary information is received and the time this information must be acted on. These concurrent activities diminish information retention. Further, while air traffic controllers usually have the status of relevant information (e.g., aircraft, flight data) continuously available on the radar display or in the flight strip bay, allowing responses based on perceptual data rather than memory data, there are frequently occasions when attention is directed away from the displays. The unforgiving possibility of automated system failure is even more dramatic. Simple electronic prolongation of visual information no longer exists, making working memory integrity essential for controlling traffic.

In a memory study of simulated communications, Loftus, Dark, and Williams (1979) obtained results similar to hundreds of studies on retention in working memory when rehearsal is prevented. They found that performance is very high at a retention interval of 0 and then declines to a stable level by about 15 sec, with minimal information being retained after 15 sec. The authors concluded that because "forgetting occurs over an interval of 15 (sec) following the initial reception of a message . . . , a message should be responded to as soon as possible after it is received" (p. 179). In addition, the authors replicated research findings (e.g., Murdock, 1961) indicating that as working memory load increases, the probability of correctly recalling information from working memory decreases. The practical implication of this finding is that "whenever possible, as little information as is feasible should be conveyed . . . at any one time. In particular, no instruction should be conveyed until 10 (sec) or so after the previous instruction has been acted upon" (p. 179).

Based on the foregoing discussion of the fragile nature of information in working memory, one might conclude that sustained attention (e.g., maintenance rehearsal) to one item of information is necessary to maintain the information in working memory. In addition to this intuitive conclusion, several studies have demonstrated that information is more volatile early in the retention interval (e.g., Dillon & Reid, 1969; Kroll, Kellicut, & Parks, 1975; Peterson & Peterson, 1959; Stanners, Meunier, Headley, 1969). These studies generally concluded that early rehearsal of information reduced the amount lost during a retention period. Klapp (1987) further elaborated that:

> A few seconds of rehearsal can largely protect (working memory) from the usual loss attributed to distraction. The potential Human Factors implications of this finding appear to have been overlooked. One would suppose that retention of information, such as directives from air traffic control, would be improved by brief rehearsal when that information cannot be used immediately. The extent to which this can lead to successful recommendations which can be implemented in practice needs to be investigated. For example, pilots might be instructed to rehearse directives which can not be implemented immediately, or transmission of additional non-emergency directives might be delayed until rehearsal (or immediate implementation) of the first directive has been accomplished. (p. 16)

Therefore, the practical implication of these studies is that if information is rehearsed immediately after it is received (early rehearsal), the process will enhance information retention in working memory (Klapp, 1987).

The influence of practice on performance is another important aspect of the dramatic attentional demands on working memory. It is well known that practice is the single most powerful factor improving the controller's ability to perform ATC tasks. Nothing is as likely to offset the frailties of working memory as will practice. The framework of "automatic" and "controlled" processing serves to help explain the influence of practice on the attentional demands of working memory (Schneider & Shiffrin, 1977; Shiffrin & Schneider, 1977).

Automatic and Controlled Processing. Considerable research has identified two qualitatively distinct ways we process and/or respond to information. These are automatic and controlled processing (e.g., Fisk, Ackerman, & Schneider, 1987; James, 1890; Kahneman & Treisman, 1984; LaBerge, 1973, 1975, 1976, 1981; Logan, 1978, 1979, 1985a, 1985b; Norman, 1976; Posner & Snyder, 1975; Schneider, Dumais, & Shiffrin, 1984; Schneider & Shiffrin, 1977; Shiffrin & Schneider, 1977). Experts and novices in any domain may well process information differently. Automatic and controlled processing can serve as a means for explaining how experienced and new controllers think and solve problems in different ways.

A well-formed representation of the stimuli in memory as a result of extensive practice is a component of automatic processing or automaticity (Schneider & Shiffrin, 1977). This extensive practice affords the development of automatic links or associations between stimulus and response that can be operated with minimal processing effort (Gopher & Donchin, 1986). The defining characteristics of automaticity are empirically well understood and documented. Automatic processing is fast, parallel (Logan, 1988a; Neely, 1977; Posner & Snyder, 1975), effortless (Logan, 1978, 1979; Schneider & Shiffrin, 1977), autonomous (Logan, 1980; Posner & Snyder, 1975; Shiffrin & Schneider, 1977; Zbrodoff & Logan, 1986), consistent (Logan, 1988a; McLeod, McLaughlin, & Nimmo-Smith, 1985; Naveh-Benjamin & Jonides, 1984), and not limited by working memory capacity (Fisk et al., 1987). It also requires no conscious awareness of the stimulus input (Carr, McCauley, Sperber, & Parmalee, 1982; Marcel, 1983), and it can be learned with extensive practice in consistent environments (Durso, Cooke, Breen, & Schvaneveldt, 1987; Fisk, Oransky, & Skedsvold, 1988; Logan, 1979; Schneider & Fisk, 1982; Schneider & Shiffrin, 1977; Shiffrin & Schneider, 1977). On the other hand, controlled processing is relatively slow, serial, mentally demanding, dependent on working memory capacity, and requires little or no practice to develop. Controlled processing is also used to process novel or inconsistent information, and essentially characterizes novice performance.

Although initial theoretical treatments viewed automaticity in terms of little or no attentional resource demands (Hasher & Zacks, 1979; Logan, 1979, 1980; Posner & Snyder, 1975; Shiffrin & Schneider, 1977), new theoretical treatments of automaticity as a memory phenomenon appear to be the most viable, particularly in terms of skill acquisition and training applications. According to the memory view, automaticity is achieved when performance is dependent on "single-step, direct-access retrieval of solutions from memory" (Logan, 1988b, p. 586). For example, an experienced controller who is familiar with the spatial layout of the ATC console visually searches for information automatically. The search goal, along with current display features, allows retrieval of prescriptive search strategies from memory. An inexperienced controller could not search automatically, because the necessary visual search strategies would not be present in memory, requiring reliance on general search skills and deliberate attention to all the potentially relevant information.

The training of automatic processing could have tremendous implications for ATC and the integrity of the controller's working memory. We have seen that the volatility of information in working memory places a tremendous burden on a controller's flight information management performance. Automaticity would allow increased information processing (e.g., parallel processing) without decrements in working memory performance. The viability of automaticity training for complex tasks, such as ATC, has been questioned by several researchers, who suggested that only "simple" tasks can be automated (e.g., Hirst, Spelke, Reaves, Caharack, & Neisser, 1980). However, Fisk et al. (1987) questioned this suggested limitation of automaticity, noting:

> Those researchers . . . do not clearly define what makes a task simple or complex. Complex tasks can be performed via automatic processing, via controlled processing, or most likely, through a combination of both processes. Simple tasks can also be performed by either automatic or controlled processing. The type of processing is not determined by the complexity (or simplicity) of a task but rather by the consistency and, if the task is consistent, the amount of practice. (p. 191)

(See Fisk et al., 1987, Fisk & Schneider, 1981, 1983, and Logan, 1988b, for a discussion of automaticity training principles and guidelines.)

However, the extent to which automaticity can lead to profitable training guidelines and recommendations that can be implemented in the complex and dynamic ATC environment is not clear and needs investigation. The identification of the ATC tasks and subtasks that would afford automatic processing and those that would afford controlled processing is a fundamental part of such an investigation.

Further research is also needed to investigate the influence of ATC automation on automatic processing. Specifically, what influence will ATC automation have on the development of overlearned (i.e., automatized) patterns of behavior, which are important for reducing the attentional demands of a controller's working memory? Another issue must be addressed. This is the concern that the cognitive structures (e.g., memory processes, conceptual knowledge) associated with overlearned patterns of behavior, which work to reduce the current load on working memory, may not be available to those controllers who "grow up" in a more automated ATC environment. The cognitive requirements of ATC will be ever changing with continued increases in ATC automation, making it difficult to reliably appreciate the nature of automatic processing in future ATC systems. How will future ATC systems afford automatic processing for the controller? One can safely conclude that the development of automaticity in future systems will be different than automaticity development in the current system.

Although there is an extensive literature on the psychology of memory and its influences on automaticity and the allocation of attention, questions still remain as to whether increased attention facilitates improved memory (Vortac, 1991). In particular, is additional attention beyond the minimum attentional threshold for a stimulus (i.e., the minimum amount of attention needed to activate a memory representation), necessary or sufficient for memory improvement? Several empirical studies (e.g., Mitchell, 1989) demonstrated that if sufficient attentional resources are available to allow the activation of a memorial process, additional attentional resources will not strengthen the activation nor improve the memory operation. Rather, the additional, unnecessary attentional resources will result in unnecessary memory loading and decreased working memory efficiency.

The previous brief discussion of attention and memory suggests that, depending on the memory processes required for a task, deliberate attention may or may not be necessary or sufficient for activation. For example, automatic processes will be activated regardless of the attentional resources available or expended. However, controlled or nonautomatic processes will not operate without the attentional resources necessary to exceed the minimum attentional threshold.

Working Memory Capacity

A number of textbooks in cognitive psychology (see Klapp et al., 1983, for a review) and human factors (e.g., Adams, 1989; Kantowitz & Sorkin, 1983; Sanders & McCormick, 1993) have proposed a single, limited-capacity system theory of working memory. This is based primarily on laboratory methods designed to measure static memory (e.g., recall of randomly presented alphanumerics, or words). Much of the ground-breaking original memory research was built around paradigms like this. The standard claim was that the maximum capacity of working memory is limited to "seven plus or minus two chunks" (Miller, 1956). This one paper has had a tremendous impact on theory and practice in memory research. A "chunk" is a single unit of information temporarily stored in working memory. This view of memory assumes that it is a single limited-capacity system and that it serves as the foundation of working memory. This standard single-system theory suggests that once working memory is filled to its five to nine chunks, maximum capacity is reached, full attention is deployed, and no further memory-involved tasks can be processed without degrading performance on concurrent tasks.

This view may be unreasonably pessimistic about human information-processing performance in situations such as strategic planning, decision making, and the processing of visual-spatial material, where extensive amounts of information are processed and retained (Chase & Ericsson, 1982; Klapp & Netick, 1988). However, it may be unreasonably optimistic in dynamic memory situations, "in which an observer must keep track of as much information as possible, when signals arrive in a continuous stream with no well-defined interval for recall" (Moray, 1986, p. 40-27). Several authors have presented data to support a multicomponent working memory system, which includes, but is not limited to, static memory (e.g., Baddeley, 1986; Brainerd, 1981; Chase & Ericsson, 1982; Hitch, 1978; Klapp & Netick, 1988; Klapp et al., 1983; Moray, 1980). For example, Baddeley (1986) described a working memory system that consists of a "central executive" that coordinates and directs the operations of two "slave" systems, the articulatory loop and the visuo-spatial "scratchpad." Essentially, these two slave systems are responsible for processing verbal and nonverbal information, respectively. Baddeley's model is very much a multiple-resource model like Wickens's (1984) model. Information on three lines of research, multiple resources, dynamic memory, and the skilled memory effect, is briefly presented next to demonstrate the dynamic nature of working memory capacity.

Multiple Resources. The literature on working memory capacity suggests that rather than a single working memory system, capable of being easily overloaded, there appear to be several systems with multiple resources, each system capable of being overloaded without interference from the other (Klapp, 1987). Multiple resource theory has been successful in describing performance in dual-task situations (Navon & Gopher, 1979; Wickens et al., 1983). For example, Klapp and Netick (1988), in examining dual-task performance in working memory, reported data suggesting that there are at least two working memory systems (i.e., auditory-verbal and visual-spatial) that differ in resource composition. Essentially, the data demonstrated that if two tasks use the same resource (e.g., auditory-verbal), interference will be reliably greater than if the two tasks use different resources (e.g., auditory-verbal vs. visual-spatial).

There are additional advantages of multiple resources theory that have potential for improving the use of memory aids so we can recall more information. Wickens et al. (1983) developed the principle of "stimulus/central processing/response compatibility." It described the optimal relationship between how information is displayed and human resources are effectively used in the form of memory codes. Displays should be designed in a format that actively helps the individual encode information into working memory. Essentially, the presentation display format should be compatible with the code used in working memory for the particular task. For example, the encoding and storage of air traffic control information is better served if it is presented in a visual-spatial format. The authors also suggested that retrieval of material from memory aids (e.g., computerized menu systems, spatially organized aids such as a "mouse") would be more effective if the resource modality needed to operate the memory aid is not similar to the storage (i.e., presentation) modality in working memory. This would reduce retrieval interference. For example, air traffic control tasks, which are heavily dependent on visual-spatial resources, may be better served by semantic-based computer menu systems or auditory-verbal systems for memory aiding.

Multiple resource theory has the potential for new approaches for improving complex and dynamic tasks such as ATC. Klapp and Netick's (1988) data suggested that in order to optimize working memory resources, tasks and subtasks need to be appropriately allocated across independent subsystems of working memory. The data also indicated that training to make the most out of task configuration may also help the management of working memory. The general guidelines offered by multiple resource theory need to be extensively investigated to determine their profitability in improving ATC tactical operations.

Dynamic Memory. Remarkably little data is available on dynamic memory in comparison to the amount of work that has been done on static memory (Moray, 1981, 1986). Dynamic memory tasks, which require operators to keep track of a continuous sequence of information with no well-defined recall intervals, are more analogous to the complex and multidimensional nature of "real-life" tasks. For example, air traffic controllers must be competent in responding to the nature of an individual aircraft under control, while concurrently "handling" the entire series of aircraft. The multidimensional nature of this task requires the controller to keep track of a large number of identified aircraft, each varying in flight data (e.g., altitude, heading, location, type), with flight data further varying along a number of values (e.g., 12,000 ft, 45 degrees north, 350 mph). Further, the number of aircraft and associated flight data are periodically updated, requiring the controller to continually acquire and forget flight information. This is done to revise the memory representation of the airspace traffic.

The research that does exist overwhelmingly suggests that dynamic memory capacity is only about three items, much less than the traditional memory capacity of seven items using a static memory paradigm (Baker, 1963; Kvalseth, 1978; Mackworth, 1959; Moray, 1980; Rouse, 1973a, 1973b; Yntema, 1963; Zeitlin & Finkleman, 1975). Based on a dynamic memory task, analogous to that of an air traffic controller, Yntema (1963) suggested three corrective solutions to reduce the severe limitations of dynamic memory capacity. First, recall performance is much better in a monitoring situation when the operator is responsible for only a few objects (e.g., aircraft) that vary on a large number of attributes (e.g., flight data), than for a large number of objects with few attributes. This recommendation is consistent with work on "conceptual chunking," which indicates that recall of a primary object or concept (e.g., aircraft) precipitates recall of associative elements or attributes (e.g., flight data) from long-term memory (Egan & Schwartz, 1979; Garland & Barry, 1990a, 1990b, 1991, 1992). Additional information on conceptual chunking is presented in a subsequent section. Second, the amount of information about each attribute (e.g., altitude, heading) has relatively little influence on dynamic memory integrity. This result is also consistent with conceptual chunking. Therefore, information precision can be increased without degrading dynamic memory performance. Third, dynamic memory performance is enhanced when each attribute value has its own unique scale. Such attribute value discriminability reduces the influence of interference due to item similarity.

Yntema's (1963) suggestions for dynamic memory enhancement warrant a note of qualification, particularly if applying them to an ATC environment. Although the conclusions are based on sound controlled laboratory experimentation, there are no data currently available that links these conclusions specifically with air traffic control. Yntema's (1963) subjects were not controllers and the task stimuli were "meaningless" to the subjects. An investigation of the applicability of these suggestions to an ATC setting is needed.

The nature of the dynamic memory tasks present in the available literature invariably involve the presentation of a time series of "random" and "meaningless" information, which the subject simply observes (i.e., monitors) (Baker, 1963; Kvalseth, 1978; Mackworth, 1959; Moray, 1980; Rouse, 1973a, 1973b; Yntema, 1963). The general finding of a limited dynamic memory capacity of approximately three items may simply be a product of these task characteristics (Moray, 1986). For example, skilled operators (e.g., air traffic controllers) who have to deal with complex and multidimensional information often exceed the three-item capacity that has been proposed. These operators process heavy information loads and are competent in recalling a considerable amount of information from their dynamic displays on demand. This superior ability may simply be a result of meaningful information processing as a result of dynamic interaction and direct manipulation of the displays. This direct manipulation (vs. monitoring) may allow the operator more meaningful encoding and retrieval strategies, which facilitate recall of the information. This explanation has definite ATC automation implications. More specifically, direct manipulation environments with motoric enactment may facilitate dynamic memory performance, while monitoring may degrade or unnecessarily restrict dynamic memory to the three item limit. This a specific concern for the future with the advent of free flight concepts as described in the RTCA concept document (Radio Technical Commission for Aeronautics, 1995).

It is tenuous at best to generalize the available dynamic memory results found in the laboratory (using meaningless material) to "real-life" dynamic environments, where operators skillfully construct the form and content of the information they need to remember (Moray, 1986). Extensive research is needed to identify the features of controllers' dynamic memory that will contribute to the development of corrective solutions and training guidelines to reduce the effects of severe memory constraints in an ATC setting. Such research is especially important with the onset of ATC automation, where the integrity of system decision making (which is based on information monitoring) is highly dependent on dynamic memory capacity. Based on work by Megan and Richardson (1979), Moray (1986) suggested that dynamic memory research may be better served if the research objectives view "the gathering of information as a cumulative process, . . . one whose outcome (is) the convolution of a data acquisition function and a forgetting function" (p. 40). Experts in many fields appear to use memory more effectively than would have been anticipated based on either the static or dynamic memory research. This may be due in part to the skilled memory effect.

Skilled Memory Effect. The intimate relationship between working memory and long-term memory provides the means to substantially increase working memory capacity beyond the traditional limits. Baddeley (1976, 1981, 1986) and his colleagues functionally described working memory as a product of several memory system components, which in combination allow for skilled tasks (e.g., reading) to exceed the traditional working memory capacity limits. Further, in a series of experiments examining memory performance as a function of practice, Chase and Ericsson (1982) demonstrated that individuals can substantially increase their working memory capacity. The authors suggested that with increased practice, working memory develops rapid-access mechanisms in long-term memory.

Over the past three decades researchers have built a solid base of empirical evidence for the "skilled memory effect" (see Chase, 1986, for a review). The literature, which covers research on a wide range of perceptual-motor and cognitive skills, generally concludes that experts in their area of expertise are able to retain information far in excess of the traditional limits of working memory (Chase, 1986; Chase & Ericsson, 1982). Based on now-classic studies with the game of chess, Chase and Simon (1973a, 1973b) theorized that for search-dependent domains like chess, domain-specific expertise can be differentiated based on how memory is organized. They suggested "that the chess master has acquired a very large repertoire of chess patterns in long-term memory that he/she can quickly recognize, although both masters and weaker players have the same (working memory) capacity" (Chase, 1986, p. 28-55).

The skilled memory effect has been replicated many times in various search-dependent domains, such as chess (Charness, 1976; Chi, 1978; Frey & Adesman, 1976; Goldin, 1978, 1979; Lane & Robertson, 1979), Go (Reitman, 1976), gomoku (Eisenstadt & Kareev, 1975; Rayner, 1958), bridge (Charness, 1979; Engle & Bukstel, 1978), and in nonsearch domains such as music (Slaboda, 1976), computer programming (McKeithen, Reitman, Rueter, & Hirtle, 1981; Schneiderman, 1976), baseball events (Chiesi, Spilich, & Voss, 1979), electronics (Egan & Schwartz, 1979), architecture (Akin, 1982), and sport (see Garland & Barry, 1990a, and Starkes & Deakin, 1984, for reviews). Research in nonsearch domains has identified "hierarchical knowledge structures" as a fundamental property of the skilled memory effect (e.g., Akin, 1982; Egan & Schwartz, 1979; Garland & Barry, 1990a, 1991). Specifically, these studies suggest experts use domain-specific conceptual knowledge to organize information, and this organization serves to facilitate storage and retrieval.

Based on the accumulated knowledge, Chase (1986) concluded that:

> The skilled memory effect is due to the existence of a vast domain-specific, long-term memory knowledge base built up by the expert with years of practice. This knowledge base can be used to serve two important mnemonic functions: (1) patterns can be used to recognize familiar situations, and (2) conceptual knowledge can be used to organize new information. (p. 28-61)

In summary, the research literature suggests that the traditional simplistic view of working memory as a single, limited-capacity system is not viable. Working memory capacity appears to be directly or indirectly related to several factors, such as the nature of the multiple working memory components (e.g., resources, conceptual organization), task parameters, meaningfulness of materials, and operator skill and experience. Despite the incredibly vast research literature on memory, Klapp (1987) asserted that a "detailed breakdown and mapping of (working) memory systems onto tasks is not yet understood" (p. 6), "largely because of our ignorance concerning the nature of the memory systems" (p. 17).

Chunking and Organization

Researchers have long recognized the principle of "chunking" as a means to expand the limits of working memory (Miller, 1956). Essentially, chunking is any operation (or operations) that combines two or more items of information into one. The resulting one item or "chunk" can then be stored as a single information unit in working memory, making available additional working memory capacity to allocate elsewhere. For example, a controller becomes familiar with the aircraft callsign TWA354 and processes it as a single chunk, requiring only one space in working memory, rather than a series of six alphanumerics, requiring six spaces in working memory. Further, a potential conflict between three aircraft—AAL348, TWA548, DAL35—will likely be organized as one chunk, rather than three, because the controller will likely not think of one without recalling the others.

Before addressing this topic, a qualification is in order to clarify the relationship between "chunking" and "organization." It is suggested that these terms refer to essentially the same processes; however, their applications are traditionally different (Klatzky, 1980). Chunking is generally associated with recent working memory storage of a relatively small number of items that will be available for immediate recall. Organization, on the other hand, is generally associated with long-term storage of a considerable amount of information. Although the terms traditionally apply to different situations, they share the underlying process of combining (organizing/chunking) two or more items of information into one. Further, because chunking is recognized as a process for the initial organization and encoding of information into long-term memory (i.e., elaborative rehearsal), it is reasonable to conclude that organization also occurs in working memory (Klatzky, 1980). In practical terms, there is relatively little substantive difference between the two terms.

In general, chunking operations can be divided into two related forms. First, chunking may be facilitated by combining items based on temporal or spatial properties—that is, combining items that occur closely in time or space. In this manner, chunking occurs without the items necessarily forming a meaningful unit (Bower & Winzenz, 1969; Huttenlocher & Burke, 1976). This sort of chunking is often referred to as "grouping" (Klatzky, 1980). Parsing is closely related to grouping. Parsing is the process of "placing physical discontinuities between subsets that are likely to reflect chunks" (Wickens, 1984, p. 222). You can improve retention of relatively meaningless information by putting gaps or breaks within the information sequence. For example, someone could recall the telephone number 516 347 0364 better than 5163470364 (Wickelgren, 1964). Loftus et al. (1979), in their study of working memory retention of air traffic control communications, reported that in certain circumstances four-digit items (e.g., 7382) were better retained when parsed into two pairs of double digits (e.g., "seventy-three, eighty-two").

Second, chunking may be facilitated if it "utilizes information from (long-term memory) to meaningfully relate many incomplete items to a single known item" (Klatzky, 1980, p. 92). The degree of the inherent meaningful relationship between the separate items is also important and can help or hinder chunking. For example, the potential conflict between AAL348, TWA548, and DAL35 allows these three aircraft to be chunked as one item (i.e., potential conflict), due to the shared meaningfulness of each being a contributor to a potential conflict.

Chunking essentially benefits two qualitatively distinct processes in working memory (Wickens, 1992). First, chunking helps the retention (i.e., maintenance) of information in working memory for a brief period of time, after which time the information is directly or indirectly "dumped." For example, controllers typically deal with a continuous flow of aircraft through their sector of responsibility. When aircraft are handed off, the associative information for that aircraft is no longer needed. Therefore it is beneficially dumped from memory. Second, chunking facilitates the transfer of information into long-term memory. Controllers must process a considerable amount of information concerning the status of several aircraft, which must be integrated and stored in long-term memory in order to initially create and subsequently revise the memorial representation (i.e., "picture") of the airspace traffic.

The psychological literature has clearly documented the contribution of organizational processes (e.g., chunking) to good memory (e.g., Ellis & Hunt, 1989). How well someone organizes material is often a clear indication of their level of expertise in any given area. Experts can take in a large quantity of task-specific information in a brief period of time and subsequently recall the information in meaningful units or chunks. Chase and Simon's (1973a, 1973b) study of chunking of stimulus information by chess experts demonstrated that experts are able to encode more information in a limited time when compared with nonexperts.

In explaining chunking behavior in the recall of task-specific stimulus information, Chase and Simon (1973a, 1973b; Simon & Chase, 1973; Simon & Gilmartin, 1973) proposed a perceptual chunking hypothesis. "Perceptual chunking" involves perception by coding the position of entire chunks or several items, by storing chunk labels in working memory, and subsequently decoding at the time of recall. Two critical features of the perceptual chunking hypothesis are that chunks are independently perceived and that recall requires decoding chunk labels in working memory. This means that heavy processing demands are placed on working memory.

Egan and Schwartz (1979), however, pointed out several problems with these critical features. First, chunk independence does not allow for global processing. For example, an air traffic control specialist can perceive the global characteristics (e.g., "a developing conflict situation") of a traffic pattern on the radar display in addition to the individual features (e.g., individual aircraft). Second, a group of display features (e.g., aircraft) may not form a functional unit or chunk, independent of other functional units. The functional units (chunks) must be context dependent. As another example, the controller in identifying and processing two concurrent potential conflict situations will form two chunks—for example, "conflict A" and "conflict B." These chunks are not independent of each other, in that the resolution of conflict A will have an influence on the resolution of conflict B and vice versa. This is due to shared airspace. In addition, the two conflict resolutions will influence and be influenced by the surrounding noninvolved air traffic. Third, some studies have shown that various interpolated tasks have no influence on recall performance of skilled chess players (Charness, 1976; Frey & Adesman, 1976). These studies strongly question Chase and Simon's position that task-specific information places substantial demands on working memory capacity.

As an alternative to perceptual chunking, Egan and Schwartz (1979; also see Garland & Barry, 1990a, 1991) proposed a conceptual chunking hypothesis, which links chunking (and skilled memory) to the organization of concepts in long-term memory. Conceptual chunking consists of a few primary features. First, skilled operators rapidly identify a concept(s) for the entire display, or segments of the display (e.g., overflights, climbing aircraft, descending aircraft, military aircraft). Second, skilled operators may systematically retrieve functional units and their elements that are related to the identified conceptual category stored in long-term memory (e.g., flights DAL1134, TWA45, UAL390, and TCA224 are elements identified as part of the conceptual category "overflights"). Third, conceptual knowledge of the display enables skilled operators to systematically search displays to verify details suggested by the conceptual category. For example, a controller is able to systematically search and detect aircraft that possess identifying flight characteristics that are consistent with the defining characteristics of the conceptual category "overflights."

Based on the available research, the conceptual chunking hypothesis appears to overcome the problems of the perceptual chunking hypothesis, by linking skilled memory and chunking to the organization of concepts in long-term memory (Egan & Schwartz, 1979; Garland & Barry, 1990a, 1991). The available data indicate that skilled operators are reliably better at recalling display features even after a brief exposure time. This superior recall performance may be based on the use of a "generate and test" process (Egan & Schwartz, 1979). This means that emphasis on processing information related to a conceptual category (e.g., potential air traffic conflict), allows skilled operators to systematically retrieve elements (e.g., the defining features of the potential conflict and the involved aircraft) that are meaningfully associated with the conceptual category. The reader may recall Yntema's (1963) research on dynamic memory, which indicated that recall performance was better in a monitoring situation when the subject was responsible for a few objects (e.g., aircraft) that vary on a number of attributes (e.g., flight data) than when subjects were responsible for a large number of objects with few attributes. These findings are consistent with conceptual chunking, in that recall of the primary object or concept (e.g., aircraft) facilitated recall of the associative elements or attributes (e.g., flight data) from long-term memory. Tulving (1962) suggested that the ability to access the whole functional unit allows for systematic retrieval of all the information within a unit or chunk. He stressed that this ability is contingent on a good organizational structure of the task-specific knowledge in long-term memory.

Ellis and Hunt (1989) noted that the question of how organization affects memory is very important and equally complex. Although memory and organization are two different processes, Ellis and Hunt suggested that the two processes are positively correlated, resulting in the assumption that "organization processes contribute to good memory." Mandler (1967) provided support for this assumption, suggesting that organization is effective because of "economy of storage." Simply, organization is similar to chunking, in that individual units are grouped into large functional units, reducing the number of items to be stored in working memory and/or long-term memory. Mandler's approach assumes that organization occurs during encoding.

In a supportive yet alternative approach, Tulving (1962) suggested that organization benefits memory because of its "effects at retrieval." Tulving agreed that the organization of information occurs at encoding. However, he stressed that the ability to access the functional units or the whole entity at retrieval facilitates memory. This ability to access the whole functional unit allows for systematic retrieval of all the information within a unit. Tulving's arguments are consistent with conceptual chunking, in that knowledge of a conceptual display would allow subjects to systematically retrieve functional units that are related to the previously identified conceptual category that has been accessed in long-term memory. In addition, conceptual knowledge of the display would enable skilled operators to systematically search the conceptual category in long-term memory to verify details suggested by the initial conceptual category.

Ericsson (1985) pointed out apparent parallels between experts' superior memory performance in their domain of expertise and normal memory for meaningful materials, such as texts and pictures. Kintsch (1974) demonstrated that a competent reader can form a long-term representation for the text's meaning very rapidly and extensively, without deliberate effort (automatic processing). In addition, pictures (e.g., spatial information) appear to be fixated in long-term memory in less than 1 sec (Potter & Levy, 1969). Those results appear consistent with the process of conceptually driven pattern recognition, which involves recognition decisions being guided by long-term memory rather than by sensory information (Ellis & Hunt, 1989).

Based on the available data, the superior perceptual skill of experts in a variety of skill domains may not involve rapidly decoding independent chunk labels from a limited-capacity working memory; rather, as Egan and Schwartz (1979) proposed, perceptual skill may be linked to the organization of task-specific concepts in long-term memory. It is suggested that expert memory performance may be more conceptual in nature, enabling skilled operators to (a) rapidly identify a concept for an entire stimulus display, (b) systematically retrieve functional units (chunks) that are related to the conceptual category stored in long-term memory through a "generate and test" process, and (c) systematically search displays to verify details suggested by the activated conceptual category. These findings and the theoretical foundations behind them re-emphasize the importance of both initial and recurrent training in any command and control environment where the situation is fluid and memory resources are in demand. Working memory will likely be used more effectively when the operator is completely up to speed in the higher order tasks and concepts. This will lead to more effective and less effortful organization in working memory.

The compatability of encoding processes with those of retrieval can have a major impact on memory organization and subsequent success or failure. Essentially, information retrieval is enhanced when the meaningful cues used at encoding are also present at retrieval. If the encoding and retrieval cues are not compatible, then memory will fail (e.g., Godden & Baddeley, 1980). For example, in the ATC setting, the flight progress strips and their manipulation serve as significant retrieval cues, because they contain essentially the same information present during initial encoding.

Although research on air traffic control memory, specifically controller memory organization and chunking behavior, is disappointingly minimal, a study by Means et al. (1988) of controller memory provides some interesting data. In an airspace traffic drawing task, controllers were presented a sector map at the end of a 30- to 45-min ATC simulation, and subsequently were instructed to group associated aircraft in the sector by drawing a circle around them. It was assumed that the aircraft groupings reflect the manner in which controllers organize airspace traffic. The findings indicated that aircraft groupings could be characterized by various kinds of traffic properties or concepts (e.g., landing aircraft, overflights, climbing aircraft, traffic crossing over a fix, etc.).

In addition, the researchers gathered data indicating that controllers who performed in a radar scenario condition (control traffic with radar and flight progress strips) tended to group aircraft based on the potential to "conflict," whereas controllers in a manual scenario condition (control traffic with flight progress strips only) tended to group aircraft based on geographical proximity. Controllers in the manual scenario condition had not controlled traffic without radar in a number of years, and therefore were less competent in controlling traffic under the experimental conditions than were the radar scenario controllers who had available the necessary displays. These data suggest that the more competent controllers tended to use higher order grouping criteria (e.g., potential conflict) than did the "handicapped" controllers, who tended to utilize simpler grouping criteria (e.g., geographical proximity).

These data are consistent with conceptual chunking in that the controllers tended to group (organize) the airspace around a number of ATC concepts and potential control problems. Further, the radar scenario controllers appeared to use more discriminating grouping criteria based on the strategic dynamics (e.g., conceptual nature) of the airspace, unlike the manual controllers, who appeared to use criteria based on simpler airspace spatial properties (e.g., the aircraft are close to one another). This suggests that the more experienced and skilled controller utilizes a larger, more discriminating conceptual knowledge base in order to control traffic. These results were consistent with the findings of Sollenberger and Stein (1995). Controllers were generally more successful in recalling aircraft in a simulation scenario based on the concept of what role they played than on what the callsigns were. Aircraft were chunked around spatiotemporal concepts. Although controllers could only recall a small percentage of the callsigns, they had little difficulty in determining what had been occurring in the airspace they had under control.

Mental Models (The Controller's "Picture"?!). Several times throughout this chapter the rather common ATC phrase *the controller's picture* was used to refer to the controller's mental representation of the airspace. At this time, an attempt is made to address this cognitive phenomenon, which seems to play such an important role in ATC memory and tactical operations.

A mental model is a theoretical construct that provides the user a framework for thinking about a complex domain of which they are a part. Mental models may be specific to a given situation (e.g., VFR traffic) or more global to the entire task domain (e.g., the entire flight sector). They may, or may not, include abstractions concerning functional relationships, operating guidelines, and systems goals and objectives (Mogford, 1991; Norman, 1986; Rasmussen, 1979; Wickens, 1992; Wilson & Rutherford, 1990). Theoretical descriptions of mental models are varied (Mogford, 1991). For example, Rouse and Morris (1986) suggested: "Mental models are the mechanisms whereby humans are able to generate descriptions of system purpose and form, explanations of system functioning and observed system states, and predictions of future system states" (p. 351). Further, Norman (1986) stated:

> Mental models seem a pervasive property of humans. I believe that people form internal, mental models of themselves and of the things and people with whom they interact. These models provide predictive and explanatory power for understanding the interaction. Mental models evolve naturally through interaction with the world and with the particular system under consideration. These models are highly affected by the nature of the interaction, coupled with the person's prior knowledge and understanding. The models are neither complete nor accurate, but nonetheless they function to guide much human behavior. (p. 46)

Research on mental models and conceptual structures in the air traffic control environment is disappointingly limited (see Mogford, 1991, for a review). However, the research that is available does suggest a connection between a controller's "picture" and the controller's understanding of, and memory for the traffic situation (e.g., Bisseret, 1970; Landis, Silver, Jones, & Messick, 1967; Means et al., 1988; Moray, 1980; Whitfield, 1979). A general conclusion of these studies is that skilled controllers, in comparison to less skilled controllers, use their picture as a supplementary display in order to enhance memory for aircraft. In addition, it is generally concluded that the quality and functionality of the controller's picture is directly related to ATC expertise.

According to Whitfield (1979), who was one of the first to study the picture systematically, the skilled controller's picture seems to use three kinds of memory: (a) static memory (e.g., sector characteristics, separation standards), (b) dynamic memory (e.g., continual updating of aircraft flight data), and (c) working memory (e.g., current status of aircraft). Further, Mogford (1991) suggested that the controller's "picture" is likely maintained in working memory, with substantial influences from "unconscious rules" stored in long-term memory. He stated that "it appears that the controller's mental model possesses various kinds of information which are reliant on different types of memory. Maps, flight plans, aircraft performance information, separation standards, and procedures are learned through training and experience and stored in memory" (p. 239).

The extent to which mental models can provide assistance with the practical problems of ATC memory enhancement remains unclear. Although the available research suggests a significant role for working memory in ATC picture development and use, research has not yet revealed empirical evidence suggesting how the controller's picture may assist in enhancing controller working memory and improving ATC tactical operations. Research on mental models in air traffic control is needed as ATC systems become more automated, forcing the controller into ever increasing levels of supervisory control. The dramatic changes with future automation will not only replace ATC technology and equipment, but will also change the way controllers conduct their job. Research is needed to investigate how increased computerization of ATC tasks influences the development of the controller's picture and its potential supporting influence on controller working memory. Hopkin (1980), in addressing this very problem, noted:

> Controllers frequently report that computer aids seem to increase the probability that they will lose the picture of the traffic under their control. This problem is the subject of research and the development of appropriate measures in air traffic control . . . in relation to concepts such as working mental models and working memory. If, as Neisser (1976) claimed, images are anticipatory phases of perceptual activity and images are plans for obtaining information from potential environments, this may provide a theoretical framework and suggest appropriate measures for evaluating the efficacy of various forms of computer assistance, particularly predictions, as aids to imagining. It could also provide hypotheses for specifying conditions when forgetting is most likely to occur. (p. 558)

An understanding of ATC mental models may prove beneficial to understanding the impact of automation on designing controller training and memory aids, because to be effective, such aids must effectively interact with the cognitive processes of the controller (e.g., Hollnagel & Woods, 1983; Moray, 1988). It is important that data and job aids be designed and presented in such a way as to work with the controllers' internal representation of the airspace rather than against it. Wickens (1992) stated:

> Within the last decade, designers of computer systems are beginning to capitalize on the fact that people have a lifetime's worth of experience in negotiating in a three-dimensional environment and manipulating three-dimensional objects (Hutchins, Hollan, & Norman, 1985). The spatial metaphor, therefore, is an important emerging concept in human-computer interaction. (p. 154)

Further, Wickens (1984) commented on a challenging and critical issue regarding advanced automation, that of design implications for computer-based data entry and retrieval systems:

> How does the computer model or understand the user's conception of the data and logic within the computer itself? Clearly the computer should organize data in a form compatible with the user's mental model. But what if different individuals possess different styles of organization? Are different organizational formats appropriate for spatial versus verbal modes of thinking, as suggested by Schneiderman (1980)? A related question concerns the assumptions that the computer should make about the level of knowledge of the user. For the same program, a computer's interaction with a novice should probably be different from the interaction with an expert user. A novice, for example, would benefit from a menu selection program in which all options are offered, since many of them are not likely to be stored in long-term memory. For the expert, this format will probably give unnecessary clutter, since the alternatives are stored and available in LTM in any case. An intriguing question from the viewpoint of systems designs is how the computer can either

> explicitly assess or implicitly deduce the level of knowledge or the format of organization employed by the user. (p. 237)

Although several researchers have suggested potential implications of mental models for both training and display design (e.g., Mogford, 1991; Wickens, 1992), Wilson and Rutherford (1989) recently asserted that "We have shown the several different interpretations of the concept (mental models) and its utility to be a weakness, which militates against the widespread use of mental models in system design" (p. 629). Obviously, further work is needed on the air traffic controller's picture.

This brief overview of the work on chunking and organization and its relevance to ATC tactical operations leads to the primary conclusion that more research is needed. In particular, research is needed in an ATC setting to better understand the conceptual structures that guide the synthesis and organization of present and future traffic situations. In support of this line of research, Whitfield (1979) suggested that a controller's mental model is required for current and future planning of the traffic situation. A further line of research is suggested by the available work on dynamic memory and conceptual organization (e.g., mental model). Perhaps the ability of controllers to exceed the traditional limits of dynamic memory (i.e., three items) is associated with the controller's conceptualization (e.g., mental model) of the ATC domain. If so, what are the features of the controller's conceptualization that may contribute to dynamic memory enhancement? How can this be used to help train and maintain controller skill? Do ATC conceptual structures fundamentally change with experience and expertise, thus facilitating the enhancement of dynamic memory and skilled memory? There are obviously more questions than answers at this point; however, with increased ATC automation, the time (although limited) is ripe for extensive investigations to address these crucial questions. Harwood, Murphy, and Roske-Hofstrand (1991) pointed out that the complexity of ATC must be recognized; otherwise, research and applications will not be useful or meaningful.

Forgetting in Working Memory

The primary reason for this chapter is to examine the relationship of working memory to controller operational errors. An FAA Administrators Task Force identified controller memory lapses (i.e., forgetting) as a significant issue related to revising and retrieving critical operational information (Operational Error Analysis Work Group, 1987). Although considerable information on forgetting is available in the psychological literature (see Klatzky, 1980, pp. 124–150 for a review), the profitable application of this material to the real-life setting of ATC is unclear. In contrast to the unforgiving nature of unintended memory failure, Hopkin (1995) noted:

> Forgetting as a boon rather than a bane has scarcely been studied at all. Yet it is not always an advantage in air traffic control to be able to recall all the details of what happened previously, as this could encourage unwarranted presumptions that any intervening changes of circumstance are trivial and that previous solutions can be adopted again, whereas it might be better to work out fresh solutions without such remembered preconceptions. Some limited guidance on how to code air traffic control information to make it more memorable can be offered, but there is no comparable practical guidance on how to code air traffic control information so that it is easy and efficient to use while it is present but is readily forgotten after it has served its purpose and there is no benefit in remembering it. Given the perennial problem of too much information in air traffic control,

recommendations on how to render the useless forgettable would have real practical value. (pp. 55–56)

Forgetting is also desirable because it provides storage space for incoming new information. The level at which information is processed plays a large role in determining how difficult it will be to remember or forget that information (Murphy & Cardosi, 1995, pp. 179–191). Thus, the nature of forgetting information in the ATC setting is paradoxical, in that it has both desirable and undesirable implications.

Essentially, research on both unintentional and intentional forgetting is necessary in order to develop aids to eliminate and/or enhance forgetting depending on the situation. The following discussion presents some of the available information on forgetting that may be applicable to the ATC setting.

Information processing models generally incorporate two mechanisms that produce memory retrieval failures. These are (a) spontaneous decay, which refers to a time-dependent process of information becoming less available over time, and (b) interference, which refers to the disruption of the memory trace due to competing activity. Considerable research effort has gone into trying to determine which of these mechanisms really drives forgetting (Card et al., 1986).

Decay. Research by Reitman (1974) initially demonstrated the separate roles of decay and interference in working memory. This research, along with others, has generally implicated time-dependent processes as being attributable to the rapid rate of decay or complete loss of information availability if the individual takes no or inefficient action to process the information for temporary short-term or permanent long-term memory. In addition to the rapid decay of information that has been actively attended to and encoded, forgetting as a result of decay is also, in part, a function of the initial level to which the material is processed. Preattentive processing of information, without higher-order encoding, will inevitably result in decay. In addressing research on the decay mechanism as a means of forgetting, Wickens (1984) stated:

> When verbal information is presented auditorily, the decay may be slightly postponed because of the transient benefits of the echoic (auditory) code. When information is presented visually, the decay will be more rapid. The consequence of decay before material is used is the increased likelihood of error. The pilot may forget navigational instructions delivered by the air traffic controller before they are implemented. . . . In fact, Moray (1980) concludes that "the task of monitoring a large display with many instruments is one for which human memory is ill suited, especially when it is necessary to combine information from different parts of the display and the information is dynamic." (p. 216)

The time-dependent decay process operates to significantly attenuate the fidelity of the memory trace (Klatzky, 1980; Wickens, 1992). The extent to which the decay process is disruptive or beneficial to the controller is situation specific. The development of techniques to optimize the decay of information seems a viable line of research. If the controller were able to reliably control the decay of information, then information management would be facilitated. The reader is referred to the section on Attention and Working Memory for additional information related to the decay mechanism of forgetting. This is clearly an area in which good display design can be beneficial. Controllers and other operators should never be forced to depend on memory with all its foibles if there is a viable way of organizing information so it is present and available when needed.

Interference. Considerable research has demonstrated that it is more difficult to retrieve an item from working memory and long-term memory if there are other similar items in the respective memory system (e.g., Conrad, 1964; Underwood, 1957). The similarity of items in memory is contingent on the memory representation of each item. For example, interference in working memory is more likely for items that sound alike (i.e., acoustic/phonetic interference). Long-term memory is more susceptible to semantic interference. That is, items (e.g., chunks) that share similar meanings are likely to share the same retrieval cues, which in turn disrupts information retrieval. Research on the interference effect, has demonstrated that much of what is commonly referred to as forgetting is simply failure to retrieve, not actual loss (e.g., decay) from memory (Card et al., 1986).

Generally, the literature recognizes three sources that may contribute to the interference effect: within-list (or information redundancy) interference, retroactive interference, and proactive interference (Wickens, 1984). *Within-list interference* is attributable to the increased similarity of items within a group that must be processed in working memory. For example, Wickens (1984) illustrated that "when an air traffic controller must deal with a number of aircraft from one fleet, all possessing similar identifier codes (AI3404, AI3402, AI3401), the interference due to the similarity between items makes it difficult for the controller to maintain their separate identity in working memory" (p. 224). Obviously, in order to alleviate within-list interference, information must be presented in a manner that reduces the information redundancy.

Retroactive interference is the detrimental effect of recently acquired information (retroactively) interfering with previously learned material (Underwood, 1957). For example, a controller may forget a newly assigned altitude of an aircraft, because an additional item of information intervened and prevented sufficient maintenance rehearsal of the new altitude and/or notation on the flight progress strip. Further, increased similarity between the item to be retained and the intervening item will increase the probability of interference.

Proactive interference is the detrimental effect of previously acquired information (proactively) interfering with recently learned material (Keppel & Underwood, 1962; Underwood, 1957). This effect may be especially profound during labor- and time-intensive situations, where there is a tendency to cognitively regress back to former firmly established ways of thinking. This is a situation where the power of long-term memory to help organize information in working memory can work against you. This creates challenges for training managers who are planning for transitions to new equipment and/or systems. Proactive interference must be considered or it could diminish the potential benefits of new technology.

A considerable amount of research has been conducted to examine the processes that reduce the effects of proactive interference, or as the literature commonly refers to it, "a release from proactive interference" (e.g., Keppel & Underwood, 1962). This phenomenon refers to the fact that if the type of stimulus material (e.g., letters, numbers) is changed from trial n to trial $n + 1$ (e.g., from numbers on trial n to letters on trial $n + 1$), then proactive interference will be reduced, resulting in a substantial decrease in forgetting of the recently acquired material (e.g., the stimulus material on trial $n + 1$) (Loftus et al., 1979). Explanations for this phenomenon are generally consistent with the following example provided by Loftus et al. (1979):

> Suppose a subject must remember two pieces of information, A and B, that are presented in close temporal proximity. To the extent that A and B may be differently encoded, they will be less confusable, and hence easier to recall. Carrying this notion over to the controller/pilot situation, it seems reasonable to expect that two pieces of numerical information will be easier to remember to the extent that they are uniquely encoded. (p. 172)

In a study of simulated communications between controllers and pilots, Loftus et al. (1979) found evidence to indicate that a "unique-encoding system" as compared to a "same-encoding system" of ATC communications led to superior memory performance. The same-encoding system referred to the current relatively standard ATC practice of transmitting virtually all numerical data in a digit-by-digit manner (e.g., the radio frequency 112.1 would be transmitted as "one, one, two, point, one"). In contrast, "an example of [the] 'unique-encoding system,' would be to encode radio frequencies in the digit-by-digit manner described above but to encode transponder codes as two pairs of double digits (e.g., '7227' would be encoded as 'seventy-two, twenty-seven')" (p. 171). This finding has definite memory implications for recall of multidimensional flight data. Loftus et al. (1979) concluded that:

> Attention is traditionally paid to the question of how transmitted information should be encoded so as to minimize errors in perception (e.g., by use of the phonemic alphabet). However, virtually no attention has been paid to the question of how information may be encoded so as to minimize errors in memory. The (unique-encoding system) represents but one possible improvement in encoding of transmitted information. Potentially, there are many others. (p. 180)

Unfortunately, several years have elapsed with little or no research to advance these findings.

In addition, previously presented information on dynamic memory is also available in support of the utility of the unique-encoding system. In particular, dynamic memory studies by Yntema (1963) and Yntema and Mueser (1960) provide the most applicable evidence. In these studies, subjects were required to keep track of a large number of objects, which varied on a number of attributes, which in turn varied on a number of unique values. These studies indicated that memory fidelity was enhanced when the attribute values each had their own unique codes (e.g., feet, speed, miles), compared to attribute values sharing common codes. For example, suppose a controller must identify and then enter the status of several aircraft along several flight data dimensions. Because the flight data are coded differently (e.g., altitude/feet, distance/nautical miles), performance will be superior if the controller deals in turn with all the relevant flight data of one aircraft before progressing to the next aircraft, rather than dealing with all the aircraft on only one flight data dimension (e.g., altitude/feet) before progressing to the next flight data dimension. The unique-encoding system appears to be a profitable means by which information can be optimally encoded, thus enhancing working memory retention and minimizing retrieval failures of critical information. Research is needed to examine the viability of such an information encoding system in an ATC environment.

Based on the available research on interference effects, Wickens and Flach (1988) suggested four ways to reduce the effects of interference on forgetting in working memory. They are:

1. "Distribute the material to be held in (working) memory over time." This will allow proactive interference from previously acquired information to be reduced.
2. "Reduce similarity between items." This is suggested since similar-looking or similar-sounding (Conrad, 1964) items lead to greater interference.
3. "Eliminate unnecessary redundancy." This suggestion is intended to reduce the effects of within-list interference.
4. "Minimize within-code interference." This suggestion is consistent with the previously presented information on code interference in working memory. For example, in the predominantly visual/spatial ATC environment, concurrent secondary tasks should minimize the use of visual/spatial codes, and instead they should utilize auditory/speech encoding (e.g., voice recognition technology). (pp. 124–126)

Directed Forgetting. As mentioned earlier, in addition to enhancing the integrity of working memory performance through the reduction of memory lapses, there are also times when the intentional "purging" of information from working memory will work to enhance memory. Hopkin (1988) asserted that intentional forgetting may be beneficial in that the "controller dealing with an immediate problem is not overburdened by recalling other problems not sufficiently similar to be helpful in solving the present one" (p. 12). Further, Hopkin (1980) noted the importance of identifying and developing ATC techniques intended to aid the controller in the forgetting of "unwanted baggage," which may prove to proactively interfere with current information processing. Such "directed forgetting" (also referred to as "motivated" or "intentional" forgetting in the cognitive literature) of information that is no longer useful would seem to be a necessary skill in a dynamic memory setting, such as ATC flight management, in which the ability to process incoming sequential information is contingent upon the availability of processing space in working memory. The available research indicates that when subjects are instructed to intentionally forget unwanted information, there are additional attention resources for dealing with concurrent tasks (e.g., E. Bjork, R. Bjork, & Kilpatrick, 1990; R. Bjork, 1972; Martin & Kelly, 1974). In addition, R. Bjork (1972) suggested that directed forgetting can be trained.

The information presented earlier on the effects of decay on forgetting is relevant to the present discussion of directed forgetting. If techniques can be identified to assist the controller in reliably controlling the decay of information, directed forgetting would be a valuable product. As mentioned previously, two qualitatively different types of rehearsal strategies are involved in working memory—maintenance and elaborative rehearsal. Short-term retention of information in working memory is achieved through maintenance rehearsal, which emphasizes the phonetic aspects (i.e., auditory, speech) of the stimuli, whereas elaborative rehearsal is important for transfer of information into long-term memory by emphasizing the semantic aspects (i.e., meaningfulness) of the stimuli and their association with the conceptual information of the controller's mental model stored in long-term memory. Because information transfer to long-term memory facilitates the undesirable effects of proactive interference (see Craik & Watkins, 1973; Glenberg, Smith, & Green, 1977), information to be retained for only a short period of time should only use phonetic maintenance rehearsal, as opposed to semantic elaborative rehearsal (Wickens, 1992). This strategy, along with directed forgetting strategies, may prove useful in enhancing directed forgetting (Bjork, 1972; Bjork et al., 1990).

Based on the available data from laboratory studies, Wickens (1984) suggested "that this technique (directed forgetting), like chunking, is a potentially valuable strategy that can be learned and subsequently employed for more efficient storage and retrieval of subsequent memory items" (p. 226). However, a note of qualification is warranted. Specifically, research is needed to determine the applicability of the laboratory findings to the ATC setting. The preceding suggestion was based on data gathered in a laboratory setting with college students (e.g., sophomores) who were required to forget meaningless information, which they had no experience in actively using and/or processing. Information is needed to determine the utility of purposefully forgetting meaningful information in a real-life, labor-intensive, time-intensive environment such as ATC. Until such data is available, instructional guidelines for the training of directed forgetting in an ATC setting will remain unrealized.

WHAT DOES THE FUTURE HOLD FOR WORKING MEMORY IN ATC?

The preceding discussion of working memory and its implications for air traffic control is by no means an exhaustive, definitive treatment of the working memory requirements of air traffic control tactical operations. Although considerable information on working memory is available (e.g., Baddeley, 1986; Klatzky, 1980), there remain more questions than answers. Working memory permeates every aspect of the human information processing system, making it virtually impossible to get a "handle" on all the parameters that define its functionality. This chapter has attempted to raise an awareness of a few of the most salient and transient characteristics of working memory and their implications for ATC. Additional areas of research that directly or indirectly influence working memory were beyond the scope of this chapter. These include, but are not limited to, long-term memory, stress, decision making, and workload.

The limiting factor in gaining a more comprehensive understanding of the working memory requirements of ATC tactical operations is the simple fact that there is relatively little human factors research on the cognitive aspects of ATC, especially on working memory. In 1980, Hopkin, in noting the importance of memory research in the ATC environment, commented that "the application of theories of memory to practical air traffic control problems must be developed more in the future" (p. 558). In calling attention to the necessity to reinterpret the air traffic controller's tasks in relation to cognitive psychology constructs, Hopkin (1995) stated:

> Some of the roles of memory in air traffic control do not fit the main theories of memory very well. Theories tend to emphasize timescales of a few seconds for short-term memory or relative permanence for long-term memory, or refer to active task performance for working memory (Baddeley, 1990; Logie, 1993; Stein and Garland, 1993). The controller relies on a mental picture of the traffic that is based on a synthesized integration of radar, tabular and communicated information, interpreted according to professional knowledge and experience (Whitfield and Jackson, 1982; Rantanen, 1994). Although a simplified form of the picture can be built in a few seconds, as is routinely done at watch handover, building the complete picture requires more processing (Craik and Lockhart, 1972) and typically takes about fifteen to twenty minutes, by which time the controller knows the full history and intentions of all current and pending traffic and can plan accordingly. (pp. 54–55)

Unfortunately, the application of psychological theories (i.e., memory theories) to practical air traffic control problems has gone largely unheeded. This is unsettling with the onset of the progression of automated systems, which will substantially alter the way in which controllers manage live traffic (Wise et al., 1991). The implications of increased ATC automation on the controller's cognitive processes are unknown. How can we gain an adequate understanding of the cognitive (e.g., working memory) requirements of advanced automation when the cognitive requirements of the current system remain elusive? Comprehensive task analyses of controllers have evolved over the years to a point today where scientists begin to understand the scope and complexity of the controller's job (Nickels, Bobko, Blair, Sands, & Tartak, 1995). There are potential human factors consequences of increasing ATC automation. These include the impact of memory aids on ATC working memory.

After considering the cognitive psychology research on the working memory system, one can safely conclude that the ATC system, given current structure and technology, will only be as efficient and reliable as the controller's working memory system. A controller's working memory directly or indirectly influences every aspect of his/her ability to control traffic. With ever-increasing amounts of ATC automation, human problem solving and other cognitive processes will change or become additional complicating factors. Researchers need a new set of cognitive performance measures to fully appreciate the consequences of automation on controller performance. Hopkin (1991c), in discussing future ATC measures of performance, noted:

> In previous studies, the most appropriate performance measures may not have been employed. In the future, better measurement tools will be needed to show the consequences of automation not only in terms of performance but also in terms of associated cognitive skills. Some cognitive effects are not currently measured at all, for example on understanding or memory, but they may be more significant than the routinely measured effects on performance. (p. 558)

Throughout this chapter considerable information has been presented emphasizing the critical importance of working memory in ATC tactical operations. Unfortunately, the available research on working memory in ATC and non-ATC settings has largely gone unnoticed in current and future ATC system design. As Hopkin noted in 1980, it is definitely time to apply existing (and new) memory research to the practical problems of air traffic control. Although there is considerable research on the frailties of working memory and ways to overcome them, there also exists a fundamental problem in making the appropriate knowledge influence the ATC system design process. For example, Hopkin (1991c) commented:

> It is not sufficient to plan and conduct research if the only products are journal articles, standards, or handbooks, essential though these are. The research evidence has to be applied and integrated into the design. Nowhere does this seem to be done satisfactorily, although its necessity is acknowledged and many attempts are made. Lack of appropriate mechanisms to apply research findings to design processes appears to be the main difficulty. This problem is linked to some uncertainty about how valid and general some of the existing data and findings are. Is all the existing evidence actually worth applying to the design process? If not, how do we determine which should be applied and which should not? What criteria could serve such a purpose? What should the balance of evidence be between previous research and current and future research? What are the best measurements to gain evidence in the form of practical advice at the design stages? How can research findings be made more acceptable to designers, so that they are more willing to

> adapt design processes of future air traffic control systems to incorporate evidence from research? (p. 555)

At present, there are no appreciable answers to these provocative questions.

For several decades, an implicit philosophy of automation has existed that adopted the assumption that maximum available automation is always appropriate (invest in hardware, not people). This philosophy has been based, in part, on the availability of increasingly sophisticated and advanced technological innovations, the need to reduce human workload, the need for increased safety of flight, and perhaps, primarily, on the assumption that the human mind (especially human memory) is similar to a silicon-based system that cannot be easily overloaded. Although automated systems have provided substantial benefits, several human factors consequences have arisen and incidents/accidents have occurred. These problems often end up by calling for human factors professionals and the aviation community to reexamine automation practices. We continue to automate without building the human factor into the design process.

There is an increasing awareness of the lack of a scientifically based philosophy of automation. This philosophy must be based in an understanding of the relative capabilities (e.g., frailties of working memory) of the controller in the system, and the circumstances under which automation should assist and augment the capabilities of the controller. What is needed is an approach that has a better philosophical base for what automation seeks to achieve and a more human-centered approach, to avoid the most adverse human factors consequences of automated systems and provide a better planned progressive introduction of automated aids in step with user needs (e.g., Garland, 1991). Such a comprehensive, scientifically based design philosophy for human-centered automation must be developed in order to avoid inevitable "one step forward and two steps backward" progression.

For the time being, the human controller, despite the limitations and constraints of the working memory system, will remain an essential part of the ATC system. Furthermore, it is suggested that with ever-increasing levels of ATC automation, the significance of the human controller in the system and the significance of the controller's working memory system should no longer be taken for granted.

In conclusion, the purpose, intent, and nature of this chapter are perhaps best reflected in ideas Levesley (1991) put forth about the way he sees the ATC system in 50 years. Levesley commented:

> What I actually predict will happen is that the lessons of the last fifty years will be repeated in the next fifty. Airlines will still prefer to spend $500 on aircraft for every $1 spent on ATC. Will the cost of potential super-systems actually prohibit their introduction, as they prove totally cost-ineffective? If I survive to the age of 93 and I fly somewhere in 2040, I suspect that there will still be a human problem solver on the ground in control of my flight, who will rejoice in the title of "the controller." And I don't think that controllers will be there because they are irreplaceable, or because the public wants someone there. I think that, with the right tools to help, the controller will still be there as the most cost effective, flexible system solution to the problem of safely guiding pilots and passengers to their destination. And *that* is what air traffic control is really all about. (p. 539)

REFERENCES

Adams, J. A. (1989). *Human factors engineering*. New York: Macmillan.
Akin, O. (1982). *The psychology of architectural design*. London: Pion.

Ammerman, H., Fligg, C., Pieser, W., Jones, G., Tischer, K., & Kloster, G. (1983). *Enroute/terminal ATC operations concept* (Report No. DOT/FAA/AP-83-16). Washington, DC: Federal Aviation Administration.

Ammerman, H., & Jones, G. (1988). *ISSS impact on ATC procedures and training* (Report No. CDRL C108). Washington, DC: Federal Aviation Administration.

Anderson, R. E. (1984). Did I do it or did I only imagine doing it? *Journal of Experimental Psychology: General, 113,* 594–613.

Backman, L., & Nilsson, L. G. (1984). Aging effects in free recall: An exception to the rule. *Human Learning, 3,* 53–69.

Backman, L., Nilsson, L. G., & Chalom, D. (1986). New evidence on the nature of the encoding of action events. *Memory & Cognition, 14,* 339–346.

Baddeley, A. D. (1976). *The psychology of memory.* New York: Basic Books.

Baddeley, A. D. (1981). The concept of working memory: A view of its current state and probable future development. *Cognition, 10,* 17–23.

Baddeley, A. D. (1986). *Working memory.* Oxford: Clarendon Press.

Baddeley, A. D. (1990). *Human memory: Theory and practice.* Boston: Allyn & Bacon.

Baddeley, A. D. (1996). Exploring the central executive. *Quarterly Journal of Experimental Psychology: Human Experimental Psychology, 49A*(1), 5–28.

Baddeley, A. D., Grant, S., Wight, E., & Thompson, N. (1975). Imagery and visual working memory. In P. M. Rabbitt & S. Dornic (Eds.), *Attention and performance V.* New York: Academic Press.

Baddeley, A. D., & Hitch, G. (1974). Working memory. In G. Bower (Ed.), *The psychology of learning and motivation: Advances in research and theory.* New York: Academic Press.

Baddeley, A. D., & Lieberman, K. (1980). Spatial working memory. In R. S. Nickerson (Ed.), *Attention and performance VIII.* Hillsdale, NJ: Lawrence Erlbaum Associates.

Baker, C. (1963). Further towards a theory of vigilance. In D. Buckner & J. McGrath (Eds.), *Vigilance: A symposium.* New York: McGraw-Hill.

Bencomo, A. A., & Daniel, T. C. (1975). Recognition latency for pictures and words as a function of encoded-feature similarity. *Journal of Experimental Psychology: Human Learning and Memory, 1,* 119–125.

Bisseret, A. (1970). Memoire operationelle et structure du travail. *Bulletin de Psychologie, 24,* 280–294.

Bisseret, A. (1971). Analysis of mental processes involved in air traffic control. *Ergonomics, 14,* 565–570.

Bjork, E. L., Bjork, R. A., & Kilpatrick, H. A. (1990, November). *Direct and indirect measures of inhibition in directed forgetting.* Paper presented at the 31st annual meeting of the Psychonomic Society, New Orleans, LA.

Bjork, R. A. (1972). Theoretical implications of directed forgetting. In A. W. Melton & E. Martin (Eds.), *Coding processes in human memory.* Washington, DC: Winston.

Bower, G. H., & Winzenz, D. (1969). Group structure, coding, and memory for digit series. *Journal of Experimental Psychology Monograph Supplement, 80,* 1–17.

Brainerd, C. J. (1981). Working memory and the developmental analysis of probability judgement. *Psychological Review, 88,* 463–502.

Broadbent, D. E. (1958). *Perception and communications.* New York: Pergamon Press.

Brooks, L. R. (1968). Spatial and verbal components in the act of recall. *Canadian Journal of Psychology, 22,* 349–368.

Brown, J. (1958). Some tests of the decay theory of immediate memory. *Quarterly Journal of Experimental Psychology, 10,* 12–21.

Card, S. K., Moran, T. P., & Newell, A. (1986). The model of human processor: An engineering model of human performance. In K. R. Boff, L. Kaufman, & J. P. Thomas (Eds.), *Handbook of perception and human performance: Volume II, Cognitive processes and performance* (pp. 45-1–45-35). New York: Wiley-Interscience.

Carr, T. H., McCauley, C., Sperber, R. D., & Parmalee, C. M. (1982). Words, pictures, and priming: On semantic activation, conscious identification, and the automaticity of information processing. *Journal of Experimental Psychology: Human Perception and Performance, 8,* 757–777.

Charness, N. (1976). Memory for chess positions: Resistance to interference. *Journal of Experimental Psychology: Human Learning and Memory, 2,* 641–653.

Charness, N. (1979). Components of skill in bridge. *Canadian Journal of Psychology, 33,* 1–50.

Chase, W. G. (1986). Visual information processing. In K. R. Boff, L. Kaufman, & J. P. Thomas (Eds.), *Handbook of perception and human performance: Volume II, Cognitive processes and performance* (pp. 28-1–28-71). New York: Wiley-Interscience.

Chase, W. G., & Ericsson, K. A. (1982). Skill and working memory. In G. Bower (Ed.), *The psychology of learning and motivation.* New York: Academic Press.

Chase, W. G., & Simon, H. A. (1973a). The mind's eye in chess. In W. G. Chase (Ed.), *Visual information processing* (pp. 215–272). New York: Academic Press.

Chase, W. G., & Simon, H. A. (1973b). Perception in chess. *Cognitive Psychology, 4*, 55–81.

Chi, M. T. H. (1978). Knowledge structures and memory development. In R. S. Siegler (Ed.), *Children's thinking: What develops?* (pp. 144–168). Hillsdale, NJ: Lawrence Erlbaum Associates.

Chiesi, H. L., Spilich, G. J., & Voss, J. F. (1979). Acquisition of domain-related information in relation to high and low domain knowledge. *Journal of Verbal Learning and Verbal Behavior, 18*, 257–273.

Cohen, S. (1981). On the generality of some memory laws. *Scandinavian Journal of Psychology, 22*, 267–281.

Cohen, S. (1983). The effect of encoding variables on the free recall of words and action events. *Memory & Cognition, 11*, 575–582.

Conrad, R. (1964). Acoustic comparisons in immediate memory. *British Journal of Psychology, 55*, 75–84.

Conrad, R., & Hull, A. J. (1964). Information, acoustic confusions, and memory span. *British Journal of Psychology, 55*, 429–432.

Cowan, N. (1995). *Attention and memory: An integrated framework.* Oxford psychology series, no. 26. New York: Oxford University Press.

Craik, F. I. M., & Lockhart, R. S. (1972). Levels of processing: A framework for memory research. *Journal of Verbal Learning and Verbal Behaviour, 11*, 671–684.

Craik, F. I. M., & Watkins, M. J. (1973). The role of rehearsal in short-term memory. *Journal of Verbal Learning and Verbal Behavior, 12*, 599–607.

Crowder, R. (1978). Audition and speech coding in short-term memory. In J. Requin (Ed.), *Attention and performance VII* (pp. 248–272). Hillsdale, NJ: Lawrence Erlbaum Associates.

Dillon, R. F., & Reid, L. S. (1969). Short-term memory as a function of information processing during the retention interval. *Journal of Experimental Psychology, 81*, 261–269.

Dosher, B. A., & Russo, J. E. (1976). Memory for internally generated stimuli. *Journal of Experimental Psychology: Human Learning and Memory, 2*, 633–640.

Durso, F. T., Cooke, N. M., Breen, T. J., & Schvaneveldt, R. W. (1987). Is consistent mapping necessary for high-speed scanning? *Journal of Experimental Psychology: Learning, Memory, and Cognition, 13*, 223–229.

Egan, D. E., & Schwartz, B. J. (1979). Chunking in recall of symbolic drawings. *Memory & Cognition, 7*, 149–158.

Eisenstadt, M., & Kareev, Y. (1975). Aspects of human problem solving: The use of internal representations. In D. A. Norman & D. E. Rumelhart (Eds.), *Exploration in cognition* (pp. 87–112). San Francisco: Freeman.

Ellis, H. C., & Hunt, R. R. (1989). *Fundamentals of human memory and cognition.* Dubuque, IA: Brown.

Endsley, M. R. (1990, June) *Situational awareness global assessment technique (SAGAT)—Air to air tactical version* (Rep. No. NOR DOC 89-58). Hawthorne, CA: Northrop Corporation.

Engle, R. W., & Bukstel, L. (1978). Memory processes among bridge players of differing expertise. *American Journal of Psychology, 91*, 673–690.

Erdelyi, M., Buschke, H., & Finkelstein, S. (1977). Hypermnesia for Socratic stimuli: The growth of recall for an internally generated memory list abstracted from a series of riddles. *Memory & Cognition, 5*, 283–286.

Ericsson, K. A. (1985). Memory skill. *Canadian Journal of Psychology, 39*, 188–231.

Federal Aviation Administration. (1989). *Air traffic control: Order No. 7110.65F.* Washington, DC: Air Traffic Operations Service, U.S. Department of Transportation.

Federal Aviation Administration. (1995). *FAA Aviation Forecasts—Fiscal Years 1995–2006, March 1995* (Report No. FAA-APO-95-1). U.S. Department of Transportation.

Finkelman, J. M., & Kirschner, C. (1980). An information-processing interpretation of air traffic control stress. *Human Factors*, 22(5), 561–567.

Fisher, D. F. (1969). *Short-term memory: An annotated bibliography supplement I.* Aberdeen Proving Ground, MD: Human Engineering Laboratories, Aberdeen Research & Development Center.

Fisher, D. F. (1971). *Short-term memory: An annotated bibliography supplement II.* Aberdeen Proving Ground, MD: Human Engineering Laboratories, Aberdeen Research & Development Center.

Fisher, D. F., & Wiggins, H. F. (1968). *Short-term memory: Annotated bibliography.* Aberdeen Proving Ground, MD: Human Engineering Laboratories.

Fisk, A. D., Ackerman, P. L., & Schneider, W. (1987). Automatic and controlled processing theory and its applications to human factors problems. In P. A. Hancock (Ed.), *Human factors psychology* (pp. 159–197). New York: North Holland.

Fisk, A. D., Oransky, N., & Skedsvold, P. (1988). Examination of the role of "higher-order" consistency in skill development. *Human Factors, 30*, 567–581.

Fisk, A. D., & Schneider, W. (1981). Controlled and automatic processing during tasks requiring sustained attention: A new approach to vigilance. *Human Factors, 23*, 737–750.

Fisk, A. D., & Schneider, W. (1983). Category and work search: Generalizing search principles to complex processing. *Journal of Experimental Psychology: Learning, Memory, and Cognition, 9*, 117–195.

Frey, P. W., & Adesman, P. (1976). Recall memory for visually presented chess positions. *Memory & Cognition, 4*, 541–547.

Garland, D. J. (1991). Automated systems: The human factor. In J. A. Wise, V. D. Hopkin, & M. L. Smith (Eds.), *Automation and systems issues in air traffic control* (pp. 209–215). Berlin: Springer-Verlag.

Garland, D. J., & Barry, J. R. (1990a). Sport expertise: The cognitive advantage. *Perceptual and Motor Skills, 70*, 1299–1314.

Garland, D. J., & Barry, J. R. (1990b, August). *An examination of chunking indices in recall of schematic information.* Paper presented at the 98th Annual Convention of the American Psychological Association, Boston.

Garland, D. J., & Barry, J. R. (1991). Cognitive advantage in sport: The nature of perceptual structures. *American Journal of Psychology, 104*, 211–228.

Garland, D. J., & Barry, J. R. (1992). Effects of interpolated processing on expert's recall of schematic information. *Current Psychology: Research & Reviews, 4*, 273–280.

Glenberg, A., Smith, S. M., & Green, C. (1977). Type 1 rehearsal: Maintenance and more. *Journal of Verbal Learning and Verbal Behavior, 16*, 339–352.

Godden, D., & Baddeley, A. D. (1980). When does context influence recognition memory? *British Journal of Psychology, 71*, 99–104.

Goettl, B. P. (1985, September 29–October 3). The interfering effects of processing code on visual memory. *Proceedings of the 29th Annual Meeting of the Human Factors Society* (pp. 66–70). Santa Monica, CA: Human Factors Society.

Goldin, S. E. (1978). Effects of orienting tasks on recognition of chess positions. *American Journal of Psychology, 91*, 659–672.

Goldin, S. E. (1979). Recognition memory for chess positions. *American Journal of Psychology, 92*, 19–31.

Gopher, D., & Donchin, E. (1986). Workload—An examination of the concept. In K. R. Boff, L. Kaufman, & J. P. Thomas (Eds.), *Handbook of perception and human performance: Volume II, Cognitive processes and performance* (pp. 41-1–41-49). New York: Wiley-Interscience.

Harper, R. R., Hughes, J. A., & Shapiro, D. Z. (1989). *The functionality of flight strips in ATC work* (Report to the U.K. Civil Aviation Authority). Lancaster, UK: Lancaster Sociotechnics Group.

Harris, J. E. (1984). Remembering to do things: A forgotten topic. In J. E. Harris & P. E. Morris (Eds.), *Everyday memory actions and absent mindedness* (pp. 71–92). London: Academic Press.

Harwood, K., Murphy, E. D., & Roske-Hofstrand, R. J. (1991). *Selection and refinement of methods and measures for documenting ATC cognitive processes* (Unpublished Draft Technical Note).

Hasher, L., & Zacks, R. T. (1979). Automatic and effortful processing in memory. *Journal of Experimental Psychology: General, 108*, 356–388.

Healy, A. F. (1975). Temporal-spatial patterns in short-term memory. *Journal of Verbal Learning and Verbal Behavior, 14*, 481-495.

Healy, A. F., & McNamara, D. S. (1996). Verbal learning and memory: Does the modal model still work? *Annual Review of Psychology, 47*, 143–72.

Hellige, J. B., & Wong, T. M. (1983). Hemisphere specific interference in dichotic listening: Task variables and individual differences. *Journal of Experimental Psychology: General, 112*, 218–239.

Hirst, W., Spelke, E. S., Reaves, C. C., Caharack, G., & Neisser, U. (1980). Dividing attention without alternation or automaticity. *Journal of Experimental Psychology, 109*, 98–117.

Hitch, G. J. (1978). The role of short-term memory in mental arithmetic. *Cognitive Psychology, 10*, 302–323.

Hollnagel, E., & Woods, D. D. (1983). Cognitive systems engineering: New wine in new bottles. *International Journal of Man–Machine Studies, 18*, 583–600.

Hopkin, V. D. (1980). The measurement of the air traffic controller. *Human Factors*, 22(5), 547–560.

Hopkin, V. D. (1982). *Human factors in air traffic control* (AGARDograph No. 275). Neuilly-sur-Seine, France: NATO.

Hopkin, V. D. (1988). *Human factors aspects of the AERA 2 program.* Farnborough, Hampshire, UK: Royal Air Force Institute of Aviation Medicine.

Hopkin, V. D. (1989). Man-machine interface problems in designing air traffic control systems. *Proceedings IEEE, 77*, 1634–1642.

Hopkin, V. D. (1991a). The impact of automation on air traffic control systems. In J. A. Wise, V. D. Hopkin, & M. L. Smith (Eds.), *Automation and systems issues in air traffic control* (pp. 3–19). Berlin: Springer-Verlag.

Hopkin, V. D. (1991b, January). *Automated flight strip usage: Lessons from the functions of paper strips.* Paper presented at the conference on Challenges in Aviation Human Factors: The National Plan, Washington, DC.

Hopkin, V. D. (1991c). Closing remarks. In J. A. Wise, V. D. Hopkin, & M. L. Smith (Eds.), *Automation and systems issues in air traffic control* (pp. 553–559). Berlin: Springer-Verlag.

Hopkin, V. D. (1995). *Human factors in air traffic control.* Bristol, PA: Taylor & Francis.

Hutchins, E. (1986). Direct manipulation interface. In D. Norman & S. Draper (Eds.), *User centered system design: New perspectives in human–computer interaction.* Hillsdale, NJ: Lawrence Erlbaum Associates.

Hutchins, E. L., Hollan, J. D., & Norman, D. A. (1985). Direct manipulation interfaces. *Human-Computer Interaction, 1*(4), 311–338.

Huttenlocher, D., & Burke, D. (1976). Why does memory span increase with age? *Cognitive Psychology, 8,* 1–31.

Jackson, A. (1989). *The functionality of flight strips* (Report to the U.K. Civil Aviation Authority). Farnborough, UK: Royal Signals and Radar Establishment.

Jacob, R. (1986). Direct manipulation. *Proceedings of the IEEE International Conference on Systems, Man, and Cybernetics* (pp. 348–359). Atlanta, GA.

James, W. (1890). *Principles of psychology.* New York: Holt.

Johnson, M. K. (1988). Reality monitoring: An experimental phenomenological approach. *Journal of Experimental Psychology: General, 117,* 390–394.

Johnson, M. K., Taylor, T. H., & Raye, C. L. (1977). Fact and fantasy: The effects of internally generated events on the apparent frequency of externally generated events. *Memory & Cognition, 5,* 116–122.

Jonides, J. (1995). Working memory and thinking. In E. E. Smith & D. N. Osherson (Eds.), *Thinking: An invitation to cognitive science* (Vol. 3, 2nd ed., pp. 215–265). Cambridge, MA: MIT Press.

Kahneman, D., & Treisman, A. M. (1984). Changing views of attention and automaticity. In R. Parasuraman & R. Davies (Eds.), *Varieties of attention* (pp. 29–61). New York: Academic Press.

Kantowitz, B. H., & Sorkin, R. D. (1983). *Human factors: Understanding people-system relationships.* New York: Wiley.

Kausler, D. H., & Hakami, M. K. (1983). Memory for activities: Adult age differences and intentionality. *Developmental Psychology, 19,* 889–894.

Keppel, G., & Underwood, B. J. (1962). Proactive inhibition in short-term retention of single items. *Journal of Verbal Learning and Verbal Behavior, 1,* 153–161.

Kinney, G. C., Spahn, M. J., & Amato, R. A. (1977). *The human element in air traffic control: Observations and analysis of the performance of controllers and supervisors in providing ATC separation services* (Rep. No. MTR-7655). McLean, VA: MITRE Corporation.

Kintsch, W. (1974). *The representation of meaning in memory.* Hillsdale, NJ: Lawrence Erlbaum Associates.

Kirchner, J. H., & Laurig, W. (1971). The human operator in air traffic control systems. *Ergonomics, 14*(5), 549–556.

Klapp, S. T. (1987). Short-term memory limits in human performance. In P. Hancock (Ed.), *Human factors psychology* (pp. 1–27). Amsterdam: North-Holland.

Klapp, S. T., Marshburn, E. A., & Lester, P. T. (1983). Short-term memory does not involve the "working memory" of information processing: The demise of a common assumption. *Journal of Experimental Psychology: General, 112,* 240–264.

Klapp, S. T., & Netick, A. (1988). Multiple resources for processing and storage in short-term working memory. *Human Factors, 30*(5), 617–632.

Klapp, S. T., & Philipoff, A. (1983). Short-term memory limits in performance. *Proceedings of the Human Factors Society, 27,* 452–454.

Klatzky, R. L. (1980). *Human memory: Structures and processes.* San Francisco: Freeman.

Koriat, A., & Ben-Zur, H. (1988). Remembering that I did it: Process and deficits in output monitoring. In M. Grunegerg, P. Morris, & R. Sykes (Eds.), *Practical aspects of memory: Current research and issues* (Vol. 1, pp. 203–208). Chichester: Wiley.

Koriat, A., Ben-Zur, H., & Nussbaum, A. (1990). Encoding information for future action: Memory for to-be-performed versus memory for to-be-recalled tasks. *Memory & Cognition, 18,* 568–583.

Koriat, A., Ben-Zur, H., & Sheffer, D. (1988). Telling the same story twice: Output monitoring and age. *Journal of Memory & Language, 27,* 23–39.

Kosslyn, S. (1981). The medium and the message in mental imagery: A theory. *Psychological Review, 88,* 46–66.

Kroll, N. E. A., Kellicut, M. H., & Parks, T. E. (1975). Rehearsal of visual and auditory stimuli while shadowing. *Journal of Experimental Psychology: Human Learning and Memory, 1,* 215–222.

Kvalseth, T. (1978). Human and Baysian information processing during probabilistic inference tasks. *IEEE Transactions on Systems, Man, and Cybernetics, 8,* 224–229.

LaBerge, D. (1973). Attention and the measurement of perceptual learning. *Memory & Cognition, 1,* 268–276.

LaBerge, D. (1975). Acquisition of automatic processing in perceptual and associative learning. In P. M. A. Rabbit & S. Dornic (Ed.), *Attention and performance V* (pp. 78–92). New York: Academic Press.
LaBerge, D. (1976). Perceptual learning and attention. In W. K. Estes (Ed.), *Handbook of learning and cognitive processes.* Hillsdale, NJ: Lawrence Erlbaum Associates.
LaBerge, D. (1981). Automatic information processing: A review. In J. Long & A. D. Baddeley (Eds.), *Attention and performance IX* (pp. 173–186). Hillsdale, NJ: Lawrence Erlbaum Associates.
Landis, D., Silver, C. A., Jones, J. M., & Messick, S. (1967). Level of proficiency and multidimensional viewpoints about problem similarity. *Journal of Applied Psychology, 51,* 216–222.
Lane, D. M., & Robertson, L. (1979). The generality of levels of processing hypothesis: An application to memory for chess positions. *Memory & Cognition, 7,* 253–256.
Lauber, J. K. (1993, July). Human performance issues in air traffic control. *Air Line Pilot,* pp. 23–25.
Levesley, J. (1991). The blue sky challenge: A personal view. In J. A. Wise, V. D. Hopkin, & M. L. Smith (Eds.), *Automation and systems issues in air traffic control* (pp. 535–539). Berlin: Springer-Verlag.
Loftus, G. R., Dark, V. J., & Williams, D. (1979). Short-term memory factors in ground controller/pilot communications. *Human Factors, 21,* 169–181.
Logan, G. D. (1978). Attention in character-classification tasks: Evidence for the automaticity of component stages. *Journal of Experimental Psychology: General, 107,* 32–63.
Logan, G. D. (1979). On the use of a concurrent memory load to measure attention and automaticity. *Journal of Experimental Psychology: Human Perception and Performance, 5,* 189–207.
Logan, G. D. (1980). Attention and automaticity in Stroop and priming tasks: Theory and data. *Cognitive Psychology, 12,* 523–553.
Logan, G. D. (1985a). Skill and automaticity: Relations, implications, and future directions. *Canadian Journal of Psychology, 39,* 367–386.
Logan, G. D. (1985b). On the ability to inhibit simple thoughts and actions: II. Stop signal studies of repetition priming. *Journal of Experimental Psychology: Learning, Memory, and Cognition, 11,* 65–69.
Logan, G. D. (1988a). Toward an instance theory of automatization. *Psychological Review, 95,* 95–112.
Logan, G. D. (1988b). Automaticity, resources, and memory: Theoretical controversies and practical implications. *Human Factors, 30*(5), 583–598.
Logie, R. H. (1993). Working memory and human–machine systems. In J. A. Wise, V. D. Hopkin, & P. Stager (Eds.), *Verification and validation of complex systems: Human factors issues* (NATO ASI Series Vol. F110, pp. 341–353). Berlin: Springer-Verlag.
Logie, R. H. (1995). *Visuo-spatial working memory. Series: Essays in cognitive psychology.* Hove: Lawrence Erlbaum Associates.
Lyon, G. R., & Krasnegor, N. A. (Eds.). (1996). *Attention, memory, and executive function.* Baltimore, MD: P. H. Brookes.
Mackworth, J. (1959). Paced memorizing in a continuous task. *Journal of Experimental Psychology, 58,* 206–211.
Marcel, A. T. (1983). Conscious and unconscious perception: An approach to the relations between phenomenal experience and perceptual processes. *Cognitive Psychology, 15,* 238–300.
Martin, P. W., & Kelly, R. T. (1974). Secondary task performance during directed forgetting. *Journal of Experimental Psychology, 103,* 1074–1079.
McKeithen, K. B., Reitman, J. S., Rueter, H. H., & Hirtle, S. C. (1981). Knowledge organization and skill differences in computer programmers. *Cognitive Psychology, 13,* 307–325.
McLeod, P., McLaughlin, C., & Nimmo-Smith, I. (1985). Information encapsulation and automaticity: Evidence from the visual control of finely timed action. In M. I. Posner & O. S. Marin (Eds.), *Attention and performance XI* (pp. 391–406). Hillsdale, NJ: Lawrence Erlbaum Associates.
Means, B., Mumaw, R., Roth, C., Schlager, M., McWilliams, E., Gagne, V. R., Rosenthal, D., & Heon, S. (1988). *ATC training analysis study: Design of the next-generation ATC training system.* Washington, DC: Federal Aviation Administration.
Megan, E., & Richardson, J. (1979). Target uncertainty and visual scanning behavior. *Human Factors, 21,* 303–316.
Miller, G. A. (1956). The magical number seven plus or minus two: Some limits on our capacity for processing information. *Psychological Review, 63,* 81–97.
Mitchell, D. B. (1989). How many memory systems? Evidence from aging. *Journal of Experimental Psychology: Learning, Memory, and Cognition, 15,* 31–49.
Mogford, R. H. (1991). Mental models in air traffic control. In J. A. Wise, V. D. Hopkin, & M. L. Smith (Eds.), *Automation and systems issues in air traffic control* (pp. 235–242). Berlin: Springer-Verlag.
Monsell, S. (1984). Components of working memory underlying verbal skills: A "distributed capacities" view. In H. Bouma & D. Bouwhuis (Eds.), *Attention and performance X* (pp. 142–164). Hillsdale, NJ: Lawrence Erlbaum Associates.

Moray, N. (1980). *Human information processing and supervisory control* (Tech. Rep.). Cambridge, MA: MIT Man–Machine Systems Laboratory.

Moray, N. (1981). The role of attention in the detection of errors and the diagnosis of failures in man-machine systems. In J. Rasmussen & W. B. Rouse (Eds.), *Human detection and diagnosis of system failures.* New York: Plenum.

Moray, N. (1986). Monitoring behavior and supervisory control. In K. R. Boff, L. Kaufman, & J. P. Thomas (Eds.), *Handbook of perception and human performance: Volume II, Cognitive processes and performance* (pp. 40-1–40-51). New York: Wiley-Interscience.

Moray, N. (1988). Intelligent aids, mental models, and the theory of machines. In E. Hollnagel, G. Mancini, & D. D. Woods (Eds.), *Cognitive engineering in complex dynamic worlds* (pp. 165–175). London: Academic Press.

Murdock, B. B. (1961). The retention of individual items. *Journal of Experimental Psychology, 62,* 618–625.

Murdock, B. B. (1968). Modality effects in short-term memory: Storage or retrieval? *Journal of Experimental Psychology, 77,* 79–86.

Murphy, E. D., & Cardosi, K. M. (1995). Human information processing. In K. M. Cardosi & E. D. Murphy (Eds.), *Human Factors in the Design and Evaluation of Air Traffic Control Systems* (Report No. DOT-FAA-RD-95-3, pp. 135–218). Washington, DC: FAA.

Narborough-Hall, C. S. (1987). Automation: Implications for knowledge retention as a function of operator control responsibility. *Proceedings of the Third Conference of the British Computer Society* (pp. 269–282). Cambridge, UK: Cambridge University Press.

National Transportation Safety Board. (1991). *Aircraft accident report. Runway collision of USAir Flight 1493, Boeing 737 and SkyWest Flight 5569 Fairchild Metroliner Los Angeles International Airport Los Angeles, California, February 1, 1991* (Rep. No. NTSB-AAR-91-8). Washington, DC: Author.

Naveh-Benjamin, M., & Jonides, J. (1984). Maintenance rehearsal: A two-component analysis. *Journal of Experimental Psychology: Learning, Memory, and Cognition, 10,* 369–385.

Navon, D., & Gopher, D. (1979). On the economy of the human processing system. *Psychological Review, 86,* 214–255.

Neely, J. H. (1977). Semantic priming and retrieval from lexical memory: Roles of inhibitionless spreading activation and limited-capacity attention. *Journal of Experimental Psychology: General, 106,* 226–254.

Neisser, U. (1976). *Cognition and reality.* San Francisco: Freeman.

Nickels, B. J., Bobko, P., Blair, M. D., Sands, W. A., & Tartak, E. L. (1995). *Separation and control hiring assessment (SACHA): Final job analysis report.* Bethesda, MD: University Research Corporation.

Nilsson, L. G., Ohlsson, K., & Ronnberg, J. (1977). Capacity differences in processing and storage of auditory and visual input. In S. Dornick (Ed.), *Attention and performance VI.* Hillsdale, NJ: Lawrence Erlbaum Associates.

Norman, D. A. (1986). Cognitive engineering. In D. A. Norman & S. W. Draper (Eds.), *User centered system design* (pp. 31–61). Hillsdale, NJ: Lawrence Erlbaum Associates.

Operational Error Analysis Work Group (1987, August). *Actions to implement recommendations of April 17, 1987.* Unpublished manuscript, Federal Aviation Administration, Washington, DC.

Peterson, L. R., & Peterson, M. J. (1959). Short-term retention of individual verbal items. *Journal of Experimental Psychology, 58,* 193–198.

Phillips, W. A., & Christie, F. M. (1977). Interference with visualization. *Quarterly Journal of Experimental Psychology, 29,* 637–650.

Posner, M. I. (1973). *Cognition: An introduction.* Glenview, IL: Scott, Foresman.

Posner, M. I. (1978). *Chronometric explorations of the mind.* Hillsdale, NJ: Lawrence Erlbaum Associates.

Posner, M. I., & Mitchell, R. F. (1967). Chronometric analysis of classification. *Psychological Review, 74,* 392–409.

Posner, M. I., & Snyder, C. R. R. (1975). Attention and cognitive control. In R. L. Solso (Ed.), *Information processing and cognition: The Loyola symposium* (pp. 212–224). Hillsdale, NJ: Lawrence Erlbaum Associates.

Potter, M. C., & Levy, E. I. (1969). Recognition memory for a rapid sequence of pictures. *Journal of Experimental Psychology, 81,* 10–15.

Rantanen, E. (1994). The role of dynamic memory in air traffic controllers' situational awareness. In R. D. Gilson, D. J. Garland, & J. M. Koonce (Eds.), *Situational awareness in complex systems* (pp. 209–215). Daytona Beach, FL: Embry-Riddle Aeronautical University Press.

Rasmussen, J. (1979). *On the structure of knowledge: A morphology of mental models in a man–machine systems context* (Rep. No. RISO-M-2192). Roskilde, Denmark: Riso National Laboratory.

Rayner, E. H. (1958). A study of evaluative problem solving. Part I: Observations on adults. *Quarterly Journal of Experimental Psychology, 10,* 155–165.

Reitman, J. S. (1974). Without surreptitious rehearsal: Information and short-term memory decays. *Journal of Verbal Learning and Verbal Behavior, 13*, 365–377.
Reitman, J. S. (1976). Skilled performance in Go: Deducing memory structures from inter-response times. *Cognitive Psychology, 8*, 336–356.
Roediger, H. L., III, Knight, J. L., & Kantowitz, B. H. (1977). Inferring decay in short-term memory: The issue of capacity. *Memory & Cognition, 5*, 167–176.
Rouse, W. B. (1973a). Model of the human in a cognitive prediction task. *IEEE Transactions on Systems, Man, and Cybernetics, 3*, 473–478.
Rouse, W. B. (1973b). *Models of man as a suboptimal controller* (NTIS Rep. No. N75-19126/2). Cambridge, MA: MIT, Ninth Annual NASA Conference on Manual Control.
Rouse, W. B., & Morris, N. M. (1986). On looking into the black box: Prospects and limits in the search for mental models. *Psychological Bulletin, 100*, 349–363.
Radio Technical Commission for Aeronautics. (1995). *Report of the RTCA Board of Directors select committee on free flight*. Washington, DC.
Sanders, M. S., & McCormick, E. J. (1993). *Human factors in engineering and design*. New York: McGraw-Hill.
Schlager, M. S., Means, B., & Roth, C. (1990). Cognitive task analysis for the real (-time) world. In *Proceedings of the Human Factors Society 34th Annual Meeting* (pp. 1309–1313). Santa Monica, CA: Human Factors Society.
Schneider, W., Dumais, S. T., & Shiffrin, R. M. (1984). Automatic and control processing and attention. In R. Parasuraman & R. Davies (Eds.), *Varieties of attention* (pp. 1–27). New York: Academic Press.
Schneider, W., & Fisk, A. D. (1982). Degree of consistent training: Improvements in search performance and automatic process development. *Perception and Psychophysics, 31*, 160–168.
Schneider, W., & Shiffrin, R. M. (1977). Controlled and automatic human processing: I. Detection, search, and attention. *Psychological Review, 84*, 1–66.
Schneiderman, B. (1976). Exploratory experiments in programmer behavior. *Journal of Computer and Information Sciences, 5*, 123–143.
Schneiderman, B. (1980). *Software psychology*. Cambridge, MA: Winthrop.
Schneiderman, B. (1983). Direct manipulation: A step beyond programming languages. *IEEE Computer, 16*(8), 57–69.
Shaffer, L. H. (1993). Working memory or working attention? In A. Baddeley & L. Weiskrantz (Eds.), *Attention: selection, awareness, and control: A tribute to Donald Broadbent*. New York: Clarendon Press.
Shiffrin, R. M., & Schneider, W. (1977). Controlled and automatic human information processing: II. Perceptual learning, automatic attending, and a general theory. *Psychological Review, 84*, 127–190.
Simon, H. A., & Chase, W. G. (1973). Skill in chess. *American Scientist, 61*, 394–403.
Simon, H. A., & Gilmartin, K. A. (1973). Simulation of memory for chess positions. *Cognitive Psychology, 5*, 29–46.
Slaboda, J. (1976). Visual perception of musical notation: Registering pitch symbols in memory. *Quarterly Journal of Experimental Psychology, 28*, 1–16.
Slamecka, N. J., & Graf, P. (1978). The generation effect: Delineation of a phenomenon. *Journal of Experimental Psychology: Learning, Memory, and Cognition, 4*, 592–604.
Smith, E. (1968). Choice reaction time: An analysis of the major theoretical positions. *Psychological Bulletin, 69*, 77–110.
Sollenberger, R. L., & Stein, E. S. (1995). A simulation study of air traffic controller situational awareness. *Proceedings of the International Conference on Experimental Analysis and Measurement of Situational Awareness* (pp. 211–217). Daytona Beach, FL: Embry Riddle Aeronautical University.
Sperandio, J. C. (1974). *Extension to the study of the operational memory of air traffic controllers* (RSRE Translation No. 518). Unpublished manuscript.
Squire, L. R., Knowlton, B., & Musen, G. (1993). The structure and organization of memory. *Annual Review of Psychology, 44*, 453–95.
Stanners, R. F., Meunier, G. F., & Headley, D. B. (1969). Reaction time as an index of rehearsal in short-term memory. *Journal of Experimental Psychology, 82*, 566–570.
Starkes, J. L., & Deakin, J. (1984). Perception in sport: A cognitive approach to skilled performance. In W. F. Straub & J. M. Williams (Eds.), *Cognitive sport psychology* (pp. 115–128). Lansing, NY: Sport Science Associates.
Stein, E. S., & Bailey, J. (1994). *The controller memory guide. Concepts from the field* (Rep. No. AD-A289263; DOT/FAA/CT-TN94/28). Atlantic City, NJ: Federal Aviation Administration.
Stein, E. S., & Sollenberger, R. (1995, April 24–27). The search for air traffic controller memory aids. *Proceedings 8th International Symposium on Aviation Psychology*, Columbus, OH, *1*(A96-45198 12-53), 360–363.

Sternberg, S. (1969). The discovery of processing stages: Extension of Donder's method. *Acta Psychologica, 30*, 276–315.

Tulving, E. (1962). Subjective organization in free recall of "unrelated" words. *Psychological Review, 69*, 344–354.

Underwood, B. J. (1957). Interference and forgetting. *Psychological Review, 64*, 49–60.

Vortac, O. U. (1991). *Cognitive functions in the use of flight progress strips: Implications for automation*. Unpublished manuscript, University of Oklahoma, Norman, OK.

Vortac, O. U., Edwards, M. B., Fuller, D. K., & Manning, C. A. (1993). Automation and cognition in air traffic control: An empirical investigation [Special issue]. Practical aspects of memory: The 1994 conference and beyond. *Applied Cognitive Psychology, 7*(7), 631–651.

Vortac, O. U., Edwards, M. B., Fuller, D. K., & Manning, C. A. (1994). *Automation and cognition in air traffic control: An empirical investigation* (Report No. FAA-AM-94-3). Oklahoma City, OK: FAA Office of Aviation Medicine Reports

Vortac, O. U., Edwards, M. B., Fuller, D. K., & Manning, C. A. (1995). *Automation and cognition in air traffic control: An empirical investigation; Final Report* (Rep. No. AD-A291932; DOT/FAA/AM-95/9). Oklahoma City, OK: Federal Aviation Administration.

Vortac, O. U., Edwards, M. B., & Manning, C. A. (1995). *Functions of external cues in prospective memory; Final report* (Report No. DOT/FAA/AM-95/9). Oklahoma City, OK: Civil Aeromedical Institute.

Welford, A. T. (1976). *Skilled performance*. Glenview, IL: Scott, Foresman.

Whitfield, D. (1979). A preliminary study of air traffic controller's picture. *Journal of the Canadian Air Traffic Controller's Association, 11*, 19–28.

Whitfield, D., & Jackson, A. (1982). The air traffic controller's "picture" as an example of a mental model. In G. Johannsen & J. E. Rijnsdorp (Eds.), *Analysis, design, and evaluation of man-machine systems* (pp. 45–52). Dusseldorf, West Germany: International Federation of Automatic Control.

Wickelgren, W. A. (1964). Size of rehearsal group in short-term memory. *Journal of Experimental Psychology, 68*, 413–419.

Wickens, C. D. (1984). *Engineering psychology and human performance*. Columbus, OH: Charles E. Merrill.

Wickens, C. D. (1992). *Engineering psychology and human performance* (2nd ed.). New York: HarperCollins.

Wickens, C. D., & Flach, D. M. (1988). Information processing. In E. L. Wiener & D. C. Nagel (Eds.), *Human factors in aviation* (pp. 111–155). San Diego, CA: Academic Press.

Wickens, C. D., Sandry, D., & Vidulich, M. (1983). Compatibility and resource competition between modalities of input, central processing, and output: Testing a model of complex task performance. *Human Factors, 25*, 227–248.

Wilkins, A. J., & Baddeley, A. D. (1978). Remembering to recall in everyday life: An approach to absent-mindedness. In M. Gruneberg, P. Morris, & R. Sykes (Eds.), *Practical aspects of memory*. London: Academic Press.

Wilson, J. R., & Rutherford, A. (1989). Mental models: Theory and application in human factors. *Human Factors, 31*, 617–634.

Wingfield, A., & Butterworth, B. (1984). Running memory for sentences and parts of sentences: Syntactic parsing as a control function in working memory. In H. Bouma & D. Bouwhuis (Eds.), *Attention and performance* X. Hillsdale, NJ: Lawrence Erlbaum Associates.

Winograd, E. (1988). Some observations on prospective remembering. In M. Gruneberg, P. Morris, & R. Sykes (Eds.), *Practical aspects of memory: Current research and issues* (Vol. 1, pp. 348–353). Chichester: Wiley.

Wise, J. A., & Debons, A. (1987). *Information systems: Failure analysis*. Berlin: Springer-Verlag.

Wise, J. A., Hopkin, V. D., & Smith, M. L. (1991). *Automation and systems issues in air traffic control*. Berlin: Springer-Verlag.

Yntema, D. B. (1963). Keeping track of several things at once. *Human Factors, 6*, 7–17.

Yntema, D. B., & Mueser, G. B. (1960). Remembering the states of a number of variables. *Journal of Experimental Psychology, 60*, 18–22.

Zbrodoff, N. J., & Logan, G. D. (1986). On the autonomy of mental processes: A case study of arithmetic. *Journal of Experimental Psychology: General, 115*, 118–130.

Zeitlin, L. R., & Finkleman, J. M. (1975). Research note: Subsidiary task techniques of digit generation and digit recall as indirect measures of operator loading. *Human Factors, 17*, 218–220.

Zingale, C., Gromelski, S., Ahmed, S. B., & Stein, E. S. (1993). *Influence of individual experience and flight strips on air traffic controller memory/situational awareness* (Rep. No. DOT/FAA/CT-TN93/31). Princeton, NJ: PERI, Inc.

Zingale, C., Gromelski, S., & Stein, E. S. (1992). *Preliminary studies of planning and flight strip use as air traffic controller memory aids* (Rep. No. DOT/FAA/CT-TN92/22). Princeton, NJ: PERI, Inc.

20

Air Traffic Control Automation

V. David Hopkin
Embry-Riddle Aeronautical University

THE NEED FOR AUTOMATION

Throughout most of the world, aviation as an industry is expanding. Far more aircraft are flying now than formerly, and still more are expected in future. Air traffic demands are notoriously difficult to predict as they are always vulnerable to powerful extraneous and unforeseeable influences beyond the control of the aviation community, but all current forecasts concur about a substantial future increase in aircraft numbers and about a continuing variety of aircraft types. As a consequence, air traffic control must seek to accommodate increasing demands for its services. Even the most efficient current air traffic control systems cannot remain as they are, because they were never designed for the quantities of air traffic now expected in the longer term, and they could not readily be adapted to cope with such increases in traffic volume. The combined sequential processes of devising, proving, and introducing major changes in an air traffic control system take several years, but to make no changes is not a practical option. Air traffic control must evolve (Wise, Hopkin, & Smith, 1991).

A relevant parallel development is the major expansion in the quantity and quality of the information available about each aircraft flight. This development has brought significant changes and benefits in the past and will bring further ones in the future, applicable to the planning and the conduct of each flight. The quality and the frequency of updating of the information about the position of each aircraft were enhanced when radar was initially introduced and then became progressively more refined, and further enhancements are expected as information becomes available from satellites, data links, and other innovations (Hopkin, 1989). In principle, it seems that practical technical limitations on data gathering for air traffic control purposes could disappear altogether because whatever information is deemed to be essential for safety or efficiency could be provided.

With limited and finite airspace, the only way to handle more air traffic in regions that are already congested is to allow aircraft to approach each other more closely in safety. Flight plans, navigational data, on-board sensors, prediction aids, and computations can collectively provide very full and frequently updated details about the state and progress of each flight, in relation to other flights or to hazards nearby and to the flight objective. The provision of high-quality information about where each aircraft is and where it is going could allow the minimum separation standards between aircraft to be reduced safely. However, the closer that aircraft are allowed to approach each other, the less is the time available to detect and respond to any emergency, and the fewer are the available options for resolving it (Hopkin, 1995).

An apparent alternative option for handling more traffic would seem to be to employ more controllers and to partition further the region of airspace for which each controller or small team of controllers is currently responsible. Unfortunately, in the regions of densest traffic where the problems of handling more traffic are most acute, this process of partitioning has often been taken already to its beneficial limits. Further partitioning becomes self-defeating and counterproductive wherever the consequent reductions in the controller's workload are outweighed by the extra work generated by partitioning, in such forms as additional coordination, liaison, communications, handovers of responsibility, and reduced experience of each flight and its history. Further partitioning would also be unwelcome in cockpits, where it would lead to extra work through additional reselections of communications frequencies. Generally the further partitioning of airspace is not a practical option. Nor is the loading of more traffic onto controllers while retaining present control methods, because dealing with current heavy traffic keeps controllers continuously busy and they could not handle much more.

The essence of the problem is therefore that each controller must become responsible for more air traffic, but without any diminution of the existing high standards of safety and efficiency in the air traffic control service, and preferably with some enhancement of them. This implies that somehow each controller must spend less time in dealing with each aircraft, with no impairment of standards (International Civil Aviation Organization, 1993). To achieve this, the controller needs help, much of which must come from automation and computer assistance (Hopkin, 1994a). As a first step, the broad human factors implications of foreseeable technical developments have to be deduced (Hopkin, 1997). There is growing awareness that cultural differences may intervene to prevent universally valid solutions to problems (Mouloua & Koonce, 1997).

AUTOMATION AND COMPUTER ASSISTANCE

This chapter covers both automation and computer assistance in relation to the air traffic controller, but does not treat these concepts as synonymous. It seems prudent to specify some distinguishing characteristics of each in relation to human factors (Hopkin, 1995).

Automation refers to functions that do not require, and often do not permit, any direct human participation or intervention in them. The human controller generally remains unaware of them unless special provision has been made to notify the controller of their occurrence or progress. The products of automation may be applied by the human controller, who nevertheless cannot normally influence the processes that lead to those products. Previous applications of automation to air traffic control have mostly been to quite simple functions that are routine, continuous, or frequently repeated. Examples of such automated functions include data gathering and storage, data compilation and correlation, the computation and presentation of summaries of data, the retrieval and updating of data, and data synthesis. Most applications of automation are universal and unselective. Some limited selective automation to accommodate different task requirements has been achieved, and this is expected to become more common. When the selectivity permits human intervention or is adaptive in accordance with the needs of individual controllers, it then constitutes computer assistance.

Computer assistance is characterized by human tasks, roles, and functions that are central in that they are the hub or focus of activities and are supported by the computer. The human controller must retain some means to guide and participate in the processes of computer assistance wherever the controller carries the legal responsibility for air traffic control events. In recent years the concept of human-centered automation has become popular (Billings, 1991). It represents a reaction against forms of computer assistance that have owed more to their technical feasibility than to user requirements, but in air traffic control computer assistance of cognitively complex human functions has always been preferred to their full automation. The latter has sometimes been proposed, as in the prevention of conflicts between flights by automated flight profile adjustments without referral to either the controller or the pilot, but a combination of formidable technical difficulties and problems of legal responsibility has so far prevented its adoption. A defining characteristic of computer assistance is that some human participation is essential in that a process or function cannot be completed without it, although the actual human role may be minimal, for example, to sanction an automated function.

Air traffic control as a whole is therefore still computer assisted rather than automated in relation to those aspects of it that involve active participation by the human controller. Without the presence of a controller there would be no air traffic control in even the most advanced current system, and evolutionary plans for future air traffic control envisage the continuing involvement of the human controller (Federal Aviation Administration, 1995). In this respect air traffic control differs from some other large human–machine systems that can function automatically (Wise, Hopkin, & Stager, 1993a). The concepts of computer assistance are being considered mostly at present for application to such complex human cognitive functions as decision making, problem solving, prediction, planning, scheduling, and the allocation and management of resources (e.g., Vortac, Edwards, Fuller, & Manning, 1993). Much of the computer assistance is selective. It does not respond identically to every individual and to every circumstance, but can adapt to aid differently particular tasks, functions, jobs, and roles of the controller, who often retains the option of dispensing with the computer assistance altogether. A characteristic of computer assistance in air traffic control, although not an inevitable property of it, is that it is intended to aid the individual controller, and comparatively few current or pending forms of computer assistance seek to aid teams of controllers or their supervisors.

In some texts, the concept of automation embraces computer assistance, and different degrees of feasible human intervention are referred to as different levels of automation (Wickens, Mavor, & McGee, 1997). A further practical categorization of air traffic control functions that are not wholly manual is between semi-manual, semi-automated, automated, and fully automated functions (Cardosi & Murphy, 1995).

TECHNOLOGICAL ADVANCES WITH HUMAN FACTORS IMPLICATIONS

Communications. At one time, spoken messages were the main form of communication between the controller and the pilot. These are being replaced by data transponded automatically or on request, which reduce both the need for human speech and the reliance on it. In principle time is saved, although not always in practice (Cardosi, 1993). Other kinds of information contained in speech may be lost, for example, those used to judge the competence or confidence of the speaker.

Radar. Radar provides a plan view of the traffic, and thus shows the lateral and longitudinal separations between aircraft in flight. The permitted separation minima between aircraft within radar coverage can usually be much less than those for aircraft beyond radar coverage, for example, in transoceanic flight. Modern secondary radars supplement the plan view with a label attached to each aircraft's position on the radar screen, showing its identity and aspects of its current state such as its flight level (height), speed, and whether it is climbing, descending, or in level flight. The changing information on the label is updated frequently and automatically, with some significant changes being signalled to the controller.

Navigation Aids. Ground-based aids that can be sensed or interrogated from aircraft mark standard routes between major centers of population. Navigation aids that show the direction from the aircraft of sensors at known positions on the ground permit computations about current aircraft track and heading. Other navigation aids can make comparisons between aircraft, and hence help the controller to maintain safe separations between them. The information available to the controller depends considerably on the navigation aids in use.

Satellites. Data derived from satellites about the location of aircraft represent an impending technical advance that can transform the accuracy of the information available about air traffic. In order to accommodate and benefit from this increased accuracy, human tasks and functions must adapt to it in terms not only of task performance but also of appropriately revised assessments of its trustworthiness and reliability.

Automatically Transponded Data. Such data are not obtained on request or as a machine response to a controller's action but are independent of controller activities. They can replace routine human actions and chores, but may also remove some adaptability and responsiveness to particular controller needs. The controller cannot know or access such data unless special provision for this is made.

Datalinks. These send data continuously or very frequently between the aircraft and the air traffic control system independently of the pilot and the controller, who may be able to tap into them for information or be presented automatically with it. The associated human factors problems are therefore centered on what information derivable from datalinks is needed by controllers under what circumstances, and on its forms of presentation.

Information Displays. Because all the information needed by the controller about the aircraft cannot be presented within the labels on a radar display without their becoming too large and cluttered and inviting label overlap, and because much of the information is not readily adaptable to such forms of presentation, there have always been further kinds of information display in air traffic control, such as maps and tabular displays. In the latter, aircraft can be listed or categorized according to flight level, direction, route, destination, or other criteria appropriate for the controller's tasks. Tabular displays of automatically compiled and updated information can be suitable for presentation as windows in other air traffic control displays.

Electronic Flight Progress Strips. These are a particular kind of tabular display intended to replace paper flight progress strips. Being electronic, they can be generated and updated automatically in ways which paper strips cannot be, and the controller must use keys instead of hand-written annotations to amend them. They have posed some human factors problems of design by revealing difficulties in capturing electronically the full functionality of paper flight strips, which are more complex than they seem (Hughes, Randall, & Shapiro, 1993). Also, the greater flexibility of electronic formats calls for some reappraisal of the desirability of adapting different flight strip formats to meet different controller needs.

Data Input Devices. Through these, the controller enters information into the computer and the system, and initiates events. The chosen input devices must be appropriate for the tasks, the displays, the communications, and the forms of feedback. Technical advances may extend the range of input devices available, for example, by introducing touch-sensitive surfaces and perhaps automated speech recognition, and they raise human factors issues about the respective merits and disadvantages of alternative input devices and their mutual compatibility within a single workspace (Hopkin, 1995).

COMPUTATIONS WITH HUMAN FACTORS IMPLICATIONS

Alerting. A variety of visual or auditory alerting signals can be provided as automated aids. They may serve as memory aids, prompts, or instructions to the controller, or may signify a state of progress or a change of state. Their intention is to draw the controller's attention to particular information or to initiate a response from the controller. They are normally triggered when a predefined set of circumstances actually arises in the course of routine recurring computations within the system. They can be distracting if employed inappropriately.

Track Deviation. To save searching, computations can be made automatically by comparing the intended and the actual track of an aircraft and by signaling to the controller whenever an aircraft deviates by more than a predetermined permissible margin. The controller is then expected to contact the pilot to ascertain the reasons for the deviation and to correct it where appropriate. The significance and degree of urgency of a track deviation depend on the phase of flight. It can be very urgent if it occurs during the final approach to landing.

Conflict Detection. This is also an aid to searching, which can facilitate judgment. Comparisons between aircraft are made frequently and automatically, and the controller's attention is drawn by changing the coding of any displayed aircraft that are predicted to infringe on the separation standards between them within a given time or distance. Depending on the quality of the data about the aircraft, a balance is struck to give as much forewarning as possible without incurring too many false alarms.

The practical value of the aid relies on computational correctness and on getting this balance right. Sometimes the position or the time of occurrence of the anticipated conflict are depicted, but a conflict detection aid provides no further information about it.

Conflict Resolution. This aid takes conflict detection a stage further. The data and computations used to identify a conflict can be related to data on other traffic in order to compute and present automatically one or more solutions to the conflict that meet all predefined criteria and rules. If more than one computed solution could be offered to the controller, the order of computer preference usually follows the same set of rules. Nominally the choice remains with the controller, who can still devise and implement his or her own solution, but controllers are trained and expected to accept the preferred computer solution in normal circumstances. It can be difficult for the controller who imposes a preferred human solution to ascertain all the factors included in the automated one, yet this is necessary either if the automation has taken account of information unknown to the controller or if the controller possesses information that is unavailable to the computer but invalidates its solution.

Computer-Assisted Approach Sequencing and Ghosting. These forms of assistance apply to flows of air traffic approaching an airfield from diverse directions, and amalgamating into a single flow approaching one runway or into parallel flows approaching two or more parallel runways. Computer-assisted approach sequencing depicts on a display the predicted flight paths of arriving aircraft. It shows directly, or permits extrapolation of, their expected order of arrival at the amalgamation position and the gaps that there will be between consecutive aircraft when they arrive there. The controller can issue instructions for minor flight path or speed changes in order to adjust and smooth gap sizes, and particularly to ensure that the minimum vortex separation standards applicable during final approach are met. Ghosting fulfills a similar function in a different way. The label of each aircraft appears normally on the radar at its position on its actual route, but it appears also as a ghost with much reduced brightness contrast in the positions on the other converging routes that it would occupy on them to arrive at the amalgamation position at the same time (Mundra, 1989). In effect, each route depicts all the traffic before route convergence as if convergence had already occurred, showing what the gaps between consecutive aircraft will be and permitting adjustment and smoothing of any gaps before the traffic streams actually merge.

Flows and Slots. Various schemes have evolved that treat aircraft as items in traffic flows and that exercise control by assigning slots and slot times to each aircraft in the flow. Separations can then be dealt with by reference to the slots. The maximum traffic handling capacities of flows can be utilized, and tactical adjustments can be minimized in the initial slot allocation by allowing for the intersection or amalgamation of traffic flows.

Traffic Flow Management. Although flow management as a concept may refer to a system that includes flows and slots, it is usually applied to larger traffic regions and to the broad procedures that prevent excessive general congestion by limiting the total amount of traffic, by diverting flows, or by imposing quotas on departures, rather than by tactically maneuvering single aircraft. Traffic flow management is more strategic than air traffic control, and precedes it. The role of traffic flow management is expected to increase (Duytschaever, 1993; Harwood, 1993)

Free Flight. The principle relies heavily on automation because most current systems do not give the controller access to all the data needed to confirm the computer calculations. Variants of free flight principles are sometimes called random routing or direct routing. The intentions are for the pilot to specify departure and arrival times, departure and arrival airports, and preferred route and flight profile, and for the computer to check and confirm that the proposed flight will not incur conflicts with other known flights and that major delays at the arrival airport will not be encountered. If all is well, the flight would be sanctioned, perhaps automatically. The flight is independent of the air traffic control route structure, and, weather permitting, would normally follow the shortest direct route, a segment of a great circle. The controllers would verify its continued safety, would deal with discrepancies between its actual and planned track, and might introduce minor modifications to enhance system capacity. Free flight could be more efficient by saving time and fuel and by allowing the optimum flight profile to be flown if known. It represents a reversion to more tactical control procedures dealing with single aircraft, at a time when most policies favor more strategic air traffic control procedures dealing with flows of traffic. It requires accurate and up-to-date information. It poses numerous human factors issues, not only of practicality and roles, but of information access and legal responsibility. Extensive planning and research effort is currently being devoted to free flight and its implications, which are not identical everywhere but depend partly on other factors such as route structures, traffic mixes and densities, and typical flight durations.

Associated Legal and International Requirements. Extensive human factors knowledge can be applied to air traffic control (International Civil Aviation Organization, 1993), and many technological advances can be matched successfully with human capabilities and limitations to further the performance of air traffic control tasks (Wise et al., 1991). Nevertheless, some practical constraints reflect the legal requirements of air traffic control or agreed international practices, procedures, and rules that must be followed in air traffic control, including the format and language of spoken communications.

Consequences for Responsibilities. When computer assistance to aid the controller is introduced, it is vital that all the responsibilities of the controller can be met fully using the facilities provided. This is a practical constraint not only on the forms of computer assistance supplied but also on their acceptability to controllers (Hopkin, 1995).

OPTIONS FOR HELPING THE CONTROLLER

The primary objective of all forms of computer assistance provided for the air traffic controller is to aid the controller's task performance (Hopkin, 1994a). The most favored forms of assistance seek to further this objective by enabling the controller to handle more traffic. The following are the main broad ways in which the controller can be assisted.

Full Automation of Functions. In this option, some functions and tasks are removed altogether from the controller. It applies particularly to the frequent and routine gathering, storage, transfer, and manipulation of data, all of which have often been automated extensively as functions, so that in many modern air traffic control systems controllers no longer spend much time on such tasks. An example is the provision of the identity of each aircraft in its label on the radar display.

Improvements to the Quality of the Data. These can be achieved in several ways. The data can become more accurate, consistent, reliable, precise, valid, trustworthy, or acceptable, for example. It is also necessary for the controller to know how much better the data are. Such knowledge can arise through training, learning from experience, or some displayed indication. A main purpose of this form of computer or machine assistance is to change the controller's behavior so that it is appropriate in relation to the actual quality of the data presented.

Reductions in the Time Needed. This option reduces the time required by the controller to perform particular functions or tasks. Several means to achieve this objective are available. The required time can be shortened by performing specific tasks or functions in less detail, less often, in condensed form, in simplified form, with less information, with fewer actions, with some parts omitted, or with data compiled more succinctly. All of these means are applied in air traffic control, with the choice depending on the forms of computer assistance that can be provided and on the ways in which they can be matched appropriately with human capabilities and limitations for the effective performance of the various tasks and functions.

Treating Aircraft as Traffic Flows. To control air traffic as flows rather than as single aircraft normally involves a change from tactical to strategic air traffic control. Among the consequences seem to be fewer human interventions, less criticality in their timing because they may often be brought forward at the behest of the controller, and more emphasis on the prevention rather than the solution of problems. Although some believe that air traffic control must evolve in this way, most forms of computer assistance that are current or soon pending are primarily tactical and applicable to single aircraft, and aids for the control of flows are generally in quite early stages of development, still lacking such fundamental human factors contributions as satisfactory codings to differentiate between the main defining parameters of traffic flows.

Sharing Human Functions With Machines. In this option, machines fulfill some aspects of functions, or they help, prompt, guide, or direct the human. In the most popular initial variant of this option, which seems attractive but actually does not work very well, the machine does much of the work and the human monitors the machine's performance. Unfortunately, the human finds it difficult to maintain concentration indefinitely in a passive role with nothing active to do. This same passivity can incur some loss of information processing and understanding, which may be tolerable in normal circumstances but becomes a liability in nonstandard ones. The introduction of any major form of computer assistance that affects the controller's tasks will change the controller's situational awareness (Garland & Hopkin, 1994) and require rematching of the human and machine databases. The human controller relies greatly on a detailed mental picture of the air traffic, which active task performance and manipulation of data help to sustain. Any forms of computer assistance that interfere with these processes may result in reported loss of clarity or detail in the controller's mental picture of the traffic.

Expanding the Range of Machine Support. A more profitable approach with longer term benefits is to employ machines to support human activities and to apply technical developments to extend and expand the forms of machine support offered, with increasing degrees of complexity. Monitoring, for which machines are generally well suited, becomes a machine rather than a human role. The machine, which gathers, stores and compiles information automatically, also collates, summarizes, selects, and presents it automatically, with timing and ordering appropriate for the tasks. It thus functions as a memory aid and prompt. Given high-quality data, machines can often make better predictions than humans, so that controllers can use them to ascertain the consequences of proposed actions before implementing them. The machine can offer solutions to problems and can aid decision making. The controller, before accepting or rejecting such assistance, needs to be able to discover which information the machine has or has not taken into account. The machine can apply far more information far more quickly than the human can. If a machine generates a preferred solution to a problem, it seems a small technical step for it to recommend that solution for human acceptance, another small technical step for it to implement that solution automatically unless the notifed controller chooses to intervene, a further small step to implement the solution automatically and then notify the controller afterward, and a final small step not to notify the controller at all. However, in human factors terms, and in legal terms, these are all big steps, the last being full automation.

A CLASSIFICATION OF HUMAN–MACHINE RELATIONSHIPS

Relationships that are feasible or will become feasible in air traffic control are listed next, to assist the recognition and categorization of those that actually occur and their comparison with alternative relationships that could apply or that would result from proposed changes (Hopkin, 1995). Relationships are listed in the approximate order in which they became or will become technically feasible. The list of relationships is expanding because technological innovations introduce new options from time to time without invalidating any of the options that already exist. The main possible human–machine relationships include the following:

1. The human adapts to the machine.
2. The human and the machine compete for functions.
3. The human is replaced by the machine.
4. The human complements the machine.
5. The human supports failed machines.
6. The human is adapted to by the machine.
7. The human and the machine function as hybrids.
8. The human and the machine function symbiotically.
9. The human and the machine duplicate functions in parallel.
10. The human and the machine are mutually adaptive to each other.

11. The human and the machine are functionally interchangeable.
12. The human and the machine have fluid and not fixed relationships.
13. The human and the machine form a virtual air traffic control world.
14. The machine takes over in the event of human incapacitation.

These are relationships. They exclude the extremes, where there is no machine but only the human and where there is no human but only the machine, neither of which seems likely in air traffic control for the foreseeable future. The corollaries of the choice of human–machine relationships, particularly in terms of options excluded and associated decisions taken, are not always realized at the time (Hopkin, 1988a).

RELEVANT HUMAN ATTRIBUTES

Some particular human characteristics with no machine equivalent have to be emphasized in relation to computer assistance, for otherwise they are not likely to feature in human–machine considerations, comparisons, or allocations of functions, as their relevance remains unrecognized, no matter how important they may actually be. Many of these characteristics are now more widely acknowledged, although sometimes they still have insufficient influence.

Common Human Attributes Related to the Workspace. Some attributes have been widely studied. Although human workload can be difficult to measure, there are always some limitations on human workload capacity that computers do not share (Costa, 1993). Human errors can be classified into distinctive categories that may be differentiated from typical machine errors and may be partly predictable because the kinds of human error that are possible often stem from design decisions about the tasks and equipment (Reason, 1993; Reason & Zapf, 1994). Human beings become tired, experience fatigue (Mital & Kumar, 1994), and need breaks and rosters and adjustments of circadian rhythms (Costa, 1991) in ways in which machines do not. Humans have social, recreational and sleep needs that rostering and work rest cycles must accommodate (Hopkin, 1982). Excessive demands may induce stress in humans, although stress and workload are complex concepts both in their causality and in their measurement (Hopkin, 1980a; Tattersall, Farmer, & Belyavin, 1991). Insufficient demands can induce boredom, the causes and consequences of which have been neglected recently in relation to air traffic control research, although most commonsense assumptions about boredom appear to be unsupported and its effects on safety are particularly obscure (Hopkin, 1980b). The optimum physical environmental conditions of the controller, in terms of heating, lighting, acoustics, airflow, temperature, humidity, radiation, and appearance, may not accord with those of the computer, but they must be met (Hopkin, 1995). The machine must adapt to human anthropometric characteristics that determine recommended reach and viewing distances and the requirements for human comfort and health (Pheasant, 1986). If computer assistance is cost-effective, manual reversion in the event of its failure must usually entail some loss of efficiency, because safety must never be sacrificed. However, the feasibility of manual reversion in the event of machine failure has too often been studied as a one-way process instead of a two-way process, to the neglect of the real difficulty that after manual reversion everyone may become too busy running the system in manual mode to spare any time to reload the machine with data after it has been repaired, prior to switching back to it.

The Context of Work. For the human, work has rewarding properties in its own right. Decisions about the nature and conditions of work affect job satisfaction, which is influenced by the level of autonomy delegated to the individual controller, by the opportunities to develop and apply particular skills, and by responsibilities and the means of exercising them, all of which can be changed greatly by computer assistance, sometimes inadvertently. Team roles and structures may also change, often because the computer assistance aids the individual controller but is introduced into contexts where much of the work has been done by teams where each controller's actions are readily observable by others in the team. This observability is crucial to the successful development of professional norms and standards, which are strong motivating forces in air traffic control, contributing to its professional ethos, morale, and camaraderie. These forces also imply the continued presence of some residual individual differences between controllers, so that controllers can to some extent develop individual styles, which are used by colleagues and supervisors to judge how good they are as controllers. These judgments in turn influence decisions on training, promotions, and career development. The exploration and understanding of the full consequences of computer assistance in air traffic control in terms of observability and the effects of its loss are becoming urgent (Hopkin, 1995).

Attitudes. To the controller, computer assistance must not only be effective, supportive, and safe; it must be acceptable. The effects of the introduction of computer assistance on attitudes have been comparatively neglected in air traffic control. Attitude formation covers the ways in which attitudes are formed and the influences on them, and the extent to which the formation of attitudes can be anticipated and therefore could be controlled and preplanned (Rajecki, 1990). Much is known, for example, from advertising and marketing studies, about how to influence and manipulate attitudes, but this knowledge has rarely been applied in air traffic control, and there is an ethical issue of whether it should be. Selective technical means to apply it have been established (Crawley, Spurgeon, & Whitfield, 1980). What would be entailed is the deliberate application to air traffic control of current evidence about the characteristics of equipment items and forms of computer assistance that improve their user acceptability, including how they work, how they feel, and how they look. Attitudes toward any change, whether favorable or not, are formed quickly, and the methods employed in the initial introduction of any change and in the earliest stages of training can therefore be crucial. Attitudes, once formed, become resistant to contrary evidence. If controllers' initial attitudes toward computer assistance are favorable, they will strive to make it work and to get the best from it, but if their initial attitudes toward the assistance are unfavorable they may become adept at demonstrating how unhelpful the computer assistance can be. The kinds of factor that can influence the acceptability of computer assistance are now becoming clearer. Among them are its effects on responsibility, on the development and applicability of skills, on job satisfaction, and on the challenge and interest of the work. Controllers generally like being controllers, and their attitudes to the work itself are often more favorable than their attitudes toward their working conditions. Perhaps some of the research effort devoted to optimizing the ergonomic aspects of the system might profitably be re-deployed on studies of attitude formation, because the evidence available from existing sources should usually be sufficient to prevent serious ergonomic errors, but without positive attitudes the computer assistance could be ergonomically optimum yet still be unacceptable to its users.

Degree of Caution. If controllers like a form of computer assistance, they can become too enthusiastic about it and too reliant on it. A positive aspect of overenthusiasm can be a dedicated effort to make the computer assistance function as effectively as possible, although controllers may try to extend its usage to assist tasks that lie beyond the original design intentions and for which its usage has not been legally sanctioned. Controllers may welcome computer assistance because it is superior to poor equipment that it replaces, but if the computer assistance is not in fact very good their positive attitudes toward the change may disguise and discourage the need for further practical improvements. Favorable attitudes may accrue for reasons that can include enhanced status or increased attractiveness, as in the case of color coding, and they can induce strong beliefs in the benefits of the system even where none can be demonstrated by objective measures of performance or safety. More sensitive and appropriate measures of the benefits of positive attitudes may relate to fewer complaints, improved collaboration, increased motivation and job satisfaction, and lower absenteeism and job attrition rates. Controllers are renowned for some skepticism toward novel forms of computer assistance. Many recall earlier forms of computer assistance, claimed to be helpful in reducing workload, which in practice had limited value and some of which actually added to the work. Controllers have learned to require more tangible evidence of promised benefits, in such forms as prior demonstration or proof.

Disagreements Between Measured Human Attributes. One justification for the employment of several measures to test the efficacy of computer assistance is the potential lack of agreement between them. This does not necessarily imply that one or other of the measures must be wrong, but the different kinds of evidence on which they draw may be contradictory. A common example in air traffic control and elsewhere concerns the replacement of monochrome coding by the color coding of displayed information. Objective measures of task performance often reveal far fewer beneficial effects of color than the glowing subjective measures would lead one to expect. The tendency to dismiss one kind of measure as spurious must be resisted. Neither kind of measure is ever complete and fully comprehensive. More probable explanations are that the performance measures fail to cover some aspects of the tasks most affected subjectively by the color coding, and that the subjective measures tap genuine benefits that are not measured objectively. Examples of the latter could include color coding as an aid to memory, as an entrenchment of understanding, and as a means to structure and de-clutter visual information, none of which might influence the chosen set of objective measures directly (Hopkin, 1994b; Reynolds, 1994).

Function Descriptions. Functions that seem identical or very similar when expressed in systems concepts can be quite different when expressed in human terms. Attempts to introduce computer assistance for human air traffic control tasks often reveal that their degree of complexity has been underestimated. The human controller who takes a decision without assistance chooses what evidence to seek, gathers it, applies rules and experience to it, reaches a decision, implements the decision, and fulfills all these functions actively, and is therefore well placed to judge whether any given change in the evidence warrants reexamination of the decision. When the computer presents a decision for the controller to accept or reject, this looks similar functionally and when described in systems concepts, but it is not. The controller does not know, and often cannot readily discover, what evidence has been taken into account in the computer decisions, whether it is correct, and what new information would invalidate it. Because the controller needs to process far less information to accept a computer decision than to reach the same decision without computer assistance, the assisted controller tends to understand and recall less about the decison and its circumstances (Narborough-Hall, 1987). In human terms, the processes of human and computer-assisted decision making often cannot be equated.

HUMAN FACTORS IMPLICATIONS OF AUTOMATION AND COMPUTER ASSISTANCE

Interface Designs. Many of the commonest human factors problems that result from automation and computer assistance in air traffic control occur under definable sets of circumstances. Making these explicit clarifies the origins of the problems, reveals the prevailing constraints, and suggests practical solutions. One of the most familar human factors problems arises when a function must be performed somehow but no machine can be devised to fulfill it. Tasks may be assigned to humans not because they do them well but because no machine can do them at all. As technology advances, this problem recedes, but in modern systems the human controller can do only what the computer allows the human to do. If the specification of the human–machine interface makes no provision for an action, it cannot be implemented no matter how correct it may be, and human attempts to implement it are liable to be ruled by the machine as invalid actions. The effectiveness of computer assistance is crucially dependent on human–machine interface designs that enable human roles to be performed. For example, the human cannot be flexible unless the human–machine interface permits human flexibility (Hopkin, 1991a).

Attributes of Speech. An artifact of computer assistance is that many dialogs formerly conducted by the controller directly with pilots or with other controllers have to be conducted through the human–machine interface so that the information in the system is updated and can be applied to automated computations. Much of the further information incidentally contained in speech, on the basis of which pilots and controllers made judgments about each other, becomes no longer available (Hopkin, 1982). Few formal studies, and no recent ones, have ever been made to quantify and describe the nature and extent of the influence of this further information on the actions of the controller and on the conduct of air traffic control, but its influence was never negligible and could become substantial, and the human factors implications of its absence could be very significant. The reduction in spoken human communications in air traffic control implies some loss of information gleaned from such attributes of speech as accents, pace, pauses, hesitancies, repetitions, acknowledgments, misunderstandings, degree of formality, standardization, courtesies, message formats, and the sequencing of message items. The basis for judgments of the speaker's confidence, competence, and professionalism is curtailed, yet these judgments may be important, particularly in emergencies. On the other hand, it is conceivable that these judgments are so often wrong that the system is safer without them. All the categories of potentially useful information lost when speech is removed should be identified, not to perpetuate speech but to determine whether surrogates are needed and what forms they could take.

Computer-Generated Workload. The basic objective of enabling the controller to deal with more aircraft implies that a criterion for the adoption of any form of computer assistance must be that it results in less work for the controller. Some previous forms of computer assistance have violated this criterion, especially when messages that are spoken also have had to be entered as data into the system, or when quite simple and standard tasks have required cumbersome keying procedures. Such forms of computer assistance not only negate their main purpose but are unpopular and can lead to counterproductive attitudes toward computer assistance in general.

Cognitive Consequences. In retrospect, some of the main cognitive consequences of various initial forms of computer assistance in air traffic control were insufficiently recognized, although they are becoming more familiar now. These consequences can be allowed for, either by designing the other human and machine functions so that certain penalties are acceptable or by redesigning the computer-assisted tasks to avoid such effects, perhaps by keeping the controller more closely involved in the control loops than the computer assistance strictly requires. The crucial influence of human cognitive functioning and information processing on the successful matching of human and machine has recently received the attention it deserves (Cardosi & Murphy, 1995; Wickens et al., 1997). Because the application of computer assistance has consistently revealed that many of the human functions that it is intended to replace or supplement are much more complex than they seem superficially to be, it has proved difficult to capture their full functionality in many forms of computer assistance in their stead. An example of more complex cognitive effects than those at first anticipated concerns the replacement of paper flight progress strips with electronic strips (Hopkin, 1991b; Vortac, Edwards, Jones, Manning, & Rotter, 1993). The paper strips seemed quite straightforward until the full complexity of their functionality became recognized. Most aspects of task performance with paper strips, of their manipulation, and of their updating are relatively easy to capture electronically, but a strip is a talisman, an emblem, a history, a record, and a separate object. Active writing on strips, annotation of them, cocking them sideways, and placement of them in relation to other strips on the board all help understanding, memory, and the building of the controller's picture. Strips collectively denote current activities and future workload and are observable and accessible to colleagues and supervisors. These and further aspects have proved much more difficult to capture fully in electronic flight strips. A recurring human factors issue is to identify which functions of paper flight strips can and should be retained electronically, which can be discarded and not replaced, and which cannot be perpetuated in electronic form yet must be retained in an alternative form.

Rules and Objectives. The preceding point about air traffic control functions also applies to some rules, which can seem quite simple until they have to be written as software, at which point they begin to look complex. There may be many exceptions to them, and considerable complexity concerning the circumstances under which one rule overrides another. The objectives of air traffic control are multiple. Not only must it be safe, orderly, and expeditious, but also cost-effective, noise abating, fuel conserving, and job satisfying, responsive to the needs of its customers while never harming the well-being of those employed in it. With so many objectives, there is much scope for their mutual incompatibility, which the rules and the relative weighting of rules attempt to resolve at the cost of some complexity.

Observability. Most forms of computer assistance have the incidental but unplanned consequence of rendering the work of the controller much less observable by others, including immediate colleagues and supervisors. Air traffic control as a team activity relies heavily on tacit understanding between controllers. Each member of the team builds expectations about the activities of colleagues and learns to rely on them. Where the activities of colleagues can no longer be observed in detail, such reliance and trust become initially more difficult to develop and ultimately impossible to build in the same way. For example, a colleague may have difficulty in detecting whether a controller has accepted or rejected a computer solution of a problem, because acceptance and rejection may both involve similar key pressings. General key-pressing activity may remain observable, but not which particular keys have been pressed. Loss of observability can make it more difficult for controllers to appreciate the skills of colleagues, to acquire new skills by observation, and to demonstrate their own accomplishments to others. A complicating factor can be reduced flexibility in nonstandard circumstances, because the options provided within the menus and dialogs available through the human–machine interface are preset.

Concealment of Human Inadequacy. Many of the forms of support that computer assistance can provide have the inherent capability of compensating for human weaknesses, to the extent that they can disguise human incompetence and conceal human inadequacy. This can become very serious if it is compounded by controllers' inability to observe closely what their colleagues are doing. If a controller always accepts computed solutions to problems, this may indeed utilize the computer assistance most beneficially, but it is impossible for others to tell from that controller's activities whether or not the controller has understood the solutions that have been accepted and their consequences, whereas in a more manual system with less computer assistance and more observability it is impossible for a controller to disguise such lack of knowledge from colleagues indefinitely. This is not an imputation on the professionalism of controllers, of which they are rightly proud. It is a statement that important safeguards that are present now may be taken too much for granted, and could inadvertently be removed by future changes introduced for other reasons.

Stress. Although there is much evidence that the problems have been exaggerated, human stress has been claimed for a long time to be associated with air traffic control, which has acquired a spurious reputation as a particularly stressful occupation (Melton, 1982). Initially stress was usually attributed to time pressures and excessive workload, coupled with responsibilities without proper means to exercise them. Computer assistance can introduce its own forms of human stress, if the controller must rely on machine assistance that is not fully trusted, must utilize forms of assistance that function too complexly to be checked, or must fulfill functions that are incompletely understood but that the controller has no power to change.

Team Roles. Computer assistance in air traffic control often changes many of the traditional roles and functions of the team, some of which may disappear altogether. This can be acceptable, provided that the full functionality of teams has been defined beforehand, so that team roles are not removed inadvertently by other events and their diminution does not come as an unwelcome surprise. The neglect of teams has been twofold. Most current and planned forms of computer assistance for air traffic control are not designed for teams, and most forms of computer assistance designed for teams in other work contexts are not being proposed for air traffic control application. Teams have many functions. They include the building and maintenance of tacit understandings, of adaptability to colleagues, and of local agreed air traffic control practices. Through team mechanisms, controllers gain and keep the trust and respect of their peers, which depend on the need for practical interactions between team members and on sufficient mutual observability of activities within the team. Computer assistance may render some of the traditional roles of the air traffic control team supervisor impractical. The future roles of supervision need planning according to known policies, and should not change by default.

Coping With Machine Failure. The functioning of many forms of computer assistance is not transparent to the controllers who use them. In particular, it is not apparent how they could fail, how they would appear if they did, or how it would be possible for the controller to discover which functions remained unaffected. This is a crucial aspect of successful human–machine matching within the system. For many kinds of failure of computer assistance, no provision has been made to inform the user that the system is not functioning normally. The controller is not concerned with the minutiae of the reasons for failure because it is not the controller's job to remedy it, but the controller does need to know the existence and ramifications of any significant failure, and especially which facilities can still be used normally.

Controller Assessment. In manual air traffic control systems, the concept of the "good controller" is familiar. The criteria for this judgment have proved elusive, but there is usually quite high consensus among colleagues familiar with their work about who the best controllers are. Computer assistance has implications for the development of this concept of the good controller because it may restrict the judgments on which it can be based. Lack of observability can make decisions about careers, promotions, and retraining seem more arbitrary. Who would be the best controller using computer assistance—one who always accepts it, one who overrides it in ways that are predictable because they are rigid, one who overrides it frequently, or one who overrides it selectively but to the occasional discomfiture of colleagues? What criteria for promotion would be acceptable as fair when controllers are expected generally to adopt the forms of computer assistance provided?

Other Air Traffic Control Personnel. There has been a long-term imbalance within air traffic control concerning the impact of automation and computer assistance on those who work within it. Almost all of the limited human factors attention has been concentrated on the air traffic controller as an individual, to the neglect of supervisors, assistants, teams and their roles and functions, and of technical and maintenance staffs. Sooner or later this imbalance will have to be redressed, but not by neglecting the controller.

IMPLICATIONS FOR SELECTION AND TRAINING

As the introduction of computer assistance progresses, at some point questions about the continuing validity of the selection procedures for controllers are bound to arise (Della Rocco, Manning, & Wing, 1991). The simplest issue is whether an ability to work well with the new forms of computer assistance should become an additional measured requirement in the selection procedure. A more complex question is whether some of the abilities for which controllers have been selected in the past no longer remain sufficiently relevant to current or future air traffic control jobs to justify their retention in selection. Issues concerning the circumstances under which it becomes necessary to adapt selection procedures are quite difficult to resolve, and the criteria for deciding when, how, and at what level intervention in these procedures becomes essential are ill-defined. Many current controllers were not originally selected for jobs like those now envisaged for them. It is not clear how far training could compensate for this discrepancy, or how far selecting different kinds of people must constitute the ultimate remedy. Perhaps modifications of the chosen forms of computer assistance could obviate many of the adjustments in selection procedures that might otherwise be needed.

Both automation and computer assistance entail some changes in what the controller needs to know. The controller's professional knowledge, much of which is gained initially through training, must match the facilities provided. The controller's training in relation to computer assistance therefore has to cover how it is designed to be used, how the human and the machine are intended to match each other, what the controller is expected to do, and what the controller needs to understand about the functioning of the computer assistance in order to work in harmony with it. Effective computer assistance also entails considerable practical training in how to access data, interrogate the computer, manipulate information, use menus and conduct dialogs, and learn all the options available through the human–machine interface. The taught procedures and instructions may have to be revised to realign the controller's actions with the computer assistance. The controller may need some human factors knowledge (Hunt, 1997). The distinction between what the controller is taught and what the controller learns may warrant reappraisal, the former referring to training content and the latter to on-the-job experience (Hopkin, 1994c).

Any changes in the machine database that affect the computer assistance of the controller always require some restoration of the optimum match between human and machine, in the form of corresponding changes in the human database that consists of the controller's knowledge, experience, skills, and professionalism. Changes may be needed to rematch the controller's situational awareness and mental picture of the traffic with the system (Mogford, 1994), taking account of the revised imagery that may have become more appropriate for the controller in the computer-assisted system (Isaac, 1994). These rematching processes begin with retraining, which obviously must accomplish the new learning required but less obviously may require the discarding of old knowledge and skills, now rendered inapplicable but thoroughly familiar through years of experience and practical application. Much less is known about how to train controllers to forget the old and irrelevant than about how to train them to learn the new, but a potential hazard, particularly under stress or high workload, is reversion to familar former habits and practices that do not apply any more. If this can happen, it must not be dangerous. Although most emphasis is on new learning, some of the most urgent practical problems concern how to make forgetting safe (Hopkin, 1988b).

Much effort is expended to ensure that all forms of computer assistance in air traffic control are safe, efficient, practical, and acceptable, but they must also be teachable. It is not sufficient for those who devise or prove a new form of computer assistance to demonstrate its successful functioning. Practical and cost-effective means must be devised to teach the new form of computer assistance to the whole workforce for whom it is intended. Learning to use it should not be laborious because that would prejudice its acceptability and raise training costs, and any difficulties in understanding its functioning will lead to its misinterpretation and misuse if they are not resolved. Training with computer assistance should always include appropriate forms of team training, work scheduling, and resource management, so that the performance of tasks with computer assistance fits snugly within all other facets of the job.

Training relies extensively on real-time simulation, as does much human factors research in air traffic control. Although real-time simulation is an essential tool, it is not a sufficient one for every purpose. A comprehensive list has been compiled of actual human factors applications in operational evaluations, many of which also cover real-time simulations. The listing distinguishes between applications that are valid, applications where simulation may be helpful if supported by external confirmatory evidence, and applications for which it is inherently inappropriate as a technique and should not be used (Hopkin, 1990). This listing is subject to modification in the light of further experience, and shares its uncertain validity with many other human factors activities concerning air traffic control. Originally validation of findings was considered essential, and it is still common in some activities such as selection procedures, but recent texts on the validation of human factors recommendations for air traffic control systems have revealed the extent of the uncertainties (Wise, Hopkin, & Stager, 1993a, 1993b), and point to the increasing difficulty of deriving independent validation criteria for human factors recommendations as systems increase in their complexity and integrality. Possible approaches are to integrate validation techniques into design processes, and to adapt certification procedures as validation tools (Wise, Hopkin, & Garland, 1994). Methods for introducing more human factors contributions into certification processes are being examined (Wise & Hopkin, 1997).

THE FUTURE

Automation and computer assistance for much current air traffic control are still confined to quite routine human functions, but their envisaged future forms will affect many cognitive functions of the controller and could change the controller's job greatly. This means that air traffic control is well placed to profit from the experience of others in contexts where computer assistance has already been applied more extensively. However, in regard to computer assistance there is a prevailing impression of responding to external events as further technical options become practicable, rather than of the implementation of broad and principled policies about what the forms of computer assistance and the resultant human roles ought to be in air traffic control.

Many attributes treated hitherto as the exclusive prerogative of humans are becoming machine attributes also. These include intelligence, adaptability, flexibility, and a capacity to innovate. The rules, insofar as they exist, about the optimum matching of human and machine when both possess these attributes are not yet firm enough to be applied now to air traffic control, although some of the issues, such as the roles of adaptive machines, are now being addressed (Mouloua & Koonce, 1997). If computer assistance reduces workload as it is intended to do, so that the controller is driven less by immediate task demands and gains more control over workload and its scheduling, perhaps excessive controller workload would occur in the future only if it was self-inflicted, because high workload could always be prevented if the controller employed the computer assistance properly. Very high workload would then signify that the controller needed further training. It would also be expected that much of the workload would become strategic rather than tactical, unless free flight became widespread.

It will become more important to understand the origins of human acceptance of computer assistance and of satisfaction with it. An incidental consequence of more widespread computer assistance could be to make air traffic control more similar to many other jobs, because its primary knowledge and skills relate more to the manipulation of a human–machine interface than to its particular application in air traffic control. Currently, most knowledge and skill as an air traffic controller do not transfer directly to other jobs. This may not remain true. Those employers who provide the best conditions of employment, the greatest satisfaction of human needs and aspirations in the workplace, and the forms of computer assistance that match human needs and responsibilities best will then attract the best applicants to their jobs, have the lowest job attrition rates, incur the lowest selection and training costs, and employ a workforce that is justifiably proud of its achievements. Such a development would expand further the human factors objectives in air traffic control.

REFERENCES

Billings, C. E. (1991). *Human-centered aircraft automation: A concept and guidelines.* (Report No. NASA TM 10385). Moffett Field, CA: NASA Ames Research Center.

Cardosi, K. M. (1993). Time required for transmission of time-critical ATC messages in an en-route environment. *International Journal of Aviation Psychology, 3*(4), 303–313.

Cardosi, K. M., & Murphy, E. D. (Eds.). (1995). *Human factors in the design and evaluation of air traffic control systems.* (Rep. No. DOT/FAA/RD-95/3). Washington, DC: Federal Aviation Administration, Office of Aviation Research.

Costa, G. (1991). Shiftwork and circadian variations of vigilance and performance. In J. A. Wise, V. D. Hopkin, & M. L. Smith (Eds.), *Automation and systems issues in air traffic control* (pp. 267–280). Berlin: Springer-Verlag, NATO ASI Series Vol. F73.

Costa, G. (1993). Evaluation of workload in air traffic controllers. *Ergonomics, 36*(9), 1111–1120.

Crawley, R., Spurgeon, P., & Whitfield, D. (1980). *Air traffic controller reactions to computer assistance: A methodology for investigating controllers' motivations and satisfactions in the present system as a basis for system design.* Birmingham: University of Aston Applied Psychology Department Report 94 (3 Vols.).

Della Rocco, P., Manning, C. A., & Wing, H. (1991). Selection of air traffic controllers for automated systems: Applications from today's research. In J. A. Wise, V. D. Hopkin, & M. L. Smith (Eds.), *Automation and systems issues in air traffic control* (pp. 429–451). Berlin: Springer-Verlag, NATO ASI Series Vol. F73.

Duytschaever, D. (1993). The development and implementation of the EUROCONTROL central air traffic control management unit. *Journal of Navigation, 46*(3), 343–352.

Federal Aviation Administration. (1995). *National plan for civil aviation human factors: An initiative for research and application.* Washington, DC: Author.

Garland, D. J., & Hopkin, V. D. (1994). Controlling automation in future air traffic control: The impact on situational awareness. In R. D. Gilson, D. J. Garland, & J. M. Koonce (Eds.), *Situational awareness in complex systems* (pp. 179–197). Daytona Beach, FL: Embry-Riddle Aeronautical University Press.

Harwood, K. (1993). Defining human-centered system issues for verifying and validating air traffic control systems. In J. A. Wise, V. D. Hopkin, & P. Stager (Eds.), *Verification and validation of complex systems: Human factors issues* (pp. 115–129). Berlin: Springer-Verlag, NATO ASI Series Vol. F110.

Hopkin, V. D. (1980a). The measurement of the air traffic controller. *Human Factors, 22*(5), 547–560.

Hopkin, V. D. (1980b). Boredom. *The Controller, 19*(1), 6–10.

Hopkin, V. D. (1982). *Human factors in air traffic control.* AGARDograph No. 275. Paris: NATO.

Hopkin, V. D. (1988a). Air traffic control. In E. L. Wiener & D. C. Nagel (Eds.), *Human factors in aviation* (pp. 639–663). San Diego, CA: Academic Press.

Hopkin, V. D. (1988b). Training implications of technological advances in air traffic control. In *Proceedings of Symposium on Air Traffic Control Training for Tomorrow's Technology* (pp. 6–26). Oklahoma City, OK: Federal Aviation Administration.

Hopkin, V. D. (1989). Implications of automation on air traffic control. In R. S. Jensen (Ed.), *Aviation psychology* (pp. 96–108). Aldershot, Hants: Gower Technical.

Hopkin, V. D. (1990). Operational evaluation. In M. A. Life, C. S. Narborough-Hall, & I. Hamilton (Eds.), *Simulation and the user interface* (pp. 73–83). London: Taylor & Francis.

Hopkin, V. D. (1991a). The impact of automation on air traffic control systems. In J. A. Wise, V. D. Hopkin, & M. L. Smith (Eds.), *Automation and systems issues in air traffic control* (pp. 3–19). Berlin: Springer-Verlag, NATO ASI Series Vol. F73.

Hopkin, V. D. (1991b). Automated flight strip usage: Lessons from the functions of paper strips. In *Proceedings of AIAA/NASA/FAA/HFS Symposium on Challenges in Aviation Human Factors: The National Plan* (pp. 62–64). Vienna, VA: American Institute of Aeronautics and Astronautics.

Hopkin, V. D. (1994a). Human performance implications of air traffic control automation. In M. Mouloua & R. Parasuraman (Eds.), *Human performance in automated systems: Current research and trends* (pp. 314–319). Hillsdale, NJ: Lawrence Erlbaum Associates.

Hopkin, V. D. (1994b). Color on air traffic control displays. *Information Display, 10*(1), 14–18.

Hopkin, V. D. (1994c). Organizational and team aspects of air traffic control training. In G. E. Bradley & H. W. Hendrick (Eds.), *Human factors in organizational design and management* (Vol. 4, pp. 309–314). Amsterdam: North Holland.

Hopkin, V. D. (1995). *Human factors in air traffic control*. London: Taylor & Francis.

Hopkin, V. D. (1997). Automation in air-traffic control: Recent advances and major issues. In M. Mouloua & J. M. Koonce (Eds.), *Human–automation interaction: Current research and practice* (pp. 250–257). Mahwah, NJ: Lawrence Erlbaum Associates.

Hughes, J. A., Randall, D., & Shapiro, D. (1993). Faltering from ethnography to design. In J. A. Wise, V. D. Hopkin, & P. Stager (Eds.), *Verification and validation of complex systems: Additional human factors issues* (pp. 77–90). Daytona Beach, FL: Embry-Riddle Aeronautical University Press.

Hunt, G. J. F. (Ed.). (1997). *Designing instruction for human factors training in aviation*. Aldershot, Hants: Avebury Aviation.

International Civil Aviation Organization (1993). *Human Factors Digest No. 8: Human factors in air traffic control* (Circular 241-AN/145). Montreal: Author.

Isaac, A. R. (1994). Imagery ability and air traffic personnel. *Aviation, Space, and Environmental Medicine, 65*(2), 95–99.

Melton, C. E. (1982). *Physiological stress in air traffic controllers: A review* (Rep. No. DOT/FAA/AM-82/17). Washington, DC: Federal Aviation Administration.

Mital, A., & Kumar, S. (Eds.). (1994). Fatigue. *Human Factors, 36*(2), 195–349 (Special issue).

Mogford, R. (1994). Mental models and situation awareness in air traffic control. In R. D. Gilson, D. J. Garland, & J. M. Koonce (Eds.), *Situational awareness in complex systems* (pp. 199–207). Daytona Beach, FL: Embry-Riddle Aeronautical University Press.

Mouloua, M., & Koonce, J. M. (Eds.). (1997). *Human–automation interaction: Research and practice*. Mahwah, NJ: Lawrence Erlbaum Associates.

Mundra, A. D. (1989). *Ghosting: Potential applications of a new controller automation aid*. (Rep. No. MW-89W00030). McLean, VA: Mitre Corporation.

Narborough-Hall, C. S. (1987). Automation implications for knowledge retention as a function of operator control responsibility. In D. Diaper & R. Winder (Eds.), *People and computers II* (pp. 269–282). Cambridge: Cambridge University Press.

Pheasant, S. (1986). *Bodyspace: Anthropometry, ergonomics and design*. London: Taylor & Francis.

Rajecki, D. W. (1990). *Attitudes*. Oxford: W. H. Freeman.

Reason, J. T. (1993). The identification of latent organizational failures in complex systems. In J. A. Wise, V. D. Hopkin, & P. Stager, (Eds.), *Verification and validation of complex systems: Human factors issues* (pp. 223–237). Berlin: Springer-Verlag, NATO ASI Series Vol. F110.

Reason, J. T., & Zapf, D. (Eds.). (1994). Errors, error detection and error recovery. *Applied Psychology: An International Review, 43*(4), 427–584 (Special issue).

Reynolds, L. (1994). Colour for air traffic control displays. *Displays, 15*(4), 215–225.

Tattersall, A. J., Farmer, E. W., & Belyavin, A. J. (1991). Stress and workload management in air traffic control. In J. A. Wise, V. D. Hopkin, & M. L. Smith (Eds.), *Automation and systems issues in air traffic control* (pp. 255–266). Berlin: Springer-Verlag, NATO ASI Series Vol. F73.

Vortac, O. U., Edwards, M. B., Fuller, D. K., & Manning, C. A. (1993). Automation and cognition in air traffic control: An empirical investigation. *Applied Cognitive Psychology, 7*, 631–651.

Vortac, O. U., Edwards, M. B., Jones, J. P., Manning, C. A., & Rotter, A. J. (1993). En-route air traffic controllers' use of flight progress strips: A graph theoretic analysis. *International Journal of Aviation Psychology, 3*(4), 327–343.

Wickens, C. D., Mavor, A. S., & McGee, J. P. (Eds.). (1997). *Flight to the future: Human factors in air traffic control.* Washington, DC: National Research Council.

Wise, J. A., & Hopkin, V. D. (1997). Integrating human factors into the certification of systems. In M. Mouloua & J. M. Koonce (Eds.), *Human–automation interaction: Research and practice* (pp. 181–185). Mahwah, NJ: Lawrence Erlbaum Associates.

Wise, J. A., Hopkin, V. D., & Garland, D. J. (Eds.). (1994). *Human factors certification of advanced aviation technologies.* Daytona Beach, FL: Embry-Riddle Aeronautical University Press.

Wise, J. A., Hopkin, V. D., & Smith, M. L. (Eds.). (1991). *Automation and systems issues in air traffic control.* Berlin: Springer-Verlag, NATO ASI Series Vol. F73.

Wise, J. A., Hopkin, V. D., & Stager, P. (Eds.). (1993a). *Verification and validation of complex systems: Human factors issues.* Berlin: Springer-Verlag, NATO ASI Series Vol. F110.

Wise, J. A., Hopkin, V. D., & Stager, P. (Eds.). (1993b). *Verification and validation of complex systems: Additional human factors issues.* Daytona Beach, FL: Embry-Riddle Aeronautical University Press.

21

Human Factors in Air Traffic Control/Flight Deck Integration: Implications of Data-Link Simulation Research

Karol Kerns
The MITRE Corporation Center for Advanced Aviation System Development, McLean, VA

The subject of this chapter is air traffic control (ATC)/flight deck integration. ATC/flight deck integration means that ground-side and air-side resources in the National Airspace System (NAS) are brought together so that the skills, knowledge, and capabilities resident in each of these elements can be shared. It starts with the proposition that there is an integrative or coordinative mechanism that makes certain components of the system act in greater concert and thereby achieve a better outcome at no greater cost.

The reason for sharing resources and developing an integrative mechanism is to be able to do a better job of meeting operational needs. Today, the primary integrative mechanism that supports the exchange of information between controllers and pilots is radiotelephone; voice messages carry virtually all air traffic clearances and flight advisory and warning information. However, years of refining the language and procedure of communications have only served to confirm the intractable nature of many of the problems inherent in the exclusive use of spoken language and simplex radio for the air–ground transfer of information. In addition, the growing demand for aviation services and the constraints on budget that both the aviation industry and the Federal Aviation Administration (FAA) face mean that the NAS will need to seek solutions in the future that make better use of existing ground-based and airborne automation systems, rather than making investments in new airport construction or expansions at major airports as a way of absorbing projected demands. Many believe that a new integrative mechanism, a digital data communications link, can help alleviate some of the communications problems in the current voice environment and more effectively couple the ground-based and airborne NAS resources in the future environment.

The purpose of this chapter is to examine how a data link should be used in the operational environment as the integrative mechanism. Operationally oriented simulation research on data link is a primary source of evidence on how human characteristics will interact with this new technology to affect the exchange of information between the ground system and the flight deck. This chapter reviews the research record on human factors of data-link communications to examine what it says about the overall effect of data link on the ATC system and how data link should be designed to reduce communications errors, enhance information transfer, and support an appropriate allocation of functions between ground-side and air-side system elements.

I begin with a brief discussion on the need for ATC/flight deck integration, summarizing the historical development of controller and pilot roles and how it impacts their workload patterns and tasks in the operational environment. This provides the motivation for the following section on defining the integration problem. The problem is multidimensional and can be analyzed from a detailed perspective of information transfer problems and needs and from a higher level perspective of system philosophy and ATC/flight deck allocation of functions. This is followed by a review of the data-link simulation literature. The chapter ends with a discussion of the implications of the research for the evolution of ATC/flight deck integration in the NAS.

WHY ATC/FLIGHT DECK INTEGRATION IS NEEDED

Controller and Pilot Roles

Today's ATC is based on a centralized, ground-based, human-intensive system with the controller having a pivotal role in planning the movement of air traffic and transmitting instructions to carry out the plan. The role of the pilot receiving ATC services is one of processing advisory information, accepting instructions, and acting on them (Billings & Cheaney, 1981). This controller/pilot role relationship has been shaped by historical trends in ATC system evolution. At present, there are signs that this role relationship is contributing to growing inefficiencies in ATC system performance. Some of the causal factors that underlie these inefficiencies include divergent air and ground operational objectives, incompatible controller and pilot tools, and workload imbalances.

Today and for the foreseeable future, pilots and controllers have primary responsibility for carrying out the operational objectives in the NAS. Pilots are responsible for transporting people and cargo in a safe and efficient manner, whereas controllers provide for the safe and expeditious movement of air traffic. As the ATC system has evolved to support these objectives and accommodate more traffic, there has been a concomitant growth in complexity in terms of the demands placed on pilots and controllers. In part, this added complexity stems from the parallel and often divergent evolution of ATC and air carrier organizational and technological systems in which controllers and pilot are embedded. Although there is a significant volume of traffic other than air carriers that further complicates the system, it is the rigorous scheduling of air carrier aircraft that causes current problems and is the driving force for the development of cockpit technology. The ATC and air carrier organizations have their own internal systems, policies, and standard operating procedures (SOPs) designed to optimize their respective operational objectives. In turn, the technological capabilities and tools used by controllers and pilots embedded in these organizations, such as information displays, databases, and flight path planning and management capabilities, have been designed to interoperate within their respective organizational systems, not between them.

Because the design of ground-side and air-side systems has been driven by internal organizational objectives and technology, neither the procedures nor the tools are well adapted for coordinated use by controllers and pilots to achieve common goals. As an example, controllers work with a two-dimensional, plan view display of traffic that is well suited to radar separation procedures and representation of vector solutions to separation and spacing problems. In contrast, SOPs and flight management systems used by airline pilots support vertical profile planning in all flight phases to manage fuel and flight schedule requirements. Consequently, vector instructions from controllers impose a high cognitive demand on pilots attempting to execute and maintain a prescribed vertical flight path.

Another aspect of the system complexity that places demands on pilots and controllers stems from the procedures they use to coordinate their activities and conduct operations in different environments. These system procedures have evolved in a piecemeal fashion, with the consequence that the allocation of responsibilities between controllers and pilots is sometimes unbalanced. Tasks and procedures have been adapted and added on, prompted by a variety of factors including new equipment, new regulations, demand for services, and in response to operational system errors, incidents, and accidents (Degani & Weiner, 1994). The net result has been a simultaneous over- and underproceduralization of system operations.

Overproceduralization reflects the system's history of gradual accretion of procedures in response to discrete situations and an inability to discriminate obsolete and situation-specific procedures. It is illustrated by the routine issuance of preferred routings and use of static flow control and airspace restrictions by controllers and traffic flow managers. These procedures, designed to manage controller workload in specific situations, also reduce system flexibility when applied indiscriminately (Degani & Weiner, 1994).

On the other hand, underproceduralization reflects a lag in the system's response to new situations. It is illustrated by variations in cockpit procedures that make it difficult for controllers to effectively time the issuance of ATC instructions so that they avoid high-workload periods in the cockpit. The lack of SOPs increases human stress and workload in environments where operations have not changed in response to expansions and traffic increases at airports (Adam, Kelley, & Steinbacher, 1994).

Air–Ground Information Exchange and Operating Environments

Notwithstanding the care and effort that have gone into designing the air–ground information exchange process, the voice communication system does not always operate as intended. Both training and operating environment characteristics contribute to failures in voice communications.

Air traffic control communications have been designed to ensure that spoken dialogues can be conducted efficiently and with minimum possibility of error or misunderstanding (Hopkin, 1988). In the current system, much information is exchanged between pilots and controllers through speech, often over noisy communication channels. To compensate for transmission over a potentially noisy channel and achieve a better comparison with the receiver's expectations than would be achievable with free language, controllers and pilots have adopted a formalized language for voice communications (Kerns, 1991). The language conventions used are called *standard phraseology* and include not only the words to be used and their sequential order, but also pronunciation, enunciation, and speed of speech. Operational communications procedures define the process for conducting the dialogue, including the cues that tell a listener when a transaction has been completed and whether a readback is required.

At present, controllers receive training only in the use of standard phraseology, not in speech rates or word enunciation. Pilot communications training is even less formally structured than that of controllers, resulting in the use of nonstandard phraseology (Adam et al., 1994). However, even if controller and pilot training were comprehensive and well structured, it would not ensure perfect conformity by controllers and pilots in the operating environment. This is because the voice channel is available to only one speaker at a time and controller and pilot workload tends to be concentrated in specific environments.

The operating environment in which an information transfer occurs is a critical factor influencing the effectiveness of air–ground information transfer. The multiple operating environments—airport, terminal, enroute, and oceanic—that compose the NAS have markedly different characteristics in terms of what information controllers and pilots expect to transfer, its criticality, and the expected timing of the response (Flathers, 1987). Busy environments, such as the airport and terminal, also tend to be communications intensive, whereas less busy environments, such as the enroute and oceanic, have fewer communications. As more transmissions are crowded onto the frequencies at busy times, the procedural steps (callsign identifications, readbacks) that assure communication are being dropped (Adam et al., 1994).

It also happens that in the busiest environments, controller and pilot duty priorities, and communications tasks are most apt to conflict. In these busy environments, controllers issue instructions constantly. They may fail to initiate lower priority advisory messages or confirm the pilot's response. Conversely, pilots who are operating in busy environments are preoccupied with external vigilance and flight control tasks. They may not wish to receive messages during high-workload periods and may fail to respond or respond in a nonstandard fashion.

DEFINING THE INTEGRATION PROBLEM

The problems that ATC/flight deck integration seeks to address are those that cannot be met by a compartmentalized ATC system and require coordination of multiple elements. The current system is compartmentalized by an overarching allocation of functions that defines the roles, procedures, and tools of controllers and pilots in a way that makes coordination between them difficult. In addition, the coordination mechanism used to communicate information between pilot and controllers in current the ATC system also has limitations. Because communication is a two-way process designed to assure mutual understanding of messages, problems in communication are not simply pilot or controller problems; they are integration problems.

The need for more effective means of information transfer between air and ground elements has been widely recognized by the human factors and aviation communities. *The National Plan for Aviation Human Factors*, a joint government/industry perspective on human factors research priorities, contains a research plan, Flight Deck/ATC Integration, that describes the problem in terms of three dimensions: management of communications errors; enhancing information transfer; and defining an appropriate allocation of functions between ground and air elements (FAA, 1990). A base of research exists in each of these areas.

Communications Errors

There is a sizable literature on communications errors. Two complementary strategies have been applied to investigate the communication process and understand the nature of the problems. One approach analyzes pilot and controller reports of incidents and operational problems associated with communications; the other analyzes tape recordings of controller–pilot communications. The first approach is useful in characterizing the information transfer process and identifying the nature of the errors and potential causal factors, whereas the second approach provides insight into the actual frequency of problems and extracts cause–effect relationships about errors occurring in routine operations.

Incident and Error Reports. Billings and Cheaney (1981) reviewed seven related studies of information transfer problems that were based on voluntary reports submitted by pilots and controllers to the Aviation Safety Reporting System (ASRS). This broad review defined the major dimensions of the information transfer problem, encompassing various methods of transfer, such as charts, manuals, electronic displays, and voice communications; the failure points in the information transfer process; and the kinds of errors between and among elements of the aviation system.

As Fig. 21.1 shows, the ASRS data suggest that, in general, visual information transfer works better than aural information transfer: 15% versus 85% of problem citations, respectively. The analysis of failure points in the transfer chain indicates that the predominant problem cited was the failure to initiate the information transfer process. Other common failure points were inaccuracies in the transmitted message and failures to transmit a message at the right time. Although the bulk of the failures were traced to the sender of the message, failures to receive the information correctly also accounted for a substantial fraction of the reports.

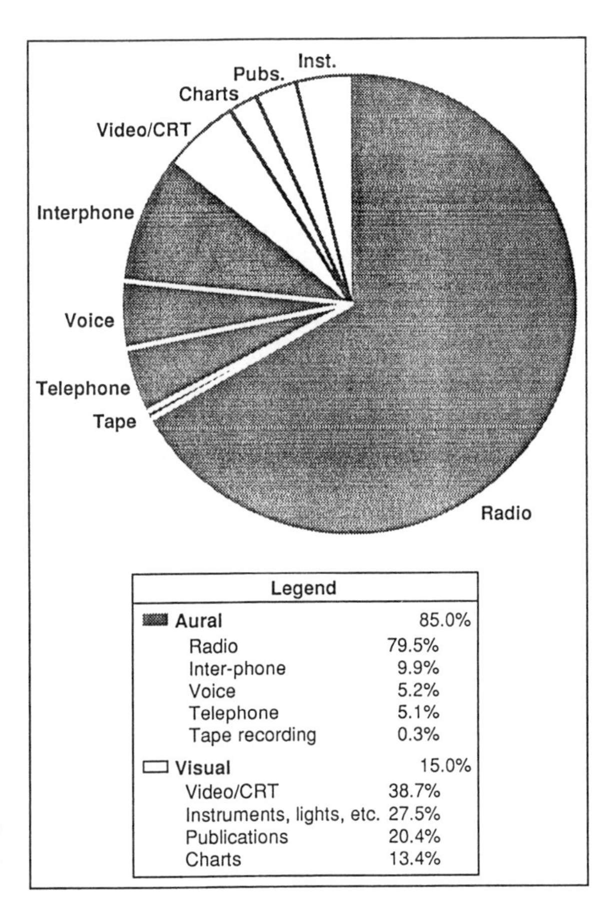

FIG. 21.1. Media involved in information transfer problems. Adapted from Billings and Cheaney (1981).

The ASRS reports that focus on air–ground information transfer deal exclusively with voice communications and emphasize aspects of human speech processing and conversational behavior that mediate communication performance (Grayson & Billings, 1981). Although much of controller–pilot communication was designed intentionally with redundant phrases or information to allow a listener to fill in what may have been missed, many ATC clearances and instructions are concise and without redundancy to minimize time on communication channels. This form of communication does not allow the listener to fill in what may have been missed. However, controller–pilot communication was also designed with safety measures, such as proper timing and readbacks, to assure that communication is taking place correctly (Adam et al., 1994). The analysis of communications problems in air–ground information transfer found that a tendency to fill in information, the expectation factor, and timing problems were implicated in many types of controller–pilot communication problems.

Misinterpretations and inaccurate communications were frequently cited problems in the ASRS reports. The causal factors included:

1. Phonetic similarities in which the words used in the message lead to confusion in meaning or in the identity of the intended recipient.
2. Transposition errors in which the sequence of numerals within the message was inaccurate.
3. Formulation errors in which messages were based on erroneous data or resulted from erroneous judgments.

In addition, transcription errors in which messages are received correctly but are recorded erroneously or entered erroneously into airborne or ground systems have been noted in related studies of navigational errors. The expectation factor also contributes to misinterpretations and inaccuracies, because pilots and controllers sometimes hear what they expect to hear. This generates what have been called "readback and hearback" errors in which respectively a pilot perceives what he or she expected to hear in an instruction transmitted by a controller and a controller perceives what he or she expected to hear in the readback transmitted by the pilot.

Another pervasive problem reported in the ASRS data was the failure to transfer traffic avoidance information and the seeming inconsistency with which information about traffic is made available. In the current system, controllers provide traffic advisories on a workload-permitting basis, which means that they are less likely to initiate traffic advisories during high-traffic periods, precisely when the pilot's need for the information is greatest. The required timing of information transfers when controllers are preoccupied with higher priority tasks accounts for failures to transmit appropriate messages, as well as untimely transmissions that are originated too late or too early to be useful to the recipient.

Voice Tapes. More recently, human factors researchers studying communications have looked to analyses of tape recordings of controller–pilot communications to better understand how often and why problems occur (Cardosi, 1993, 1994; Morrow, Lee, & Rodvold, 1993). Overall, the results of the analyses of voice tapes complement the results derived from the incident reports. The kinds of communications problems identified in the tapes analyses consist of procedural deviations, which include partial and missing readbacks, and inaccuracies, which include readback and hearback errors and callsign confusions. Although the incident data provide detailed information on errors in which pilots misunderstood controllers or controllers failed to initiate transfers of traffic avoidance information, the tapes analyses balance this perspective with data on error rates in routine communications and procedural deviations in which pilots do not follow standard communication procedures (Morrow et al., 1993).

Results from studies of voice tapes from the terminal, enroute, and tower environments found a consistently low incidence of communications errors. A readback error rate of less than 1% was observed in the terminal (Morrow et al., 1993), enroute (Cardosi, 1993), and tower (Cardosi, 1994) tapes. The rate of procedural deviations was much higher than the error rate but consistent for the terminal and en route environments. Morrow et al. (1993) observed a rate of deviations that ranged between 3% and 13% in the terminal environment, whereas Cardosi (1993) observed a rate that ranged between 3% and 12% for partial readbacks, acknowledgments without readback, and missing acknowledgments in the enroute environment. However, data from the tower environment indicate wider variation in procedures and a higher incidence of reduced acknowledgments. These results show that the responses to nearly three-quarters of the tower messages (72%) comprised partial readbacks, other forms of reduced acknowledgments, and missing acknowledgments.

The terminal and enroute studies also found that errors and procedural deviations increased as clearances increased in complexity. In these environments, the complexity of clearances ranged from one to two pieces of information up to five or more pieces in a single transmission. Again, the tower study results were not consistent with those of the terminal and enroute studies. In the tower environment, message complexity seemed to have little effect on communications error rate, and messages tended to be more complex overall (Cardosi, 1994). The author cited a number of reasons for this inconsistency. Compared to other operating environments, transmissions in the tower environment tend to be more predictable and more often reflect standard procedures (e.g., standard arrival routes and departure procedures). Moreover, inasmuch as the individual pieces of information contained in tower clearances are logically grouped by the pilot and therefore not independent, the study data may overestimate the complexity level and memory burden of the messages.

Although pilots frequently view controller speech rate as a cause of miscommunication, speech rate was not observed to be higher in transactions with errors (Cardosi, 1994; Morrow et al., 1993). Instead, the study results suggest that higher speech rates are a response to communication failures caused by the complex messages. During busy periods, controllers issue longer, more complex messages in an attempt to minimize use of the radio frequency and ensure timely delivery of messages. These complex messages place a greater demand on the pilot's memory and are more likely to be misunderstood, resulting in incorrect readbacks or requests for repetition (Cardosi, 1993). The controller's response when correcting or clarifying a message has a higher speech rate than when presenting the original message (Cardosi & Boole, 1991).

Information Transfer

Studies of voice communications errors offer guidance on improving information transfer between controllers and pilots. Billings and Cheaney (1981) concluded that both human factors (distraction, forgetting, failure to monitor, nonstandard procedures, and phraseology) and system factors (unavailability of traffic information, ambiguous procedures, and high workload) contribute to information transfer deficiencies. Cardosi

(1993) and Morrow et al. (1993) emphasized that complex messages overtax pilots' working memory, and nonstandard procedures reduce the time devoted to checking mutual understanding and therefore interfere with collaboration between controllers and pilots.

These investigators recommend remedies such as wider use of memory aids including written and verbal techniques, and better training including training in collaborative principles. Instructing controllers and pilots to keep messages short and make more disciplined use of scratch-pad notes can reduce the memory burden. Training programs can be developed to ensure good communication and illustrate why shortcuts in phraseology and procedures can increase transaction time and induce errors. Recurrent training can also encourage habitual use of established principles and methods of voice communication.

Studies of voice communications also point out the urgent need for alternative means of transferring information that is now communicated exclusively by voice (Billings & Cheaney, 1981; Adam et al., 1994). Unlike voice radio, the data-link communications medium can transmit coded, digital data to individual addressees. Data-link system and transaction status is monitored through a built-in feedback path or protocol that automatically verifies the integrity of the message reaching the addressee and provides information to the sender concerning responses, interruptions, or failures. Once received, message data can be stored for future reference and formatted for easy access by the user. These features of data link can be expected to alleviate problems induced by user interaction with the voice radio system at nearly all stages of the communication process.

Off-loading some of the voice radio message traffic onto another communication link will alleviate miscommunication problems caused by congestion. In addition, data-link system capabilities such as discrete addressing of messages, preformatted messages and standard protocols, and preservation of information can minimize radio-based problems like callsign confusion, overlapping transmissions, procedural deviations, and memory lapses (Kerns, 1991; Morrow et al., 1993). Failures and inconsistencies in communications, such as the transfer of traffic avoidance information, can also be alleviated by reduced frequency congestion or use of digital broadcast media for transmission of the traffic information.

Perhaps the greatest potential for improving information transfer via a data link will come from exploiting device-independent message standards and coding that allow flexible representation of information and direct interface of the data to automation systems. At the most basic level, standardization supports computer aiding to simplify the human's communication subtasks such as message formulation, communication system monitoring and error checking, and message logging. At a more sophisticated level, standardization supports transfers of data between aircraft and ground automation systems to ensure common databases and consistent solutions, without requiring the human to recode and reenter information.

Counterbalancing the potential improvements in information transfer are significant design challenges. Many of the basic principles and methods of effective voice communication also apply to data link, but safety measures for data link must address different albeit analogous design and procedural issues. Visual display and manual control of transmitted information add load to the human's busy visual information-processing channel. There is evidence that pilots are already at the limits of visual channel capacity in some operating environments (Billings, 1978). Even though research suggests that a visual display may be less prone to misinterpretation than an acoustic display, visual perception is still susceptible to the effects of expectations and errors.

The potential kinds of errors in visual perception also are predictable; confusion is no longer between information that sounds like other information when spoken but between alphanumeric characters that look alike or are physically adjacent (Billings & Cheaney, 1981; Morrow et al., 1993; Hopkin, 1988).

In addition to issues of channel loading and display formats for a visual mode of presentation, another important area of design concerns the impact of the digital communications medium on failure points in the transfer process. Data-link system capabilities can be exploited to standardize the content and procedures used in the ATC/flight deck segment of the information transfer path. However, new procedures will be needed to ensure coordination within controller teams and flight crews when the communications medium is silent and less readily observable by multiple operators. Furthermore, the availability of a second communications link will introduce greater complexity in procedures by requiring the users to switch between links during operations and to keep track of two systems to obtain information that was previously available on a single voice radio party line. As regards advanced applications of data link, automation issues and the definition of appropriate roles for the pilot and controller vis-à-vis these computer-computer information exchanges constitute important human factors problems.

Allocation of Functions Between ATC and Flight Deck

Enhancing information transfer and reducing communications errors are necessary conditions for effective integration of air and ground system resources but they are not sufficient. For the ATC/flight deck integration to achieve the desired outcome, the issue of allocation of functions is central to controllers and pilots endorsing and accepting current and emerging developments. Greater use of aircraft and ground automation in system operations may radically, if only implicitly, alter the current allocation of functions (FAA, 1990). There is a literature on allocation of functions that provides insight on the goals that underlie alternative allocations of functions, the requisite distribution of information between ATC and the flight deck, and user acceptance issues.

Some of the original research on allocation of functions and ATC/flight deck integration examined a distributed management philosophy as an alternative to a centrally managed ATC system. By providing pilots with the "big picture" of the traffic situation through a cockpit display of traffic information (CDTI) and the ability to participate in local traffic management, it was thought that controller and pilot workloads could be balanced and information transfer could become more efficient. Kreifeldt (1980) summarized a body of work on CDTI that explored alternative allocations of functions between the ground and the air. Experiments with CDTI concepts showed that the most preferred mode of distributed management was one in which both controllers and pilots participated: Controllers issued a sequence order, and pilots managed their own position in the sequence from that point on. The results also showed that interaircraft spacing and spacing variability with a mix of equipped and unequipped aircraft was at least as good in the distributed as it was in the centralized management condition. Finally, distributed management resulted in reduced controller workload, and although pilots reported a higher visual workload with CDTI, they still preferred distributed management. Kreifedlt concluded that using a CDTI pilots can maintain separation better than controllers. An even greater benefit of CDTI is its potential ability to cope with unplanned contingencies in which the speed and quality of a pilot's response are essential. These properties are difficult to achieve with the inherently long delay times of an air–ground loop.

More recent work on the allocation of functions extends the scope of the integration problem to formally consider the next level of integration, including the role of the air carrier's aeronautical operational control (AOC) center on the airside of the system and the role of traffic flow management (TFM) on the ground side. The FAA, Eurocontrol, and the civil aviation authority in Germany all have long-term projects underway that are attempting to exploit existing and emerging technology such as aircraft collision avoidance systems and data link to support alternative function allocations and define complementary capabilities for automation resources.

Sorenson, Miller, Simpson, and Murray (1992) summarized recent technology developments and opportunities to meet operational needs through the integration of aircraft flight management system (FMS) and AOC capabilities with air traffic management (ATM) capabilities, that is, the combination of ATC and TFM functions. In this paper, alternative allocations of functions were described for specific flight phases and airspaces. By adapting the allocation of functions in specific flight phases and airspaces, the FMS/AOC/ATM integration applications seek to balance controller and pilot workload and achieve greater flight efficiencies. Two example applications taken from the oceanic airspace, where current communications performance and surveillance capabilities are limited, envision pilot use of the FMS in conjunction with a CDTI for (a) station keeping to maintain spacing with respect to nearby aircraft and (b) maneuvering to execute control actions such as passing and climbing with respect to nearby traffic. In contrast, airspace capacity and frequency congestion are limiting factors in terminal airspace. FMS/AOC/ATM integration applications for this environment envision use of FMS-based approach and departure procedures to take advantage of FMS precision guidance and flight-path conformance without the need for progressive controller instructions and conformance advisories.

Requirements for such an integrated system include information sharing to ensure that airborne and ground-based automation systems generate compatible solutions when predicting flight times and paths, and development of SOPs to ensure that there is no role ambiguity or diffusion of responsibility for decisions and outcomes. The development of these SOPs and the overall philosophy of human–technology system relationships that govern them are key areas for human factors input.

REVIEW OF DATA-LINK SIMULATION STUDIES

Although the data-link was originally conceived of as a mechanism for pilot–controller communications, recent developments in the concept for an aeronautical data-link system recognize the need to broaden the scope of digital communications and include air–ground exchanges of weather, surveillance, and navigation information. Accordingly, the aeronautical data-link system is being developed to support both addressed communications, which direct information between specific pairs of users, and broadcast communications, which transfer information intended for simultaneous receipt by multiple users (FAA, 1994). Pilot–controller communications and flight information services (FIS) constitute two classes of communications targeted for early implementation. In the longer term, data link will be used for advanced air traffic management communications, which involve computer–computer information exchanges among ATC computers, avionics, and AOC computers.

For decades, operational use of data link has been the subject of simulation studies (Kerns, 1991). The vast majority of the research conducted to date has focused on initial use of data link for controller–pilot communications. Approaching the problem from both controller and pilot perspectives, these separate but similar studies have already produced consistent findings that have been captured in human factors recommendations and requirements documents for data-link systems (Human Factors Task Force, 1992; Society of Automotive Engineers [SAE], 1994). The following sections summarize the simulation results in terms of system-level effects that obtain across specific applications and operating environments and design-dependent effects that illustrate how features of the procedures, user–system interface, and application influence data-link effectiveness.

ATC System-Level Effects

Communications Efficiency. Congestion on the voice radio frequency has been related to a number of information transfer problems (Lee, 1989b). One of the measures used to estimate the impact of data link on frequency congestion is the amount of time controllers spend in voice communication tasks when data link is available. Talotta et al. (1990) looked at how much time controllers spent on the voice channel under three levels of data-link equipage. In this study, enroute controllers used data link to issue radio frequencies and altitude assignments; other communications were conducted via voice, regardless of aircraft equipage. Relative to a voice communications baseline, this study found a 28% reduction in controller time spent on the voice channel when 20% of the aircraft under control were data-link equipped and a 45% reduction when 70% of the aircraft were equipped. Off-loading routine communications to the data link not only creates additional capacity on the voice channel to encourage better procedural discipline (full readbacks) but also reduces the possibility of missed or blocked transmissions. Comparable reductions in radio frequency utilization were also reported for terminal controllers (Blassic & Kerns, 1990; Talotta et al., 1992a, 1992b). Again, reductions were a function of data-link equipage in the traffic scenario. In two recent studies that simulated highly congested enroute and terminal environments (Data Link Benefits Study Team, 1995, 1996) the results indicated that voice radio usage by controllers dropped dramatically as data link equipage reached 90% of the air traffic. Controllers reduced voice channel usage by as much as 84% in enroute airspace and by as much as 70% in the terminal airspace. In addition, the studies showed that controllers continued to use data link as traffic density and volume increased.

For equivalent scenarios, study results also show that a dual-media voice and data-link system requires fewer total transmissions than an all-voice system. Three studies (Blassic & Kerns, 1990; Hinton & Lohr, 1988; Talotta et al., 1990) compared transaction counts in scenarios with and without data link. Hinton and Lohr (1988) found that with increasing levels of data-link capability in the aircraft the number of data-link transmissions did not increase as rapidly as voice transmissions decreased for comparable flight scenarios. Consistent with this finding are data reported by Talotta et al. (1990). This study found an overall decrease in total voice and data-link transmissions with increasing levels of data-link equipage in the traffic. Again, the drop in voice transmissions that was observed as data-link equipage increased was greater than the increase in data-link transmissions. Blassic and Kerns (1990) produced similar results in a terminal environment simulation in which data link was used to transmit messages containing terminal information and expected approach procedures and for repetitively used ATC instructions. They compared the number of instructions issued in voice-only and data-link conditions. Compared to voice baseline, the number of instructions fell by 8.5% when 30% of the aircraft were data-link equipped and by 12% when 80% of the aircraft were equipped. All three studies attributed the reduction in total transmissions to fewer missed calls and fewer repetitions of information. Results from several studies support this attribution (Data Link Benefits Study Team, 1995, 1996; Knox & Scanlon, 1991; Lozito, McGann, & Corker, 1993). These studies found that use of data link for routine ATC messages in conjunction with voice radio as a backup and for nonroutine communications resulted in fewer messages to correct, repeat, or clarify a clearance as compared to the voice-only baseline.

From the flight crew perspective, there is additional evidence of time efficiency with data link as compared to voice communication. Talotta et al. (1990) and Uckerman and Radke (1983) compared the total time spent by flight crews on communication tasks using voice and using data link. Both studies found that time for crews to process messages was somewhat shorter with data link than with voice. When the data-link system is interfaced to flight management functions of the aircraft, the time savings is even more pronounced. Waller (1992) and Knox and Scanlon (1991) compared time required to receive and enter data into the subsystems of the airplane using voice communication and using a data-link system that allowed the crew to transfer data directly into the airplane flight guidance system or the FMS without reentering the information. Both studies demonstrated a substantial time savings for the crew when using the data-link communication process.

Speed and Timing of Communications. The simulation literature documents a marked difference in the speed and timing of communications when data link is used as compared to voice. These studies indicate that both total transaction time and the timing of specific events in the communication process are altered.

Total transaction time represents the entire time span when the controller would be concerned with a given communication. With a data-link system, it includes the inherent transmission delays, uplink and downlink, and the time required for a crew response. Simulation results indicate that, on average, total transaction time was twice as long for data link as it was for voice. In an enroute simulation, Talotta et al. (1990) estimated an average total transaction time of 21 sec for data link as compared to 10 sec for voice, whereas Waller and Lohr (1989) estimated an average total transaction time of 19 sec for data link and 8 sec for voice based on a full-mission simulation.

Data taken from the terminal controller's perspective confirmed that the viability of data link in the terminal environment will depend on total transaction time (Talotta et al., 1992a, 1992b). The 1992 study showed that as total data-link transaction time increased up to an average of 37 sec, controllers reverted to the voice channel. The final control sectors were most sensitive to longer delays. Aggregating across a set of subjective and objective problem indicators measured in the study, the investigators suggested that transactions times in excess of 22 sec would seriously limit the usability of data link in the departure and final approach sectors. Communication strategies also changed as a function of transaction time. Results showed that departure controllers, faced with increasing but not unworkable delays, tended to issue more complex messages—messages containing two or more instructions.

Data from several studies consistently estimate an average delay of 10 sec for crews to access and respond to a data-link message (Kerns, 1991), a performance that is comparable to estimates of processing times for crews to listen and respond to a voice message (Cardosi & Boole, 1991). The data-link simulations also suggest a relationship between flight phase and crew response time. Two studies (Diehl, 1975; Waller & Lohr, 1989) found that crew response times tended to be shorter and less variable during the arrival phase as compared to the enroute and departure phases. Similarly, Van Gent et al. (1994) found that crew response times decreased significantly from the oceanic through the cruise and descent phases. In this study, the results also suggest that longer message lengths in the oceanic and cruise phases may partially account for the differences in response times.

Other findings consistently reported in the literature indicate that although the sequence of procedural steps in the flight deck communication process is altered with data link, controllers will not necessarily be aware of the change because the timing of the execution of instructions is comparable for data link and voice. In simulations with both single-pilot and two-pilot crews, pilots initiated maneuvers to comply with an instruction before a response was dispatched to the controller (Hinton & Lohr, 1988; Parker, Duffy, & Christenson, 1981; Rehman & Mogford, 1994; Waller & Lohr, 1989). However, even with the changed sequence of steps on the flight deck, results from Rehman and Mogford (1994) suggest that controllers would not necessarily perceive any difference because they received a display of the pilot's data-link response before any path change was evident on their situation display. Findings from Lozito et al. (1993) further indicate that the time required to complete execution of ATC instructions did not differ significantly for voice and data-link crews.

Evidence from controller simulations suggests that they too are altering the sequence of procedural steps for communications with data link. With voice communications, controllers are required to obtain readbacks or some other acknowledgment from the pilot. With data link, Talotta et al. (1990) and Blassic and Kerns (1990) observed that controllers did not wait for a display indication of crew acknowledgment before issuing subsequent instructions or performing other tasks. Similarly, when Rossiter, Wiseman, Connolly, and Morgan (1975) evaluated a control-by-approval mode of messages transmission, in which the controller initiated transmission of computer-generated messages, they found that controllers did not wait for crew acknowledgment before beginning to process the next instruction.

Workload. Across the entire body of simulation studies, there is little evidence of impacts on overall controller or pilot workload as a result of using data link. The evidence on workload shows, however, that the visual mode of information presentation and the delay performance of the communication system can result in measurable shifts in controller and pilot workload.

A general effect of data link documented in the simulation studies is a redistribution of workload: Visual and manual workload increase, whereas auditory and speech workload decrease. Groce and Boucek (1987) documented an impact of data link on copilot visual task load as a result of data link and a minimal impact on pilot visual tasking. For the copilot this increase was largely offset by a corresponding decrease in auditory task load as a result of decreased radio transmissions. The minimal increase in pilot visual tasking noted in this study tended to reach an overload state for brief intervals when monitoring of data-link information overlaid the normal instrument scanning tasks. Perceived workload data from Waller and Lohr (1989) support the same pattern of effects. When operating in the copilot role and therefore handling ATC communications, crewmembers reported a reduction in workload with data link. But when operating in the pilot role, crewmember ratings of workload were mixed. Pilots with more experience reported reduced workload with data link, whereas pilots with less experience reported increased workload.

A study by Talotta et al. (1992a) that investigated the effect of delay performance of data link found that controller estimates of perceived workload in the terminal environment increased as a function of increasing delays in data-link test conditions. Delays in excess of an average 22 sec total transaction times produced a significantly higher level of workload when compared to voice.

Although the reduction in controller speech task load is well documented in the literature (Blassic & Kerns, 1990; Talotta et al., 1990, 1992a, 1992b), a corresponding increase in controller visual and manual workload as a result of the visual mode of communications has not been measured. It is interesting to speculate that the absence of an increase in perceived workload could mean that controller's visual channel is not yet at capacity.

Implications of Party-Line Information. In the voice communication environment, pilots use information overheard on the common radio frequency to develop their understanding and a representation of the current operating environment. Because some information will be discretely addressed with data link and therefore not available to all users operating in an airspace, researchers have undertaken to (a) characterize important party-line information elements, (b) analyze the role of this information in pilot decision making, and (c) develop implications for application of data link.

Two studies were conducted to identify important party-line information elements and assess their accuracy and relevance to specific flight operations (Midkiff & Hansman, 1993; Pritchett & Hansman, 1994). Both studies surveyed pilots and found that specific information elements related to traffic and weather were rated as critical. Although pilots rated party-line information as highly important, they perceived it to be only moderately available and accurate (Pritchett & Hansman, 1994). Survey results also indicated that the importance of party-line information appeared to be greatest for operations near or on the airport. However, in a related simulation experiment to investigate pilot use of party-line information Midkiff and Hansman (1993) found that the ability to assimilate and use party-line information was lowest during high-workload periods. This result indicates that the party-line monitoring task may be shed in high-workload situations. It also suggests that party line does not constitute a reliable mechanism for delivering critical information.

Simulation results also recommend use of alternative means of information transfer and presentation to enhance or replace party-line information as a source of weather and traffic information. Weather and traffic information are spatially oriented data, accessed simultaneously by multiple users, and may be transferred more effectively via data-link broadcast media and presented in a graphical display format. Studies by Lee (1991) and Wanke and Hansman (1990) investigated a graphical format for presentation of wind shear information. Results of both studies showed that graphical displays improved avoidance of wind shear by pilots when compared to a voice presentation. Lee also found that flight crews provided only with conventional ATC transmission of weather information had difficulty discriminating conditions conducive to microburst events from less hazardous wind shear events and that real-time updates of the data-linked information contributed to improved situation awareness for microburst events.

Traffic information is already being provided to air carrier aircraft as part of the traffic advisory and collision avoidance system (TCAS); however, not all TCASs contain a graphical display showing the positions of proximate air traffic. Currently, the FAA is developing a lower cost version of a traffic information display for general aviation aircraft (FAA, 1994b). Future TCAS implementations will likely evolve toward a standard configuration that includes a graphical format.

Design-Dependent Effects

Operational Communications, Flight Crew, and Controller Procedures. The relative utility of alternative prototcols for conducting the controller–pilot communications dialogue via data link has been examined in several studies. Talotta et al. (1988) investigated use of data link as a confirmation of voice transactions, whereas Cox (1988) reported on use of data link as a prenotification of a voice transaction. Other studies used data link to replace voice transactions. Overall, the simulation results generally favor protocols that minimize switching between media to complete a specific transaction (Kerns, 1991). The difficulty in synchronizing the timing of the two media has led pilots (Cox, 1988; Eurocontrol, 1986; Hinton & Lohr, 1988) and controllers (Talotta et al., 1988) to judge cross-media protocols as unnecessarily complex. Instead, the study results recommend that data link be used to replace selected voice communications. The findings also support consistency between procedures used in conducting communications over the two media (Kerns, 1991). More specifically, the research indicates that like voice communication procedures, data-link communication will require controller and pilot operational acknowledgment of clearance requests and ATC instructions to assure that communication is taking place correctly (Talotta et al., 1988).

Although standard communication protocols and phraseology can be assured through automated procedures in a data-link communication environment, the addition of this medium also allows for flight crew and controller team responsibilities and procedures to be altered. Traditionally, procedures for two-pilot crews assign responsibility for handling ATC communications to the copilot or pilot not flying. However, in the voice radio environment, both crew members monitor the radio frequency and simultaneously have access to messages. With data link there will be a need to develop new procedures that provide feedback to the pilot flying and thereby ensure mutual understanding of information. Research results on two-pilot crews have recommended a data-link procedure that requires a verbal communication between pilots before responding to a data-link message (Hahn & Hansman, 1992; Lee, 1989a; Lozito et al., 1993; Waller & Lohr, 1989). Speech generation technology has also been investigated as a mechanism for ensuring crew understanding of data-link messages, and this work is discussed in the following section. Furthermore, a recent study of pilot procedures highlights the need for new procedures in a dual-media communications environment. In this study McGann, Morrow, Rodvold, and Mackintosh (in press) found that a single pilot adapted voice communication procedures to be compatible with the sequential nature of data link thus causing voice performance to deteriorate in the dual media environment relative to the voice-only environment.

Data link may also enable some alternative allocations of responsibility and changes in controller team procedures. The availability of a second communication link will permit an assistant controller to take on some of the communications tasks during busy periods. A study by Talotta and Shingledecker (1992) compared data-link effectiveness with single-controller and two-controller teams operating combined approach and final control positions in terminal airspace. In this study, controller teams devised alternative strategies for allocating responsibilities. A prototypical allocation of responsibilities observed in the study was that the assistant controller used data link to issue instructions to aircraft on entry into the airspace and at outer fixes, whereas the primary controller used voice and handled turns to final and approach clearances. Controller ratings indicated that a team was judged as being capable of producing higher overall capacity than a single controller at the combined sector. Controllers also reported that the two-controller team was capable of increasing capacity only when voice communication was supplemented by data link. Results from a later study in which the terminal radar controller received assistance from a supervisor or handoff controller, confirmed that the control team can successfully share tasks and increase sector capacity when data link is used to permit simultaneous communication with multiple aircraft (Data Link Benefits Study Team, 1996).

Similarly, research on enroute operations (Data Link Benefits Study Team, 1995; Shingledecker & Darby, 1995) examined how changes in the allocation of controller team responsibilities affected sector capacity. Study results show that when teams of three controllers used a combined data-link and voice radio communication system they were able to provide ATC services that improved enroute sector efficiency and productivity. These effects were reflected in reduced aircraft ground delay, flight time, and flight distance in comparison to the current operational environment using only voice communications. Data on the duties and tasks of each control position showed that the addition of data link not only resulted in a shift in the distribution of air–ground communication tasks among controllers it also produced a shift in responsibilities in other sector tasks. While the radar controllers continued to perform all voice radio communications, the two assistant controllers, the data controller and the tracker, sent data-link messages. Along with the data-link communications tasks, the assistant controllers also took increased responsibility for monitoring the traffic situation and making control decisions. Moreover, as the level of workload sharing increased on a variety of sector tasks, the radar controllers were able to devote more time to overseeing and directing the team's activities. Under voice-only conditions sharing of planning and decision making tasks with the assistant controllers is limited largely because the radar controller is heavily occupied by communications responsibilities. The radar position's significant involvement with aircraft communications tasks tends to hinder interactions required for team direction. Under data link conditions, the radar controllers reported a higher frequency of directing the team's actions and all three positions reported a greater level of team interaction.

User–System Interaction

In my earlier review of the simulation literature (Kerns, 1991), I observed that most of the researchers had tended to investigate display/control capabilities related to one side of the (two-way) communication process, depending on their operational perspective. Controller-oriented studies focused on capabilities for generating and sending outbound messages and on the design of display feedback to permit monitoring the progress of information exchanges. Alternatively, pilot-oriented studies focused on capabilities for display of received messages and for generating responses. Considerable progress has been made since that review to develop design philosophies and principles for two-way message handling for controllers and pilots. Many of the design principles have general applicability to both controller and pilot work environments; others address unique flight deck or controller operational requirements.

Automation. The level of automation to be applied in designing data-link message sending and receiving functions has received a good deal of attention in the literature. Controller-oriented studies manipulated levels of automation in message generation and sending (Rossiter et al., 1975; Talotta et al., 1988, 1989, 1992b). Results of these studies indicated that controllers should retain manual control over the transmission of data-link messages and that receipt of pilot acknowledgments should automatically update the ground-system database.

For both controllers and pilots, message composition appears to work best when it is computer assisted. Menu-driven message composition has been the most widely implemented interface style in controller and pilot simulations (see Figs. 21.2 and 21.3). Users build messages by selecting from a menu of predefined messages and message elements that have been stored in their data-link system. Although alternative mechanisms for message composition will be needed in exceptional situations, simulation research suggests that considerable input error potential exists when controllers compose messages using less automated styles such as command entries (Blassic & Kerns, 1990; Talotta et al., 1992b).

A key automation issue for the flight deck system design concerns the use of automatic versus crew actions to acknowledge messages and transfer message data to other aircraft subsystems. Pilot-oriented simulations have examined alternative levels of automation and crew procedures in this process. Waller (1992) evaluated a semi-automatic process that allowed the pilot to acknowledge the message to ATC and copy the data into a standby area of the data-link display. A second action was then required to copy data to the appropriate flight subsystem. Results from this study indicated that allowing the pilot to transfer data directly from the data-link system to flight management functions without reentry of the information was highly beneficial as long as the flight crew remained in control of data input into the subsystems of the airplane. At the same time, the study pilots also recommended that certain messages such as radio frequency assignments should only require a single input to dispatch an acknowledge and transfer the data. Knox and Scanlon (1991) compared "Roger" and "Roger/Enter" options that either dispatched a response to ATC or dispatched the response and inserted data into the airplane's flight control or flight management system. In this study, the combined Roger/Enter option was used 86% of the time for tactical messages, and results suggested that the single-input transfer process greatly reduced crew workload. Hahn and Hansman (1992) also used a semiautomatic process and obtained similar results when they compared automated FMS programming of data-linked clearances with manual FMS programming of data-link and voice clearances. Their findings indicate that crews were better able to detect errors in messages with the use of automated FMS programming. Crew comments noted that the automated programming allowed them to focus their time and effort on the highest level of interpretation; rather than focusing on isolated words and numbers.

In the Knox and Scanlon (1991) study, several crews suggested that they would have liked the option to transfer data to the FMS as provisional modifications prior to dispatching a response to ATC. Lozito et al. (1993) examined a message preview and direct entry option. In the simulated data-link interface, it was possible for the pilot to automatically load information from the message before or after dispatching an acknowledgment. If the pilot previewed and directly entered message data into an aircraft subsystem, the data-link system would automatically send an acknowledgment. The pilot could also elect to acknowledge the message without first exercising the preview and direct entry option. This study reported longer response times than those previously found in simulations. A mean acknowledgment time of about 21 sec was observed for data-link messages, over twice as long as the average 10-sec response times observed in previous studies. Longer response times notwithstanding, data-link crews in this study

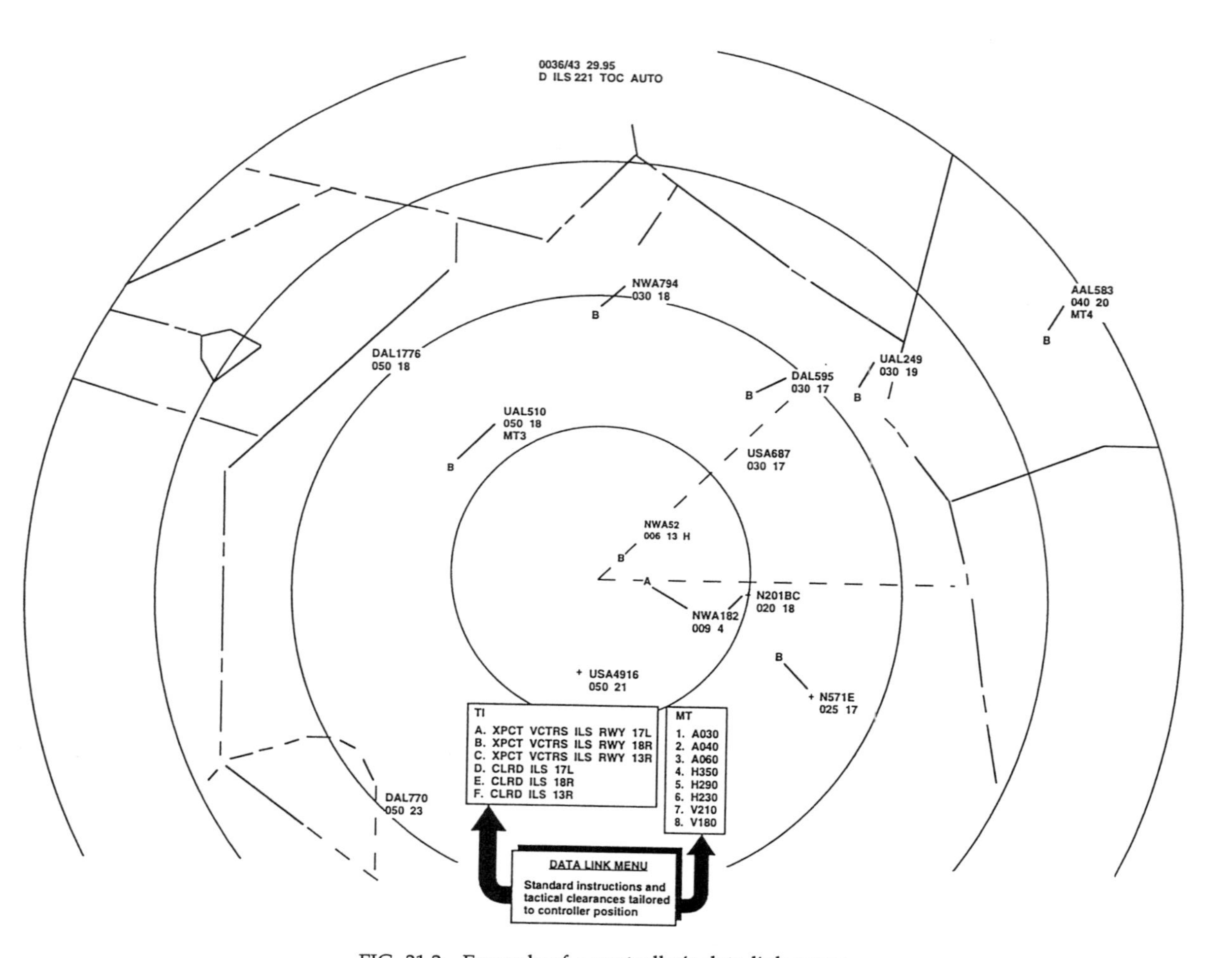

FIG. 21.2. Example of a controller's data-link menu.

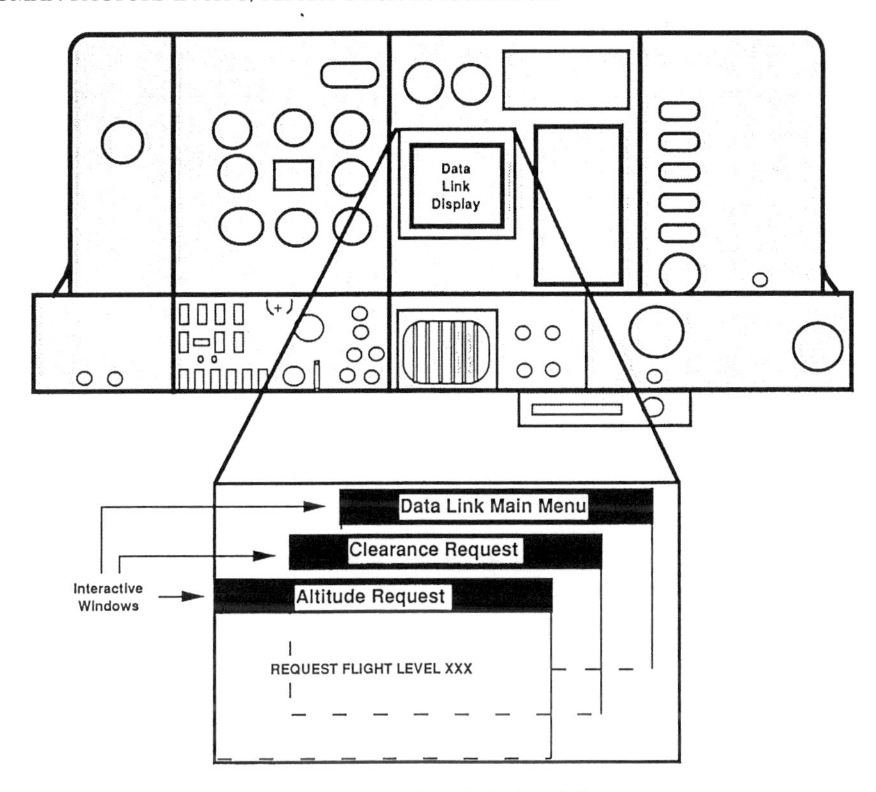

FIG. 21.3. Example of a pilot's data-link menu.

had fewer communications errors, performed more concurrent tasks, and took the same amount of time to comply with the messages when compared to the voice crews.

Display Surfaces and Locations. Determining the best display surfaces and locations for supporting data-link functions in the controller workstation and the cockpit has been the subject of several simulations. Controller-oriented simulations (Rossiter et al., 1975; Talotta et al., 1988, 1989) evaluated the desirability of integrating data-link message data with aircraft data tags on the situation display as compared to grouping all message and transaction data in lists. Results consistently indicate that the integrated presentation is most efficient for controller monitoring of data-link transactions, but that dedicated lists of active and completed data-link transactions are also desirable as redundant displays.

Pilot-oriented simulations (Rehman & Mogford, 1994; Van Gent, Bohnen, & Jorna, 1994) evaluated alternative display locations comparing forward- and aft-mounted data-link displays. Not surprisingly, the forward-mounted display locations resulted in shorter response times. When data link shared a display with other flight management functions, this too resulted in shorter crew response times compared to a dedicated data-link display (Van Gent et al., 1994).

Display Modes and Formats. The benefits and drawbacks of alternative display modes and formats, including speech, graphics and textual presentations, have been investigated in a number of flight deck studies (Diehl, 1975; Hahn & Hansman, 1992; Hilborne, 1975; Lee, 1991; Rehman & Mogford, 1994; Waller, 1992). Overall, the results of these investigations favor redundant presentations of information such as text with speech and text with graphics. At a minimum, a textual format appears to be required for presentation of most messages. However, the addition of speech and graphical formats enhances individual and team comprehension of messages by allowing human access through multiple attentional resources and at multiple levels of message interpretation.

Audio annunciation of data-link messages using speech generation technology has been investigated in a number of flight deck simulation studies (Diehl, 1975; Groce & Boucek, 1987; Hilborne, 1975; Rehman & Mogford, 1994; Waller, 1992). This presentation technique has promise for offsetting some of the added demand data link places on the crew's visual and manual processing resources and for supporting crew coordination in a data-link environment. The findings consistently show that (a) flight crews judge speech output to be desirable as a redundant display option for data-link messages, but (b) the addition of speech output increases crew response times substantially when compared to visual displays alone. Related research has also shown that comprehension of spoken messages takes longer than reading text of a similar complexity, with the amount of time increasing as a function of decreasing speech quality (Baber, cited in Stanton, 1993). A number of design and procedural issues were also noted in the research results. Speech output can be very intrusive, and can disrupt current task activity. Specifically, it was noted that automatic activation of speech output can disrupt flight crew tasks. Even with a manual control, in order to allow both crew members to orient themselves, the speech presentation should be preceded with a sound or phrase, as is the case with current voice communications from controllers to pilots, which start with a callsign. Some studies also reported crew problems because of speech rate and quality. Research on comprehension of synthetic speech indicates that human listeners focus more attention at lower levels of interpretation, such as isolated words, when quality is poor and therefore have less spare capacity for understanding the intent of the message (Pisoni, Nusbaum, & Greene, 1985; Baber, cited in Stanton, 1993). However, the additional processing demand and cognitive effort imposed by synthetic speech can be decreased with training and practice. Perhaps the most critical design issue identified in these studies is the potential for speech output to interfere with concurrent voice communications on the ATC radio frequency.

Graphical and textual display formats have also been compared to voice presentations (see Fig. 21.4 for examples of a graphical display formats). Hahn and Hansman (1992) found that a graphical presentation of data-link message improved the ability of crews to detect errors in clearances when compared to a textual or voice presentation. The advantage was most pronounced for detection of erroneous clearances into weather. Despite the improved error detection performance with the graphical format, flight crews in this study felt that a combined textual and graphical presentation was desired to support the full range of decision tasks (e.g., strategic evaluation, extraction of detailed flight parameters) that are required for compliance with clearances. Like the improved listening comprehension that results from higher quality speech, this study suggests that graphical format allowed the crew to focus attention at a strategic level. Results from Lee (1991) and Wanke and Hansman (1990) also support the advantage of graphical formats for presentation of weather information.

Data-Link Applications

According to the research, successful application of data link to ATC/flight deck information exchanges depends on the operating environment in which the exchange occurs and the type of information contained in the message. Figure 21.5 depicts the

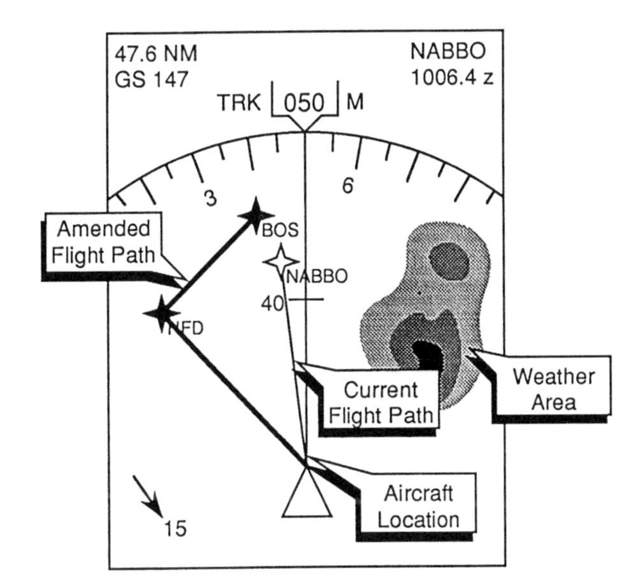

FIG. 21.4. Example of graphical ATC route amendment and weather displays.

areas that controller- and pilot-oriented research has identified as suitable for data-link application (Kerns, 1991). As the figure shows, data link is generally more acceptable in less busy operational environments and flight phases, such as predeparture and enroute.

In the terminal environment, pilots and controllers have long expressed cautions and reservations over use of data link, although until recently simulation research in the terminal environment was limited (Kerns, 1991). Early studies of data link reported that pilots judged most applications related to local and ground control functions unacceptable for data link. The landing clearance was deemed acceptable as a data-link instruction, depending on when it was given (Hilborne, 1975; Hinton & Lohr, 1988).

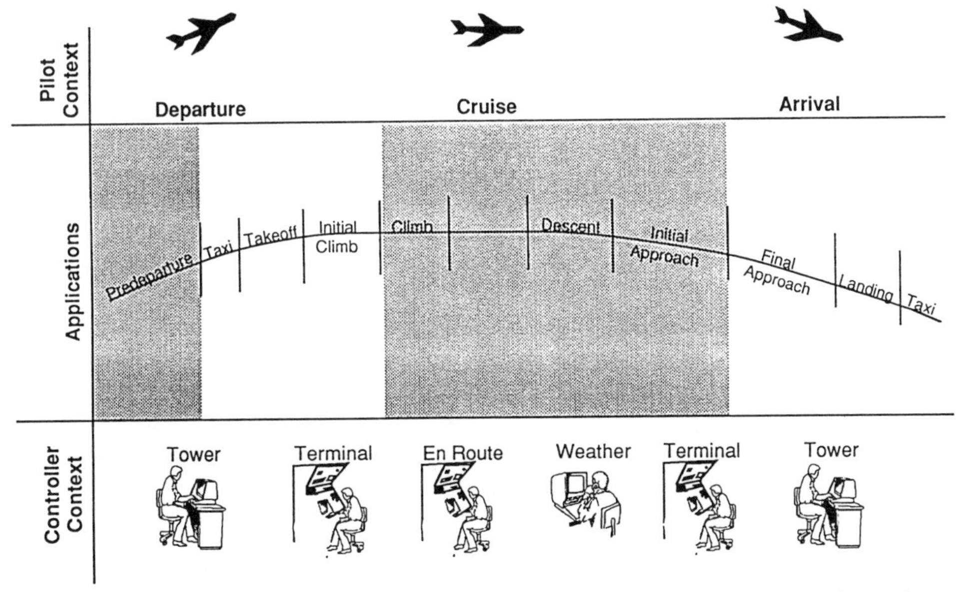

FIG. 21.5. Data-link application areas with convergent, controller and pilot, evidence of effectiveness and acceptability. From Kerns (1991).

More recently, controller-oriented simulation studies have begun to fill in more of the details concerning the types of messages that are acceptable and to establish the boundaries of the operational envelope for using data link in the current terminal environment. These studies suggest that data link is most appropriate for longer, repetitive messages, such as routine information on terminal operating conditions and expected approach procedures, which tie up the voice channel (Blassic & Kerns, 1990; Talotta & Shingledecker, 1992a). Results also indicate that in general, data link would not be suitable for time-critical instructions, such as turning aircraft onto final or dealing with missed approaches (Talotta & Shingledecker, 1992b). Thus, it appears that the timing of message delivery is a driving factor for data-link use during approach operations. Messages that can be prepared and issued in advance of final approach and landing operations should be acceptable for data link in terminal airspace.

Early operational experience with data link in the predeparture environment has also been instructive in the design of data-link applications. Since 1991, the FAA and several airlines have been using a data link to alleviate severe congestion problems on clearance delivery frequencies and to improve the transfer of involved, often lengthy predeparture clearance (PDC) instructions to pilots. Controllers issue PDCs digitally to participating airlines through the airline communications network. In turn, the airline dispatch office has responsibility for actual delivery of PDCs to flight crews, typically using the company data link and a cockpit display or printer. This first operational experience with digital PDCs received broad support from the participants (Moody, 1990); it also served to document some of the weaknesses in the indirect method of controller–pilot communication and highlighted a number of human factors issues associated with data-link communications (Drew, 1994). Incident reports on PDC cite procedural deviations in which crews failed to obtain their PDC or received the wrong PDC. Under the current PDC delivery system, voice communications are used to verify that the PDC has been issued correctly; however, this procedure is not standard across airports. Another class of incidents relates to problems in interpreting the PDC information. These reports indicate that crews have difficulty interpreting some cockpit display formats used for PDC. They cannot find critical information quickly and accurately, and they cannot easily detect changes in their usual routes.

On balance, the advantages of a digital PDC in terms of reduced frequency congestion and a clearer, persistent visual presentation of messages outweigh the drawbacks in terms of the limited feedback to verify clearance delivery and the display formatting problems in the initial system. This has encouraged the FAA and the airlines to begin experimenting with a digital automatic terminal information service (ATIS; Kuhl & Berry, 1989). The ATIS is also a lengthy, involved message. Controllers, working in airport traffic control towers, record the ATIS and broadcast it continuously over dedicated radio frequencies. It gives local weather conditions, runways in use, and other airport advisories. Pilots are required to listen to the ATIS during approach to and before departure from an airport. A field evaluation of a prototype digital ATIS application conducted at two airports produced uniformly positive responses from pilots, although controllers identified a number of required enhancements for the system they used to generate the ATIS (Aeronautical Radio, 1992). Flight crew comments indicated that a digital ATIS afforded them greater flexibility to plan approaches early and improved message clarity over the voice broadcast. On the other hand, controllers found that the initial system placed a significant demand on them to coordinate the dual tasks of composing the text version and recording the voice version of the ATIS. Future refinement of the digital ATIS application will incorporate speech generation technology to allow the controller to create text and voice messages with a single procedure.

Apart from considerations of the operating environment, there appears to be an intrinsic advantage in applying data link to the transfer of weather information because of the graphical formatting capability (Lee, 1991; Wanke, Chandra, Hansman, & Bussolari, 1990; Wanke & Hansman, 1990). Although the graphical format is undoubtedly the most efficient means for representing this information, the time criticality of bringing hazardous weather situations to the crew's attention also places stringent demands on transmission system performance and crew alerting capabilities. Moreover, operational issues associated with controller awareness of and responsibility for delivery of hazardous weather advisories as well as verification of the crew's intent following notification have yet to be resolved.

Summary

From the review of the simulation research, there is ample evidence that data link is a useful adjunct to voice communications from a number of perspectives. Table 21.1 summarizes some of the advantages and disadvantages of using data link for ATC/flight deck information transfer. Data link improves the overall efficiency of the communications system by reducing the frequency of communications failures and consequently the number of attempts required for successful information transfer. In simulations, this effect is primarily attributable to the availability of a clearer, usable representation and a persistent, storable reference of message content, although in actual operations errors caused by noise or blocked transmissions would also be avoided, as would some message formulation and transcription/data transfer errors.

TABLE 21.1
Generalized Advantages and Disadvantages of Using Data Link for ATC/Flight Deck Information Transfer

Advantages
Increased efficiency and communications system capacity.
Relief from memory burden of lengthy, involved messages.
User pacing of communications tasks allows for more effective multitasking.
Increased timeliness of message delivery (e.g., where frequencies are congested).
Greater consistency in procedures and message content (protocol and application standards).
Display flexibility allows for more efficient and accurate assimilation of information.
More efficient and accurate information transfer to other ATC and flight deck subsystems.
Disadvantages
Longer transmission delays; utility seen as limited for time-critical instructions.
Less flexibility to handle unanticipated or rapidly changing conditions.
Visual format seen as disruptive in visually busy environments.
Design Issues
Message generation efficiency and accuracy improved or reduced?
Workload peaks dampened or exacerbated?
Team/crew shared awareness and understanding of information enhanced or restricted?
ATC and flight deck situation information comprehensive and consistent?
Overall efficiency of operations increased or decreased?

User perceptions and the performance of data link indicate that its superior precision and accuracy are generally obtained at the cost of speed in the information transfer. The research findings show that delay factors associated with message generation and transmission times account for longer total transaction times with data link than with voice. The time required for message interpretation and acknowledgment is comparable for the two media, although accuracy is improved with data link. Simulation results also reveal that, within limits, controllers and pilots can effectively adapt to the added delay by performing other tasks concurrently and adjusting the timing of their communications. Because of such adaptations, execution of ATC instructions seems to take about the same amount of time regardless of the communications medium. However, the delay factors associated with message generation appear to limit data link's utility in rapidly changing conditions, whereas transmission delays limit its utility for time-critical instructions.

The simulation research clearly documents the redistribution in controller and pilot workload that accompanies data link communications: Visual and manual workload increase, whereas auditory and speech workload decrease. This redistribution takes on operational significance in specific environments and for specific classes of information. In environments where severe frequency congestion currently exists, the controller's auditory and speech workload can reach an overload state. In such environments, data link helps to achieve more timely performance within acceptable workload limits by providing an additional channel for message generation and transmission, especially for repetitive messages issued to each aircraft. Conversely, the pilot's visual and manual resources are already heavily loaded in some environments. In these situations, data link increases perceived workload and could inappropriately interrupt and disrupt visual scanning and flight management tasks.

Greater consistency in procedures and message content along with display flexibility are important advantages of data link. Although the data-link system inherently supports greater procedural consistency in the ATC/flight deck segment of the transfer, the silent communication process requires extra measures, such as cockpit and controller team coordination procedures, voice generation technology, and display layouts, that ensure access and understanding of information by multiple operators. The research further shows that redundant display formats widen the band of information available and improve the user's access to the particular features and data that are most compatible with the mental representation of the situation and task requirements. Operationally, both analytic and holistic processing of information are combined in many of the user's tasks (Wickens, 1984, p. 121). The simulation research highlights some specific classes of information, such as weather, traffic, and route, that are most likely to show benefits of more efficient and accurate assimilation when presented in spatially oriented, graphical formats.

Taken as a whole, the research indicates that data link allows controllers and pilots to devote more attention to critical communications functions such as interpreting, evaluating, and formulating messages. It does this by offering them relief from many of the overhead functions, such as repetitive message preparation, transcription of data to preserve information for future reference, and entry of data to provide input to other systems. The ability to automate these overhead functions while retaining human involvement in critical functions not only yields greater efficiency in operations by eliminating redundant transcription and data entry tasks; it also has great potential for preventing data entry errors, and when coupled with flexible display formatting it should reduce the opportunity for errors of interpretation.

IMPLICATIONS FOR THE EVOLUTION OF ATC/FLIGHT DECK INTEGRATION

Human Role in Air Traffic Management Communications

The data-link simulation research has direct application to air–ground information transfers associated with the safe and efficient management of air traffic. In this regard, the human role in the data-link communications process that emerges from the simulation research is one of management by approval. This means that controllers and pilots will have final authority to approve the transfer of information to each other and to their automation systems. Consistent with the philosophy of management by approval is the notion of provisional approval also mentioned in the literature, wherein the human operator can delegate final approval of flight plan clearances and amendments to the automation system within the bounds of specific operating constraints and parameters. In addition to provisional approval, the application of management by approval to air–ground information transfer should also include a range of approval options that enable the human operator to retain regular and meaningful involvement in the process. For example, human selection of specific messages for automatic transfer would be consistent with this philosophy, as would human selection of operating parameters that govern provisional approval of information transfers. Finally, use of automation to evaluate incoming messages and formulate preliminary approvals or identify potential constraint violations would constitute another variant consistent with this philosophy.

Negotiation and collaborative decision making are central to most advanced air traffic management communications concepts. Consistent with the previous research are several plausible ways of conducting the negotiation in a data-link communications environment. The flexibility of voice communications could be exploited to work out a mutually agreed-on decision, which in turn could be converted to a data-link message by the pilot or controller, depending on the situation and the available system capabilities. Additionally, data link could be used to make information on situation constraints available for common access by controllers and pilots prior to negotiation, or data link could be used in concert with automation for the iterative exchange of candidate flight path modifications. Controllers and pilots would manage the data-link process using the various approval options described earlier.

Alternative Allocations of Functions

The strategic direction for the NAS operation is called air traffic management (ATM). Among other characteristics, ATM assumes a shift away from today's ground-based, tactical ATC operations toward a more cooperative arrangement in which NAS users routinely participate in flight planning and airspace management decisions. It is expected that closer cooperation will require the direct exchange of digital messages between airborne and ground automation systems to ensure common databases and common solutions to route planning and flight scheduling problems.

Simulation results on data-link applications confirm that although the addition of a digital communications link can address many information transfer problems, this capability alone will not be sufficient to address user acceptance issues. In concert with data link, alternative allocations of functions between ATC and the flight deck will be required to take full advantage of data-link capabilities for information presentation and retrieval. In particular, successful applications of data link in terminal area operations will depend on the coordinated use of automation, procedures, and communications to better balance controller and pilot workload.

Convergent Evolution of ATC and Flight Deck Functionality and Information

As user participation increases and air and ground system elements focus more on solving a common problem, the separate air and ground systems will begin to more closely resemble each other (Wiener, 1988). Closer resemblance between systems will mean that the cognitive and information-processing demands on controller and pilot and their awareness of the situation will depend entirely on the ability to obtain, interpret, and assimilate information provided by displays (Houck, cited in Flach, 1994). Although the definitions of comprehensive and coherent situation representations for controllers and pilots may never be identical, it is reasonable to assert that there will be greater commonality in terms of the types of information represented and the most efficient formats.

Research on information transfer failures (Billings & Cheaney, 1981) and the importance of party-line information (Pritchett & Hansman, 1994) reveals that the current voice delivery mechanism is both unreliable and inefficient. The research further shows that in terms of the mental effort and attention required to access and recode information into a usable representation of the situation, digital transfers of traffic, route, and weather information promise to improve both controller and pilot situation awareness.

REFERENCES

Adam, G. L., Kelley, D. R., & Steinbacher, J. G. (1994). *Reports by airline pilots on airport surface operations: Part 1. Identified problems and proposed solutions for surface navigation and communications* (MITRE Tech. Rep. No. MTR94W60). McLean, VA: MITRE Corporation.

Aeronautical Radio, Inc. (1992, December). *Automatic terminal information service (ATIS): Field evaluation* (Prepared for ARD-270, Federal Aviation Administration). Washington, DC: U.S. Department of Transportation, Federal Aviation Administration.

Billings, C. E. (1978). Human factors associated with profile descents. *ASRS Fifth Quarterly Report* (NASA Tech. Memo TM-78476). Moffett Field, CA: NASA Ames Research Center.

Billings, C. E., & Cheaney, E. S. (1981). *Information transfer problems in the aviation system* (NASA Tech. Paper 1875). Moffett Field, CA: NASA Ames Research Center.

Blassic, E., & Kerns, K. (1990). *Controller evaluation of terminal data link services: Study 1* (MITRE Tech. Rep. No. MTR90W215). McLean, VA: MITRE Corporation.

Cardosi, K. (1993). *An analysis of en route controller–pilot voice communications* (Rep. No. DOT/FAA/RD-93/11).Washington, DC: U.S. Department of Transportation, Federal Aviation Administration.

Cardosi, K. (1994). *An analysis of tower (local) controller–pilot voice communication* (Rep. No. DOT/FAA/RD-94/15).Washington, DC: U.S. Department of Transportation, Federal Aviation Administration.

Cardosi, K., & Boole, P. (1991). *Analysis of pilot response time to time-critical air traffic control calls* (Rep. No. DOT/FAA/RD-91/20).Washington, DC: U.S. Department of Transportation, Federal Aviation Administration.

Cox, M. E. (1988). The Mode S data link: Experimental work and possible future applications in western Europe. In *Proceedings of the AIAA/IEEE 8th Digital Avionics Systems Conference* (pp. 695–703). San Jose, CA: American Institute of Aeronautics and Astronautics.

Data Link Benefits Study Team (1995). *User benefits of two-way data link ATC communications: Aircraft delay and flight efficiency in congested en route airspace* (Rep. No. DOT/FAA/CT-95/4). Washington, DC: U.S. Department of Transportation, Federal Aviation Administration.

Data Link Benefits Study Team. (1996). *Benefits of controller–pilot data link ATC communications in terminal airspace* (Rep. No. DOT/FAA/CT-9613). Washington, DC: U.S. Department of Transportation, Federal Aviation Administration.

Degani, A., & Weiner, E. (1994). *On the design of flight deck procedures* (NASA Contractor Rep. No. 177642). Moffett Field, CA: NASA Ames Research Center.

Diehl, J. M. (1975). *Human factors experiments for data link, interim report No. 6: An evaluation of data link input/output devices using airline flight simulators* (Rep. No. FAA-RD-75-160). Washington, DC: U.S. Department of Transportation, Federal Aviation Administration.

Drew, C. (1994, March). PDC's: The problems with pre-departure clearances. *ASRS Directline, 5*, 2–6.

EUROCONTROL Experimental Centre. (1986). *Real time simulation for the evaluation of SSR mode s ground/air data link* (Controller interface; Eurocontrol Experimental Centre Rep. No. 196). Bretigny, France: Author.

Federal Aviation Administration. (1990, November). *The national plan for aviation human factors* (Draft Report).Washington, DC: U.S. Department of Transportation, Federal Aviation Administration.

Federal Aviation Administration. (1994). *The aeronautical data link system operational concept.* Washington, DC: U.S. Department of Transportation, Federal Aviation Administration.

Flach, J. M. (1994). Situation awareness: The emperor's new clothes. In M. Mouloua & R. Parasuraman (Eds.), *Human performance in automated systems: Current research and trends* (pp. 241–248). Hillsdale, NJ: Lawrence Erlbaum Associates.

Flathers, G. W. (1987). *Development of an air ground data exchange concept: Flight deck perspective* (NASA Contractor Rep. No. 4074). Hampton, VA: NASA Langley Research Center.

Grayson, R. L., & Billings, C. E. (1981). Information transfer between air traffic control and aircraft: Communication problems in flight operations. In C. E. Billings & E. S. Cheaney (Eds.), *Information transfer problems in the aviation system* (NASA Technical Paper 1875). Moffett Field, CA: NASA Ames Research Center.

Groce, J. L., & Boucek, G. P. (1987). *Air transport crew tasking in an ATC data link environment* (SAE Tech. Paper 871764). Warrendale, PA: SAE International.

Hahn, E. C., & Hansman, R. J., Jr. (1992). *Experimental studies on the effect of automation on pilot situational awareness in the datalink ATC environment* (SAE Tech. Paper 922022). Warrendale, PA: SAE International.

Hilborne, E. H. (1975). *Human factors experiments for data link, final report* (Rep. No. FAA-RD-75-170). Washington, DC: U.S. Department of Transportation, Federal Aviation Administration.

Hinton, D. A., & Lohr, G. A. (1988). *Simulator investigation of digital data link ATC communications in single-pilot operations* (NASA Tech. Paper 2837). Hampton, VA: NASA Langley Research Center.

Hopkin, D. V. (1988). Air traffic control. In E. Wiener & D. Nagel (Eds.), *Human factors in aviation* (pp. 639–662). San Diego, CA: Academic Press.

Human Factors Task Force. (1992). *Human factors requirements for data link* (RTCA Paper No. 754-92/SC169-194). Washington, DC: RTCA.

Kerns, K. (1991). Data-link communication between controllers and pilots: A review and synthesis of the simulation literature. *International Journal of Aviation Psychology, 1*(3), 181–204.

Knox, C. E., & Scanlon, C. H. (1991). *Flight tests with a data link used for air traffic control information exchange* (NASA Tech. Paper 3135). Hampton, VA: NASA Langley Research Center.

Kreifeldt, J. G. (1980). Cockpit displayed traffic information and distributed management in air traffic control. *Human Factors, 22*(6), 671–691.

Kuhl, F. S., & Berry, D. L. (1989). *Automatic terminal information service (ATIS) via data link: Evaluation of a controller's workstation* (MITRE Tech. Rep. No. MTR89W240). McLean, VA: MITRE Corporation.

Lee, A. T. (1989a). *Display-based communications for advanced transportation aircraft* (NASA Technical Memorandum 102187). Moffett Field, CA: NASA Ames Research Center.

Lee, A. T. (1989b). Human factors and information transfer. In *Proceedings of the Second Conference, Human Error Avoidance Techniques* (pp. 43–48). Warrendale, PA: SAE International.

Lee, A. T. (1991). Aircrew decision-making behavior in hazardous weather avoidance. *Aviation, Space, and Environmental Medicine, 62*, 158–161.

Lozito, S., McGann, A., & Corker, K. (1993). Data link air traffic control and flight deck environments: Experiment in flight crew performance. In R. E. Jensen & D. Neumeister (Eds.), *Proceedings of the Seventh International Symposium on Aviation Psychology* (pp. 1009–1015). Columbus, OH: Ohio State University.

McGann, A., Morrow, D., Rodvold, M., & Mackintosh, M. (in press). Mixed-media communication on the flight deck: A comparison of voice, data link, and mixed ATC environments. *International Journal of Aviation Psychology.*

Midkiff, A. H., & Hansman, R. J., Jr. (1993). Identification of important "party line" information elements and implications for situational awareness in the datalink environment. *Air Traffic Control Quarterly, 1*(1), 5–30.

Moody, J. C. (1990). *Predeparture clearance via tower workstation: Operational evaluation at Dallas/Ft. Worth and Chicago O'Hare airports* (MITRE Tec. Rep. MTR90W108). McLean, VA: MITRE Corporation.

Morrow, D., Lee, A., & Rodvold, M. (1993). Analysis of problems in routine controller-pilot communications. *International Journal of Aviation Psychology, 3*(4), 285–302.

Parker, J. F., Duffy, J. W., & Christensen, D. G. (1981). *A flight investigation of simulated data link communications during single-pilot IFR flight: Volume I. Experimental design and initial tests* (NASA Contractor Rep. No. 3461). Hampton, VA: NASA Langley Research Center.

Pisoni, D. B., Nusbaum, H. C., & Greene, B. G. (1985, November). Perception of synthetic speech generated by rule. *Proceedings of the IEEE, 73*(11), 1665–1676.

Pritchett, A. R., & Hansman, R. J., Jr. (1994). *Variations in party line information requirements for flight crew situation awareness in the datalink environment* (Rep. No. ASL-94-5). Cambridge, MA: Department of Aeronautics and Astronautics, MTI.

Rehman, A. J., & Mogford, R. H. (1994). *Airborne data link study report* (Draft Tech. Rep.). Washington, DC: U.S. Department of Transportation, Federal Aviation Administration.

Rossiter, S., Wiseman, R., Connolly, M, & Morgan, T. (1975). *The controller–computer interface with an air-ground data link, Volume II* (Rep. No. FAA-RD-75-133). Washington, DC: U.S. Department of Transportation, Federal Aviation Administration (NTIS No. AD-A017369).

SAE Aerospace Recommended Practice. (1994). *Human engineering recommendations for data link systems* (ARP-4791). Warrendale, PA: SAE International.

Shingledecker, C. A., & Darby, E. R. (1995). Effects of data link ATC communications on teamwork and sector productivity. *The Air Traffic Control Quarterly, 3*(2), 65–94.

Sorenson, J., Miller, J., Simpson, R., & Murray, J. (1992). Opportunities for integrating the aircraft FMS with the future air traffic management system. *37th Annual Air Traffic Control Association Conference Proceedings* (pp. 534–543). Arlington, VA: ATCA, Inc.

Stanton, N. (1993). Speech-based alarm displays. In C. Barber & J. M. Noyes (Eds.), *Interactive speech technology: Human factors issues in the application of speech input/output to computers* (pp. 45–54). London: Taylor & Francis.

Talotta, N. J., Shingledecker, C., Zurinskas, T., Kerns, K., Marek, H., Vancampen, W., & Rosenberg, B. (1988). *Controller evaluation of initial data link air traffic control services: Mini-study 1, Volume I* (Rep. No. DOT/FAA/CT-89/25, I). Washington, DC: U.S. Department of Transportation, Federal Aviation Administration.

Talotta, N. J., Shingledecker, C., Zurinskas, T., Kerns, K., & Marek, H. (1989). *Controller evaluation of initial data link air traffic control services: Mini-study 2, Volume I* (Rep. No. DOT/FAA/CT-89/14, I). Washington, DC: U.S. Department of Transportation, Federal Aviation Administration.

Talotta, N. J., Shingledecker, C., & Reynolds, M. (1990). *Operational evaluation of initial data link en route services, Volume I* (Rep. No. DOT/FAA/CT-90/1, I). Washington, DC: U.S. Department of Transportation, Federal Aviation Administration.

Talotta, N. J., & Shingledecker, C. (1992a). *Controller evaluation of initial data link terminal air traffic control services: Mini-study 2, Volume I* (Rep. No. DOT/FAA/CT-92/2, I). Washington, DC: U.S. Department of Transportation, Federal Aviation Administration.

Talotta, N. J., & Shingledecker, C. (1992b). *Controller evaluation of initial data link terminal air traffic control services: Mini-study 3, Volume I* (Rep. No. DOT/FAA/CT-92/18, I). Washington, DC: U.S. Department of Transportation, Federal Aviation Administration.

Uckerman, R., & Radke, H. (1983). *Evaluation of an airborne terminal for a digital data link in aviation* (Deutsche Forschungs-und Versuchsanstalt fr Luft-und Raumfahrt-FB83-05). Braunschweig, Federal Republic of Germany: DFVLR Institute for Flight Guidance.

Van Gent, R. N., Bohnen, H. G., & Jorna, P. G. (1994). *Flight simulator evaluation of baseline crew performance with three data link interfaces* (Draft Tech. Report). Washington, DC: U.S. Department of Transportation, Federal Aviation Administration.

Waller, M. C. (1992). *Flight deck benefits of integrated data link communication* (NASA Tech. Paper 3219). Hampton, VA: NASA Langley Research Center.

Waller, M. C., & Lohr, G. W. (1989). *A piloted simulation of data link ATC message exchange* (NASA Tech. Paper 2859). Hampton, VA: NASA Langley Research Center.

Wanke, C. R., & Hansman, R. J., Jr. (1990, May). *Operational cockpit display of ground-measured hazardous windshear information* (Rep. No. ASL-90-4). Cambridge, MA: MTI.

Wanke, C. R., Chandra, D., Hansman, J. R., & Bussolari, S. R. (1990, November). *A comparison of voice and datalink for ATC amendments and hazardous wind shear alerts.* Paper presented at the 4th International Symposium on Aviation and Space Safety, Toulouse, France.

Wickens, C. D. (1984). *Engineering psychology and human performance.* Columbus, OH: Sijthoff & Noordhoff.

Wiener, E. L. (1988). Cockpit automation. In E. L. Weiner & D. C. Nagel (Eds.), *Human factors in aviation* (pp. 433–459). San Diego, CA: Academic Press.

V

AVIATION OPERATIONS AND DESIGN

22

Human Factors of Functionality and Intelligent Avionics

John M. Hammer
Hammer & Associates, Norcross, GA

The usability of systems is determined by interface functionality and presentation. Much research in human factors has concentrated primarily on presentation—the surface aspects of the interface. Today, interface presentation and interaction have greatly improved to the extent that navigating the interface is much easier. Understanding the functionality of the interface from its presentation can still be quite difficult, and this functionality has a significant impact on usability. This problem is termed the *human factors of functionality*. It complements the area of mental models, which studies how users adapt and understand system functionality (Kieras, 1990; Rouse, Salas, & Cannon-Bowers, 1992). This chapter deals predominantly with how system functionality should be designed to adapt to users.

How does functionality influence usability? First, the functionality partly determines the tasks that the user performs and, in some cases, how these tasks are performed. The functions require inputs, and there are a variety of ways that inputs can be organized conceptually. The concepts presented determine in part how the user thinks about the interface. For example, the degree of control level automation determines whether the user is continuously involved in the task or only intermittently involved. The amount of sensing available and the information integration determine whether the user must gather information from one subsystem source and enter it into another subsystem. The degree to which the system understands its own functions, their applicability, and the user's goals can have a potentially tremendous impact on the usability of functions (Hammer & Small, 1995; Rouse, Geddes, & Curry, 1987; Rouse, Geddes, & Hammer, 1990).

This chapter argues that a system that understands itself and the user is the next revolutionary step in avionics architecture. It covers some human factors problems in avionics functionality and describes a revolutionary avionics architecture that we feel can address these problems.

PROBLEMS WITH EXISTING SYSTEMS

It is widely accepted that automation does not eliminate human–machine interaction problems (Wiener & Curry, 1980). Instead, these problems are displaced or transformed from one type of problem to another. Typically, physical interaction problems are transformed into cognitive problems that are concerned with understanding the system. The following describes some problems that are covered in more detail elsewhere in this volume.

Automation Modes Work at Cross-Purposes

The modes of various automated systems can be set to work at cross-purposes with each other or the crew. The many possibilities include modes that are not designed to be combined and modes that inhibit other modes or aircraft capabilities. For example, in the crash of China Airlines Flight 140, the crew mistakenly activated an automatic go-around without realizing it (Mecham, 1994). The autopilot tried to increase power and gain altitude while the crew attempted to maintain power and reduce altitude. Because the crew controlled the elevators and automation controlled the horizontal stabilizer, the nature of the conflict was not apparent. The aircraft eventually attained an unrecoverable state and then crashed. If the aircraft had been flown completely manually to a landing or fully automatic in a go-around, no accident would have occurred.

Completely Automatic or Completely Manual Subsystems

Automation is now sufficiently capable that some subsystems could be completely automated, except when there is a malfunction that requires manual operation. Depending on the design philosophy, there may not be any intermediate levels of automation. In some ways, this simplicity is attractive. Only two modes—on or off—reduces training requirements. On the other hand, manual operation could impose a significant workload on the pilot who may be unfamiliar with operations that are usually automated and require no intervention. Some subsystems on the aircraft have multiple levels of automation. One example is flight path control, which can be done manually or automatically by relatively simple autopilots or by sophisticated flight management systems. The levels of automation available in flight control are probably an artifact of the history in which they were introduced. In designing new systems, some consideration needs to be given to how the intermediate automation should work.

Function Cannot Be Understood from Display

Interpreting system functions from displays can be difficult. There are several causes for this. First, finding the relevant information is difficult because most displays include more information than is relevant at any one time. Second, the consequences of many courses of action are not displayed (often, the automation does not know the consequences). Frequently, automation is so complex that displaying more complete information about what it is doing would worsen the information overload. At the same time, displaying limited information makes it difficult to understand the automation.

Although display shortcomings can be considered at fault, it is also possible that the functions themselves are too complicated to display, given the current understanding of display design. In fact, that is probably the case today, as our ability to conceive and implement functions has far exceeded our ability to display their state or consequences. This situation is not likely to improve in the future, given the minimal amount of research on this problem.

Automation Compensates for Worsening Failure While Complicating Recovery

Automation can also compensate for failures, but this compensation may mask a problem until it becomes severe. An example is the China Air Flight 006 where the autopilot compensated for reduced thrust from one engine (Wiener, 1988). When the autopilot was disengaged, the crew was apparently unaware of the extent to which the autopilot had compensated for the engine failure. Disengaging the autopilot removed this compensation. The crew did not reapply it, and the aircraft rolled into a dive that lost 30,000 of its 40,000 ft of altitude. The automation did not understand its own limitations or how authority should be transferred to the crew.

Surprising Engagement or Disengagement of Automation

At times, automation fails to engage as expected. The pilot will configure the aircraft and attempt to engage automation, but it refuses. Frequently, the automation checks some condition that must be true before it may be engaged. Because the pilot does not have access to the precondition, it is sometimes impossible to understand why engagement fails. Conversely, automation may change modes or disengage due to similar tests, and for the same reasons, the automation is difficult to understand. This problem is not confined to on/off engagement. Sometimes, the automation will engage but do something entirely unexpected.

Unnatural Behavior

Automation sometimes operates the aircraft in ways that are unlike human pilots. For example, if a human pilot wanted to achieve a particular altitude and speed at a particular navigation fix, the aircraft would be flown to achieve those goals a few miles in advance of passing through the fix. A flight management system, on the other hand, would attempt to achieve the goals exactly at the fix. The crew needs to learn two ways of doing something: the natural way and the automation way.

Conclusion

So much control has been delegated to automation that it would be reasonable to consider it a member of the crew, at least for the purposes of discussing its interface. As such, it could be evaluated with respect to its cockpit resource management (CRM) skills (Foushee & Helmreich, 1988). Although it is unusually precise, it cannot explain its actions, nor is it aware of the interaction problems it causes. A human with such poor skills would be sent back to CRM training to improve these skills. Automation granted near-human levels of authority should also have near-human interaction skills.

HOW FUNCTIONS ARE DESIGNED—THE CURRENT PRACTICE

There are at least two views on how avionics functionality is designed today. In commercial air transport, the new design is based on the most recent design with whatever minimal changes are necessary or desirable. Change is minimized for several reasons. First, it reduces design cost and pilot retraining cost when pilots change aircraft. Second, minimal change means minimal opportunity to introduce major problems in the interface. Third, it allows vendor components to be reused with little change, which reduces the time to market.

The disadvantage to minimal change is that it traps the design at a local maxima. In the case of both computer hardware and software, there will be enormous strides in technological capabilities and concepts. Minimizing the introduction of change reduces the ability to make improvements. The greatest improvements possible would probably come from a rethinking of the functionality.

The second way that functionality would be designed would be through task analysis and related system analysis. Task analysis starts with a mission and decomposes it into a sequence of user tasks and attributes that form a model of the user's overall activity over time. This fine-grained description of user activity should provide insight to the designer as to the consequences of particular functionality choices.

Problems with Task Analysis

From the designer's perspective, the task analysis does not tell one how to design. Instead, it is intended to show the consequences of design. Furthermore, whatever design feedback is given is probably useful only for local optimization. This is not bad in and of itself, but it implies that the assumed functionality on which the task analysis is built is itself not necessarily questioned by the task analysis. There is the same possibility of being trapped at a local functionality maximum as in evolutionary modification.

The attributes associated with individual tasks are often meant to further describe human performance and the demands on human performance. The difficulty here is that it is often difficult to justify the particular values assigned to the attributes. The results of the analysis can be no better than the inputs, as represented in these attributes. This approach is becoming increasingly problematic as the attributes used shift from physically observable measurements to unobservable cognitive concepts. A related problem, exemplified by the issue of pilot workload, concerns the dependency of the attributes on precise characteristics of the mission situation. For example, in a tactical air to air engagement, we might expect that pilot workload might be influenced more by the caliber of the opponent than with anything else. In general, workload might depend more on the situation than the tasks being performed (Suchman, 1987). If this is the case, and it seems to be true in the case of pilot workload, the wisdom of basing a workload analysis on tasks seems questionable. One might well base workload measurement on factors that exert more influence. This example raises the question of whether tasks are a suitable basis for answering the questions that are supposed to be answered by a task analysis.

The conceptual distance between function design and task analysis may be too far for any connection to be made. The concepts manipulated in task analysis are rather distant conceptually from the design of functions. Because of this, conclusions drawn in task analysis may indicate problems in function design but not solutions, at least for macroscopic design problems.

Task analysis in practice seems to be performed only on a relatively small number of missions due to the cost of doing each one. If the task analysis incorporates many mission specifics, the results of the task analysis are really influenced more by the mission. If so, the resulting conclusions may be mission artifacts more than anything else. This, compounded with the limited number of missions studied, could result in a design for a mission that is never actually flown in the fielded system.

Impact of Conceptual Organization on Interface Functions

The concepts used to organize an interface influence its usability. Even something as simple as a route may be thought of either as a sequence of points or as a sequence of segments. A point perspective is more appropriate for arrivals at a particular point at a particular time. A segment perspective is more appropriate for considerations such as fuel usage and speed. Of course, the system may offer both point and segment perspectives due to functional requirements. This third option, perhaps to be labeled creeping featurism, presents additional functional complexity to the user. The system becomes overconstrained, in the sense that the user may express more constraints than are physically possible to be realized. For example, once a start time and speed are specified, the end time is determined for a given segment. The choice of point, segment, or both perspectives influences how routes are viewed, what inputs and outputs (categorically) are available/required, and what functions are available to the user.

The functionality determines what an input is and what a consequence is. In a tightly coupled system such as an aircraft, there are many system variables that one can choose to control. However, only a subset of these can be controlled, whereas the others are determined by the values chosen by the controlled set. For example, to reach a particular point, a pilot may control fuel consumption by selecting the most efficient climb rate or may control trajectory by flying directly to that point.

Choosing Among Concept Spaces

One important decision from a human factors standpoint for the avionics software designer is designing a conceptual organization for an interface. Conceptual organization means the objects that are manipulated, the possible manipulations, and the behavior that is manifested. This design choice is not merely selection of compatible components but rather a choice among incompatibles. Consider the example of mechanically and electronically steered radar antennas. An electronically steered antenna can be pointed much more rapidly. The slow pointing of a mechanically steered antenna so constrains the feasible patterns of movement that the interface concepts are tied to the movement patterns. The interface to an electronically steered antenna is not so constrained. Traditional concepts from mechanically steered antennas could be used, or the problem could be viewed entirely differently as a resource allocation problem (e.g., keep track of targets with these properties, search this sector if there are any remaining resources). Even if a resource allocation approach is taken, there are a tremendous number of potential conceptual organizations to be considered by the designer.

In general, entirely electronic systems controlled by avionics computers have an enormous design flexibility in how the functionality of the user interface is organized. This problem may in fact be insurmountable in the following sense. The traditional approaches to effective user–system operation include selection, training, human factors design, and aiding. Design is the principle focus of this chapter so far, yet it is unclear how a suitable solution can be found in this enormous design space. Indeed, the designer would have to be wise to be able to predict how a novel system would be used and what the effects would be of various types of possible functionality on system performance. Possibly, there is no solution to these problems that can be applied during design, at least as design is practiced now.

To conclude, functional design is conceptually difficult and perhaps intractable given the large amount of functionality to be hosted in modern automation. There is already more capability than human operators typically utilize. The next section describes an alternative approach to perfecting functionality. With aiding and training, perhaps functionality does not need to be perfect; maybe it just needs to understand itself, the user, and the situation.

A STRUCTURE FOR A SOLUTION

For the last decade, my colleagues and I have been investigating a new concept termed the *intelligent interface.* The intelligent interface goes beyond traditional interfaces, whatever their surface form, in that it contains intelligent functions that are intended only to help the user. The specific functions could include any of the following: managing displayed information, watching for hazards, adaptively executing tasks on behalf of the crew, assessing the situation, and recommending responses. Because these functions are intended to be intelligent, they employ models of the user and situation that provide an intelligent, richer description.

Will an intelligent interface help to remedy the problems described earlier? It could be argued that automation is the problem and that intelligent automation may compound the problem. One description of the intelligent interface is a system to help the pilot use the systems on the aircraft. Obviously, increasing the level of automation could worsen the automation-related problems. This chapter discusses how intelligent interfaces should be designed to be successful.

There are several differences between intelligent interfaces and traditional automation. An intelligent interface contains models that enable it to be aware of some of the conditions that lead to the automation defects described previously. A second difference is in the automation philosophy by which the intelligent interface is designed. Its sole purpose is to help the pilot fly the aircraft. Although this claim could be made of traditional automation, there is a significant difference. Traditional automation helps the pilot operate the aircraft by taking tasks away from the pilot. These tasks are automated, and the pilot monitors this automation and makes occasional commands to that automation. The intelligent interface tries to keep the pilot in charge of tasks while supporting pilot decision making.

An Overview of Intelligent Interface Processing

The heart of the issue is the depth of processing of the intelligent interface that causes it to avoid problems of traditional automation. In terms of system design, it is important to distinguish the depth and functionality of an intelligent interface from a traditional interface. A traditional interface is an interconnect between the user and the automation. Typically, there are no changes made between the inputs/outputs of the automation and those made by the user. If the automation needs a target airspeed, the user will enter a target airspeed into the interface by typing, adjusting a knob, adjusting a slider on a mouse-based graphical user interface (GUI), and so on. The format may vary, but the information content is not changed by the traditional interface.

In an intelligent interface, the distance from user input to avionics input is much larger, and there is a considerable functionality between the user and the traditional automation. To continue the example, further intelligent processing will be done once the speed has been entered into the interface. First the speed will be examined for possible hazards. This will involve bounds checking on the speed itself as well as examination of the impact of speed changes on other hazards that are currently being monitored. The speed changed will be interpreted in terms of previously identified user plans, and if there is a significant change, the displays themselves might be configured. The intelligent interface attempts to determine the meaning and consequences of the speed change before passing the change to the traditional avionics system.

None of this description of additional processing describes the depth or intelligence of the intelligent interface. For example, in evaluating the speed for hazard, the monitoring might consider the aircraft configuration (flags, speed brakes and spoilers, gear) to avoid damage to the aircraft. It might consider location to avoid speeds over 250 knots near terminal airspace. It might consider the weather so that the aircraft is not flown too fast in turbulence. It will consider the flight plan to determine what impact speed changes might have on it. Even these various checks on airspeed could be themselves fairly elaborate. For example, flying too fast with the gear down might cause further consideration of whether the gear can in fact be retracted (or have they been damaged or perhaps must be left down). The diagnosis of the problem may be either "slow down" or "raise gear." Intelligent processing can determine which recommendation is appropriate.

THE INTELLIGENT INTERFACE LEVEL

Architecturally, the intelligent interface occupies a significant place between the traditional system and the user (Fig. 22.1). There are two general types of inputs to the intelligent interface. First, as discussed there, are the inputs that are ultimately bound for the traditional avionics system. The second general type of input is that needed explicitly by the intelligent interface.

To understand this second category, consider the role of the intelligent interface. The intelligent interface serves as an aid or assistant to the user. In this role, the expected form of interaction is approximately that of what would be expected of two human

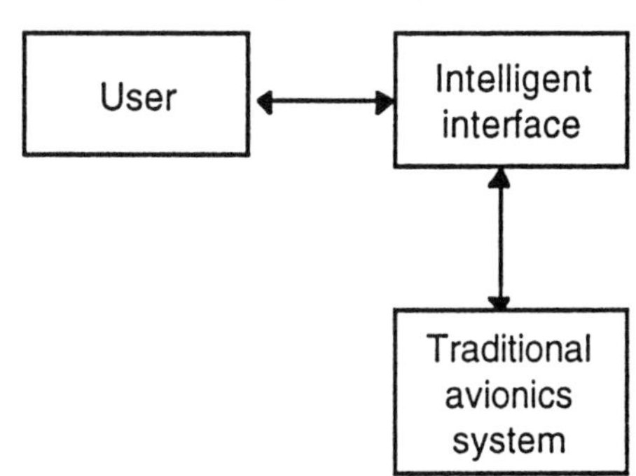

FIG. 22.1. Intelligent interface stands between the user and the traditional system.

users who as a crew must interact to operate a system. Thus, we would expect a communication about intentions—what each of the agents is doing, expects to do, and what the other is doing. We would also expect certain information to be highlighted or emphasized, as in "look at the fuel transfer system—see that growing imbalance." Although traditional automation may have some idea of what is worth emphasizing, there is never a claim that such emphasis is either intelligent or complete. Closely related to the topic of intention is that of permission and authority. Intentions may be communicated not merely to share information and update user situation awareness, but also to seek permission or concurrence for execution of certain actions.

The remainder of this chapter discusses the components and structure of an intelligent interface. The intelligent interface has models of crew intentions, crew information needs, and aircraft hazards. These models are intended to be intelligent enough to avoid many of the problems associated with traditional avionics.

INTENTIONALITY

One pointed criticism of modern avionics is its minimal or nonexistent understanding of what is happening overall. In other words, the avionics has at best a limited model that is useful in understanding what the flight crew is attempting to do, what responses by the crew are appropriate, or even the situation in which the crew finds itself. Virtually all communication between humans takes advantage of or even depends on a contextual model. Because the avionics lacks such a model, communication with it seems difficult. For example, extra communicative acts are required because there is no contextual model to fill in the gaps.

The reason for the minimal contextual model within the avionics is that the data representations within the avionics are intended primarily to support the avionics itself in its automatic control of the aircraft. Despite claims that another automation philosophy drives design, a detailed audit of the purposes for which each datum is represented would show a predominant bias toward supporting automation rather than the user. The reason for this bias is primarily that of the organizational forces in which the avionics software designer practices. Unless there are strong forces to the contrary, the avionics software design and representations will support the needs of the avionics software itself rather than those of the user.

As envisioned earlier, current automation uses control limits as an approximation to authority limits. These limits are typically quite modest in the coverage, at least with respect to robustness. Thus, one finds that the limits err by being too aggressive or too conservative, with the aforementioned China Air an example of aggressiveness (in that the engine balance was never announced) and failure to engage as an example of conservativeness.

Content of a Model of Intentions

The representations of a model of intentions depend on what uses are made of it to make contextual decisions about information, authority, hazards, and so forth. For example, to make decisions about information needs, the model must be able to recognize situations when a particular piece of information is and is not needed. Naturally, to accomplish this the model depends on the structure of the domain, particularly the situations that occur, and the user actions that can be taken as well as the structure of the information—its meaning, breadth, and resolution.

The model of intentions is typically based on a structure that is similar to a task analysis that describes the missions that might be expected to occur. This generality is important because the intentional model should cover—to the extent possible—all possible situations so that there is no gap in the functional coverage of the intelligent interface. There are a number of differences between the structures used for manual task analysis, as practiced during design, and intentional models that are used for real-time decision aiding. First, a manual task analysis uses tasks that are identified by the designer during analysis. A model of intentions must have a structure that permits online task recognition by a computer. The feasibility of recognition is a primary concern of the architect of the intentional model. For example, distinguishing between a touch-and-go practice landing and a real landing in advance is impossible. Only after the fact can the two be distinguished, and after-the-fact intentional structures are somewhat less useful. The model is more useful if the recognized intentions are temporally leading or at least concurrent indications of activities, rather than trailing indications (Geddes, 1989).

The intentional model should be made an active component of the intelligent interface. It should react to the situation and activate or deactivate elements of the model structure to keep the model as an accurate description of what is happening. In recognizing transitions in the situation, it is advisable not to rely primarily on the passage of time, unlike a task analysis conducted for design purposes. In other words, the task should be recognized based on what is happening rather than on what happened previously. Those who have organized models temporally, although the models seem attractive, have found that models often get stuck in particular states when an out-of-the-ordinary turn of events occurs. Time should be used only when it truly is the mechanism that activates and deactivates elements. Both temporal processing and situational processing have found places in current models, although situational elements tend to be generally more descriptive than temporal elements.

A hierarchical intentional model is employed for several reasons. First, a hierarchical model can represent the situation at several different conceptual levels. As such, it can describe multiple intentions for a low-level action. For example, a military aircraft could jettison fuel either to reduce weight for landing or to send a visual signal when the jettisoned fuel is ignited with afterburners. Second, a hierarchical model may be able to represent a situation at a high level when model limitations prevent a low-level description. For example, determining that the pilot is attempting to land may be possible, but determining the runway may not.

The intentional model is also both descriptive and prescriptive. The descriptive model represents what the pilot is attempting to do. The prescriptive model represents a recommendation from decision aids to the pilot about what should be done. Although these two models are processed separately, they share a common representation to facilitate communication in the intelligent interface. The descriptive and prescriptive models can synergistically aid the pilot. When the pilot begins something new, the descriptive model can recognize it. This description is then specialized by the prescriptive model and displayed to the pilot. The result is a system that anticipates the information the pilot needs.

Our experience has been that the intentional model, shared as described earlier, has had a profound influence on the architecture of the intelligent interface. Designers, once exposed to the power of the model, tend to make heavy use of the model in functional processing. It is difficult to appreciate how significant such a model can be in an intelligent interface. Because we as humans use an intentional model, it is difficult for us to appreciate the significant impact such a model makes on software without seeing it firsthand.

INFORMATION MODELS

Although modern avionics possesses considerable data, it has little idea of which data are actually information (data of value) to the pilot. As a result, the avionics is limited in its ability to change the information on the displays. Intelligent information management automatically selects the displayed information, its level of emphasis, and its format. There are several reasons for the importance of intelligent information management in a modern crewstation. First, there is a tremendous volume of data available. Current trends in data communication and data storage aboard aircraft promise to increase this volume. Second, most of the data is without value at any one particular time, although presumably all of it can be of value under some circumstances. As a result, the user can spend a considerable amount of time and effort selecting the appropriate data for display (Small & Howard, 1991).

Representational Needs in Information Modeling

The three representational needs in information modeling are, in order of importance, information need, emphasis, and format (Fig. 22.2). Information need is modeling the information that is relevant in current and near-term situations. Information emphasis is determining which selected information should receive increased display emphasis. Information format modeling is determining how information that is relevant and possibly emphasized should be displayed. The range of display choices includes such dimensions as location, size, color, shape, symbology, and modality.

Information Need. The foremost modeling question is, what gives rise to information requirements? The most obvious requirement for information is task execution, which is described in the intention model. Indeed, one of the traditional uses of task analysis models was to determine information requirements. Both task analysis and intentional models are top-down methods of determining information requirements. Given such a model, information requirements are associated with tasks or intentions.

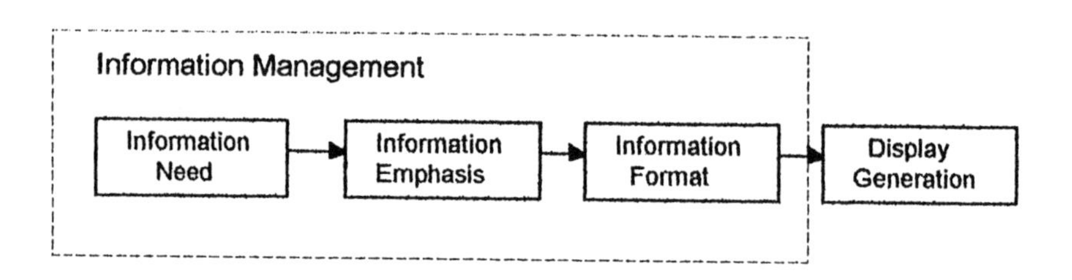

FIG. 22.2. Information management processing sequence.

The second source of information requirements is significant situational changes, or *events*. Events are detected by an assessment module that finds the few significant state changes among the many insignificant changes. Events are bottom-up sources of information because they are unanticipated within the intentional structure and are the result of noticing changes to low-level data. Because information requirements arise from two sources—events and intentions—combining the information requirements into a single representation of the pilot's complete needs is important to select the right information to display.

There are dozens of dimensions that can be used to describe information. Some examples include the priority, the use (warning, control, checking), and the type (navigation, weather, system). Starting with these dimensions makes the problem of automatic information selection and formatting seem extraordinarily difficult. Our experience has been that most of these dimensions are not useful in information management. What has been more practical is to work backward from the decisions to the inputs necessary for those decisions (i.e., need, emphasis, format).

Information arising from intentions and events can be in conflict, in that there may be more information required than fits within the available display area. Further, information requirements arising from a single type of source, such as intentions, can be in conflict with each other. Fundamentally, information selection is a resource allocation problem, and resource allocation usually means that there is competition for resources. Concepts such as priority and importance are essential to resolving these conflicts optimally.

Event-based information may or may not be of interest to the pilot at the moment the event is detected. One way to determine whether an event is of interest is to map it onto changes in capabilities of the aircraft: thrust, sensing, navigation, fuel, and so forth. In the intention model, interest in changes in the capabilities can be expressed. Determining whether there is any interest is simply a matter of looking at the capability concerns of the active tasks.

Information Emphasis. The emphasis function causes certain display elements to be given perceptual attributes that cause their salience to be higher. The exact attributes changed are determined in the format decision, which is discussed later. The emphasis decision merely decides what information should be emphasized, but not how it should be done. The remainder of this discussion on emphasis concentrates on how this decision can be made.

There are numerous reasons why information should be emphasized. One reason is the extreme consequences of not taking into account the information content. An example is failure to complete some item on a checklist, such as setting flaps for takeoff and landing. Another reason is doctrine. For example, current commercial air transport practice is to call out the altitude at 100-ft intervals during landings (in the United States). This information is already available on the altimeter; it is emphasized by a voice callout. In addition to emphasis, this procedure also presumably increases altitude awareness for one crew member beyond what would otherwise be the case. Emphasis is also required by unusual or urgent events to alert the crew to the unusual conditions and to secure a prompt response to the urgent event.

Correspondingly, representation of the need for emphasis can be included in several places within the models discussed thus far. The most frequent source of emphasis is in significant events. Typically, if an event's information is important enough to display (i.e., to change the displays in favor of this information), then emphasis is also required. Intentions can also serve as a convenient structure on which to associate the need to emphasize, particularly with regard to information that is emphasized due to policy. A third source, not unlike the first, is the error monitor (discussed later), which monitors for hazardous situations and produces events to notify the pilot.

Information Format. The final decision to be made about information is the display format or modality. This includes selection of modality (aural, visual, both) and the display element to use (bar chart, digital, tape, etc.). The motivation behind these decisions is to configure the most directly visible perceptual aspects of the displays to convey information to the user in a way that it is most suitable for its intended use.

This process is most akin to the traditional human factors display design process. In fact, it could be considered an online version of the same. Any of the criteria used in conventional display design are potential candidates for the online version. Examples include:

The accuracy with which the information must be perceived.

Whether rates of change of the displayed information are needed.

Whether the information is to be used for a checking or control.

Recent research by Shalin and Geddes (1994) has shown considerable performance improvements by adapting the information format to the task.

The final shape of the modality selection depends to a great deal on the display flexibility available to it. For example, if there are few display capabilities for varying display of altitude, there is little need to consider it during design. A highly capable information manager can place heavy demands on display generation. In practice the display programming effort has been at least as large as the information management effort. Display flexibility is less of a restriction in selection and emphasis of information because virtually all display generators have some way to perform both of these functions.

Conclusion

It is worth making a few points about visibility and appreciation for various types of functionality. Selection of information is a highly visible function. Its changes are immediately apparent on the display system, and if correct, they immediately give a positive impression. Emphasis, which consists of highlighting visual displays, is less apparent than selection, and formatting is less apparent than either of them. The practical implication of this difference in functionality, or rather this perception of functionality, is that some consideration needs to be given during requirements generation to the perceived value of the various functions.

One criticism of traditional automation is that it takes too much authority and awareness away from the pilot. At first glance, the same claim could be leveled at information management because it controls displays automatically. There are several reasons why this claim does not hold up under scrutiny. The first reason is that the information manager is intended to improve the pilot's situation awareness by showing information that the pilot would have selected anyway. The pilot does miss out on the reasoning that went into display selection, but the displays selected should make this reasoning evident.

The second reason is that the pilot always has the authority to override the information manager. When this happens, the displays are under manual exclusive pilot control until certain conditions are met. Conditions might be that a certain amount of time has elapsed or the situation has changed significantly. The best condition to use is still an active research topic.

A third reason is that the behavior of the information manager can be adjusted to the domain in a way that does not diminish the pilot's authority. For example, display selection may remind the pilot by placing an icon at the edge of a screen. Automatically replacing one display with another would be reserved for the most immediate and serious problems. One approach to this essential problem is to develop a model of how crew members share information and would perform this task for each other (Zenyuh, Small, Hammer, & Greenberg, 1994).

ERROR MONITORING

Aviation has adopted several approaches to the problem of human error in the cockpit: selection, training, human factors design, and accident investigation. Selection tests candidates before and during training to select those most likely to succeed. Training attempts to reduce error by instilling correct practices and knowledge in pilots, and by practicing unusual or dangerous situations in simulators or with instructors so that pilots will be prepared for them should they really occur. Human factors design attempts to design crewstations and tasks to eliminate error-prone characteristics. Accident investigation is a feedback loop that investigates accidents and incidents to identify defects in all of the approaches.

These various approaches combine as a layered defense against human error. Selection, training, and human factors design operate before flight and investigation after an accident or incident. One layer that has received less than its due is concurrent error monitoring that takes place during flight, especially as implemented in an intelligent interface.

Currently, concurrent detection of errors is implemented by redundant human operators and the somewhat limited practices implemented in today's avionics. One justification for multiple crewmembers, human air traffic controllers, and shared radio channels is as a check on human errors. To some extent, traditional avionics had limited checks for errors.

In traditional software, including avionics software, the perspective on error detection is virtually always that of the software itself. Software detects those errors that affect its processing and functionality, not necessarily those that represent errors of the pilot. The reason for this is that what error processing is present is embedded within the functional context of the avionics. In other words, the purpose of the avionics is to accomplish some control function, and error processing is possible only to the extent that it fits within the functional context. From an organizational context, the budget for error processing is controlled by those who seek increased functionality.

For example, consider the error of commanding the aircraft to land at the wrong airport. The avionics could check this command for plausibility before passing it to a lower level control loop. It is only within the context of automation that checks are performed on commands. If the automation is turned off or the pilot flies manually, the checks in automation are not made.

Some avionics, such as the ground proximity warning system, have a functional purpose that is solely oriented toward error detection. There are several problems with this type of functionality in today's traditional avionics. First, there is too little of it. Second, it is not very intelligent. Third, it lacks independence from the functional aspects of traditional avionics. Finally, it does not consider consequences.

Comprehensive Coverage

A number of authorities have advocated an electronic cocoon around the aircraft. As long as the aircraft was operated within the safety of this cocoon, the crew would be free to do whatever it wanted (Wiener & Curry, 1980). However, the crew would be alerted as the aircraft drew near the edge of the cocoon and prevented from unintentionally leaving the cocoon. We are still far from achieving this goal.

The goal of a complete, airtight safety cocoon seems to be theoretically unachievable in the sense that one could convincingly demonstrate that an aircraft could never be operated unsafely. There are simply many ways to fail. A more practical approach is to enumerate a large number of unsafe situations and actions and then prepare software to detect each one of them.

Intelligent Monitoring

Traditional ground proximity warning systems (GPWS), which have yielded a reduction in accidents, have often been criticized for an excessive false alarm rate. Consider the information available to a GPWS unit. The desired altitude for a warning is keyed into the unit. A radar altimeter measures the altitude directly underneath the aircraft. The aircraft's position can be measured by GPS/GLONASS and INS systems, and it would not be difficult to install ground altitude data on CDROM for the area over which the aircraft is expected to operate. Using these data recently made available, a true cocoon could be established, at least with respect to the problem that GPWS is intended to prevent.

The point is that to make an intelligent decision about the need for an alarm requires access to many sources of information, not just one. It is easy to build an alarm system that provides many false alarms, and then rely on the pilot to sort out the true alarms. The delirious effects of excessive false alarms on human performance have been known for some time. Our contention is that more sophisticated processing of more inputs should reduce the false alarm rate and thus improve response to warnings.

Independence from Traditional Avionics Functionality

To be successful, error detection should be functionally independent of the traditional avionics. There are several reasons for this. First, the purposes of traditional avionics and error monitoring are dissimilar. To embed error monitoring within traditional avionics is to limit monitoring to those situations that are recognizable from within the traditional avionics perspective (i.e., the data it stores). Second, an error monitor must have data structures and models that meet its primary purpose of detecting and thus avoiding errors. From an object-oriented perspective, the separation of error monitoring from traditional avionics would be to give first-class status (i.e., object status) to errors.

Finally, error monitoring should not depend on whether functions are enabled in traditional avionics.

Consequences Are an Important Structural Orientation

To consider GPWS again, its processing has no concept of consequences. Of course, the designers knew that flight below a certain altitude could have most severe consequences. However, none of that consequential reasoning is present in the GPWS unit itself. It merely compares the radar altitude to the threshold and sets off an alarm if the threshold is transgressed. As a result, GPWS can be considered to cause many false alarms, at least when evaluating the true state of the aircraft with respect to the distance to the ground. In other words, if the aircraft continues on its current trajectory, how far is it from the edge of the cocoon? The GPWS has no representation about cocoon borders.

The point is that a situation is a hazard only if the potential consequences are severe. Evaluating errors requires a structural orientation toward consequences within the monitor. Other approaches that have been tried include omission of prescribed actions and human error theory. Experience with the omission of actions is that the severity of the error is usually unknown without other information about consequences. Human error theory can suggest what might be done about repairing the error (e.g., omission or repetition errors are somewhat self-diagnosing) or explain why it happened. Understanding the cause for an error may be useful for the designer or the pilot (in a debrief), but it serves little purpose in alerting the pilot to a serious error (Greenberg, Small, Zenyuh, & Skidmore, 1995).

CONCLUSION

A high-level architecture for an intelligent interface has been described. The description represents a family of solutions, not an individual solution. The model structures described provide a sufficient framework for dealing with the problems of automation. One key property of the intelligent interface is that it increases the level of intelligence in the avionics to correspond more nearly with the authority already granted. Historically, the intelligent interface represents the next generation of automation that is built on the current layers of flight management systems and autopilots. The purpose of the intelligent interface is to support the pilot's decision making. This differs from the purpose of traditional automation, which is to automate tasks for the pilot.

System engineering becomes an essential effort for any system constructed with an intelligent interface. To build an intelligent interface component requires a thorough understanding of the purpose, benefits, and employment of each subsystem component to be installed on the aircraft. This understanding is a necessary part of the system engineering because knowledge engineering about the subsystem is necessary. The questions asked include:

> What are the effects of using the subsystem in each of its modes on the aircraft and environment? This is aimed at producing a device level model of the subsystem.
>
> When is it appropriate to use the subsystem?

How does the subsystem interact with other subsystems on the aircraft, especially with regard to the previous questions?

When would using the subsystem be inappropriate or erroneous (as opposed to ineffectual or irrelevant)?

It is widely suspected that those who construct new systems do not fully understand all the ramifications and implications of what they are designing. Answering these questions will challenge the designers of traditional avionics.

Those who have participated in the design of an intelligent interface have found that the scrutiny given the traditional avionics design can produce a more purposeful product. During design, a number of intelligent interface models are constructed of how the entire system will be used from the pilot's perspective. This model building can yield benefits by improving the design as well as incorporating the intelligence interface functionality. For example, I was once preparing the knowledge base for an information manager that was to select from one of several available displays. It used information requirements that were associated with intention structures, and it picked the best display by matching its information display capabilities to the information requirements that had been accumulated from all active intentions.

While I was debugging the knowledge base, I noticed that some displays were never chosen and that other displays were frequently chosen. Naturally, this was assumed to be a fault of the knowledge base, as it was under development. After close observation of the display selection algorithm, I came to the conclusion that the algorithm and knowledge base were correct. The problem was in the displays themselves. Some displays lacked elements that were always demanded. Other displays seemed to support situations that would never occur. To fix the problem, new display designs were prepared. The point of this example is that evaluation of the information content of displays was made possible only by computing a match of displays to situations. Although it would certainly be possible to prepare a written argument that the displays are well designed, computation was a more compelling proof.

The strength of this approach lies in the executable nature of the knowledge. It is not merely that the knowledge can then be applied via execution to produce simulations of the effects of the subsystems along with the associated knowledge. As such, it represents a powerful system engineering capability that is especially useful to those who are responsible for the overall technical project administration. To succeed, those developing this type of system require the support of management to get answers to knowledge engineering questions. These answers are not always simple to obtain but can benefit both the design and the operation of complex systems.

REFERENCES

Foushee, H. C., & Helmreich, R. L. (1988). Group interaction and flight crew performance. In E. L. Wiener & D. C. Nagel (Eds.), *Human factors in aviation* (pp. 189–227). San Diego: Academic Press.

Geddes, N. D. (1989). *Understanding human operator's intentions in complex systems*. Unpublished doctoral thesis, Georgia Institute of Technology, Atlanta, GA.

Greenberg, A. D., Small, R. L., Zenyuh, J. P., & Skidmore, M. D. (1995). Monitoring for hazard in flight management systems. *European Journal of Operations Research, 84*, 5–24.

Hammer, J. M., & Small, R. L. (1995). An intelligent interface in an associate system. In W. B. Rouse (Ed.), *Human/technology interaction in complex systems* (Vol. 7, pp. 1–44). Greenwich, CT: JAI Press.

Kieras, D. E. (1990). The role of cognitive simulation on the development of advanced training and testing systems. In N. Fredrickson, R. Glaser, A. Lesgold, & M. G. Shafto (Eds.), *Diagnostic monitoring of skill and knowledge acquisition* (pp. 51–73). Hillsdale, NJ: Lawrence Erlbaum Associates.

Mecham, M. (1994, May 9). Autopilot go-around key to CAL crash. *Aviation Week & Space Technology*, pp. 31–32.

Rouse, W. B., Geddes, N. D., & Curry, R. E. (1987). An architecture for intelligent interfaces: Outline of an approach to supporting operators of complex systems. *Human-Computer Interaction, 3*(2), 87–122.

Rouse, W. B., Geddes, N. D., & Hammer, J. M. (1990). Computer-aided fighter pilots. *IEEE Spectrum, 27*(3), 38–41.

Rouse, W. B., Salas, E., & Cannon-Bowers, J. (1992). The role of mental models in team performance in complex systems. *IEEE Transactions on Systems, Man and Cybernetics*, 22(6), 1296–1308.

Shalin, V. L., & Geddes, N. D. (1994, October). Task dependent information management in a dynamic environment: Concept and measurement issues. *Proceedings of the IEEE International Conference on Systems, Man and Cybernetics*, pp. 2102–2107. San Antonio, TX.

Small, R. L., & Howard, C. W. (1991, October). A real-time approach to information management in a pilot's associate. *Proceedings of the Tenth Digital Avionics Systems Conference*, pp. 440–445. Los Angeles, CA.

Suchman, L. A. (1987). *Plans and situated actions*. Cambridge: Cambridge University Press.

Wiener, E. L. (1988). Cockpit automation. In E. L. Wiener & D. C. Nagel (Eds.), *Human factors in aviation* (pp. 433–461). San Diego: Academic Press.

Wiener, E. L., & Curry, R. E. (1980). Flight-deck automation: Promises and problems. *Ergonomics, 23*, 955–1011.

Zenyuh, J. P., Small, R. L., Hammer, J. M., & Greenberg, A. D. (1994). Principles of interaction for intelligent systems. The human–computer crew: Can we trust the team? *Proceedings of the Third International Workshop on Human-Computer Teamwork*, Cambridge, UK.

23

Weather Information Presentation

Tenny A. Lindholm
The National Center for Atmospheric Research
Research Applications Program

Before we can fully relate the aviation user's weather information needs to the function or task at hand, both now and in the future, we must comprehend and contrast the differences between the current aviation environment and whatever is envisioned for the future. It follows that we begin this chapter with a description of current aviation weather information available to users and then provide a vision of the future air traffic control system and associated weather information. The human factors and display design for this future system can then considered from a functional point of view, allowing user needs to evolve in the proper context. Each element and user within the National Airspace System (NAS) will be considered, as well as the implications of their interactions within the system. The goal is to develop a true system-level understanding of weather information display needs, because users and their needs for aviation weather to support decision making are so varied. The approach taken here insures that the functional interactions of the air traffic control system and its allocation of weather display capabilities will be well understood as the system is modernized.

AVIATION WEATHER DISSEMINATION—CASE STUDIES

On the afternoon of August 2, 1985, a wide-bodied jetliner crashed short of Runway 17L at Dallas-Fort Worth Airport, with considerable loss of human life. The only indication of a hazard to the flight crew was a moderate to severe rain shower just to the right of the approach course. The National Transportation Safety Board (NTSB) listed as a probable cause the occurrence of a small, short-lived but severe downburst now widely known as a microburst (Fugita, 1986).

A number of years ago, a South American jetliner crashed while on approach to New York's John F. Kennedy Airport during marginal ceiling and visibility conditions. The aircraft arrived in the terminal area with just enough fuel and reserves to complete a normal sequence to landing. After unplanned and lengthy holding delays, the aircraft crashed about 10 miles short of the runway from fuel starvation. The NTSB identified the flight crew's lack of awareness of evolving weather impact to normal sequencing as a contributing factor.

In 1988, a jetliner crashed on departure from Detroit with, again, considerable loss of life. The NTSB identified the flight crew's failure to properly configure the aircraft flaps and leading edge devices as the probable cause of this accident. However, cockpit voice transcripts clearly indicate confusion by both pilots as they tried to convert encoded alphanumeric weather data to a graphic portrayal on a map in order to make the information more usable on departure. These actions could have contributed to flight crew distraction while completing checklist actions. In fact, the cockpit voice recorder revealed that the captain of this flight remarked, "Not now, I'm weathered out," in response to updated alphanumerics just prior to departure (Sumwalt, 1992).

Finally, a medium-sized jetliner crashed in 1991 just south of the Colorado Springs Airport after encountering a severe roll to the right and immediate dive to the ground. The NTSB was unable to agree on a probable cause for this accident.

For the major air carriers, the NTSB reports that 35.6% of all accidents between 1983 and 1988 were weather related (NTSB, 1989a). For general aviation, 26.1% of all accidents, and 38.1% of all fatal accidents, were weather related (NTSB, 1989b). Aside from the obvious economic and societal costs associated with these numbers, improved weather information can potentially save NAS operators literally hundreds of millions of dollars annually through elimination of needless ground holds for weather, unnecessary diversion of aircraft and associated system-wide disruption, more efficient routing, and better planning for ground operations and terminal sequencing. The FAA stated that 80% of all delays greater than 15 min are caused by weather, resulting in an "economic loss" of $1 billion per year (FAA, 1992b). Airspace planning can and should become more strategic. In order to realize these benefits, however, airspace system designers must address the following top-level user needs relative to aviation weather:

1. The pilot (the ultimate end user) needs accurate, timely, and appropriate information. The flight crew in our first example case study required precise information regarding the occurrence, position, and intensity of a weather phenomenon that is very localized and short-lived. A weather detection and dissemination system for this type of hazard should meet these needs and nothing else, for this phenomenon requires immediate evasive action by the pilot. There is no time for interpreting a complex display.

2. The system or structure that supports the ultimate end user—operations, meteorology, air traffic control—needs accurate, timely, and appropriate information. In our second case study, the NTSB verified the existence of an information void confronting this crew. The crew was unaware of the developing weather situation to the point that proper fuel management was relegated to a very low priority. The information needs for those supporting airspace operations are quite different, as we show later, from those of the end user.

3. The weather information must be presented in an optimal form that makes it quickly and unambiguously usable as a decision aid. This requires the system developer to understand the extrinsic as well as the cognitive aspects of each user's task. In our third case study, experienced pilots were bewildered with a long, complex alphanumeric teletype message that was trying to describe a complex, three-dimensional graphic. Pilots think in four dimensions (the fourth being time); decision support information, in most cases, should be presented similarly.

4. A mechanism should be in place to develop new weather "products" or refine current ones to address new aviation hazards as they are identified. Our final case study left a field of questions unanswered regarding the possible existence of essentially clear air, terrain-induced, extremely severe wind phenomena. History has shown that the scientific community and aircraft operators can work extremely well together to precisely define new hazards, determine how to detect or predict them, and get appropriate information to the end user in time to initiate a proper response (Mahapatra & Zrnic, 1991).

In summary, to best serve the aviation weather user community, weather observations and forecasts must improve, aviation weather information dissemination must improve, and users of aviation weather must be trained in its proper use, given their functional role. These goals translate into human factors issues that will challenge human factors researchers and practitioners alike.

HUMAN FACTORS CHALLENGES

The use of weather information by the pilot and other users is regulated heavily, and in some cases is mandated in terms of source, format, validity, and geographic orientation. Unfortunately, weather forecasting is an inexact science, and aviation weather "products" in the past have been lacking in credibility. The combination of regulated use and incredible weather products has created a situation where different classes of users have different expectations, and even individuals within a particular class will indicate differing needs. This creates a human factors challenge in that, in this inexact environment, the system developer must probe deeply into the cognitive use of such information from varied sources. That is, individual users will, through personal experiences, establish perceptual biases on how weather information currently affects their behavior. The difficulty, of course, is transitioning the user from this experiential-based behavior to one that is rule-based for time-critical weather encounters.

This general observation translates into a host of other issues facing the human factors community, which include:

1. This is a crucial first step: Identify who the users are and what function (that is, the complete task) they perform. End users typically use weather information as a decision aid and are not, generally, meteorologists. To the system developer, this means information presentation must be tailored to exact needs and criticality. Users that are not end users fulfill a number of roles, such as traffic management, weather information provider, and air traffic control. We precisely define classes of users later.

2. Identify how weather information is to be used, both in today's system and that of the future. The system developer must establish a realistic vision of how the air traffic control system will permit operations in the future automated environment, and how classes of users will be permitted to interact with the environment. Obviously, we can only predict the future the best we can, but the user cannot be expected to elicit his or her needs within an unknown system.

3. How should the weather information be displayed to various classes of users? Too much information, or improper display of needed information, is dangerous. The example case study where the flight crew was required to decode alphanumerics to

obtain needed information is an excellent instance of improper information transfer, and other examples abound.

4. Carefully identify exactly what information is needed. We need to think of weather end users as process controllers at a top level (Wickens, 1984), who need information to manage and control the flight process. Approaching information needs through a functional breakout using systems engineering analysis techniques will objectively identify information needs in a top-down fashion, separating such subtasks as course tracking and controlling the flight process. Conceptually, consider placing aviation weather users along a continuum that indicates relative closeness to actual aircraft operation. For example, a meteorologist at a regional National Weather Service (NWS) forecast center would be placed on one end of the continuum, and a pilot would be placed on the opposite end. In general, users fall onto the continuum according to how close they are to actual aviation operations (National Research Council [NRC], 1994). The closer the user is to actual operations, the less is the amount of analytical detail the user needs to aid the decision process. In other words, the operators require decision aids, and the weather information providers require analytical detail. Above all, give the user no more than what is needed.

5. How does the system developer integrate new weather technologies into current airspace system functions? Two issues are crucial in terms of new technology insertion. First, the airspace management and control system is slowly evolving, whereas weather information technology is in the midst of a revolution. Second, users for the most part have never had the kind of weather products that are about to be introduced—extremely high resolution (spatially and temporally) and accurate—and they do not know how to use them to the best benefit.

6. Finally, how can a sound human systems engineering approach integrate with needed scientific research to produce more advanced aviation weather products to handle not-yet-defined atmospheric hazards? We have seen revolutionary success with the concepts of concurrent engineering in large system development. A similar hand-in-hand approach to air traffic control and aviation weather product development would help insure user needs are addressed throughout the development process.

When we talk about integrating user needs with the development process for an air traffic control function, which includes advanced weather information, we are trying to capture the concept of shared situational awareness among all national airspace users. If we can cause our transfer of weather information to represent a common mental model of the weather situation across the spectrum of aviation users, we can enhance decision making in a cooperative way. That is, decisions are mutually arrived at based on the same perception of system state. This is a top-level goal; we need to address how this goal filters down into lower functional levels within the system, and how do functions, information needs, and goals interrelate at the lower levels. These broad questions suggest the need for a top-down, systems engineering approach to user needs as they relate to function. This concept will permeate the chapter.

We approach these questions in greater detail as this chapter unfolds.

TRANSFORMATION OF AVIATION WEATHER

I stated earlier that the aviation weather system is in the midst of a revolution, whereas the national airspace system is slowly evolving to higher levels of automation. In a contradictory sense, the aviation weather revolution will evolve also, to supply information when and where appropriate to support the NAS evolution. The revolutionary aspect of change relates to the precision, timeliness, system integration, display capabilities, and above all accuracy of weather information provided to NAS users, and implementation of these changes will begin in the very near term (McCarthy & Serafin, 1990, p. 4). It is important to summarize these phased changes to weather information dissemination so we can properly address their implications to the various classes of users.

Gridded Data

The heart of the emerging aviation weather system is known as the Aviation Gridded Forecast System (AGFS) and is being validated and fielded by the National Weather Service. The AGFS is a national, four-dimensional database consisting of atmospheric variables of most interest to aviation—wind speed and direction, clouds and other impediments to visibility (from temperature and relative humidity), turbulence, and icing. In addition, gridded radar mosaics from a national Doppler weather radar network will provide three-dimensional views of convective systems. The fourth dimension provides forecast information.

The greatest impact to the aviation user is that one will now have the data to support weather products in all three spatial dimensions. That is, the user can now "slice and dice" the atmosphere anywhere over the United States and view graphic depictions of the aviation impact variables (AIVs) that are route and altitude specific to the user's particular needs. The concept of a national database also lends itself well to frequent updates from varied sources, greatly impacting accuracy of weather information and forecasts.

Observations—Density, Timeliness

In an effort to increase user credibility in aviation weather forecasting, a major revolution is occurring in the sensing of hazardous and operationally significant weather. In 1990, spacing of the most basic observation (balloon radiosonde measurements) was about 250 miles. A number of new sensing mechanisms are planned to increase atmospheric sampling by several orders of magnitude. Some of these include inflight sampling and data link, automated surface observing systems, wind profiler networks, and the Doppler weather radar network. These data will be used to provide users better current information and will also be used to increase the accuracy of weather forecasts in three spatial dimensions (McCarthy, 1991).

Temporal and Spatial Resolution

The present observing system was designed to provide information about large-scale weather phenomena that shape the evolving nature of weather conditions. However, the weather events of most interest to aviation are of much smaller scales—tens of kilometers and less than 1 hr in duration. With greater computing capability, increased observation density, and the AGFS, spatial and temporal resolution will increase to better meet the needs of aviation users. For example, the mechanisms that create and propagate turbulence are very small, on the order of a few kilometers or less. Because of increased temporal resolution, the NAS user can expect more frequent forecasts based on updated data, and potentially better decision aids for preflight and enroute operations.

Forecasting Capabilities

Given much higher resolution input data on the state of the atmosphere, it does not necessarily follow that forecasting capabilities will improve. Forecasters are faced with the same dilemma we want to avoid with other users within the NAS, and that is data overload. As part of the aviation weather revolution, the scientific community (FAA, 1992a; McCarthy, 1993) is concurrently developing the algorithms and automated processes to transform the huge amount of incoming raw data into the aviation impact variables, contained in the national four-dimensional AGFS, that will support graphics portrayal to the user. An integral part of this effort is an ongoing verification program that documents accuracy of the resulting information and recommends improvements to the algorithms to enhance accuracy.

The impact to the user will almost be immediate in terms of better forecasts and increased resolution. The accuracy and resolution improvements will continue for many years as driven by the needs of the user.

New Advanced Weather Products

The concept of "weather product" has emerged, and will continue as user needs evolve. To illustrate, let us explore an example. A significant weather advisory describing the potential for moderate icing (very similar to the alphanumeric advisory presented to the pilots in our third case study) might be issued to a pilot in the following encoded format:

> WA OR CA AND CSTL WTRS FROM YXC TO REO TO 140 SW UKI TO 120W FOT TO 120W TOU TO YXC LGT OCNL MDT RIME ICGICIP FRZLVL TO 180. CONDS SPRDG EWD AND CONT BYD 08Z.

First, this advisory is very difficult to read and understand. It requires the pilot to plot the corners of the affected area on a flight chart, then observe if his route of flight passes through the volume. Second, once the pilot does the plotting exercise, he observes that the affected volume encompasses a three-state area up to 18,000 ft. Finally, when compared to actual icing encounters, he might find the area of actual icing conditions might be only 25 miles square. So, when we consider a weather product, we think of meteorological information tailored to route and altitude, that is spatially and temporally accurate, and that is presented in a display concept that is appropriate to the user.

ADVANCED AVIATION WEATHER AND THE NATIONAL AIRSPACE SYSTEM

Now we need to address the evolving NAS by describing a vision of the future at a very top functional level. In what sort of system will the end user control his or her process, and what weather information needs will confront the non-end-user?

NAS Evolution

A quick observation is in order before we focus on the future NAS. Weather information will, for some time to come, have to support the current, largely human-directed and workload-intensive NAS structure. A tremendous human factors challenge exists with this task, because we can expect the growth of air traffic to continue, with little assistance to address the problem for the human in the loop. Weather information to the aviation user will have to overcome years of incredulity and perceptions, plus be presented to the user such that it helps with task accomplishment, before we can expect any derived benefits to accrue.

> Today's aviation weather system provides imprecise information, covers huge geographic areas, and often overpredicts or entirely misses adverse weather conditions. When weather conditions are marginal or rapidly changing, the safety, efficiency and capacity of aviation operations are compromised. (McCarthy, 1993, p. 1)

Today's aviation weather information is basically data rich—it requires some understanding of meteorology, is difficult to interpret relative to a particular situation, is not very timely, and generally is not route or altitude specific. We find the primary end user—the pilot—faced with a number of choices as to where to obtain weather information. The information is given in a fairly standard format, usually alphanumeric and compatible with typical data manipulation and transmission schemes such as a teletype. It often provides a textual description of weather hazards, clouds, winds, and other information of interest to the pilot over a much larger geographical area than needed, plus terminal weather conditions for departure and arrival. Some hard-copy or computer graphics are available, but generally these products require the pilot to take extraordinary efforts to obtain them. Little information, other than verbal updates, is available to the pilot in flight. This situation is true for the commercial as well as the general aviation pilot. For the ground user involved with air traffic control and management, we find a better situation in that computer graphics are more prevalent. They are on separate displays from the primary workstation, however, and require mental integration with the four-dimensional process being controlled. A considerable amount of data is available on paper, which many times has to be transferred manually to a graphics display prior to use. Information is routinely updated on the order of every 12 hr, except for hazardous weather conditions, which can be updated as often as necessary. This description is necessarily brief, but paints a picture of the weather information system that will be replaced gradually over the next decade.

The current weather system essentially feeds a manually operated NAS. Pilots are responsible for avoiding hazardous weather conditions appropriate for their operation and type of aircraft, and they do so in a tactical way using a see-and-avoid concept or airborne weather radar. Automatic weather updates are not routinely provided to the pilot. Controllers maintain an overall awareness of hazardous weather conditions that might impact their area of responsibility, but are not required to separate aircraft from weather. Strategic planning of routes due to weather does occur, but many times it is based on an incomplete picture of the weather state. As a result, traffic is usually organized in an "in-trail" structure, requiring aircraft to fly at less-than-optimal altitudes, speeds, and routings.

NAS modernization will introduce automated functions that will transition the controller from tactical control to strategic traffic management. The pilot is transformed into a systems monitor who will no longer personally interact with air traffic control. These things will, of course, occur in carefully orchestrated stages over many years, and the aviation weather system must match the needs of each stage. A vision of the NAS of the future shows aircraft separation being maintained by satellite communications; computer routing of aircraft to permit direct and optimal routings; extensive use of satellite and data-link communications for flight management and weather information; highly interactive and graphic displays for process management; and, overall, strategic planning being the rule instead of the exception. As the NAS evolves, so must the preciseness and informational content of the aviation weather supporting it.

Users of Aviation Weather

The aviation weather users of today will change as the NAS changes, but generally the functions performed will remain intact, possibly performed by computer or defined by other roles in the system. That is why it is so important to assume a task or functional orientation when allocating weather information needs. The following discussion of aviation weather users is by no means exhaustive, but illustrates conceptually the broad spectrum of user needs being addressed.

Airline and military users cover the type-of-user spectrum, from meteorologist to end user, or pilot. Weather information needs, similarly, span the spectrum from raw data to decision aids, and careful consideration of these needs can literally make the difference between high payoff or miserable failure of the informational impact to operations.

Functions within the broad area of air traffic control also require diverse approaches to defining needs. Floor controllers within ARTCC sectors and Terminal/Radar Approach Control (TRACON), for example, are concerned with tactical control and separation of traffic in smaller geographical areas. Relative to weather, this function is perhaps limited by today's incomplete and imprecise weather information picture—that is, better information and presentation might expand the strategic planning role of these users. The ARTCC Traffic Management Unit (TMU) and Air Traffic Control System Command Center (ATCSCC), on the other hand, are primarily concerned with strategic planning and traffic flow management from a national level. Weather information needs tend to be at a top-level for the most part, but also needs to include detailed and precise descriptions of weather conditions at key, or pacing, airports across the country. The Air Traffic Control Tower (ATCT) is much like the pilot—aids to support quick and accurate decision making, and not data, are clearly needed.

Conceptually, a completely different set of needs is represented by the many information producers and providers of today's aviation weather information, and these functions will continue to exist for the foreseeable future. These users need varying shades of data, because they are predominantly highly trained in meteorology and their primary role is to transform atmospheric data into sometimes rigidly defined aviation weather products. There is a continuum here also. For example, meteorologists in the National Weather Service (NWS) and Aviation Weather Center (AWC) rarely interact with aviation operators and/or end users. They generate the current and forecast weather products for passive transmission to other users and providers. On the other hand, Automated Flight Service Station (AFSS) specialists are primarily concerned with communicating weather information via briefings to pilots, both on the ground and airborne, in response to specific requests. Additionally, there are numerous commercial weather service providers that disseminate weather information to end users in the form of hard-copy graphics, text, verbal briefings, and computer graphics.

Perhaps the most important point that can be made relative to a human systems approach to aviation weather systems development is this: Each step of the way, we are never certain how this new capability will impact a person's job performance. We mention briefly that more precise, complete weather information might create a more efficient method of planning air traffic—that is, make planning more strategic—but we will never really know until a near full-capability prototype is evaluated operationally. As we set the stage for this important process, it is instructive to again consider the concept of user classifications.

CLASSES OF USERS

Recall the concept introduced earlier that places users along a continuum based on their relative closeness to actual aviation operations. This is how the system designer should look at classifying users according to information needs. This is, of course, a good starting point for the entire process to be introduced shortly.

We've been using the term *end user* throughout to refer loosely to the pilot operating within the NAS. The end user represents one end of the continuum. The function represented here is a seeker and consumer of information to support strategic and tactical decision making. This implies, from a workload management viewpoint, that this information should be immediately useful with little or no mental capacity required to make it match the situation or transform it from data to useful information. It should be presented in such a way that it already matches the temporal and spatial situation driving its need. It is a decision aid. Users in this class are, of course, the flight crew, and some air traffic control users such as ATCT specialists and traffic management personnel in the TMUs and at the ATCSCC.

Meteorologists who transform the huge amount of atmospheric data into usable aviation products fall on the other end of the continuum. They use the finite, absolute detail to form mental and physical models of what is happening to generate specific weather events, and form products to convey information in specific formats to other users. They need structure and organization to the large amount of data they must assimilate. For this reason, data formatting must fit the assimilation process they mentally use to forecast weather. As an example, because atmospheric trends are so important to forecasting, meteorologists will frequently ask for weather data and graphics to be presented in a movie loop, or time-sequenced images. This aids the meteorologist in visualizing the developing weather pattern.

The users that fit between the extremes on the continuum are the information conduits to other users. These users include meteorologists (when filling their briefing role), AFSS specialists, ARTCC sector supervisors and controllers, airline dispatchers, and any other users who, as part of their function, are required to convey an understanding of the weather state to another consumer of information. It follows that, based on a particular function, a single user's needs can vary considerably in the course of performing a task.

HUMAN FACTORS SYSTEM DESIGN ISSUES

We have mentioned a number of theoretical human factors principles and constructs, and identified at a top level some of the human factors challenges facing the system designer of aviation weather products and displays. Now we need to address them more within a formal human factors framework, but still in the context of aviation weather, while leading into a discussion of the process needed to address them properly.

Paradigm Fixation

Users are not always right. They cannot anticipate with complete accuracy how they will use new technology. I call this phenomenon *paradigm fixation,* and it occurs whenever new technology introduces a new informational environment for the user. As a designer, one must build a system that users will want when it gets here, not build the system they want as they see things today (Lewis & Rieman, 1993). In practice, lengthy operational exposure during development is the only way to fully understand how new information will be assimilated into the current task structure. Even then, there is a nagging doubt that operational usage did not expose one or more crucial information needs or exercise every critical decision path.

We mentioned the "evolution" of automation within the NAS versus the "revolution" in aviation weather sensing, forecasting, and display. The natural tendency, for some time to come, will be for the user to continue business as usual. This presents a real difficulty to the system designer, who must somehow elicit the future weather information needs from a user who is working within the constraints of an NAS that is evolving in small but sure steps. Even in a rapid prototyping environment, where a near fully capable system is exercised in an operational role, lack of confidence in and lack of time to be creative with a new source of information, and the comfort of the current task structure, will probably result in invalid user feedback.

Validation, Verification, Evaluation

Closely related to the preceding discussion, and absolutely crucial to the solution of any human engineering problem, is the issue of validation/verification. Validation is the process of assessing the degree to which a test or other instrument of measurement does indeed measure what its supposed to (Hopkin, 1993). Woods and Sarter (1993) go further in saying that validation as an iterative evaluation should be an integral part of system design rather than something tacked on at the end. It should help the designer improve the system and not simply justify the resulting design. Validation as a process should provide "converging evidence on system performance." Verification, on the other hand, is the process of determining the truth or correctness of a hypothesis (Reber, 1985), or for this context, it should explore how far major system elements, such as software, hardware, and interfaces, possess the properties of theories, or confirming appropriateness by gathering environmental information informally (Hopkin, 1993).

These two general concepts can be further placed into the context of aviation weather products. There is an operational *validation* that must occur, and a continuous meteorological *verification* to measure and document how accurately algorithms describe meteorological phenomena. Obviously these two tasks go hand in hand. I would suggest another concept is needed to complete the triad—*evaluation.* Evaluation is a means to determine how well initial goals have been achieved (Hopkin, 1993). However, evaluation may also reach conclusions about feasibility, practicality, and user acceptance. I mention each—validation, verification, evaluation—separately only to relate the task at hand to a formal process, when indeed elements of all three should be integrated into the design process, and they probably have to occur iteratively.

As always, certain "social issues" need to be considered in any val/ver/eval (validation/verification/evaluation) process. For aviation weather, or any verifiable information that can directly impact aviation safety, one must address the type of evaluation that is acceptable to users and public, how much time and money should be used to test, and, relatedly, when is it good enough to place in the public's hands verses current, not-as-reliable information. Finally, what level of security and reliability will the public demand from the system? These questions will impact how extensive an operational evaluation is permitted prior to "complete" verification (Wise & Wise, 1993), and will certainly have an effect on the perceived value of an evaluation.

A final issue—technical usability—is absolutely crucial and needs to be an integral part of evaluation. Basically, technical usability refers to traditional human factors issues—display design, domain suitability, human–machine interface match to cognitive problem solution. If the system is difficult for the user in any way, evaluation results will be confounded and difficult to parse. The val/ver/eval process should try to eliminate user annoyances or system usability issues as soon as possible in order to keep the user focused on true operational utility.

To paraphrase past experience: "An ATC system can only be validated in operation" (Smoker, 1993, p. 524).

THE "CRITERION PROBLEM"

The "criterion problem" directly relates to some of the issues already identified in conjunction with evaluating the utility of a weather information system against a set of goals. The definition of the "criterion problem" as given by Fitts is the problem of validating and verifying procedures and equipment against a goal, purpose, or set of aims (Fitts, 1951). Three related problems arise in the context of NAS modernization (Harwood, 1993):

1. The NAS goal is to provide safe, expeditious, and orderly flow of traffic. The goal of NAS modernization (and aviation weather improvement) is to enhance NAS safety, capacity and efficiency. The problem here is to establish objective, concise measures of success that represent consensus of the user community.
2. Lack of knowledge of task structure of individual and controller teams in current and future ATC environments. The system developer must consider the resulting ATC environment after each incremental change to the NAS on the way to full modernization, and the resulting user needs.
3. There is a requirement for sensitive criterion measures when transitioning from old to new systems, to maintain continuity and safety. The system developer must be sensitive to consequences of the new system for controller task performance. The question becomes, "when is it good enough for testing, and implementation?"

There are no answers to the questions raised by the "criterion problem." A systematic approach to the evaluation phase of development, to include extensive user and customer involvement and agreed-upon criteria for success that are goal-oriented, will help. The following discussion on task structure should provide some guidance on relating task performance to user needs and evaluation of system utility.

TASK STRUCTURE

This section summarizes some of the literature relating task to validation and evaluation, and places this knowledge in the context of aviation weather information. It is so very important for the system developer to understand the physical and cognitive processes involved with user task accomplishment. During system evaluation, the system developer should be as familiar with the task as the user so that meaningful observations of how the system is being used can be made. Kantowitz (1992) suggested that external validation has three components, or characteristics: A validation process must be representative of subjects, of variables, and of setting. This means evaluation should occur on the job with real users. To extend this basic rule of thumb further, development and evaluation of complex human–machine systems require not only an adequate representation (prototype) of the user interface, but also an appropriate understanding and representation of the task environment.

Going further, validation sometimes identifies unanticipated interactions between the user, work environment, system, and outside environments, creating a need for redesign or resulting in suboptimal system performance. Extending Kantowitz' suggestion, the "TEST" model identifies variables and interactions that have to be addressed or controlled in design and validation. "TEST" is an acronym for *task, environment, subject* (that is, don't use highly skilled, or test subjects; use normal people in normal work environment, which includes normal lighting, fatigue, and stress), and *training* (i.e., you must train, but the user is not fully trained in use of the system on task, and system performance will improve on a learning curve). Measures of effectiveness are system performance, operator performance, workload reduction, skill acquisition, and development of individual differences and strategies (Jorna, 1993).

A definition of a *task* is, the act of pursuing a desired goal through a set of operations by utilizing the potentials of the available system (Sanders & Roelofsma, 1993). I would supplement this definition by suggesting there is a hierarchy of "subtasks" that are somehow dependent on each other for accomplishment. These dependencies might be predicated on simply subtask accomplishment, or possibly on information about the state of the environment, and are where true informational needs of the user come from.

Closely related to the task structure is the concept of a mental model, which very much guides task accomplishment by virtue of its regulative function on activity and a reflection of the world on which the subject acts (Dubois & Gaussin, 1993). Most definitions of a mental model include such words as symbolic; relation-structure of the system it imitates (Waern, 1989); parallel to reality; and knowledge of the potential characteristics of a part of the external world (Payne, 1991). In general, the system developer should strive to understand the user's mental model of either the real world the user is attempting to describe, or the prescribed task structure (as the user perceives it), and match the user interface and system software structure to the model. This has many implications to future use, to include training and eventual user acceptance. A mismatch would explain why certain users, when given a new tool to incorporate into job accomplishment, initially have difficulty in assimilating it into their task, and why a suitably long evaluation period is necessary. A simple example will further illustrate this point. The AFSS Preflight position is responsible for generating and giving preflight weather briefings to pilots that are route and altitude specific. The structure and content of this briefing is regulated and rigid. AFSS specialists have a very specific mental model of this structure, which guides the seeking of information to maintain awareness and develop specific briefings. A weather display system must match this structure in terms of how the user interfaces with the system and the type of weather information presented. If it does not, severe training and acceptance penalties will surely result.

Often, a revolutionary decision aid or information source, such as aviation weather, is introduced operationally that fundamentally changes the current task accomplishment or structure, or even eliminates the need to perform a particular task. In the interest of improving overall system performance, the mental model justifiably should be adapted to accommodate the new capability. That is, the system developer accepts the fact that a fundamental change in doing business is necessary, and so is willing to accept (perhaps) a significant training burden. Keep in mind the following, from Stager (1993): "A design requirement that is often overlooked is that the information provided at the interface must support the acquisition, maintenance, and updating of a *valid* mental model for the user" (p. 105). Ensure that the process defines a valid model of system operation for the user.

As the NAS modernizes, we would expect increasing levels of task automation to occur, to include the processing, display, and impact of weather information on system decision making. This situation introduces the question of how tasks are allocated between human and machine, or task sharing (Hancock, 1993; Kantowitz & Sorkin, 1987). There seems to be universal agreement that functions or tasks should be performed by human and machine together. This implies that automation should provide an effective decision aid to the human, and not always make crucial decisions for the operator. By taking this approach, the operator is physically and cognitively entrained in the system operation, enhancing overall situation awareness. I don't believe this to be a principle that applies in every case; however, I believe it especially applicable to aviation weather information use by NAS users. In either case, if the concept of task allocation is used, a top-down understanding of the task is essential.

To summarize, in the same sense of a knowledge-based system development, the designer must more than observe, but really understand each action of the user, and what cognitive process and mental model the user exercises as he or she invokes each task. This understanding also has direct application to the graphical user interface, display, and software structure design.

DISPLAY AND INFORMATION TRANSFER ISSUES

There is a huge body of literature on computer–human interface and display design. I cover here only some top-level principles that have been validated through research or operational experience and that have relevance to weather information presentation.

A very important principle that follows from the previous discussion on task structure is this: Insure compatibility of analog display with the orientation of human mental representation. Remember that most aviation users are spatially oriented. For example, digitized aircraft altitude requires significant mental processing to transform it into the analog conceptual representation and adds error (Grether, 1949). The same applies to display movement—make route and height depictions congruent with the real world (e.g., east-to-west is shown right-to-left).

The concept of perceptual schema—the form of knowledge or mental representation that people use to assign stimuli to ill-defined categories, a general body of knowledge about a perceptual category, developing from perceptual experience with examples rather than a strict listing of features (Wickens, 1984)—is important when applied to information supporting effective air traffic control. Posner and Keele (1968) suggested that there are two components—a general representation of the mean, and some abstract representation of the variability. We find that experienced controllers and pilots have developed schema relative to system states that potentially have significant impact to system operation, such as weather hazards. Posner and Keele's research suggests that variability must be addressed directly in training and display design, and not just the prototypical case. For example, the system developer should fully understand the system "outlier" states important to the user so that training and display design properly highlight them.

There is some advantage to top-down, context-driven processing when the user seeks information. That is, have the user work from top-level displays down to the level of detail needed, rather than just flash a "chunk" of information on the screen in response to a request. This is because there is strong research evidence that says human processing generally follows this model (Wickens, 1984). Closely related is the concept of holistic processing (Wickens, 1984), which "describes a mode of information processing in which the whole is perceived directly rather than as a consequence of the separate analysis of its constituent elements" (p. 164). According to Navon (1977), this does not mean that perceptual analysis of the whole precedes analysis of the elements, but rather suggests that the conscious perceptual awareness is initially of the whole and that perception of the elements must follow from a more detailed analysis. Generally, aviation users seek to process stimuli in a holistic way. The reason for this is to relieve demands on short-term memory (the "whole" sticks in short-term memory better than an enumeration of its parts). If a detail in the whole is important, it must be highlighted in some way.

Why do weather graphics seem so important to aviation users? Pure human factors research suggests that providing information along one dimension—text, color, quality—and then expecting the human to make an absolute judgment about the stimulus is very difficult. When more dimensions are added, research suggests less information is transmitted along each dimension, but more overall information is transmitted. This lessens the demands on the human; a graphic decision aid uses many dimensions for this reason. It also matches the structure of the mental model used by the user (Wickens, 1984). With graphics, it is important that the image given to the user or pilot be right the first time. Research suggests that subjects place an undue amount of diagnostic weight to the early stimulus, called *anchoring*. Subsequent sources of evidence are not given the same amount of weight, but are used only to shift the anchor slightly in one direction or another (Wickens, 1984). Also, research suggests the number of cues has a negative impact on response accuracy. This means cues must be informative, not so salient that it overrides the information content, reliable, and limited to those that are truly diagnostic of the situation you want to convey. Once again, provide the user no more information than is needed (Wickens, 1984).

The implication is that information and displays must be unambiguous, context driven, and require little mental interpretation, that the structure of the graphical user interface and display must match the user's model and task structure, and that the information transmitted must be accurate the first time. A goal, of course, is to enhance user awareness of the weather state without negatively impacting workload.

WORKLOAD AND TIME SHARING

Users within the NAS have learned to be efficient at time sharing and allocating limited attentional resources. The current air traffic control system and weather displays have had an important role in this learning process, because they almost require the user to develop work-around strategies to be effective. Generally, research on workload suggests there is a trade-off between maintaining situational awareness and minimizing workload through automation. The trick is to optimally allocate workload between human and computer aiding without removing the human from the cognitive control loop. The user's primary task is flying or controlling air traffic. Information and decision aids are provided to make this task easier, and they must be provided so they can be used without the user having to devote excessive attentional resources and mental processing. All of the preceding principles and concepts really relate to this one, top-level goal.

We must think of users within the NAS as process controllers in varying degrees. That means they spend a lot of time attempting to predict future environmental states to support decision making. Computers and automation can have tremendous impact on the current NAS via computer aiding (Wickens, 1984) and anticipating future goals and system responses based on states. The cognitive element of task performance can perhaps best be left to the user.

THE PROCESS—ADDRESSING THE ISSUES

Model of System Development

I introduce now a model of system development that addresses the issues we have previously identified and discussed relative to aviation weather presentation. It is fairly generic, not very profound, but carries with it a practical human factors focus that includes many of the necessary elements we have been talking about. Please refer to Fig. 23.1.

As can be seen, the model describes a circular, iterative process that includes numerous opportunities for the user to actively influence system design. In the upper left corner, a concept enters the process and exits as a specification at various stages in its life while the process continues to operate on future iterations of the concept. In other words, this is a living process that really never ends. For the weather products, we include scientific assessment and research to address meteorological verification and systematic product improvement. User needs are addressed through assessment, validation of strawman capability and rapid prototype capability, and feedback into science and engineering development. The engineering function performs system integration, software development, human factors assessment, and display design. This model represents true concurrent engineering with built-in preplanned product improvement.

Obviously, this concept of system development is an ideal that could be applied to just about anything. It particularly works well with aviation weather product development because we are essentially building a national capability from the foundation up. The many opportunities for user involvement in the process are not just an ideal situation, but rather an absolute necessity, for we are unsure exactly what weather information and capability will demonstrate true economic benefit. We also address the difficulty of supporting a NAS in evolution, because the process is circular. And we overcome the user's tendency to fixate on current task structures with lengthy operational evaluations and iteration.

Perhaps one difficulty we still need to address is how the system developer gets an objective handle on user tasks, functions, and the information needs to support their efficient accomplishment. The user's help is again invaluable, along with analytical tools to help structure the investigation.

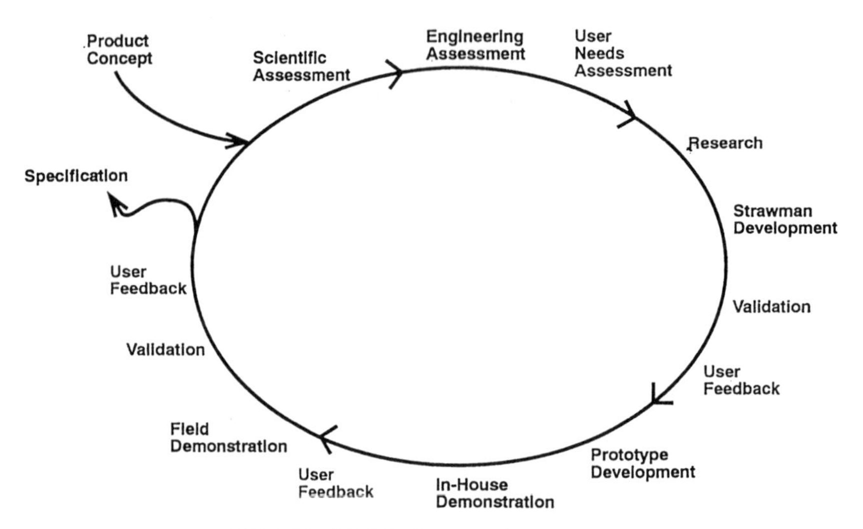

FIG. 23.1. Model of system development.

Systems Engineering Model

There are a number of tools available that can be used to model the functions of a system, but actually what is important is the process or way of thinking and not the idea that one must use a formal tool. The example in Fig. 23.2 (Bachert, Lindholm, & Lytle, 1990) is from the integrated computer aided manufacturing definition (IDEF) modeling language, which has the same structure and basis as the SAINT or MicroSAINT modeling packages. What is important is that the analysis focuses on the functional hierarchy of the system and that it proceeds in a top-down fashion to insure a one-to-one mapping from top-level to lower level functions. As one identifies each activity and breaks it down into lower level activities, interrelationships begin to emerge

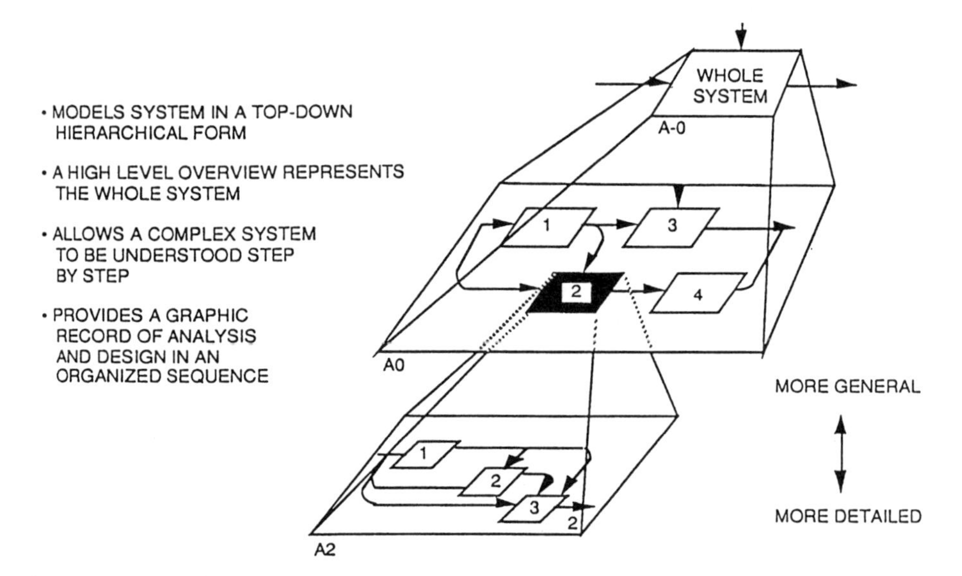

FIG. 23.2. Systems engineering model.

between activities, and the resources and information needed to support each activity become more specific. The method establishes a source document for tracking information requirements. It becomes an easy matter, for example, to identify resource need differences as the functions begin to evolve with the NAS. Also, with new and most changed systems, a task analysis using IDEF and tested with SAINT or MicroSAINT can be used to uncover shortcomings in the old system that need changing. Interrelationships between tasks and information resource needs will clearly emerge, from top to bottom, to support decision making (Sanders & Roelofsma, 1993).

Rapid Prototyping

Rapid prototyping has no value unless it is rapid. Incremental capability improvements need to be in front of the user as soon as possible. The development process, really, shows two phases of rapid prototyping. First, an early capability is demonstrated in-house to gather early user feedback, which might actually occur several times. Second, a near-full-capability prototype is evaluated in a operational field demonstration over a suitably long period of time.

The idea of quick exposure is important, but keep in mind an apparent paradox associated with rapid prototyping. You must field the system in order to evaluate its effect on the user and his or her function. In order to do this the system must be "realized" at various levels, such as the operational and display concept, the software, and the human interface. Once the designer is committed to a design represented by a particular realization, acceptance of change becomes more difficult. This means evaluation questions asked late in the design process are very narrow, and tend to be focused on how to show that the system could work as opposed to finding out what the contextual limits of the system are and identifying the impact of the new device on the joint cognitive system and operational processes (Woods & Sarter, 1993). How to decouple the effects of different levels of realization from actual system effects during the evaluation is difficult, as mentioned previously in the context of the user interface. For example, things like system crashes or improper contrast and colors will illicit responses about the system that are not of interest, or unfairly harsh, and miss entirely the issues the evaluator is trying to resolve (Woods & Sarter, 1993).

I want to emphasize the idea of early and continuous user involvement throughout the weather product development cycle. User exposure and familiarity with the various new weather capabilities are really the only way to overcome fixation with current paradigms and identify benefits derived from previously un-thought-of ways of doing business. Users are creative with new capabilities, and can provide the system designer with insight that will never emerge from an engineering design team. On the negative side, attempts to incorporate user input late in the development process is much more difficult and expensive than building in user needs from the beginning. The $4 billion aviation weather modernization program is full of examples of creative user input, derived from actual operational experience, that demonstrate collectively a huge potential impact to airspace safety and efficiency.

By emphasizing early and often operational demonstrations using prototypes, the system developer maintains a task (function) orientation throughout development. The process of development using a task orientation might go like this: First, write descriptions of all the tasks and circulate to users for comment (understand their cognitive process and resource or information needs); second, rough out an interface design and produce a scenario for each of the tasks; third, represent the scenarios using storyboards, never taking the process out of the context of the task. The "cognitive walkthrough" is sometimes helpful in understanding the mental processes being used in accomplishing a particular task (Lewis & Rieman, 1993). The task orientation is essential; however, there are some cautions:

1. As a developer, you cannot cover every task the user is expected to perform, and without a top-down, functional approach you will probably miss the cross-task interactions (Lewis & Rieman, 1993). These functional and informational interactions are very important, because these will tend to identify areas where duplication of tasks will occur, and efficiencies will arise out of better planning.
2. Do not fragment tasks when evaluating new information capabilities with users. Field tests using operational tasks are better than laboratory tests using fragmented tasks. You need the complete task (Lewis & Rieman, 1993).
3. Use process data as opposed to bottom-line data. Process data are observations of what the test users are doing and thinking as they work through the tasks. Bottom-line data give a summary of what happened: How long did users take, were they successful, were there any errors? When evaluating in the field, process data validate the task process and even identify better ways of doing the task (Lewis & Rieman, 1993).

My bottom line is: Do not shortcut the rapid prototyping process—to do so will introduce rapid prototyping risks associated with too little training and lack of familiarity with the new weather capability, thereby inhibiting user opportunity to develop new strategies that integrate performance evaluations and workload management (Jorna, 1993).

User Needs

At this point, we can begin to speak definitively about aviation weather needs in today's air traffic control environment. In actual practice, the process described in Fig. 23.1 has been very helpful in merging scientific discovery with what current and future aviation systems need in terms of weather information. We now bring all we have discussed in this chapter together to identify a starting point for pinpointing top-level user needs. Operational evaluations using prototypes will then have a departure point from which detail can be identified for a particular class of user.

In general, critical issues in designing any user interface are: Sufficient information is available for each step in the decision sequence, the information is accessible within the time available for the decision step, and the information is presented within the context of the overall task structure (Clare, 1993). These are good rules to follow when developing a weather information and display system for the end user who requires a decision aid. To generalize for all classes of user, a weather information dissemination system must (Tonner & Kalmbach, 1993):

1. Make the job easier.
2. Be easy to work with.
3. Be compatible with neighboring systems.
4. Not lull controllers (users) into false sense of security.
5. Keep the user in the decision loop, in charge, well informed.
6. Be reliable.

Situation Awareness

Weather impacts air traffic control and operations more than any other state variable. Although reference to a weather information system will be accomplished by the user to support decisions, a display system must double as a continuous provider of information about the weather state both now and in the future. Further, proper information about the weather must be provided across the spectrum of users to avoid perceptions of contradictory state information and promote cooperative decision making. When we want to simply give all users the same information in the proper context, with the goal of all users perceiving the same situational state, we are enhancing their shared situational awareness (SA).

The concept of shared SA is fairly simple—obviously, controllers and pilots cannot be expected to arrive at the same decision about, say, routing if both are depending on conflicting information on location and severity of enroute thunderstorms. Endsley (1988) defined SA as the perception of the elements in the environment within a volume of time and space, and the comprehension of their meaning and projection of their status in the near future. Pew (1994) offered a concise definition of a situation: A situation is a set of environmental conditions and system states with which the participant is interacting that can be characterized uniquely by its priority goals and response options. Keywords for the system developer are perception, time and space, projection into the near future, system state, priority goals, and response options. We must be sure of the user's perceptions of the information and display we provide, so that it supports the proper decision. The information must be location specific, and must provide some predictive element of the system state. Finally, the system developer must be absolutely sure of the user's goals and what options are driven by perceptions. These points are absolutely essential to maintain and enhance SA, and the weather information system should be expected to do just that.

Here is one more point relative to SA. Weather information can be provided in three basic formats: graphics, usually derived from gridded weather aviation impact variables; icons showing location and spatial extent; and text. In general, the best format for the end user is defined by the amount of interaction with the information that the user can be expected to accomplish. For example, with three-dimensional graphics the user can interact with and "slice-and-dice" the graphic as desired. With text and most icons, the user receives the information and acts with no interaction or further inquiry. Clearly, more user interaction is good for SA, but only to the extent that workload is impacted or there is diminished value in terms of task accomplishment. However, the icon is a good indication of a state variable that needs immediate attention, and text is a provider of background information that requires no action. So, again we are forced to determine how the information relates to the task.

Needs by Class of User

Based on extensive operational prototype experience with advanced weather products and display concepts, here are some verified top-level user needs.

Given NAS users' previous experience with an outdated, sometimes inaccurate aviation weather system, validated data and information perhaps take on the highest priority to spur user confidence in the consistent good quality of information. Before users actually make strategic traffic flow decisions based on advanced weather products, and real benefits are derived, the information and decision aids will have to prove their utility. In general, we can also state that users who are not meteorologists (or whose tasks do not require them to interpret raw atmospheric data) need information and display concepts that require little or no interpretation. By this, I mean most weather information will be presented in the form of decision aids to these users, and that some data will always be required to support certain functional needs.

In general, our end user (the pilot) needs highly focused information rather than data, and he or she needs decision aids relevant to the immediate situation rather than general advisories. Weather observations should be transformed into visualizations and decision aids that facilitate direct inferences and immediate action by the pilot. When I distinguish between weather information and decision aids, I am suggesting the concept of a hierarchy or pyramid of data and information, stratified by increasing direct relevance to aeronautical operations (NRC, 1994). Temporal and spatial needs are defined by the fact that the systems and phenomena of most interest to aviation and many other activities are of small scale—tens of miles in size and often less than an hour in duration. Weather products include variations in wind speed and direction, clouds and other impediments to visibility, turbulence, icing, and convective systems such as thunderstorms (NRC, 1994).

User information and presentation concepts are always tailored to task. For this reason, the meteorologist (or, if you will, the weather product generator function) requires considerable atmospheric data that is properly formatted to enhance and quicken its utility. Sensed and measured data should be presented as close as possible in the same format as the user's mental model of the forecasting process. For example, graphic looping is very useful for visualizing atmospheric processes in change and in large measure aiding the forecasting task. Decision aids in the form of simple graphics or icons are probably not very useful in this environment.

Display concepts must meet these general needs:

1. Aviation impact variables or state-of-the-atmosphere variables (SAVs) for meteorologists and decision aids must be presented in easily comprehended formats, with visualizations.
2. The presentations must be available for flight planning through remote computer access, and must be an integral part of the information for controllers.
3. The system must provide for user interaction with three-dimensional, route-specific, vertical cross sections so pilots can easily study alternatives and make informed decisions. The same weather information, perhaps formatted differently, must be given to distributors of weather information, such as the AFSS.
4. The information must be provided to the cockpit while airborne. Much is yet to be learned on the exact products and format to provide, and the supporting communications or data-link infrastructure; however, we certainly know the top-level needs well enough to begin a user needs investigation through our rapid prototyping process.

PERSPECTIVES ON THE PROCESS

The process described in Fig. 23.1 works very well in the development of any user-intensive system, and it really works for defining advanced weather products. Since the mid-1980s, we have collected a considerable amount of experience bringing together needs from all classes of users within the NAS, and this work will continue for some time. The most important ingredient for success is extensive user involvement from the beginning. As always, there are potential pitfalls, so it is fitting that I conclude this chapter with some of the more crucial lessons learned from exercising this process.

Advanced weather products represent a new information technology. With the introduction of any new process, source of information, or task structure, the system developer should temper user-stated needs with observations from actual operational experience. The user will initially state his or her needs from the perspective of how his or her task was performed in the past. By all means note and consider all user input, but give extra attention to contradictions between observed needs and user-stated needs. Always be aware that users initially have difficulty in using new information sources that are revolutionary in terms of content. Mistrust, difficulty in fitting into current task structure, and the inherent delay involved in formulating new structures that use the new information are all valid reasons for this.

Use caution with user feedback that comes from a laboratory experiment or an environment that is different from an operational setting where the user has had the opportunity to use the products for a significant period of time. The displays and advanced weather technology should be exercised with the complete task and under identical physical conditions as those in the operational setting: that is, lighting, stress, interaction with other functions, workload surges, decision making, and planning expectation (Smoker, 1993, p. 524).

Because most weather products will be presented as decision aids for users, be aware of some difficulties with this form of information, as identified by Wickens (1984). The complexity of the aid can make it difficult to use, because many times it depends on the ability of the user to "divide and conquer," or divide the overall decision into its constituent components for analysis in the context of the given aid. This method of processing can alter the nature of the problem to the point that it is difficult to evaluate the success of decision aid—that is, would the decision have been better with or without it.

Relative to the display of complex weather graphics, there are two very interrelated aspects of display design that must be evaluated with users in an integrated way—display functionality and display concept. The complex weather graphics that will be presented require some means for the user to manipulate, interact with, "slice and dice," zoom and pan, very easily and intuitively. The display concept refers to the structure built into the software that defines how the user interacts with the entire display. The concept or structure should strike a match with the user's mental model or defined task structure as a starting point.

Do not permit such basic human factors issues such as color/contrast, usability, display clarity, and functionality to confound the results of attempting to evaluate the weather product for utility. Work these difficulties out carefully in-house prior to introducing the system to operational use.

Similarly, carefully verify the accuracy of meteorological information in parallel with your validation effort. The introduction of inaccurate and invalid weather products during rapid prototyping will very quickly destroy the credibility of the entire demonstration.

And finally, as a broad issue to guide implementation, airspace users and the service providers should agree to implement future airborne and ground systems and improved standards and procedures simultaneously to ensure incremental benefits throughout the transition period (ATA, 1994). Advanced weather products are in a sense perishable; that is, if benefits are not shown quickly, support for better aviation weather will be lost. And without an NAS that is structured to use better weather information, benefits will be difficult if not impossible to show.

REFERENCES

Air Transport Association. (1994, April 29). *Air traffic management in the Future Air Navigation System.* (White paper)

Bachert, R. F., Lindholm, T. A., & Lytle, D. D. (1990, October 1–4). *The training enterprise: A view from the top* (SAE Technical Paper Series 901943). Presented at the Aerospace Technology Conference and Exposition, Long Beach, CA.

Clare, J. (1993). Requirements analysis for human system information exchange. In J. A. Wise, V. D. Hopkin, & P. Stager (Eds.), *Verification and validation of complex systems: Human factors issues* (p. 333). Berlin: Springer-Verlag.

Dubois, M., & Gaussin, J. (1993). How to fit the man-machine interface and mental models of the operators. In J. A. Wise, V. D. Hopkin, & P. Stager (Eds.), *Verification and validation of complex systems: Human factors issues* (p. 385). Berlin: Springer-Verlag.

Endsley, M. R. (1988). Design and evaluation for situation awareness enhancement. *Proceedings of the 32nd Annual Meeting of the Human Factors Society, 1*, 97.

Federal Aviation Administration. (1992a, April). *A weather vision to support improved capacity, efficiency and safety of the air space system in the twenty-first century* (FAA Code ASD-1). Washington, DC: Author.

Federal Aviation Administration. (1992b, November). *Proposed aviation weather system architecture* (FAA Code ASE-400). Washington, DC: Author.

Fitts, P. M. (Ed.). (1951). *Human engineering for an effective air-navigation and traffic control system.* Washington, DC: National Research Council.

Fugita, T. T. (1986). *DFW microburst.* Chicago: Satellite and Mesometeorology Research Project (SMRT).

Grether, W. F. (1949). Instrument reading I: The design of long-scale indicators for speed and accuracy of quantitative readings. *Journal of Applied Psychology, 33*, 363–372.

Hancock, P. A. (1993). On the future of hybrid human-machine systems. In J. A. Wise, V. D. Hopkin, & P. Stager (Eds.), *Verification and validation of complex systems: Human factors issues* (p. 73). Berlin: Springer-Verlag.

Harwood, K. (1993). Defining human-centered system issues for verifying and validating air traffic control systems. In J. A. Wise, V. D. Hopkin, & P. Stager (Eds.), *Verification and validation of complex systems: Human factors issues* (pp. 115–129). Berlin: Springer-Verlag.

Hopkin, V. D. (1993). Verification and validation: Concepts, issues, and applications. In J. A. Wise, V. D. Hopkin, & P. Stager (Eds.), *Verification and validation of complex systems: Human factors issues* (pp. 9–33). Berlin: Springer-Verlag.

Jorna, P. G. A. M. (1993). The human component of system validation. In J. A. Wise, V. D. Hopkin, & P. Stager (Eds.), *Verification and validation of complex systems: Human factors issues* (pp. 295–298). Berlin: Springer-Verlag.

Kantowitz, B. H. (1992). *Selecting measures for human factors research. Human factors psychology.* Amsterdam: North-Holland.

Kantowitz, B. H., & Sorkin, R. D. (1987). Allocation of functions. In G. Salvendy (Ed.), *Handbook of human factors.* New York: Wiley.

Lewis, C., & Rieman, J. (1993). *Task-centered user interface design.* Boulder CO: Textbook published as shareware.

Mahapatra, P. R., & Zrnic, D. S. (1991). Sensors and systems to enhance aviation safety against weather hazards. *Proceedings of the IEEE, 79.*

McCarthy, J. (1991, June 24–26). The aviation weather products generator. *American Meteorological Society, 4th International Conference on Aviation Weather Systems*, Paris, France.

McCarthy, J. (1993, March 2–4). *A vision of aviation weather system to support air traffic management in the twenty-first century.* Presented to Flight Safety Foundation 5th Annual European Corporate and Regional Aircraft Operators Safety Seminar, Amsterdam, the Netherlands.

McCarthy, J., & Serafin, R. J. (1990, November 19–22). An advanced aviation weather system based on new weather sensing technologies. *Proceedings, 43rd International Air Safety Seminar*, Rome, Italy. Arlington, VA: Flight Safety Foundation.

National Research Council. (1994, March). *Weather for those who fly*. Prepared by the National Weather Service Modernization Committee, Commission on Engineering and Technical Systems, National Research Council. Washington, DC: National Academy Press.

National Transportation Safety Board. (1989a). *Annual review of aircraft accident data, U.S. air carrier operations. Calendar year 1987*. Washington, DC: Author.

National Transportation Safety Board. (1989b). *Annual review of aircraft accident data, U.S. general aviation. Calendar year 1987*. Washington, DC: Author.

Navon, D. (1977). Forest before trees: The presence of global features in visual perception. *Cognitive Psychology, 9*, 353–383.

Payne, S. J. (1991). A descriptive study of mental models. *Behaviour and Information Technology, 10*, 3–21.

Pew, R. W. (1994). Situation awareness: The buzzword of the '90s. *CSERIAC Gateway, V*(1), 2.

Posner, M. I., & Keele, S. W. (1968). On the genesis of abstract ideas. *Journal of Experimental Psychology, 77*, 353–363.

Reber, A. S. (1985). *The Penguin dictionary of psychology*. Harmondsworth, England: Penguin Books.

Sanders, A. F., & Roelofsma, P. H. M. P. (1993). Performance evaluation of human–machine systems. In J. A. Wise, V. D. Hopkin, & P. Stager (Eds.), *Verification and validation of complex systems: Human factors issues* (p. 316). Berlin: Springer-Verlag.

Smoker, A. (1993). Simulating and evaluating the future—Pitfalls or success. In J. A. Wise, V. D. Hopkin, & P. Stager (Eds.), *Verification and validation of complex systems: Human factors issues* (p. 524). Berlin: Springer-Verlag.

Stager, P. (1993). Validation in complex systems: Behavioral issues. In J. A. Wise, V. D. Hopkin, & P. Stager (Eds.), *Verification and validation of complex systems: Human factors issues* (pp. 99–114). Berlin: Springer-Verlag.

Sumwalt, Captain R. L. III. (1992, January). Weather or not to go. *Professional Pilot*, pp. 84–89.

Tonner, J. M., & Kalmbach, K. (1993). Contemporary issues in ATC system development. In J. A. Wise, V. D. Hopkin, & P. Stager (Eds.), *Verification and validation of complex systems: Human factors issues* (p. 492). Berlin: Springer-Verlag.

Waern, Y. (1989). *Cognitive aspects of computer supported tasks*. Chichester, England: John Wiley & Sons.

Wickens, C. D. (1984). *Engineering psychology and human performance*. Columbus, OH: Charles E. Merrill.

Wise, J. A., & Wise, M. A. (1993). Basic considerations in validation and verification. In J. A. Wise, V. D. Hopkin, & P. Stager (Eds.), *Verification and validation of complex systems: Human factors issues* (p. 88). Berlin: Springer-Verlag.

Woods, D. D., & Sarter, N. B. (1993). Evaluating the impact of new technology on human–machine cooperation. In J. A. Wise, V. D. Hopkin, & P. Stager (Eds.), *Verification and validation of complex systems: Human factors issues* (pp. 133–158). Berlin: Springer-Verlag.

24

Human Factors in Aviation Maintenance

Colin G. Drury
State University of New York at Buffalo

THE MAINTENANCE AND INSPECTION SYSTEM

Before human factors techniques can be applied appropriately in any system, the system itself must be well understood by the human factors engineers. The following description of aviation maintenance and inspection emphasizes the philosophy behind the system design, and the points where there is potential for operator error.

An aircraft structure is designed to be used indefinitely provided that any defects arising over time are repaired correctly. Most structural components do not have a design life, but rely on periodic inspection and repair for their integrity. There are standard systems for ensuring structural safety (e.g., Goranson & Miller, 1989), but the one that most concerns us is that which uses engineering knowledge of defect types and their time histories to specify appropriate inspection intervals. The primary defects are cracks and corrosion (which can interact destructively at times), arising respectively from repeated stretching of the structure from air or pressure loads, and from weathering or harmful chemicals. Known growth rates of both defect types allow the analyst to choose intervals for inspection at which the defects will be both visible and safe. Typically, more than one such inspection is called for between the visibility level and the safety level to ensure some redundancy in the inspection process. As the inspection system is a human/machine system, continuing airworthiness has thus been redefined by the design process from a mechanical engineering problem to a human factors one. Inspection, like maintenance in general, is regulated by the Federal Aviation Administration (FAA) in the United States, the Civil Aviation Authority (CAA) in the United Kingdom, and equivalent bodies in other countries. However, enforcement can only be of following procedures (e.g., hours of training and record keeping to show that tasks have been completed), not of the effectiveness of each inspector. Inspection is also a complex sociotechnical system (Taylor, 1990), and as such, can be expected to exert stresses on the inspectors and on other organizational players (Drury, 1985).

Maintenance and inspection are scheduled on a regular basis for each aircraft, with the schedule eventually being translated into a set of workcards for the aircraft when it arrives at the maintenance site. Equipment that impedes access is removed (e.g. seats, galleys).

The aircraft is cleaned, and access hatches are opened. Next comes a relatively heavy inspection load, to determine any problems (cracks, corrosion, loose parts) that will need repair. During inspection, each of these inspection findings is written up as a nonroutine repair (NRR) item. After some NRRs are repaired, an inspector must approve or "buy back" these repairs. Thus, the workload of inspectors is very high when an aircraft arrives, often necessitating overtime working, decreases when initial inspection is complete, and slowly increases toward the end of the service due to buybacks. Much of the inspection is carried out in the night shift, including routine inspections of aircraft between the last flight of the day and first flight of the next on the flightline.

Maintenance can be performed either in parallel with inspection, or following the raising of an NRR. Much maintenance is known to be required in advance of inspection and can thus be scheduled before the aircraft arrives. In contrast to this scheduled maintenance, response to an NRR is considered unscheduled. At present unscheduled maintenance represents a large and increasing fraction of the total repair activity, primarily due to the aging of the civil fleet. In 1990, the average age of jet transport aircraft in the United States was 12.7 years, with over a quarter of the aircraft more than 20 years old (Bobo, 1990). From 1980 to 1988, as the aircraft fleet increased by 36%, the maintenance costs increased 96%.

HUMAN FACTORS ANALYSIS OF THE MAINTENANCE AND INSPECTION SYSTEM

One early and thorough analysis of the inspection function (Lock & Strutt, 1985) used logical models of the process and field observations to understand potential errors within the system. It is still the case that inspection and maintenance tasks need to be analyzed in more detail than the preceding systems description if human factors techniques are to be brought to bear in a logical fashion. At the level of function description, Tables 24.1 and 24.2 give a generic function listing for the activities in inspection and maintenance. Note that not all "inspection" activities are performed by

TABLE 24.1
Generic Task Description of Inspection

Function	*Visual Example*
Initiate	Read and understand workcard. Select equipment. Calibrate equipment.
Access	Locate area on aircraft. Move to worksite. Position self and equipment.
Search	Move eyes (or probe) across area to be searched. Stop if any indication.
Decision	Reexamine area of indication. Evaluate indication against standards. Decide whether indication is defect.
Respond	Mark defect indication. Write up nonroutine repair (NRR). Return to search.
Buyback	Examine repair against standards. Sign off if repair meets standards.

TABLE 24.2
Generic Functions in Aircraft Repair

Function	*Tasks*
Initiate	Read and understand workcard. Prepare tools, equipment. Collect parts, supplies. Inspect parts, supplies
Site access	Move to worksite, with tools, equipment, parts, supplies.
Part access	Remove items to access parts. Inspect/store removed items.
Diagnosis	Follow diagnostic procedures. Determine parts to replace/repair. Collect and inspect more parts and supplies if required.
Replace/repair	Remove parts to be replaced/repaired. Repair parts if needed. Replace parts.
Reset systems	Add fluids supplies. Adjust systems to specification. Inspect adjustments. Buyback, if needed.
Close access	Refit items removed for access. Adjust items refitted. Remove tools, equipment, parts, unused supplies.
Respond	Document repair.

a person with the title of "inspector." Examples are transit checks, "A" checks, and avionics diagnostics, which are often performed by an aviation maintenance technician (AMT), also known as a mechanic. Each of the functions listed has different human factors considerations as critical elements. Some, such as *search* in Inspection, depend critically on vision and visual perception. Others, such as *site access* in Repair, are motor responses where human motion and motor output are critical.

In principle it is possible to proceed through each function and task, listing the major human subsystems involved, the error potential of each, and the design requirements for reducing these errors. Indeed, the first part of this exercise has been performed for inspection by a team working for the FAA's Office of Aviation Medicine, on the basis of field observations of many different inspection tasks (Drury, Prabhu, & Gramopadhye, 1990). The error mechanisms of interest in these systems were enumerated and studied by Latorella and Drury (1992) and Prabhu and Drury (1992). Drury (1991) provided an overview of these error studies and included error breakdowns of the inspection function originally developed for the *National Plan for Aviation Human Factors* (FAA, 1991). As an example of the listing of possible errors, Table 24.3 shows those for the initiate function of Inspection.

Mere listing of possible errors is often less useful than classifying errors into the behavioral category or stage of human information processing involved. Examples of error classification schemes abound, such as Hollnagle (1989), Reason (1990), and Senders and Moray (1991), depending on the use to be made of the data. An example in Table 24.4, also from Drury (1991), shows the error framework of Rouse and Rouse (1983) applied to one task (1.8) from Table 24.3.

As a technique for structuring the systematic application of human factors to aircraft inspection and maintenance, the error approach suffers from a fundamental flaw: In

TABLE 24.3
Sample of Aircraft Maintenance and Inspection Errors by Task Step for the Initiate Task

Task	*Error(s)*
1.1 Correct instructions written.	1.1.1 Incorrect instructions. 1.1.2 Incomplete instructions. 1.1.3 No instructions available.
1.2 Correct equipment procured.	1.2.1 Incorrect equipment. 1.2.2 Equipment not procured.
1.3 Inspector gets instructions.	1.3.1 Fails to get instructions.
1.4 Inspector reads instructions.	1.4.1 Fails to read instructions. 1.4.2 Partially reads instructions.
1.5 Inspector understands instructions.	1.5.1 Fails to understand instructions. 1.5.2 Misinterprets instructions. 1.5.3 Does not act on instructions.
1.6 Correct equipment available.	1.6.1 Correct equipment not available. 1.6.2 Equipment is incomplete. 1.6.3 Equipment is not working.
1.7 Inspector gets equipment.	1.7.1 Gets wrong equipment. 1.7.2 Gets incomplete equipment. 1.7.3 Gets nonworking equipment.
1.8 Inspector checks/calibrates equipment.	1.8.1 Fails to check/calibrate. 1.8.2 Checks/calibrates incorrectly.

such a complex system the number of *possible* errors is very large and effectively innumerable. In human factors methodology it is usual to make use of existing error data, if the system has been in operation long enough, in order to prioritize the errors. However, for aviation maintenance, the error data collection systems are not particularly useful.

Currently, error reports are primarily used for documenting error situations for administrative purposes by internal or external regulatory agencies. All of these reporting systems suffer from a number of problems when regarded as feedback or corrective mechanisms at the systems level. First, they are driven by the external event of a problem being detected: If the problem is not detected, the error is not captured. In flight operations, in contrast, there are selfreporting mechanisms that capture a broader range of error events. In principle, these can be used by maintenance and inspection personnel, but in practice this is rarely the case.

Second, the feedback of digested error data to users is not well human-factored. Often the data is merely compiled rather than digested, so that mechanics or inspectors

TABLE 24.4
Example of Possible Errors for Task Step of Calibrated NDI Equipment (adapted from Drury, 1991)

Level of Processing	*Possible Errors*
1. Observation of system state	Fails to read display correctly.
2. Choice of hypothesis	Instrument will not calibrate: inspector assumes battery too low.
3. Test of hypothesis	Fails to use knowledge of NiCads to test.
4. Choice of goal	Decides to search for new battery.
5. Choice of procedure	Calibrates for wrong frequency.
6. Execution of procedure	Omits range calibration step.

must search large amounts of data with little reward. They rarely have the opportunity for this; thus, much of the data collection effort is wasted on the users.

Third, error reports in maintenance and inspection produced for administrative purposes are typically concerned with establishing accountability for an error and its consequences, rather than understanding the causal factors and situational context of the error. This type of information is not appropriate for use as performance feedback to inspectors or maintenance personnel, nor is it helpful information for error-tolerant system design. Error-reporting schemes are developed from within an organization and therefore vary greatly among organizations. The framework of these error reporting schemes is event driven and developed iteratively; thus, additions are made only with the occurrence of a new error situation. To a large extent, the information recorded about a situation is constrained by the format of the error reporting scheme. For example, in one error-reporting scheme, the reviewer is required to attribute the error to some form of human error unless the situation can be described as an "act of God" (Drury, 1991). Analysis of the data collected by such a scheme will invariably find the human at fault, rather than working conditions, equipment, procedures, or other external factors. This biased representation has serious implications for error prevention, especially considering that equipment design and job aiding have been found to be more efficacious than selection or training approaches in error prevention (Rouse, 1985). To alleviate the difficulties of inconsistency, and provide an appropriate and useful structure for error data collection, an error-reporting scheme should be developed from a general theory of the task and the factors that shape how the task is performed. Principally, the behavioral characteristics of the operator, but ideally also organizational environment, job definition, workspace design, and the operator's physical, intellectual, and affective characteristics should be considered.

At present, work is under way to produce much better error data so as to guide human factors interventions. Allen and Marx (1994) proposed the Maintenance Error Decision Aid system, in which aircraft maintenance and inspection personnel will self-report errors in a format compatible with human factors analysis methods. This project should provide the bridge between systems interpretation in terms of error taxonomies for inspection (e.g., Latorella & Drury, 1992) and practical interventions across the whole maintenance and inspection system.

A CLASSIFICATION OF HUMAN FACTORS INTERVENTIONS IN MAINTENANCE AND INSPECTION

If the aim of applying human factors to aircraft inspection and maintenance is to improve both human performance and human well-being, then any interventions should address human–system mismatches, either potential or actual. Direct interventions can be logically only of two types: changing the operators to better fit the system, and changing the system to better fit the operators. The former are personnel subsystem interventions, whereas the latter are hardware/software interventions. (In terms of the SHELL model of ICAO, 1989, these would be classified as liveware and hardware/software/environment, respectively.) In addition to such direct interventions, there are examples of system-level actions designed to enable system participants to understand, evaluate, and facilitate change within the system.

Since the public concern for maintenance and inspection human factors arising from the Aloha Airlines incident in 1988, there have been ongoing programs to identify and tackle human factors issues in this field, led by the FAA. The function breakdown of the necessary activities (Tables 24.1 and 24.2) and the classification into systems-level, personnel/hardware, and software interventions forms a convenient framework for presentation of the literature describing these efforts. It also helps to point out where programs exist, and hence helps guide future research and application.

Table 24.5 presents a merging of the function descriptions from Tables 24.1 and 24.2, in the order expected when an inspection activity discovers the defect that must be repaired. Scheduled maintenance activities would generally start at the *initial maintenance* function and omit the *inspection buyback* function. The entries in Table 24.5 provide the framework for presentation of current interventions. In parallel to this effort have been research efforts, for example, aimed at understanding error mechanisms (Latorella & Drury, 1992; Prabhu & Drury, 1992) and speed–accuracy tradeoff (Drury & Gramopadhye, in Section 5.3.4, Galaxy Scientific Corporation, 1993) in inspection.

HUMAN FACTORS ACTIONS AND INTERVENTIONS

This section provides additional detail on the entries in Table 24.5, showing the human factors considerations in each project. System-level actions are treated first to provide additional system overview information.

TABLE 24.5
Classification of Interventions for Human Factors in Maintenance and Inspection

	System Level Actions	
	Development of human factors audit programs	
	CRM Analysis of maintenance and inspection	
	CRM Training for maintenance and inspection	
	Hangar-floor ergonomics programs	
	Characterization of visual inspection and NDI	
	Error analysis and reporting systems	
	PENS system for audit	
	Human factors guide	
	Function-Specific Interventions	
Function	*Personnel Subsystem*	*Hardware/Software Subsystem*
Initiate inspection		Workcard redesign
Inspection access		Restricted space changes
Search	Visual search training	Task lighting design
Decision	Feedback for decision training	
	Individual differences in NDI	
Inspection response		Computer-based workcards
Initiate maintenance		
Maintenance site access		
Diagnosis	Diagnostic training	ITS* computer-based job aid
Maintenance part access		
Replace/repair	International differences	
Reset system		
Inspection buyback	International differences	
Close access		
Maintenance response		

*Intelligent Tutoring System.

Sociotechnical Systems Analysis. Within such a complex system, highly technical, labor-intensive, and highly regulated, there is still considerable room for alternative organizational designs. In a comprehensive sociotechnical system (STS) analysis of many maintenance/inspection organizations, Taylor (1991) found three quite different models of which group was in control of work assignments. Although individuals were usually highly motivated and conscientious in their work, communications patterns between groups and between shifts was in need of improvement. The benefits of organizational changes that move decision making closer to the work point have already been demonstrated in improved aircraft availability and fleet performance in a military context (Rogers, 1991).

Crew Resource Management Training. The preceding STS analysis suggested the need for improved communication procedures. Hence a project was undertaken to provide crew resource management (CRM) training within the maintenance and inspection function on one airline and measure its results (Taylor, 1993). CRM has already been applied successfully to reduce crew coordination errors in flight crews (Heimreick, Foushee, Benson, & Russini, 1986). A training course was developed and presented to over 500 management personnel in an airline maintenance organization. As a result, significant improvements were measured in job-relevant attitudes using a standard questionnaire. Safety-related measures of system performance also improved, but this effect could not be tied exclusively to the CRM training. Measurements continue on this project to establish the long-term effects of CRM training.

Development of Human Factors Audit Programs. The need for an ergonomics/human factors evaluation system has been apparent for some time, and audit programs have been developed (e.g., Drury, 1990) to provide a rapid overview of the factors likely to impact human/system mismatches at each workplace. In the aircraft inspection context, there is no fixed workplace, so that any audit program has to start with the workcard as the basic unit rather than the workplace. Such a system was produced in conjunction with two airline partners (Lofgren & Drury, 1994) and tested for both large airliners and helicopters. The system was tested for reliability, and modified where needed, before being validated against human factors expert judgments. Significant agreement was found. The system can be used from either a paper data collection form (with later data entry), or directly from a portable computer. The computer is used to compare the data collected against appropriate standards and to print out a report suitable for use in an existing airline audit environment. The report allows the airline to focus available change resources on major human/system mismatches. Currently, a similar program is under development for maintenance activities to complement the existing inspection audit.

Hangar-Floor Human Factors Programs. The change process in ergonomics typically involves small groups of users and human factors specialists performing analysis, redesign, and implementation on the users' own workplaces. At one airline partner, implementation was performed using the analyses already carried out as part of the restrictive space project (see Access section), with good results. An existing methodology (Reynolds, Drury, & Broderick, 1994) is currently being adapted for use at that partner airline to provide a more systematic model using the audit program (described earlier) for analysis, rather than the particular measures relevant to restrictive spaces. This intervention process has so far (as of mid-1994) trained a team of inspectors in human factors applied to the aviation hangar environment. This team has analyzed several jobs and is in the process of developing ergonomic redesign of several systems. To ensure compatibility with other human factors efforts in aviation the SHELL (software/hardware/environment/liveware) model used by the International Civil Aviation Organization (ICAO) is used as a framework for the ergonomics training.

Characterization of Visual Inspection and Nondestructive Inspection (NDI). The process of inspection is, like other human activities, error prone. Ultimately, inspectors can make two errors (Drury, 1991):

Type 1: Reporting an indication that is not a defect (false alarm).

Type 2: Not reporting an indication that is a defect (miss).

However, all of the processes within inspection (Table 24.1) can contribute to these errors, so that a detailed error analysis is required. Over the years there have been attempts to quantify inspection reliability, so that models of crack growth can be combined with detection probabilities to optimize inspection intervals. Two recent studies of human plus equipment performance in eddy-current inspection for cracks have been undertaken. The first, Spencer and Schurman (1994), evaluated inspectors at nine facilities, and established probability of detection (POD) curves against crack size for each. There were significant differences between facilities, much of which were accounted for by differences in calibration and probing techniques. The second study, Murgatroyd, Worrall, and Waites (1994), used computer-simulated signals in a laboratory setting, finding no effects of a degraded inspection environment, but again large individual differences between inspectors. Such individual differences were also studied in laboratory experiments, reported in later discussions.

Error Analysis and Reporting Systems. The error characterization work in inspection has continued in the broader context of maintenance (Allen & Marx, 1994). In one airline, maintenance, towing, pushback, and servicing errors accounted for over $16 million over a 3-year period, with the majority of errors being procedural. Most common errors were fitting of wrong parts and incorrect installation, plus not securing the aircraft after repair. To obtain better data for feedback and system improvements, the authors have formed a consortium of several airlines and one manufacturer to develop the Maintenance Error Design Aid (MEDA). This is a multilevel reporting and scenario-based analysis system, which is at the time of this writing being tested within the consortium.

Audit System for Regulators. In addition to the ergonomics audit (described earlier), the concept of auditing has a long history in the regulatory environment, which provides an additional source of feedback to the maintenance and inspection system. Layton and Johnson (1993) reported on a job aid for these FAA inspectors, based on a pen computer. This system, Performance Enhancement System or PENS, contains most of the relevant federal aviation regulations in its database, as well as details of aircraft operators and their aircraft. Thus the FAA inspectors can rapidly enter heading data into a report, and can both rate and comment on the performance of the person being observed. An ongoing field evaluation of this system in all nine FAA regions is showing evidence of considerable cost savings (Layton & Johnson, 1993).

Human Factors Guide for Aviation Maintenance. With so much research and development activity taking place on human factors in maintenance and inspection, there is an obvious need to get usable information into the hands of nonspecialists within the system. Since 1992, a guide has been under development to codify the human factors principles, techniques, and findings into a guide for use by system participants such as managers and supervisors of maintenance and inspection. As Parker (1993) showed, an industry survey was used to determine appropriate structure, topics and level of the information. In 1993 first draft of this guide was produced in hard copy form, parts of which were used in team training for the human factors interventions noted above. Based on review and feedback a second hard-copy version is now being produced (M. Maddox, editor) with a hypertext version to follow.

Workcard Redesign. Because existing workcards were often found to be unsatisfactory from a human factors viewpoint, a project was undertaken to show how they could be improved. The first phase of this project (Patel, Drury, & Prabhu, 1993) used the human factors literature to determine principles of information design applicable to workcards, and to design new workcards embodying these principles. These new workcards were developed as job aids for two distinct types of inspection. For a C-check, which is a heavy inspection conducted infrequently, inspectors need detailed guidance on what defects to expect and which areas to search. For the more frequent A-checks, the inspection is typically the same every day (or more accurately, every night) so that a layered information system is needed. Here a checklist provides procedural and sequence information to prevent procedural errors, and more detailed information is available behind the checklist for reference as needed. Evaluation of the C-check prototype showed highly significant improvements when inspectors rated workcard design (Patel, Drury, & Lofgren, 1994). The design techniques developed for this project have now been extended to many other workcards by the airline involved.

Restrictive Space Changes. Many aircraft inspection tasks must be performed in restrictive spaces due to airframe structural constraints. A study at an airline partner measured the effect of restrictions on postural accommodations (e.g., movements), perceived discomfort, and perceived workload (TLX). It was found possible to differentiate between good and poor workspaces using these measures, and to use the findings to initiate countermeasures in the form of improved access equipment (Reynolds, Drury, & Eberhardt, 1994). A classification scheme for restricted spaces was developed to assist in this work, and has been tested using laboratory simulations of inspection tasks (Reynolds, Drury, Sharit, & Cerny, 1994).

Visual Search Training. A comprehensive series of projects used a workstation-based visual inspection simulator (Latorella et al., 1992) to test many hypotheses about improvement of inspection training. For visual search training, both improvements in defect conspicuity and improvements in search strategy were sought (Drury & Gramopadhye, 1992). Current inspection training procedures are largely either classroom based, covering theory and regulation, or on-the-job practice. Neither technique is most appropriate to the skills required in inspection, particularly the search skills. One experiment tested a technique of practice on a visual-lobe testing task and showed that this practice transferred to search performance for both the same and perceptually similar defects. The second experiment evaluated both performance feedback and cognitive feedback as techniques for improving search strategy and performance. It was found (Drury & Gramopadhye, 1992) that the two types of feedback have different effects, so that both may be needed to obtain the best results.

Task Lighting Design. To perform the inspection task effectively, the inspector must be able to detect the indication (e.g., crack or corrosion), which is often a difficult visual task. As search performance depends on detection off the optic axis, good lighting is extremely important to enhance the conspicuity of indications. Lighting can range from ambient, through portable, to personal (e.g., flashlights), but together these must provide illumination of the structure of sufficient quantity and quality to give a high probability of detection. Using the existing hangar of an airline partner, detailed lighting surveys were carried out, and the results were used to determine the need for improvement. A multifactor evaluation of alternative light sources was performed, and a methodology was developed to allow airlines to specify particular devices that will supply adequate lighting and meet other safety and portability criteria (Reynolds, Gramopadhye, & Drury, 1992).

Feedback Training for Decision. Using the same eddy current simulator as described by Latorella et al. (1992), Drury and Gramopadhye (1992) compared the different techniques available to help train inspectors to make complex, multifactorial judgments. In decision training, the experiments showed that an active training program significantly improved the number of correct decisions made on multi-attribute indications, whether the inspector was given specific standards in training or had to develop a template during training (Gramopadhye, Drury, & Sharit, 1993). Thus, it is more advantageous to train inspectors to make complex judgments about indications with many attributes (e.g., for corrosion these could be area, depth, severity) if the inspector is actively involved in each decision, rather than passively watching another inspector make the decision.

Individual Differences in Inspection. As noted in the crack detection studies just discussed, there are large differences in performance between inspectors. This has been known for many years in the industrial inspection literature (e.g., Drury & Wang, 1986; Gallwey, 1981). Because of the possibility of selection tests for inspectors, Thackray (1995) ran a series of experiments to find correlates of performance on a simulated NDI task. The task chosen was the NDI task detailed by Latorella et al. (1992), which simulated eddy current inspection of lap splice joints on an aircraft fuselage. Thackray found significant correlations between different aspects of performance and a number of pretest measures, of which the best predictor was mechanical aptitude. The full implications of these findings have yet to be integrated into either aircraft inspection practice or the industrial inspection literature.

Computer-Based Workcards. Lofgren and Drury (1994) described an implementation of improved workcards (discussed earlier) as a Hypertext program on a portable computer. The relevance to the response function of inspection was the automatic generation of much of the information needed on the NRR forms. Computer-based delivery of workcard information to the mechanic has been tested in a military context (Johnson, 1990), but the hypercard system developed here used the human factors guidelines for information design derived earlier. There are obvious savings from having an easily updated electronic delivery system, but it must also meet the inspectors' needs. In a direct evaluation against both original and improved paper-based workcards for part of an A-check, Lofgren and Drury found overwhelming support for the computer-based system over the original. It should be noted, however, that about 80% of the improvement was also seen for the improved paper-based workcards. Clearly, it is a good strategy to implement changes to the existing system without waiting for the benefits of electronic delivery.

Diagnostic Improvements: Intelligent Tutoring Systems for Training and Job Aiding. The costs of incorrect diagnosis in aircraft systems are high. If the wrong unit is removed then there is a cost of the test process for a good unit, as well as the cost of a delay until the malfunctioning unit is found. Training in fault diagnosis is thus a critical skill in ensuring both the effectiveness and efficiency of the maintenance system. Johnson, Norton, and Utsman (1992) showed how computer-based training has evolved into intelligent tutoring systems (ITS), in which models of the instructor and trainee are included in the software. Thus, in addition to system logic and data, the program for instruction contains person-models that allow more appropriate feedback and branching. An ITS was developed for the environmental control system of a Boeing-767-300 (Johnson et al., 1992), usable both as a job aid and a training device. An evaluation of this system, using 20 AMTs, compared the ITS with instructor-led instruction by comparing performance on a post-training examination (Johnson, 1990). No significant performance differences were found, showing that the system was at least as effective as the much more expensive instructor-led training.

As technology evolves to allow use of portable computer systems at the work point, the basic logic and interface of such an ITS can become a useful job aid. Particularly when interfaced with the central maintenance computer of a modern aircraft, it can support improved diagnosis techniques. Indeed, in the military, Johnson (1990) showed that a diagnosis task is dramatically improved in speed and accuracy with the use of a portable-computer-based job aid. Aircraft are now designed with on-board maintenance computer systems, so that the hardware support for such tasks is in place. Human factors in design of the interface and logic are still required to ensure usability.

International Differences in Inspection. The organization of the inspection/repair/buyback process is different in the United States and the United Kingdom. A study of these differences (Drury & Lock, 1992) showed that integration between inspection and repair was emphasized in the United Kingdom, while organizational separation of these functions was considered desirable in the United States. Recent work (parallel to the preceding program) at an airline (Scoble, 1994) showed that it is possible to better integrate the repair and buyback functions with the inspection process within the existing United States context.

Human Factors in Airways Facilities Maintenance. Throughout this chapter we have considered only the maintenance and inspection activities associated directly with aircraft and aircraft components. However, the techniques used here are equally applicable in other maintenance contexts within the aviation community. Examples would be maintenance of ground vehicles, of NDI equipment, and of ground-based equipment.

In the United States, the FAA has established a program to apply human factors techniques to Airways Facilities, such as radars, VORs (Very High Frequency Omni-directional Radio ranges), and the control centers with which they are associated (D. Wagner, private communication, 1995). The objective of this program is to ensure that equipment, systems procedures, and organizational concepts maximize human productivity, improve training concepts and methods, reduce stressful work environments, and minimize errors. All projects must have operational sponsorship to ensure smooth transition from research through to deployment in a busy operational environment. Projects have been started in operation control center design, GPS (global positioning system) validation, and producing design guides.

An earlier project (Jones & Jackson, 1992) applied many of the intelligent tutoring systems developed for airline maintenance to an airways facilities environment. This advanced technology training system used the MITT Tutor, developed by Galaxy Scientific Corporation to develop a trouble shooting training program for the air traffic control beacon interrogator (ATCBI-4). The trainee was able to interact with a model of the ABI-4 and solve problems using various diagnostic procedures. The program allows access to flow diagrams and oscilloscope traces, while monitoring trainee progress and errors.

FUTURE CHALLENGES FOR HUMAN FACTORS IN MAINTENANCE AND INSPECTION

The function- and task-based approach detailed in this chapter was introduced to put human actions, and particularly human error, into a systems context of ensuring continuing airworthiness. In this way the potential for human factors interventions can be seen, alongside those of the physicists and engineers who specify the inspection intervals and who design the equipment for defect detection and repairs. The need for human factors effort is clear, as it continues to be in flight operations. Maintenance and inspection error shows itself in spectacular system failures with depressing regularity.

As will be clear from the review of both system level studies and function-specific interventions in the previous section, there is a variety of valid work being done to bring human factors techniques into a domain neglected for far too long. These are not the only efforts, just those for which specific references can be cited. In a number of airlines, human factors is being introduced: error reporting in one, human factors audits in another, new forms of work organization in a third. In addition, aviation regulatory authorities beyond the FAA and CAA already mentioned are analyzing maintenance human factors in aircraft accidents. ICAO is pushing its concept of a human factors model (the SHELL model) beyond the cockpit into the hangar. Indeed, it has already mandated human factors training for all aircrew; can a mandate for ground crew be far behind?

Lest we conclude that we have applied human factors in a comprehensive manner, reference to Table 24.5 shows just how spotty is our coverage of the essential functions. We can list referenced interventions in only about a third of the cells of this table, and only a single intervention in most cells. Compared to the literature on human factors in flight operations, we have barely begun. Some of the cells of Table 24.5 can be covered with small extensions from other cells. Thus, the redesigned workcards for inspection should be applicable almost *in toto* to the initiate maintenance function. Similarly the restricted space studies and improved task lighting go beyond inspection.

However, 10 years after the Lock and Strutt study, and 5 years after the FAA's large involvement, we still need more work at both the systems level and with demonstration projects at the function level. We need to move as well from retrofitting existing systems to designing out some of the error-prone situations in new systems. Already new aircraft can be designed with anthropometric models in the CAD (computer-assisted design) system much publicized for the Boeing 747 (*Aviation Week and Space Technology*, 1993). Such an intervention should prevent the creation of future restricted space problems. But we also need to be designing human interfaces for new NDI systems, using task analytic techniques for new methods of repairing composites, applying STS design ideas to new contract repair stations, and helping design new computer systems for job control and workcard presentation.

The aviation industry has made itself into an extremely safe transportation system, but more is always demanded. As long as there are people in the system, there will be the potential for those errors that arise from human–system interaction. Human factors has far to go to ensure that the next level of system safety is reached.

ACKNOWLEDGMENTS

The author's work reported here was performed under a contract from the Federal Aviation Administration, Office of Aviation Medicine (Dr. W. Shepherd), through Galaxy Scientific Corporation (Dr. W. L. Johnson).

REFERENCES

Allen, J., & Marx, D. (1994, November). Maintenance error decision aid project. *Proceedings of the 8th FAA/OAM Meeting on Human Factors in Aviation Maintenance and Inspection: Trends and Advances in Aviation Maintenance Operations*, Alexandria, VA, November 16–17, 1993, 101–116.

Aviation Week and Space Technology. (1993, November 22). Boeing 777 design targets gate mechanic (p. 60).

Bobo, S. (1990). Communication and transfer of non-destructive inspection information. *Final Report, Second Federal Aviation Administration Meeting on Human Factors Issues in Aircraft Maintenance and Inspection—Information Exchange and Communications*, 151–166. Washington, DC: Office of Aviation Medicine.

Drury, C. G. (1985). Stress and quality control inspection. In C. L. Cooper & M. J. Smith (Eds.), *Job stress and blue collar work* (Vol. 7, pp. 113–129). Chichester, England: John Wiley.

Drury, C. G. (1990). The ergonomics audit. In E. J. Lovesey (Ed.), *Contemporary Ergonomics* (pp. 400–405). London: Taylor & Francis.

Drury, C. G. (1991, September). Errors in aviation maintenance: taxonomy and control. *Proceedings of the Human Factors Society 35th Annual Meeting*, San Francisco, CA, *1*, 42–46.

Drury, C. G., & Gramopadhye, A. K. (1992). Training for visual inspection: Controlled studies and field implications. *Proceedings of the Seventh FAA Meeting on Human Factors Issues in Aircraft Maintenance and Inspection, Atlanta, GA*, 135–146.

Drury, C. G., & Lock, M. W. B. (1992). Ergonomics in civil aircraft inspection. *Contemporary Ergonomics 1992* (pp. 116–123). London: Taylor & Francis.

Drury, C. G., Prabhu, P. V., & Gramopadhye, A. K. (1990, October). Task analysis of aircraft inspection activities: methods and findings. *Proceedings of the Human Factors Society 34th Annual Meeting, Santa Monica, CA*, 1181–1185.

Drury, C. G., & Wang, M. J. (1986). Are research results in inspection tasks specific? *Proceedings of the Human Factors Society 30th Annual Meeting, 1*, 393–397.

Federal Aviation Administration. (1991). *National Plan for Aviation Human Factors*. Washington, DC: U.S. Department of Transportation.

Galaxy Scientific Corporation. (1993). *Human Factors in Aviation Maintenance, Phase 2: Progress Report* (Rep. No. DOT/FAA/AM-93/5). Springfield, VA: National Technical Information Service.

Gallwey, T. J. (1981). Selection task for visual inspection on a multiple fault type task. *Ergonomics, 25*, 1077–1092.

Goranson, U. G., & Miller, M. (1989). Aging jet transport structural evaluation programs. *Proceedings of the 15th ICAF Symposium: Aeronautical Fatigue in the Electronic Era*, Jerusalem, Israel, 319–353.

Gramopadhye, A., Drury, C. G., & Sharit, J. (1993, October). Training for decision making in aircraft inspection. *Proceedings of the 37th Annual Human Factors and Ergonomics Society Meeting, Seattle, WA, 1*, 1267–1271.

Heimreick, R. I., Foushee, H. C., Benson, R., & Russini, R. (1986). Cockpit management attitudes: Exploring the attitude–performance linkage. *Aviation, Space and Environmental Medicine, 57*, 1198–1200.

Hollnagel, E. (1989). The phenotype of erroneous actions: Implications for HCI Design. In G. R. S. Weir & J. L. Alty (Eds.), *Human–computer interaction and complex systems*. London: Academic Press.

International Civil Aviation Organization. (1989). *Human Factors Digest No. 1, Fundamental Human Factors Concepts* (Circular No. 216-AN/131). Montreal, CA: International Civil Aviation Organization.

Johnson, W. B., Norton, J. E., & Utsman, L. G. (1992, October). Integrated information for maintenance training, aiding and on-line documentation. *Proceedings of the 36th Annual Meeting of the Human Factors Society*, 87–91. Atlanta, GA: The Human Factors Society.

Johnson, W. S. (1990). Advanced technology training for aviation maintenance. *Final Report of the Third FAA Meeting on Human Factors Issues in Aircraft Maintenance and Inspection*, 115–134, Atlantic City, NJ.

Jones, J. A., & Jackson, J. (1992, August). Proficiency training systems for airway facilities technicians. *Human Factors Issues in Aircraft Maintenance and Inspection: Science, Technology, and Management: A Program Review, Report of a Meeting*. Atlanta, GA: Galaxy Scientific Corporation.

Latorella, K. A., & Drury, C. G. (1992, August). A framework for human reliabiity in aircraft inspection. *Proceedings of the Seventh Federal Aviation Administration Meeting on Human Factors Issues in Aircraft Maintenance and Inspection*, Atlanta, GA, 71–82.

Latorella, K. A., Gramopadhye, A. K., Prabhu, P. V., Drury, C. G., Smith, M. A., & Shanahan, D. E. (1992, October). Computer-simulated aircraft inspection tasks for off-line experimentation. *Proceedings of the Human Factors Society 36th Annual Meeting, 1*, 92–96. Santa Monica, CA: Human Factors Society.

Layton, C. F., & Johnson, W. B. (1993, October). Job performance aids for the flight standards service. *Proceedings of the 37th Annual Meeting of the Human Factors Society, Designing for Diversity*, Seattle, WA, *1*, 26–29. Santa Monica, CA: Human Factors Society.

Lock, M. W. B, & Strutt, J. E. (1985). *Reliability of in-service inspection of transport aircraft structures (CAA Paper 85013)*. London: Civil Aviation Authority.

Lofgren, J., & Drury, C. G. (1994, November). Human factors advances at Continental Airlines. *Proceedings of the 8th FAA/OAM Meeting on Human Factors in Aviation Maintenance and Inspection: Trends and Advances in Aviation Maintenance Operations*, Alexandria, VA, 117–138.

Murgatroyd, R. A., Worrall, G. M., & Waites, C. (1994). *A Study of the Human Factors Influencing the Reliability of Aircraft Inspection* (Rep. No. AEA/TSD/0173). Warrington, England: AEA Technology.

Parker, J. F. (1993). A human factors guide for aviation maintenance. *Proceedings of the Human Factors and Ergonomics Society 37th Annual Meeting*, Seattle, WA, *1*, 30–33.

Patel, S., Drury, C. G., & Lofgren, J. (1994). Design of workcards for aircraft inspection. *Applied Ergonomics, 25*(5), 283–293.

Patel, S., Drury, C. G., & Prabhu, P. (1993, October). Design and usability evaluation of work control documentation. *Proceedings of the 37th Annual Human Factors and Ergonomics Society Meeting*, Seattle, WA, 1, 1156–1160.

Prabhu, P., & Drury, C. G. (1992, August). A framework for the design of the aircraft inspection information environment. *Proceedings of the Seventh Federal Aviation Administration Meeting on Human Factors Issues in Aircraft Maintenance and Inspection*, Atlanta, GA, 83–92.

Reason, J. (1990). *Human error*. Cambridge, England: Cambridge University Press.

Reynolds, J. L., Drury, C. G., & Broderick, R. L. (1994). A field methodology for the control of musculoskeletal injuries. *Applied Ergonomics 1994, 25*(1), 3–16.

Reynolds, J. L., Drury, C. G., & Eberhardt, S. (1994, November). Effect of working postures in confined spaces. *Proceedings of the 8th FAA/OAM Meeting on Human Factors Issues in Aircraft Maintenance and Inspection: Trends and Advances in Aviation Maintenance Operations*, Alexandria, VA, 139–158.

Reynolds, J. L., Drury, C. G., Sharit, J., & Cerny, F. (1994, October). The effects of different forms of space restriction on inspection performance. *Proceedings of Human Factors and Ergonomics Society 38th Annual Meeting, 1*, 631–635.

Reynolds, J. L., Gramopadhye, A., & Drury, C. G. (1992, August). Design of the aircraft inspection/maintenance visual environment. *Proceedings of the Seventh Federal Aviation Administration Meeting on Human Factors Issues in Aircraft Maintenance and Inspection*, Atlanta, GA, 151–162.

Rogers, A. (1991). Organizational factors in the enhancement of military aviation maintenance. *Proceedings of the Fourth International Symposium on Aircraft Maintenance and Inspection* (pp. 43–63). Washington, DC: Federal Aviation Administration.

Rouse, W. B. (1985). Optimal allocation of system development resources and/or tolerate human error. *IEEE Transactions Systems Man & Cybernetics, SMC-15*(5), 620–630.

Rouse, W. B., & Rouse, S. H. (1983). Analysis and classification of human error. *IEEE Transactions Systems, Man & Cybernetics, SMC-13*(4), 539–549.

Scoble, R. (1994, November). Recent changes in aircraft maintenance worker relationships. *Proceedings of the 8th FAA/OAM Meeting on Human Factors in Aviation Maintenance and Inspection, Trends and Advances in Aviation Maintenance Operations*, Alexandria, VA, 45–48.

Senders, J. W., & Moray, N. P. (1991). *Human error: Cause, prediction, and reduction*. Hillsdale, NJ: Lawrence Erlbaum Associates.

Spencer, F., & Schurman, D. (1994, November). Human factors effects in the FAA eddy current inspection reliability experiment. *Proceedings of the 8th FAA/OAM Meeting on Human Factors in Aviation Maintenance and Inspection, Trends and Advances in Aviation Maintenance Operations*, Alexandria, VA, 63–74.

Taylor, J. C. (1990, October). Organizational context for aircraft maintenance and inspection. *Proceedings of the Human Factors Society 34th Annual Meeting, 2*, 1176–1180.

Taylor, J. C. (1991). Maintenance organization. *Human Factors in Aviation Maintenance—Phase 1: Progress Report* (Rep. No. DOT/FAA/AM-91/16, 15–43). Washington, DC: Office of Aviation Medicine.

Taylor, J. C. (1993). The effects of crew resource management (CRM) training in maintenance: An early demonstration of training effects on attitudes and performance. *Human Factors in Aviation Maintenance—Phase Two: Progress Report* (Rep. No. DOT/FAA/AM-93/5, 159–181). Springfield, VA: National Technical Information Service.

Thackray, R. (1995). Correlates of individual differences in non-destructive inspection performance. *Human Factors in Aviation Maintenance—Phase Four, Volume 1 Program Report* (Rep. No. DOT/FAA/AM-95/14, 117–133). Springfield, VA: National Technical Information.

25

Human Factors in U.S. Civil Aviation Security[1]

Ronald John Lofaro
Federal Aviation Administration

This chapter may well be very different from others in this text. The first basic reason is that Civil Aviation Security in the United States is the responsibility of the Federal Aviation Administration (FAA). This is by law—in the main, Public Law 101-604, The Aviation Security Improvement Act of 1990. Basically, it sets up a Civil Aviation Security System, with the FAA as the focal point. Second, although there are other organizations that are involved in aviation security, such as the Federal Bureau of Investigation (FBI), air carriers, and aviation security companies, it is still the FAA that acts both as a "glue" and in a cooperative, oversight capacity. But the major reason that this chapter differs is that much of the information and work in aviation security are not able to be made public, as they can be sensitive, "for official use only," or classified data. The basic rule of thumb is this: No data that could afford a possible terrorist (or any other who may pose a threat to civil aviation) accurate insights into either the capabilities or the limitations of the civil aviation security system can be made public. Because the aviation security system comprises the devices, equipment, procedures, and personnel involved in keeping the flying public safe from sabotage and terrorism, it is obvious that this chapter cannot contain certain levels of detail. Although this is regrettable, as much of security can be novel, interesting, and exciting to the human factors practitioner, the reason for a rather broad-based approach is clear. In fact, every time you fly, there is reason to be thankful that those details about the security system that could result in a successful seizure or bombing of an aircraft are not made available to the public.

The FAA, through its Associate Administrator for Civil Aviation Security, read and edited this chapter before it went to press. This FAA approval is a legal requirement. What this chapter does contain is a history and overview of the current security system, followed by current issues and human factors efforts, and some concluding thoughts on the future.

[1]The view and opinions expressed in this chapter are solely those of the author. They do not necessarily represent the views or policies of the FAA or of any FAA organizational unit.

A few more caveats: There is no effort made in this chapter to go into international aviation security. This is because—as anyone who has flown overseas, to Canada, or Mexico knows—there are very stringent rules and procedures for nondomestic air travel. These procedures, the work that produced them, and the associated technology are even more closely guarded than domestic data. Although the FAA is also the U.S. government agency that has responsibility for international security, the cross-links to other governments and their security agencies make any attempt to go into such things an essentially impossible effort. However, much of domestic security does involve "lessons learned" from the procedures and equipment on the international side of the security equation.

The final "reader beware" is that today's geopolitical situation is highly volatile. The number of known terrorist groups is very large. The number of disgruntled, hostile people—some of whom are not mentally stable—may be incalculable. All of these can and do pose significant threats to aircraft and passengers. There is a never-ending ebb and flow of political situations today. Each new action (viz., Cuba, Haiti, Bosnia, or Rwanda) brings a new set of factors into play, and possibly another set of potential terrorists ready to act. What this chapter presents is based on the current assessments of what are seen as the major concerns for civil aviation security, that is, hijacking and bombing. These may not be operative in years to come, or even for very long.

Before the readers wonder why, or if, they should proceed any further in this chapter, there are some stable factors:

1. The technology used in detecting weapons and explosives is mainly in a highly developed form. Although there are refinements and advances being worked on, there may well be no quantum leaps in this arena. Still, the example of GPS (global positioning system) and its impact on military and commercial aviation operations comes to mind, so we must be alert and not place too high a confidence level here.
2. The human factors issues in security are basically those of the person-in-the-system. These issues are generically similar to those encountered in any "human-in-the-loop" system.

CIVIL AVIATION SECURITY

Overview

The major threat to civil aviation security has changed dramatically in the last decade. The previous danger was that of hijacking. In an effort to counter the hijacking threat, Congress enacted Public Law (PL) 93-366 in 1974. Title I of this law is known as the Anti-Hijacking Act of 1974, and Title II, the Air Transportation Security Act of 1974. PL 93-366 became one of the three bases of the current U.S. Civil Aviation Security Program. It also significantly changed the Federal Aviation Act of 1958, another of the three bases for U.S. civil aviation security. The 1958 act was, and still remains, the principal statute for U.S. civil aviation.

The FAA's role in aviation security further expanded in 1985 following the hijacking of TWA Flight 847 in the Middle East. However, there soon followed a shift to the far deadlier threats of sabotage by bombings and armed terrorist attacks. It became clear that the United States and its aviation industry were prime targets of terrorists.

In response, the international aviation industry initiated efforts to upgrade its security systems and procedures. Unfortunately, the terrorist tactic of sabotage culminated in the bombing of Pan Am Flight 103 over Lockerbie, Scotland, in December 1988. The destruction of this Boeing 747 and the killing of 270 people deeply affected America. The U.S. government undertook a major effort to learn what had happened, to identify corrective measures, and to implement enhanced security programs.

The May 1990 report of the President's Commission on Aviation Security and Terrorism found that "current aviation security systems are inadequate to provide such protection" and gave over 60 recommendations on improving security. Congress reacted swiftly. In November, the President signed PL 101-604, the Aviation Security Improvement Act of 1990—the third of the three legal bases referred to. Title I of this law deals with aviation security and Title II with U.S. response to terrorism affecting Americans abroad. Most of the recommendations of the President's Commission were implemented in this new law. Along with establishing several federal security positions, it made the federal government directly responsible for managing the airport security program and for establishing plans and policies in cooperation with other airport security organizations.

As one of its responses to this Congressional mandate, the FAA rapidly expanded the Aviation Security R&D Service, housed at the FAA Technical Center in New Jersey. As part of the Aviation Security R&D Service, the Aviation Security Human Factors Program (ASHFP) was established in 1991. In general, the ASHFP seeks to develop guidelines, specifications, and certification criteria for human performance with aviation security systems. A major emphasis is on improving human–machine interface and the human input to decision making. The research and development being done are aimed at methods to improve and assess the human and system performance and operational effectiveness. In this way, both system and operator performance can be strengthened.

Terrorism

Generally, terrorism is seen to encompass the use of physical force or psychological intimidation against innocent targets. The objectives of such methods are the social, economic, political, and strategic goals of the terrorist or terrorist organization.[2]

The U.S. Department of State defines terrorism as "premeditated, politically motivated violence perpetrated against noncombatant targets by sub-national or clandestine agents, usually intended to influence an audience" (Title 22; U.S. Code 2656(d)). According to the Department of State view, "the term non-combatant target is interpreted to include, in addition to civilians, military personnel who at the time of the incident are unarmed and/or not on duty." The Department of State also considers "as acts of terrorism attacks on military installations on armed military personnel when a state of military hostilities does not exist at the site, such as bombings against U.S. bases in Europe, the Philippines, or elsewhere."

Terrorism has proven to be a cost-effective tool that violates all law. In the late 1960s, terrorism became a constant occurrence on the international scene. During the 1970s and 1980s, the threat posed to the U.S. and U.S. aviation increased, as the U.S. Congress Office of Technological Assessment (OTA, 1991) documented:

---

[2]See *Patterns of Global Terrorism: 1990* (1991). For latest sources on the definitional forms see, for instance, Yonah (1990–1991, 1991).

> The statistics of both domestic and international incidents are startling. During the decade of the 1970s, the total number of incidents worldwide was 8,114 people killed and 6,902 injured. The most targeted victim during the 1970s was the business community, with a total of 3,290 incidents recorded.
>
> The next decade was even more intensive in scope and destructive force. In 1980, 2,755 attacks were registered and their number increased to a record high of 4,422 in 1989, a 16-percent increase over the previous year. The 1980s saw a grand total of 31,426 incidents, with 70,859 killed and 48,849 injured, reflecting a lethality trend of more attacks designed to kill random victims. The 1980s ended with the destruction of Pan Am 203 over Lockerbie. The 1990s have already seen the World Trade Center (NYC) bombing.

For some surveys on terrorism, see Yonah and Cline (1982), Yonah (1987), and Yonah and Foxman (1989, 1990).

Technology

The U.S. Office of Technology Assessment (OTA, 1991) identified one direction for countering domestic terrorism. Their 1991 report said that, although it may be impossible to end terrorism, we can try to reduce the vulnerabilities (and, thereby, the likely number of terrorist incidents) to the greatest degree possible, consonant with a free and open society. OTA holds that many terrorist acts, particularly those against transportation systems and against highly visible fixed sites, can be deterred or prevented by using technological tools in conjunction with antiterrorist and anticriminal methods. Although not an all-pervasive panacea (there is no "silver bullet"), technology is and will continue to be a highly useful tool in the ongoing battle. In fact, it may well play a far greater role in the future than it does today.

Terrorism, Technology, and Human Factors

In a follow-on to their 1991 effort, the U.S. OTA (1992) published a report that gave guidance for using technology to combat terrorism and for structuring aviation security. Chapter 5 of this report is entirely devoted to the findings of OTA on the necessary human factors roles in civil aviation security. It was stressed that human factors in the implementation of civil aviation security have not received the attention they deserve. The OTA also indicated that technology can enhance, but not replace, the capabilities of the personnel in any security system.

In summary, the OTA findings on FAA human factors, while somewhat critical of the FAA's (then) capability, emphasized that great strides can be made. Specifically, passenger profiling and airline operating or managing practices in safety were targeted as human factors research and development areas that needed FAA attention. As discussed later, the FAA and the ASHFP were cognizant of and responsive to the OTA findings and recommendations.

THE U.S. CIVIL AVIATION SECURITY PROGRAM AND SYSTEM

As already stated, U.S. civil aviation security is the responsibility of the FAA. The Federal Aviation Administration, working with the airport operators and air carriers, has established domestic security requirements that address the many threats to civil aviation.

The U.S. Civil Aviation Security Program developed by the FAA is based on a combination of interrelated security measures and resources for safeguarding the civil aviation industry and the traveling public. Under the U.S. program, the federal government, primarily through the FAA, is responsible for:

- Identifying and analyzing threats to security.
- Prescribing security requirements.
- Coordinating security operations.
- Enforcing the appropriate federal air regulations (FARs).
- Directing law enforcement activities under the governing statutes and regulations.

The airport operators are charged with providing a secure operating environment for the air carriers. Specifically, they are responsible for ensuring that:

- Responsive security programs and emergency action plans are maintained.
- Air operations areas (AOA) are restricted and protected.
- Law enforcement support is provided to respond to various security threats.
- Physical security measures for the airport are provided.

The air carriers are responsible for the most visible security measures for safe and secure travel, which include:

- Screening passengers with metal detectors and with x-ray equipment and inspecting their carry-on articles.
- Securing baggage and cargo.
- Protecting the aircraft.
- Maintaining responsive security programs.

[*Note*: Air carriers generally contract with private security firms to perform most of these functions. Nevertheless, the air carriers are held accountable by the FAA for the effectiveness of the screening operation and can be assessed penalties for noncompliance.]

It is important to be aware that other federal agencies (the FBI, the Department of State, the Immigration and Naturalization Service, etc.) are also linked in the arena of civil aviation security. Much of the interchange and support these agencies provide to the FAA is in the sharing of intelligence data. Thus, we can see that civil aviation security in the United States is an interlocking system where the air carriers and the airport operators are joined to the FAA, under PL 101-604.

During the 1990s, FAA Aviation Security Research and Development Service emphasized a systems approach to the myriad of problems that arise from forestalling and combatting terrorist (be they internal or foreign) attacks. The problems involved simply with the coordination and orchestration of a system that involves many federal agencies, a formidable number of air carriers, security companies, and airport employees, and the FAA's own Civil Aviation Security organization are apparent and often staggering. This chapter refrains from any attempt to go further than a condensed consideration of the U.S. Civil Aviation Security program and the human factors thereof.

This completes the overviews of the terrorist threat, the role of the FAA, other government agencies, and the air carriers, plus an attempt to clarify why much of what is happening in U.S. civil aviation security cannot be put into print for "public consumption." We next examine the role of human factors and of the FAA's Aviation Security Human Factors Program. We then proceed to a critical look at where human factors finds itself in security as well as where it could well be directing more, or new, efforts and resources.

HUMAN FACTORS

The author is in agreement with Meister (1989), who sees human factors as an umbrella discipline encompassing a broad range of interest areas, although this has also led to a degree of fragmentation. *Human factors* (and nowhere in this chapter is an effort made to raise or resolve the human factors vs. human factors engineering issue) is used to denote human performance in a human–machine system; the variables that effect and affect performance; the human elements involved in the design, development, testing, and operation of the equipment in a system; and, finally, the system itself.

To date and from personal experience in the FAA, the author must also agree with Meister (1989) that human factors research has been either lacking or not vigorously applied in system design and development. However, recent developments, FAA-wide and in security human factors, may foreshadow both the actualization of a more complete systems perspective and the embedding of human factors in the FAA research and development that deal systems design, development, and acquisition.

In civil aviation security, the goals of human factors research must be those well articulated by Meister (1989): to describe, predict, and control human performance in the system context. Only by achieving these goals can civil aviation security guaruntee the flying public safety from terrorist acts.

Aviation Human Factors

In aviation, there are many events that can be seen as human factors landmarks. More than 30 years ago (1976), the FAA and NASA established the Aviation Safety Reporting System (ASRS). This was followed by the International Civil Aviation Organization (ICAO) 1986 Resolution A-26-9, which addressed human factors and flight safety. The U.S. Congress then passed two public laws, the Aviation Safety Act of 1989 and the Aviation Security Improvement Act of 1990. These laws focused on the role of human factors in aviation. Further, PL 101-604 mandated optimizing the performance of the human in the (security) system, another of Meister's key concepts. The total security system can thus be optimized, with increased safety for the flying public.

Finally, the FAA in 1990 took the lead in producing a *National Plan for Aviation Human Factors*. This document now serves as one reference for indicating the needed aviation human factors research and development (R&D) in the 1990 to 2000 time frame.

Human Factors in the FAA

The FAA has been struggling with the discipline of human factors for approximately 15 years. This statement is not to be construed as meaning that the FAA either had, or in many ways now has, (a) embedded human factors into the agency's daily operations, climate, and culture or (b) a unified FAA-wide human factors program. The FAA has been questioned on its efforts in human factors by publications such as *Aviation Week and Space Technology*; by Congress through the Government Accounting Office (GAO) and OTA; and by prominent members of the Human Factors and Ergonomics Society.

In 1988, partially in response to outside influences (PL 100-591; the Aviation Safety Research Act), and partly due to increased internal awareness and to a top-level FAA commitment to exploring the role and function of human factors, the FAA created the position of Chief Scientific Advisor for Human Factors. Congress had mandated the FAA, in concert with NASA, industry, and academia, to develop a multiyear (10) plan for aviation safety research and development that emphasized human factors. The major accomplishment of the first person to hold that FAA post (Dr. H. Clayton Foushee, formerly of NASA-Ames Research Center, now at Northwest Airlines) and of his staff was the initial draft of the *National Plan for Aviation Human Factors*. During Foushee's tenure, the FAA Human Factors Coordinating Committee was established to approve and coordinate both FAA-funded and internal human factors projects.

Within the FAA during the late 1980s, and continuing, human factors was represented by (a) the FAA Civil Aeromedical Institute (CAMI), (b) the Human Resources Organization, and (c) a portion of the Research and Development Division (ARD) of the Systems Engineering Organization. There were no clear-cut boundaries. Some difficulties in responsibility, authority, and resources were encountered. In 1991, the Aviation Security Research and Development Service established its own Human Factors (for civil aviation security only) Program at the FAA Technical Center.

The years 1989 through 1994 saw some significant efforts by the FAA in human factors. There was R&D, based on FAA funding and technical guidance, in crew resource management, cockpit display technology, data link, and air traffic control. Some of the human factors R&D were done by such organizations as NASA-Ames and Langley, Boeing, and the John A. Volpe National Transportation Systems Center (VNTSC). Some parts were done by universities such as the University of Illinois, Embry-Riddle Aeronautical University, and the University of Colorado at Boulder. In all cases, the FAA provided funding and a measure of cooperative oversight and direction. Various results and products were the result. Still, there existed a perception that the FAA had no integrated human factors program and vision. Rather, the FAA, through a limited number of its technical staff, was seen as working on a broad spectrum of projects that had some human factors aspects. Further, human factors had not found its way into the arena where it must be a recognized and effective player: systems design, development/assessment, and acquisition, with resultant deployment and implementation.

The state of affairs was somewhat compounded by the FAA (an agency formerly with 52,000 personnel and now, due to downsizing, approximately 46,000 employees), having approximately 10 to 20 (estimates vary) recognized and experienced human factor practitioners. Some of these resided in the human resources organizations where their efforts were mainly directed at FAA internal employee issues (such as job satisfaction, hiring, and training) and FAA internal organizational development and effectiveness.

However, events of late 1993 and 1994 may herald significant changes. In October 1993, FAA Administrator Hinson signed FAA Order 9550.8, Human Factors Policy. This order specifies that "human factors shall be systematically integrated into the planning and execution of the functions of all FAA elements and activities associated with systems acquisitions and systems operations."

The order recognizes the importance of human factors. It calls for a systems approach to human factors and points out the need to institutionalize human factors principles within the FAA. Some steps in these directions are underway. A prototype course in the use of human factors in the planning and acquisition of aviation systems is being developed. It is hoped that this course, the result of an FAA-internal request, will lead to a series of courses for the requesting organization and will serve as a paradigm for future courses in other FAA organizations, such as security.

Another set of developments includes, first, the rewriting of a handbook for FAA managers. This rewrite includes the role of human factors in programs and systems. Second is an FAA document to be used as a guide for the inclusion of human factors in all programs and system developments. This includes embedding human factors in systems specifications and requirements as well as the assessment of the human factors work in a program/system. This document is currently specific to one major FAA organization, but has the potential for modification and use across the FAA. The FAA Human Factors Coordinating Committee (HFCC) was revitalized and significantly expanded in late 1993. This expansion placed FAA personnel with decision-making power on the HFCC, making the committee much more than a recommending and oversight body. To date, the new HFCC has been more proactive and has taken the lead in both resurrecting the National Plan for Aviation Human Factors and attempting to embed human factors in the FAA culture and climate.

The only cloud, and it is more a thunderstorm cell, on the horizon of human factors in the FAA is an ongoing downsizing, with reorganization, of the agency. The FAA, in concert with the Team for Re-Inventing Government, had agreed in 1993/1994 to considerably downsize the agency by 1999. To date (1997), this has resulted in early retirement by many long-term employees in the technical as well as the managerial ranks. There are signs that all available methods (early retirement; buy-outs; possible reduction in force [RIF] procedures) will continue to be used to achieve the number of FAA employees that can remain in the downsized FAA.

A major reorganization occurred in late 1994, and has led to an instability and state of flux where programs are "up for grabs" as to future existence and funding. In point of fact, all too many programs are in competition to even retain their current levels of personnel and funding. The 1994 reorganization merged aquisition and research. It created a large aviation research organizational element, and placed aviation security R&D in/under this organization. Further, the reorganization expanded both the staff (by shifting existing FAA personnel) and the monies controlled by the FAA's Chief Scientific Advisor for Human Factors. Although the results of these actions remain to be played out, human factors does not have a history of being seen as vital to the FAA mission and is relatively "new" to the FAA. Therefore, there still remains the distinct possibility that human factors may not actually grow, or may even be "distributed" across the FAA so thinly that it ceases to have any impact on decisions or operations. Additionally, at the time of this (re-)writing, the U.S. Congress proposed to slash the FAA's R&D funds in fiscal year 1996 from $143 million to $116.19 million (*Aviation Week and Space Technology*, July 3, 1995). Since the FAA's human factors work is a part of their R&D effort, such a drastic cut does not bode well. Time will tell.

The author's purpose in writing this chapter, and including citations from certain reports somewhat critical of the FAA's human factors efforts, is not to provide yet another critique. Another critique is neither needed nor productive. The citations used were selected for their relevance to aviation security human factors and to provide the reader with both a frame of reference and some very recent history regarding the observations of oversight and other commissions. Further, such reports have given a large measure of direction to the FAA's ASHFP, which as of this writing is barely 3 years old. The author, as a former member of the ASHFP, must bear some responsibility for its efforts.

Finally, as the disclaimer at the beginning of this chapter states, this chapter does not necessarily reflect the views or policy of the FAA or of any of its organizational units. That disclaimer can be relied on.

THE FAA CIVIL AVIATION SECURITY HUMAN FACTORS PROGRAM

The FAA ASHFP, in various documents, defines its view of human factors research and development and the role of the ASHFP in civil aviation security. Some of these documents are not available to the public. However, the remainder of the chapter attempts to capture the gist of current ASHFP efforts.

The main drivers for the FAA ASHFP, established in 1991 and, until late 1993, consisting of one scientist, were, are, and continue to be:

1. The funding levels (OST; FAA; Congressional) for the ASHFP. It must be noted that, in 1991, Congress "zeroed out" funding the ASHFP. Happily, monies were internally reprogrammed to keep the program afloat and progressing.
2. The requirements that FAA Headquarters, through the (now) Associate Administrator for Civil Aviation Security, establishes for the ASHFP.
3. Whatever other new requirements that the Aviation Research section places on the ASHFP.

These drivers are, in part, related to the U.S. Congress Office of Technology Assessment (OTA) Reports (1991, 1992); Public Law 101-604, The Aviation Security Improvement Act of 1990; and a GAO Report (1994) assessing the FAA's efforts in implementing PL 101-604.

However, PL 101-604 is the real impetus, as it legally mandates the FAA to perform a wide range of activities in security. A representative statement is that the FAA "shall review issues relating to human performance in the aviation security system with the goal of maximizing such performance."

Further, the FAA Administrator is tasked to provide recommended guidelines and to prescribe appropriate changes to existing procedures to improve such performance. The FAA Act of 1958 was amended to add a program of accelerated research, development, and implementation requirements. The human factors aspects of this program were to include "both technological improvements and ways to enhance human performance."

New FAA emphasis on human factors stems from the realization that any system functions as well, or as poorly, as the operators and maintainers of that system. An interesting and hopefully instructive analogy comes from cockpit technology. Beginning in the late 1970s and continuing, there have been quantum leaps in cockpit automation. Auto-throttle, auto-land, on-board computers ("flight directors") coupled to flight controls, and more have been developed culminating in the so-called *glass cockpit* and the *electronic aircraft*, where cathode ray tubes (CRTs) have replaced the dials and gauges and control of the plane is often a computerized function. A major rationale for such development was the belief that cockpit automation would do away with pilot/crew error. What also has resulted is a new and dangerous class of error: automation errors. Such errors include, but are not limited to, incorrect input to flight director computers, incorrect use or interpretation of flight director data, and lack of coordination and control of the on-board automated systems. There now is a realization that human–machine interface issues as well as system–system interface issues were not well recognized, defined, and researched prior to installation of the automation. As aviation security systems, such as x-ray detection devices, continue to be developed and refined, the FAA is working to ensure that the humans who operate the systems will be trained and tested with state-of-the-art devices to maximize their performance.

The FAA's ASHFP does research and development to define, develop, and implement FAA objectives and to develop methods and products to improve human performance consistent with a broad system integration perspective. Current and future technologies that have significant impact on human performance in aviation security are evaluated and improved. The detection devices and the security systems of which they are a part will be carefully analyzed and evaluated as to their displays, the training necessary to operate them, and the optimum operator/device interface and procedures. Finally, as detection and training systems are combined with each other and as different detection systems are integrated into new systems, the operator and system function issues of integration, training, and procedures will be carefully studied.

These human factors activities cross technology, application, and organizational boundaries. FAA workforce human factors awareness and training needs will be established and the requisite training developed and delivered. The FAA Technical Centers Explosive Detection System (EDS) programs will have in-house human factors personnel matrixed to their work. This effort involves collaborations with government agencies, airlines, security companies, equipment suppliers, and the aviation industry as a whole.

FAA ASHFP: Major Projects

There now are three major areas of human factors security research and development in the FAA. There are the preboard x-ray screening checkpoint, domestic passenger profiling, and human–systems integration. Let us examine these three areas, the rationale for each, and the current/planned work in each.

The Advanced Screening Checkpoint

The preboard x-ray screening checkpoint, the place that all air travelers remember for instructions such as "lay your bags flat on the conveyer belt" and "please put all metallic items in your pockets onto the tray," is now part and parcel of domestic flying. The older of us remember the days when such was not the case, but the 1970s and 1980s have made the checkpoint part of federal law for all passengers with or without carry-on baggage, seeking to board a flight. Under the rubric of screening checkpoint, human factors encompasses the selection, training, and performance assessment of security company personnel who operate the x-ray devices.

Briefly, the U.S. Congress has mandated that the FAA enhance airport security by improving the performance of the checkpoint screening process. Congress specifically recommended that human factors engineering technologies be brought to bear to improve the state of the airport security system.

A Department of Transportation (DOT) task force on aviation security concluded that human performance is the *critical* (italics added) element in the screening process. The realization of the criticality of human performance in the security area was based on the fact that, although engineering evaluations could readily determine the physical capabilities of screening equipment, the reasons why an object might pass undetected during checkpoint screening are complex and go beyond equipment capability. As a result, the limiting factor in determining the effectiveness of the screening process was seen to be the ability of screeners to correctly recognize and identify threat objects embedded with x-ray images.

An FAA ASHP project that deals with x-ray screening is called the Advanced Screening Checkpoint. Before you proceed to your gate to emplane, you will pass through what is called a screening checkpoint. This consists of an x-ray machine for your carry-on baggage and a "portal" (magnetometer) that you walk through. The portal will alarm if there are metallic objects on your person. As two backups, there are, first, the hand-held magnetometers ("wands"), which are passed near your arms/legs/torso. This is to resolve any portal alarm that still occurs after you have emptied your pockets of keys, coins, and the like. The second backup in resolving screeners' concerns about objects in your luggage that may be threats (weapons and the like) is to open and examine your luggage. The human factors here are:

1. The selection, training, and performance proficiency of the x-ray screening personnel.
2. The same concerns for the "wanding" personnel.

It must be noted that at a checkpoint there is a team of security personnel to do various functions. The team's membership can change from day to day; it usually ranges from two to six, depending on passenger flow and other nonoperational factors. The team members rotate through the various checkpoint (e.g., wanding, x-ray) positions during their shift. In fact, they rotate many times per shift, due, in part, to the issue of sustained vigilance on the x-ray machine and, in general, to keep them familiar and proficient in all aspects of the checkpoint. Rotation also provides a degree of relief from possible boredom and inattention due to repeatedly doing the same limited set of actions. However, not all members of the team will be used as x-ray screeners, as some persons are not seen as adequate to the task. This introduces new issues on vigilance, training (assumed equivalent for all screeners), and performance. Thus, although we must not lose sight of the human subsystem (team) that works a checkpoint, we must be aware that there are operational realities that can lead to team function/disfunction problems:

1. The human interface and operational issues with the various devices used in a checkpoint.
2. The function of the entire checkpoint, seen as a system.

Additionally, there is an overlay that affects all these factors, that of new and emergent technology in x-ray and other detection devices and in training devices. Extensive work is being done on other methods to detect threat objects, such as vapor analysis of the emanations from a piece of baggage; mmwave devices to detect threat objects; and pattern recognition of threat objects using neural nets (neurocomputers) "grown" from an adaptive fuzzy system and linked to an x-ray (or other type) baggage scanner. We cannot go into further detail on such work as it could enable potential terrorists to devise countermeasures. Suffice it to say that as new technology comes forward, the same set of human factors will apply and may well be extended as some of the emerging technologies are integrated into a composite detection system.

The FAA ASHP, from its beginning in 1991, has made x-ray screener selection training and proficiency evaluation one of its main thrusts. A systems concept—the Screener Proficiency Evaluation and Reporting System (SPEARS)—was laid out by Dr. Mitchell Grossberg of the FAA. Industry was asked to begin or continue development of equipment that could be used in the actual system.

The *Commerce Business Daily* published, in 1991 and 1993, the FAA's functional capabilities for any candidate SPEARS device(s). Candidate SPEARS training components were to have the following capabilities:

1. That of automatically training and operationally testing individual airport security screeners in using x-ray display equipment to detect and identify diverse concealed weapons, explosives, and other dangerous devices ("threats").
2. That of scoring and managing multiple training lessons which provide initial or advanced student with immediate constructive feedback based on training performance. The training lesson sequence shall progress students through increasingly difficult tasks in detecting and identifying diverse x-ray threat images.
3. That of assessing trainee performance in relation to identified detection tasks and automatically recording the assessment results in an assumable, secure database. This database will be capable of evaluating individual and group analysis at each screening site and from remote locations.
4. That of having training consoles which are simultaneously operable at multiple airport screening sites. Such consoles shall be capable of use in proficiency-based screener selection, training, recurrent training, and operational testing.
5. That of using training and operational test materials which are representative of current threat objects as well as being efficiently modifiable to include new/current threat images and FAA-disseminated reference information and procedural guidelines.

SPEARS real-time operation components for testing screener vigilance and proficiency of threat detection and identification were to have the following characteristics:

- Fictional threat images can be automatically and unpredictably (for screeners) inserted into baggage x-ray images at screening checkpoints.
- Screeners receive immediate feedback on their decisions about such images while the prevailing flow rate of passengers and bags through the checkpoint is not disrupted.
- At least 300 different threat images shall be conveniently supervisor-selectable for insertion. The image transition and position orientation for each bag will be controllable.
- Images shall be selected for automatic occurrence either individually or in series, based (in part) on time of day, checkpoint activity, or screener identity.
- Screener proficiency for identified fictional images shall be automatically measurable recordable, and accessed to a secure database. This database on screener performance shall be capable of evaluation analyses to determine individual and group proficiencies.

- The images that result in screener false alarms will be identified and recorded in a database file for analyses.

From mid-1993 and continuing, the ASHFP has been involved in a very extensive set of both laboratory and field-based evaluations of candidate SPEARS devices. It is foreseen that a certification standard and process will begin for SPEARS devices in the future.

In late 1994, the SPEARS effort was broken out so as to separate the x-ray screener off-line training/testing from the online testing that uses fictional threat image projection. The off-line training effort is now called STEARS, for Screener Training Evaluation and Reporting System. At this time, there are three candidate STEARS devices that are undergoing lab and field-based assessments. It is likewise foreseen that a certification standard and process may begin for STEARS devices in the future.

X-Ray Screener Selection. In the area of x-ray screener selection, there are legal (and political) issues with the FAA being proactive. However, the need to select x-ray screener candidates with the highest potential for successful performance is not open to question. A review was made by the FAA and VNTSC personnel of the only existing selection test in use; one major air carrier used it. This review and other considerations led to the FAA awarding a multiyear (1993–1995) aviation security grant to Embry-Riddle Aeronautical University (E-RAU).

The E-RAU work is directed at developing the taxonomy of underlying abilities and traits (Fleishman & Quaintance, 1984) needed as one basis for any instrument used in the selection of x-ray baggage screeners. E-RAU work encompasses:

1. An exhaustive literature search, focusing on methods for the identification of underlying abilities/traits, with extension and emphasis on use in x-ray screener selection was done. Not only methods were considered, but also those underlying abilities and traits that have been shown to be critical and essential for "success" as a baggage screener or, by extension, other positions that require similar cognitive, pattern recognition, vigilance, and information processing.
2. Concomitantly, subject-matter-expert (SME) input identifying current, critical, and essential x-ray screener abilities and traits was generated via modified Delphi techniques (Lofaro, 1992).
3. A job task analysis (JTA), focused on the cognitive aspects of x-ray screening, was accomplished. A final technical report, "Development of Decision-Centered Interventions for Airport Security Checkpoints," was done for the FAA by Klein Associates, Inc., in Fairborn, Ohio, in 1994.
4. The tests (and the test development literature) that are candidates for inclusion, or modification leading to inclusion, in a prototype battery were analyzed to determine whether or not the underlying abilities/traits they purport to assess are those that correspond to the ones generated via the SME workshops, the JTA, and the literature review. The result was an initial matrix of cognitive, perceptual/psychomotor, and personality abilities and traits that are seen as essential to, and typical of, a successful x-ray baggage screener. This matrix was used in the selection of two COTS (commercial, off-the-shelf) tests to be used in the prototype test battery.
5. Specifications for the hard- and software needed to computerize the selected subtests was developed. A test plan for the development and validation of a prototype test battery was developed.
6. An FAA review and progress meeting was held on all these efforts.

As a result of this review and some additional areas where the E-RAU team had begun exploratory work in during 1993/1994, the E-RAU effort took some new directions for the remainder of 1994 and 1995. To briefly synopsize:

1. Continue with an expansion and sharpening of the Delphi workshop data in order to do scaling; this would be done through an x-ray screener questionnaire, to be administered across many airports and security companies. The goal was a sample size of 1,300 plus.
2. As a result of some of the 1993/1994 E-RAU work, one security company at one airport was to institute a trial series of monetary, organizational, and procedural changes for its x-ray screeners. E-RAU would track any changes in morale, job turnover, and so on that can be considered a result of these changes.
3. E-RAU personnel began to develop a small database on the ergonomics of screening and checkpoint workstations; this would continue.
4. The two COTS tests selected would be used in a predictive validation study for the prototype battery.
5. Quantifiable x-ray screeners performance measures, based on false image projection (see earlier discussion, The Advanced Screening Checkpoint), would be developed. These measures will become the basis for a job sample-type component of the prototype battery.
6. Finally, in late 1994, E-RAU and the ASHFP began discussion on R&D to ascertain the traits, capabilities, and performance of screeners using computer tomography-based (CAT scan) x-ray baggage screening systems.

X-Ray Screener Training and Performance Assessment. Performance assessment of x-ray screeners goes beyond assessing training performance, although that will be integral to SPEARS. A significant technological step forward has been the development of the ability to, electronically and from a remote station, insert what appears to be a threat object (a "false image") into an image of a bag that an x-ray screener is scanning in real time and online. The image of the threat object is "false" only in the sense that the object is not actually in the baggage. The appearance of the object in the x-ray image is veridical to what the screener would see if the object were physically present in the baggage. This development can allow x-ray screeners to be confronted with what they would see if the actual threat object were in the piece of luggage, but without the danger of having a real threat object present.

Before we go further, let us be sure to cover a major point: X-ray screeners are regularly tested by the FAA, the air carriers, and their security company, while they are working. This is done by inserting one of the FAA-approved test objects into a piece of baggage and ascertaining if the screener identifies it. All test objects used are nonfunctional, although they appear to the screener to be "the real thing." Testing screeners this way is highly labor-intensive and a screener can reasonably expect to be tested relatively infrequently, in terms of the number of baggage that the screener sees in any given period of time. The ability to project a "false" threat image to a screener allows for increased levels of performance evaluation while significantly decreasing the number of persons physically needed to do the testing.

Further, the presentation rate of test objects to a screener can be raised to where the vigilance level is positively impacted. All screeners also can be presented with a full panoply of the most current and likely threat objects known.

The current training that x-ray screeners receive is based on FAA-approved, ATA-developed materials, augmented by OJT. The training materials used are in need of real revision, using an accepted and standard process, such as ISD, for such revision. Further, an expansion of the topics is needed to include emphasis on the current technology and weaponry available to today's terrorist. Any such revision involves many considerations, because the FAA has regulatory and oversight functions for security. Cost-effectiveness and possible mandatory standards for training effectiveness are but some of the issues with any attempt at revising the existing training.

The ASHFP is both doing and exploring several training projects. One of these involves taking the existing training from the candidate SPEARS devices and installing it on lap-top computers, then ascertaining, via a field test, if any training effect is significant and comparable to the effect produced by the candidate SPEARS device itself. The lap-top versions will offer tremendous economy and portability and require almost no storage or presentation space. Another project will explore a computerized, multimedia approach to SPEARS and other security training. This project may also go into the redevelopment and upgrading of the existing FAA/ATA screener training.

Domestic Passenger Profiling

A second major FAA ASHFP project is in the arena of passenger profiling. In the last GAO Report (1994), it was recognized that more work on passenger profiling is needed. An OTA report (1992) had found that research and development on profiling should be done by the FAA. This report also discusses, in some detail, the role of humans in passenger profiling. It is instructive to consider some of what was said:

> There are two general approaches to operational profiling. One compares passenger demographic and other background data (age, sex, nationality, travel itinerary, etc.) to historic or recent intelligence-derived "threat profiles." The other is based on the examiner's psychological assessment of the passenger, taking into account nervousness, hostility, or other suspicious characteristics.
>
> Airline passenger profiling, in most cases, must be fast (and consequently cursory) enough so as not to impose excessive delays. In other security contexts, such as screening for the "insider threat" profile within an organization where time is not so critical, much more detailed background data and questioning is possible. A different, although overlapping, form of profiling is used by law enforcement and investigatory agencies. Given pertinent data and evidence from a crime scene or threat, experts compile a profile of likely social, psychological and physical characteristics of the criminal. However, much of the work and methodology could be transferred from one of the broad profiling regimes to the other. (U.S. Congress OTA, 1992, p. 27)

Passenger profiling techniques can run the gamut from a passive mode (gathering data to ascertain if more stringent security measures are warranted) to active modes of interrogation and search. As an example, profiling can be done by using behavioral indices, such as valid "soft" signs, to identify a potential terrorist. Nervousness comes to mind, but "white knuckle" flyers exist, so obviously, it is not a single sign but a constellation of such behavioral cues that needs to be to be identified and validated.

There are technology and software possibilities involving fuzzy (logic) systems and neural nets, growing a neural net from a fuzzy system's data. One avenue of research can be in pattern recognition, such as developing a database of digitalized facial images of known terrorists and using a TV (scanning) camera linked to a computer with a neural net based on known terrorist facial images for the identification of terrorists, even if disguised. Some of these possibilities have been considered and may be undergoing some development.

Passenger profiling efforts have a long history. In fact, the FAA in the late 1960s and early 1970s had played a pioneering role in profiling during the large number of airline hijacking attempts to go to Cuba. Today, the ASHFP is working to reestablish a more proactive FAA effort in domestic profiling research and development. There are a myriad of legal (e.g., civil liberties, privacy) and cultural issues involved. Happily, space precludes any presentation or discussion of these. At this time, the FAA has a focus on doing domestic profiling research and development in a two-stage effort, with one concurrent and related project.

Manual, Domestic Passenger Profiling. A 1993/1994 FAA effort with one major air carrier worked on developing a manually done and passive profiling instrument. The goal was to develop a prototype instrument that could be used to identify domestic passengers judged *not* to represent a terrorist threat. The initial profiling data elements were identified through contributions from FAA Civil Aviation Security personnel and Northwest Airlines Security personnel. These elements were evaluated and refined at a subject-matter expert (SME) workshop using Federal Bureau of Investigation (FBI), Immigration and Naturalization Service (INS), Customs, security personnel from three airlines, and FAA personnel. A worksheet and scoring procedures for the manual use of the profiling elements in an operational setting were developed so that domestic passengers could be profiled by airline or other personnel.

A field test of the feasibility of the manual domestic passive profiling (MDPP) worksheet was conducted at a domestic airport in early 1994, with a resultant FAA Report published in late 1994. Further work with other air carriers is planned.

The manual effort served other purposes. One was to examine what available data were of use in developing a profile: what available data were not of use, and what currently nonavailable data were seen by the experts as having strong value in profiling. The other purpose was to see if it was possible to proceed further, with the "further" being an automated passive system. The answer to that question was "yes," and therefore the FAA ASHFP approved a grant proposal to develop such a system.

Automated Domestic Passenger Profiling. FAA-funded and directed research and development, which began in late 1994, was scheduled to proceed as follows:

1. A comprehensive search and review of all pertinent literature pertaining to terrorist methods, profiling, and attempts at automation of the profiling system was conducted. This search will be extended to include not only profiling as it pertains to U.S. Civil Aviation Security, but also profiling as performed by other law enforcement organizations or foreign governments.

2. An SME panel will obtain consensus from SMEs regarding those items contained in a computerized reservations system that they believe indicate facts about a passenger, useful for making a determination that the passenger poses no threat to civil aviation and to obtain opinions as to those items in terms of which are the relevant items and which are informational elements that are not now in a typical computerized reservations system but are potentially useful for profiling. Finally, the SME panel developed a weighting system for each of these identified items. The panel will draw

its membership from experts in the aviation security field as well as other law enforcement agencies familiar with the profiling technique. A modified Delphi methodology will be used to develop SME consensus (Lofaro, 1992). Unlike other derivations of the Delphi technique, this methodology relies heavily on direct, facilitated interaction between investigators and SME. This approach is selected primarily because of the complexity of the objectives and the requirement to maintain a high level of focus for the group consensus.

3. Opinions and evaluations from a broader sampling of experts than just the SME panel will be obtained through a broad-based survey. Survey techniques also provide the capability of assessing a greater geographic dispersion of respondents. The survey will be sent to a broader sampling of experts in the aviation field and will be compared to the results from the SME panel to assess any possible effects of such a limited subject expert population. The data generated by the SME workshop as well as informational elements identified in the literature search will form the basis of the survey instrument.

4. A computer program will be developed to extract as much of the relevant data as possible from the air carriers reservation system and process it according to the scoring method identified by the SME panel.

This research and development project will determine how to enhance the ability of security personnel to accurately profile passengers, thus better assessing threats to domestic aircraft associated with particular passengers. This profiling assessment will be done by using information contained in a computerized reservation system. The profiling will consist of a mathematical score resulting from analysis of the combined data elements. Domestic locations will be selected for field testing. The focus will be on using the profile score to eliminate low-risk passengers from additional special scrutiny.

Active Profiling: "Dupe" Card. International passenger profiling uses active interrogation. One rationale is that a passenger may have become a "dupe." A dupe is one who, although not fitting known terrorist profiles or descriptions, has been duped by a terrorist group into carrying baggage that has had a threat object placed in it without the person's knowledge. Such a person thus may not be the subject of active and intense security measures. In this way, an innocent passenger can actually carry a threat, such as a bomb, on board a plane. This is the scenario that resulted in the destruction of Pan Am Flight 103 over Lockerbie, Scotland. The FAA wants to ascertain the operational suitability of active (written and/or oral) interrogation, patterned after some international procedures, in identifying potential dupes on domestic flights. The ASHFP supports this by cooperative FAA/Industry research and development. Contingencies requiring administration of a domestic airline passenger dupe card could arise under conditions of raised threat to domestic air travel. An ASHFP project has begun work to develop and field-test a written dupe questionnaire (card) for administration to domestic passengers prior to boarding their flight. At this time, a prototype dupe card with instruction and screening template has been developed. Although such a card will be "self-administered" in that passengers are expected to complete it by themselves, airline personnel will be needed to review and process the completed cards.

The initial FAA research and development (early 1995) will be a field-based gathering of data, and analyses to ascertain the operational suitability and resources needed to administer, either on a contingency or full-time basis, a domestic airline passenger questionnaire that will determine whether the passenger is likely to have been duped into carrying unknown and potentially restricted baggage contents onto a domestic flight.

Research and development on other active profiling techniques, such as actual interrogation to resolve inappropriate dupe card responses, and manual or domestic profile data that indicate if additional security measures may be needed and other problematic profiling issues, are planned by the ASHFP.

HUMAN–SYSTEMS INTEGRATION

The final area, human–systems integration (HSI), is the newest ASHFP thrust area with only a few efforts now underway, all in initial stages.

Canine (K-9) Olfactory Detection: Training and Assessment

Currently, the FAA has a large number of dog handling teams available within the United States. The dogs are trained by the Department of Defense (DoD) in explosive and other detection; the dogs are maintained by local law enforcement agencies with the commitment to serve FAA needs first. This is both convenient and cost-effective for the FAA. However, the down side is that the FAA loses direct control of how the dogs, the handlers, and the dog–handler teams (systems) are trained and maintained. Therefore, two critical elements that need FAA attention are the initial training and evaluation along with the annual recertification of each dog–handler team.

In fiscal year 1994 (FY 94), there were significant changes in the FAA K-9 program. In a series of reviews and meetings, it was seen that there was a need to reevaluate and redirect this program by:

1. Identifying effective training and testing methods for dogs, handlers, and the dog–handler team.
2. Assessing baseline performance standards in order to develop recertification performance standards.
3. Developing a field quality control procedure to routinely verify and validate canine performance after initial FAA training.
4. Scientifically identifying and characterizing the detection thresholds of the canines for explosives and certain other chemical compounds.

The ASHFP will support the FAA K-9 program by doing research and development on:

1. Identifying effective training and testing methods for dogs, handlers, and the dog–handler team.
2. Developing procedures for assessing baseline—and recertification—performance standards for dogs, handlers, and dog–handler teams.

Human Factors Training for FAA Personnel

FAA Order 9550.8 (10/27/93), *Human Factors Policy*, specifies that human factors are to be systematically integrated into the planning and execution of the functions of all FAA elements and activities associated with systems acquisition and systems operations. In order to prepare FAA personnel to effectively implement such a policy, courses of instruction in human factors, both general and specific to the mission of the specific FAA organizational unit to be trained, are required.

The author was a member of a five-person FAA working group charged with developing the first FAA-designed human factors course to be given to FAA personnel. A prototype of this human factors training course was developed. It will now be delivered to sample groups, the working group included, for critique and revision. It is expected that the development process, as well as the actual training itself, will be used as both paradigm and (partial) content base for the future development of other specific (to include security-specific) human factors courses.

For security human factors training, the ASHFP's first step will be to do a needs analysis within the FAA Civil Aviation Security Organization, followed by a survey to sharpen and extend the data in the needs analysis. At this point and using the prior course development work (now in progress) and materials, security-specific courses can be designed, validated, and delivered.

Passenger "Wand" and Physical Searches

Functional requirements for passenger searches with hand-held detectors have been developed. This work will include identifying screener characteristics that enhance or degrade hand-held detection searches and the sensitivity thresholds of "wands." Data from these efforts will be used in designing and assessing the current and future checkpoints and in designing and assessing emerging detection/integrated technology systems.

The human factors work and analyses needed for the current ASHFP efforts are complex. Doing aviation security human factors seems analogous to Sisyphus and his rock or Hercules cleaning the Augean stables. However, the future can only hold more—much more.

THE FUTURE

The same conditions and pressures that will shape much of the future of the role human factors in civil aviation security have determined its present role. These are:

1. The "threat" probability, that is, the current geopolitical situation as to who may be arrayed "against" the United States—their commitment to, as well as their capability for, terrorist actions.
2. The FAA mandate to regulate and enforce all domestic, civil aviation security.
3. The air carriers' economic status.
4. The civil aviation security companies' economic status.
5. Funding.

To an extent, the ASHFP will also be driven by emerging detection technology.

All major air carriers have their own security organizations and all, save one, have security companies to do the actual security detection work. No air carrier has a dedicated security human factors staff. The same is true of security companies. However, in some air carriers and security companies, there may be one (or more?) persons who have human factors expertise and do some human factors work, usually in the areas of training, personnel selection, and OD/OR. The occasional human factors person does not make for an organizational commitment.

The air carriers, the security companies, and the airport operators are not set up for, nor resourced (personnel and money) to perform, human factors research and development. However, it is also not their responsibility, and in the past they have been extremely helpful in providing information, support, and even some of their personnel to the ASHFP in its research and development work. In fact, much of the FAA's security research and development work would not have been possible without the help of the air carriers and security companies.

The FAA, since 1991, has had an Aviation Security Human Factors Program. This program works with air carriers, security companies, and other FAA organizational units to provide expertise and research results across a large variety of areas. Because the FAA has both a mandated involvement and funds, it is unlikely that the air carriers or security companies will go forward on their own. Rather, they will continue to work with the FAA, providing the FAA with data and support—such as access to employees, access to operational settings, and active cooperation for FAA field testing of the performance of security personnel, procedures, and equipment. Such industry support allows for "value-added" for everyone, while allowing air carriers and security companies to compete in the market place. Or, put another way, the research and development funding for security human factors comes, in the main, from the FAA, thus allowing the industry to use its monies for their operations.

What are the areas in civil aviation security that seem likely for continued, or new, efforts? Are there "targets of opportunity?" The answer follows the break-out of the standard basic human factors areas of:

1. Performance.
2. Selection.
3. Human–machine interface.
4. Training.
5. System performance and integration.
6. (Certain) organizational issues.

Let us examine three relevant areas from this list. A major set of interrelated issues in aviation security revolve around measurement, analysis, and evaluation—surely no surprise. The portions of aviation security that are most in need of identification, clarification, and definitive work are (a) security personnel training, (b) performance (of all system components and of the security system itself), and (c) organizational and management policy and practices, both FAA and industry.

What follows is prescriptive. It offers no evaluation of the current states of affairs in aviation security human factors, but rather attempts to present what processes and procedures can profitably be applied. An effort is made to show where work has not been done but needs to be, with analogies to other areas in aviation human factors and training such as crew resource management (CRM) and advanced qualification programs (AQP). The tripartite break-out used (training, performance, and OD) is somewhat artificial. Therefore, there will be overlap. The order of the three areas is also arbitrary and not meant to show priority.

Security Personnel and Training

Training, in the author's view, has too often served both as a (false) panacea and a resultant whipping boy for performance deficiencies. And, in too many instances, training has served as the "fix" of choice for performance concerns that would have been more correctly addressed through redesign or retrofit of equipment and devices. But, in a bottom-line (read: "dollars") world, training is usually much less expensive, and most assuredly quicker, than equipment redesign or retrofit.

An anecdotal example comes to mind. While the author was a participant-observer in a CRM course at an airline training center, the word came that 25 of Airplane Type X had been purchased from a recently defunct airline. The purchasing airline was quite happy, as this particular type was an answer to some of its needs. However, it turned out that among the 25 copies, there were many cockpit display/instrumentation variants—and at least eight of these were seemingly major ones. The "fix"? It was not to standardize the cockpits, but to develop "differential" training.

In security, the training of screening checkpoint personnel, individually, and as a team, is not extensive. In fact, team training does not exist. The current x-ray screener training was not developed through a structured process, such as the ISD (instructional systems development) process called out by the FAA SFAR 58 advanced qualification program (AQP) for aircrew. There have been no changes in this initial screener training for well over 3 years and no validation of the efficiency or efficacy of the training as regards performance. Further, the training is specific to position and has no team component.

Further, neither methodology nor rating system exists for evaluating the screening checkpoint team. Team training does not exist for personnel who work as a team. This may remind some readers of the late 1970s and early 1980s in terms of the cockpit crew (team). Such aviation human factors efforts as CRM training, which has been extended to dispatch, flight attendants, and may be extended to air traffic control specialists (ATCS), could contribute to the training and team performance arena of aviation security.

Thus, the area of training the x-ray screener and the x-ray checkpoint team is in need of extensive effort. The FAA SPEARS project will work on some level of online proficiency and recurrency testing, in which each online test is also seen as a training event. (The reader is left to pursue this rationale.) At this time, all security companies have extensive training and training-related costs. These are high because the turnover rates, as has been mentioned, are so high, creating a situation where many are trained, but few last very long on the job. Further, the attrition rates in the training itself, be it the initial phase or the on-the-job training (OJT), are tremendous. Add to the never-ending training costs the price of required background and fingerprint checks for job applicants and one gets a feel for the price that the security companies currently pay: a price that better selection and better training could reduce. Both on an individual and a team level, training is an area that has need of basic human factors research and development.

Human and Systems Performance

The signal–sensory detection paradigm (SDT; Green & Swets, 1974) is commonly used—or proposed—to evaluate x-ray screener performance and, by analogy, the security checkpoint for carry-on baggage. Probability of detection (PD) and false alarm (FA) rate are often used to "judge" particular x-ray devices as to their efficiency and feasibility, and are used as indicators of x-ray screener proficiency. The PD must be very high and the FA rate low enough so as not to impede passengers who have a flight to board. The "throughput rate" of baggage, and passengers, must be maintained.

A closer examination shows some difficulty in the application of the SDT 2 × 2 matrix. The first, minor difficulty is this: A hit (a screener correctly identifies a threat object in baggage), a miss (a screener does not identify a threat object which is present), and a false alarm (a screener saying a threat object is present when none is there) are straightforward. But what does a correct rejection (saying no threat object is present and none is) mean operationally? Is not a hit any time the screener is correct? Even more difficult is that x-ray screeners are trained to make one of three decisions: there is no threat object in the baggage ("pass"); there is definitely a threat object present ("stop and hold" the bag in the x-ray compartment and get a law enforcement person to the checkpoint); and "not sure," that is, there may be or there may not be a threat object in the baggage. In this case, the screener asks to take the suspect bag to the side and then conducts a physical search.

In the first two cases, the screener has declared a surety that a threat is, or is not, present in the baggage. There remains the prior concern that a "correct rejection," in operational terms, is equivalent to a "hit." However, what if "not sure, open the bag" is the x-ray screener's response? If the screener is wrong—that is, no threat is there—is that a false alarm? If the screener is correct, is that a hit? If so, we must combine categories of responses. But, operationally, do we want to hold it against screeners when they seek to resolve uncertainty by a hand search—and call it a false alarm if nothing is there?

Screeners are actively discouraged from having high false alarm rates—it slows the throughput and can be seen as a negative reflection on the screener's ability and proficiency. However, many, if not most, bags are filled with a staggering variety of items, all to be looked at and a decision reached by the screener in a very short time. So, we find three decisions with six outcomes open to a screener, at least two of which are problematic in scoring. Beta (β), ROC curves, and d prime (d′) statistics are used, but these are based on the 2 × 2 matrix. More importantly, on the operational and motivational levels, the decision to inspect the bag must not be seen negatively. Screeners must be encouraged to resolve their uncertainty by inspection—at the same time they must be better trained and tested to lower the occurrences of uncertainty. This is a problem, both organizational and theoretic, that requires new work.

The X-Ray Screener: Vigilance and Workload

As has been said, there are many issues of performance and measurement that need study. The person working the x-ray scanning device typically works 20 to 30 min before rotating to another position. The rationale for this is not any direct, controlled studies, either in a laboratory environment or in the operational milieu. Rather, the rationale goes back to work on vigilance, which may or may not generalize. Certainly the airport environment, especially during high-workload times and with the consequences of even a single failure-to-detect, needs extensive work. There has been no workload research and development for any single member of the security checkpoint team, or for the team. This is a situation that needs a remedy, especially because the aviation industry has recently produced many crew workload measures (TLX for one) that can have application.

System Integration

There are two separate, vital areas here. The first is the integration of human factors in the development of detection systems, be this a system's actual components, the detection system as a unit, or the aviation security system itself. The second area is the use of human factors in integrating the components and function of any security system, from an x-ray checkpoint, with its team of personnel, the x-ray devices, and portals, to an airport, with its complex system (or systems), to the aviation security system.

Two Israeli security persons (Lewis & Kaplan, 1990) articulated what a security system entails:

> When talking about security systems, we are talking about five elements: 1. *intelligence or information*, 2. *personnel*, 3. *equipment or technology*, 4. *procedures* (which divide into routine and emergency), and, finally 5. *evaluation*—drilling, *exercising, and auditing the system.* A security system is as good as it will perform the day, the hour or the minute it is required. These elements are inter-related. You can have the best equipment in the world, but if the people running it are not adequately trained, the system will fail.
>
> Procedures for implementing these five elements are also necessary. [emphasis added] (p. 38)

Keeping this definition in mind, let us look at system design and integration. Because human factors is the discipline that can act as a bridge between the behavioral and the physical in systems development, it must be used in precisely that manner. It seems simplistic to view (continue to view?) human factors as an afterthought whose function is to enable a box to be checked. As human factors practitioners want to believe, the role of human factors is well recognized and accepted. Meister (1989) presented a survey result that indicated that this view of human factors and the real world were somehow dichotomous. The ability of human factors to influence system development varies from environment to environment. As an example, the U.S. Army MANPRINT (manpower personnel integration) effort has become stronger and more effective. Yet it has taken MANPRINT almost 10 years to achieve this. In aviation security, the system concept is much cited but, perhaps, not nearly as much used. It remains for the human factors program and practitioners, in aviation security, to put life into Meister's blueprint.

Organizational Issues

Within the Aviation Security Industry. The U.S. Congress OTA (1992) indicated some of the issues involved with human resources/organizational development and security: The FAA mandates certain positions in an airline's organizational structure, such as a security director for the airline and security coordinators at each airport, but airline management practices and philosophy usually fall outside the scope of FAA regulatory authority. OTA found that the effect of airline operating or management practices on airline safety, and changes in those practices, were rarely addressed in FAA safety analyses. The FAA Human Factors Plan cites the influence of management "culture" on human performance as one area where basic research is needed. If the organizational "climate" (i.e., working conditions, wages, management, organizational culture, etc.) does not allow an individual to perform at his or her peak, it may not matter how well he or she is trained or how well designed the technology is.

Plainly put, human resources, in the OD area, are rarely addressed by airlines security organizations or security companies. The single example that the author is aware of where this is not the case is that already alluded to as a result of the E-RAU work—and that is on a test basis, with only one security company, at one site.

The selection, training, and performance of security personnel, in the airport environment, not at corporate headquarters, are fertile research and development areas. The management practices and procedures as they have impact on security personnel performance and satisfaction may be even more in need of research and development. In the current economic climate and using industry forecasts of the airline's economic health through 2000, none of this is likely—unless, that is, the FAA provides money and/or personnel to the industry.

The focus of this chapter has been the x-ray checkpoint screener/team. Basically, these people receive minimum wage and minimum benefits. Turnover is high (say 25% or more a year) to extraordinarily high (6% per month).

Thus, the last line of defense—the preboard security team—is poorly paid and poorly rewarded in other ways. What is the impact on morale and performance? Performance has been discussed as to evaluation and baseline. As to morale, there are, not unexpectedly, problems.

Within the FAA. The FAA had officially embraced and espoused total quality management (TQM). The various FAA aviation security organization members, as is true with the remainder of the Agency, have no formal or informal training in human factors. PL 101-604 speaks to human factors (as does PL 100-591) and to the FAA's needs and responsibilities. This chapter has tried to make clear the many and varied components of human factors work needed by the FAA.

EPILOGUE

To return to Meister (1989, p. 45): "Human factors has three goals: to describe, to predict, and to control the performance of the human in the system structure. These goals are no different from those of any other scientific discipline, but their implications for human factors are distinctive." These *are* the goals for aviation security human factors.

REFERENCES

Fleischman, E. A., & Quaintance, M. K. (1984). *Taxonomies of human performance*. New York: Academic Press.

Government Accounting Office. (1994). *Report to Congressional Committees: Aviation security: Additional action needed to meet domestic and international challenges* (Rep. No. GAO/RCED 94-38). Washington, DC: Author.

Green, D. M., & Swets, J. A. (1974). *Signal detection theory and psychophysics*. New York: Krieger.

Kaempf, G., Klinger, D., & Wolf, S. (1994). *Development of Decision-Centered Interventions for Airport Security Checkpoints* (Final Tech. Rep. prepared under contract DTRS-57-93-C-00129). Fairborn, OH: Klein Associates, Inc. (Available from Klein Associates, Inc., 582 E. Dayton-Yellow Springs Rd., Fairborn, OH 45324-3987)

Lewis, A., & Kaplan, M. (Eds.). (1990). *Terror in the skies*. Jerusalem: Hemed Press.

Lofaro, R. J. (1992). *A small group Delphi paradigm. Human Factors Society Bulletin, 35*(2).

Meister, D. (1989). *Conceptual aspects of human factors*. Baltimore, MD: Johns Hopkins University Press.

U.S. Congress Office of Technology Assessment. (1991, July). *Technology against terrorism: The federal effort* (Report No. OTA-ISC-481). Washington, DC: U.S. Government Printing Office.

U.S. Congress Office of Technology Assessment. (1992, January). *Technology against terrorism: Structuring security* (Report No. OTA-ISC-511). Washington, DC: U.S. Government Printing Office.

Yonah, A. (Ed.). (1987). *The 1986 annual on terrorism*. Dordrecht, the Netherlands: Martinus Nijhoff.

Yonah, A. (Ed.). (1990–1991). *Terrorism: An international resource file, 1989 Index, and 1990 Index*. Ann Arbor, MI: University of Michigan.

Yonah, A. (Ed.). (1991). *Terrorism: An international resource file, 1970–1989 Bibliography*. Ann Arbor, MI: University of Michigan.

Yonah, A., & Cline, R. S. (Eds.). (1982). Worldwide chronology of terrorism—1981. *Terrorism: An International Journal, 6*(2), 107–388.

Yonah, A., & Foxman, A. H. (Eds.). (1989). *The 1987 annual on terrorism*. Dordrecht, the Netherlands: Martinus Nijhoff.

Yonah, A., & Foxman, A. H. (Eds.). (1990). *The 1988–1989 annual on terrorism*. Dordrecht, the Netherlands: Martinus Nijhoff.

26

Aviation Incident and Accident Investigation

Sue Baker
Civil Aviation Authority, Gatwick, UK

Any discussion of aviation-related incident and accident investigation invariably prompts a number of questions, many of which raise fairly fundamental issues about the nature and purpose of the investigation process. For example, should time and resources be expended on the investigations of incidents rather than focusing all the effort on the major aviation accidents? What is the underlying purpose of investigations and who should conduct them and, if a full-scale field investigation is conducted, what benefits can be gained from this as against a more limited and less resource-intensive "desk-top" enquiry? One of the aims of this chapter is to attempt to answer these questions and to consider, in some detail, the practice and process of investigation in the aviation sphere. The information on which this chapter is based is drawn from firsthand experience of the investigation of air traffic control (ATC) related incidents and accidents in the United Kingdom, but it seems reasonable to assume that the points raised have a general application extending beyond the ATC area or any one state.

In order to convey an insight into what incident investigation is and what it does, it may be helpful to consider what incident investigation is *not*. First and foremost, it should not be an exercise in the apportioning of blame. The individual does not work in a vacuum. Mistakes are made in the context of the system, and unless the system itself is considered during an investigation, the whole process is likely to be of dubious value. Blaming and/or punishing an individual serves no valuable function for the person concerned. All this would do is to maintain the status quo and therefore the circumstances under which further errors may occur, doing little or nothing to rectify shortcomings or prevent future occurrences. A report published by the UK Air Accident Investigation Branch in 1990 illustrates the point. In the accident in question, a BAC 1-11 had been inadvertently fitted with the wrong sized windscreen retaining bolts during maintenance. At around 17,000 ft the affected windscreen separated from the aircraft, and in the ensuing depressurization the pilot was partially sucked through the gap. The accident gave rise to a number of human factors concerns regarding the maintenance procedures, which are outside the scope of this chapter. However, what is of direct interest here is the manner in which this, admittedly highly unusual, situation was handled by one of the air traffic controllers involved. Specifically, doubts were cast on the quality of the training received by the controller in incident and accident handling. The recommendations of the subsequent report, backed up by data from other, less serious occurrences, led to a reappraisal and reconfiguration of emergency training for controllers within the United Kingdom. The point is that data from this accident, together with those gathered from other occurrences, pointed, not to a negligent or blameworthy controller, but rather to a system deficiency that needed to be rectified if further similar events were to be avoided. This is not to suggest that, on occasions, the major casual factor in an incident or accident will not be the actions performed by the individuals involved. However, to ignore the system-related factors is to ignore the opportunity to make system-based improvements to the whole organization and its staff and not just to provide a stopgap, quick fix on an individual level. Undoubtedly, problems will be discovered and individual errors found, but these need to be viewed as chances for improvement, not opportunities for punishment. Somewhat paradoxically, perhaps, incident and accident investigation need not only discover information on deficiencies in the system. The lessons learned from the successful handling of an incident or accident are equally valuable. A recent investigation involving an aircraft with engine problems seeking a diversion for a speedy landing was skillfully and expeditiously handled by a trainee controller who had recently undergone a period of training in the handling of emergencies in accordance with the recommendation made in the BAC 1-11 accident report. It is important that successful performance be given as much "publicity" as inadequate performance, not only for its motivating effect but also because it illustrates that improvements can be made to existing systems.

Incident investigation is not then a justification for punishment. Equally, it is not, or at least should not be, simply an academic data-gathering exercise. The collection and analysis of data, together with the knowledge gained and conclusions drawn about individual and system performance and problems, should be undertaken with the aim of improving flight safety. It is for this reason that the provision of accurate and adequate feedback on the lessons learned from investigations is so vitally important.

It has now become a truism that incidents and accidents tend to be the result of a chain of causal events and/or contributory events. To look at these chains is to describe the system, not the individual. Human factors input is of value because it can be one of the ways in which the scope of error causation is extended from the so-called person at the sharp end, to a consideration of the wider aspects underlying the organisation and its structure and function.

It has been suggested (International Civil Aviation Organization [ICAO] Circular 247-AN/148, 1993) that this extension of emphasis is shifting the "blame" for incident causation from the individual who perpetrated the visible error to decisions made at management level. The logical extension of this, it is suggested, is a failure to recognize or accept individual culpability or responsibility, because the onus for all incidents and accidents could be firmly laid at the door of management. This line of argument misses the point.

Individuals do make errors, sometimes without any evident predisposing factors in the system. There is also little doubt that on some, fortunately rare, occasions, individuals or groups will deliberately violate rules and procedures (Reason, 1989). Such a situation, once discovered, obviously requires remedial action. However, remediation at the individual level is only going to prove of limited value. At best it may help the individual mend his or her ways, but it is likely to do little in terms of future prevention in more general terms. The individual exists in the system, and overlooking the possibility of system-based antecedents in error occurrence is to overlook the opportunity to take more far-reaching preventative measures. The major point has to be, however, that the investigator should not approach the investigation with preconceived ideas regarding the causal factors nor attempt to validate some existing, perhaps prematurely formed, hypothesis or pet theory. The presence on the team of more than one human factors specialist also helps to ensure that the conclusions reached are not a function of one individual's perspective. The opportunity to discuss incidents and accident data with peers and to "bounce" ideas off colleagues goes some way toward preventing an idiosyncratic approach.

INCIDENTS VERSUS ACCIDENTS

The decision to investigate incidents as well as accidents is not one that can be taken lightly. The investigation of incidents and accidents is a specialized, resource-intensive activity. It is reasonable to ask whether the end justifies the means in terms of any benefits gained when seen against the outlay of resources. It has been asserted that "an incident investigation can often produce better accident prevention results than can an accident investigation" (ICAO Circular 240-AN/144, 1993). If this assertion is true, and many investigators believe that it is, it still remains to justify the cost and effort involved in the investigation of incidents.

The first and most obvious reason for investigating incidents, as well as accidents, is, quite simply, that there are more of them. This allows the investigators to build up a wider picture of the problems encountered and also to gain an understanding of any trends. A database developed in this way gives the investigator a baseline from which to assess whether subsequent occurrences are a further indication of a known problem or are an unfortunate "one-off." The more data available, the firmer is the basis on which conclusions and decisions can be made.

From the human factors perspective, the behavior manifested by the individuals or groups involved in incidents may not differ greatly from that observed in accident scenarios. Granted, the gravity of an accident will add another dimension to the situation in which the controller or pilot finds him- or herself, but, generally, speaking, the cognitive failures, problems in decision making, communications breakdown, distraction, and all the other factors that contribute to the sum total of behavior in an accident will also be present in incidents. Because the major reason for investigation is the promotion of lessons learned to prevent future similar occurrences, knowledge gathered before an accident occurs can be seen to justify the effort and resources expended.

It could possibly be argued that a thorough investigation of a small number of accidents would yield data of such "quality" that decisions could be made on the basis of this small, but detailed, data set. It is certainly true that generalizable lessons can be learned from accident investigations, but it is also true that the focusing of attention on such a limited set of occurrences may overlook the opportunities offered by incident investigation to have prevented such accidents in the first place. Nor does homing in on a limited number of instances provide the type of overall picture of system health that can be gained through more numerous, but still rigorous, incident investigations.

DATA QUALITY

Although the need to investigate the human factors aspects of incidents and accidents is gaining wider acceptance, there is still a degree of apprehension in some quarters resulting from the perception of human factors findings as "speculative" and the assessment of human factors data as being of a lower order of credibility than more "physical" data such as instrument readings, cockpit voice recordings (CVR), engine damage, or even body parts. While human factors data are viewed in this light, reports are likely to present an incomplete account of the antecedents of incident and accidents. What is worse, human factors issues left uninvestigated and unaddressed can form no part of the lessons learned for the future. Although it is true that the evidence associated with human factors findings may not be as tangible, in some respects, as the other data just described, investigation of human factors issues is invaluable in shedding light, not only on *what* occurred but also on *why* it occurred, especially when the event involved human error and not just mechanical failure.

A full-scale investigation, looking at all aspects, including human factors, can be seen as providing the optimum opportunity for the collection of good quality data. In most aviation incidents a wide range of information sources is available to the investigators, such as radiotelephony (RTF) recordings and transcripts, video recordings of radar displays, and controller and pilot reports. When coupled with visits to the units concerned and face-to-face interviews with the personnel involved, a picture is gained of the whole context in which an incident or accident actually occurred. This broader picture is essential to a system-based approach, allowing for a consideration of each of the factors that contributed to the occurrence and, equally important, the interaction among them.

DATA COLLECTION

Incident and accident investigation is, by definition, post hoc, involving a reconstruction of the events, actions, and decisions that took place at the time. The accuracy of the reconstruction will depend to a great extent on the quality of the data gathered and the relevance of the questions asked. It is particularly difficult to conduct an analysis later from data that have not been specifically gathered for the purpose. There are a number of aspects to data collection. Some of these are considered next.

Who Collects the Data?

So far in this chapter, the assumption has been made that human factors data will be collected by a human factors specialist, although this may not necessarily be the case. Indeed, it has been argued that "most accidents and incidents are investigated by investigators who are trained as 'generalists'" and that the human factors investigators need not be "physicians, psychologists, sociologists or ergonomists" (ICAO Circular 240-AN/144, 1993, p. 14). This attitude is particularly unfortunate in a climate in which greater efforts are made to look closely at each aspects of the system. It is highly unlikely that anyone would suggest that the engineering or avionics side of an investigation could be conducted by a generalist. The generalist approach is certainly not the case in the United Kingdom, where a human factors specialist is an integral part of the investigation team, at least where ATC-related events are concerned.

To accept the principle that anyone with training can conduct human factors investigations is to denigrate the role of human factors in investigations and is also likely to lead to the collection of data of a lower quality than might otherwise have been achieved. Many of the issues arising from the investigation of incidents and accidents are essentially in the realm of psychology. In this one can include questions of decision making, problem solving, perception, attention, and so on. To this one can add equipment design and ergonomic aspects. These are specialist areas, an understanding of which is not easily acquired without an appropriate educational background. There are other areas of expertise, however, that possibly need to be developed on the job. These would relate to the specific details of the investigative process, such as the role of the various interested parties, or the legal and licensing aspects, coupled with at least a broad familiarization with aviation and air traffic control. Whether skill in investigation is an art or a science and whether some individuals have a particular facility in this area is open to debate. The ideal situation would be for a potential human factors investigator to come to the task already armed with a background and experience in the human factors field as a basic requirement. To this can be added on-the-job training in those aspects of the task not already acquired. It would seem logical to develop a multidisciplinary team of investigators, each with his or her own area of specialization, which can be enhanced by familiarisation training in the tasks performed by their colleagues. This cross-fertilization would serve to facilitate the working of the team and the data-gathering process.

What Data Are Collected?

Reference has already been made to the data sources available to aviation incident and accident investigators. The data of most interest to each member of the investigation team will depend to some extent on the area of specialization of the particular team members. From the human factors point of view, transcripts and recordings of RTF communication will be essential, as will written reports from the perspective of the individual controller and/or pilots concerned. This allows the investigators to appreciate the background to an incident or accident and prepares the way for the later stages of the investigation, that is, unit visits and face-to-face interviews. From the point of view of the human factors investigator, this background information is invaluable because it allows an opportunity to draw on the expertise of team colleagues who will probably be more familiar with the specific aviation-related aspects of the event. As a result of this preparation, the human factors investigator is in a better position to frame human factors questions relevant to the context in which the incident or accident occurred.

OTHER METHODS OF DATA COLLECTION

The thesis presented in this chapter is that the optimal method of conducting the human factors side of incident and accident investigation is for the human factors specialist to be presented as an integral part of a team of experts, each with possibly different, but complementary, areas of expertise. There are, however, other means of gathering data, and some of these are discussed here.

Checklists

It would be possible to provide a nonspecialist with a checklist against which human factors data could be gathered. The data collected, however, would, in all probability, be more a function of the nature of the comprehensiveness of the checklist items than any real, in-depth, understanding of the occurrence in question. The checklist approach has a number of disadvantages:

1. The data are likely to be rather "coarse grained" in that they would not reflect the contributory factors in any great detail.
2. The data would be limited to the contents of the checklist, rather than reflecting the nature of the specific incident or accident.
3. Although the checklist may be useful for noting the more tangible items of record such as hours worked, weather conditions, and so on, the approach would not lend itself as readily to an understanding of the less evident data that are vital to an investigation. In this category one might include the more cognitive aspects of the performance displayed by the individuals concerned in the event, which are, arguably, best investigated by the human factors specialist.
4. A standardized checklist approach is also less likely to pick up on the more generic issues involved, which may not be immediately apparent.
5. If the data initially gathered are prone to the shortcomings already mentioned, this will have serious implications for any subsequent uses to which those data may be put. If the right questions are not asked at the outset, it could prove difficult, if not impossible, to retrieve necessary information at a later stage. Attempts have been made in the past to conduct human factors analyses of incidents and accidents from occurrence reports. Many of these attempts have been flawed by virtue of the fact that the source material has not been collected from a human factors perspective.

Self-Reporting Schemes

A further means of gathering data without the necessity for a full-scale field investigation is to enlarge the scope of the self-reports completed by personnel involved in the occurrence. Currently, in the United Kingdom, ATC personnel involved in incidents and accidents are subject to a mandatory reporting program and complete a report that covers the incidents as they saw it, including such aspects as time on shift, shift start, equipment serviceability, and so on. This could be extended to include additional subjective data such as perceived workload, distractions, and more detailed data on the nature of the incident itself. However, asking the individuals concerned to, in effect, conduct a human factors analysis of their own behavior is fraught with problems. First, there is the question of reporter bias, which always needs to be taken into account when individuals report on their own behavior. The incident is, naturally enough, going to be described from one viewpoint, which may or may not accord with events as they actually occurred. In addition to the more obvious memory problems, individuals involved in incidents or accidents are likely to try to make sense of what may otherwise appear to be an illogical situation. If we accept the premise that, generally speaking, individuals will not deliberately make errors, then a situation in which an error occurs may be, almost by definition, a meaningless situation. Consequently, individuals placed in the situation of having to report on their own errors may attempt to present the event in a much more rational light. This is not to suggest that they are necessarily lying, but rather that they are attempting to understand a situation or an error for which they have no rational explanation.

Confidential Reporting Schemes

In addition to more formal incident reporting programs, a number of states also run confidential reporting schemes that allow individuals to air their concerns on aviation safety in a confidential manner. Such schemes, such as CHIRP in the United Kingdom, CAIR in Australia, and the ASRS scheme in the United States, are valuable in drawing attention to human factors issues, often before the problems reported have manifested themselves in incidents or accidents. However, it has to be borne in mind that data gathered via these schemes is not of the same type as that gathered during an incident investigation. Reporters are very much a self-selected group, motivated by their view of events to report in a more public forum but, for whatever reason, unable or unwilling to utilize the more formal reporting channels. These reports are therefore likely to be even more prone to reporter bias and the problems mentioned earlier than the other methods already described, although the very act of reporting can serve a cathartic function for the reporters. There is also the question of how far such reports can be progressed through the system, because there could well be a conflict between verifying the veracity of the reports and adhering to the stated pledge to maintain confidentiality. These caveats do not, however, denigrate the value of these schemes in providing an educational function for other pilots or controllers by which they can learn from the mistakes of others. They also serve as a useful "barometer" of the current state of aviation safety as perceived by those actually doing the job.

International Aspects

In the United Kingdom, in excess of 1,200 Air Traffic Services (ATS)-related incidents and accidents are reported per annum. All of these are investigated in order to determine the causal and contributory factors and to put in place remedial measures as appropriate. Of the total, around 60 are the subject of a full field investigation involving a close examination of all the available data together with site visits to the ATC facilities involved and interviews with relevant members of staff. Resource allocation means, of necessity, that decisions have to be taken regarding the selection of occurrences to be investigated at this level, and priority is normally given to those assessed as involving the most risk. From a human factors perspective, this may not necessarily be the best criterion, but an examination of the events investigated over a 7-year period would suggest that the existing human factors database is fairly representative of the problems inherent in United Kingdom ATC. The scope of the field investigation is such that a comprehensive picture is obtained of the event and its causation and mechanisms are in place to allow the feedback of lessons learned. However, even 60 events per annum represents a relatively small data set from which to draw conclusions and make recommendations. Therefore, the availability of other confirmatory data is highly desirable. Communication between investigation agencies from different states is a valuable source of information. The exchange of information and ideas can only serve to strengthen the quality of the investigative process generally. Attempts are in hand to achieve some form of commonality in data bases to facilitate the transfer of information. With the current state of technology this is an achievable aim. What is likely to prove more difficult, however, is achieving some commonalty in the method and process of incident investigation, including the human factors side. Different states vary in the methods adopted and the number and scope of the events covered, with attention often being focused on accidents at the expense of the seemingly less serious, but nevertheless important, incidents. However, if common terminology and taxonomies can be agreed on for the storage of investigation data, this would go some way toward overcoming the disadvantages of differences in method. It has already been suggested in this chapter that data can be viewed as varying in quality depending on the manner in which they are collected and by whom. As liaison between investigation bodies improves and data transfer among them becomes easier and more frequent, it will be important that not all data are regarded as "equal" and that, ideally, the sources of data are specified when data transfer occurs.

INVESTIGATION FRAMEWORK

It is important during an investigation that care is taken to ensure that no relevant information is overlooked in the data-gathering process. For this reason it is sometimes proposed that investigators adopt a particular model as an investigative framework. Many of these are not models in the accepted sense of the term; they have little or no predictive capability and are, at best, a set of guidelines which can be used to inform the investigative process. In essence, they tend to represent explicit statements of the good practice that any investigator worth the name should be utilizing. Models may serve as a structure in which the nonspecialist can collect data. However, it could be argued that they have only limited utility in the most important aspects of investigation—namely, the evaluation, prioritization, and interpretation of the data. It is in these areas that the specialist investigator comes into his or her own.

The problem of bias has already been mentioned from the perspective of the reporter. However, investigators can have biases too, and it is essential that the investigator be aware of the danger of bringing a biased approach to a particular investigation or set of circumstances and also of forming hypotheses before the relevant data have been sifted and analyzed. The decisions as to what data are relevant in the causal chain and what can be safely left out of the equation is an exercise of judgment that forms one of the most important aspects of the investigation process. Any specialist may tend to see things in terms of his or her own field of specialization and interpret the data accordingly.

The point is that there would be a number of facets to the overall picture of the incident. Any facet ignored, or allowed to dominate to the detriment of the rest, is going to produce an outcome in terms of the feedback from the accident that is biased and essentially flawed. The construction of multidisciplinary teams of specialists working together helps to militate against bias and may also prevent the formation of premature hypotheses on the part of any one investigator.

FEEDBACK

The investigation of incidents and accidents has a number of phases—background preparatory work, initial data gathering, evaluation, prioritization, and interpretation. These are all essential elements in the investigative process, but are, of themselves, only subgoals in fulfilling the overall aim of investigation, that is, the prevention of future incidents and accidents. The fulfillment of this aim demands the provision of clear, logical, and above all practicable feedback. An incident or accident from which no lessons are learned is an incident or accident that has been wasted in air safety terms.

The point has been made in relation to the provision of feedback that "it is probably worth making a distinction between safety regulation (i.e. the establishment of goals and standards) and safety management (which relates to the day to day application of those goals and standards in the operational environment)" (Baker, 1992, p. 248). There are a number of target areas to which feedback from incidents and accidents can, and should, be addressed. First and foremost this involves the providers of the services in question, in this case air traffic control. The individuals and their management involved in the occurrence have a need and a right to be informed of the findings of any investigation. Those responsible for safety regulation and the setting of safety standards also need firsthand information on the state of health of the systems they are regulating. Incident investigation is a reactive process that indicates, post facto, that whatever safeguards were in place have failed in some way because an incident or accident has occurred. However, as already stressed, the reactive nature of incident investigation does not preclude its having an additional, proactive role. One of the major ways in which this proactive role can be realized is in the provision of input to research activities. Investigation of incidents and accidents can, and should, provide quite detailed information on each facet of the system under investigation. The expertise that a human factors specialist brings to the investigation of behavior and performance can, for example, be invaluable in informing the development of interfaces and decision support aids for both the pilot and the controller. The very fact that the investigation process concentrates on events in which the system has broken down serves to illustrate those areas demanding of most attention and helps to focus on those aspects of the task where research could most usefully be targeted. It is essential that lessons learned are effectively fed forward to those in a position to make decisions regarding the development and procurement of future systems. A knowledge of past problems and inadequacies should help in the development of the sort of informed opinion that can ask the appropriate, and often awkward, questions of the system designers and equipment retailers to ensure that problems identified with past systems are not perpetuated in the future.

SYSTEM DEVELOPMENT AND EVALUATION

It has been argued that the development and evaluation of new systems should, necessarily, involve a comparative study of the performance of the old system by comparison with the new (Baker & Marshall, 1988), in order that the advantages of the new system can be more readily and clearly demonstrated. This comparative approach, although desirable, is time-consuming and expensive. Expediency and cost often require that only the new, developmental system is tested and adequate evaluation of the pros and cons of the new system vis-à-vis the old is frequently omitted. However, during the investigation process much can be learned about the relative merits and demerits of existing systems as they are perceived by the end user. As a result, it should be possible to indicate to system designers the strengths and weaknesses of current systems and to indicate those areas that need improvement and those that function well. The success of this process does, however, require a symbiotic relationship between the human factors investigator and the designers and evaluators of systems. The problems inherent in this approach have been pointed out elsewhere; for example, Baker and Marshall (1988) made the point that "However desirable a co-operation between designers and human factors experts might be, human factors specialists are still not sufficiently involved in the design phase with the result that, often, the anticipated benefits from the system in question are not spelled out in any clearly testable way. Typically, psychologists are simply requested to validate or demonstrate the *advantages* of a new system" (p. 83). In the United Kingdom at least, very few ATC-related incidents can be traced directly to problems related to inadequate or less than optimal equipment. Rather, poor equipment and facilities tend to be implicated as contributory, not causative factors. Nevertheless, investigations do reveal areas in which equipment development is needed. A good deal of attention has been focused, for example, on alerting systems that inform the pilot or controller of an impending collision. This is well and good and very necessary if airborne collisions are to be avoided. However, relatively less attention has been focused on the development of systems that aid the planning and decision making aspects of the ATC tasks, that is, to prevent the development of situations in which conflicts arise in the first place. The investigation of the human factors aspects of incidents and accidents can be helpful here in highlighting those aspects of the planning and decision making process most in need of support.

Feedback is not, however, restricted to ergonomic and equipment-related issues. The adoption of the system approach discussed earlier facilitates the gathering of information on all aspects of ATC functioning. Human factors recommendations that ensue from investigations can range from fairly basic "quick fixes" to more far reaching issues involving, for example, such aspects as training or the role of management. In Reason's terms (Reason, 1989), both the "active" and "latent" failures in the system need to be addressed, and careful sifting and analysis of the information gathered from investigations can reveal, not only those areas in which failures have already occurred and errors been made, but also those aspects of the system that if left unaddressed could well lead to problems in the future. The existence of these generic problems, which may not have manifested themselves directly in an incident or whose connection to an occurrence may seem somewhat tenuous, is often difficult to demonstrate. This is one area where the advantages of incident as well as accident investigation are most evident. It may be difficult to demonstrate, say, on the basis of one accident, that a particular problem exists. However, if it can be shown that similar conclusions have been reached as a result of the more numerous incident investigations, the case for a closer examination of the problem and perhaps the initiation of research will be greatly strengthened.

CONCLUSION

The role of human factors in incident and accident investigation has received increased attention in recent years. Even so, the extent to which human factors considerations are taken into account during the investigation process varies from state to state. This chapter has focused on the investigation of civil ATC-related incidents and accidents in the United Kingdom, where a human factors specialist is routinely included as part of a multidisciplinary investigation team.

The motivation for conducting investigations extends beyond discovering the cause of any one incident or accident. The main focus has to be on lessons learned with a view to the prevention of similar incidents or accidents in the future. The greater the volume of information that can be gathered, the more complete is the picture that can be gained and the firmer is the basis for any recommendations for future improvements. The additional knowledge gained from investigating incidents, in addition to less frequent accidents, is invaluable in compiling the overall picture.

The collation and analysis of data, together with the compilation of reports and recommendations arising from a specific incident or accident, is not the end of the story. An incident or accident from which no lessons are learned is a wasted event. There has to be practicable and accurate feedback and that feedback has to be acted on. It is therefore essential that efficient mechanisms exist, not only to disseminate information to those individuals and/or organizations where it can do most good in terms of prevention, but also to monitor that the feedback has been utilized.

A successful investigation demands a balanced approach to the problem. Each of the team of experts involved in the investigation will have his or her own area of expertise, none of which should be allowed to assume undue priority and importance in the investigative process. The underlying causal factors in incident and accident occurrence will vary, however. Accidents involving engine failure as the root causal factor, for example, will give rise to different findings with different emphases than those in which training or ground equipment are primarily implicated.

The inclusion of human factors as a potential issue in incident and accident occurrence has come fairly late on the investigative scene. However, to ignore the human factors aspects of these events will, almost inevitably, lead to an unbalanced and incomplete picture in attempting to determine not only what happened, but why.

REFERENCES

Air Accidents Investigation Branch. (1990). *Report on the accident to BAC One-Eleven, G-BJRT over Didcot, Oxfordshire on 10 June 1990.* Department of Transport, HMSO, London.

Baker, S. (1992). The role of incident investigation in system validation. In J. A. Wise, V. D. Hopkin, & P. Stager (Eds.), *Verification and validation of complex systems: Human factors issues* (p. 248). NATO ASI Series.

Baker, S., & Marshall, E. (1988). Evaluating the man-machine interface—The search for data. In J. Patrick & K. D. Duncan (Eds.), *Training, human decision making and control* (p. 83). Amsterdam: North-Holland.

International Civil Aviation Organization. (1993). *Human factors, management and organization* (ICAO Circular 247-AN/148, Human Factors Digest No. 10). Montreal, Canada: Author.

International Civil Aviation Organization. (1993). *Investigation of human factors incidents and accidents* (ICAO Circular 240-AN/144, Human Factors Digest No. 7). Montreal, Canada: Author.

Reason, J. T. (1989). The contribution of latent human failures to the breakdown of complex systems. *Philosophical Transactions of the Royal Society (London), B, 327,* 475–484.

27

Forensic Aviation Human Factors [Accident/Incident Analyses for Legal Proceedings]

Richard D. Gilson
University of Central Florida

When someone has an accident[1] with an aircraft, automobile, or with any other system or device, people want to know why. The primary motivation[2] for this is the *prevention* of their own and others' involvement in similar mishaps. Humans do make mistakes with machine systems, and are listed as the probable cause of accidents more often than not.[3] But who actually is to blame, why, and how to avert a reoccurrence may not be obvious or easy to uncover. A closer look is often in order through forensic human factors, particularly in such complex systems as aviation.

First there must be an agency review of investigatory findings. If there are no other definitive causes of a mishap such as a mechanical failure, then the operator is likely to become the focus of blame. If the circumstance is truly of an individual's making, then, beyond the determination of personal culpability, further human factors analyses may not be warranted. However, if external factors could have actually *induced* the behavior, then the problem may not be a one-time occurrence and further analyses are called for. Experts in human factors have long recognized that designs, procedures, and training (beyond mere talent) affect the safe and efficient use of systems. Thus, human decisions made by systems designers, and ultimately by the manufacturers, also influence outcomes and may contribute to or principally underlie the problems.

Courts of law serve as one of several means for determination of cause and blame, and may spawn potential remedies. Because those involved in the judicial process

[1]An accident or mishap refers to an undesirable, unintentional, and unexpected event. However, used herein, an accident (or mishap) is *not* an unforeseeable random event, totally without cause, as is an act of God. Forensics presumes and seeks out underlying causes or contributing factors and assumes that future care could make such accidents preventable.

[2]Certainly there are other less altruistic motivations. The public has almost a morbid interest in aviation accidents, some people involved have mercenary pursuits, and even others may seek vengeance or absolution.

[3]Human error as the primary cause of aviation accidents is frequently put at anywhere from 60% to 80%, depending on the source.

typically have no training in human factors or in aviation, forensic human factors experts specializing in aviation are needed to analyze, explain, and give probabilistic opinions, so that judges and juries can make *informed decisions*. Although litigation is considered by many as a negative, a positive outcome may be to help prevent further occurrences.

INTRODUCTION

Opening

An airplane crashes and someone dies. Federally mandated investigations are launched to determine the probable cause(s) and to uncover any regulatory violation(s). Recommendations for rule changes or penalties may follow. These are prospectively aimed at preventing similar mishaps. Civil legal actions may also follow but are more retrospective as they look for specific blame and seek compensation for losses.

Both avenues, governmental and civil, pursue recourses/remedies but from different perspectives. Governmental investigations are typically public, fact finding, and directed at categorical reporting and statistical manipulation. Usually only recurrent accidents evoke response to the underlying issues in question, human or machine, and then only after there is some general consensus that a fix is needed and proper.

Civil actions, however, are typically private, cloistered, argue the meaning of specific facts already known, and focus on alleged errors. The very adversarial nature of the "civil" process places the individual pilot and system(s) under extreme scrutiny to force justification of their respective actions and design. As each matter is settled either by mediation or by verdict, the precedent becomes a deterrent for similar incidents.

Taken in tandem, both approaches provide real-world feedback about practical human factors issues, with regard to the merits of a system design and its use. Clearly, these postcrash analyses are the last resort, coming too late for those involved, but are preferable to inaction if a real problem exists.

Government agencies do have the weight of law behind mandated changes, but must wait to amass statistical evidence and to make a persuasive case concerning the relatedness of what appear to be similar crashes. The process of proposed rule making has to overcome many checks and balances designed to guard against unnecessary fixes or unwanted regulation that might arise from a truly isolated mishap, that is, a sampling error.

Civil lawsuits, on the other hand, evaluate single cases, but carry no regulatory authority. Indeed, the fact that many final judgments in civil lawsuits do not eventually become law suggests governmental dismissal of the cause as an act of God or as a unique event generally unlikely to reoccur (e.g., preventable carelessness). It is also possible that there were flaws in the final judgment because of a lack of knowledge or distortion of the meaning of facts about aviation or human factors. The extent to which the judgment was valid, yet not made regulatory, may serve as warning to designers and users alike that certain actions or inaction may carry the added risk of penalties.

Human factors experts in aviation can assist in all these processes, from postcrash analysis and through recommended remedies. However, because it is the civil legal system that most often (vs. government agencies) retains human factors experts for aviation forensic services, this chapter focuses primarily on civil proceedings in an attempt to provide the reader with an understanding of the issues involved.

Forensics

The terminology of forensics deserves some explanation. First, the word *forensics* stems from forum, the marketplace or assembly place of an ancient Roman city, which formed the center of judicial or public business. As a noun, it implies oratory, rhetoric, and argumentative discourse, such as might be used in a debate. This use implies an advocacy position.

On the other hand, as an adjective, forensics is usually associated with judicial inquiries or with evidence given in courts of law. For example, forensics medicine is considered a science that deals with the relation and application of medical facts to legal problems, as in the role of "Quincy," the fictitious forensics pathologist in the long-running television series. In it, as in real life, an expert's opinion becomes legally admissible evidence, given under oath. The implications of the testimony are argued by the attorneys and ultimately are used in legal judgments.

These differences in meaning, I believe, sparked a controversial position statement and subsequent brouhaha over the value of forensics human factors in *Ergonomics in Design*. In an editorial, Daryle Jean Gardner-Bonneau (1994a) insinuated that forensics in human factors was less than a worthwhile scientific pursuit by taking the unusual action of deliberately removing "Formative Forensics" from the masthead of *Ergonomics in Design*. Gardner-Bonneau's complaint appears to be that forensics human factors has "little science" and that human factors/ergonomics professionals as expert witnesses are "often called on to consider isolated aspects of a case and render judgements based on limited information . . . [thus] the waters of forensics practice, from [the] editor's perspective, are simply too murky to play in" (p. 3). Howell (1994), in a published article within that same issue entitled "Provocations: The human factors expert mercenary," also supported this view by suggesting that "the credibility of the human factors discipline [was being questioned] . . . as a result of our rapidly growing presence in the forensics area" (p. 6). It seems that both writers are considering forensics as a noun, suggesting a narrow, almost quarreling/advocacy view for the benefit of one side—hardly the basis of an unbiased scientific inquiry and analysis.

A fusillade of letters offered rebuttal in the subsequent issue of *Ergonomics in Design*. In one letter by Richard Hornick (a former president of the Human Factors Society, and chair-elect of the Human Factors and Ergonomics Society, Forensics Professional Group), the editorial position was severely criticized as potentially damaging to that society as well as to individual forensics practioners (Hornick, 1994). Hornick argued that judicial scrutiny and challenges to an expert's opinion(s) "far exceed those [scientific peer reviews] that occur in academia and industry . . . [and that the] legal arena provides a powerful tool to correct flaws [in product/workplace design and as a defense against wrongful suits or unfair claims]" (p. 4). Other letters in the issue claimed that forensics work "mediate[s] scientific findings and practical requirements . . . for real-world problems" (Beaton, 1994, p. 5) and that "forensics work drives you back to the basics of the subject" (Corlett, 1994, p. 5). Editor Gardner-Bonneau responded in her own letter (Gardner-Bonneau, 1994b) that "that human factors analysis *can* [her emphasis] have [value] in forensics work . . . [and that she] encourage[s] submissions that emphasize the application of human factors methodologies and techniques to the analysis of forensics cases" (p. 35). Finally, Deborah A. Boehm-Davis, then President of the Human Factors and Ergonomics Society, also took the time to write (Boehm-Davis, 1994) to "make it clear that the statements made in the editorial *do not* [her emphasis] represent the Human Factors and Ergonomics Society's position on the practice of forensics. The

Society encourages the practice of sound forensics and sponsors a technical group that supports human factors/ergonomics professionals who work in this field" (p. 35).

The intent here is to use forensics as a subspecialty of human factors relating scientific principles/facts to the analysis of a specific accident or mishap. At issue is the determination of what happened, why it happened, and how to help prevent reoccurrence of the underlying problem (if indeed there is one, beyond an "act of God"). This does not mean merely rendering professional views and expert opinions based on theory and knowledge. But rather, as Senders (1994) advocated, it should involve field studies and empirical data, wherever possible. This approach is no different from using the procedures, investigative techniques, and findings/knowledge of other disciplines and subspecialties to forensics, such as forensics ballistics, forensics chemistry, forensics psychiatry, or the like.

Aviation Human Factors

Human factors itself is relatively new as an applied science, originally often aimed at problems or hazards in aviation systems. Forensic human factors is newer yet in courts of law, and applies to a number of domains in addition to aviation. Typically, interest is in the discovery of potential misuse or danger of products from the legal standpoint of liability,[4] for example, in the context of human factors design-induced errors, procedural flaws, or inadequate training or warnings. Behavior with products may range from what is normally prudent, to what should be reasonably expected (even if wrong), to what may be even deliberate risk-taking activities.

Because the law has always focused on people with problems, the emergence of human factors in the legal system was probably inevitable and should provide a valuable service. Ideally, beyond the determination of fault, the legal system is designed to be one of a number of means to change future behavior (whether that of the manufacturer, seller, or user). But keep in mind in order to change behavior (by design, prevention, procedures, warnings, training, etc.), the problem must be well understood. Evidence regarding aviation operations and the even lesser known field of human factors is often foreign to the courts. Evaluation of what went wrong in these contexts often must be explained by expert testimony as part of the evidence.

Judges and juries (as well as attorneys and litigants) expect more than a reconstruction of the events and an analysis of the crash dynamics. They want a reliable determination of the *why* (of behavior), along with the *what* (happened). Details about how the design was intended to be used and how it actually was used contribute salient evidence that often tilts the balance in decisions. Were errors the result of the individual or because of the design itself? Were there attentional lapses, slowed reactions, inaccurate perceptions, wrong expectancies, hazardous activities stemming from deliberate behavior by the individual involved? Or were the problems induced by defective design, inadequate protections/warnings, deficient instructions/training, or even improper integration with other systems?

The determination of the problem and restitution for it occurs at many stages in legal proceedings, such that most potential lawsuits do not come to full fruition. Even before an accident ever occurs the manufacturer may reevaluate and modify a design for better usability and safety using an outside human factors expert. After an accident,

[4]Liability in this context generally suggests failure to exercise ordinary or reasonable care in the manufacture or use of a product.

but before a lawsuit, a manufacturer or insurance company may ask questions of a human factors expert about its culpability or its relative exposure. During a lawsuit, settlement questions arise requiring expert opinions.

As a goal, specific human factors input into the original design stages should help to insure that a product or system is easy and safe to use. This is done in automobile manufacturing as it fits changes into the competitive marketplace, creating maximum user preference and minimal user problems. In actuality, automobile manufacturing changes are constantly being rethought—responding to evolving market forces. With mega-volume sales and yearly changes in design, some automobile manufacturers actually have groups of full-time human factors practitioners on staff. These efforts have paid off lately, with the successful selling of safety (once thought impossible) as well as convenience, economy, and comfort. Similarly, with high per-unit cost of a commercial or military aircraft, large companies such as Boeing and others also use staffs of human factors experts throughout the design and modification stages. Also, in commercial aviation and beyond, there is the pull from sales potentials and the incentives of profit. The consumer drives the marketplace and spawns product developments with purchasing power. New developments are proudly announced and advertised with gusto.

In contrast, general aviation manufacturers, with annual sales in the hundreds and dwindling, are forced to continue to use specialized designs often 40 years old (before human factors became separate from engineering design). Employment of full-time human factors professionals is not a practical consideration, although their engineering staffs may have been updated with human factors courses. Fortunately, with so few changes being made, there is plenty of experience with what works and what modifications are needed. Moreover, there are a number of mechanisms in place to provide feedback regarding problems with existing designs. These include FAA Airworthiness Directives, Service Difficulties Reports, Manufacturer's Service Bulletins, NASA's Aviation Safety Reporting System, various National Transportation Safety Board (NTSB) recommendations, exchanges in technical and trade literature, insurance company publications, and a variety of magazines for pilots, aircraft owners, mechanics, fixed-base operators, and so on. These are evidence that today's general aviation suffers from remedies that come by way of push, not pull. The push comes primarily from government agencies with new regulations and the threat of penalties, an unnatural force for change. Fixes, if deemed appropriate, are disseminated in drab technical literature, published as advisories, or mandated by regulation. Thus, without strong market forces and with few accidents, changes in general aviation are traditionally slow to emerge, and these, for the most part, are evolutionary or regulatory refinements, not revolutionary leaps ahead.

Despite all this, accidents do occur, ratcheting up demands for narrow analyses and remedies. News media reports, followed by NTSB Accidents Reports (or Military Safety Boards), manufacturers' mishap reports, and legal proceedings, typically help illuminate what happened, but not necessarily why. Certainly they do not incorporate specialized human factors considerations. Unfortunately, almost any aircraft accident creates intense public interest (in comparison to the automobile accident) and the call for hurried fixes.

Detractors of this media hysteria abound. Some denounce it as creating a "tombstone mentality" delaying the evaluation of the efficacy/soundness of a design until damage, injury, or death has occurred. Others maintain that an atmosphere of fear of litigation has paralyzed the industry, causing the demise of general aviation. Their argument is that advances in technology become *prima facie* evidence of design shortcomings, halting even the discussion of proposed modifications by a "circle the wagons" mentality. Even others condemn the process as relinquishing important complex design decisions to lay juries and to inconsistent tort law mandates, while dividing any joint efforts by those most directly involved, namely, the manufacturers and the users.

Rightly or wrongly, all this means that human factors experts currently have (with a few exceptions) their greatest input into aviation via postaccident analyses in response to questions raised by the legal system. Although this may change, with aviation product liability reform that limits liability to 18 years after manufacture, it is more probable that there will only be a shift from blame of older aircraft involved in accidents to blame of component manufacturers, maintenance facilities, and service organizations (private and public). In any event, human factors analyses will still be required.

For now forensics aviation human factors experts, working within the system, can bring about thoughtful application of scientific knowledge and analysis to specific problems. Their purpose should be to educate juries who are called upon to make legal judgments. Despite the fact that plaintiff and defense experts may disagree, the didactic process is helpful in revealing most sides of what are often complex issues. Imperfect as it may be, this process provides checks and balances of both sides, ironically in much the same way as the open literature provides checks and balances for scientific disputes.

Even without the prospect of civil exposure or penalties, progressive manufacturers will continue to improve their product line, including enhancements in safety. Companies seek human factors experts for design assistance, and for positive marketing advantages, such as better panel layouts, more efficient controls, improved seating comfort, reduced cabin noise, and so on. At the same time, however, such experts always should be aware of the goal of designing a reasonably safe product including consideration of foreseeable hazards, alternative design features against those hazards, and evaluation of such alternatives (Sanders & McCormick, 1993). Sanders and McCormick (1993) also pointed out that human factors design and evaluation often are not straightforward, with a note of caution that alternatives may introduce other hazards. For example, if warnings are chosen as a means of highlighting potential hazards, then there may be a problem of warning overload, diluting the importance of vital information.

Of course, the best solution is to create enough sales in general aviation to enlarge the feedback loop so market forces dominate again. That would create the incentive for people to buy airplanes again in large enough quantities so that human factors research and input during design and manufacture could make aviation systems, easier, safer, and more reliable to use. Better designs should beget more sales, and more sales will inevitably bring about better designs, spiraling the market outward, not inward as it has been for far more than a decade. Moreover, with more aviation activity in the mainstream of public life, distorted views should also diminish. For example, public interest in discovering the why of almost any one of the nearly 40,000 lives lost each year through traffic accidents is disproportionate to the scrutiny given to almost any aviation mishap (admittedly there are number of reasons, rational and otherwise). Certainly, however, a life lost on the highway is not worth less then a life lost in the air. The goal obviously should be to responsibly reduce all risk as much as possible.

However, the law recognizes, and the public should too, that there is a trade-off between risk and benefit, and that no system or product is absolutely safe or reliable. Human factors can and should help reduce risk. For now, this is realistically only possible within the systems in place, regardless of their imperfections.

Forensic Human Factors

As specifically applied to aviation, it seems appropriate to consider forensic human factors as a broad view of inquiry beyond (i.e., not limited to) the courts of law and including occurrences that may surface in a variety of ways.[5] Regardless of how problems become known, external analyses of human behavior with aviation systems, whether involving accidents, mishaps, incidents, or errors, are often extensive. They utilize various forms of investigative techniques, and now include human factors analyses as part of the whole process.

Unfortunately, aviation is nearly unique in modern societies—that is, outside military mission environments. It demands high levels of performance with the penalty for error being extensive damage, injury, or death. As the old saying goes, mistakes in aviation are often unforgiving. Yet, mistakes with actual equipment/systems, whether highlighted by incidents, mishaps, or accidents, may provide the truest forum for evaluation of human factors designs, procedures, and training methodology. As such, forensics provides invaluable feedback to correct problems.[6]

Accordingly, forensics inquiries in various disciplines are indispensable ingredients in aviation accident investigation. However, human factors analyses are often inductive in nature and are hampered by loss of evidence, whether by death of those involved, by memory losses of those injured, by deliberately ambiguous statements, or even by rationalizing or intentional misstatements.

Nonetheless, forensic human factors in aviation legitimately takes its place as a method for integrating remaining evidence, specific facts, and circumstances to determine what, and more importantly, why human factors is more often than not the primary cause of an accident. Ideally, such analyses focus specifically on the objective of how to make systems/people operate safely and efficiently together in future designs, procedures, and training. (Often as a result of forensic human factors work in aviation design or training, remedies are recommended.)

According to many, forensic human factors in aviation started with aviation's first fatality, Lt. Selfridge, a U.S. Army pilot killed after crashing in a Wright Flyer in the early days of aviation. That accident analysis led to the search for preventative solutions, such as the use of helmets in the early days of flying, eventually developing into the overall consideration of methods for survivability and crashworthiness. Similarly, many non-combat-related accidents during World War II brought human factors to the forefront with design changes and classic texts.

What has grown from these seeds is the stimulus for applied experimental research and true evaluation of products in environment. What remains is the need to evaluate risk before the accident, although headway in this area has been made through a

[5]For example, National Transportation Safety Board (NTSB) accident reports; National Aeronautics and Space Administration/Aviation Safety Reporting System (NASA/ASRS) incident reports; Federal Aviation Administration (FAA) Service Difficulty Reports (SDRs), FAA Condition Difficulty Reports (CDRs), and FAA enforcement actions; incidents made known in the technical literature and commercial magazines, books, or videos; through manufacturer's field reports; insurance company's databases; pilot organization publications, such as the Aircraft Owners and Pilot Association (AOPA *Pilot*), the AOPA Air Safety Foundation (AOPA/ASF), the Experimental Aircraft Association (EAA), the Flight Safety Foundation (FSF), etc.; professional organizations such as the *Forum* of the International Society of Air Safety Investigators (ISASI); military investigatory agencies and publications; and the list goes on and on.

[6]If forensics is the sole technique to initiating fixes, it properly deserves the criticism derisively insinuated by the moniker "tombstone mentality."

large-scale adaptation of the "critical incident" in databases, such as the NASA Aviation Safety Reporting System (ASRS), and the FAA Service Difficulty Reports (SDR).

Forensic Aviation Human Factors

For decades statistical trends have illustrated continuing improvements in aircraft accident rates, primarily due to system/equipment reliability. Human fallibilities have lessened also, but at a slower rate. Proportionately, then, the contribution of the human factor in accidents has grown compared to aircraft faults, underscoring the importance of understanding why. The following illustrates the growing importance of human factors.

According to a January 3, 1994, editorial in *Aviation Week and Space Technology*,

> Human factors engineering ... is required if the air transport industry is to achieve an acceptable level of safety in the next century. Human error is the cause of the vast majority of civil aircraft accidents ... [projecting from] the current rate of hull losses, there could be a major accident every 10 days by 2006. (Editorial, 1994, p. 66)

Perhaps the complication of sophisticated systems mitigates the benefits of improved designs and training. When infrequent and unanticipated faults do occur, they become blurry enigmas [engendering vague responses] for human supervisors, an example being what occurred during the "Three Mile Island" nuclear incident. Forensic human factors experts can reveal underlying areas of inadequacies in perception, attention, memory, situation awareness, and so forth, adding evidence to an accident investigator's reconstruction of the physical evidence and background data.

HOW HUMAN FACTORS CONTRIBUTES TO AVIATION FORENSICS

Many postaccident investigations now include aviation human factors consultants and experts in addition to traditional accident reconstruction experts.[7] Aviation human factors forensics experts can serve to explain human behavior (right or wrong) with the equipment people use and often innocently depend upon. In most cases, the question is, why were the errors made? Could the errors have been avoided? Was the behavior typical—to be expected within the specific set of circumstances—or was it improper? Were there contributory factors involved such as fatigue, stress, intentional misuse, or even alcohol or drugs? Could the design be more error tolerant, and if so what are the trade-offs?

Attorneys want an analysis of human and design-induced errors, both for their own understanding and to provide a credible approach to initiate or defend a lawsuit. Were errors the result of the individual or because of the design itself? Human errors might include attentional lapses, slowed reactions, inaccurate perceptions, risk-taking activities, and wrong expectancies stemming from inadequate situation awareness. Design-induced errors might include problems induced by defective design, hidden hazards, inadequate protection or warnings, deficient instructions or training, or even improper integration with other systems.

[7]The International Society of Air Safety Investigators (ISASI) has a Human Factors Working Group to address "issues arising from examining human factors in accident investigations"; see McIntyre (1994).

Juries expect more than a reconstruction of the events and an analysis of the crash dynamics. They want a reliable determination of the *why* (of behavior) along with the *what* (happened). Details about how the design was intended to be used and how it actually was used contribute salient evidence that often tilts the balance in their decisions.

Postaccident Investigative Procedures

Scientists serving as forensics experts initially spend much of their time reviewing a mountain of diverse information from various sources, the aftermath of most aircraft accident investigations involving fatalities. The task then shifts to selecting and piecing together what appear to be causal factors into a cohesive theory of what happened and why, together with supporting evidence. Often this is followed by an active search for other sources of information to fill in inevitable gaps. This search can consist of flight demonstrations or even specific experiments carried out to verify or to test a theory.

The following are three descriptions of accidents involving aviation human factors forensics analyses. As you can see, although they were successful in identifying human errors, design defects, and operational deficiencies, the outcomes may be to some readers far from definitive. Most accidents do not cleanly fall into any particular category; indeed, most often there are multiple causal and contributing factors leading to a fatal crash. Beyond the presentation of evidence, the dynamics of legal proceeding brings into play the personal, social, and economic factors of those directly involved. This is a mix whose outcome may be baffling to some, yet it is a true reflection of real life. In all fairness, the process seems to hit the mark most of the time.

Human Error Exemplar. A 71-year-old pilot was the sole occupant of a high-performance single-engine airplane proceeding into night IMC (instrument meteorological conditions) conditions from Raleigh, NC, to his home base located at a South Carolina Airport. This was the fifth flight leg of a 12-hr business day. Although the evidence indicates that he had flown at least one coupled ILS (instrument landing system) approach on that day, other information indicated that he had little overall night IFR (instrument flight rules) flight experience. He commenced an NDB 21 instrument approach with the autopilot heading and pitch command engaged for track and descent. The pilot stopped the descent using the altitude hold function at 1,200–1,300 ft, well above minimum descent altitude (MDA), after air traffic control (ATC) issued a low-altitude alert (the airplane was off course, low for that position, and in the clouds). The pilot then started a climb for a missed approach by pulling back against the altitude hold. The autopilot resisted the altitude change by counter trimming, eventually reaching full nose down trim. After several minutes of holding 40–50 pounds of control yoke back force [according to subsequent test flights], the autopilot was disconnected electrically by pulling the trim and autopilot circuit breakers and even momentarily turning off the master switch, but the pilot never retrimmed the aircraft manually. The down force remained unabated, as evidenced by panting sounds transcribed from the ATC audiotape. Confusion of both the pilot and ATC controllers as to the source of problem led to the interpretation that the autopilot was stuck on and pitching down because of a runaway "hardover" condition.

The radio transmissions are dramatic. "I'm in trouble I was using my autopilot . . . and I can't get it off . . . ah autopilot is ah all is hung the trim I I'm fighting it like a bastard trying to keep the thing up . . . (panting sounds)" ". . . I pulled every circuit breaker I can find . . . Negative [answering to ATC suggestion to turn off/on the master switch] I moved the master switch (unintelligible) turned it on." [There was no indication of an attempt to manually retrim, or to use the autopilot malfunction checklist that calls for retrim.]

For the next 32 mins, the airplane was vectored and flown erratically while various "solutions" were tried including resetting the autopilot circuit breaker pitch command, but while leaving the trim circuit breaker still pulled. This configuration pitted the autopilot clutch force (about 20–30 pounds) against the airplane maximum down trim, leaving about 20 pounds of force on the control yoke. Eventually, at between 700–1,200 ft just below the clouds (according to weather reports, radar data, and intermittent communications), the pilot maneuvered to within sight of the Chapel Hill Airport on a base leg approach. In preparation for landing, he apparently disengaged the autopilot and the powerful out-of-trim condition reappeared in full force. With obvious control difficulty and distraction, he overshot the airport and reengaged the autopilot pitch control to abate the out-of-trim forces. Efforts to maneuver for another visual approach were convoluted but finally led back to the airport. However, when the pilot apparently again turned off autopilot for landing, this time clearly exhausted, he lost control and crashed to his death. There was no factual evidence of any actual airplane, autopilot, or system interface malfunction or failure prior to ground impact.

This pilot's confusion and "mental set" with the autopilot were consistent with his past behavior during other episodes. Previously he had tried to troubleshoot the autopilot in the air, on one occasion by climbing above the clouds to VFR (visual flight rules) conditions "on top" and resetting the autopilot "to see if it would do it again." On another occasion, he reportedly "played with the switches" and presumably the circuit breakers. On even another occasion, while using the autopilot for a coupled ILS approach, he apparently lost control to the point where the passenger (a VFR pilot) briefly took over control. Yet there were "no mechanical problems" as demonstrated by the fact that the second coupled ILS approach was completed without difficulty. In each of these incidents, there was no mention by anyone on board that the flight manual (the FAA-approved Airplane Flight Manual Supplement, AFMS) was ever referred to and no evidence that postflight maintenance or instruction was ever sought. Apparently, this pilot had become highly dependent on the autopilot for flying. At the same time he clearly misunderstood how the autopilot system worked, how to test it, or how to disengage it, and he made no efforts to learn more.

This pilot exacerbated his problem by his failure to use basic operational procedures, that is, aviate first, so as to concentrate all his mental and physical resources on landing, then troubleshoot when safely on the ground. By dividing his attention and by not concentrating fully on flying the aircraft, as evidenced by his haphazard ground track, he undoubtedly prolonged the flight, leading to the eventual crash. By attempting to troubleshoot while flying, he added to his mental workload and created more control problems; for example, by turning off the master switch, he also turned off the cabin lights, making controlled flight at night all the more difficult. He was distracted from compliance with ATC clearances (aimless headings, unauthorized altitude changes, disregarded low-altitude alerts), which more than likely eliminated his opportunity to land safely with a surveillance approach at the Raleigh-Durham airport. The transcript showed that controllers remarked among themselves their reservations about the pilot's ability to handle his aircraft safely.

In his fatigued state this pilot forgot the basics of flying manually. If he had concentrated on flying, he might have simply tried manual trim with the trim wheel, thereby removing the pitch down force and the problem. To the contrary, he flew desperately for more than 30 min simply because he did not neutralize the out-of-trim forces. Tragically, he could have retrimmed the airplane at any time and landed normally.

The complaint alleged that the autopilot malfunctioned and that the disconnect was defectively designed. Further, it was alleged that the aircraft manufacturer was negligent for choosing this autopilot for this airplane. This case was settled before it went to trial, even before the expense of many depositions. Attorneys on all sides were finally convinced that the primary cause was the pilot's misuse of the autopilot system and his failure to comply with the FAA-approved Airplane Flight Manual Supplement.

Design Error Exemplar. A medium-size twin-engine helicopter departed from an oil platform in the North China Sea, with five people on board including two pilots, one American, one Chinese. The American pilot was flying on the right side (the PIC position in this helicopter). Shortly after liftoff at about 200 ft above the water, one engine experienced what is described as a turbine burst. The engine blades and housing came apart with a loud bang, a fire flash, black smoke, and debris that fell from the right side of the aircraft, all heard or seen by witnesses on the oil platform. The cockpit voice recorder (CVR) indicated the Chinese word "fire" and the master fire warning light presumably illuminated along with a 250-Hz continuous tone. The red fire light was located in a master warning panel (a "four-pack" design capable of showing fire, caution, and left or right engine failure). The crew responded according to prescribed procedures first by pushing the master fire light/switch to silence the alarm sound (the master fire light remains "on"). The fire suppression procedures then call for the pilot to reach overhead to find the lighted "T" handle signifying the affected engine. The CVR indicates that the American pilot did this, pulling back the lighted "T" handle, which in turn retarded the engine lever, shut off the fuel supply, and discharged the contents of the fire extinguisher bottle into the engine. All this took about 15.2 sec, perhaps long, but it includes the element of surprise and perhaps confusion caused by a language barrier between the two pilots.

Just before (1.6 sec) a statement by the American pilot "pulling number two," an alternating 550/700-Hz tone signaled an engine out. The sound itself does not convey the engine number. This alarm sound is designed to trigger when N_1 (rpm) drops below 59%, accompanied by the illumination of a warning light indicating the affected engine. The actual affected engine was number 1 (on the left side). Yet the pilot was responding as if the fire was in the number 2 engine (on the right side). By the time the engine-out alarm sounded, the pilot may have been looking overhead, or his response may have been already mediated and in motion. Therefore, even if he did see the number 1 engine-out warning light, he may have merely silenced the sound by pushing the light/switch and did not mentally process the number at that point in time. In any event, 8.5 sec after the irreversible action of "pulling number two" (probably after realizing that the number 1 engine-out light had come "on") came the last words on the CVR, an expletive phrase, "f—ing number two was the wrong engine." With the loss of all power, and the low altitude, a power-off autorotation was not possible and the helicopter crashed into the sea 5 to 7 sec later, killing all on board.

What looks at first like a case of human error is more complicated. The postcrash investigation revealed that the turbine burst had sent shrapnel through the fire wall between the two engines, creating holes large enough for light to go through. The infrared-light fire detectors in both engine compartments were activated, in turn signaling *both* "T" handle alarms. The right-side American pilot apparently saw the closest number 2 "T" handle lighted (which visually overlaps the number 1 "T" handle from the pilot's vantage point) and presumably he thought it was the engine signaled by the master fire alarm. Evidently, the pilot did not look further because he found what he was directed to look for by the procedures, a lighted "T" handle.

If the pilot had been alerted to validate the affected engine by means other than the "T" handles, he could have done so. Or if the pilot had been directed to look at both "T" handles before responding, he could have done so. Time was available. In retrospect, without any response, power for continued single-engine flight was not only possible but would have been fairly routine (both engines power the main and tail rotors through the same shafts). Moreover, the engine compartment fire in all likelihood could have been contained well beyond the time needed for validation of the affected engine (via other instruments) and a shutdown. Other precautionary actions were available as well, for example, a return to the platform or deployment of floats and a water landing.

Documents in the aircraft manufacturer's possession, obtained by subpoena, indicated that other turbine bursts like this (but not of this consequence) were known to have occurred before in this model aircraft/engines. However, there was no indication in the operational procedures or in other information available to pilots (pilot publications, alerts or advisories, service letters or bulletins, airworthiness directives, etc.) that both "T" handles could be lighted, one as a false alarm, which if responded to would lead to a total power loss. The manufacturer had designed and installed a "five-pack" master warning panel in the subsequent model helicopter, showing both the left and right fire alarms. (Note that "five-packs" were state-of-the-art at the time for comparable helicopters from other manufacturers.) If a turbine burst sets off both master fire alarms (side by side in the pilot's direct view), then both would be acknowledged by the pilot before shutting down an engine, thereby alerting the pilot to check before further action. This design was never offered as a retrofit to the prior model involved in this crash.

The complaint alleged that the engine was flawed and that the helicopter's warning system design misled the pilot to shut down the wrong engine, resulting in the death of the pilots and the passengers. Further, it was alleged that this event was foreseeable based on past similar incidents. The jury decided that helicopter operator was not in any way at fault because its pilot had been misled to shut down the wrong engine.

System Error Exemplar. An ILS (instrument landing system) approach by the pilot of a high-performance single-engine airplane to the North Bend Airport, Oregon, resulted in a crash nearly 3 miles beyond the threshold of the runway. The final approach to Runway 4 was initiated from an assigned holding pattern at 4,000 ft, about 3,000 ft high. The final approach, as recorded on radar, was well aligned along the localizer (LOC) course up to the last radar point near the landing threshold for Runway 4. However, the final descent path was always extremely high above the glideslope (GS) and continued beyond the missed approach point (MAP) in a descent. The crash site was along the extended LOC course at an elevation just above the decision height (DH), in an area enshrouded in clouds according to a nearby witness at the time of the occurrence.

Why did this acknowledged careful pilot overfly the entire airport into higher terrain beyond? The evidence suggests that he may have been confused by information on the approach chart and by the procedures required after the approach clearance. With respect to the approach chart, he may have mistaken the distance from the final approach fix to the North Bend Vortac (6.3 nautical miles [nm] away) as the distance to the North Bend Airport (only 2.5 nm away). Both have the same three-letter identifier, *OTH*, which would be entered as such into the airplane's LORAN (long-range navigation) receiver. With a 3,900-ft altitude leaving the holding pattern over the final approach fix and only 2.5 nm straight ahead to the runway, even a rapid descent would place this airplane high over the touchdown zone, as it did. With the actual position unknown to the pilot, he continued ahead in an unknowingly futile pursuit of the narrow (1.4 degree) glideslope beam down to just above decision height (DH) at the crash site. Oddly, given the particular offset location of the Vortac in relation to the airport, the pilot never read less than 2.5 nm even near the crash site. Under the high workload and stress of a single-pilot IFR (instrument flight rules) operation, this pilot might have misinterpreted his location as always being outside the outer marker (2.5 nm from the airport).

Why wasn't this pilot prompted by cockpit indications as to his true position and to his misinterpretation? First, flying the ILS itself does not depend on identifying geographical locations, such as the outer marker (OM) or the middle marker (MM). The ILS simply depends on intercepting and following the LOC (left–right) course and the GS (up–down) course down to the decision height (DH, 262 ft). At this point a decision to land or execute a missed approach is made dependent on whether the runway environment is in sight. Notably, this pilot did identify the outer marker as the holding fix during probably five or six times around the racetrack pattern, while waiting for the weather to improve. However, when cleared for the ILS 4 approach, the outer marker was no longer relevant and he apparently focused his attention on the descent needed to acquire the glideslope, straight ahead. Second, because the airplane was in fact well above the glideslope at the point of crossing what should have been the actual geographic location of the missed approach point (MAP), cockpit identification of it would not be possible (although the location was clear on ATC radar). Cockpit indications of the ILS missed approach point depend on being "on" glideslope and at decision height altitude. Neither had been reached because of nonstandard procedures during the approach. Thus, the pilot, unaware, continued to descend beyond. Finally, timing the approach from the outer marker (and using speed for distance), although not required for an ILS, could have provided a clue about location, but not if the pilot thought he had 6.3 nm to go (vs. the actual 2.5 nm).

Undoubtedly, both parties were confused by the approach procedure. The holding pattern (and altitude) had been verbally assigned by ATC (it was not depicted on the approach chart), and the inbound portion of the hold was aligned precisely along the final approach course toward the airport. Normally, except for actually depicted holding patterns with altitude profiles on the approach chart, the procedure turn or radar vectoring is required to provide the maneuvering room for a descent to the glideslope intercept altitude (here at 1,300 ft). When the approach clearance was originally issued, the pilot was turning directly inbound, making the procedure turn appear superfluous, except for the excessive altitude. ATC communication transcripts did not reveal any discussion or concern by the two controllers on duty about the unusual radar path/altitude, nor did they offer radar assistance. On radar, it was apparent that the point-by-point position/altitude was very high along the approach path and unlikely to result in a successful landing on Runway 4. There were a total of 12 radar "hits" covered about 4 miles up to the runway threshold. Two of the specific "hits" pinpointed the airplane almost exactly over the actual MAP and over the runway touchdown zone, both extremely high. Strangely, even the prolonged appearance of the airplane itself on radar could have signified that something was wrong. Final approaches to Runway 4 in this area are not even seen by ATC, because they are usually well below radar coverage.

The air traffic control system, one that was intentionally designed to be redundant by depending on the vigilance of both controllers and pilots, is precisely the reason why this type of ILS accident does not occur more often. The *Airman's Information Manual*[8] (AIM) states, "The responsibilities of the pilot and the controller intentionally overlap in many areas providing a degree of redundancy. Should one or the other fail in any manner, this overlapping responsibility is expected to compensate, in many cases, for failures that may affect safety." In this case, a breakdown of redundancy did occur, with several opportunities to avert this crash lost, some available only to the controllers. Notably, this was the second ILS 4 approach to the North Bend Airport for this aircraft. A similar sounding aircraft was heard by the ear witness about 30 min earlier very close to the crash site, just about the time of this airplane's first missed approach.

The complaint alleged that despite the obvious pilot errors, the ATC failed to warn of the danger. This litigation entailed extensive discovery through various investigations, depositions, and expert's reports, which are now a requirement in some federal cases. A settlement took place just before trial.

CIVIL PROCEEDINGS IN AVIATION LAWSUITS

Expert Witnesses

Clearly, the preceding examples involve varying degrees of complexity even for those of us involved in aviation and in human factors. Obtaining a working knowledge of aviation takes years of study, and even more so for human factors. There are universities offering 4-year degrees devoted to aviation and aerospace, and other universities offering doctoral degrees in human factors, exemplifying the level of effort required. There are difficult concepts to grasp, deep and eclectic knowledge to acquire, unique nomenclature to learn, and of course ubiquitous acronyms in both aviation and human factors to know. However, over the years human factors has used an organization concept using software (ideas), hardware (things), environmental factors (surroundings), and liveware (people), or SHELL, a concept developed by Edwards (1972) and Hawkins (1984). It can provide a way of systematically organizing factual data as they relate to human performance and interactions. The NTSB report addresses each of these areas, except software, to standardize data collection. It is for an expert's ability to integrate information and from it to render a reasonable opinion that he or she is called.

Consider, if you will, how convoluted cases might be decided by juries or judges unknowledgeable in aviation. Consider also the effects of those judgements on the aviation industry, if they are made in isolation from complete information and real-world operations. A fair hearing requires facts fairly presented and an *understanding* of the significance of those facts in context by those making decisions. A fair decision is ultimately of benefit to everyone.

---

[8]*Airman's Information Manual* (AIM), Pilot/Controller Roles and Responsibilities, paragraph 400.

An expert is someone who has been determined by the court to have superior or special knowledge in a particular field or domain, obtained by education, by specialized training, and by personal experience. The expert serves two functions, to educate those involved in the case and to offer opinions, if asked (by either side).

Armed with extensive knowledge, an expert's educational function is to present background information to juries, judges, and attorneys on both sides in sufficient detail so that they can understand the facts that are presented in the case. This means providing a "foundation" of underlying concepts, customary understandings in the field, typical occurrences, common expectancies, and so forth as they relate to the case specifics.

Armed with special knowledge about the case, the expert, if asked, also can explain what he or she thinks may be a basis for evidence and also can give opinions that may serve as fact(s) to support legal judgments (of juries or judges). Unlike a lay witness, who can testify only about personal knowledge of facts, a court-appointed expert may offer opinion(s) to be considered as facts. Experts can also suggest likely scenarios and suggest reasonable expectancies based on their unique expertise. These opinions must have reasonably testable objectivity, which is why a scientific foundation is necessary before the explanations of facts and opinions are presented. A vigorous cross-examination serves to reinforce credible opinions, if responded to professionally (see Surosky, 1993, for techniques.)

Before any testimony is given, experts first must be qualified by the court concerning their expert status and about their relevance to the particular case. Unless stipulated by all parties, an expert's qualifications and purpose can be challenged in the proceedings. Proof of "superior knowledge" as an aviation forensics expert is supported by professional certification (Ph.D., ATP, Board Certification, etc.), publications (patents, operational designs, etc.), and experience (prior court qualifications, current and past employment in human factors and aviation, honors, awards, operational experience, etc.).

Even with authoritative knowledge about the subject matter, the expert must have specific knowledge about the case. This usually requires extensive reading and intimate familiarity with the facts, always with the potential that opinions may change as new evidence becomes available. The latter creates a delicate balance between the timing of information reviewed and the need to formulate opinions—the longer the time available, the better. For example, sometimes preliminary opinions change as new information arrives, occasionally to the detriment of the hiring attorney. If the expert is serving only as a background consultant to the attorney in the case, not as an expert declared to the other side, then any apparent weakness can remain confidential, often leading to an out-of-court settlement. However, if the expert is declared as such, then the opinion(s) become available to both sides, under the rules of discovery. At this stage, the earlier the hiring attorney knows, the better, because a lot of "lawyering" is strategy. If the expert becomes strangely unavailable to the other side, it may change the dynamics of a settlement. Therefore the initial understanding of the case issues, the level of commitment expected, and a clear agreement that opinions do change engenders a professional relationship between the expert and the attorney, keeping unnecessary expenses and embarrassment from arising.

Process

Typically a case for an expert starts with a phone call from an attorney with a brief description of the facts (and perhaps the contentions) of the case. The request is normally to review the facts and circumstances of a specific mishap in order to determine human factors contributions/causation, focusing on system use, interface design, procedures, training, human error(s), and so forth. Sometimes the request is to analyze a specific design or an associated warning as to the likelihood of human errors (misunderstanding and misuse) or designed-induced error(s). Occasionally, there is a request to specifically comment on the opinions offered by an opposing expert(s).

Without commitment on either side, follow-up material is usually sent for an initial review. Most of the time the minimum information is the National Transportation Safety Board (NTSB) factual or preliminary report. (Notably, litigation usually follows the lengthy investigative process by the NTSB that results in the final NTSB Factual Report. The supporting documentation produced by the NTSB investigators is destroyed.)

The next contact is usually by telephone from the expert, with his or her verbal assessment of preliminary opinions. Based on that conversation, if there is an agreement by the attorney that the expert might help the case and if the expert is willing to take on the case, then there is some level of commitment. There are many reasons at this point for a noncommitment: The issue is not in the expert's area, the problem not of sufficient interest to the expert, the expert's opinion is adverse to the attorney's opinion, there are conflicts that might create a potential bias, incompatible scheduling demands, and so on. Before an agreement, it is appropriate to indicate what the general nature of testimony might entail in the area of human factors design (e.g., display and information processing), and the human factors of flight operations (e.g., flight procedures, pilot behavior, pilot expectancies from air traffic control, etc.). It is also appropriate to indicate areas of nonexpertise; there will be no testimony outside of an expertise, such as in crashworthiness or in accident reconstruction. Most often testimony is in the form of expectancies, likelihoods, or probability that something did occur or that should have been reasonably foreseeable for the manufacturer or user.

The level of commitment of an expert might be as a background consultant (available for questions from the hiring attorney) or as a declared expert available for deposition and trial. With such an agreement a full analysis gets underway with material and sent by the attorney. Information also may be requested by the expert including other information that is available, or information that needs to be obtained, such as depositions, proposed test, experiments, and so forth. The following are some typical materials that may be available through the discovery process (each party in the lawsuit must inform the other of documents they have, in a process known as discovery).

- Factual data
  - National Transportation Safety Board (NTSB) factual report
  - FAA ATC plots of the flight path and ATC communications
  - Cockpit voice recorder (CVR) communications
  - Photographs of the accident site and the wreckage
  - VFR charts or IFR charts (as appropriate)
- Aircraft information
  - Pilot operating handbook (POH)
  - Weight and balance information
  - Airframe, engine, propeller (or other component) logbooks
- Pilot/crew information
  - FAA pilot certificates and medical records

Pilot logbooks and training records

Weather information

Weather reports and forecasts

Testimony

Statements of eye or ear witnesses

Opposing expert reports

Depositions taken by attorneys on all sides

This mass of material arrives at various times, in sundry order, and at times by the box load. How is it organized? How is it verified, given natural errors in assembling and recording so much information, the different perceptions by people seeing or hearing the same thing, and even outright fabrication of the truth by people with real self-interests? (Presumably, misrepresentations are an uncommon experience for most experts coming from a scientific setting.)

The organizational headings just listed work well for referencing material in most cases, but verification of conflicting data is more difficult. In general, the pattern of evidence can be viewed in a statistical context with occasional "outliers." The preponderance of the evidence centers around the median of the data and should anchor an opinion, unless there are extraordinary and explainable reasons for doing otherwise. One unsubstantiated conflict should raise a question and deserves consideration, but without corroboration it should not unduly sway an expert's view. Most often crash witness statements are in conflict and can be checked against one another using something akin to law enforcement techniques. Law enforcement officers often interview people separately to highlight discrepancies and to look for converging evidence. This is not to say that an apparent discrepancy should be dismissed as untrue or just wrong. However, in real life most events have some basis, right or wrong, and even the smallest troublesome detail may lead may to a breakthrough of understanding. But there should be caution for relying too much on any one piece of evidence; sometimes the underlying basis will never be known.

After a preliminary review, if there are gaps in existing evidence that need to be filled before a suitable human factors analysis can be completed, then it is appropriate, if not expected, for an expert to actively seek additional facts (in coordination with the attorney). Information might be available by data searches, additional interrogatories or depositions, through videotaped observations, or even from modest experiments. Experts who should seek additional data that might be available, but do not do so, are open to spirited questioning by the other attorneys. They will be suspicious that such passive behavior can be manipulated and therefore is not appropriate for an expert, or that this is really indicative of the expert "not asking the question, because of fear of getting the answer (an unwanted one)." It is true that regardless of the findings, all such pieces of evidence are subject to discovery rules, as they should be in the search for truth. Generally, just like scientific research, if the theory is correct, then new data will fit. The converse is also true, signaling the need for a new approach and for serious discussions with the attorney. It is safe to say that there are no attorneys that want to hear something negative about their side of the case, but they probably would all agree it is best to know the down sides of a case before a deposition is taken (questioning by the other party's attorney) and certainly before written or oral testimony at trial (written reports are now required as testimony at certain federal trials).

Plaintiff-Defense Approaches

Although this chapter addresses forensics, direct reference to legal principles has been avoided up to now, because the primary focus of the book is on aviation and human factors. However, some discussion of legal issues is necessary to explain the legal context of forensics, but is given here in a compilation of words of others, because the author is not an attorney.

> Products liability is the legal term used to describe a type of civil lawsuit. In court, usually before a lay jury, an injured party (the plaintiff) seeks to recover damages for personal injury or loss of property from a manufacturer or seller (the defendant) because the plaintiff believes that the injuries or damages resulted from a defective product. Products liability falls under case law where each new court decision adds, changes, clarifies, or sometimes obscures the prior legal precedents. Civil actions are unlike criminal proceedings, where there is a presumption of innocence with proof of guilt required beyond reasonable doubt, well beyond the 50% mark. In civil actions judgements can be made on the preponderance of the evidence, that is on a 50% tilt or on an apportionment of blame among parties, by percentages. Just as there are few absolutes in life, there are few absolutes in the law, therefore what is "reasonable" and "what is likely" are often heard.

Kantowitz and Sorkin (1983, pp. 629–630, 633) stated emphatically that:

> There is no such thing as a perfectly safe product. . . . Instead, there must be a balance between the potential harm a product may cause and the benefits to society of a plentiful supply of products. . . . As laws and judicial interpretations evolve, the definition of a reasonably safe product changes. . . . In a product liability lawsuit an injured party—the plaintiff—brings suit against a manufacturer or seller—the defendant—who has provided the allegedly defective product. . . . There is no absolute standard for unreasonable danger. . . . Expert witness testimony is often used to establish the degree of danger associated with a product. But even experts disagree. It is not unusual to find the human factors specialist testifying for the plaintiff at odds with the specialist for the defendant.

In the prior accident exemplars, the legal complaints ask who has responsibility, usually a euphemism for monetary damages. In the case of the autopilot, it was alleged that the autopilot manufacturer allowed a defective design and that the aircraft manufacturer was negligent for choosing this autopilot for this aircraft. Surosky (1993, p. 29) stated:

> A product defect is one that makes it unreasonably dangerous for the user, or dangerous to an extent beyond what and ordinary user might contemplate. The basis for product liability is usually a defective design, defective manufacturing, or failure to warn of the hazards of an inherently dangerous product. . . . Contributory negligence is the failure on the part of the injured party to exercise ordinary care in self-protection, where such carelessness (along with any negligence by the defendant) is a direct cause of injury.

The settlement included some payment made by the autopilot manufacturer, but there was no payment made by the airplane manufacturer.

In the case of the helicopter turbine burst, it was alleged that the engine had a known manufacturing defect and that the helicopter manufacturer knew about this defect but failed to design a proper warning for the failure. The complaint was essentially one of product liability. The engine was flawed, and the helicopter design misled the pilot to shut down the wrong engine, resulting in the death of the pilots and passengers. Further, it was alleged that this event was foreseeable based on past similar events. Kantowitz and Sorkin (1983, pp. 632–633) stated:

> The first step in establishing product liability in cases where no express warranty or misrepresentation is involved is to prove that the product was defective. . . . Product defects arise from two sources. First, a flaw in the manufacturing process may cause a defect. This results in a defective product that does not meet the manufacturer's own quality-control standards. . . . Second, a product may have a design defect. . . . The manufacturer is liable for any defects in a product, including those that were unknown at the time of manufacture [but defects that were reasonably foreseeable (e.g., probable errors by the operator) in normal use or even from misuse].

The jury held the engine manufacturer was responsible for the crash and awarded damages.

In the case of the instrument approach crash, it was alleged that the air traffic controllers had unique knowledge about the ensuing danger (via radar) but failed to warn of the pilot of the danger. Further, even though it was admitted that the pilot made errors, it was contended that the pilot was, in part, misled into those errors by the placement of government-installed navigational aids, ambiguous charts, and procedures. In effect, the allegations were of negligence on the part of the government. Kantowitz and Sorkin (1983, pp. 630–631) stated:

> To establish negligence, the plaintiff must prove that the conduct of the defendant involved an unreasonably great risk of causing damage. The plaintiff must prove that the defendant failed to exercise ordinary or reasonable care. A manufacturer has a legal obligation to use new developments that will decrease the level of risk associated with his or her product. Failure to keep abreast of new technology can be grounds for negligence. . . . There is no absolute standard for "unreasonably great risk." It is quite possible that the same defendant would be judged negligent in one court and innocent in another court. . . .

According to Joseph Nall, a lawyer member of the NTSB, as quoted by Barlay (1990, p. 125):

> In many instances, the public can sue the government because of the Federal Tort Claims Act, but there is an exemption called Discretionary Function to contend with. A government agency cannot be sued for making the wrong judgement. But if a government employee is negligent, the government is liable.

Clearly, there was not a single cause here, but errors made by both the pilot and the ATC controllers representing a system breakdown. This case resulted in a substantial monetary out-of-court settlement.

Ethics

Regardless of an expert witness's experience, most attorneys give him or her their own "standard lecture" before a deposition or before trial. In virtually all such lectures, they start out with "tell the truth." Charles Smith (personal communication, May 3, 1994), an attorney in Dallas, even wrote out his 36-item list of basic rules, applicable to expert and lay witnesses alike, irrespective of their importance or the role they will play in the case. Smith's rule number one is:

> 1. *Tell the truth.* This is more than a copybook maxim; it is rule of self-preservation. Always assume that the examining counsel is supporting himself on his professional ability, and that this includes the ability to make a witness who is playing fast and loose with the truth very uncomfortable.

True integrity comes from within, and is not foisted on us from the outside. But in reality it is better to heed former President Ronald Reagan's famous words, "trust but verify." It has been suggested that, "it would be preferable to have nonpartisan experts called by the court itself . . . divorcing expert testimony from the source of compensation [the contesting parties] therefore, as close as possible, ensuring the impartiality of the witnesses involved" (Dunham & Young, 1995, p. 446). Others disagree. J. A. Wise (personal communication, April 13, 1994) argued that experts on both sides serve an important check and balance function for each other, making it easier for others to judge the merits of their positions and their credibility.

Regardless of the court's approach, there is a "Darwinian-like" selection for experts. As an expert's portfolio of cases builds up, all past depositions and testimony, readily available to any attorney, become a potential source for questions. If there are changes in views or inconsistences, rest assured a competent attorney will be happy to point them out at trial. Poor or incomplete explanations degrade credibility. Lost credibility means the expert will no longer be employed as such. Because it is impossible to anticipate every question, particularly on cross-examination, the most sensible approach is to follow the rules telling the truth according to a consistent approach well supported by scientific facts. If for some reason a new scientific revelation occurs, the expert should be prepared to be deeply questioned and have to have ample justification for it. Sharp questions will be forthcoming in cases to follow.

Unemotional advocacy of the truth, not for any side of a lawsuit, provides an enduring foundation for any expert. The attorneys are the advocates for their clients, and accordingly it is they who win or lose cases. Experts are only one ingredient in an assembly of facts ultimately argued by the attorneys. For experts it is often difficult to tell who won or who lost, as the prior examples illustrate. Oftentimes with cross claims and legal maneuvering the true litigants become lost, plaintiffs take on defensive positions, and defense strategies take a decided offensive flavor or may even become plaintiffs against yet another defendant. A party to a lawsuit who might appear to be obviously at fault or almost fault-free may become ancillary to or a target of the case because of its available resources (known as deep pocket). If there is comparative negligence and one party cannot pay damages, in some states the other party must pick up the entire bill[9] even if that party is only 1% at fault. To stay out of the fray, experts would be well advised to set a uniform fee structure based on time worked, not on the importance of the action (e.g., appearance in court). They definitely should not base fees on the outcome of a case (e.g., contingency fees) because this can be construed as having a financial interest in the outcome. This time-based fee structure helps ensure a dispassionate analysis and a fair hearing for the facts.

---

[9]The doctrine of *joint and several liability.*

HOW AVIATION FORENSICS CONTRIBUTES TO HUMAN FACTORS

Real-World Feedback

Aviation forensics provide real-world feedback about human behavior involving sophisticated designs in high stress environments. Aircraft, in many ways, are like caricatures of other complex devices in our society, emphasizing, if not exaggerating, the need and role of human factors in the design of such systems. Flying demands rapid understanding and integration of assorted information, while extracting rigorous performance that is unforgiving of errors. It is rare in our daily activities that the consequence of "underachievement" is so high. Death can be the penalty for errors, and, many errors can be made in aircraft.

When something goes wrong with an aircraft resulting in a crash, aerodynamicists, reconstructionists, and crashworthiness experts converge on the evidence to analyze and render their opinions. Similarly, McIntyre (1994), Chairman of the ISASI Human Factors Working Group, urged that human factors investigation specialists be included in all accident and incident investigations. McIntyre (1994, p. 18) went on to state "that the predominant factor in aviation mishaps [is]—the Human Factor. . . . Human factors data should be subject to the same rules of evidence that are applied to other mishap data. Findings must relate to factors which were significant in the mishap scenario, and be integrated with other factual evidence."

Because of the intimate relationship between human operator and machine in aircraft, the human factors of aircraft design is constantly being tested, and in the case of a crash, both the user and the design are literally on trial. What human factors professionals learn from accidents is that nothing is perfect, and what is reasonable usually prevails in jury decisions, despite reports of extremes. In most cases users and designers could have or should have anticipated the results. In particular, fixes should be considered if problems are expected in normal use, if there is potential misuse, or if there could be undesirable interactions with other systems or in unusual environments. It is unrealistic, of course, to think that all errors can be eliminated, but many errors can be anticipated initially or after an accumulation of service experience. The key is that isolated "acts of God" must be separated from carelessness (which can be prevented by the user or designer). The difficulty is that with so few similar events occurring (fortunately by virtue of a low accident rate), the discrimination of "acts of God" from carelessness is not statistically amenable and is not entirely clear, but must be based on what is judged most probable (more than less). The courtroom is where many issues of human factors issues are practically judged, for better or worse, tilting the balance for the future of the discipline in the consumer area.

Consider as an example, the incorporation of standardized control shapes such as the flaps switch (a control surface profile) and the landing gear handle (a tire form) in order to reduce gear-up accidents. Laboratory testing by Jenkins (1947) suggested that control shapes can provide unique discrimination characteristics, but would such findings would hold up under the stress of actual flying and divided attention? Certification criteria for standardization of switch shapes were set for new airplane designs. Statistical evidence indicated fewer control-confusion accidents, but the issue is not entirely clear. There are interacting variables and conditions specific to a given airplane design that raise questions as to the true cause of any one accident. Can the placement of switches negate the benefits of discriminable shape, or should the shape overcome the effects of placement? Ultimately, acceptance or rejection of a human factors design or procedure goes beyond statistics and the imposition of certification criteria to decisions reached by mainstream juries, who are convinced by intelligent and motivated people battling over the individual facts of the case. Human factors experts help by explaining the advantages and disadvantages of various designs, procedures, and training and by giving probabilities founded on accepted theory and fact, not on possibilities speculated by attorneys. Such decisions may have far-reaching implications not only for design, recommendations, and regulations, but for human factors teachings as well.

Clearly, accident analysis is not a desirable way to collect data and a sample of one is not scientifically reliable. However, even considering extensive certification trials and testing, an individual design in an operational environment could never receive such scrutiny in any other circumstance. Moreover, accidents do happen and the real-world feedback can provide invaluable information for human factors analyses. If nothing else, legal decisions serve to identify and highlight errors or potential errors, and they may set into motion regulatory actions or set implied standards for the use of aviation products and systems. The NTSB, NASA/ASRS, FAA, manufacturers, and training organizations all use accident data to help detect and prevent human and design-induced errors that occur or could occur in aviation incidents/accidents. Similarly, human factors professionals should use these data and forensic analyses to assess the validity of human factors theory and principles.

Murphy's Law

Many suggest aviation product liability law is in disarray. Perhaps this is an overstatement, but there are conflicting messages that legal circles send to the human factors community. One of the biggest "bones of contention" is the legal concept of strict liability, which disconnects the issue of blame, risk, and error. Human factors aviation experts usually help explain why certain behavior occurred, versus what was expected, so that juries can judge what is reasonable. Strict liability judgements appear to some to bypass reason.

With regard to blame, strict liability[10] allows for damage recovery without the need to show negligence. It is applicable if a product is found defective and unreasonably dangerous,[11] regardless of the care in design, in manufacturing, or in the preparation of instructions, given that the user had not mismaintained, misused, or abused the product. In essence, manufacturers are held responsible to analyze their products for any inherent danger and are in the best position to foresee risk that reasonably might arise with their normal use.

> This [strict liability] theory of recovery says that if a product is defective, and the defect makes it unreasonably dangerous, and that defect causes and injury or damage to someone who has not misused or abused the product, and the product is in substantially in the same condition as it was when it left the manufacturer's hands, then regardless of how much care the manufacturer used in designing and building the product, the manufacturer

[10]In effect, *negligence* tests the conduct of the defendant; *strict liability* tests the quality of the product; and *implied/expressed warranty* tests the performance of the product against representations [made by the seller].

[11]It appears that in some states the fact in itself of death, injury, or damage is evidence of a dangerous defect.

> is still liable for money damages to the injured or killed person or his family. (Wolk, p. 166)

Of course it must be decided just what constitutes misuse.

With regard to risk, Holahan (1986) indicated:

> The roots of the product liability dilemma do not lie in aviation, but in the social-political attitude that has been growing in America for the last 25 years. Americans want guarantees of a riskless society. They demand that products be absolutely safe, even when they are misused. And when they have accidents, some one has to pay for them. . . . Our obsession for a riskless society has permeated the judicial system to the point where it has been instrumental in overturning our former negligence-based tort law, where once cases were judged on the basis of the comparative negligence of the parties involved. Replacing the negligence standard has been the doctrine of *Strict Liability*, which hold the manufacturer liable for whatever he did even though no negligence was involved. . . . Strict liability says "honest mistakes don't count any more" and the doctrine gives rise to the allowance of evidence (in most states) which judges yesterday's designs by today's technical know-how. (p. 78)

With regard to the prevention of error, "Murphy's law" and its ubiquitous corollaries are always written in the negative: "If it can go wrong, it will go wrong." Murphy's laws never suggest that it couldn't go wrong. It is inevitable that humans will err and that machines will break (Sanders & McCormick, 1987). Nickerson (1992, p. 341) made the point that "There is no such thing as a risk-free existence, and the attempt to eliminate risk entirely can produce only disappointment and frustration." Nickerson (1992, p. 333) also stated that "No one with a rudimentary understanding of probability believes that zero-risk policies are attainable."

With the understanding that zero errors are impossible, human factors has developed specific strategies for reducing likelihood or consequence of human errors (Sanders & McCormick, 1987). Designs can be fail-safe (or error tolerant) to reduce the consequence of error without necessarily reducing the likelihood of errors. There are even exclusion designs that eliminate certain errors or that minimize the possibility of most errors. Both approaches make the reality of errors more acceptable, although still unwanted.

Preventing Human and Design-Induced Errors

Product liability law is primarily in place to recover damages resulting from product defects. It is not in place to provide solutions for mishaps or to prevent reoccurrences. Some attorneys argue that the logical outcome of adverse judgements will be to prod manufacturers into designing safer products. Ideally, such decisions favorable to the plaintiff should engender amendments, modifications, or the redesigning of products to be safer or easier to use. Other attorneys counter with the argument that a fusillade of frivolous lawsuits diminishes resources and incentives for attempting design changes. The "Catch 22" is that product changes can be used in court against a manufacturer as *prima facie* evidence of a past problem, in effect, admitting a defect. Because changes also increase the risk of new as yet unanticipated problems arising, the effect may be the opposite of the desired outcome, that no changes are made unless forced.

Although the intricate dynamics of change are beyond the scope of forensic human factors in aviation, it should be clear the court system is neither the solution nor the problem. Most law is written in the limited negative, "thou shall not," leaving the individual empowered to act creatively within those limits. The leading edge of aviation has been advanced by stellar avionics, not by legal inquiries into the aftermath of a crash. Accordingly, the courts rightfully should remain on the trailing edge as a tool to restore for transgressions, and if possible to point out areas in need of further inquiry.

The crux of a lawsuit is the judgment of "what is reasonable." A legal action usually will prevail or not depending on whether the design in question is needed, reasonable to use, and prudently safe in the eyes of an average juror. For example, a product may be potentially dangerous but needed and reasonable to use, such as an airplane or even a knife, if its benefits outweigh the risk of use. On the other hand, it may not be defendable if at the time of manufacture safer alternative(s) were feasible and were not substantially outweighed by other factors such as utility and cost. Problems/errors that might be encountered in normal intended use or even foreseeable misuse must be considered. The latest developments, particularly with respect to safety, should be incorporated into new designs and, if possible, available as modifications for prior designs. For example, a potential danger that is not obvious and that cannot be prevented needs a warning that is obvious, easily understood, and heeded. (There is no warning about the possibility of a cut from a sharp knife, because it is a patently an obvious danger.)

Standing behind the meeting of government standards, in and of itself, is usually not a successful defense, because such standards represent what is minimally acceptable as safe. Most jurors at least want an explanation of why the company chose only to meet those standards. Usually jurors want to hear the logic behind the plaintiff's theories. They then carefully judge supporting evidence for those positions, such as behavioral expectancies, prior incident or mishap surveys, design recommendations based on human factors research findings, or even specific experiments/demonstrations definitively showing proof. Interestingly, the dual goals of human factors, ease of use and safety, are generally disassociated in legal arguments. In most cases the safety of a product often has little to do with how easy it is to use, although it seems that good designs are usually simple ones.

Many in aviation blame the legal system for the near demise of general aviation, whose sales have dropped by more that 90% Draconian since the heydays of 1978. Arguments range from single-event law suits imposing judgments against manufacturers to liability insurance skyrocketing new airplane prices to beyond reach of nearly everyone. Defense advocates say this stifles innovation (where changes are a tacit admission of a design inadequacy) and sales. Others argue a steady improvement in the accident rates is evidence that legal process is enhancing safety. Plaintiffs point to cases forcing changes in design that might not have been addressed otherwise, such as improved seat and restraint designs, better component reliability, detection of structural defects, and refinements for control stability.

Regardless of one's position, legislative reform has started. So far its effect is unknown. The fallout of the General Aviation Revitalization Act has yet to be measured, except for the decision by Cessna management to resume single-engine airplane production. This act is essentially one of repose. After 18 years the airplane and its components have proven themselves by law, barring lawsuits (immunity from product liability actions) against the manufacturer.

But questions abound. Undoubtedly, part replacement, overhaul, or modification starts a new 18-year period, but does this also restart the calendar for the subsystem that part goes into, such as the engine or a flight control system? Will the design of such a subsystem be removed from immunity for a design-defect lawsuit, because the replaced part depends on and effectively revitalizes the original design? Will interacting subsystems also be affected? What happens if the pilot's operating handbook (POH) is updated or replaced with revised information (e.g., performance figures, checklists, warnings, etc.); does it affect liability for the manual or even for the aircraft it represents or both? Will legal sights be refocused on other defendants not covered by this federal statute, such parts and component manufacturers, maintenance facilities, the FAA/ATC, fixed-base operators (FBOs), fuel distributors, training organizations, and flight instructors or designated pilot examiners?

The act also does not bar claims for injury or damage to persons or property on the ground, only for those who were on board (presumably who voluntarily have accepted a degree of risk). Further, it does not protect a manufacturer who knowingly misrepresents to the FAA or who conceals a defect which could cause harm. Therefore even beyond the 18 years, accident victims or their families still will have rights and legal recourse. But the argument may shift to what is a hidden defect (and a failure to warn), what is a misrepresentation, and of course when is a defect harmful or "unreasonably dangerous." Obviously these could be future points of contention.

The fact that the General Aviation Revitalization Act legislates only for general aviation aircraft, with fewer than 20 seats, and not used in scheduled passenger-carrying or military operations, raises even other legal questions. It has been predicted that there will be challenges to this law on many grounds. One compelling argument is that the law seems to favor a small subset of the consumer product industry. Indeed, if the concept of repose is correct, then some contend that it should be applied to all product areas including lawn mowers, automobiles, and household appliances.

Finally, there may be other higher issues to be resolved, namely, complaints that tort law,[12] under the jurisdiction of state systems and courts, supersedes and makes irrelevant federal authority (FAA) to set standards for safety. Such legal precedents, in effect, set different standards that vary from state to state, governing aircraft that cross state boundaries. Moreover, judgments are made by lay juries, for the most part unknowledgeable in aviation, rather than by federal agencies set up to represent the aviation industry. Others argue that these federal agencies are inadequate and that it is improper to use as a defense the fact that the design was FAA certified.

It is unlikely these mixed messages will disentangle in the near future. However, many advocate that placing aviation accident cases under federal jurisdiction, along with all the other federal regulations that govern aviation, is the correct action. Thus, one suggestion is to place aviation under federal tort law. Clearly, the involvement of lawyers in aviation is not going away. Therefore, it seems reasonable that aviation experts, including those in human factors (the highest risk area of aviation), remain involved in the legal process and outcomes. Forensic aviation human factors is one way.

Issues remain for human factors experts to explain in the context of aviation accidents that have happened and will happen. Humans (both designers and product users) have not yet discovered all the ways to make mistakes. Therefore, what happened and why will continue to direct inquiries and future research. Human error comes from various sources and in various forms, spanning the range from accident-prone products (low degree of avoidability) to accident-prone people (low degree of alertness). Whether a product (aviation or otherwise) induces an error or fails to prevent one, or whether the

[12]Product liability tort law essentially refers to civil (vs. criminal) proceedings that allow for compensation to an injured party as the result of a wrongful act or neglect that directly causes injury to a person or property.

error is self-induced by commission or omission, mistakes and blame for them will continue to be argued in court. The people may change but the problems remain the same.

REFERENCES

Barlay, S. (1990). *The final call: Why airline disasters continue to happen* (p. 125). New York: Pantheon Books.

Beaton, R. J. (1994, July). Letters from our readers. *Ergonomics in Design*, pp. 4–5.

Boehm-Davis, D. A. (1994, July). HFES responds. *Ergonomics in Design*, p. 35.

Corlett, E. N. (1994, July). Letters from our readers. *Ergonomics in Design*, p. 5.

Dunham, C. W., & Young, R. D. (1995). *Contracts, specifications, and law for engineers* (2nd ed., p. 446). New York: McGraw-Hill.

Editorial. (1994, January 3). *Aviation Week and Space Technology*, p. 66.

Edwards, E. (1972). Man and machine: Systems for safety. In *Proceedings of the British Airline Pilots Association Technical Symposium* (pp. 21–36). London: British Airline Pilots Association.

Gardner-Bonneau, D. J. (1994a, April). Comment from the editor. *Ergonomics in Design*, p. 3.

Gardner-Bonneau, D. J. (1994b, July). The Editor responds. *Ergonomics in Design*, p. 35.

Hawkins, F. H. (1984). Human factors education in European air transport operations. In *Breakdown in human adaptation to stress: Towards a multidisciplinary approach* (Vol. 1, for the Commission of the European Communities). The Hague: Martinus Nijhoff.

Holahan, J. (1986, January 1). Product liability. *Aviation International News*, p. 78.

Hornick, R. J. (1994, July). Letters from our readers. *Ergonomics in Design*, p. 4.

Howell, W. C. (1994, April). Provocations: The human factors expert mercenary. *Ergonomics in Design*, p. 6.

Jenkins, W. (1947). The tactual discrimination of shapes for coding aircraft-type controls. In P. Fitts (Ed.), *Psychological research on equipment design* (Research Report 19). Columbus: Ohio State University Army Air Force, Aviation Psychology Program.

Kantowitz, B. H., & Sorkin, R. D. (1983). *Human factors: Understanding people–system relationships*. New York: Wiley.

McIntyre, J. A. (1994, September). Perspectives on human factors: The ISASI perspective. *Forum, 27*(3), 18.

Nickerson, R. S. (1992). *Looking ahead: Human factors challenges in a changing world*. Hillsdale, NJ: Lawrence Erlbaum Associates.

Sanders, M. S., & McCormick, E. J. (1987). *Human factors in engineering and design* (6th ed.). New York: McGraw-Hill.

Sanders, M. S., & McCormick, E. J. (1993). *Human factors in engineering and design* (7th ed.). New York: McGraw-Hill.

Senders, J. W. (1994, April). Warning assessment from the scientist's view. *Ergonomics in Design*, pp. 6–7.

Surosky, A. E. (1993). *The expert witness guide for scientists and engineers*. Malabar, FL: Krieger.

Wolk, A. A. (1984, October). Points of law: Product liability—Aviation's nemesis of conscience (a personal opinion). *Business and Commercial Aviation*, p. 166.

Author Index

A

B

C

D

E

F

G

H

I

J

K

L

M

N

O

P

Q

R

S

T

Y

Z

Subject Index

A

B

C

E

F

G

H

I

Q

R

S

T

U

V

W